Wetlands: Ecology, Conservation and Management

Volume 10

The recognition that wetlands provide many values for people and are important foci for conservation worldwide has led to an increasing amount of research and management activity. This has resulted in an increased demand for high quality publications that outline both the value of wetlands and the many management steps necessary to ensure that they are maintained and restored. Recent research and management activities in support of conservation and sustainable development provide a strong basis for the book series. The series presents current analyses of the many problems afflicting wetlands as well as assessments of their conservation status. Current research is described by leading academics and scientists from the biological and social sciences. Leading practitioners and managers provide analyses based on their vast experience.

The series provides an avenue for describing and explaining the functioning and processes that support the many wonderful and valuable wetland habitats, such as swamps, lagoons and marshes, and their species, such as waterbirds, plants and fish, as well as the most recent research directions. Proposals cover current research, conservation and management issues from around the world and provide the reader with new and relevant perspectives on wetland issues.

Tatiana Lobato-de Magalhães • Kevin Murphy
Marinus L. Otte • Eugenio Molina-Navarro

World Atlas of Freshwater Macrophytes

Dicotyledonous species I (Acanthaceae – Menyanthaceae) - Volume 1

Springer

Tatiana Lobato-de Magalhães
Natural Sciences Faculty
Autonomous University of Queretaro
Santiago de Querétaro, Mexico

Marinus L. Otte
Dept of Biological Sciences
North Dakota State University
Fargo, ND, USA

Kevin Murphy
University of Glasgow
Glasgow, UK

Eugenio Molina-Navarro
Dpto. de Geología
University of Alcalá
Alcalá de Henares, Spain

ISSN 1875-1261 ISSN 1875-127X (electronic)
Wetlands: Ecology, Conservation and Management
ISBN 978-3-031-52748-7 ISBN 978-3-031-52749-4 (eBook)
https://doi.org/10.1007/978-3-031-52749-4

This Springer imprint is published by the registered company Springer Nature Switzerland AG
The registered company address is: Gewerbestrasse 11, 6330 Cham, Switzerland

Paper in this product is recyclable.

Preface

This book is the first volume of a series of five forming a World Atlas of Freshwater Macrophytes. The need for such an Atlas became apparent from the work of an informal international research network dedicated to advancing macroecological knowledge of freshwater macrophytes, set up by Kevin Murphy in 2017. This has to date produced a number of well-cited publications, collectively authored by network collaborators, which together form the primary underlying material for this Atlas (e.g., [7, 79–81, 104, 105]). During this work, it became clear that the knowledge base of macrophyte distribution and associated ecological information, at world scale, was extremely scattered in the literature and often difficult to access, hence the decision to produce an Atlas that seeks to bring together such knowledge in an easily accessible location.

We define vascular macrophytes as being vascular plants large enough to see with the naked eye, which actively grow permanently or periodically submerged below, floating on, or up through the water surface of inland freshwater or brackish waterbodies [104]. This is a group of plants of very varied age, taxonomic origin, and functional characteristics, though all have, by definition, at least some populations with an obligate requirement for inland open water conditions to complete their life cycle. A good example is the rare Neotropical endemic *Isoetes jamaicensis* Hickey with a range restricted to ephemeral pools in open xerophytic scrublands of Jamaica. The dormant corms of this species only sprout in years when enough seasonal rainfall has occurred to fill the pools [53]. Our definition excludes terrestrial plant species that occur in wetland systems, such as Brazilian várzea floodplain grasslands (e.g., [25, 76, 92]). These species are adapted to tolerate periodic flooding but usually do not have a physiological requirement for inundation as an essential part of their population survival strategy (e.g., [114]).

The majority of the 92 vascular plant families that contain macrophyte species also have other species that are non-aquatic, and most of the c. 3544 accepted macrophyte species also have populations that occur outside of inland fresh or brackish water bodies (though often closely associated with water, for example, as riparian populations). Even in those families which entirely comprise obligate macrophytes (notably Podostemaceae, Nymphaceae and Hydrostachyaceae), it is possible to find examples of species which have such populations. A good example is provided by the species of Podostemaceae (e.g., *Inversodicraea bosii* (C. Cusset) Rutish. & Thiv, *I. cristata* Engl.), which live in riparian spray-zone habitats, as well as in the water itself in the cascades and waterfalls of fast-flowing tropical rivers.

Macrophytes are generally a rather young group of plants in evolutionary terms, and it appears that the speed with which they have successfully managed to move into an aquatic lifestyle at least in part reflects this. Although the oldest species for which a reliable speciation age estimate has been made is *Lycopodiella inundata* (L.) Holub, at 110 million years ago, Ma, most macrophyte species evolved much more recently. Of the approximately 560 macrophyte species for which speciation age data are available [80], 73% evolved in the Pleistocene, Pliocene, or late Miocene, and a few species are even younger. The evidence suggests that at least two species (*Nymphaea loriana* Wiersema, Hellq. & Borsch and *Ranunculus schmalhausenii* Luferov, both high-latitude northern hemisphere plants) arose in the post-glacial Holocene, via rapid hybridization events, and have probable speciation ages <0.01 Ma [13, 15].

There is evidence that the survival strategies (sensu [50]) of most macrophyte species encompass a strong element of stress tolerance (e.g., [9, 101]). However, it is likely that the very varied environmental conditions experienced by different macrophyte species have resulted in the evolution of a wide range of survival strategies across the macrophytes as a group, given that they may experience pressures on their survival ranging from extreme disturbance (e.g., in fast-flowing rocky rivers, occupied by Hydrostachyaceae and Podostemaceae) to highly competitive conditions in shallow eutrophic lakes occupied by mixed stands of submerged and floating species [104].

Overall, it is clear that the vascular macrophytes are a very diverse group of plants, united by a common life requirement for aquatic conditions, which also provide vital support for other aquatic biota (e.g., [85, 109]) in inland waterbodies. In this Atlas, we have tried to bring together available information for aquatic macrophytes, which is otherwise widely scattered through the literature, highlighting aspects of their ecology, including range size, endemism, world rarity, ecozone/macroregional occurrence, ploidy state, species age, nuisance state, uses, and endangered status, as a resource to inform and assist the management and conservation of these remarkable plants. Although similar books exist for local regions (e.g., [62, 73, 117, 139]), none provides detailed information for macrophytes at global scale, so this atlas provides a novel resource on the distribution and ecology of macrophytes around the world.

Santiago de Querétaro, Mexico Tatiana Lobato-de Magalhães

Glasgow, UK Kevin Murphy

Fargo, ND, USA Marinus L. Otte

Alcalá de Henares, Spain Eugenio Molina-Navarro

Introduction to the Atlas Series

In this Anthropocene era, life on Earth is going through unprecedented changes driven by human activities. As the human population has grown faster than ever over the past century, so has the demand for resources. Cities cover an estimated 1% of the Earth's surface and are home to about 56% of the human population, while 38% of the land is used for agriculture [49, 126, 165]. As a result of land use changes such as providing shelter for humans, food production, infrastructure, and industrial activities, distributions of species have been significantly altered. For example, crops and livestock have been intentionally and widely distributed outside their original ranges, while agricultural practices have unintentionally provided habitat for less desirable species [23]. Surface and subsurface hydrology have been significantly altered to secure water supplies for consumption and food production, moving water from places where it is stored naturally to places where humans need it, drying out natural bodies of water, while increasing water levels of others, changing natural fluctuations in water levels, and creating new bodies of water as well as new pathways for migration. Global change through altered land use changes is driving climate change. Species distributions are changing rapidly, and extinctions of species occur at a very high rate [27, 38].

These changes on a global scale pose unprecedented challenges to environmental management. As we come to understand how human health and welfare are intrinsically linked with environmental quality and provision of ecosystem services, we also realize that we lack the information needed to determine environmental management needs and approaches properly. Not only do we lack knowledge about the numbers of species on earth, but also about their distributions. In addition, what we know is biased toward species and places that are easily accessible [40, 56]. For example, it is much easier to study plants on land than underwater, which is one of the reasons why terrestrial plants have been studied in more detail than aquatic plants [89, 144]. Terrestrial plants can be observed by walking the surface of the earth. Specialists, as well as hobbyists and citizen scientists (e.g., [140]), have contributed to our knowledge of terrestrial, and to a lesser extent aquatic plants. It is generally easy anywhere in the world to find a regional flora with images for identification of terrestrial plants, and in recent decades, electronic databases of terrestrial plants for use on mobile devices have been developed. But when it comes to aquatic plants, the story is different. To study aquatic plants, one must get into the water and under the water's surface. Some species float at least partially at the surface and can be easily observed, but others are entirely submerged, often not just obscured from view by turbidity or reflections on the surface of the water but also by floating and emergent plants. Not to mention the fact that safety is a concern in deeper water and special equipment is needed to venture under the surface. Of course, this does not mean that extensive data on aquatic plants do not exist. The data are there, but less accessible than for terrestrial plants.

This World Atlas of Freshwater Macrophytes for the first time brings together available data on global distributions of those vascular aquatic plants which occur in inland freshwater and brackish waterbodies. It is based particularly on five recent publications on the world distributions of >3500 species of aquatic macrophytes in 453 genera from 92 families ([79–81, 104, 105]). Aquatic macrophytes here are defined as "aquatic photosynthetic organisms, large enough to see with the naked eye, that actively grow permanently or periodically submerged below, floating on, or up through the water surface" of inland freshwater or brackish waterbodies [28, 104]. That definition would include macrophytic algae (such as green filamentous algae and charophytes, as well as liverworts and mosses that have aquatic populations), but we do not include these non-vascular plants here, in part because their distributions at world scale are very incompletely known, though the situation is improving [26]. Our definition emphasizes the importance of the presence of water as an essential part of the population survival strategy (sensu [50, 51]) of freshwater macrophytes. This, in turn, means that many "obligate" wetland plants (following the US definition of wetland plants for assessment of wetlands for legal purposes, [152]) are included in this group, but "facultative" wetland plants are not. Facultative wetland plants are tolerant of temporary waterlogging and submersion but do not require that to complete their life cycles [114]. All aquatic macrophytes are wetland plants, but the reverse is not so. This restrictive definition of "aquatic macrophyte" is a pragmatic approach, allowing us to deal exclusively, in this Atlas, with those plants which are most closely associated with inland brackish and fresh waterbodies. The most widely accepted definition of wetlands follows the Ramsar Convention, which reads: "wetlands are areas of marsh, fen, peatland or water, whether natural or artificial, permanent or temporary, with water that is static or flowing, fresh, brackish or salt, including areas of marine water the depth of which at low tide does not exceed six metres" [120]. Plants that occur in those wetlands then are "wetland plants." If this project had included all species of plants associated with wetlands according to the Ramsar definition, it would have included many more species, for example, from floodplains along rivers and coastal marine environments down to six metres: a very ambitious undertaking. However, upon completion, we hope that this Atlas will provide a starting point for a broader-scale effort to catalogue and map the world distributions of all wetland plants, both marine and freshwater.

Purpose of the Atlas

This Atlas provides invaluable information to anybody interested in freshwater macrophytes, from citizen scientists to professional scholars and environmental managers. It also provides a framework for further study, particularly near the boundaries of distributions. The grid size used for this Atlas is 10 degrees of latitude by 10 degrees of longitude, which means that, for example, one grid cell contains several of the smaller countries of Europe. Two grid cells cover Madagascar. It shows, therefore, accurate information about global distributions of aquatic macrophytes and a rough indication of the outer boundaries of their distributions. Still, it lacks information suitable for assessment of detailed changes in plant distributions, for example, driven by a changing climate [122]. For such studies, details of distributions at a much finer scale are needed, but across large swathes of the planet such fine-scale data are simply not available, and this Atlas serves to identify areas that need further investigation. Of particular interest are the boundaries of global distributions where significant changes in plant distributions in response to changes in land use and climate are likely to occur, but also inside distributions, such as regions prone to desertification. The North American Plains, for example, may see more frequent and prolonged droughts in the west [63] but are understudied. One example is the state of North Dakota, where large areas of the western half have not been surveyed. The only flora of North Dakota to date was compiled during the 1950s by O.A. Stevens, who spent his life surveying the counties of eastern North Dakota on foot or by bicycle because he did not own a car [148]. Similar examples exist from around the world, particularly where human population densities are low, and access is difficult, such as the vast taigas and tundras of North America and Eurasia, remote islands in the world's oceans, the Sahara Desert, and the rain forests of South America, Africa, and Asia [70, 84].

Structure of the Atlas

The Atlas consists of five volumes, with one or more chapters providing an overview of topics relevant to plant distribution, followed by detailed descriptions of the distributions of each species within families. Each book will include an introduction followed by a chapter or more on select topics related to the subject matter of plant distributions, such as endemism, rarity, and invasiveness, then by descriptions of species within plant families and species in alphabetical order. The first two volumes contain data on plants that were until relatively recently considered to belong to the "Dicots," Dicotyledonae. However, that is not a monophyletic group and is now divided into the monophyletic Eudicots ("true" dicots), while several families, including the Nymphaeales, are now outside that group [65, 141]. For the purpose of this book, which is not about systematics but about distributions, we have separated the Monocots from all other vascular families, which we will refer to by the older term "Dicots." Volumes three and four will focus only on Monocots, and the final volume will encompass a final round of Monocot species as well as the Lycopodiophyta, Pteridophytes, and Gymnosperms. The latter is based on an initial dataset from Murphy et al. [104], [79–81, 105]. Finally, we include information from the Pacific Ocean islands of Oceania, and the sub-Antarctic islands and Antarctic peninsula (which collectively contain only extremely small areas of inland waterbodies suitable for macrophyte colonization) in a separate chapter of the final volume.

Volumes Included in This Atlas

Vol. 1: Dicotyledonous species I

- Aquatic macrophytes and wetlands
- Global distribution of macrophytes
- Species descriptions by families in alphabetical order

Vol. 2: Dicotyledonous species II

- Macrophyte endemism and rarity
- Species descriptions by families in alphabetical order

Vol. 3: Monocotyledonous species I

- Genetic diversity and global ploidy distribution of macrophytes
- Species descriptions by families in alphabetical order

Vol. 4: Monocotyledonous species II

- Vectors of long-distance movements and distribution of invasive macrophyte
- Species descriptions by families in alphabetical order

Vol. 5: Monocotyledonous species III, Lycopodiophyte, Pteridophyte, and Gymnosperm species

- Macrophyte distributions on Pacific Ocean islands of Oceania, and the sub-Antarctic islands and Antarctic peninsula
- Species descriptions by families in alphabetical order
- Summary of all families

Introduction to Volume 1

This first Volume of the World Atlas of Freshwater Macrophytes includes two chapters describing where in general freshwater macrophytes occur. Chapter 1 discusses how all freshwater macrophytes are wetland plants, but not all wetland plants are freshwater macrophytes. Why is that? What are the similarities and differences between the two groups of plants? Chapter 2 addresses the question of which overarching factors determine global distributions of freshwater macrophytes. For example, there are differences in animal and tree species diversity between Eurasia and North America at the same latitudes, but is that also true for freshwater macrophytes? This is followed by Chapter 3, which describes the global distributions of >700 individual species of aquatic macrophytes, based on geographical gridcell areas of 10° latitude by 10° longitude. These species are listed alphabetically within 33 dicotyledonous families (from 18 orders of the class Magnoliopsida), which are also listed alphabetically. The families (with order in brackets, and number of macrophyte species given per family) included in this volume are as follows:

Acanthaceae (Lamiales):	15 species
Amaranthaceae (Caryophyllales):	6 species
Apiaceae (Apiales):	66 species
Apocynaceae (Gentianales):	1 species
Araliaceae (Asterales):	8 species
Asteraceae (Apiales):	60 species
Balsaminaceae (Ericales):	1 species
Boraginaceae (Boraginales):	12 species
Brassicaceae (Brassicaceae):	29 species
Burmanniaceae (Dioscoreales):	5 species
Cabombaceae (Nymphaeales):	6 species
Campanulaceae (Asterales):	44 species
Caryophyllaceae (Caryophyllales):	2 species
Ceratophyllaceae (Ceratophyllales):	4 species
Convolvulaceae (Solanales):	4 species
Crassulaceae (Saxifragales):	9 species

Droseraceae (Caryophyllales):	1 species
Elatinaceae (Malpighiales):	29 species
Euphorbiaceae (Malpighiales):	3 species
Fabaceae (Fabales):	24 species
Haloragaceae (Saxifragales):	69 species
Hydatellaceae (Nympheales):	13 species
Hydroleaceae (Solanales):	4 species
Hydrostachyaceae (Cornales):	22 species
Hypericaceae (Malpighiales):	2 species
Lamiaceae (Lamiales):	23 species
Lentibulariaceae (Lamiales):	103 species
Limnanthaceae (Brassicales):	4 species
Linderniaceae (Lamiales):	10 species
Lythraceae (Myrtales):	103 species
Malvaceae (Malvales):	3 species
Melastomataceae (Myrtales):	6 species
Menyanthaceae (Asterales):	69 species

Key to Atlas Species Entries

Macrophytes definition: "vascular aquatic photosynthetic organisms, large enough to see with the naked eye, that actively grow permanently or periodically submerged below, floating on, or up through the water surface" of inland freshwater or brackish waterbodies [28, 104]. The definition emphasizes the importance of the presence of water as an essential part of the population survival strategy (sensu [50]) of freshwater macrophytes. Here we included plants in brackish inland waterbodies as well as freshwater ones. This Atlas encompasses world distributions of more than 3500 species of aquatic macrophytes.

Synonyms: Homotypic synonyms are usually provided, mainly following those allocated to the species by Plants of the World Online [118]. Synonyms labelled "nom. illeg", "nom superfl.", or "not validly publ." are not given here, while infraspecific synonyms are generally avoided. An exhaustive list of synonyms is not given in most cases. Hybrids and infraspecific taxa are not included in the Atlas.

World ecozones: The Atlas primarily covers macrophyte species present in the six terrestrial world ecozones, which primarily provide areas of inland freshwater and brackish waterbodies, potentially open to occupancy by macrophytes. These are the Palaearctic (PAL), Nearctic (NEA), Neotropics (NEO), Orient (OR), Australasia (AUS), and Afrotropics (AFR) (Fig. 1). Records from the Pacific islands of Oceania (OCE), and mainland Antarctica together with the sub-Antarctic islands of the Southern Ocean, such as Kerguelen (collectively ANT), are noted where relevant to world distribution in individual species entries but are excluded from range estimates because of the extremely small areas of inland freshwater or brackish waterbodies present in these ecozones. To provide some examples, figures available for inland water area (comprising rivers, lakes, floodplains, reservoirs, wetlands, and inland brackish systems) (http://data.un.org/Data.aspx?d=FAO&f=itemCode%3a6680) suggest a value of just 1 km^2 for Samoa, 3 km^2 for Tonga, 30 km^2 for New Caledonia, and 91 km^2 for the Solomon Islands. A listing of macrophyte species present in each of the five main island groups of Pacific Oceania, as well as Antarctica, and the sub-Antarctic islands, is provided in Volume 5. For a comprehensive listing of 10 × 10° (latitude × longitude) gridcells covering each of the six main ecozones, see Appendix A7 [104].

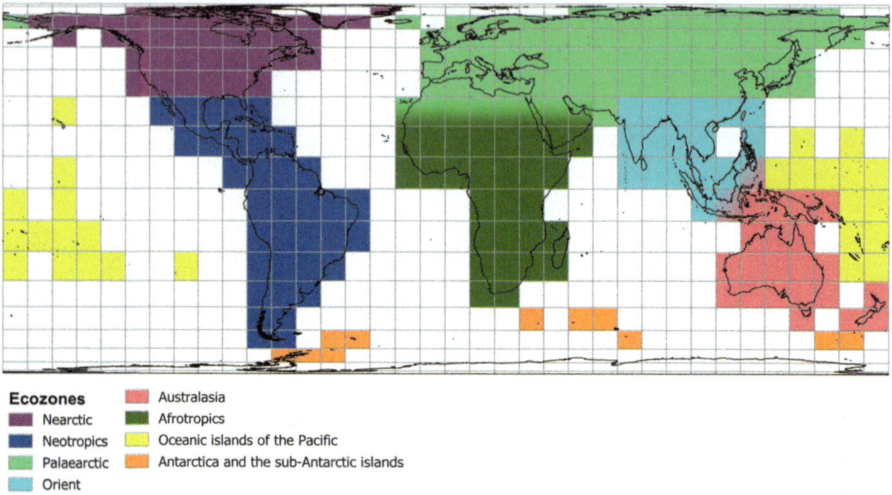

Fig. 1 Map of the world ecozones divided into 10 × 10° gridcells

Ecozone macroregions (for full details, codes, locations and gridcell coverage of each, see [104, 105]):

Palaearctic (PaW Western Palaearctic; PaC Central Palaearctic; PaE Eastern Palaearctic)

ARC	Arctic	PaC S – V; PaE DD – KK
CH	China, Mongolia	PaC L, PaC P; PaE R – T; PaE X – Z
W EUR	Western Europe (incl. Azores)	PaW G, PaW H, PaW I (western half: 10–15°E); PaW M, PaW N (northern half: 45–50°N), PaW O (northern half: 45–50°N)
E EUR	Eastern Europe	PaW I (eastern half: 15–20°E); PaW J – L; PaW P – R; PaW X
N EUR	Northern Europe	PaW A – F
MED	Mediterranean	PaW S – U; PaW N (southern half: 40–45°N); PaW O (southern half: 40–45°N); PaW V, W
NAA	North Africa/ Arabia/Canaries)	PaW Y, Z; PaW AA – EE (all 20–30°N)
UR/ W SIB	Urals/ Western Siberia	PaC A – G; Pac H (western half: 80–85°E)
E SIB	Eastern Siberia	PaC H (eastern half: 85–90°E); PaE A – M
FER/J	Far Eastern Russia/Japan	PaE N – Q; PaE U – W; PaE AA – CC
C ASIA	Khazakhstan/ Iran/ Hindu Kush	PaC I – K; PaC M, N; PaC O (northern half: 35–40°N); PaC Q (20–30°N); PaC R (20–30°N, incl. eastern Oman)

Nearctic (NeaN Northern North America; NeaS Southern North America)

ARC CAN	Arctic-Canadian Shield	NeaN A – K; NeaN P – Z; NeaN AA – EE
W NAM	Western North America	NeaN L – O; NeaS A – C; NeaS I
E NAM	Eastern North America	NeaS D – H; NeaS L – N
SW NAM	Southwestern North America	NeaS J, K

Neotropics (NeoC Central America; NeoN Northern South America; NeoS Southern South America)

MG MX	Mega Mexico I	NeoC A, B; NeoC C (northern half: Florida, 25–30°N); NeoC K
MES AM	Meso America	NeoC C (southern half: Cuba, Yucatan, 20–25 °N); NeoC D – J
NWSA	North Western South America	NeoN C, NeoN D (northern half: 5–10°N); NeoN F (western half: 75–80°W)
AMZ	Amazon	NeoN D (southern half: 0–5°N); NeoN E, NeoN F (eastern half: 70–75°W); NeoN G, H; NeoN L (northern half: 10–15°S); NeoN M
CHA	Chacoan	NeoN I, J; NeoN N; NeoS B (eastern half: 60–65°W), NeoS F, G
PAR	Paranaense	NeoN O; NeoS C, D
AND	Andean	NeoN K; NeoN L (southern half: 15–20°S); NeoS A, NeoS B (western half: 65–70°W); NeoS E; Neo S H – K

Dividing line with Neotropic region, actually runs through gridcells NeoC A and NeoC B in N. Mexico, through gridcell NeaS L (excluding SW corner of gridcell, which comprises northwestern Texas), then following 30° N line of latitude along south edge of eastern gridcell NeaS L, then NeaS M and NeaS N through northern Florida, and N of Bahamas (see [43]). For the purposes of this exercise, the dividing line was taken as corresponding with 30°N line of latitude.

Orient (Or)

IND	Indian Subcontinent	Or A – C; Or G, H; Or L, M
ICH	Indochina	Or D – F; Or I, J
SUP	Sunda Shelf and Philippines	Or K; Or N – S

Australasia (Aus)

NG	New Guinea and associated islands to Sulawesi and Lesser Sunda Is.	Aus E – H
NA	Tropical to subtropical Northern Australia (Northern Territory, The Kimberley, Queensland)	Aus I – K; Aus N, O (northern half of both: 20–25°S)
WA	Western Australia	Aus L, M; Aus Q, R
ESA	Eastern and Southern Australia, including Tasmania	Aus N, O (southern half of both: 25–30°S); Aus P; Aus S – U; Aus W
NZ	New Zealand	Aus V; Aus X, Y

Afrotropics (AfrN northern Afrotropics; AfrS southern Afrotropics)

SAH/AR	Sahara/ Sahel/S. Arabian Peninsula/ Horn of Africa	AfrN I – P; AfrN W
FOR	African Forest Zone	Afr N Q – U; Afr S A, B
EA	East African Savannah and Highlands	Afr N V; Afr S C
ZAM	Zambesian Region	Afr S D – F; AfrS J
SDES	African Southern Deserts	AfrS H, I
CAPE	Cape Region	AfrS L, M
MAD	Madagascar	AfrS G, K

In addition to the six ecozones above, within which macrophyte records were mapped on a gridcell basis, macroregions are broadly defined for the two remaining ecozones where gridcells were not used, owing to the extremely small areas of freshwater present on the Pacific oceanic islands, sub-Antarctic islands, and Antarctic Peninsula, making up the parts of these ecozones where freshwater macrophytes can exist. However, macrophyte records from within these broad macroregions were collated at macroregion scale, so they are described below.

Oceanic islands of the Pacific (Oce)

SW Oce	South Western Oceania (Melanesia, including New Caledonia and Norfolk Is.)	approx. lat. 0–30°S; long. 160–175°E
SC Oce	South Central Oceania (incl. Tuvalu, Fiji, Tonga, Samoa and adjacent islands)	approx. lat. 5–30°S; long. 175°E – 170°W
SE Oce	South Eastern Oceania (Cook Is., Line Is., Polynesia, Pitcairn Is., Easter Is.)	approx. lat. 0–30°S; long. 105–170°W)
Mic Oce	Micronesia (incl. Palau, Mariana Is., Marshall Is., Kiribati)	approx. lat. 0–20°N; long. 130–180°E
Haw Oce	Hawaii	approx. lat. 20–30°N; long. 150–160°W

Antarctica and the sub-Antarctic islands (Ant)

SG Ant	South Georgia/ South Orkney/ South Sandwich/ South Shetland Is.	approx. lat. 55–65°S; long. 30–60°W
PE Ant	Prince Edward Is.	approx. lat. 45–50°S; long. 35–40°E
CKH Ant	Crozet/ Kerguelen/ Heard Is.	approx. lat. 45–55°S; long. 50–75°E
MCA Ant	Macquarie/ Campbell/ Auckland Is.	approx. lat. 50–55°S; long. 155–170°E
Ant P	Antarctic Peninsula	approx. lat. 65–70°S; long. 60–70°W

Range area (grAH km^2): Area of inland freshwater and brackish water potentially open to occupancy by macrophytes, as sum of gridcell aquatic habitat area (grAH: derived from satellite imagery data) values for each species, for whole world, or for individual ecozones. Due to seasonal and year-on-year variations in within-gridcell areas of inland water, these values are only approximations and for that reason are rounded to the nearest hundred km^2 here. Because of this rounding, ecozone totals may not sum exactly to the given world total grAH for each species. See [104, 105] for full details.

Ecozone origin-status: E: ecozone-endemic, i.e., natural occurrence in only a single world ecozone; N: native in >1 world ecozones; I: introduced, naturalised, or invasive in one or more ecozones outside the native range of the species; E(I) endemic in a single ecozone but also present in one or more additional ecozones via human interference vectors (anthropochory). See [80] for details.

World rarity status: VR: very rare: occupancy range <1% total world grAH; R: rare, occupancy range 1– 5% total world grAH; C: common, occupancy range >5% total world grAH. See [81] for details.

Species world distribution maps: For each species entry, the world distribution maps show species occurrence per gridcell within the six primary target world ecozones, on an azimuthal equal-area projection, within 5 range bands of grAH (km^2) occurrence. The Equator (zero° latitude) and Greenwich Meridian of zero° longitude are shown in thicker line size.

Life form(s): In vascular macrophytes, life form relates to the position of plant foliage relative to the water surface [140], usually classified into four main groups (Fig. 2):

I. **Emergent plants (EM)**: Plants normally erect and standing above the water surface, but some species tolerate submergence; all produce aerial reproductive structures.
II. **Floating-leaved rooted plants (FR)**: Plants produce surface-floating leaves that differ in morphology from any submerged leaves present; produce floating or aerial reproductive structures and are rooted, attached to, or closely associated with submerged substrate.

Fig. 2 Life forms of macrophytes. Image copyright: JJ Herrera-Gallegos, with permission

III. **Submerged plants (SU)**: Plants with most or all foliage permanently submerged (some may also produce emergent or floating leaves), and rooted, attached to, or closely associated with submerged substrate, produce floating, aerial, or submerged reproductive structures.

IV. **Free-floating plants (FF)**: Plants not rooted in, attached to, or closely associated with the submerged substrate, floating on or below the surface, produce floating, aerial, or rarely submerged reproductive structures.

Note: In *Genlisea* (Lentibulariaceae), usually with EM or SU life-form or both, the "roots" are actually heavily modified underground leaves which act as traps for subterranean microorganisms, mainly protozoa, which the plant digests. In *Salvinia* (Salviniaceae: FF life form), modified hair-like submerged leaves dangle into the water from the surface-floating fronds and act as roots, taking up nutrients from the water column.

Many macrophyte species may possess more than one life form in some of their populations, or at different points in their life cycle. A question mark against the life form means that possession of this life form is rarely observed or uncertain for the species.

Where relevant, additional information is given about the plant's life form: primarily whether it shows tree, shrub, or subshrub habit (a subshrub is a plant that bears hibernating buds on persistent shoots, sometimes woody, near the ground; a hydrosubshrub usually has these structures underwater); climbing habit; rheophytic habit (direct attachment to rock substrates); epiphytic habit (growing on other plants); or some combination of these (information mainly sourced from Plants of the World Online; POWO [118]. Where such information is not given, the species is an annual/ perennial herbaceous plant.

Ploidy state: H: haploid; D; diploid; P: polyploid; DP: species with haploid or diploid and polyploid races. See [79] for details.

Speciation age (Ma): age of speciation, Ma, millions of years ago, providing an indication of species old enough (speciation age older than end of the Miocene, 5.3 Ma) to have had their world distribution potentially influenced by ancient vicariance factors associated with changes in sea level and tectonic land movement. See [80] for details.

Populations also occur in terrestrial, wetland and/ or riparian habitats: many macrophyte species occurring in inland freshwater or brackish waterbodies also have populations that occur in other habitats, usually but not always associated with watercourses or other inland waters.

Long-distance movement vectors: natural and human-related potential vectors of long-distance propagule movement: BIRD (ectozoochorous or endozoochorous transport, primarily between ecozones, by migratory and wandering bird species); HYDROCHORY (potential long-distance hydrochorous movement of propagules, only noted for transboundary waterways crossing between ecozones, and indicating sending and receiving ecozone); ANTHROPOCHORY (human-interference effects facilitating long-distance movement of propagules within or between ecozones); ?AV (potential ancient vicariance influences on global distribution because the speciation age of the species is old enough (before the end of the Miocene, ≥5.3 Ma) for it to have been potentially affected by tectonic land movement or sea level change). A question mark against a vector means that full details are not established, but there is some evidence, or potential, for the species utilizing the vector concerned. See [80] for details.

Nuisance status and uses: notes on aquatic weed or other nuisance (e.g., toxicity) status of the species; and uses made of it by humans, e.g., medicinal, horticultural, aquarium trade, phytoremediation, ornamental, food, forage).

Endangered status (IUCN): global conservation status of species as given by IUCN Red List of Threatened Species [60]: LC: Least Concern; DD: Data Deficient; NT: Near Threatened; VU: Vulnerable; EN: Endangered; CR: Critically Endangered. Occasionally other sources are also cited if IUCN designation has not been made.

NA: no data available.

Methodology Used for Compilation of Atlas Data

Full details of the methods used in collecting, collating, and analyzing the data used to compile the Atlas species entries are provided by [104, 105] and [79–81]. Here, we summarize the main approaches used in compiling the datasets. For a detailed explanation of the criteria for including or excluding species in the world list of macrophytes, see Appendices A1 and 5 of [104].

Post-1950 world geospatial records were collated for all known vascular aquatic macrophyte species. Native, introduced, naturalized, and invasive records were all included, but not records known to be for plants growing in cultivated locales such as crops and ornamental parks or gardens. Records were primarily extracted from the Global Biodiversity Information Facility world species distribution database [47]. The data were supplemented by information from primary datasets held by the authors, and records from other online geospatial plant species distribution information resources covering large areas of the planet (notably Flora of China, Flora of North America, e-Monocot, CJB African Plant Database, IUCN Red List, www.plantarium.ru, Moscow Digital Herbarium, Checklist of the Flora of the Caucasus, Flora do Brazil, Flora Zambesiaca, and Cátalogo de Angiospermas Acuáticas de México), as well as numerous sources providing more local distributional coverage (for the full list of sources, see Appendix A4 of [104]). The primary dataset was updated regularly from its first publication in 2019 (Appendix A2 of [104]), with the version used to compile Vol. I of the Atlas revised up to April 2023. Records were included from inland freshwater and brackish waterbodies, but not coastal or marine low-salinity aquatic habitats such as coastal salt-marsh channels, and the Upper Baltic Sea, nor highly-saline or highly-alkaline ("soda lake") inland waterbodies such as the Great Salt Lake (Utah, USA), and Lake Bogoria (Kenya). Records were collated from sites located in 238 10 × 10° (latitude × longitude) gridcells, located between 80°N and 60°S, covering the six major global land ecozones which primarily provide macrophyte habitat, PAL, NEA, NEO, OR, AUS, and AFR (see Appendix A3 of [104] for full details of gridcell locations in each ecozone). Detailed records from the two ecozones which only provide very limited areas of freshwater habitat: Antarctica and the Pacific islands of Oceania were excluded, but a summary of the macrophyte floras of these global regions is provided in Vol. 5 of the Atlas. Some species were excluded from the database for other reasons, which are outlined in detail in Appendix A5 of Murphy et al. [104]. In brief these grounds for exclusion included the following:

(i) Inaccurate or impossible records were ignored if other evidence, such as original site location information on voucher specimen sheets, could not be found to correct the error.
(ii) Fossil records and records derived from cultivated locales such as Botanic Gardens.
(iii) Some species were omitted from the study because they lack post-1950 geospatial records (and any corroborative information on more-recent occurrences, from the IUCN Red List or other reliable sources), or because available information about their location(s) was too vague to allow them to be correctly spatially positioned in one or more world gridcells.
(iv) A few macrophyte species were identified with ranges entirely restricted to areas outside the target ecozones. These are species endemic to parts of Oceania (e.g., the Hawaiian Islands) or the sub-Antarctic islands (see Appendix A6 of [104]). These are included in this Atlas in Volume 5.

A single species record, in a single year during the study period, within a gridcell unit was considered the minimum needed for a positive result for a given species in that unit of the planet surface. For the Atlas, in providing an estimate of the range of each species, within gridcells and ecozones, and across the world, the "extent of occurrence" (broadly following the concept used in the IUCN Red List) was calculated as the total area of inland water bodies present in the gridcell(s) in which a given species is present. The gridcell aquatic habitat area (grAH, km^2) occupied by inland waterbodies such as lakes, reservoirs, rivers, and canals (see details in [104]) for gridcells supporting individual species was then summed to provide an overall estimate of range for the species at ecozone and global scales. Due to seasonal and year-on-year variations within gridcell areas of inland water, these range values are only approximations and for that reason are rounded to the nearest hundred km^2 in Atlas species entries. The limitations and advantages of this approach to assessing macrophyte range size are further discussed in [104, 105], and [79–81], but an important point to note here is that the range estimates represent only potential areas of inland water that macrophytes might occupy. In actuality, the area occupied by macrophytes within the grAH of an individual gridcell will almost always be smaller than the value given here, because depth, exposure, and other physico-chemical limitations will usually restrict the area open to macrophyte colonization. The lowest extremes of this approach to measuring macrophyte range size are seen in some desert gridcells, especially in the Sahara, with tiny grAH values, representing lakes and pools present in isolated oases, but probably offering most of these usually shallow areas of open water as potential habitat for aquatic macrophyte growth. At the other extreme, the gridcells containing the Nearctic Great Lakes have extremely high grAH values, but most of this water is unsuitable as macrophyte habitat primarily due to depth of the lakes, and also high wave disturbance impacts on open stretches of lake shore with very large wind and wave fetch, preventing macrophyte growth.

The cartography of the grAH values for each species was produced with QGIS (version 3.16). First, a preliminary map was designed, adding the required layers (continents, gridcells with a sample of grAH data, equator and Greenwich meridian) and assigning them a North Pole Lambert azimuthal equal area projection. The Arctic and Antarctic 10 × 10° gridcells without data at this level of representation were removed to improve map's visualization. The authors then agreed on a final map design, including a blue palette legend with four classes.

Subsequently, grAH data was cleaned and sorted into different spreadsheets, one per subsection in Chapter 3. The "Join" function in QGIS was used to incorporate the grAH data into the gridcells shapefile attribute table, thus preserving the entire species name regardless the number of characters. Then, a sequence of three Phyton scripts was created with the invaluable help of Nikolai Shurupov (see acknowledgements) to read the data spreadsheets, create a list of species' names, find those in the gridcells shapefile attribute table, and plot a map per species with the selected design. The scripts were run within the Python console included in QGIS for each data spreadsheet, and a quality control was then carried out to ensure that the script was reading the right data column when producing the maps, plotting some of them manually. For each species in each ecozone, information was collated on its "origin status," as endemic to a single ecozone, E; native to ≥ one ecozone, N; introduced/invasive (i.e., present in the ecozone via anthropochorous "human interference" transfer vectors acting into the ecozone, I; or endemic to one ecozone but transferred to ≥ one additional ecozone by anthropochorous vectors, E(I) origin-status [80]. For I and E(I) species only, a note was also made of the home ecozone(s) in which the plant is native. We also made a note of the ecozone macroregions (defined in Appendix A2 of Lobato-de Magalhães et al. [80]) and also summarized in the Key to Species Entries in the Atlas) within which each I or E(I) species occurred. Primary sources for assigning origin status included Plants of the World Online [118], and major national floras such as those mentioned above, as well as other publications providing more geographically limited information on macrophyte species ranges and origins (listed in Appendix A1 of [80]). Full details are also given in that Appendix of the approaches used to assess:

(i) Co-occurrence and potential ecozones of origin for N and I species
(ii) Inter-ecozone hydrochory vector status for macrophyte species per gridcell
(iii) Speciation age and Late Quaternary glaciation effects potentially influencing macrophyte world distribution
(iv) Bird vector status for long-distance movement of macrophyte species

We include in the Atlas information on the ploidy state of macrophyte species, where available. Chromosome number change (e.g., polyploidy or whole-genome duplication; dysploidy or change in single chromosome numbers) is a class of common genetic mutations in plant species with complex effects on plant ecological responses and several evolutionary implications [153]. Ploidy state ("cytotype") has been determined, initially by chromosome counts and more recently by techniques such as flow cytometry [124], though variation in DNA content among cytotypes is still unknown for many aquatic plant species, especially in tropical regions [121]. Collation of ploidy information was undertaken from a total of 532 sources (see Appendix A of [79]), commencing with the database underpinning [125], who included >800 angiosperm macrophyte species in their world dataset, and complemented by information derived from Floras, scientific papers, and unpublished compiled databases that review in detail data for chromosome number and ploidy level of each species. Our study covers non-angiosperm vascular species of aquatic clubmosses, horsetails, ferns, and fern-allies (but not non-vascular macrophytes, the macroalgae and bryophytes) as well as angiosperm macrophyte species. A substantial part of the relevant ploidy information was found in local journals, dissertations, books, as well as non-English language publications.

Ploidy state was characterized in three classes, following [162] and [41]:

(i) Haploid/Diploid (H, D), species having solely haploid or diploid (or both) populations
(ii) Polyploid (P), species exhibiting various levels and forms of polyploidy
(iii) Mixed ploidy (DP), species showing "other" ploidy, i.e., with both haploid/diploid, and polyploid populations (or variants showing, for example, agmatoploidy or dysploidy) occurring in different parts of their range

For uses of macrophyte species, we utilized the species name and keywords such as "plant use," "ethnobotany," "medicine," "folk medicine," "bioactive," "phytoremediation," "ornamental" to search into databases such as PubMed, Web of Science, and Google Scholar to include scientific publication (e.g., [14, 18, 61, 68, 72, 87, 96, 160]) as well as common uses mentioned in the web (e.g., aquarium, ornamental, folk medicine). We included each chapter's website links and publication Digital Object Identifier (DOI) links. To build a database of photographs, we collected pictures from research colleagues with permission and on the iNaturalist website [58], considering only materials with CC0 and CC-BY licenses. All photos were confirmed by at least one specialist in aquatic macrophytes.

List of Image Contributors

Alexander Ebert
Andrey Banyshenko
Andrey Efremov
Anh Tuan Dang-Le
Diana Davey
Ehoarn Bidault
Eusebio Roldán Félix
Gary Ervin
Greg Holland
Greg Neill
James Bailey
Jens Christian Schou
Joanne Ling
João Tiago Rocha Tavares
José Marfori
Juan José Herrera Gallegos
Karina Fidanza
Keith A. Ward
Ken Harris
Lindy Fung
Marco Schmidt
Michael Kennedy
Nicholas P. Tippery
Nick Helme
Olga Chernyagina
Paulo Gonçales Carvalho
Riana Fourie
Robert Davis
Santiago Mailhos
Sidinei Magela Thomaz
Steve Maldonado Silvestrini

Subhajit Roy
Tatiana Malchinski
Tian-Chuan Hsu
Timothy Rudman
Yair Ur
Yanghoon Cho
Yoannis Domínguez
Zhizhong Li
Zig Madycki

Many users from iNaturalist, photos with CC0 and CC-BY licenses (authors listed in the photographs for all CC-BY), images taken by these licenses in July–August 2023.

Acknowledgments

We thank the book series editor Max Finlayson for supporting this book idea; Débora Orsini Lobato for her support in building the species utilities' database, particularly for Neotropical species, and reviewing citations and references; all image contributors, particularly Andrey Efremov, who sent us an enormous number of photos for the Atlas, and for requesting from his colleagues to send us images, permission forms, and making pictures available through CC0 and CC-BY licenses in iNaturalist, which made it possible to include many species images in this book. Special thanks to Nikolai Shurupov, PhD student at the University of Alcalá Geography Unit, who helped E. Molina-Navarro construct the program code to create thousands of maps for the Atlas, and Juan José Herrera Gallegos for the beautiful watercolor of macrophyte life forms.

Contents

About the Authors

Tatiana Lobato-de Magalhães is a research professor at the Universidad Autónoma de Querétaro (UAQ), QRO, Mexico, and has specialized in wetland research and education for more than 10 years, with a strong focus on ecology, landscape genetics, and distribution of aquatic plants. Her work includes research projects and teaching across the Americas, including the Amazon and Southern Brazil Highlands. She is an associate editor of an international book series (Wetland Ecology, Management, and Conservation) and scientific journals (Wetlands, Aquatic Botany); International Chapter Chair at the Society of Wetland Scientists (2016–2025); and Lead of the Wetland Environments Specialist Group of the Commission on Ecosystem Management of the IUCN. She published about 20 peer-reviewed scientific articles and the books "Wetlands for Remediation in the Tropics" and "Insights into Sustainable Management of Water and Wetlands in the Sierra Gorda Biosphere Reserve, Mexico." Tatiana is recognized by the National System of Researchers in Mexico and certificated as a Professional Wetland Scientist by PCPSWS. She was awarded the Academic Merit Medal by the UAQ, José Mariano Mociño Medal by the Mexican Society of Botany, Distinguished Foreign Scholar by the International Association for Landscape Ecology, and Research Fellow Wetland Ambassadors by the Society of Wetland Scientists. [https://lobatomagalhaes.weebly.com/]

Kevin Murphy is a senior lecturer (retired) at the University of Glasgow, Scotland. His research emphasizes plant and freshwater ecology, particularly the role of vegetation in the biodiversity support functioning of aquatic ecosystems, and the macroecology of freshwater macrophytes. He has extensive experience in applied ecological work on sustainable development issues in a range of temperate, tropical, and sub-tropical aquatic and wetland habitats (primarily in Europe, Africa, and South America), plus applied studies of vegetation management and eutrophication management in aquatic habitats worldwide. He was involved in a series of studies in South America, concentrating on the ecology of the Rio Paraná, one of the world's largest regulated river systems. After retiring from the University of Glasgow, he set up an informal international research network in 2017 dedicated to advancing macroecological knowledge of freshwater macrophytes. This has to date produced a number of well-received publications collectively authored by network collaborators.

Marinus L. Otte is a professor at North Dakota State University, Fargo, North Dakota, USA. Originally from the Netherlands, Marinus has studied wetlands around the world, particularly in The Netherlands, Ireland, Kyrgyzstan, China, Taiwan, Mexico, South Carolina, North Dakota, and Minnesota. His research has focused on biogeochemistry, plant ecophysiology, and metal tolerance in wetland plants, as well as constructed wetlands for water quality improvement and restoration of wetlands. He has been the editor-in-chief of the scientific journal *WETLANDS* since 2012, which has provided him with the widest overview possible of current wetland science. He has been the recipient of two Chinese Academy of Sciences President's International Fellowship Initiative Awards (2016, 2019) and was awarded a Fulbright Specialist Award in 2021. His recent activities focus on watershed-level issues with water availability and quality and implications for indigenous people.

Eugenio Molina-Navarro is an associate professor at the University of Alcalá, Madrid, Spain, where he belongs to the Research Group in Water, Climate and Environment (ACMA). He teaches in the BSc in Environmental Sciences and in the MSc in Geographic Information Technologies. Eugenio has developed his academic career with a multidisciplinary perspective, focused on answering scientific and management questions regarding water availability and quality, particularly regarding the prospective impacts of climate change. Eugenio's career is also internationally oriented, highlighted by a 3.5-year postdoc at Aarhus University (2015–2019). He has participated in 25 research projects and contracts, leading four of them, and is co-author of more than 60 scientific publications, including 35 WOS-indexed papers. He is also associate editor of the journal *Limnetica*. Besides, Eugenio is co-author of more than 60 conference communications and has peer-reviewed more than 70 manuscripts. He has supervised 17 BSc and 3 MSc theses and is supervising two PhD theses in progress.

Chapter 1
Aquatic Macrophytes and Wetlands

Abstract This chapter provides a definition of the wetland types where the aquatic macrophytes described in this Atlas occur. There are also wetlands where plant species occur that are not considered aquatic macrophytes within the context of this Atlas, e.g., intertidal estuarine and marine wetlands. The chapter also mentions some of the main human interference impacts on macrophyte diversity in different wetland types.

Keywords Aquatic macrophyte · Wetland classification · Wetland types

The definition used in this Atlas to describe aquatic macrophytes is "aquatic photosynthetic organisms, large enough to see with the naked eye, that actively grow permanently or periodically submerged below, floating on, or up through the water surface of inland freshwater or brackish waterbodies" [28, 104]. Although there are non-vascular plants, such as charophytes, that meet this definition we do not include them in the Atlas, mainly because of a lack of accurate distributional data for these species at global scale. Aquatic macrophytes occur in wetlands. Wetlands in turn are defined as "areas of marsh, fen, peatland or water, whether natural or artificial, permanent or temporary, with water that is static or flowing, fresh, brackish or salt, including areas of marine water the depth of which at low tide does not exceed six metres" [120]. This means that all aquatic macrophytes described in this Atlas occur in wetlands, but there are also wetlands where plant species occur that are not considered aquatic macrophytes within the context of this Atlas. That relationship deserves further consideration.

Classification of wetlands used around the world goes back to [37]; see also [44] (Fig. 1.1). It recognizes five major wetland systems: marine (seas and oceans), estuarine (for example, river deltas or mangroves; Fig. 1.2), riverine (rivers; Fig. 1.3), lacustrine (lakes), and palustrine (for example, marshes; Fig. 1.4 and swamps; Fig. 1.5). By the definition of aquatic macrophytes used in this Atlas, all plant species that occur solely in marine systems are excluded. The systems are subdivided into subsystems, for example, limnetic and littoral within the lacustrine systems.

T. Lobato-de Magalhães et al., *World Atlas of Freshwater Macrophytes*, Wetlands: Ecology, Conservation and Management 10,
https://doi.org/10.1007/978-3-031-52749-4_1

Wetland and deep-water habitats classification

System	Subsystem	Class
Marine	Subtidal	Rock bottom, unconsolidated bottom, aquatic bed, reef
	Intertidal	Aquatic bed, reef, rocky shore, unconsolidated shore
Estuarine	Subtidal	Rock bottom, unconsolidated bottom, aquatic bed, reef
	Intertidal	Aquatic bed, reef, streambed, rocky shore, unconsolidated shore, emergent, scrub-shrub, forested
Riverine	Tidal	Rock bottom, unconsolidated bottom, streambed, rocky shore, unconsolidated shore, aquatic bed, emergent
	Lower perennial	Unconsolidated bottom, rocky shore, unconsolidated shore, aquatic bed, emergent
	Up perennial	Rock bottom, unconsolidated bottom, rocky shore, unconsolidated shore, aquatic bed, emergent
	Intermittent	Streambed
	Unknown perennial	Unconsolidated bottom
Lacustrine	Limnetic	Rock bottom, unconsolidated bottom, aquatic bed
	Littoral	Rock bottom, unconsolidated bottom, rocky shore, unconsolidated shore, aquatic bed, emergent
Palustrine	--	Rock bottom, unconsolidated bottom, aquatic bed, unconsolidated shore, moss-lichen, emergent, scrub-shrub, forested
Constructed		reservoirs, ponds, canals, drainage channels, treatment wetland, educational/recreational wetland

Not included   Maybe included   Probably included

Fig. 1.1 Wetlands and deep water habitat classification. Modified from Federal Geographic Data Committee [44]

Fig. 1.2 Estuarine wetlands, mangrove on Preguiças River near Lençois Maranhenses, Maranhão State (Brazil). Image copyright: T. Lobato-de Magalhães

Fig. 1.3 Riverine wetland, Amazon River floodplains, Santarém, Pará State (Brazil). Image copyright: T. Lobato-de Magalhães

Fig. 1.4 Emergent palustrine wetland, temporary marsh, Huimilpan, Querétero State (Mexico). Image copyright: T. Lobato-de Magalhães

Fig. 1.5 Forested palustrine wetland, swamp, Camécuaro, Michoacán State (Mexico). Image copyright: T. Lobato-de Magalhães

The estuarine system is subdivided into subtidal and intertidal. All plant species limited to subtidal subsystems are again excluded from this Atlas, but species from some wetlands in the intertidal subsystem are included. Subsystems are then divided into classes based on their substrates and local hydrology. Plant species from wetlands in the intertidal subsystem classes included in this Atlas are aquatic beds and emergent wetlands, but others are not. Another further division is the distinction between habitats within systems, for example, between areas that are flooded only occasionally (temporarily flooded), seasonally, semi-permanent, or permanently flooded only in the deepest parts. For a more in-depth study of wetlands, see Mitsch et al. [93]. In addition, wetlands are constructed for many purposes (e.g., reservoirs, ponds, canals, drainage channels: Murphy and Eaton [100], Ali et al. [8], Sabbatini and Murphy [134], Bini et al. [12], Bottino et al. [16], Molina-Navarro et al. [94]; Figs. 1.6 and 1.7), including treatment wetlands, sponge cities and 'wetland parks' ([77]; Fig. 1.8). Constructed wetlands can mimic any type of wetland but may also be designed to look unlike anything found in nature. For example, constructed wetlands can be designed to vertically force water through the system so that contact with the substrate is optimized to remove pollutants from water. Regardless, constructed wetlands contribute to macrophyte diversity and distribution and compensate for some wetland losses in many parts of the world.

Fig. 1.6 Itaipú reservoir, Paraná River (Brazil). Image copyright: S. M. Thomaz, with permission

Fig. 1.7 Pareja Limno-reservoir (Spain). Image copyright: E. Molina-Navarro

Fig. 1.8 Qionghai Luojishan National Scenery Spot wetland park in Xichang, Sichuan (P.R. China). Image copyright: M.L Otte

Human activities are typically centred near wetlands [111]. Usually, shallow wetlands are the locations where crops can be grown during periods of drought, particularly in temporary and seasonally flooded wetlands (e.g., Nile and Amazon river floodplains), especially in warmer parts of the world. This means that the impacts of conversion of wetlands to agricultural land have been greatest in shallow wetlands [11, 155] and that the distributions of aquatic macrophytes of temporary wetlands have been impacted by human activities more than those of permanent wetlands [20]. Also, small and temporary wetlands (e.g., vernal pools and prairie potholes in North America) are particularly important to maintain macrophyte diversity and landscape connectivity ([21, 78]). Yet, that does not mean that human activities do not impact species of deep, permanent wetlands. Closed-basin wetlands are often particularly prone to increasing eutrophication (and sediment accumulation, leading to loss of open water area: a good example being Lake Chad in West Africa), particularly during recent decades [167, 168]. Some deeper wetlands can accumulate salts, mainly in hotter climates, producing, for example, soda and saline lakes e.g., in East Africa and Australia). However, there are many examples where deep lakes (for example, almost every one of the deep Scottish lochs, e.g., Loch Lomond, Murphy et al. [102]; and Lake Tanganyika on the Tanzania-Congo border) are oligotrophic and do not accumulate salts. On the other hand, there are also shallow lakes which certainly accumulate salts (e.g., Lake Naivasha in Kenya, Lake Cuitzeo in Mexico).

It is probably an accurate statement that overall human activities are diminishing the areas of habitats suitable for colonisation by aquatic macrophytes in wetlands around the world. On the scale of the gridcells used in this Atlas (ten degrees longitude by ten degrees latitude), the global distributions of aquatic macrophytes may have stayed the same over the past century. However, the frequency of occurrences within each gridcell will likely have diminished for many species. Most macrophyte species that show decreased range are rare endemics, vulnerable to habitat loss and degradation. On the other hand, introduced macrophyte species (~30% of global macrophyte species) probably increased their gridcell occurrence across the world in the past 100 years (some of them, the aggressively invasive ones, quite substantially, but some non-invasive species. e.g., *Erythranthe guttata*, have also shown major range extensions). The lack of older data for the occurrence of macrophyte species in many regions of the planet makes it challenging to go into more detail for many species. It is hoped that this Atlas will encourage further surveys of aquatic macrophytes within gridcells worldwide.

Chapter 2
Global Diversity and Distribution of Aquatic Macrophytes

Abstract This chapter describes the global diversity and distribution of 3543 species within the 453 genera and 92 plant families that contain vascular macrophytes across the eight world ecozones in which these plants are present. The chapter highlights 12 high-diversity families (each having >100 macrophyte species), and discusses range distribution (broad, intermediate, narrow and very narrow world ranges), latitudinal distribution and ecological boundaries of macrophyte distribution. Small global ranges are the norm for most macrophyte species. Macrophyte diversity is significantly related to latitude, showing a strong Rapoport Effect, though less so with other physical geographical drivers.

Keywords Aquatic macrophyte · Biodiversity · Rapoport effect · Global distribution

2.1 World Macrophyte Species Pool: Family Composition and Species Distributions

Geospatial-position records were initially collated by Murphy et al. [104] for 3457 species within 456 genera, and 93 families of macrophytes. Those numbers are now updated, to 2023, as 3541 species within 453 genera, and 92 families of vascular macrophytes. Monocotyledonous plants (Liliopsida) form the largest set of species within the world vascular aquatic flora. The current taxonomic breakdown is club-mosses and horsetails: six species (0.17% of total); ferns and fern-allies: 214 species (6%); conifers: two species (0.06%); dicots: 1520 species (44%); and monocots: 1757 species (50%).

The world macrophyte species pool is dominated by 12 high-diversity families, each having >100 macrophyte species. Except for Isoetaceae, these families all include >1 genera containing aquatic macrophytes (Table 2.1). Cyperaceae is the family with the greatest number of macrophyte species worldwide, though many of these are multiple-habitat species, which also occur in non-aquatic conditions.

T. Lobato-de Magalhães et al., *World Atlas of Freshwater Macrophytes*, Wetlands: Ecology, Conservation and Management 10,
https://doi.org/10.1007/978-3-031-52749-4_2

Table 2.1 Numbers of genera and species per family for macrophyte families with ≥100 macrophyte species worldwide (excluding Oceania and Antarctica endemic species) and per ecozone

Family	Genera per family	Species per family	Ecozone: Species/family					
	–	–	Nearctic	Neotropics	Australasia	Palaearctic	Afrotropics	Orient
Poaceae	70	237	91	145	78	87	81	90
Podostemaceae	53	367	1	143	2	7	111	108
Cyperaceae	35	554	196	323	156	218	163	180
Araceae	25	159	27	61	22	28	21	89
Alismataceae	18	107	36	67	22	31	15	24
Plantaginaceae	15	144	41	54	34	48	39	39
Hydrocharitaceae	13	112	22	26	32	40	45	47
Lythraceae	8	103	18	38	16	27	40	32
Eriocaulaceae	7	135	9	77	10	13	21	32
Potamogetonaceae	6	113	38	33	28	57	19	32
Lentibulariaceae	2	103	16	49	39	9	21	17
Isoetaceae	1	130	30	56	11	25	14	7

Comparing between the six world ecozones, which primarily provide inland waterbody habitats suitable for macrophyte growth, the Neotropics (NEO) has the greatest number of species per family for nine families (Cyperaceae, Poaceae, Podostemaceae, Plantaginaceae, Eriocaulaceae, Araceae, Lentibulariaceae, Isoetaceae and Alismataceae), while the Afrotropics (AFR: Lythraceae), Orient (OR: Hydrocharitaceae), and Palaearctic (PAL: Potamogetonaceae) have one apiece. The Nearctic (NEA) and Australasia (AUS) have generally lower species diversity within these 12 families. With one exception the high-diversity families all have moderate to high numbers of species occurring in every ecozone. However, Podostemaceae only has large numbers of species present in sub-tropical to tropical areas, especially in the Neotropics, and to a lesser extent in the Afrotropics and Orient. At least one species of Podostemaceae occurs in all three of the other world ecozones that primarily support macrophyte growth (Nearctic, Palaearctic and Australasia), while there is one report of an unnamed species of this family from New Caledonia in the Pacific Oceania ecozone. Unsurprisingly, the family is completely absent from the Antarctic ecozone. In the Nearctic this family is represented solely by *Podostemum ceratophyllum* Michx., which has a restricted distribution in the eastern and southern USA. Podostemaceae species are slightly better represented in the Palaearctic and Australasia.

At the other extreme, 16 families are represented in the dataset by single macrophyte species. These are: Lycopodiaceae (*Lycopodiella inundata* (L.) Holub), Polypodiaceae (*Microsorum pteropus* (Blume) Copel.), Podocarpaceae (*Retrophyllum minus* (Carrière) C.N.Page), Cupressaceae (*Taxodium distichum* (L.) Rich.), Philydraceae (*Philydrum lanuginosum* Banks and Sol. ex Gaertn.), Rapateaceae (*Spathanthus bicolor* Ducke), Restionaceae (*Platycaulos mahonii* (N.E.Br.) H.P.Linder and C.R.Hardy), Cannaceae (*Canna glauca* L.), Zingiberaceae (*Hedychium coronarium* J.Koenig), Droseraceae (*Aldrovanda vesiculosa* L.), Portulacaceae (*Neopaxia australasica* (Hook. f.) Ö. Nilsson), Balsaminaceae (*Hydrocera triflora* (L.) Wight and Arn.), Apocynaceae (*Oxystelma esculentum* (L. f.) Sm.), Pedaliaceae (*Trapella sinensis* Oliv.), Rosaceae (*Comarum palustre* L.), and Molluginaceae (*Corrigiola litoralis* L.).

Regarding species range distributions, there are substantial differences between the proportions of the total world macrophyte species pool with broad (very common), intermediate (common), narrow (rare) and very narrow (very rare) world ranges.

- 55 species (2% of total) are very common at world scale (broad range: >50% of world total gridcell aquatic habitat (grAH) area within the six main world ecozones included in the study);
- 1159 species (33%) have common occurrence (intermediate range: 5–50% of world grAH area); and
- 1092 species (31%) are classed as rare at world scale (narrow range-size, 1–5% of world grAH); while 1235 species (34.9%) are very rare (very narrow range: present in <1% of total world grAH).

The total range areas of at least 1400 species (mostly with broad or intermediate ranges) which have macrophyte populations are likely to be inflated, because these plants also occur in wetland, riparian or terrestrial environments (and these populations cannot be separated out from their world total range areas on currently available information). Examples include *Ranunculus repens* L., *Cyperus rotundus* L., *Juncus effusus* L. and *Echinochloa colona* (L.) Link. A similar range-inflation effect is likely for macrophyte species which are also grown as food crops, or otherwise cultivated, e.g., rice: *Oryza sativa* L. (Poaceae). Unfortunately, it is not feasible to check every record given by GBIF [47] and the other sources used in the world distribution study for information on habitat or cultivation status (even if provided, which often is not the case, from the sub-samples of records examined) of these common to very common species. There are simply too many records to make this possible: in July 2018, for example, GBIF held 365,622 occurrence records for *O. sativa* alone. This point should be borne in mind when considering the range data calculated here for such macrophyte species: their true aquatic (and non-cultivated, where relevant) ranges are certainly overestimated. Having said that, obligate macrophytes are predominant among the ten broadest-range macrophyte species. In order of global range size these are *Stuckenia pectinata* (L.) Börner (range area 71.6% of total world grAH), *Ceratophyllum demersum* L. (71.1%), *Potamogeton pusillus* L. (68.9%), *Rorippa palustris* (L.) Besser (68.4%), *Echinochloa crus-galli* (L.) P.Beauv. (67.9%), *Persicaria lapathifolia* (L.) Delarbre, *Phragmites australis* (Cav.) Trin. ex Steud., *Hippuris vulgaris* L. (65.9%) L. (65.7%), *Juncus bufonius* L. (65.7%) and *Ranunculus trichophyllus* Chaix ex Vill. (65.2%). These plants either have true world-cosmopolitan distributions (e.g, *C. demersum*), or they tend to have distributions centred on northern temperate latitudes (e.g., *H. vulgaris*).

At the other extreme, 121 species have tiny world ranges, individually <0.1% of world grAH, and usually limited to a single 10 × 10° gridcell. The families to which these extreme narrow-range species mainly belong are Araceae (19 species: all either *Cryptocoryne* or *Lagenandra*), Aponogetonaceae (12 species), Menyanthaceae (11 species), Isoetaceae (eight species), and Podostemaceae and Haloragaceae (six apiece). Of these extreme narrow-range ecozone-endemic macrophyte species 19 have world range <1000 km^2 grAH (0.03% of total world grAH). No fewer than 14 of these are endemic to the Cape macroregion of AFR, a well-known global hotspot for plant diversity. The narrowest ranges are seen for *Marsilea latzii* (0.004%: around 100 km^2, in the North Australia macroregion of AUS), *Noveloa longifolia* and *Carex michoacana* (both 0.017%: around 500 km^2, in the Meso America macroregion of NEO), and *Utricularia helix* (0.02%: around 600 km^2, in the Western Australia macroregion of AUS.

It should be noted that range size as measured by total area of inland fresh and brackish water present within the gridcell(s) in which a species occurs (i.e., grAH) does not represent the actual area of occupancy (AOO: [104], following the definition used by IUCN) for the species, which will always be smaller, sometimes considerably so, since not all parts of most inland waterbodies are actually open to macrophyte colonization (due to factors such as depth, exposure, substrate suitability, pollution etc.). The macrophyte species with probably the smallest known AOO is *Polypleurum longicaule* M.Kato, just 100 m^2, with its only known population located in a single waterfall in northeastern Thailand (www.iucnredlist.org/species/194789/8901730).

2.2 Global Macrophyte Latitudinal Distribution

Macrophyte total species (α-diversity) per gridcell shows a noisy but significant non-linear relationship with latitude, generally following the Rapoport effect [105] (Fig. 2.1a). Maximum values of number of species occur in latitude band 10–30° absolute: i.e., tropical to subtropical conditions, both North and South of the Equator, dropping off slightly in the Tropics closest to the Equator, and steadily declining through the temperate zone in both hemispheres, sub-Arctic and sub-Antarctic zones, with lowest values seen in the Arctic/ Antarctic. A rather similar pattern is seen for the γ-diversity-latitude curve (Fig. 2.1b), based on total number of macrophyte species present in each 10° latitude band ("species pool per latitude band") from the Equator to 80° absolute (within the six target ecozones.

A previous study showed evidence that latitude and landscape features (inland waterbody area, total land cover) are associated with gridcell diversity more than altitude [104] (Fig. 2.2). There was no great change across the lower latitudes, with inland waterbody area and total land cover more influential below 60° latitude. For richness patterns inland waterbody area is important at the low end of its gradient whereas total land cover is influential across the entire length of the gradient.

2.3 Distributions and Diversity Within Six Major Ecozones of the World

Most macrophyte species are limited to a single ecozone, with only a few occurring in all six ecozones (Fig. 2.3), but this pattern is less clear at genus level. Although, again, most macrophyte genera show limited occurrence, in one to four ecozones, there are a substantial number which have broad ecozone distribution (with 100 genera occurring in all six ecozones), a fact which has long been known (e.g., [28, 48]). However, it remains the case that more macrophyte genera (62%) have "narrow" distributions (occurring in no more than half the world's six ecozones that primarily support macrophyte populations) than those with broader distributions.

Many macrophyte species (excluding those few which exhibit broad-range worldwide distributions and are usually present in all ecozones: for example, *Stuckenia pectinata*) show global patterns of distribution which tend to be restricted to a single area of the planet. In size these range from the tight distributions seen for very narrow-range species such as *Marsilea capensis* A. Br. with a range area of

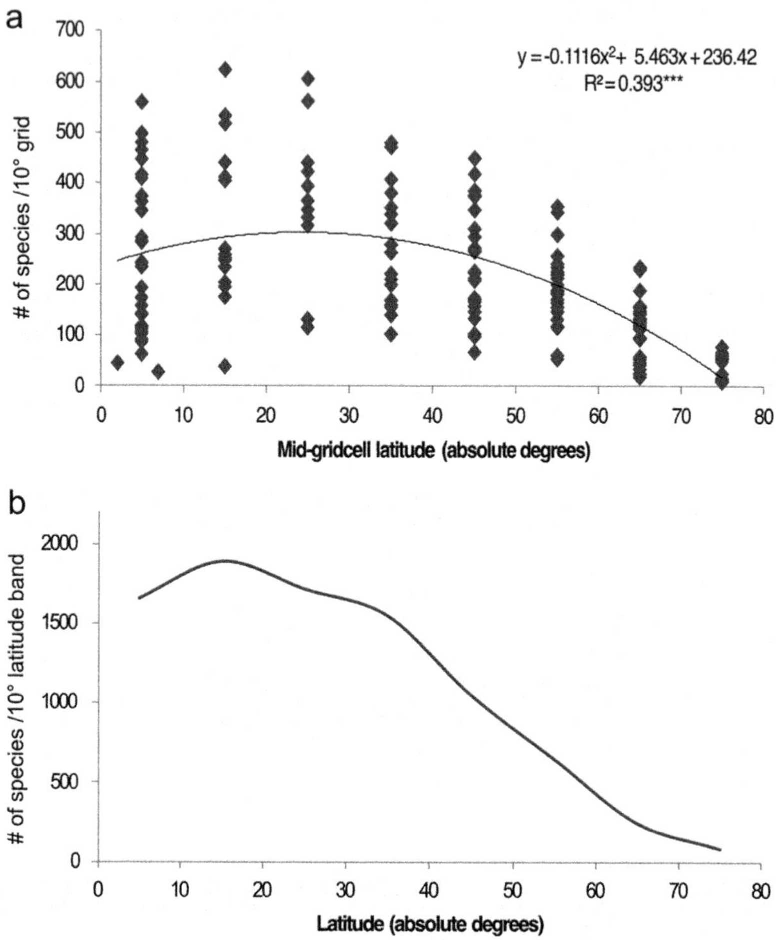

Fig. 2.1 (**a**) Global latitudinal distribution of macrophyte α-diversity: measured as S_T per gridcell (non-arid gridcells only), (**b**) Global latitudinal distribution of macrophyte γ- diversity (as total number of species present per 10° latitude band)

0.77% of total world grAH), all in southern Africa; to intermediate-range distributions spanning more than one ecozone, but still clearly occupying a single area of the planet's surface. A good example of a plant with the latter distributional pattern is the narrow-range species *Bolbitis heteroclita* (Presl.) Ching (5.11% world grAH) which has records scattered across Australasia and the Orient (as well as some records from islands of Oceania located close to the Pacific margin of these two ecozones).

In contrast some species show more disjunct distributions. Perhaps the clearest examples are plants which occur solely in the Neotropics and Afrotropics, with a substantial trans-oceanic distance separating them across the Atlantic, e.g.,

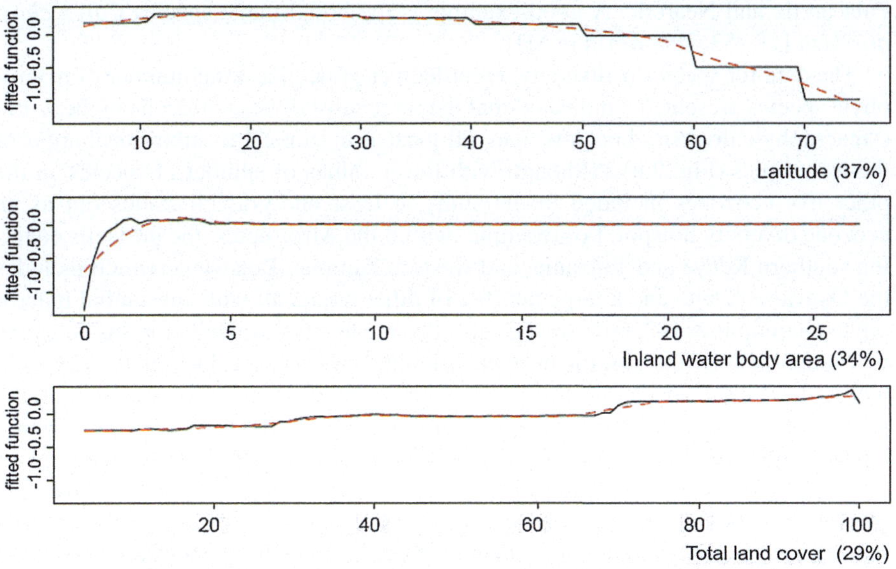

Fig. 2.2 Boosted Regression Tree (BRT) plots for total of species predicted from three spatio-environmental physical variables, with % variance explained across the range of each variable. (Modified from Ref. [104])

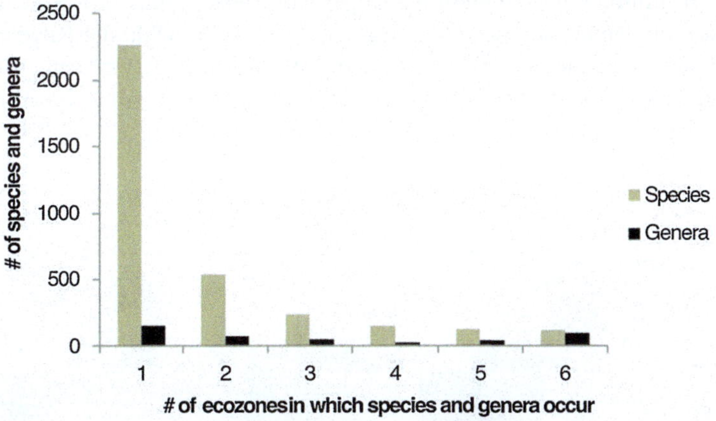

Fig. 2.3 Species and genera by ecozone occurrence

Phyllanthus hyssopifolioides Kunth (7%, world grAH) which has the bulk of its records in the Neotropics but is also introduced in West Africa.

A third principal type of global distribution pattern is seen for northern hemisphere high-latitude species which have circumboreal distributions (with a range of near-continuous to more disjunct patterns evident) around the planet through the

Palaearctic and Nearctic. A good example is the broad-range species *Carex chordorrhiza* L.f. (53.64% world grAH).

The data for species α-diversity, recorded per gridcell as total number of macrophyte species present, S_T, indicated that the three most diverse gridcells in the world ("macrophyte diversity hotspots") are all located in tropical to subtropical areas of the Neotropics (Fig. 2.4). Although with lower values of number of species, in the other five ecozones included in the study at least one gridcell stands out as an ecozone diversity hotspot. For example, within the Afrotropics, the gridcells covering southern Kenya and Tanzania, and western Zambia / Botswana (which includes the Okavango Delta and a large number of other rivers, all with substantial macrophyte communities [42, 66]) are the most diverse hotspot gridcells for this ecozone.

Comparison of the significance of differences in mean values of the diversity measure total of species (Table 2.2), across the ecozones, confirmed that the Neotropics has significantly higher gridcell diversity than all other ecozones, which did not differ significantly amongst themselves. The richest gridcell on Earth for macrophyte diversity covers Brasilia/Goiás (Brazil: 10–20°S; 40–50°W), with 628 species. This lies within the Brasilian cerrado area, known to be one of the planet's major biodiversity hotspots for a range of different biota [69]. The Neotropics also had, by a large margin, the richest ecozone total species γ-diversity ("regional species pool") of the six world ecozones examined: 1615 species. Next are the Palaearctic (1031 species) and Orient (991 species). However, if we standardise γ-diversity by total gridcell area per ecozone (to give values as number of species present per million km^2) the picture changes somewhat (Table 2.2). Now the Orient comes out top, Neotropics second and Australasia third, while the Palaearctic is in the last place on this metric. When the exercise is repeated, considering total inland fresh and brackish waterbody area per ecozone, it is interesting to see how well

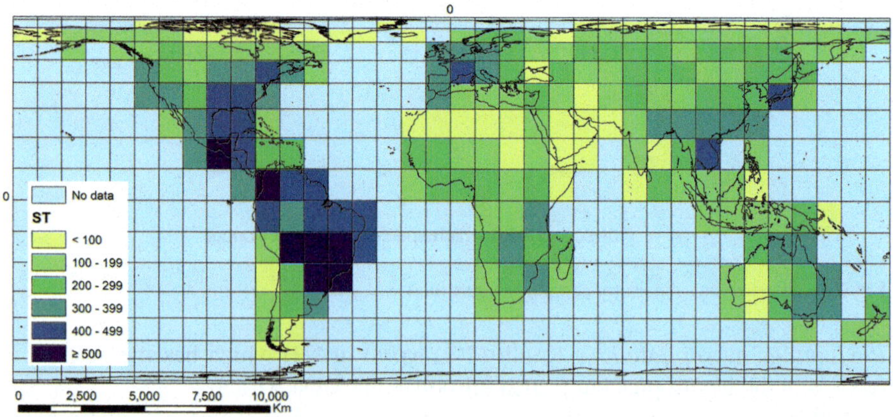

Fig. 2.4 World distribution of gridcell diversity hotspots, measured as S_T: total number of macrophyte species per gridcell (source: [104]). Position of Greenwich Meridian (0° longitude) and Equator (0° latitude) marked as "0". Scale bar is indicative only, referring to average longitudinal distance on the map

Table 2.2 World macrophyte biodiversity hotspots: total ecozone species pool (γ-diversity; γ-diversity corrected for ecozone total gridcell area per M km^2, γ-diversity corrected for ecozone total inland waterbody (grAH) area per M km^2); mean (± SE) gridcell); all-species α-diversity (mean values sharing a superscript letter label in common do not differ significantly: ANOVA with Tukey's mean separation test); and richest "hotspot" 10 × 10° (latitude × longitude) gridcell per ecozone measured as S_T (total number of macrophyte species recorded per gridcell). Values shown are from the updated 2023 database held by the authors, and hence differ from the values given in the equivalent table in [104]

	Neotropics	Nearctic	Palaearctic	Afrotropics	Orient	Australasia
Ecozone total gridcell area (M km^2)	36.9628	29.4476	63.7281	37.1328	21.5414	23.4116
Ecozone total inland waterbody area (grAH: M $km^{2)}$	0.4142	1.0549	1.0418	0.3543	0.1704	0.1061
Ecozone total species pool (γ-diversity)	1615	930	1034	915	991	824
Ecozone γ-diversity per M km^2 total gridcell area	43.7	31.6	16.2	24.64	46.0	35.2
Ecozone γ-diversity per M km^2 total inland waterbody area (grAH: M km^2)	3899	882	993	2582	5815	7766
Ecozone mean (± SE) α-diversity per gridcell (S_T)	348 (± 29.1)^a	169 (± 21.2)^b	146 (± 10.2)^b	128 (± 17.8)^b	211 (± 28.8)^b	191 (± 21.2)^b
Gridcell location of richest diversity hotspot (α-diversity per gridcell, S_T)	625 Brasilia/ Goiás: 10–20°S; 40–50°W	479 Georgia/ Kentucky/ Florida panhandle: 30–40°N; 80–90°W	418 France: 40–50°N; 0–10°E	347 Southern Kenya/ Tanzania: 0–10°S; 30–40°E	410 Indochina: 10–20°N; 100–110°E	377 Northern Queensland (Cairns/ Cape York): 10–20°S; 140–150°E

Australasia does. Despite having a rather small area of inland waters open to macrophyte colonisation (the dataset excludes inland saline lakes, of which there are many in Australia, because these do not support macrophyte growth) AUS still has over 800 macrophyte species, although a high proportion of these are introduced [80].

Most of the gridcells at the low end of the macrophyte diversity range, with $S_T \leq 15$ species per gridcell ("diversity coldspots") occur in the Sahara and Arabian Deserts, immediately north or south of the Tropic of Cancer, which divides latitude band 20–30°N into the Palaearctic and Afrotropic ecozones. The other two macrophyte diversity coldspots with very low S_T (each with 9 species per gridcell) are both Arctic: comprising the gridcells covering the Yamal Peninsula in Siberia, Russia, and the Boothia Peninsula of northern Nunavut, Canada.

2.4 Boundaries of Macrophyte Distribution

The limits of global distributions of species are determined by the upper and lower limits of conditions favourable for their survival, growth and reproduction. There may be many conditions that must be met for any one species to exist in any given place, but on a global scale, extremes of temperature are a dominant factor.

2.5 Boundaries Determined by Temperature

Freshwater macrophytes, like all other vascular plants, have a latitudinal distribution from the Arctic, with a limit of about 80°N (e.g., *Ranunculus hyperboreus* occurs as far north as 82.5°N, on Ellesmere Island in northern Canada), to the Antarctic, down to about 70°S on the Antarctic Peninsula, where the presence of water is limited by the length of the season when liquid water exists. However, the number of species found at these latitudinal extremes in the sub-Arctic and Arctic (here taken as comprising gridcell records north of 60°N), compared with Antarctica and the sub-Antarctic islands show huge differences. There is just one macrophyte species (*Deschampsia antarctica*) present in the continent of Antarctica, south of 60°S (and none at all south of 70°S), rising to 17 if the sub-Antarctic islands (some of which, e.g., Kerguelen, actually lie at latitudes lower than 60°S) are included in the ecozone. However, at least five of these species were introduced by human activities, bringing them from ecozones outside ANT. In contrast, the sub-Arctic and Arctic zone north of 60°N supports 387 macrophyte species, with 96 of those also occurring north of latitude 70°N. This difference between the high latitude parts of the planet is explained by the isolation of Antarctica from the other continents; essentially land surrounded by ocean, whereas the Arctic is an ocean surrounded by land, and part of it (especially northern Scandinavia) experiences a warming influence produced by the North Atlantic Drift. There is much land between 40 and 75°N, providing a large area of inland waterbodies for macrophytes to live in, and for pathways for dispersal, but very little between 40 and 70°S. That isolation, dating back to the Cretaceous [71] also means that migration of species between the nearest land to the north down to Antarctica, other than through human intervention, is very unlikely, though a few macrophyte species have managed to make the trip using natural vectors [80].

There are, however, also differences in macrophyte species richness between the regions north of 70°N when comparing between PAL (the northern Canadian islands and the ice-free coastal part of eastern Greenland) and NEA (northern Siberia and the Russian Arctic islands, between 50 and 160°E). All 96 species of the high Arctic macrophyte flora occur in the PAL section of this Arctic region, but only 26 in the NEA section. The difference is almost certainly related to the fact that parts of the PAL section remained uncovered by ice during the Last Glacial Maximum, hence offering a potential glacial refuge for high latitude macrophyte species, whereas this was much less the case in Arctic NEA ([105, 136]; https://www.qpg.geog.cam.ac.uk/lgmextent.html).

It should be noted that, unlike the situation regarding I-origin species in ANT, anthropochory has played only a very limited role in influencing the high latitude macrophyte flora of PAL and NEA, with <2% of the 387 species of the sub-Arctic to Arctic regions of these ecozones being of I-origin from ecozones other than PAL or NEA. One example is the presence of Neotropical *Elodea densa* (Planch.) Casp. (Hydrocharitaceae) in thermal pools in arctic Iceland, which is thought to originate either from deliberate introduction, or via propagule escape from hobbyist aquaria [163].

Migration of species may also be limited due to the presence of mountain ranges, not so much as a physical barrier, but because mountains are colder at higher altitudes. Löffler [82] describes how the predominant orientation of mountains in particular plays a role in migration and evolution of species to explain why animal species richness in the Americas is higher than in Europe and Asia north of the Himalaya. This argument was also used by Huntley [57] regarding differences in richness of species of trees. However, this does not appear to be the case for aquatic macrophytes, many of which rely on migratory bird vectors for long-distance dispersal, and hence are largely unaffected by mountain range orientation (e.g., [35, 36, 80]). North of 30 °N, 1031 macrophyte species occur in Europe and northern Asia, of which 318 are endemic to that region, whereas in North America there are 930 species with 196 endemics. We can only speculate about how changes in climate will affect distributions of aquatic macrophytes, but it is evident that vertical distributions of a range of other organisms in mountain ranges are changing rapidly [59].

2.6 Boundaries Expanded by Human Activities

Humans have willingly and unwillingly dispersed plants as long as people have walked the earth [80]. Willingly, primarily for the purpose of agriculture, unwillingly due to the presence of unwanted species among the seeds of crops [23]. Humans move aquatic plants across the globe via various modes of transportation, most recently by Internet-faciitated trade in aquarium plants [55, 86, 113]. For, example about 13% (at least 100 species: see Chapter 3) of the dicot macrophyte species covered in this Volume are available for sale through the aquarium trade, over the Internet, with few or no barriers to international movement via the mail (though the sale of some species known to cause invasive weed problems is illegal in certain countries, an example being *Crassula helmsii* in the UK). Escape of propagules from aquarium systems owned by hobbyists is now a major vector of long-distance macrophyte movement out of their native ranges. Global dispersal of freshwater macrophytes by human activities is discussed by [80] and will be further elaborated on in Volume 4 of this Atlas.

2.7 Other Factors Influencing Global Macrophyte Distribution and Diversity

Prior to the work of Murphy et al. [104], on which this chapter is broadly based (species numbers have been updated in the light of revisions to the underlying database, between 2019 and 2023, but the analyses summarised here used the 2019 version of the dataset, and are as published by Murphy et al. [104]), aquatic macrophytes were generally assumed to show broad world distributional ranges (e.g., [74, 75, 137]). However, these studies were based on analysis of small subsets of the global macrophyte species pool. Santamaría's study, for example, was based on a sample of <1% of the total number of macrophyte species identified by Chambers et al. [28]. There is now good evidence that the majority (about 66%) of vascular macrophyte species have small global ranges (defining "small" as present in ≤5% of the estimated total area of freshwater and brackish inland waterbodies, potentially available for macrophyte colonisation, across the Earth), and we examine that evidence in this chapter. We also examine the issues of macrophyte global rarity and endemism in more detail in Volume 2 of the Atlas series.

Numerous studies undertaken during the past century (relevant older studies include [19, 48, 52, 140]), have shown the importance of several factors, acting at various scales from local to global, in influencing macrophyte diversity and assemblage patterns. Latitude and altitude (elevation above sea level: a.s.l) are good examples of relevant spatio-environmental drivers (e.g., [39, 64, 149]). Other identified drivers are factors connected with hydrological regime, including annual evapotranspiration and flood patterns: especially flood pulse duration, size, and frequency [42, 154, 156]; alkalinity/acidity [99, 157]; and land use influences (particularly nutrient inputs from agriculture and other sources influencing trophic conditions in inland waters: e.g., [2, 90]).

Other studies focused on more general terms at drivers of macrophyte diversity and distribution in different types of waterbodies: lakes, rivers and wetlands (e.g., [10, 91, 103, 119, 127, 128, 132, 138, 151]). In addition, several studies have examined how a range of different driving factors may influence macrophyte diversity and assemblage at large scales, for example across Europe, and regions of Africa and the Neotropics (e.g., [3–5, 22, 29, 30, 42, 66, 95, 150, 158]), or at global scale (e.g., [7, 105]).

Several other factors, related to movement of macrophyte propagules, are also thought to be important in this context (see [80]: discussed in more detail in in Volume 4 of this Atlas). For example, the role of both endo- and ectozoochorous vectors in influencing the movement of macrophyte propagules (over both short and longer distances), primarily via migratory birds, is becoming increasingly well understood as a factor helping explain macrophyte global distributions (e.g., [1, 33, 35, 36, 45, 116, 147]). Historic (late Quaternary) climate-change impact has also been shown to predict the current distribution of high-latitude narrow-range endemic species, both in plants and also some less-mobile animal species (e.g., [98, 105, 108, 136]), while older planetary-scale phytogeographical vicariance factors also undoubtedly played a role in determining current macrophyte global distributions, at least for some species (e.g., [31, 32, 75, 161, 169]). A good example is *Pistia stratiotes* L. (Araceae), an ancient species (first known from seeds dated to the Late Oligocene/Early Miocene, some 20–18 million years ago: [123]) which is thought to be native to both the Neotropics and Afrotropics, and with a lineage which may stretch back to Gondwana, according to IUCN (http://www.iucnredlist.org/details/168937/0). However, studies which have used molecular estimates of divergence time to compare closely related aquatic macrophyte taxa that show discontinuous intercontinental distributions (e.g., [75]), do not generally support ancient vicariance as an explanation for macrophyte distribution. These studies have generally found divergence times that are "far too recent to implicate continental drift as a major determinant of discontinuous distributions in aquatic plants" [75]. Nevertheless at least 347 macrophyte species are old enough (speciation ages estimated as before the end of the Miocene, 5.3 million years ago: Ma, or earlier) to have been influenced by ancient vicariance speciation [75], associated with geographical land movement and sea level change [80]; also discussed in more detail in Volume 4 of this Atlas).

There is also growing evidence that species niche breadth is a strong indicator of global range-size for many organisms, including aquatic macrophytes. Specialist-strategy species [50] with narrow niches tend to have limited planetary distributional ranges, while broad-niche species often have much wider distributions [6, 17, 24, 67, 83, 88, 97, 115, 129–131, 133, 142, 143, 159].

2.8 Conclusions

In line with the general comments of [136], who referred to a range of biota other than macrophytes, the evidence summarised in this chapter strongly supports the assertion that small global ranges are the norm for most macrophyte species. The data also show that macrophyte diversity has a significant relationship with latitude, though less so with other physical geographical drivers.

Chapter 3
Species Descriptions, Distribution Maps and Images

Abstract This chapter covers 33 dicotyledonous families (listed in alphabetic order: Acanthaceae to Menyanthaceae), containing 760 species of aquatic macrophytes. For the species of each family, the information is provided to a standardized format, with species listed in alphabetic order. Each species entry describes the overall distribution, range area, origin within each ecozone where it is present (i.e., endemic to the ecozone, native to the ecozone and at least one other, or introduced to the ecozone via one or more anthropochorous vectors), rarity at global scale, life form, ploidy state, speciation age, long-distance vectors of species distribution, actual or potential uses, and global endangered status (as listed by the IUCN Red List). Some data are unavailable for some species, and the entries indicate where this is the case (NA: not available). The chapter also includes global distribution maps for every species and pictures of almost all species.

Keywords Aquatic macrophyte · Dicotyledonous species · Distribution · Diversity

In this chapter the species of the Magnoliopsida (dicotyledons) in the families Acanthaceae to Menyanthaceae are listed in alphabetical order. Within each family, the species are listed within genera, again in alphabetical order. For each family the results are briefly summarised, describing overall distribution, rarity and endangered status. For families with more than 15 species, those numbers are expressed as percentages.

Quick Links to Families (Order):

T. Lobato-de Magalhães et al., *World Atlas of Freshwater Macrophytes*, Wetlands: Ecology, Conservation and Management 10,
https://doi.org/10.1007/978-3-031-52749-4_3

7. Balsaminaceae (Ericales)
8. Boraginaceae (Boraginales)
9. Brassicaceae (Brassicales)
10. Burmanniaceae (Dioscoreales)
11. Cabombaceae (Nymphaeales)
12. Campanulaceae (Asterales)
13. Caryophyllaceae (Caryophyllales)
14. Ceratophyllaceae (Ceratophyllales)
15. Convolvulaceae (Solanales)
16. Crassulaceae (Saxifragales)
17. Droseraceae (Caryophyllales)
18. Elatinaceae (Malpighiales)
19. Euphorbiaceae (Malpighiales)
20. Fabaceae (Fabales)
21. Haloragaceae (Saxifragales)
22. Hydatellaceae (Nympheales)
23. Hydroleaceae (Solanales)
24. Hydrostachyaceae (Cornales)
25. Hypericaceae (Malpighiales)
26. Lamiaceae (Lamiales)
27. Lentibulariaceae (Lamiales)
28. Limnanthaceae (Brassicales)
29. Linderniaceae (Lamiales)
30. Lythraceae (Myrtales)
31. Malvaceae (Malvales)
32. Melastomataceae (Myrtales)
33. Menyanthaceae (Asterales)

3.1 ACANTHACEAE (Lamiales)

The Acanthaceae have two genera containing a total of 15 freshwater macrophyte species, which have a generally sub-tropical to tropical world distribution. Ten species occur in OR, nine in NEO, six in AUS, four in AFR, three in PAL and two in OCE. Six of the 15 species are endemic (E or E(I) origin status) to OR, two to AFR and one to NEO. Anthropochory vectors have quite strongly influenced the world distribution of Acanthaceae macrophyte species, with I-origin (introduced or invasive) records for four species in each of NEO and AUS, two each in NEA and OCE, and one each in AFR and OR. One species (*Hygrophila polysperma*) is a cause of major concern as an aquatic weed in at least one world ecozone. The world rarity findings for species of the Acanthaceae are VR 7%, R 53%, and C 40%. In terms of endangered status, the Acanthaceae have 67% of their species not assessed (NA) or DD, and 33% have LC status.

Quick Links to Genera

1. *Hygrophila*
2. *Justicia*

***Hygrophila abyssinica* (Hochst. ex Nees) T. Anderson**

Synonyms: *Hemigraphis abyssinica* (Hochst. ex Nees) C.B.Clarke, *Polyechma abyssinicum* Hochst. ex Nees, *Synnema abyssinicum* (Hochst. ex Nees) Bremek.

Range area (grAH km^2): AFR: 145000
Ecozone origin-status: AFR: E
World rarity status: R
Ecozone-endemic macroregion occurrence: AFR: FOR, ZAM
Life form(s): EM
Ploidy state: NA
Speciation age (Ma): NA
Populations also occur in terrestrial, wetland and/ or riparian habitats: YES
Long distance movement vectors: NA
Nuisance status and uses: no nuisance issues; no known uses
Endangered status (IUCN): LC

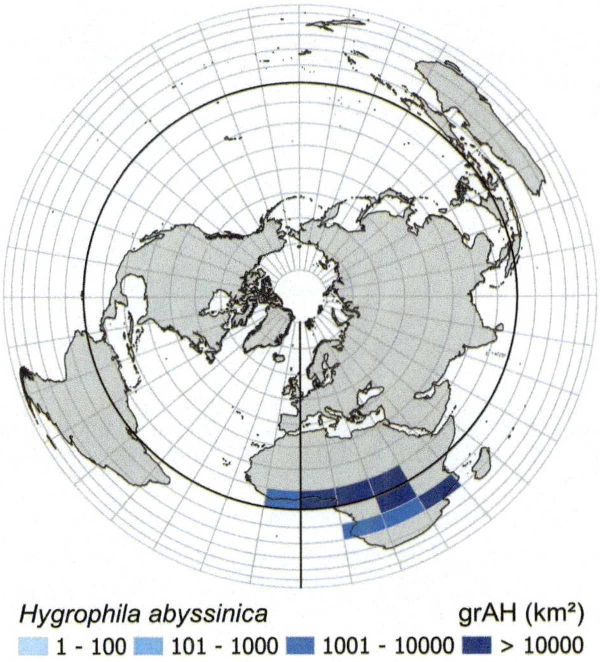

***Hygrophila auriculata* (Schumach.) Heine**
Synonyms: *Barleria auriculata* Schumach.

Range area (grAH km^2): PAL: 86000; OR: 53200; AFR: 293600
Ecozone origin-status: PAL: N; OR: N; AFR: N
World rarity status: C
Ecozone-endemic macroregion occurrence: not endemic to any world ecozone
Life form(s): EM (sometimes a subshrub)
Ploidy state: NA
Speciation age (Ma): NA
Populations also occur in terrestrial, wetland and/ or riparian habitats: YES
Long distance movement vectors: NA
Nuisance status and uses: no nuisance issues; folk medicine
Endangered status (IUCN): NA

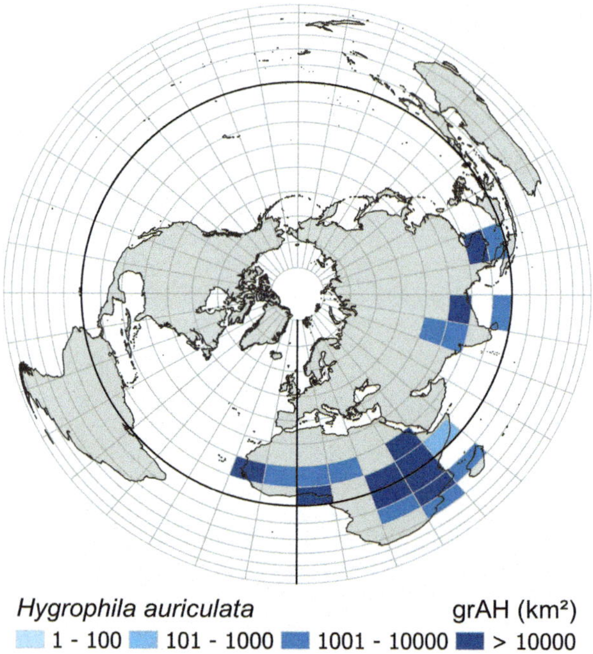

Hygrophila auriculata (Schumach.) Heine
Image copyright: iNaturalist CC0

***Hygrophila balsamica* (L.f.) Raf.**
Synonyms: *Cardanthera balsamica* Benth. ex C.B.Clarke, *Ruellia balsamica* L.f., *Synnema balsamicum* (L.f.) Alston

Range area (grAH km^2): OR: 61900
Ecozone origin-status: OR: E
World rarity status: R
Ecozone-endemic macroregion occurrence: OR: IND
Life form(s): EM
Ploidy state: NA
Speciation age (Ma): NA
Populations also occur in terrestrial, wetland and/ or riparian habitats: YES
Long distance movement vectors: NA
Nuisance status and uses: no nuisance issues; aquarium trade, bioactive
Endangered status (IUCN): LC

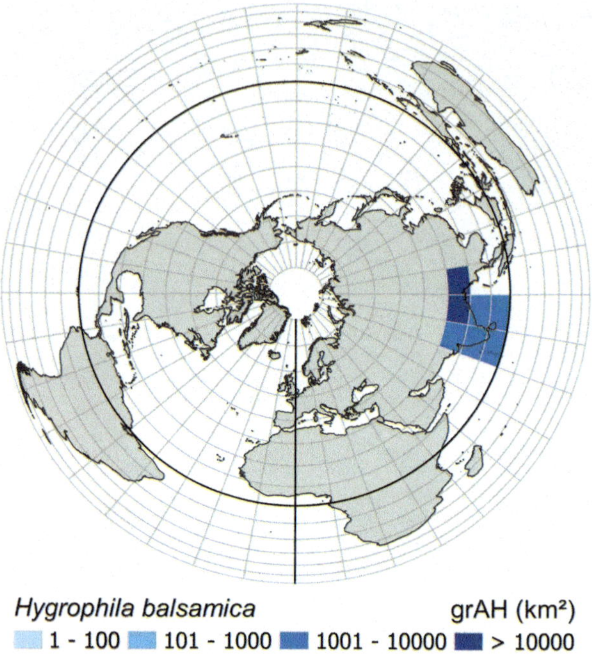

Hygrophila balsamica (L.f.) Raf.
Image copyright: iNaturalist CC0 @nuwan (Nuwan Jayawardana)

***Hygrophila biplicata* (Nees) Sreem.**
Synonyms: *Adenosma biplicata* Nees, *Synnema biplicatum* (Nees) J.B.Imlay

Range area (grAH km^2): OR: 51900
Ecozone origin-status: OR: E
World rarity status: R
Ecozone-endemic macroregion occurrence: OR: IND, ICH
Life form(s): EM
Ploidy state: NA
Speciation age (Ma): NA
Populations also occur in terrestrial, wetland and/ or riparian habitats: YES
Long distance movement vectors: NA
Nuisance status and uses: no nuisance issues; no known uses
Endangered status (IUCN): NA

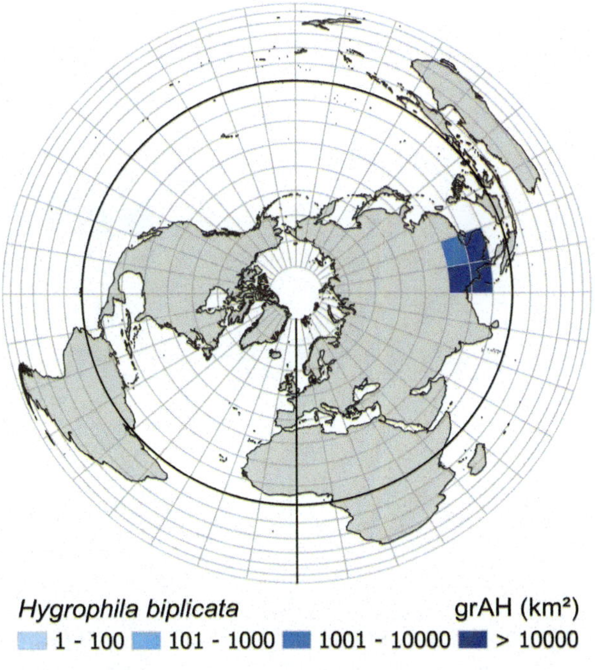

***Hygrophila corymbosa* (Blume) Lindau.**
Synonyms: *Nomaphila corymbosa* Blume

Range area (grAH km^2): NEA: 12800; NEO: 5200; OR: 226000; present in OCE
Ecozone origin-status: NEA: I; NEO: I; OR: E(I); OCE: I
World rarity status: R
Ecozone-endemic macroregion occurrence: OR: IND, ICH
Life form(s): EM
Ploidy state: NA
Speciation age (Ma): NA
Populations also occur in terrestrial, wetland and/ or riparian habitats: YES
Long distance movement vectors: ANTHROPOCHORY
Nuisance status and uses: low to no nuisance issues; aquarium trade, medicinal potential
Endangered status (IUCN): NA

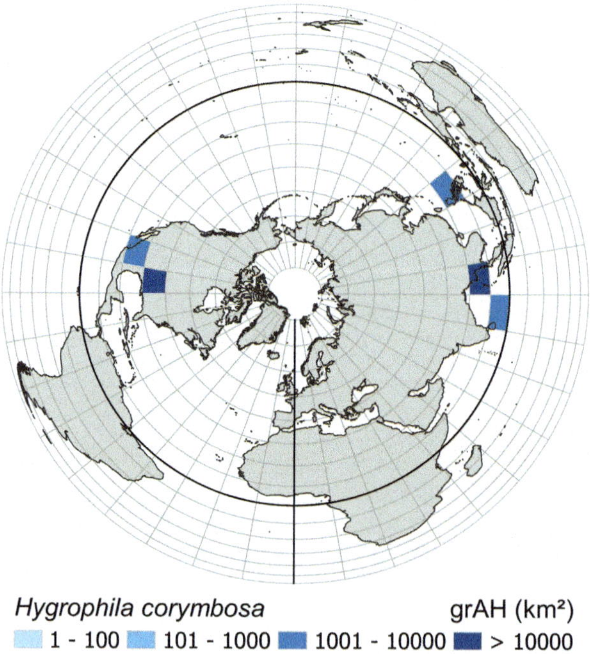

Hygrophila corymbosa (Blume) Lindau.
Image copyright: iNaturalist CC-BY @alisonnorthup

Hygrophila costata **Nees**
Synonyms:

Range area (grAH km^2): NEA: 25800; NEO: 301200; OR: 18100; AUS: 9700
Ecozone origin-status: NEA: N; NEO: N; OR: I; AUS: I
World rarity status: C
Ecozone-endemic macroregion occurrence: not endemic to any world ecozone
Life form(s): EM (hydrosubshrub or subshrub)
Ploidy state: NA
Speciation age (Ma): NA
Populations also occur in terrestrial, wetland and/ or riparian habitats: YES
Long distance movement vectors: ANTHROPOCHORY
Nuisance status and uses: serious environmental weed in Queensland; aquarium trade, folk medicine
Endangered status (IUCN): NA

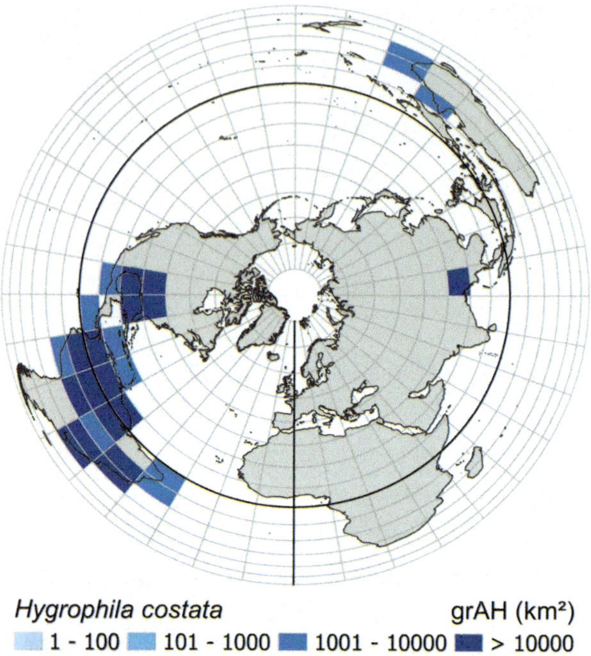

Hygrophila costata Nees

***Hygrophila difformis* (L.f.) Blume**

Synonyms: *Cardanthera difformis* (L.f.) Druce, *Ruellia difformis* L.f.

Range area (grAH km^2): NEO: 13800; OR: 48600; AUS: 10400
Ecozone origin-status: NEO: I; OR: E(I); AUS: I
World rarity status: R
Ecozone-endemic macroregion occurrence: OR: IND, ICH
Life form(s): EM, SU
Ploidy state: NA
Speciation age (Ma): NA
Populations also occur in terrestrial, wetland and/ or riparian habitats: YES
Long distance movement vectors: ANTHROPOCHORY
Nuisance status and uses: low to no nuisance issues; aquarium trade, medicinal, folk medicine
Endangered status (IUCN): LC

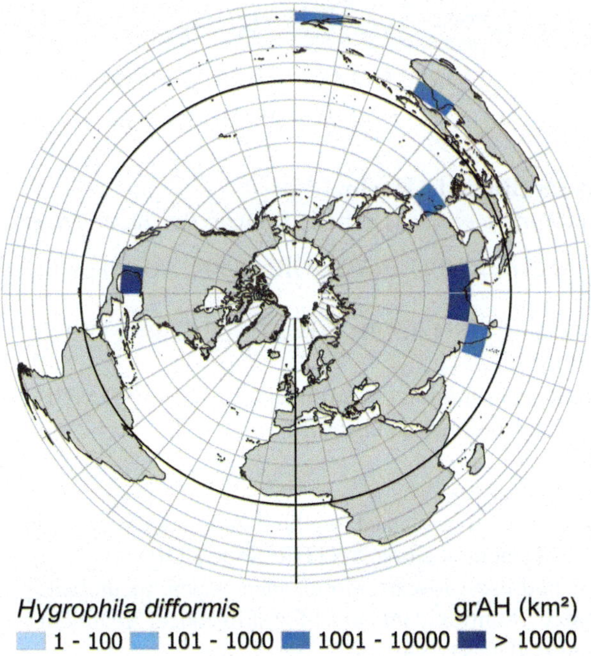

Hygrophila difformis (L.f.) Blume
Image copyright: iNaturalist CC0

Hygrophila erecta (Burm.f.) Hochr.
Synonyms: *Dyschoriste erecta* (Burm.f.) Kuntze, *Ruellia erecta* Burm.f.

Range area (grAH km^2): NEO: 36700; OR: 98900; AUS: 11800; AFR: 6000; present in OCE
Ecozone origin-status: NEO: I; OR: E(I); AUS: I; AFR: I; OCE: I
World rarity status: R
Ecozone-endemic macroregion occurrence: OR: ICH, IND, SUP
Life form(s): EM (sometimes a subshrub)
Ploidy state: NA
Speciation age (Ma): NA
Populations also occur in terrestrial, wetland and/ or riparian habitats: YES
Long distance movement vectors: ANTHROPOCHORY
Nuisance status and uses: low to no nuisance issues; medicinal: potential medical usage as source of antioxidants used for skin cancer treatment
Endangered status (IUCN): NA

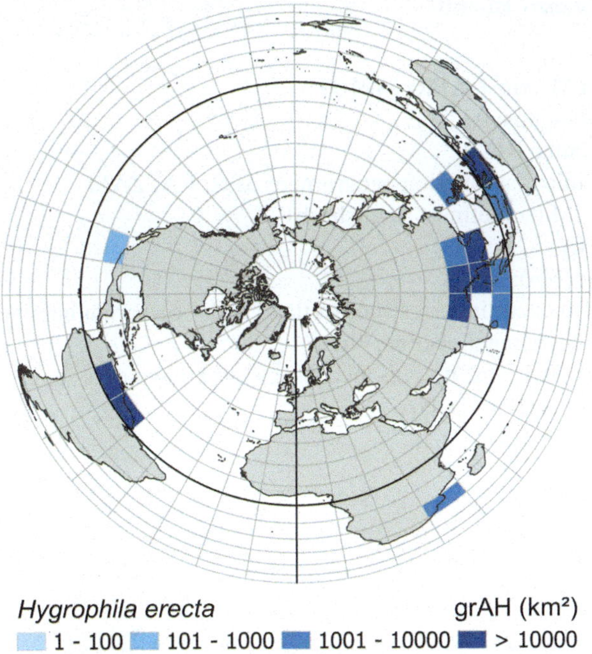

Hygrophila erecta (Burm.f.) Hochr.
Image copyright: iNaturalist CC0 @pcbose

Hygrophila linearis **Burkill**
Synonyms:

Range area (grAH km^2): AFR: 140100
Ecozone origin-status: AFR: E
World rarity status: R
Ecozone-endemic macroregion occurrence: AFR: EA, ZAM
Life form(s): EM (sometimes a subshrub)
Ploidy state: NA
Speciation age (Ma): NA
Populations also occur in terrestrial, wetland and/ or riparian habitats: YES
Long distance movement vectors: NA
Nuisance status and uses: no nuisance issues; no known uses
Endangered status (IUCN): NA

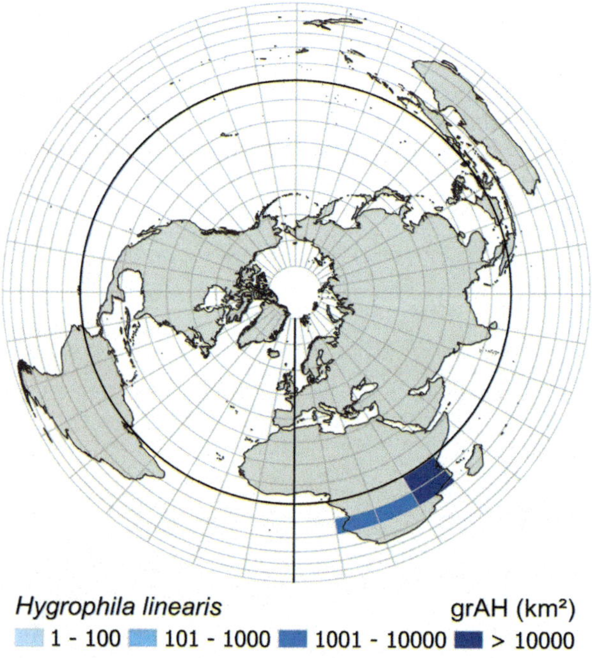

Hygrophila polysperma (Roxb.) T.Anderson

Synonyms: *Hemiadelphis polysperma* (Roxb.) Nees, *Justicia polysperma* Roxb.

Range area (grAH km^2): PAL: 41400; NEA: 21000; NEO: 24100; OR: 117800; AUS: 14000

Ecozone origin-status: PAL: N; NEA: I; NEO: I; OR: N; AUS: I

World rarity status: C

Ecozone-endemic macroregion occurrence: not endemic to any world ecozone

Life form(s): EM

Ploidy state: P

Speciation age (Ma): NA

Populations also occur in terrestrial, wetland and/ or riparian habitats: YES

Long distance movement vectors: ANTHROPOCHORY

Nuisance status and uses: major aquatic weed in at least one world ecozone outside home range [54, 80]; aquarium trade, human food, medical uses

Endangered status (IUCN): LC

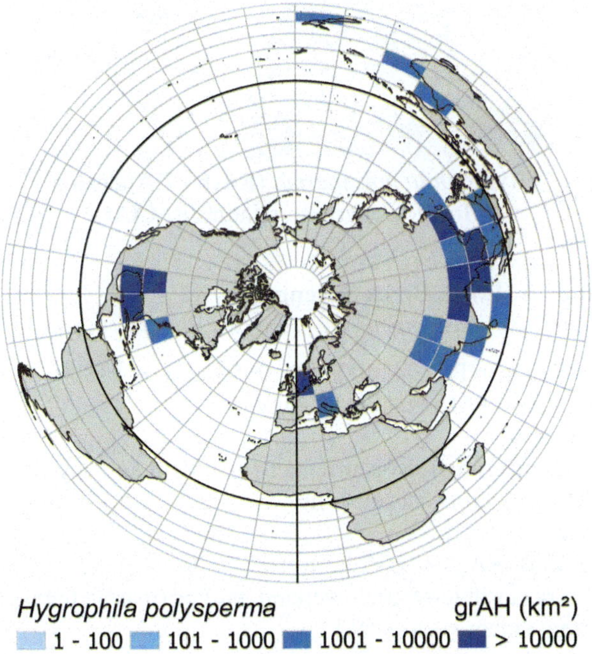

Hygrophila polysperma (Roxb.) T.Anderson
Image copyright: iNaturalist CC0

***Hygrophila ringens* (L.) R. Br. ex Spreng.**
Synonyms: *Dipteracanthus ringens* (L.) Abeyw., *Ruellia ringens* L.

Range area (grAH km^2): PAL: 41400; OR: 99500; AUS: 51800
Ecozone origin-status: PAL: N; OR: N; AUS: N
World rarity status: C
Ecozone-endemic macroregion occurrence: not endemic to any world ecozone
Life form(s): EM (sometimes a subshrub)
Ploidy state: NA
Speciation age (Ma): NA
Populations also occur in terrestrial, wetland and/ or riparian habitats: YES
Long distance movement vectors: NA
Nuisance status and uses: no nuisance issues; no known uses
Endangered status (IUCN): NA

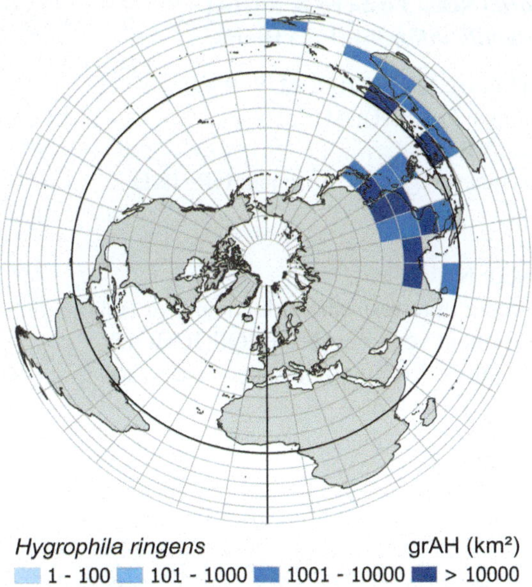

Hygrophila ringens (L.) R. Br. ex Spreng.

Hygrophila triflora **(Nees) Fosberg & Sachet**
Synonyms: *Hygrophila difformis* (L.f.) Blume

Range area (grAH km^2): OR: 18100; AUS: 9300
Ecozone origin-status: OR: E(I); AUS: I
World rarity status: VR
Ecozone-endemic macroregion occurrence: OR: IND
Life form(s): EM
Ploidy state: NA
Speciation age (Ma): NA
Populations also occur in terrestrial, wetland and/ or riparian habitats: YES
Long distance movement vectors: ANTHROPOCHORY
Nuisance status and uses: low to no nuisance issues; aquarium trade
Endangered status (IUCN): NA

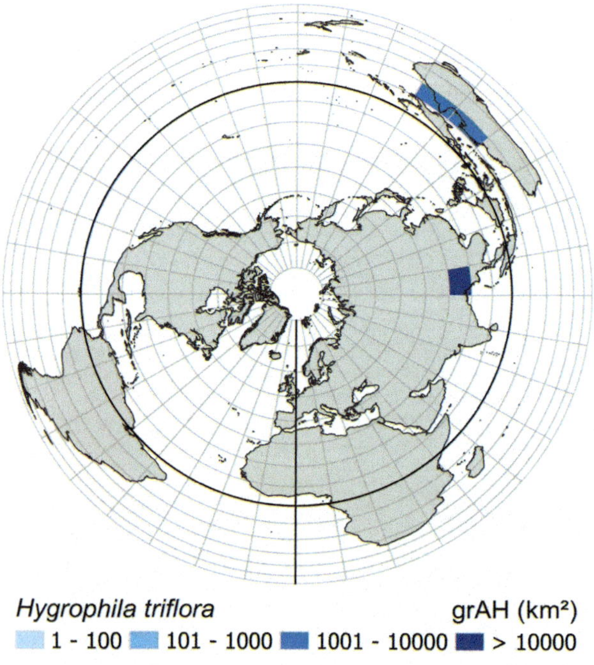

***Justicia americana* (L.) Vahl**

Synonyms: *Dianthera americana* L.

Range area (grAH km^2): NEA: 323000; NEO: 35600
Ecozone origin-status: NEA: N; NEO: N
World rarity status: C
Ecozone-endemic macroregion occurrence: not endemic to any world ecozone
Life form(s): EM
Ploidy state: D
Speciation age (Ma): NA
Populations also occur in terrestrial, wetland and/ or riparian habitats: YES
Long distance movement vectors: NA
Nuisance status and uses: low to no nuisance issues; horticulture, water garden plant, medicinal potential
Endangered status (IUCN): LC

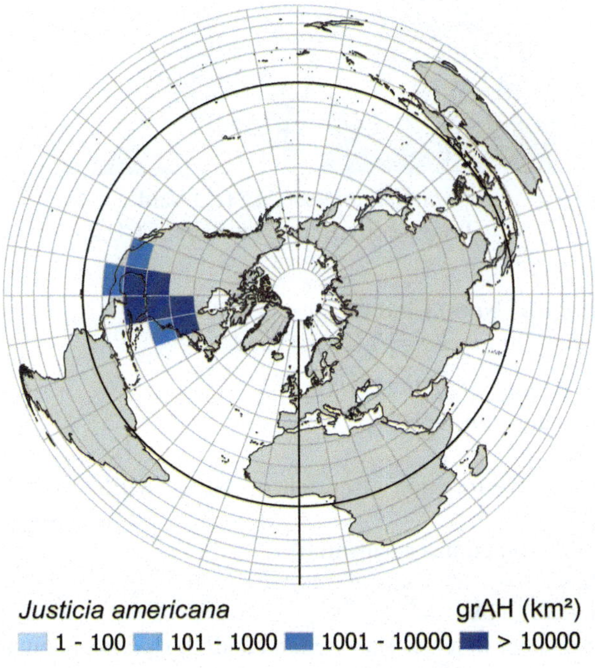

Justicia americana (L.) Vahl
Image copyright: iNaturalist CC0

Justicia laevilinguis (Nees) Lindau
Synonyms: *Dianthera laevilinguis* (Nees) Lindau

Range area (grAH km^2): NEO: 272100
Ecozone origin-status: NEO: E
World rarity status: C
Ecozone-endemic macroregion occurrence: NEO: MES AM, NWSA, AMZ, CHA, PAR
Life form(s): EM (sometimes a subshrub)
Ploidy state: NA
Speciation age (Ma): NA
Populations also occur in terrestrial, wetland and/ or riparian habitats: YES
Long distance movement vectors: NA
Nuisance status and uses: no nuisance issues; folk medicine
Endangered status (IUCN): NA

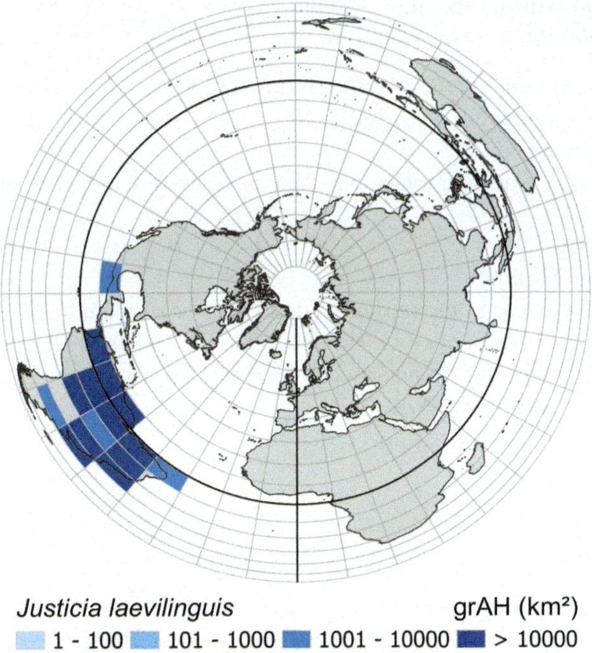

Justicia laevilinguis (Nees) Lindau

Justicia ovata **(Walter) Lindau**
Synonyms: *Dianthera ovata* Walter

Range area (grAH km^2): NEA: 25800; NEO: 33200
Ecozone origin-status: NEA: N; NEO: N
World rarity status: R
Ecozone-endemic macroregion occurrence: not endemic to any world ecozone
Life form(s): EM
Ploidy state: NA
Speciation age (Ma): NA
Populations also occur in terrestrial, wetland and/ or riparian habitats: YES
Long distance movement vectors: HYDROCHORY (NEA to NEO)
Nuisance status and uses: no nuisance issues; no known uses
Endangered status (IUCN): NA

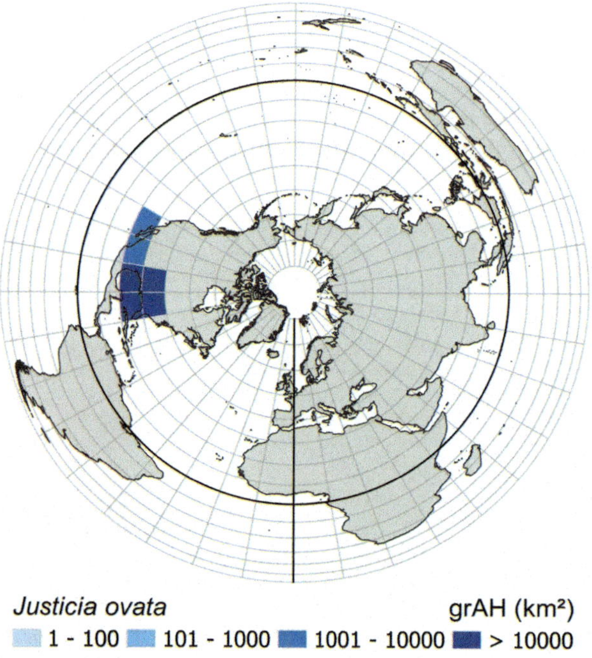

Justicia ovata (Walter) Lindau
Image copyright: iNaturalist CC0

3.2 AMARANTHACEAE (Caryophyllales)

The family Amaranthaceae has two genera containing a total of six freshwater macrophyte species, generally with a warm-temperate to tropical world distribution, with four species occurring in NEO, two each in PAL, OR, NEA, OCE and AUS, and one in AFR. Three species are endemic (E or E(I) origin status) to NEO and one to NEA. Anthropochory vectors have strongly influenced the world distribution of the freshwater macrophytes of the Amaranthaceae, with I-origin (introduced or invasive) records for two species in each of OCE, PAL, OR, NEA and AUS, and one in AFR. *Alternanthera philoxeroides* and *Alternanthera ficoides* are major invasive aquatic weeds in at least one world ecozone. The world rarity findings for species of the Amaranthaceae are one species VR, one R, and four C. The endangered status of four of the species has not been assessed (NA) or DD, while two have LC status.

Quick Links to Genera
1. *Alternanthera*
2. *Centrostachys*

***Alternanthera aquatica* (Parodi) Chodat**
Synonyms: *Mogiphanes aquatica* D.Parodi

Range area (grAH km^2): NEO: 155100
Ecozone origin-status: NEO: E
World rarity status: R
Ecozone-endemic macroregion occurrence: NEO: AMZ, CHA, PAR
Life form(s): EM

Ploidy state: NA
Speciation age (Ma): NA
Populations also occur in terrestrial, wetland and/ or riparian habitats: YES
Long distance movement vectors: NA
Nuisance status and uses: no nuisance issues; aquarium trade
Endangered status (IUCN): NA

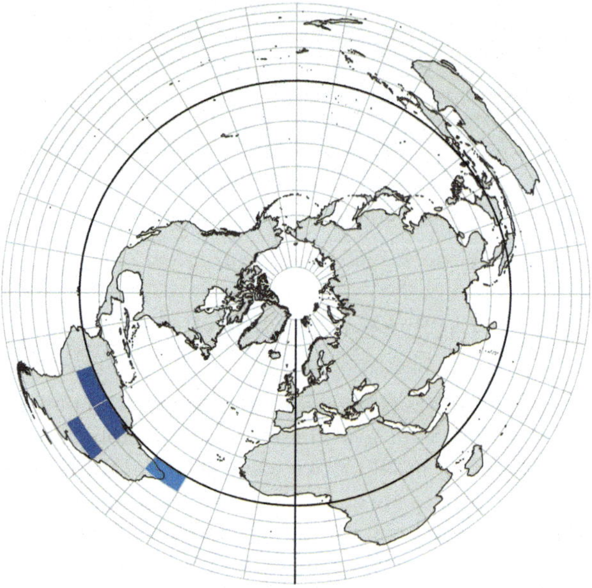

Alternanthera ficoidea **(L.) P.Beauv.**

Synonyms: *Achyranthes ficoidea* (L.) Lam., *Bucholzia ficoidea* (L.) Mart., *Gomphrena ficoidea* L., *Steiremis ficoidea* (L.) Raf., *Telanthera ficoidea* (L.) Moq., *Illecebrum ficoideum* (L.) L., *Paronychia ficoidea* (L.) Desf.

Range area (grAH km^2): PAL: 22400; NEO: 356400; OR: 98600; AUS: 65600; AFR: 63600; present in OCE
Ecozone origin-status: PAL: I; NEO: E(I); OR: I; AUS: I; AFR: I; OCE: I
World rarity status: C
Ecozone-endemic macroregion occurrence: NEO: MG MX, MES AM, NWSA, AMZ, CHA, PAR
Life form(s): EM (sometimes a subshrub)
Ploidy state: NA
Speciation age (Ma): NA
Populations also occur in terrestrial, wetland and/ or riparian habitats: YES

Long distance movement vectors: ANTHROPOCHORY
Nuisance status and uses: low or no nuisance issues; medicinal, garden plant
Endangered status (IUCN): NA

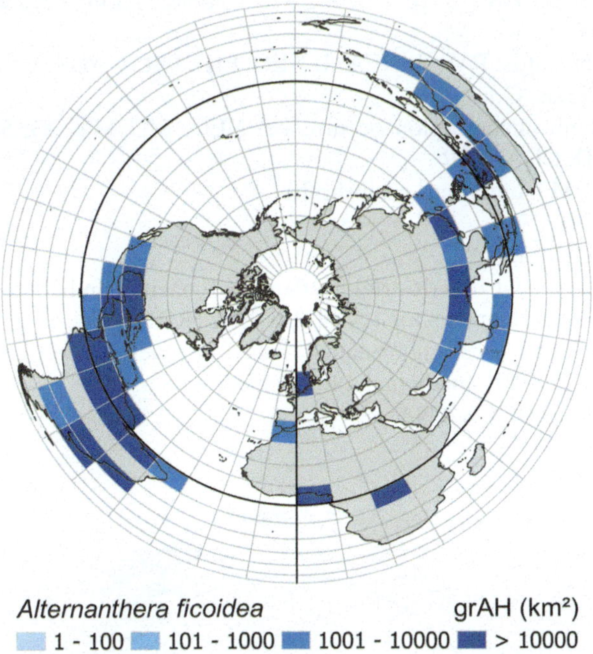

Alternanthera ficoidea (L.) P.Beauv.
Image copyright: iNaturalist CC-BY @yunglunlin (Yung-Lun Lin)

***Alternanthera philoxeroides* (Mart.) Griseb.**

Synonyms: *Achyranthes philoxeroides* (Mart.) Standl., *Bucholzia philoxeroides* Mart.

Range area (grAH km^2): PAL: 70200; NEA: 38500; NEO: 320100; OR: 95200; AUS: 60200

Ecozone origin-status: PAL: I; NEA: I; NEO: E(I); OR: I; AUS: I

World rarity status: C

Ecozone-endemic macroregion occurrence: NEO: MG MX, MES AM, NWSA, AMZ, AND, CHA, PAR

Life form(s): EM (sometimes a subshrub)

Ploidy state: DP

Speciation age (Ma): NA

Populations also occur in terrestrial, wetland and/ or riparian habitats: YES

Long distance movement vectors: ANTHROPOCHORY

Nuisance status and uses: major aquatic weed in several world ecozones outside home range [54, 80]; medicinal, phytoremediation uses

Endangered status (IUCN): NA

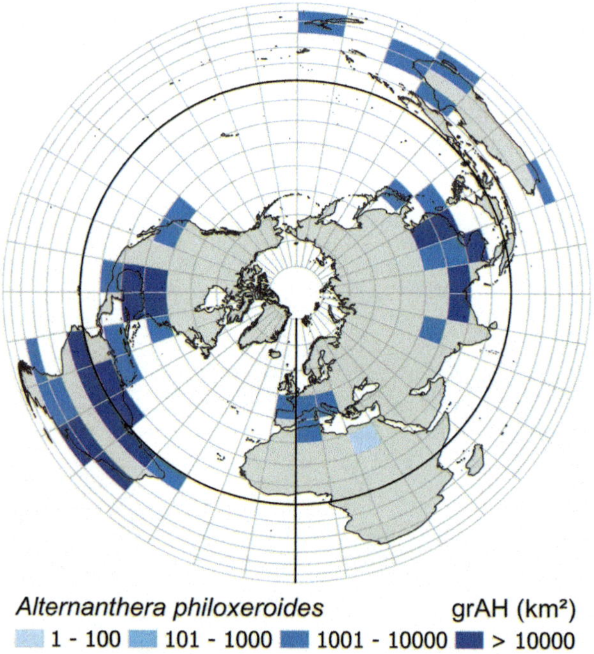

Alternanthera philoxeroides (Mart.) Griseb.
Image copyright: G. Ervin, with permission

Alternanthera sessilis **(L.) R.Br. ex DC.**
Synonyms: *Achyranthes sessilis* (L.) Besser, *Gomphrena sessilis* L., *Illecebrum sessile* (L.) L.

Range area (grAH km^2): PAL: 104600; NEA: 25800; NEO: 300400; OR: 144100; AUS: 102000; AFR: 352100; present in OCE
Ecozone origin-status: PAL: N; NEA: I; NEO: N; OR: N; AUS: N; AFR: N; OCE: I
World rarity status: C
Ecozone-endemic macroregion occurrence: not endemic to any world ecozone
Life form(s): EM (sometimes a subshrub)
Ploidy state: DP
Speciation age (Ma): NA
Populations also occur in terrestrial, wetland and/ or riparian habitats: YES
Long distance movement vectors: ANTHROPOCHORY
Nuisance status and uses: major aquatic weed in at least one world ecozone both within and outside home range [54, 80]; some medicinal and food uses, aquarium trade
Endangered status (IUCN): LC

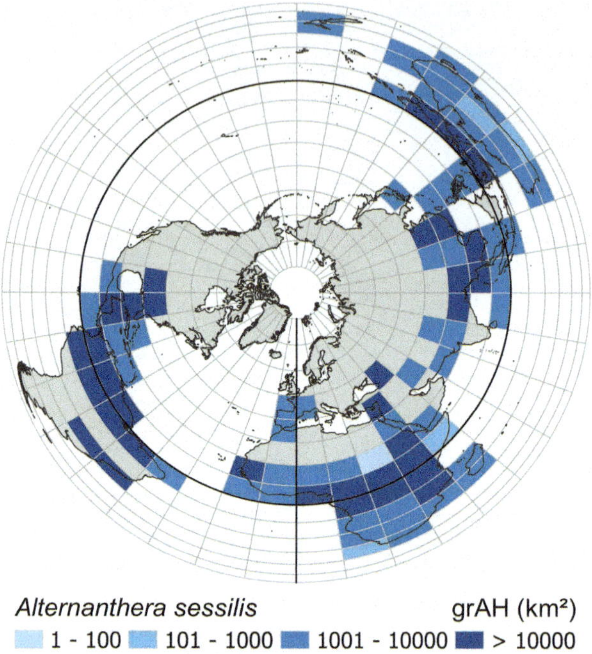

Alternanthera sessilis (L.) R.Br. ex DC.
Image copyright: A. Efremov, with permision

Alternanthera tetramera **R.E.Fr.**
Synonyms:

Range area (grAH km^2): NEO: 21100
Ecozone origin-status: NEO: E
World rarity status: VR
Ecozone-endemic macroregion occurrence: NEO: CHA
Life form(s): EM (sometimes a subshrub)
Ploidy state: NA
Speciation age (Ma): NA
Populations also occur in terrestrial, wetland and/ or riparian habitats: YES
Long distance movement vectors: NA
Nuisance status and uses: no nuisance issues; no known uses
Endangered status (IUCN): NA

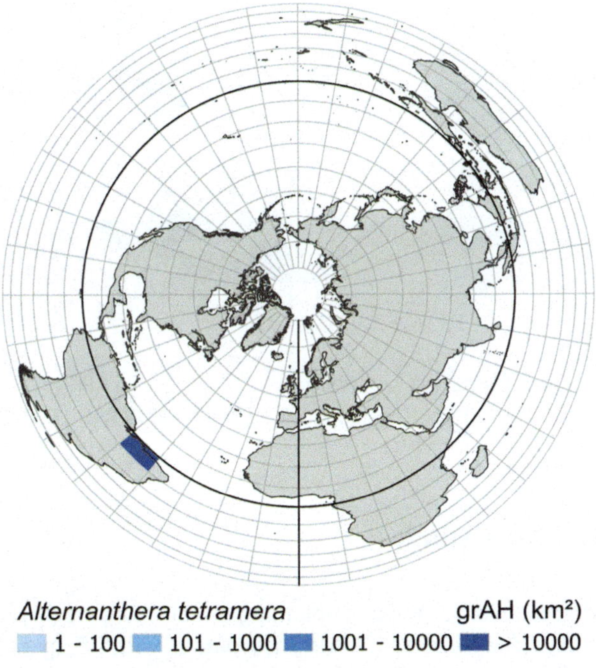

Centrostachys aquatica **(R.Br.) Moq.**
Synonyms: *Achyranthes aquatica* R.Br

Range area (grAH km^2): PAL: 137300; OR: 96600; AFR: 259600
Ecozone origin-status: PAL: N; OR: N; AFR: N
World rarity status: C
Ecozone-endemic macroregion occurrence: not endemic to any ecozone
Life form(s): EM (sometimes a subshrub)

Ploidy state: NA
Speciation age (Ma): NA
Populations also occur in terrestrial, wetland and/ or riparian habitats: YES
Long distance movement vectors: NA
Nuisance status and uses: no nuisance issues; bioactive
Endangered status (IUCN): LC

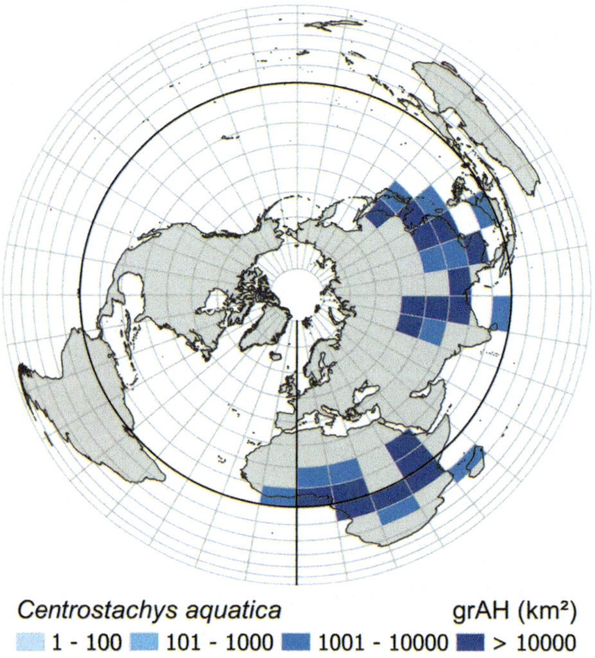

3.3 APIACEAE (Apiales)

The family Apiaceae has 15 genera containing a total of 66 freshwater macrophyte species. They have a widespread world distribution with 35 species occurring in NEA, 28 in NEO, 27 in PAL, 12 in AUS, nine in OR, five in AFR and four in OCE. There are none in ANT. Ecozone-endemic macrophytes (E or E(I) origin status) of Apiaceae show high occurrences in PAL (19 species), with a further 14 in NEA, 10 in NEO, three in AUS and one in AFR. Anthropochory vectors have played only a limited role in influencing the world distribution of species of the Apiaceae, except in AUS, where seven of the 12 Apiaceae species present are introduced, while there are six I-origin species in NEA. All four Apiaceae species in OCE are introduced, three in NEO and two in OR, PAL and AFR. Macrophyte species of Apiaceae rarely or never cause nuisance and none are reported as a cause of major aquatic weed concerns in any world ecozone. The world rarity findings for species of the Apiaceae are VR 38%, R 38%, and C 24%. In terms of endangered status, the Apiaceae have a large proportion of endangered species: 56% NA/DD, 32% LC and 12% assessed in the higher risk categories (NT/VU/EN/CR) though none are in the highest, Critically Endangered, category.

Quick Links to Genera

1. *Apium*
2. *Berula*
3. *Caropsis*
4. *Carum*
5. *Centella*
6. *Cicuta*
7. *Cynosciadium*
8. *Eryngium*
9. *Lilaeopsis*
10. *Limnosciadium*
11. *Oenanthe*
12. *Oxypolis*
13. *Peucedanum*
14. *Ptilimnium*
15. *Sium*

***Apium graveolens* L.**

Synonyms: *Carum graveolens* (L.) Koso-Pol., *Celeri graveolens* (L.) Britton, *Helosciadium graveolens* (L.) Rojas Acosta, *Seseli graveolens* (L.) Scop., *Sium graveolens* (L.) Vest

Range area (grAH km^2): PAL: 341500; NEA: 127000; NEO: 208000; OR: 54300; AUS: 38500; AFR: 31000; present in OCE
Ecozone origin-status: PAL: E(I); NEA: I; NEO: I; OR: I; AUS: I; AFR: I; OCE: I
World rarity status: C
Ecozone-endemic macroregion occurrence: PAL: MED, NAA, W EUR, E EUR, UR/ W SIB, C ASIA, CH, FER/J
Life form(s): EM
Ploidy state: D
Speciation age (Ma): 4.9
Populations also occur in terrestrial, wetland and/ or riparian habitats: YES
Long distance movement vectors: ANTHROPOCHORY
Nuisance status and uses: no nuisance issues; human food (cultivated celery) and medical purposes
Endangered status (IUCN): LC

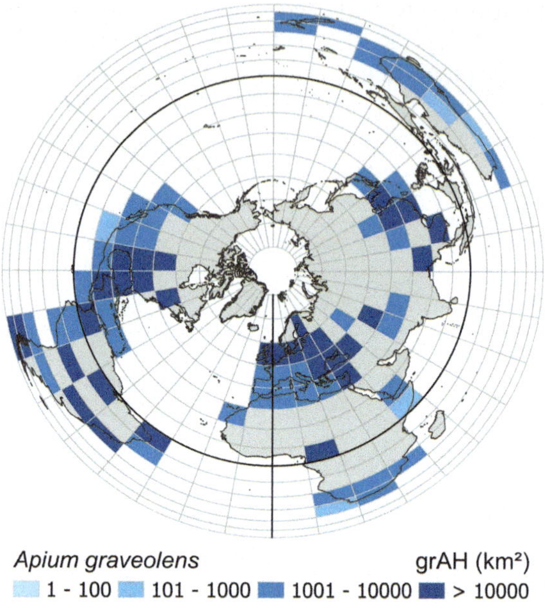

Apium graveolens L.
Image copyright: J.C. Schou, with permision

Apium inundatum **(L.) Rchb.f.**
Synonyms: *Helosciadium inundatum* (L.) Koch

Range area (grAH km²): PAL: 89500
Ecozone origin-status: PAL: E
World rarity status: R
Ecozone-endemic macroregion occurrence: PAL: MED, W EUR, N EUR
Life form(s): EM, SU
Ploidy state: D
Speciation age (Ma): NA
Populations also occur in terrestrial, wetland and/ or riparian habitats: YES
Long distance movement vectors: NA
Nuisance status and uses: no nuisance issues; some use for human food and medical purposes
Endangered status (IUCN): LC

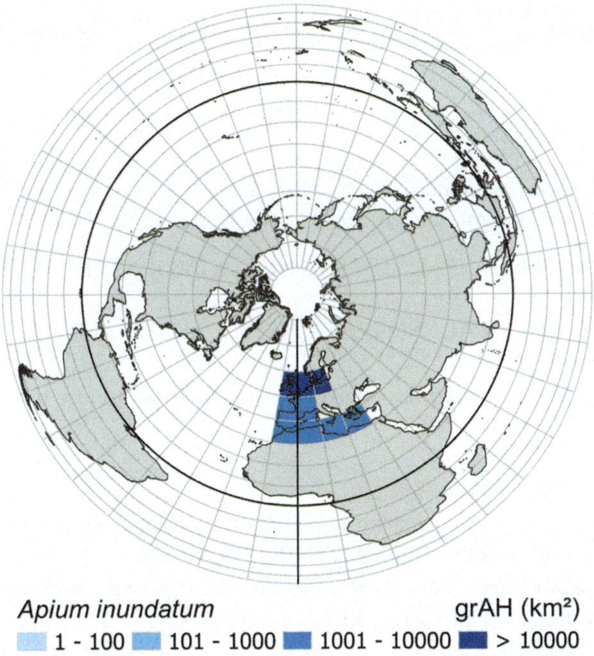

Apium inundatum (L.) Rchb.f.
Images copyright: J. C.Schou, with permission

***Apium nodiflorum* (L.) Lag.**
Synonyms: *Helosciadium nodiflorum* (L.) Koch

Range area (grAH km^2): PAL: 187200; NEA: 57300; NEO: 53800; AUS: 30400; AFR: 114800
Ecozone origin-status: PAL: N; NEA: I; NEO: I; AUS: I; AFR: N
World rarity status: C
Ecozone-endemic macroregion occurrence: not endemic to any world ecozone
Life form(s): EM
Ploidy state: D
Speciation age (Ma): NA
Populations also occur in terrestrial, wetland and/ or riparian habitats: YES
Long distance movement vectors: ANTHROPOCHORY
Nuisance status and uses: low to no nuisance issues; some uses in folk medicine, and limited usage for phytoremediation purposes.
Endangered status (IUCN): LC

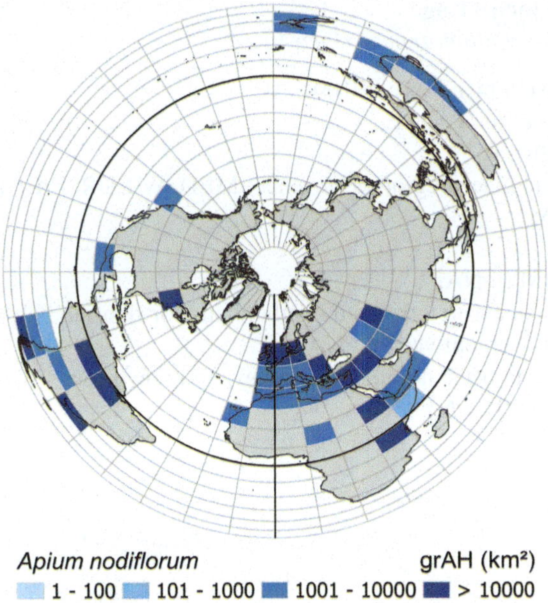

Apium nodiflorum (L.) Lag.
Image copyright: iNaturalist CC0

***Apium repens* (Jacq.) Lag.**
Synonyms: *Helosciadium repens* (Jacq.) Koch

Range area (grAH km^2): PAL: 90100
Ecozone origin-status: PAL: E
World rarity status: R
Ecozone-endemic macroregion occurrence: MED, W EUR, E EUR, NAA
Life form(s): EM
Ploidy state: D
Speciation age (Ma): NA
Populations also occur in terrestrial, wetland and/ or riparian habitats: YES
Long distance movement vectors: NA
Nuisance status and uses: no nuisance issues; no known uses
Endangered status (IUCN): VU

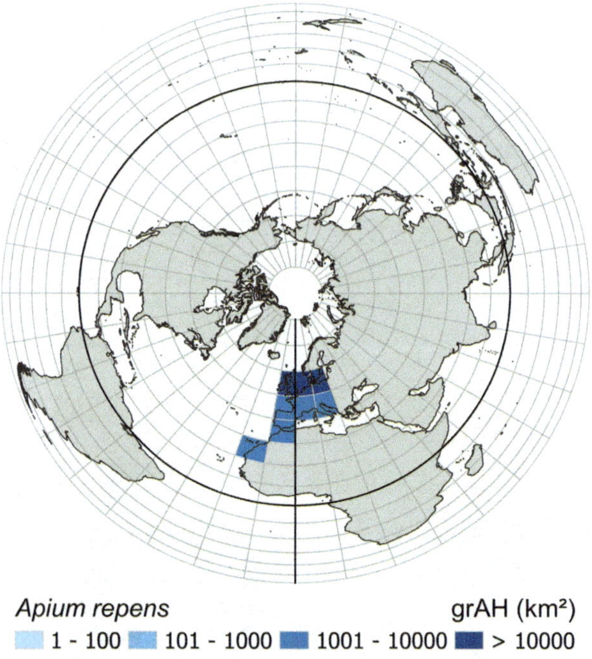

Apium repens (Jacq.) Lag.
Image copyright: iNaturalist CC0

***Berula erecta* (Huds.) Coville**
Synonyms: *Siella erecta* (Huds.) Pimenov

Range area (grAH km^2): PAL: 320600; NEA: 300600; NEO: 25900; OR: 19900; AUS: 19100; AFR: 181700
Ecozone origin-status: PAL: N; NEA: N; NEO: N; OR: N; AUS: I; AFR: N
World rarity status: C
Ecozone-endemic macroregion occurrence: not endemic to any world ecozone
Life form(s): EM
Ploidy state: D
Speciation age (Ma): NA
Populations also occur in terrestrial, wetland and/ or riparian habitats: YES
Long distance movement vectors: NA
Nuisance status and uses: no nuisance issues; phytoremediation, human food: the plant is edible (though there are some reports that it can be toxic to people and animals), folk medicine
Endangered status (IUCN): LC

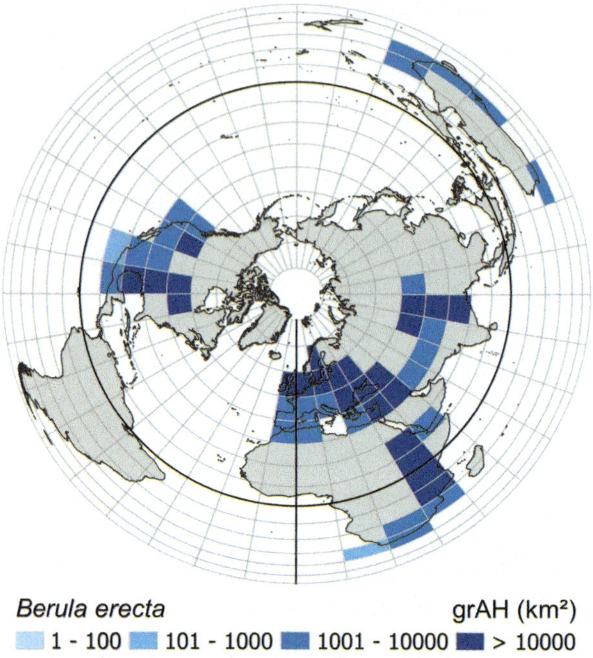

Berula erecta (Huds.) Coville
Image copyright: J.C. Schou, with permision

***Caropsis verticillatoinundata* (Thore) Rauschert**
Synonyms: *Ptychotis verticillatoinundatum* (Thore) M.Hiroe, *Sison verticillatoinundatum* Thore, *Thorella verticillatoinundata* (Thore) Briq.

Range area (grAH km^2): PAL: 19300
Ecozone origin-status: PAL: E
World rarity status: VR
Ecozone-endemic macroregion occurrence: PAL: MED, W EUR
Life form(s): EM
Ploidy state: NA
Speciation age (Ma): NA
Populations also occur in terrestrial, wetland and/ or riparian habitats: YES
Long distance movement vectors: NA
Nuisance status and uses: no nuisance issues; no known uses
Endangered status (IUCN): NA

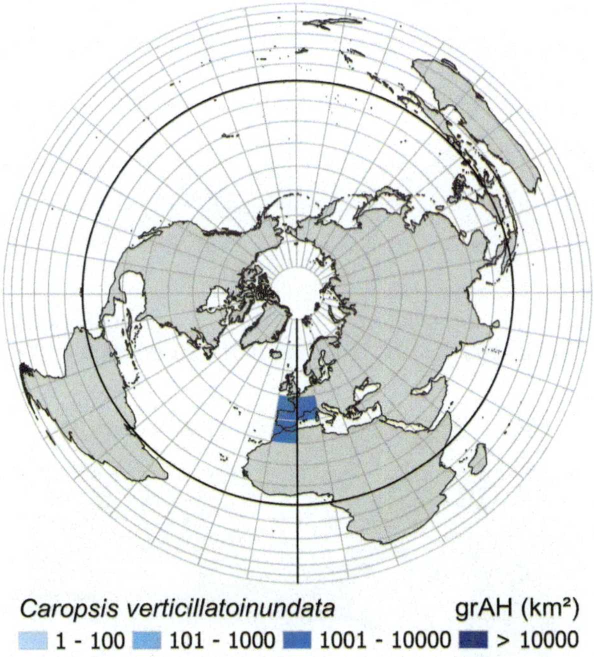

***Carum verticillatum* (L.) W.D.J.Koch**
Synonyms: *Trocdaris verticillata* (L.) Raf.

Range area (grAH km^2): PAL: 46600
Ecozone origin-status: PAL: E
World rarity status: R
Ecozone-endemic macroregion occurrence: PAL: MED, W EUR
Life form(s): EM

Ploidy state: D
Speciation age (Ma): NA
Populations also occur in terrestrial, wetland and/ or riparian habitats: YES
Long distance movement vectors: NA
Nuisance status and uses: no nuisance issues; food, folk medicine
Endangered status (IUCN): LC

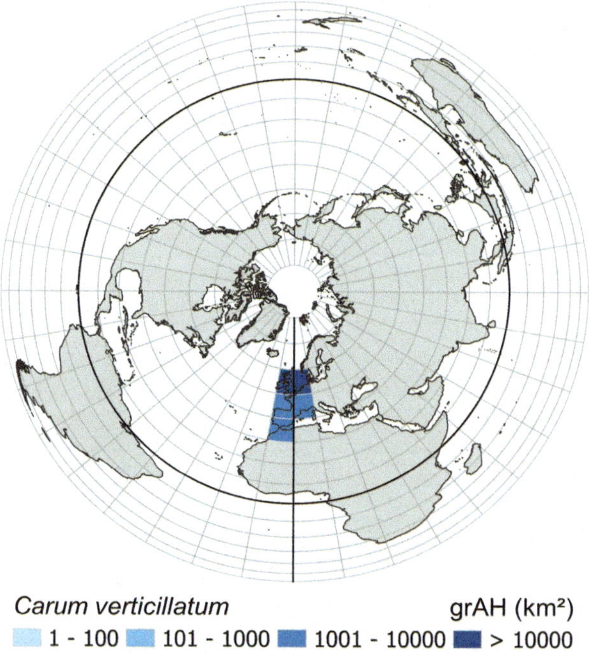

Carum verticillatum (L.) W.D.J.Koch
Image copyright: iNaturalist CC0

***Centella asiatica* (L.) Urb.**
Synonyms: *Hydrocotyle asiatica* L.

Range area (grAH km²): PAL: 61500; NEA: 34000; NEO: 216200; OR: 141800; AUS: 82400; AFR: 284600; present in OCE
Ecozone origin-status: PAL: N; NEA: I; NEO: I; OR: N; AUS: N; AFR: N; OCE: I
World rarity status: C
Ecozone-endemic macroregion occurrence: not endemic to any world ecozone
Life form(s): EM
Ploidy state: DP
Speciation age (Ma): NA
Populations also occur in terrestrial, wetland and/ or riparian habitats: YES
Long distance movement vectors: ANTHROPOCHORY
Nuisance status and uses: no nuisance issues; medical use as skin cream ingredient especially in India, and in traditional Chinese medicine systems, locally used for human food
Endangered status (IUCN): LC

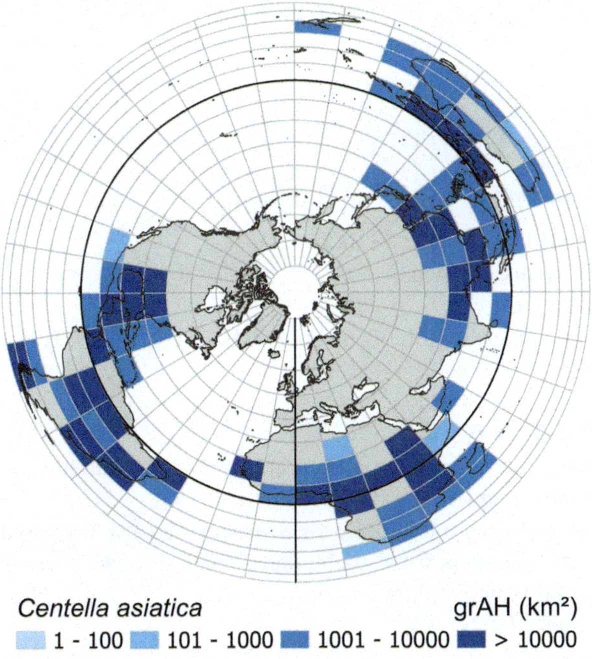

Centella asiatica (L.) Urb.
Image copyright: iNaturalist CC0

Cicuta bulbifera **L.**
Synonyms: *Keraskomion bulbiferum* (L.) Raf.

Range area (grAH km^2): NEA: 773000
Ecozone origin-status: NEA: E
World rarity status: C
Ecozone-endemic macroregion occurrence: NEA: ARC CAN, E NAM, W NAM
Life form(s): EM
Ploidy state: D
Speciation age (Ma): 4.0
Populations also occur in terrestrial, wetland and/ or riparian habitats: YES
Long distance movement vectors: NA
Nuisance status and uses: no aquatic weed issues; not used for trade or consumption. The entire plant is toxic to humans and livestock, particularly the roots, stem bases, and young shoots
Endangered status (IUCN): LC

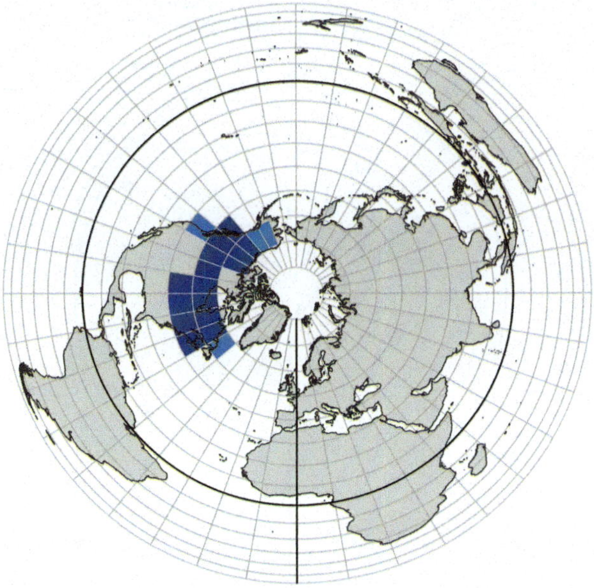

***Cicuta douglasii* (DC.) J.M.Coult. & Rose**
Synonyms: *Sium douglasii* DC.

Range area (grAH km²): NEA: 11700
Ecozone origin-status: NEA: E
World rarity status: R
Ecozone-endemic macroregion occurrence: NEA: ARC CAN, W NAM, SW NAM
Life form(s): EM
Ploidy state: P
Speciation age (Ma): 2.3
Populations also occur in terrestrial, wetland and/ or riparian habitats: YES
Long distance movement vectors: NA
Nuisance status and uses: no aquatic weed issues; no uses (extremely toxic to livestock and humans)
Endangered status (IUCN): LC

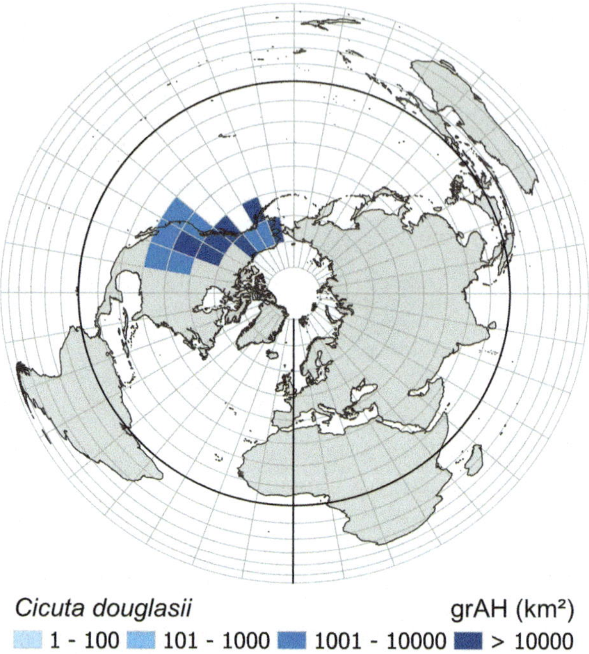

Cicuta douglasii (DC.) J.M.Coult. & Rose
Image copyright: iNaturalist CC0

***Cicuta maculata* L.**
Synonyms:

Range area (grAH km^2): NEA: 637800; NEO: 69700
Ecozone origin-status: NEA: N; NEO: N
World rarity status: R
Ecozone-endemic macroregion occurrence: not endemic to any world ecozone
Life form(s): EM
Ploidy state: D
Speciation age (Ma): 2.3
Populations also occur in terrestrial, wetland and/ or riparian habitats: YES
Long distance movement vectors: NA
Nuisance status and uses: no aquatic weed issues; medicinal uses
Endangered status (IUCN): LC

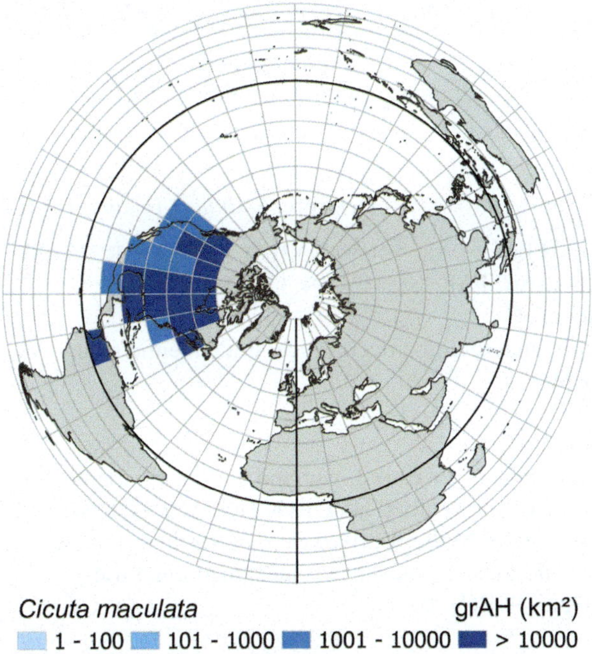

Cicuta maculata L.

***Cicuta virosa* L.**
Synonyms: *Cicutaria virosa* (L.) Delarbre, *Selinum virosum* (L.) E.H.L.Krause

Range area (grAH km^2): PAL: 856500; NEA: 402100; OR: 25100
Ecozone origin-status: PAL: N; NEA: N; OR: N
World rarity status: C
Ecozone-endemic macroregion occurrence: not endemic to any world ecozone
Life form(s): EM
Ploidy state: DP
Speciation age (Ma): 9.0
Populations also occur in terrestrial, wetland and/ or riparian habitats: YES
Long distance movement vectors: HYDROCHORY (PAL to OR); BIRD: *Anas platyrhynchos* (Northern Mallard); *Anas acuta* (Northern Pintail); ?AV
Nuisance status and uses: no nuisance issues; medicinal uses
Endangered status (IUCN): LC

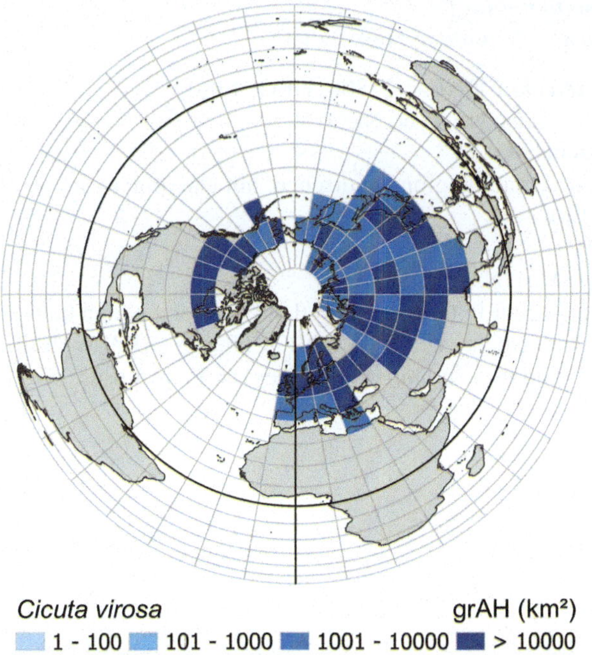

Cicuta virosa L.
Image copyright: J. C.Schou, with permission

***Cynosciadium digitatum* DC**
Synonyms: *Oenanthe digitata* Nutt. ex DC.

Range area (grAH km^2): NEA: 25800; NEO: 13800
Ecozone origin-status: NEA: N; NEO: N
World rarity status: R
Ecozone-endemic macroregion occurrence: not endemic to any world ecozone
Life form(s): EM
Ploidy state: H
Speciation age (Ma): NA
Populations also occur in terrestrial, wetland and/ or riparian habitats: YES
Long distance movement vectors: HYDROCHORY (NEA to NEO)
Nuisance status and uses: no nuisance issues; no known uses
Endangered status (IUCN): NA

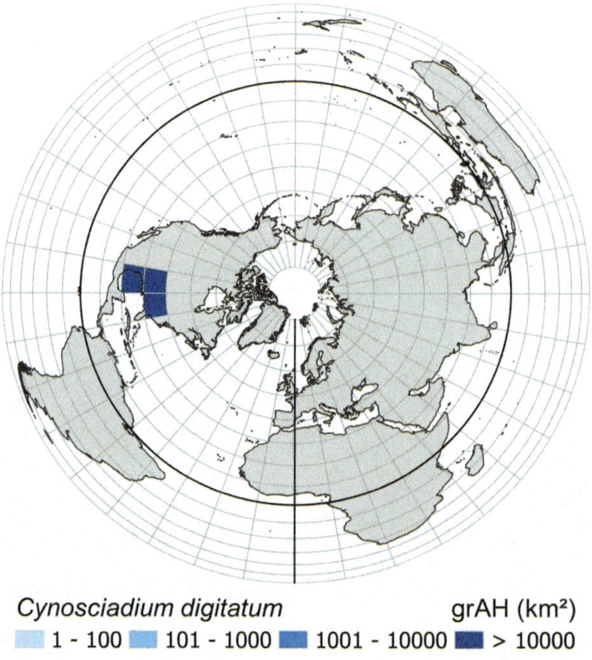

Cynosciadium digitatum DC
Image copyright: iNaturalist CC0

***Eryngium aquaticum* L.**
Synonyms:

Range area (grAH km^2): NEA: 21200; NEO: 10300
Ecozone origin-status: NEA: N; NEO: N
World rarity status: R
Ecozone-endemic macroregion occurrence: not endemic to any world ecozone
Life form(s): EM
Ploidy state: D
Speciation age (Ma): NA
Populations also occur in terrestrial, wetland and/ or riparian habitats: YES
Long distance movement vectors: HYDROCHORY (NEA to NEO)
Nuisance status and uses: no nuisance issues; medicinal and ornamental uses, horticulture: water garden plant
Endangered status (IUCN): NA

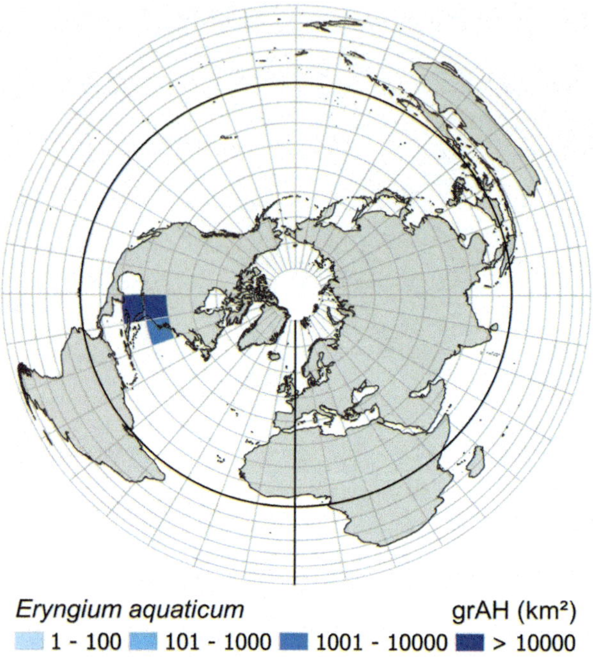

Eryngium aquaticum L.
Image copyright: iNaturalist CC0

***Eryngium aristulatum* Jeps.**
Synonyms:

Range area (grAH km^2): NEA: 11000
Ecozone origin-status: NEA: E
World rarity status: VR
Ecozone-endemic macroregion occurrence: NEA: W NAM, SW NAM
Life form(s): EM
Ploidy state: P
Speciation age (Ma): NA
Populations also occur in terrestrial, wetland and/ or riparian habitats: YES
Long distance movement vectors: NA
Nuisance status and uses: no nuisance issues; horticulture: water garden plant
Endangered status (IUCN): NA

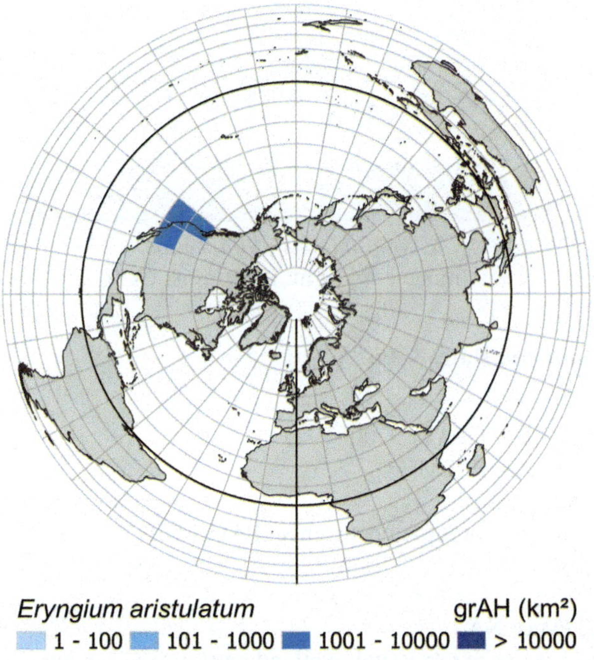

Eryngium aristulatum Jeps.
Image copyright: iNaturalist CC0

Eryngium atlanticum Batt. & Pit.
Synonyms:

Range area (grAH km^2): PAL: 5200
Ecozone origin-status: PAL: E
World rarity status: VR
Ecozone-endemic macroregion occurrence: PAL: MED
Life form(s): EM
Ploidy state: D
Speciation age (Ma): NA
Populations also occur in terrestrial, wetland and/ or riparian habitats: YES
Long distance movement vectors: NA
Nuisance status and uses: no nuisance issues; no known uses
Endangered status (IUCN): NT

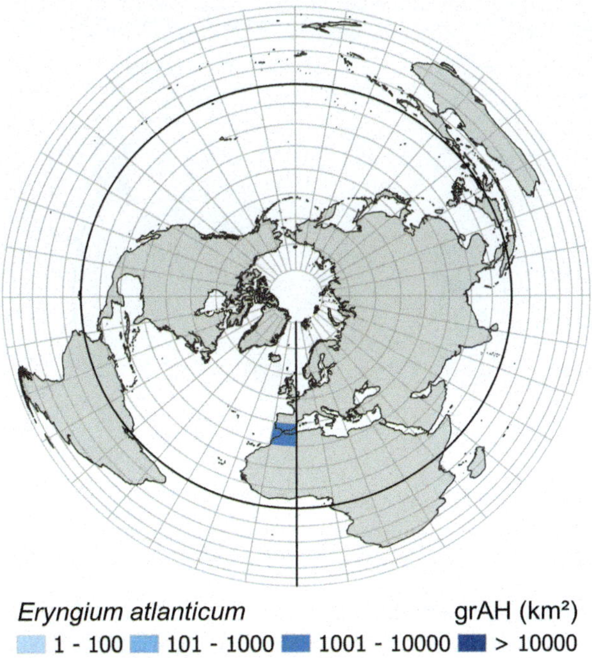

***Eryngium cervantesii* F. Delaroche**
Synonyms: *Eryngium viviparum* Cerv. ex F.Delaroche

Range area (grAH km²): NEO: 19600
Ecozone origin-status: NEO: E
World rarity status: VR
Ecozone-endemic macroregion occurrence: NEO: MG MX, MES AM
Life form(s): EM
Ploidy state: NA
Speciation age (Ma): NA
Populations also occur in terrestrial, wetland and/ or riparian habitats: YES
Long distance movement vectors: NA
Nuisance status and uses: no nuisance issues; no known uses
Endangered status (IUCN): NA

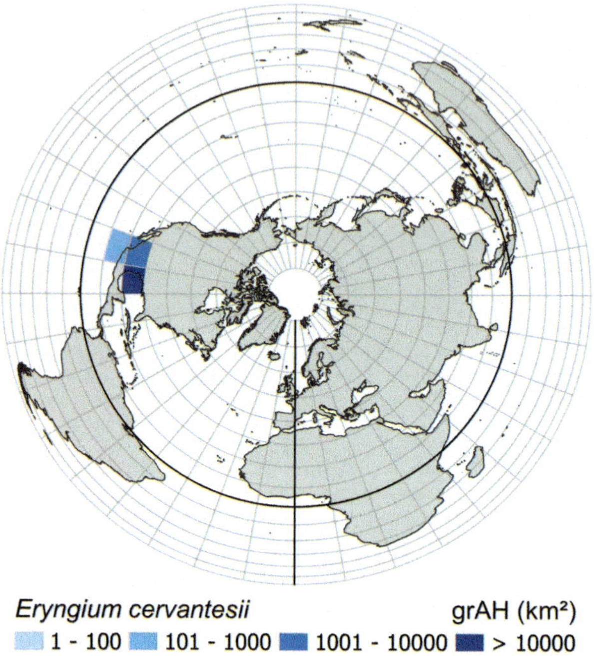

***Eryngium corniculatum* Lam.**
Synonyms:

Range area (grAH km^2): PAL: 20500
Ecozone origin-status: PAL: E
World rarity status: VR
Ecozone-endemic macroregion occurrence: PAL: MED, W EUR
Life form(s): EM
Ploidy state: D
Speciation age (Ma): NA
Populations also occur in terrestrial, wetland and/ or riparian habitats: YES
Long distance movement vectors: NA
Nuisance status and uses: no nuisance issues; no known uses
Endangered status (IUCN): LC

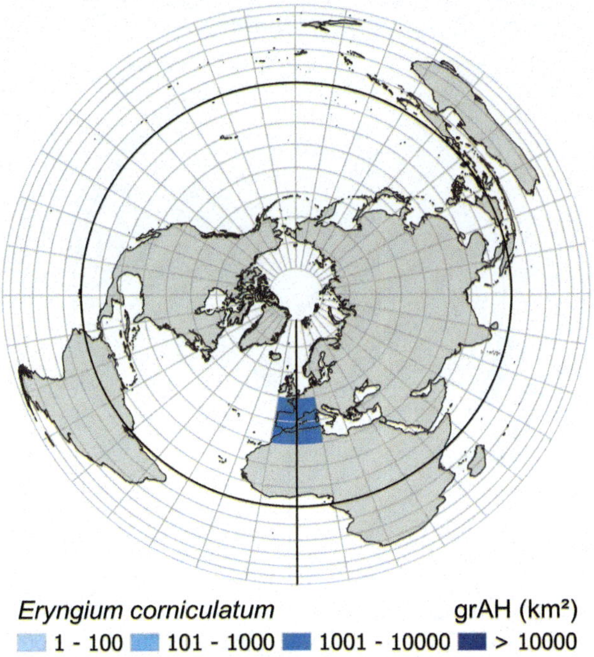

Eryngium coronatum Hook. & Arn.
Synonyms: *Eryngium coronopifolium* Decne.

Range area (grAH km^2): NEO: 97600
Ecozone origin-status: NEO: E
World rarity status: R
Ecozone-endemic macroregion occurrence: NEO: CHA, PAR
Life form(s): EM
Ploidy state: D
Speciation age (Ma): NA
Populations also occur in terrestrial, wetland and/ or riparian habitats: YES
Long distance movement vectors: NA
Nuisance status and uses: no nuisance issues; medicinal potential
Endangered status (IUCN): NA

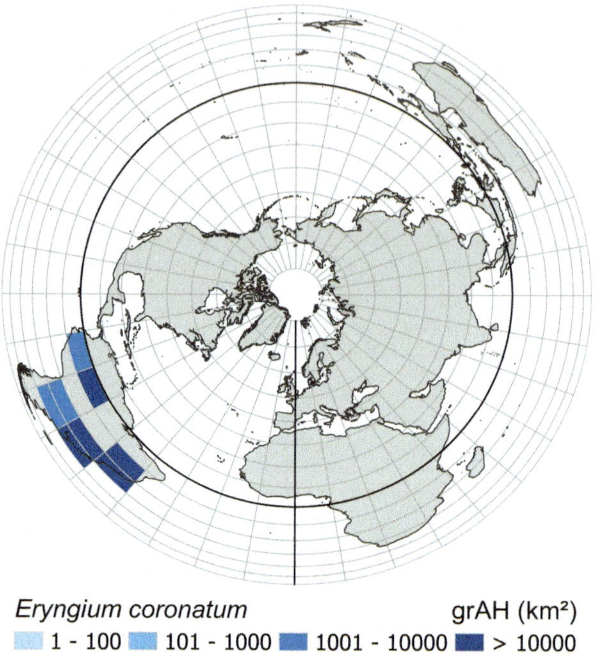

Eryngium coronatum Hook. & Arn.
Image copyright: iNaturalist CC-BY @romigaleota

Eryngium ebracteatum **Lam.**
Synonyms:

Range area (grAH km^2): NEO: 142400
Ecozone origin-status: NEO: E
World rarity status: R

Ecozone-endemic macroregion occurrence: NEO: MES AM, NWSA, CHA, AND, PAR
Life form(s): EM
Ploidy state: D
Speciation age (Ma): NA
Populations also occur in terrestrial, wetland and/ or riparian habitats: YES
Long distance movement vectors: NA
Nuisance status and uses: no nuisance issues; medicinal and food uses
Endangered status (IUCN): NA

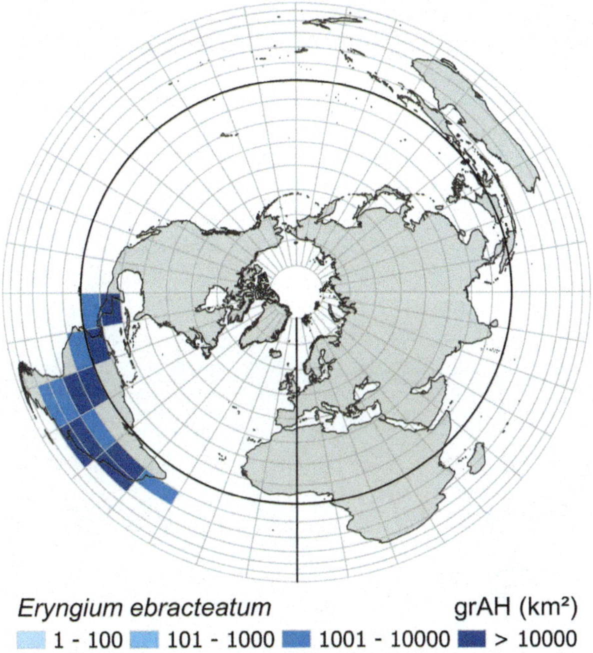

Eryngium ebracteatum Lam.
Image copyright: iNaturalist CC0

Eryngium eburneum **Decne.**
Synonyms:

Range area (grAH km^2): NEO: 47900
Ecozone origin-status: NEO: E
World rarity status: R
Ecozone-endemic macroregion occurrence: NEO: CHA, PAR
Life form(s): EM
Ploidy state: D
Speciation age (Ma): 4.0
Populations also occur in terrestrial, wetland and/ or riparian habitats: YES
Long distance movement vectors: NA
Nuisance status and uses: no nuisance issues; medicinal, dyeing and fibre uses
Endangered status (IUCN): NA

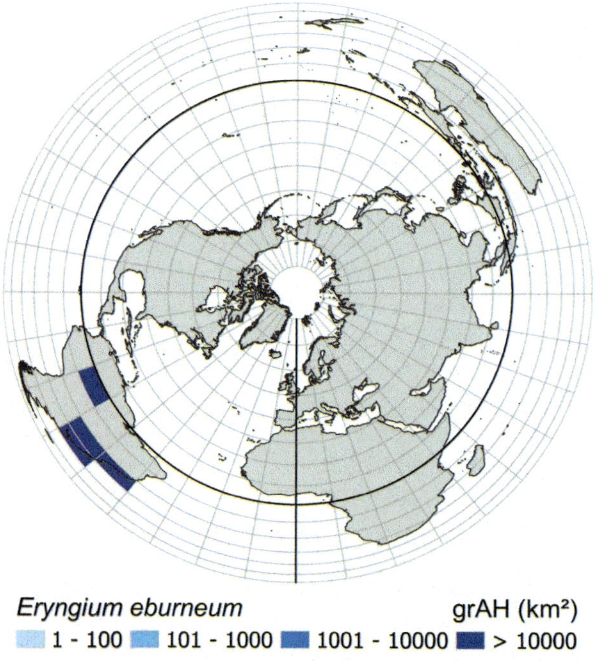

Eryngium eburneum Decne.

***Eryngium floribundum* Cham. & Schltdl.**
Synonyms:

Range area (grAH km^2): NEO: 67600
Ecozone origin-status: NEO: E
World rarity status: VR
Ecozone-endemic macroregion occurrence: NEO: AMZ, CHA, PAR
Life form(s): EM
Ploidy state: D
Speciation age (Ma): NA
Populations also occur in terrestrial, wetland and/ or riparian habitats: YES
Long distance movement vectors: NA
Nuisance status and uses: no nuisance issues; medicinal potential
Endangered status (IUCN): NA

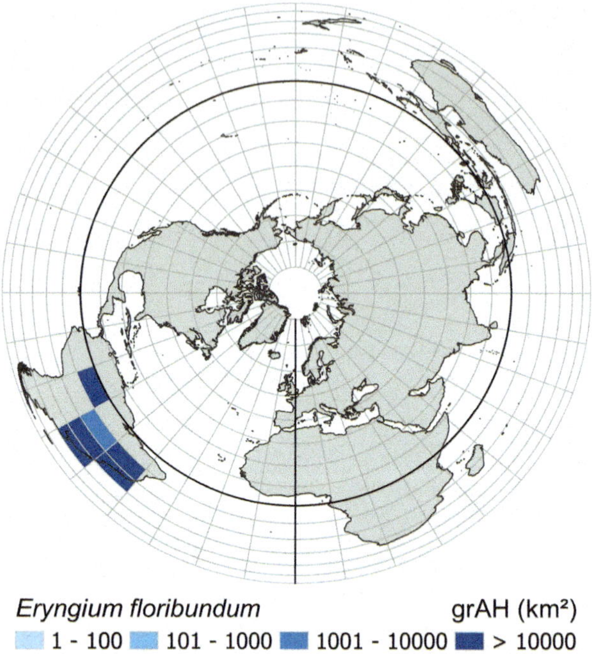

Eryngium floribundum Cham. & Schltdl.
Image copyright: T. Lobato-de Magalhães

***Eryngium fluitans* M.E. Jones**
Synonyms:

Range area (grAH km^2): NEO: 12000
Ecozone origin-status: NEO: E
World rarity status: VR
Ecozone-endemic macroregion occurrence: NEO: MG MX, MES AM

Life form(s): EM
Ploidy state: D
Speciation age (Ma): NA
Populations also occur in terrestrial, wetland and/ or riparian habitats: YES
Long distance movement vectors: NA
Nuisance status and uses: no nuisance issues; bioactive
Endangered status (IUCN): NA

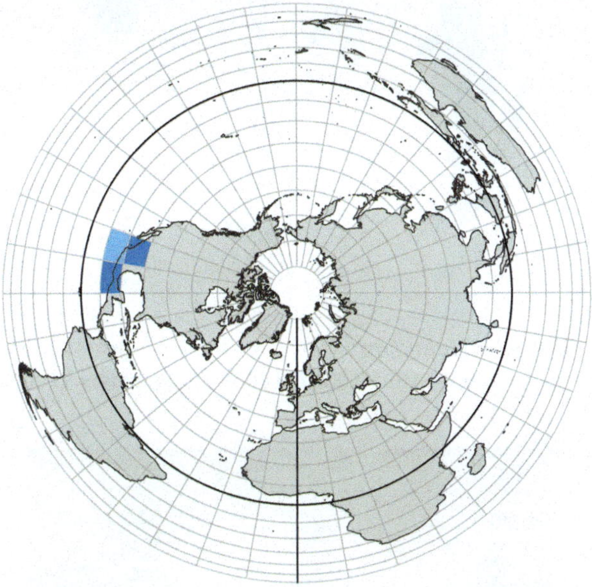

***Eryngium foetidum* L.**
Synonyms:

Range area (grAH km²): PAL: 22700; NEA: 13000; NEO: 311900; OR: 83100; AUS: 21100; AFR: 34800; present in OCE
Ecozone origin-status: PAL: I; NEA: I; NEO: E(I); OR: I; AUS: I; AFR: I; OCE: I
World rarity status: C
Ecozone-endemic macroregion occurrence: MG MX, MES AM, NWSA, AND, AMZ, CHA, PAR
Life form(s): EM
Ploidy state: D
Speciation age (Ma): NA
Populations also occur in terrestrial, wetland and/ or riparian habitats: YES
Long distance movement vectors: ANTHROPOCHORY
Nuisance status and uses: no nuisance issues; medicinal, phytoremediation and food uses
Endangered status (IUCN): NA

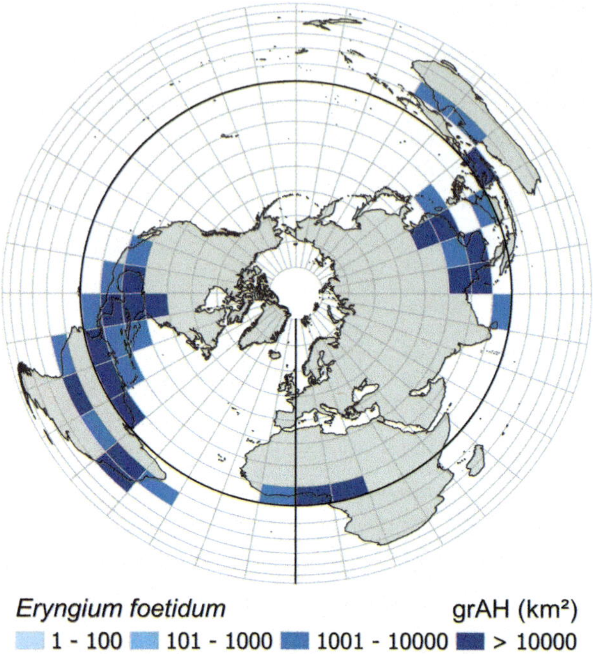

Eryngium foetidum L.
Image copyright: iNaturalist CC0

Eryngium galioides **Lam.**
Synonyms:

Range area (grAH km^2): PAL: 10200
Ecozone origin-status: PAL: E
World rarity status: VR
Ecozone-endemic macroregion occurrence: PAL: MED
Life form(s): EM
Ploidy state: D
Speciation age (Ma): NA
Populations also occur in terrestrial, wetland and/ or riparian habitats: YES
Long distance movement vectors: NA
Nuisance status and uses: no nuisance issues; no known uses
Endangered status (IUCN): EN

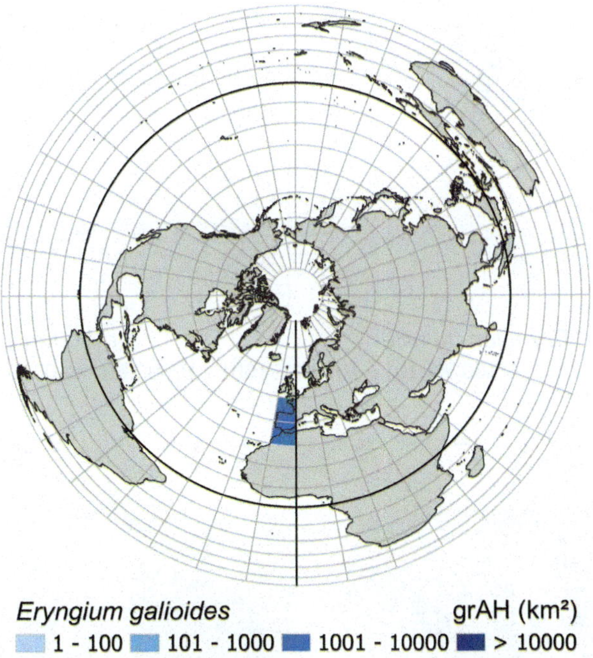

Eryngium humile **Cav.**
Synonyms:

Range area (grAH km^2): NEO: 45600
Ecozone origin-status: NEO: E
World rarity status: R
Ecozone-endemic macroregion occurrence: NEO: MG MX, MES AM, NWSA, AND
Life form(s): EM
Ploidy state: D

Speciation age (Ma): NA
Populations also occur in terrestrial, wetland and/ or riparian habitats: YES
Long distance movement vectors: NA
Nuisance status and uses: no nuisance issues; folk medicine uses
Endangered status (IUCN): NA

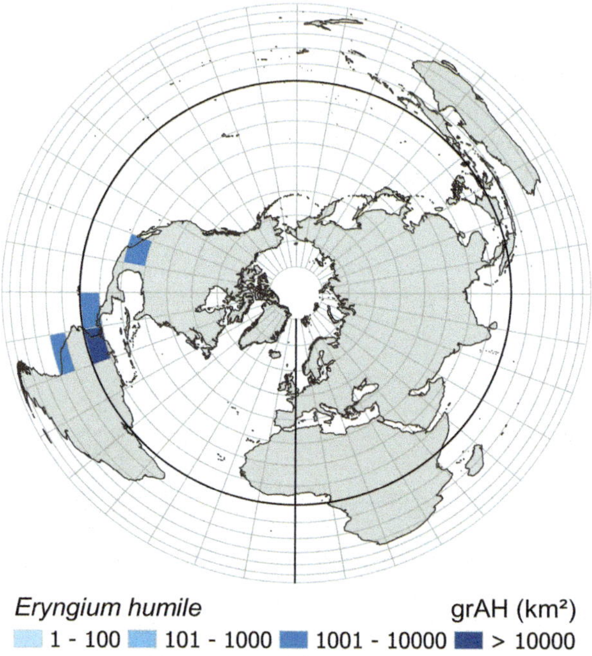

Eryngium humile Cav.
Image copyright: iNaturalist CC0

Eryngium maroccanum **Pitard**
Synonyms:

Range area (grAH km^2): PAL: 5200
Ecozone origin-status: PAL: E
World rarity status: VR
Ecozone-endemic macroregion occurrence: PAL: MED
Life form(s): EM
Ploidy state: NA
Speciation age (Ma): NA
Populations also occur in terrestrial, wetland and/ or riparian habitats: YES
Long distance movement vectors: NA
Nuisance status and uses: no nuisance issues; no known uses
Endangered status (IUCN): NT

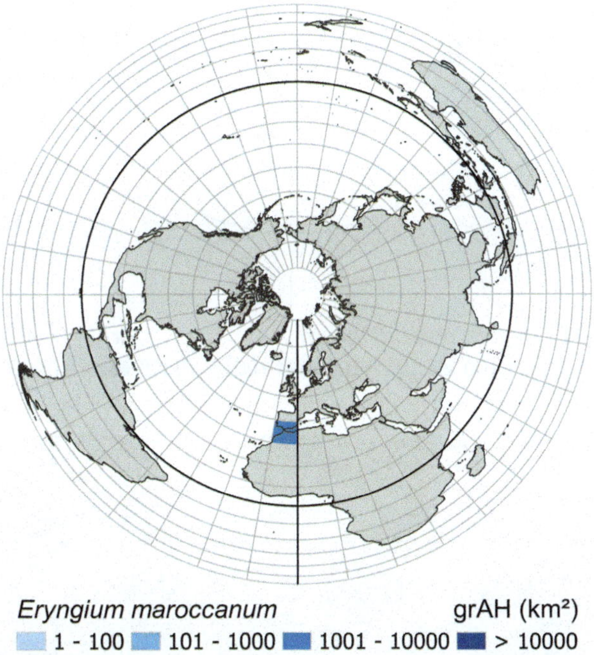

Eryngium nasturtiifolium **Juss. ex F.Delaroche**
Synonyms:

Range area (grAH km^2): NEA: 3500; NEO: 56000
Ecozone origin-status: NEA: N; NEO: N
World rarity status: R
Ecozone-endemic macroregion occurrence: not endemic to any world ecozone
Life form(s): EM
Ploidy state: NA

Speciation age (Ma): NA
Populations also occur in terrestrial, wetland and/ or riparian habitats: YES
Long distance movement vectors: NA
Nuisance status and uses: no nuisance issues; folk medicine
Endangered status (IUCN): NA

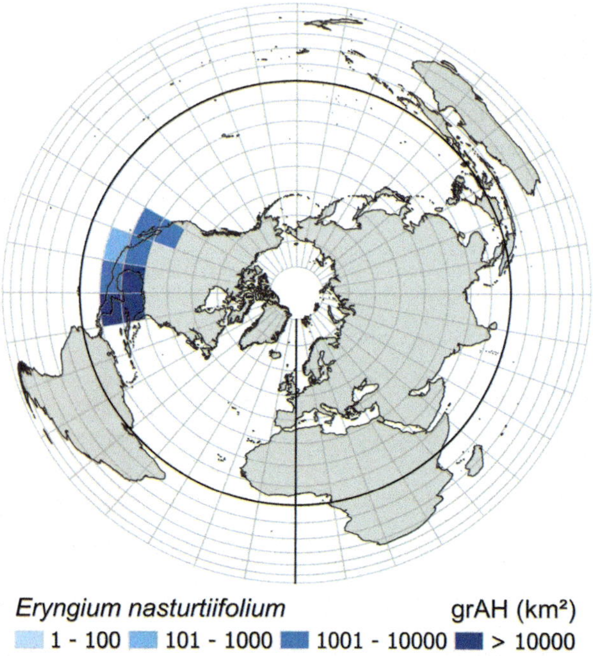

Eryngium nasturtiifolium Juss. ex F.Delaroche
Image copyright: iNaturalist CC-BY @alexiz

Eryngium paniculatum **Cav. & Dombey. ex F.Delaroche**
Synonyms:

Range area (grAH km^2): NEO: 58200
Ecozone origin-status: NEO: E
World rarity status: R
Ecozone-endemic macroregion occurrence: NEO: AND, CHA, PAR
Life form(s): EM
Ploidy state: P
Speciation age (Ma): NA
Populations also occur in terrestrial, wetland and/ or riparian habitats: YES
Long distance movement vectors: NA
Nuisance status and uses: no nuisance issues; garden plant, also medicinal and food uses
Endangered status (IUCN): NA

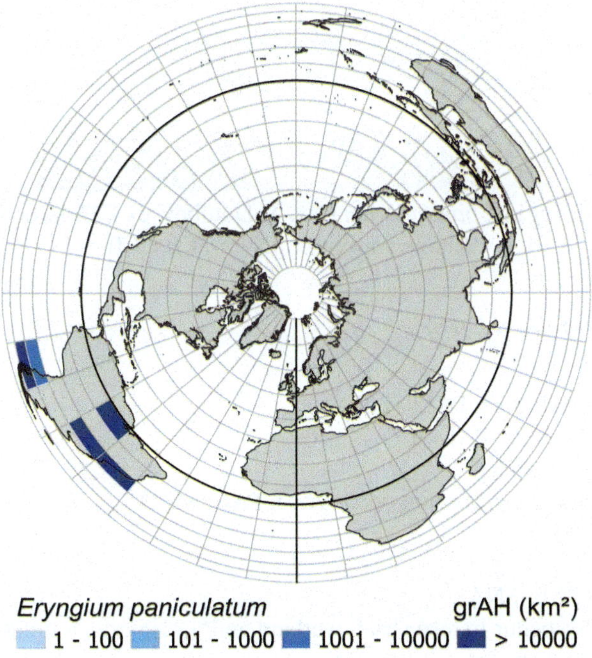

Eryngium paniculatum Cav. & Dombey. ex F.Delaroche
Image copyright: iNaturalist CC0

***Eryngium variifolium* Coss.**
Synonyms:

Range area (grAH km^2): PAL: 5200
Ecozone origin-status: PAL: E
World rarity status: VR
Ecozone-endemic macroregion occurrence: PAL: MED
Life form(s): EM
Ploidy state: D
Speciation age (Ma): NA
Populations also occur in terrestrial, wetland and/ or riparian habitats: YES
Long distance movement vectors: NA
Nuisance status and uses: no nuisance issues; horticulture: garden plant
Endangered status (IUCN): VU

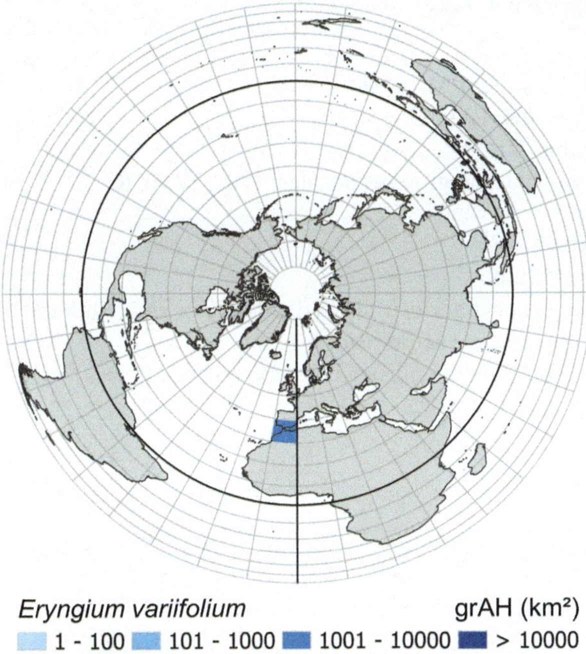

Eryngium variifolium Coss.
Image copyright: iNaturalist CC0

***Eryngium viviparum* J. Gay**
Synonyms:

Range area (grAH km^2): PAL: 5000
Ecozone origin-status: PAL: E
World rarity status: VR
Ecozone-endemic macroregion occurrence: PAL: W EUR
Life form(s): EM
Ploidy state: NA
Speciation age (Ma): NA
Populations also occur in terrestrial, wetland and/ or riparian habitats: YES
Long distance movement vectors: NA
Nuisance status and uses: no nuisance issues; no known uses
Endangered status (IUCN): EN

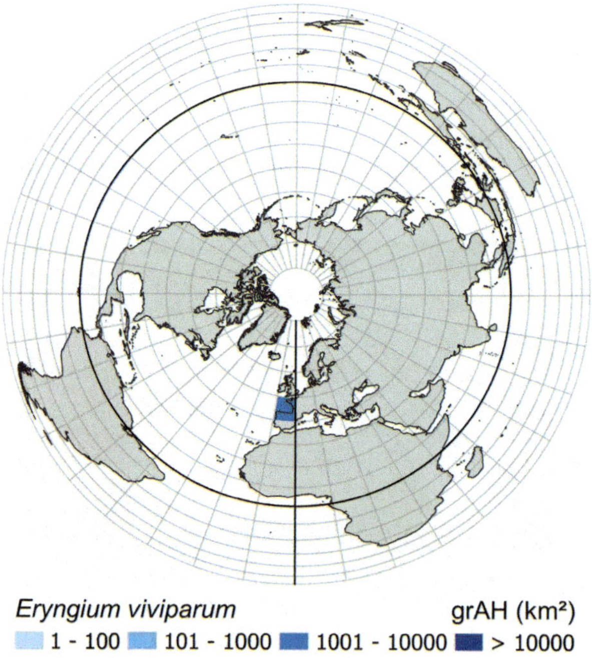

Eryngium viviparum J. Gay
Image copyright: iNaturalist CC0

***Lilaeopsis brasiliensis* (Glaz.) Affolter**
Synonyms: *Lilaeopsis minor* (A.W.Hill) Pérez-Mor.

Range area (grAH km^2): NEO: 51500; AUS: 5400
Ecozone origin-status: NEO: E(I); AUS: I
World rarity status: R
Ecozone-endemic macroregion occurrence: NEO: CHA, PAR
Life form(s): SU
Ploidy state: NA
Speciation age (Ma): 2.0
Populations also occur in terrestrial, wetland and/ or riparian habitats: NO
Long distance movement vectors: ANTHROPOCHORY
Nuisance status and uses: no nuisance issues; phytoremediation, aquarium trade
Endangered status (IUCN): NA

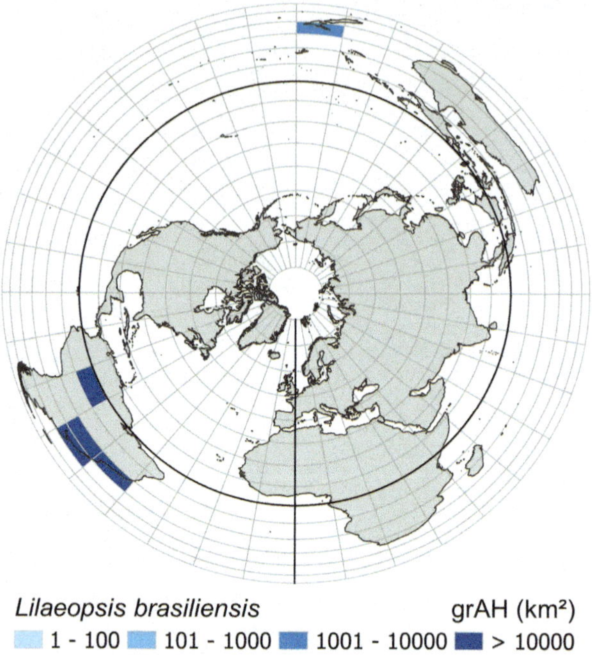

Lilaeopsis brasiliensis (Glaz.) Affolter

Lilaeopsis carolinensis **J.M.Coult. & Rose**
Synonyms: *Crantzia carolinensis* (J.M.Coult. & Rose) Chodat

Range area (grAH km^2): PAL: 5000; NEA: 25800; NEO: 60600
Ecozone origin-status: PAL: I; NEA: N; NEO: N
World rarity status: R
Ecozone-endemic macroregion occurrence: not endemic to any world ecozone
Life form(s): EM, FR, SU
Ploidy state: D
Speciation age (Ma): 0.8
Populations also occur in terrestrial, wetland and/ or riparian habitats: NO
Long distance movement vectors: ANTHROPOCHORY; HYDROCHORY (NEA to NEO)
Nuisance status and uses: no nuisance issues; aquarium trade
Endangered status (IUCN): NA

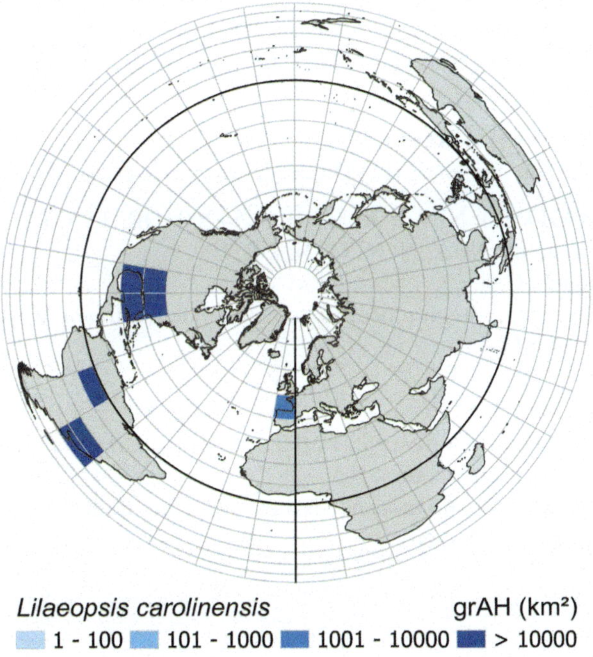

Lilaeopsis carolinensis J.M.Coult. & Rose

Lilaeopsis chinensis **(L.) Kuntze**
Synonyms: *Crantzia chinensis* (L.) Druce, *Hydrocotyle chinensis* L.

Range area (grAH km^2): NEA: 93000; NEO: 54200; OR: 15700
Ecozone origin-status: NEA: N; NEO: N; OR: N
World rarity status: C
Ecozone-endemic macroregion occurrence: not endemic to any world ecozone
Life form(s): EM, SU
Ploidy state: D
Speciation age (Ma): 3.0
Populations also occur in terrestrial, wetland and/ or riparian habitats: NO
Long distance movement vectors: NA
Nuisance status and uses: no nuisance issues; aquarium trade
Endangered status (IUCN): NA

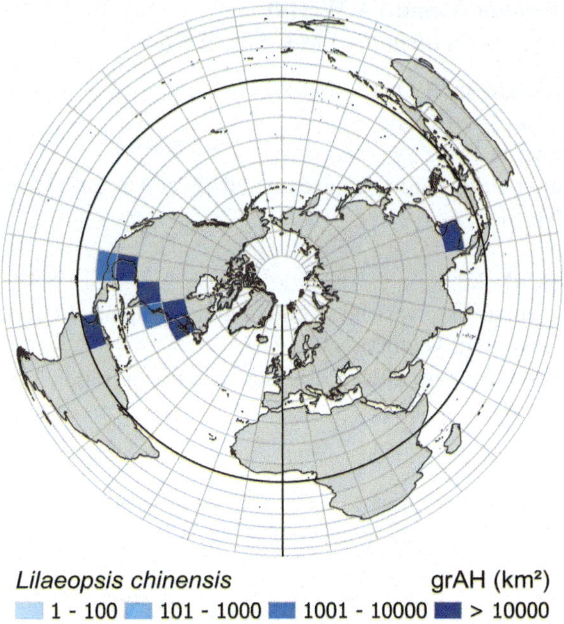

Lilaeopsis chinensis (L.) Kuntze

Lilaeopsis macloviana **(Gand.) A.W. Hill**
Synonyms: *Lilaeopsis occidentalis* J.M.Coult. & Rose

Range area (grAH km²): NEO: 121500
Ecozone origin-status: NEO: E
World rarity status: R
Ecozone-endemic macroregion occurrence: NEO: NWSA, CHA, AND
Life form(s): EM, SU
Ploidy state: NA
Speciation age (Ma): 2.5
Populations also occur in terrestrial, wetland and/ or riparian habitats: NO
Long distance movement vectors: NA
Nuisance status and uses: no nuisance issues; aquarium trade
Endangered status (IUCN): NA

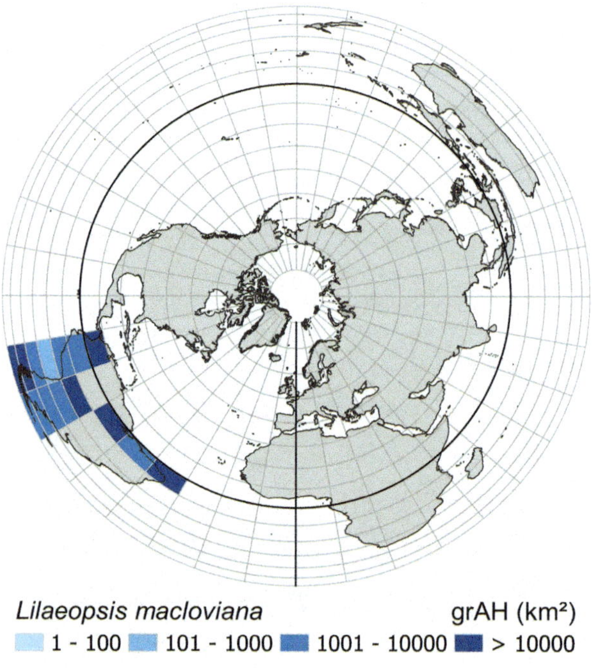

Lilaeopsis macloviana (Gand.) A.W. Hill
Image copyright: iNaturalist CC0

Lilaeopsis masonii **Mathias & Constance**
Synonyms:

Range area (grAH km^2): NEA: 1000
Ecozone origin-status: NEA: E
World rarity status: VR
Ecozone-endemic macroregion occurrence: NEA: W NAM
Life form(s): EM, SU
Ploidy state: DP
Speciation age (Ma): NA
Populations also occur in terrestrial, wetland and/ or riparian habitats: NO
Long distance movement vectors: NA
Nuisance status and uses: no nuisance issues; no known uses
Endangered status (IUCN): NA

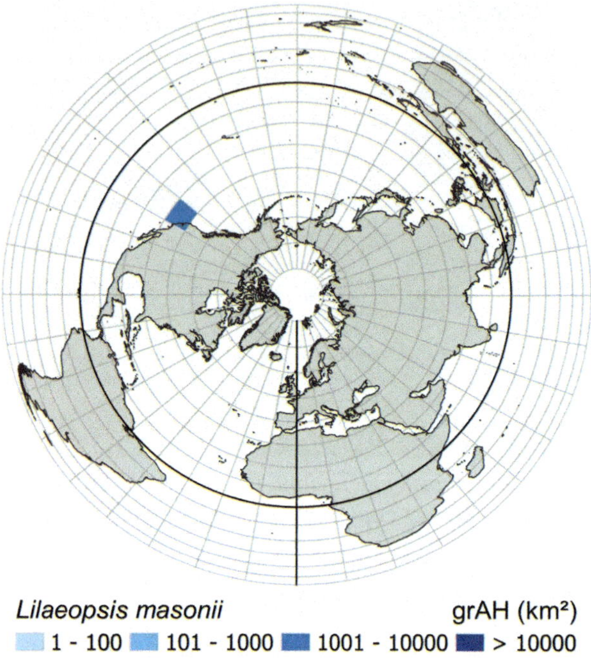

***Lilaeopsis mauritiana* G.Petersen & Affolter**
Synonyms:

Range area (grAH km^2): AFR: 2000
Ecozone origin-status: AFR: E
World rarity status: VR
Ecozone-endemic macroregion occurrence: AFR: MAD
Life form(s): SU, ?EM,
Ploidy state: D
Speciation age (Ma): 2.0
Populations also occur in terrestrial, wetland and/ or riparian habitats: NO
Long distance movement vectors: NA
Nuisance status and uses: no nuisance issues; aquarium trade
Endangered status (IUCN): NA

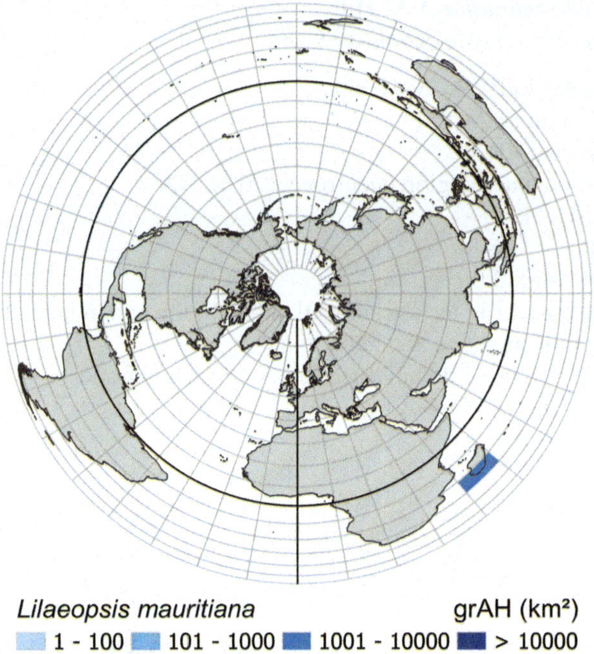

Lilaeopsis mauritiana G.Petersen & Affolter
Image copyright: iNaturalist CC0

Lilaeopsis novae-zelandiae **A.W.Hill**
Synonyms: *Crantzia novae-zelandiae* Gand.

Range area (grAH km^2): AUS: 16900
Ecozone origin-status: AUS: E
World rarity status: VR
Ecozone-endemic macroregion occurrence: AUS: NZ
Life form(s): SU, ?EM,
Ploidy state: P
Speciation age (Ma): 3.0
Populations also occur in terrestrial, wetland and/ or riparian habitats: NO
Long distance movement vectors: NA
Nuisance status and uses: no nuisance issues; aquarium trade
Endangered status (IUCN): NA

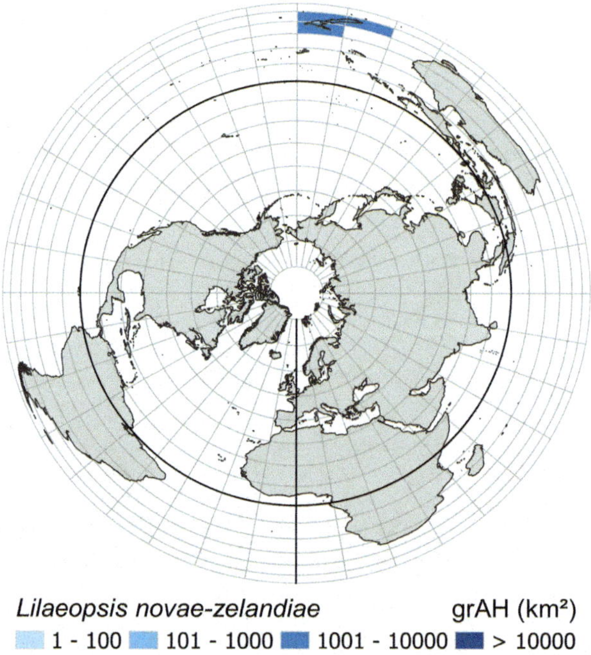

Lilaeopsis novae-zelandiae A.W.Hill
Image copyright: iNaturalist CC0

***Lilaeopsis occidentalis* J.M. Coult. & Rose**
Synonyms: *Crantziola occidentalis* (J.M.Coult. & Rose) Koso-Pol.

Range area (grAH km^2): NEA: 33200; NEO: 19100
Ecozone origin-status: NEA: N; NEO: N
World rarity status: R
Ecozone-endemic macroregion occurrence: not endemic to any world ecozone
Life form(s): SU, EM,
Ploidy state: P
Speciation age (Ma): 2.8
Populations also occur in terrestrial, wetland and/ or riparian habitats: YES
Long distance movement vectors: HYDROCHORY (NEA to NEO)
Nuisance status and uses: no nuisance issues; occasional food use
Endangered status (IUCN): LC

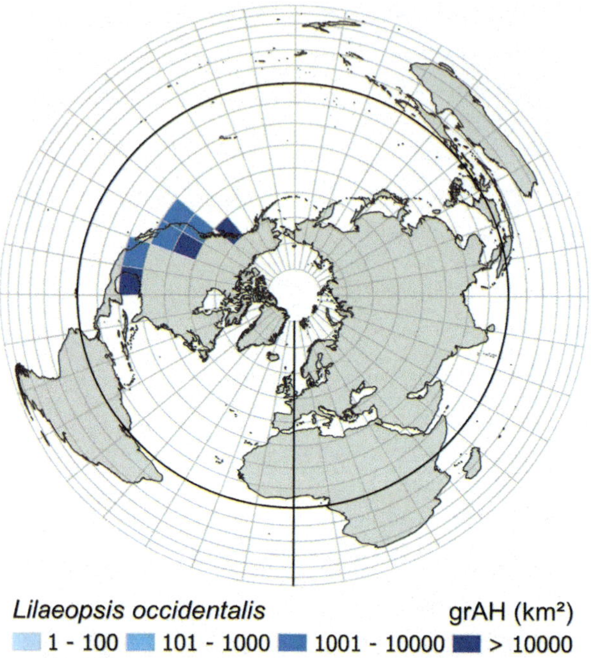

Lilaeopsis occidentalis J.M. Coult. & Rose
Image copyright: iNaturalist CC-BY @kueda

Lilaeopsis polyantha **(Gand.) H.Eichler**
Synonyms: *Crantzia polyantha* Gand., *Lilaeopsis australica* A.W.Hill

Range area (grAH km^2): AUS: 22900
Ecozone origin-status: AUS: E
World rarity status: VR
Ecozone-endemic macroregion occurrence: AUS: ESA, WA
Life form(s): SU, EM,
Ploidy state: NA
Speciation age (Ma): NA
Populations also occur in terrestrial, wetland and/ or riparian habitats: YES
Long distance movement vectors: NA
Nuisance status and uses: no nuisance issues; aquarium trade
Endangered status (IUCN): NA

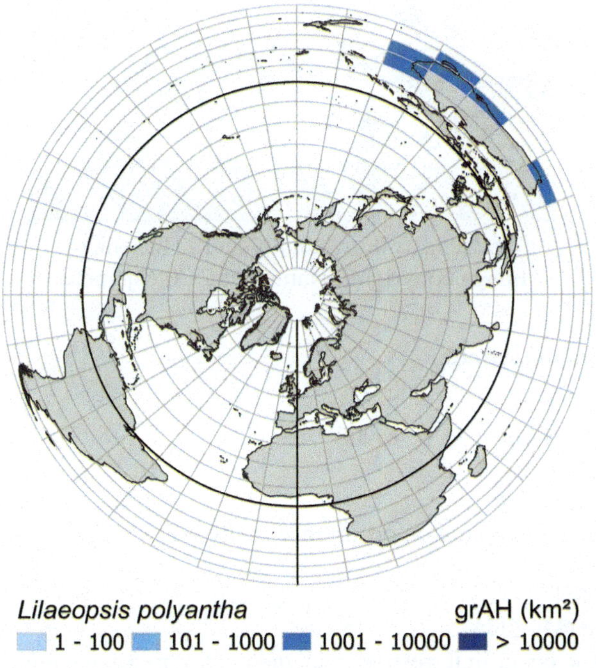

Lilaeopsis polyantha (Gand.) H.Eichler

***Lilaeopsis ruthiana* Affolter**
Synonyms:

Range area (grAH km^2): AUS: 16900
Ecozone origin-status: AUS: E
World rarity status: VR
Ecozone-endemic macroregion occurrence: AUS: NZ
Life form(s): SU
Ploidy state: NA
Speciation age (Ma): NA
Populations also occur in terrestrial, wetland and/ or riparian habitats: NO
Long distance movement vectors: NA
Nuisance status and uses: no nuisance issues; aquarium trade
Endangered status (IUCN): NA

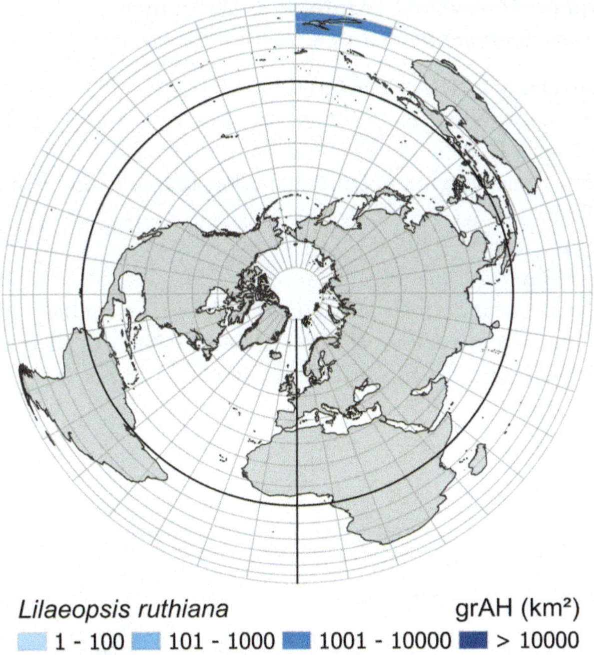

Lilaeopsis ruthiana Affolter
Image copyright: iNaturalist CC0 @john_barkla (John Barkla)

Limnosciadium pinnatum **(DC.) Mathias & Constance**
Synonyms: *Cynosciadium pinnatum* DC., *Oenanthe pinnata* (DC.) Koso-Pol.

Range area (grAH km^2): NEA: 12800; NEO: 13800
Ecozone origin-status: NEA: N; NEO: N
World rarity status: VR
Ecozone-endemic macroregion occurrence: not endemic to any world ecozone
Life form(s): EM
Ploidy state: NA
Speciation age (Ma): NA
Populations also occur in terrestrial, wetland and/ or riparian habitats: YES
Long distance movement vectors: HYDROCHORY (NEA to NEO)
Nuisance status and uses: no nuisance issues; no known uses
Endangered status (IUCN): NA

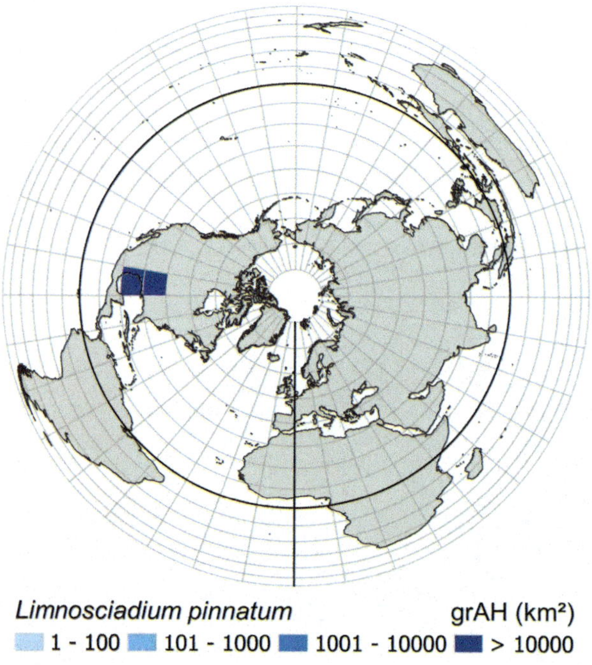

Limnosciadium pumilum **(Engelm. & A. Gray) Mathias & Constance**
Synonyms: *Limnosciadium pinnatum* (DC.) Mathias & Constance

Range area (grAH km^2): NEA: 12800; NEO: 13800
Ecozone origin-status: NEA: N; NEO: N
World rarity status: VR
Ecozone-endemic macroregion occurrence: not endemic to any world ecozone
Life form(s): EM

Ploidy state: NA
Speciation age (Ma): NA
Populations also occur in terrestrial, wetland and/ or riparian habitats: YES
Long distance movement vectors: HYDROCHORY (NEA to NEO)
Nuisance status and uses: no nuisance issues; no known uses
Endangered status (IUCN): NA

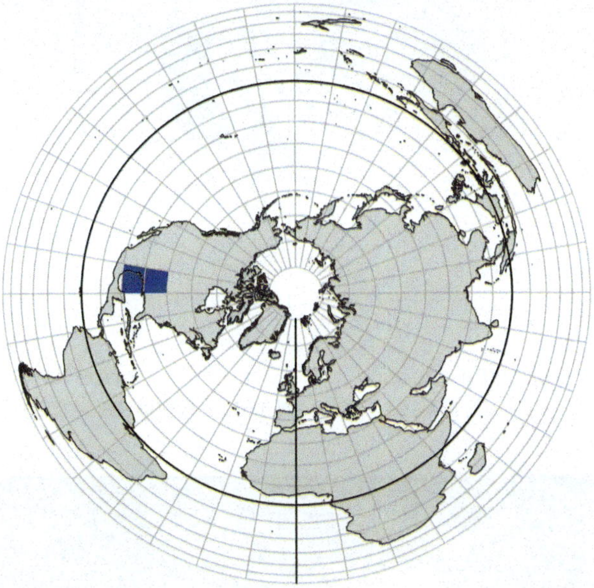

***Oenanthe aquatica* (L.) Poir.**
Synonyms: *Phellandrium aquaticum* L.

Range area (grAH km^2): PAL: 447900; AUS: 7400
Ecozone origin-status: PAL: E(I); AUS: I
World rarity status: C
Ecozone-endemic macroregion occurrence: PAL: MED, W EUR, N EUR, E EUR, UR/ W SIB, CH, E SIB
Life form(s): SU
Ploidy state: DP
Speciation age (Ma): NA
Populations also occur in terrestrial, wetland and/ or riparian habitats: NO
Long distance movement vectors: ANTHROPOCHORY; BIRD: *Spatula querquedula* (Garganey); *Anas platyrhynchos* (Northern Mallard)
Nuisance status and uses: no nuisance issues; bioactive
Endangered status (IUCN): LC

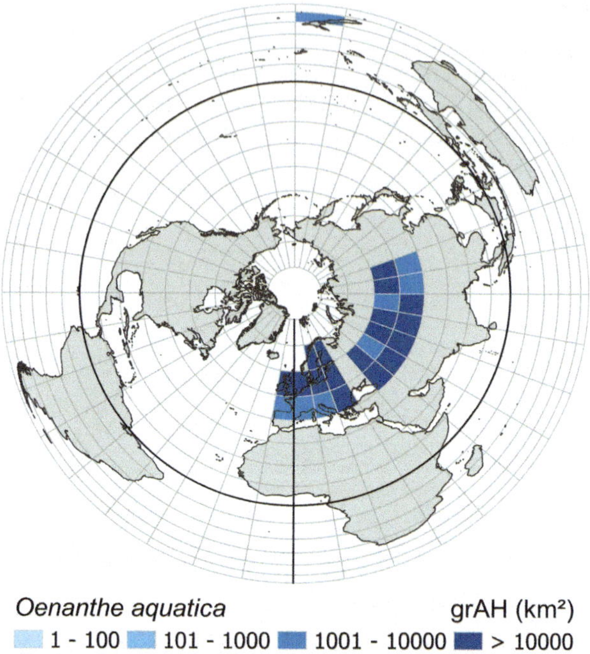

Oenanthe aquatica (L.) Poir.
Image copyright: J. C.Schou, with permission

Oenanthe crocata **L.**
Synonyms:

Range area (grAH km^2): PAL: 79500
Ecozone origin-status: PAL: E
World rarity status: R
Ecozone-endemic macroregion occurrence: PAL: MED, W EUR, N EUR
Life form(s): EM
Ploidy state: D
Speciation age (Ma): NA
Populations also occur in terrestrial, wetland and/ or riparian habitats: YES
Long distance movement vectors: NA
Nuisance status and uses: no aquatic weed issues; homeopathic medicine, highly toxic to humans and livestock
Endangered status (IUCN): LC

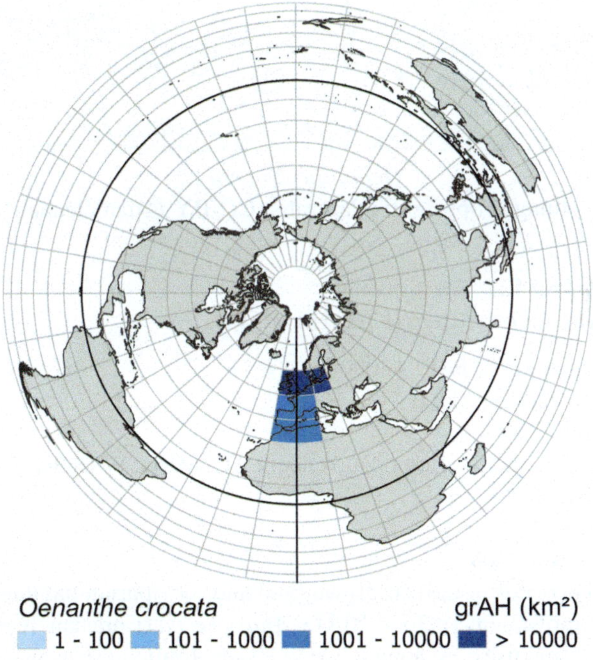

Oenanthe crocata L.

***Oenanthe fistulosa* L.**
Synonyms: *Selinum fistulosum* (L.) E.H.L.Krause

Range area (grAH km^2): PAL: 130000
Ecozone origin-status: PAL: E
World rarity status: R
Ecozone-endemic macroregion occurrence: PAL: MED, W EUR, N EUR, E EUR
Life form(s): SU, EM
Ploidy state: D
Speciation age (Ma): NA
Populations also occur in terrestrial, wetland and/ or riparian habitats: YES
Long distance movement vectors: BIRD: *Anas crecca* (Eurasian Teal)
Nuisance status and uses: no aquatic weed issues; limited use as water garden plant
Endangered status (IUCN): LC

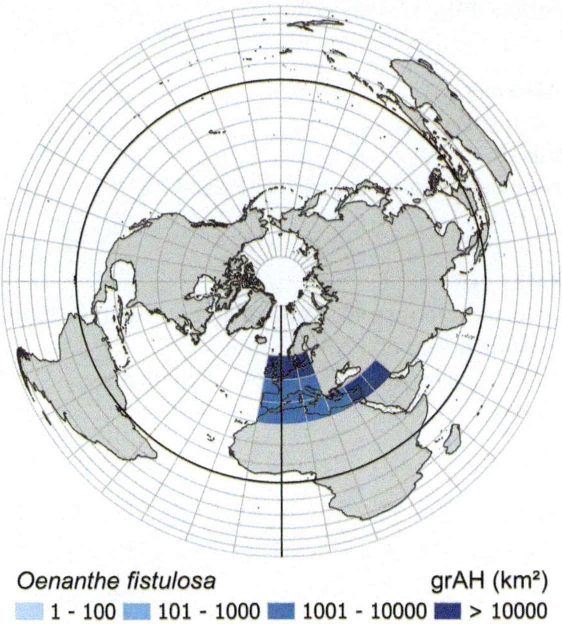

Oenanthe fistulosa L.
Image copyright: J. C.Schou, with permission

***Oenanthe fluviatilis* (Bab.) Coleman**
Synonyms:

Range area (grAH km^2): PAL: 36400
Ecozone origin-status: PAL: E
World rarity status: R
Ecozone-endemic macroregion occurrence: PAL: W EUR
Life form(s): SU, EM
Ploidy state: D
Speciation age (Ma): NA
Populations also occur in terrestrial, wetland and/ or riparian habitats: NO
Long distance movement vectors: NA
Nuisance status and uses: no nuisance issues; medicinal use
Endangered status (IUCN): NT

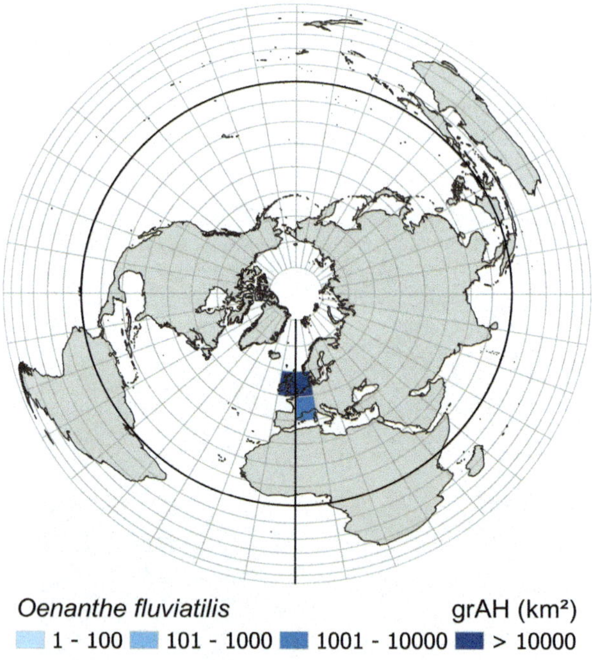

Oenanthe fluviatilis (Bab.) Coleman
Image copyright: J. C.Schou, with permission

***Oenanthe javanica* (Blume) DC.**

Synonyms: *Cyssopetalum javanicum* (Blume) Turcz., *Dasyloma javanicum* (Blume) Miq., *Falcaria javanica* (Blume) DC., *Sium javanicum* Blume

Range area (grAH km^2): PAL: 212100; NEA: 21000; OR: 85800; AUS: 47100; present in OCE
Ecozone origin-status: PAL: N; NEA: I; OR: N; AUS: N; OCE: I
World rarity status: C
Ecozone-endemic macroregion occurrence: not endemic to any world ecozone
Life form(s): EM
Ploidy state: DP
Speciation age (Ma): NA
Populations also occur in terrestrial, wetland and/ or riparian habitats: YES
Long distance movement vectors: ANTHROPOCHORY
Nuisance status and uses: no aquatic weed issues; folk medicine uses, especially in China, Korea, and Japan, limited food use in Asia
Endangered status (IUCN): LC

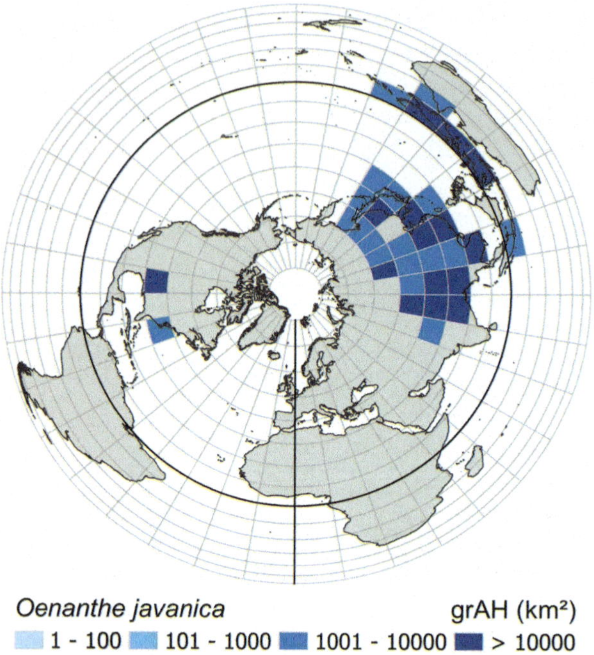

Oenanthe javanica (Blume) DC.
Image copyright: A. Efremov, with permission

***Oenanthe sarmentosa* C. Presl ex DC.**
Synonyms:

Range area (grAH km²): NEA: 54400
Ecozone origin-status: NEA: E
World rarity status: R
Ecozone-endemic macroregion occurrence: NEA: ARC CAN, W NAM, SW NAM
Life form(s): EM
Ploidy state: P
Speciation age (Ma): NA
Populations also occur in terrestrial, wetland and/ or riparian habitats: YES
Long distance movement vectors: NA
Nuisance status and uses: no nuisance issues; harvested from the wild for local use as a food, medicine and source of materials
Endangered status (IUCN): LC

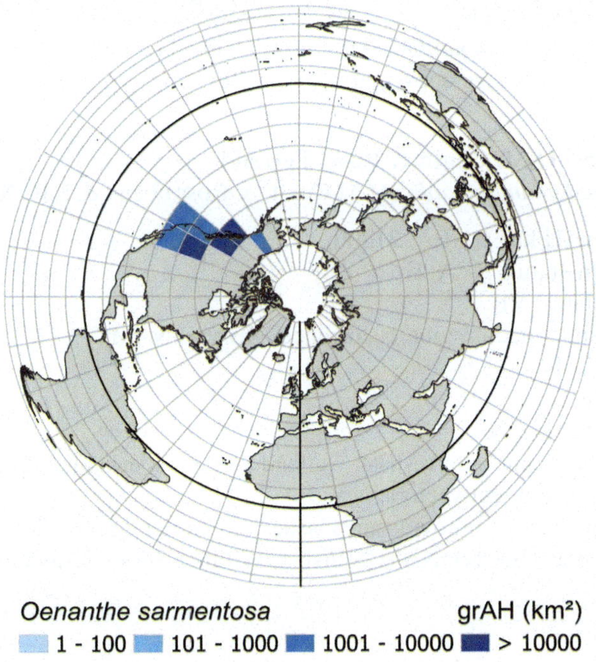

Oenanthe sarmentosa C. Presl ex DC.
Image copyright: iNaturalist CC-BY @aiva

***Oxypolis canbyi* (J.M. Coult. & Rose) Fernald**
Synonyms: *Tiedemannia canbyi* (J.M.Coult. & Rose) Feist & S.R.Downie

Range area (grAH km^2): NEA: 21200
Ecozone origin-status: NEA: E
World rarity status: VR
Ecozone-endemic macroregion occurrence: NEA: E NAM
Life form(s): EM
Ploidy state: D
Speciation age (Ma): NA
Populations also occur in terrestrial, wetland and/ or riparian habitats: YES
Long distance movement vectors: NA
Nuisance status and uses: no nuisance issues; no known uses
Endangered status (IUCN): NA (US Fish & Wildlife Service: Endangered in USA)

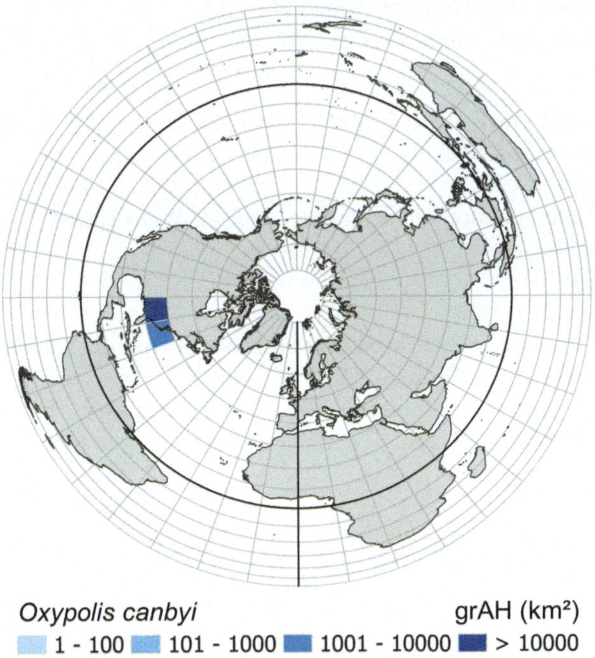

Oxypolis canbyi (J.M. Coult. & Rose) Fernald
Image copyright: A. Ebert, with permission

Oxypolis fendleri **(J.M. Coult. & Rose) Fernald**
Synonyms: *Archemora fendleri* A.Gray, *Tiedemannia fendleri* (A.Gray) J.M.Coult. & Rose

Range area (grAH km^2): NEA: 11800
Ecozone origin-status: NEA: E
World rarity status: VR
Ecozone-endemic macroregion occurrence: NEA: W NAM, SW NAM
Life form(s): EM
Ploidy state: DP
Speciation age (Ma): 7.0
Populations also occur in terrestrial, wetland and/ or riparian habitats: YES
Long distance movement vectors: ?AV
Nuisance status and uses: no nuisance issues; no known uses
Endangered status (IUCN): NA

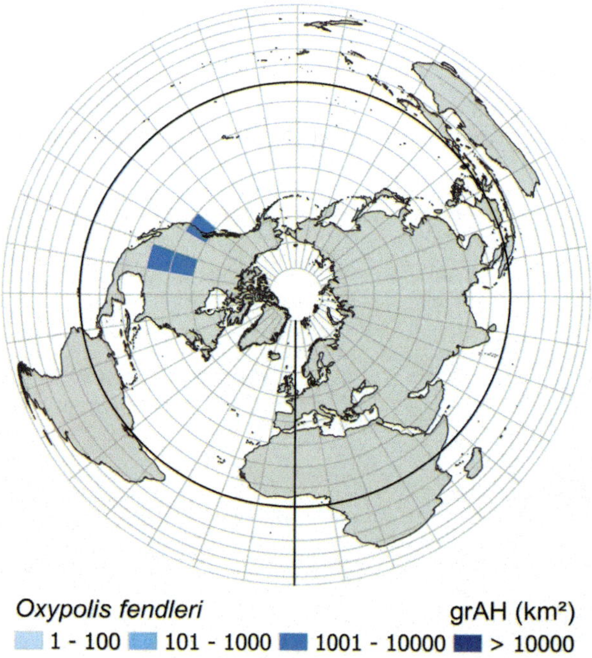

Oxypolis fendleri (J.M. Coult. & Rose) Fernald
Image copyright: iNaturalist CC0 @suejanssen

***Oxypolis filiformis* (Walter) Britton**
Synonyms: *Tiedemannia filiformis* (Walter) Feist & S.R.Downie

Range area (grAH km^2): NEA: 13000; NEO: 10300
Ecozone origin-status: NEA: N; NEO: N
World rarity status: VR
Ecozone-endemic macroregion occurrence: not endemic to any world ecozone
Life form(s): EM
Ploidy state: D
Speciation age (Ma): NA
Populations also occur in terrestrial, wetland and/ or riparian habitats: YES
Long distance movement vectors: HYDROCHORY (NEA to NEO; NEO to NEA)
Nuisance status and uses: no nuisance issues; no known uses
Endangered status (IUCN): NA

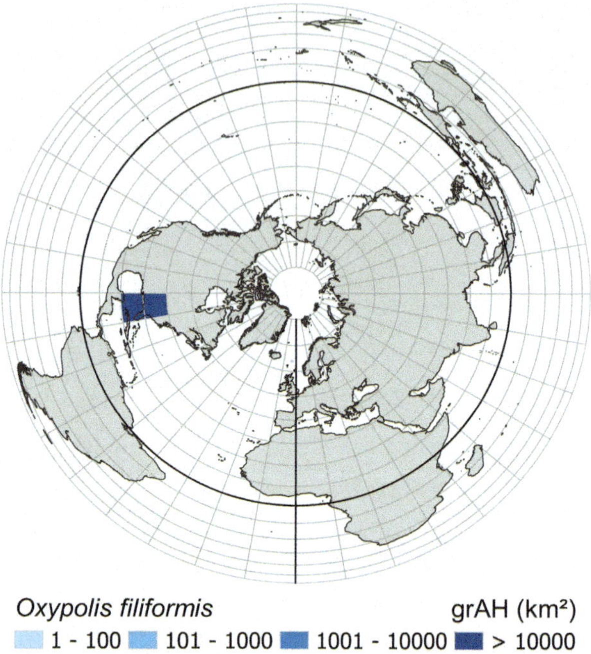

Oxypolis filiformis (Walter) Britton
Image copyright: iNaturalist CC-BY @jayhorn

Oxypolis greenmanii Mathias & Constance
Synonyms: *Tiedemannia filiformis* subsp. *greenmannii* (Mathias & Constance) Feist & S.R.Downie

Range area (grAH km^2): NEA: 13000
Ecozone origin-status: NEA: E
World rarity status: VR
Ecozone-endemic macroregion occurrence: NEA: E NAM
Life form(s): EM

Ploidy state: NA
Speciation age (Ma): NA
Populations also occur in terrestrial, wetland and/ or riparian habitats: YES
Long distance movement vectors: NA
Nuisance status and uses: no nuisance issues; no known uses
Endangered status (IUCN): NA

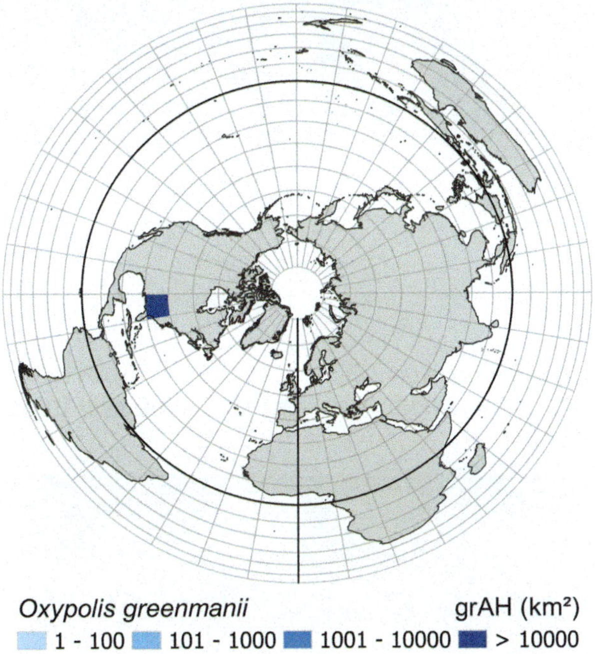

***Oxypolis occidentalis* (Roxb.) T.Anderson**
Synonyms:

Range area (grAH km^2): NEA: 23100
Ecozone origin-status: NEA: E
World rarity status: VR
Ecozone-endemic macroregion occurrence: NEA: W NAM, SW NAM
Life form(s): EM
Ploidy state: P
Speciation age (Ma): 8.0
Populations also occur in terrestrial, wetland and/ or riparian habitats: YES
Long distance movement vectors: ?AV
Nuisance status and uses: no nuisance issues; no known uses
Endangered status (IUCN): LC

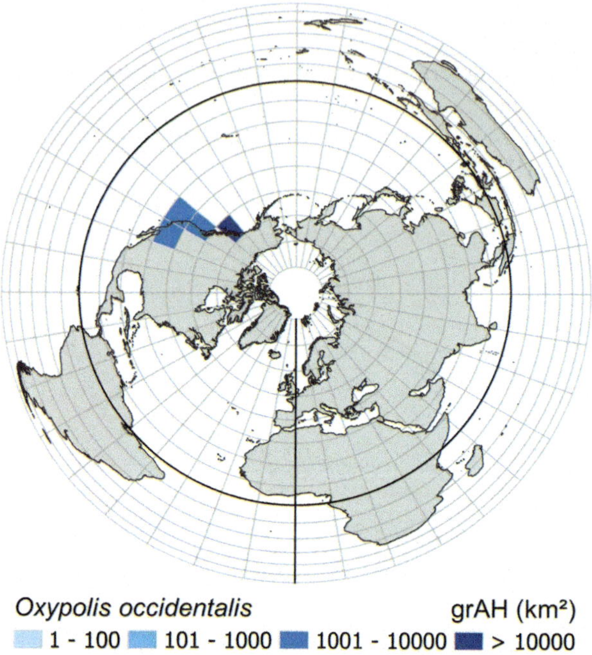

Oxypolis occidentalis (Roxb.) T.Anderson
Image copyright: A. Ebert, with permission

***Oxypolis rigidior* (L.) Raf.**
Synonyms: *Archemora rigidior* (L.) DC., *Oenanthe rigidius* (L.) Crantz, *Pastinaca rigidior* (L.) Spreng., *Sium rigidius* L., *Tiedemannia rigidior* (L.) J.M.Coult. & Rose

Range area (grAH km^2):NEA: 359300
Ecozone origin-status: NEA: E
World rarity status: C
Ecozone-endemic macroregion occurrence: NEA: W NAM, E NAM
Life form(s): EM
Ploidy state: P
Speciation age (Ma): 4.5
Populations also occur in terrestrial, wetland and/ or riparian habitats: YES
Long distance movement vectors: NA
Nuisance status and uses: no aquatic weed issues; no known uses (poisonous to mammals)
Endangered status (IUCN): LC

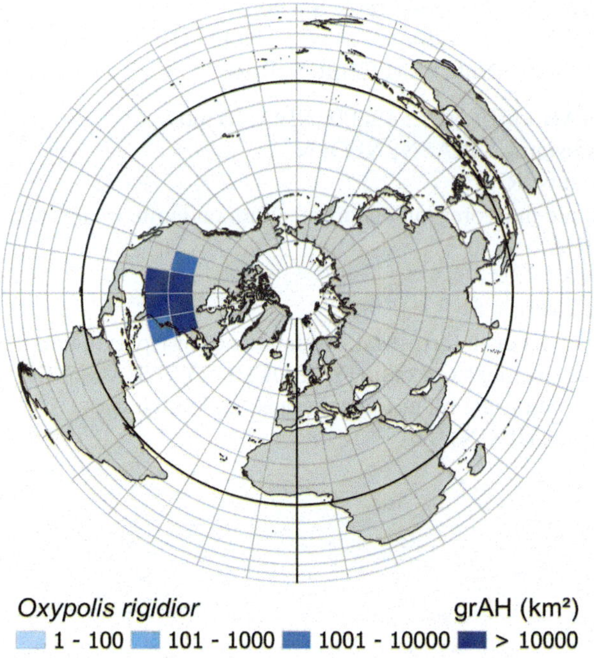

Oxypolis rigidior (L.) Raf.

***Oxypolis ternata* (Roxb.) Fosberg & Sachet**
Synonyms: *Archemora ternata* (Nutt.) Nutt., *Peucedanum ternatum* Nutt., *Tiedemannia ternata* (Nutt.) J.M.Coult. & Rose

Range area (grAH km^2): NEA: 13000; NEO: 10300
Ecozone origin-status: NEA: N; NEO: N
World rarity status: R
Ecozone-endemic macroregion occurrence: not endemic to any world ecozone
Life form(s): EM
Ploidy state: NA
Speciation age (Ma): 4.5
Populations also occur in terrestrial, wetland and/ or riparian habitats: YES
Long distance movement vectors: HYDROCHORY (NEA to NEO)
Nuisance status and uses: no nuisance issues; no known uses
Endangered status (IUCN): NA

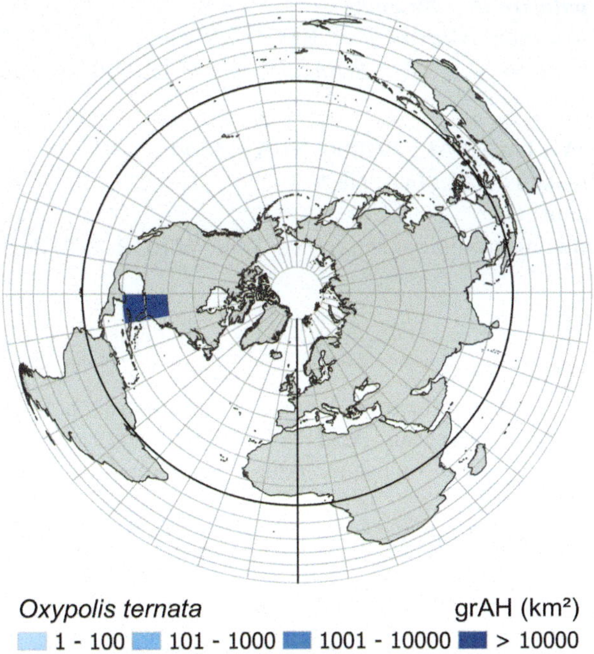

Oxypolis ternata (Roxb.) Fosberg & Sachet
Image copyright: iNaturalist CC0 @dryopteris (John Kees)

Peucedanum palustre **(L.) Moench**

Synonyms: *Thyselium palustre (*L.) Raf., *Calestania palustris* (L.) Koso-Pol., *Peucedanum sylvestre* (L.) DC., *Selinum palustre* L., *Thysselinum palustre* (L.) Hoffm.

Range area (grAH km^2): PAL: 397000; NEA: 15500
Ecozone origin-status: PAL: E(I); NEA: I
World rarity status: C
Ecozone-endemic macroregion occurrence: PAL: W EUR, N EUR, E EUR, UR/ W SIB, E SIB
Life form(s): EM
Ploidy state: D
Speciation age (Ma): NA
Populations also occur in terrestrial, wetland and/ or riparian habitats: YES
Long distance movement vectors: ANTHROPOCHORY
Nuisance status and uses: no nuisance issues; occasional use in wild flower seed mixes, medicinal uses
Endangered status (IUCN): NA

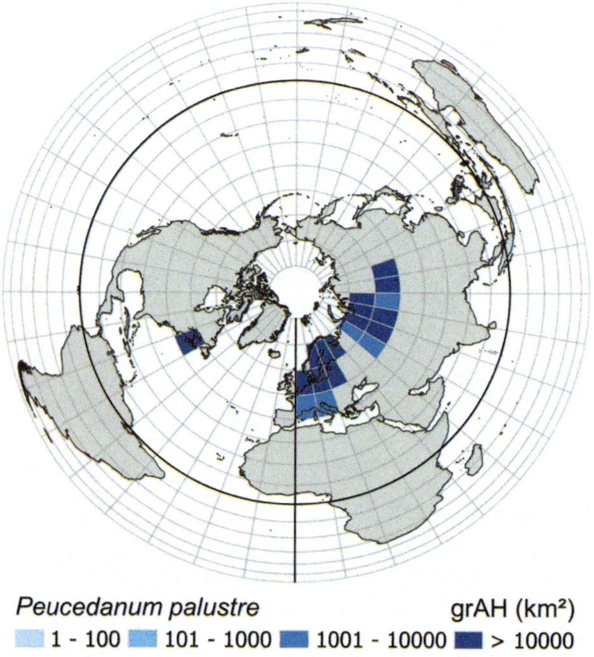

Peucedanum palustre (L.) Moench
Image copyright: J. C.Schou, with permission

Ptilimnium capillaceum **(Michx.) Raf.**
Synonyms: *Ammi capillaceum* Michx., *Discopleura capillacea* (Michx.) DC., *Sison capillaceum* (Michx.) Spreng.

Range area (grAH km^2): NEA: 90200; NEO: 24100
Ecozone origin-status: NEA: N; NEO: N
World rarity status: R
Ecozone-endemic macroregion occurrence: not endemic to any world ecozone
Life form(s): EM
Ploidy state: D
Speciation age (Ma): 0.1
Populations also occur in terrestrial, wetland and/ or riparian habitats: YES
Long distance movement vectors: HYDROCHORY (NEA to NEO)
Nuisance status and uses: no nuisance issues; no known uses
Endangered status (IUCN): NA

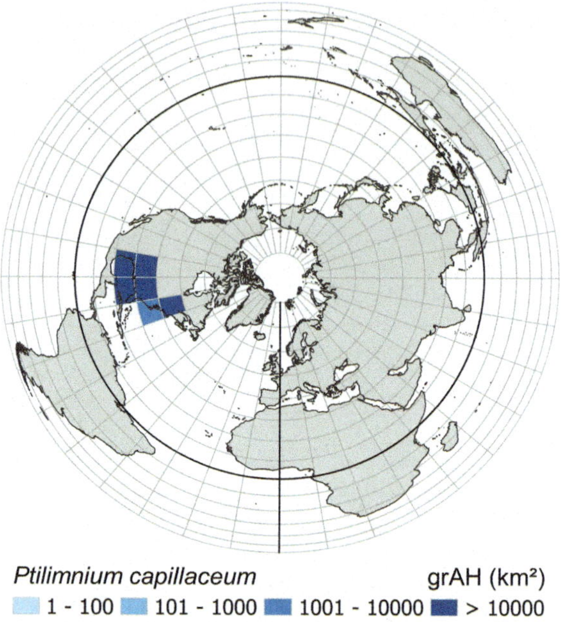

Ptilimnium capillaceum (Michx.) Raf.

***Ptilimnium costatum* Raf.**

Synonyms: *Ammi costatum* Elliott, *Discopleura costata* (Elliott) Steud., *Sison costatus* (Elliott) Eaton

Range area (grAH km^2): NEA: 25800
Ecozone origin-status: NEA: E
World rarity status: VR
Ecozone-endemic macroregion occurrence: NEA: E NAM
Life form(s): EM
Ploidy state: D
Speciation age (Ma): 2.8
Populations also occur in terrestrial, wetland and/ or riparian habitats: YES
Long distance movement vectors: NA
Nuisance status and uses: no nuisance issues; no known uses
Endangered status (IUCN): NA

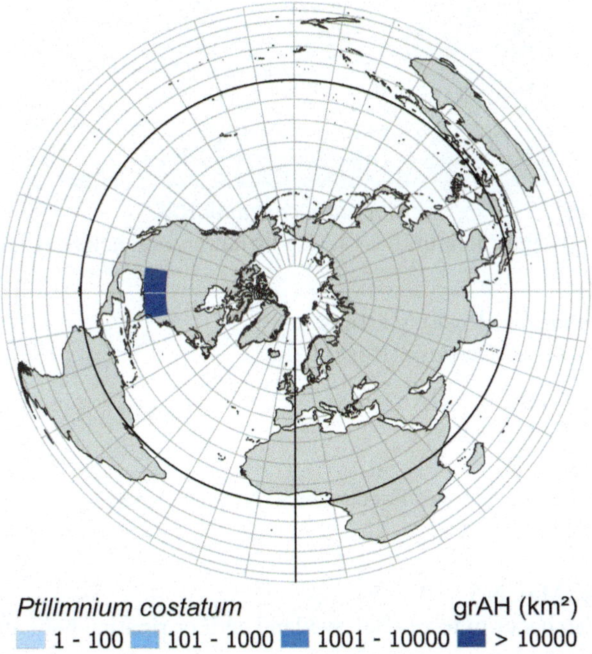

Ptilimnium costatum Raf.

Ptilimnium nodosum **(Rose) Mathias**
Synonyms: *Harperella nodosa* (Rose) Rose

Range area (grAH km^2): NEA: 34000
Ecozone origin-status: NEA: E
World rarity status: R
Ecozone-endemic macroregion occurrence: NEA: E NAM
Life form(s): EM
Ploidy state: H
Speciation age (Ma): NA
Populations also occur in terrestrial, wetland and/ or riparian habitats: YES
Long distance movement vectors: NA
Nuisance status and uses: no nuisance issues; no known uses
Endangered status (IUCN): NA

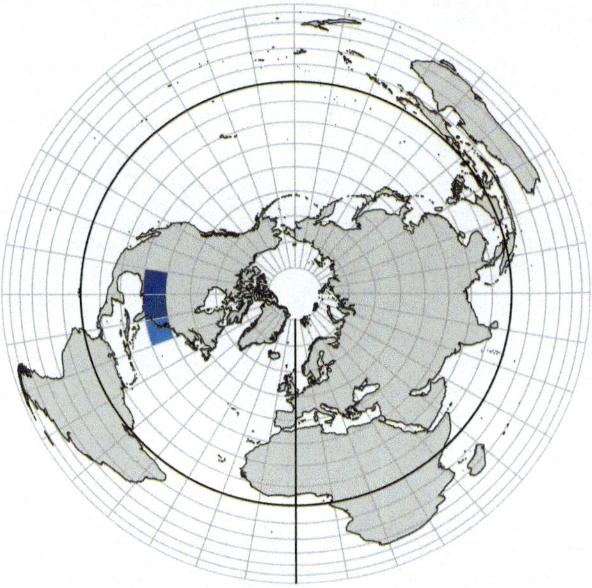

Ptilimnium nuttallii **(DC.) Britton**
Synonyms: *Discopleura nuttallii* DC.

Range area (grAH km^2): NEA: 12800
Ecozone origin-status: NEA: E
World rarity status: VR
Ecozone-endemic macroregion occurrence: NEA: E NAM
Life form(s): EM
Ploidy state: H
Speciation age (Ma): 3.6
Populations also occur in terrestrial, wetland and/ or riparian habitats: YES
Long distance movement vectors: NA
Nuisance status and uses: no nuisance issues; no known uses
Endangered status (IUCN): NA

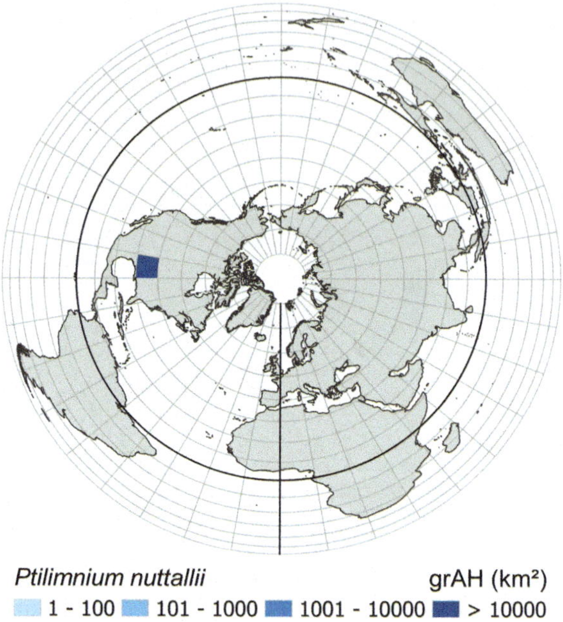

Ptilimnium nuttallii (DC.) Britton

***Sium latifolium* L.**

Synonyms: *Cicuta latifolia* (L.) Crantz, *Coriandrum latifolium* (L.) Crantz, *Drepanophyllum latifolium* (L.) Koso-Pol., *Pimpinella latifolia* (L.) Stokes, *Selinum sium* E.H.L.Krause

Range area (grAH km^2): PAL: 410300
Ecozone origin-status: PAL: E
World rarity status: C
Ecozone-endemic macroregion occurrence: PAL: MED, W EUR, N EUR, E EUR, UR/ W SIB, C ASIA, E SIB, CH
Life form(s): EM
Ploidy state: DP
Speciation age (Ma): 3.1
Populations also occur in terrestrial, wetland and/ or riparian habitats: YES
Long distance movement vectors: NA
Nuisance status and uses: no nuisance issues; some medicinal uses, food
Endangered status (IUCN): NA

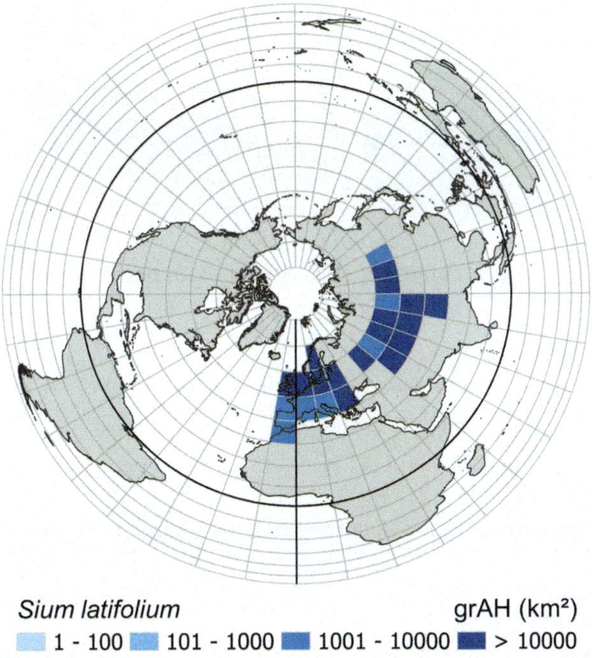

Sium latifolium L.
Image copyright: J. C.Schou, with permission

Sium medium Fisch. & C.A. Mey.
Synonyms: *Drepanophyllum medium* (Fisch. & C.A.Mey.) Koso-Pol.

Range area (grAH km^2): PAL: 128200; OR: 19900
Ecozone origin-status: PAL: N; OR: N
World rarity status: R
Ecozone-endemic macroregion occurrence: not endemic to any world ecozone
Life form(s): EM
Ploidy state: D
Speciation age (Ma): 3.2
Populations also occur in terrestrial, wetland and/ or riparian habitats: YES
Long distance movement vectors: NA
Nuisance status and uses: no nuisance issues; no known uses
Endangered status (IUCN): NA

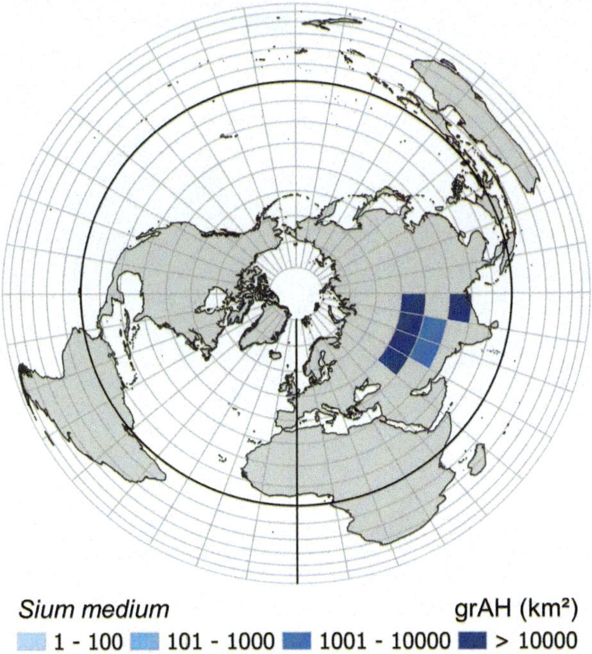

Sium sisarum L.

Synonyms: *Apium sisarum* (L.) Calest., *Carum sisarum* (L.) Baill., *Pimpinella sisarum* (L.) Jess., *Selinum sisarum* (L.) E.H.L.Krause, *Seseli sisarum* (L.) Crantz

Range area (grAH km^2): PAL: 202900; AUS: 12800
Ecozone origin-status: PAL: E(I); AUS: I
World rarity status: C
Ecozone-endemic macroregion occurrence: PAL: N EUR, E EUR, MED, UR/ W SIB, C ASIA, CH, FER/J
Life form(s): EM
Ploidy state: D
Speciation age (Ma): 0.5
Populations also occur in terrestrial, wetland and/ or riparian habitats: YES
Long distance movement vectors: ANTHROPOCHORY
Nuisance status and uses: no nuisance issues; sometimes used as a root vegetable, for example in Iran
Endangered status (IUCN): LC

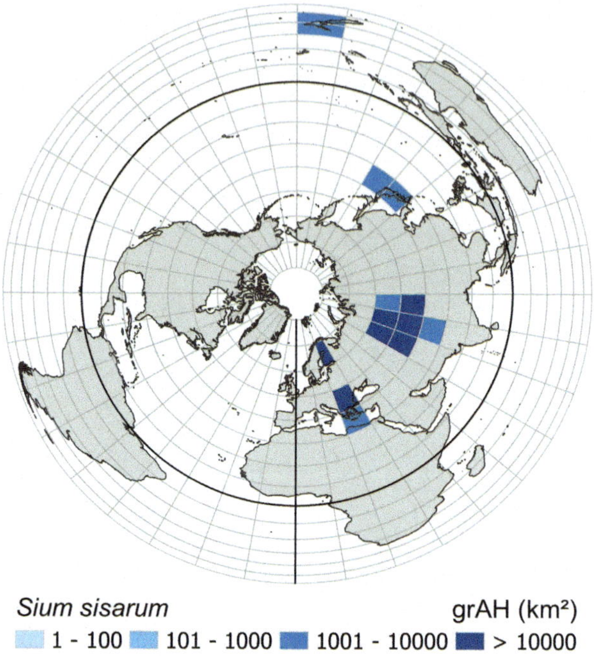

***Sium suave* Walter**
Synonyms:

Range area (grAH km²): PAL: 219400; NEA: 769600; NEO: 10300; OR: 2600
Ecozone origin-status: PAL: N; NEA: N; NEO: N; OR: N
World rarity status: C
Ecozone-endemic macroregion occurrence: not endemic to any world ecozone
Life form(s): EM
Ploidy state: DP
Speciation age (Ma): 3.1
Populations also occur in terrestrial, wetland and/ or riparian habitats: YES
Long distance movement vectors: HYDROCHORY (NEA to NEO)
Nuisance status and uses: no nuisance issues; the roots are edible, and were widely eaten by Native American people. The plant was also used for medicinal purposes
Endangered status (IUCN): LC

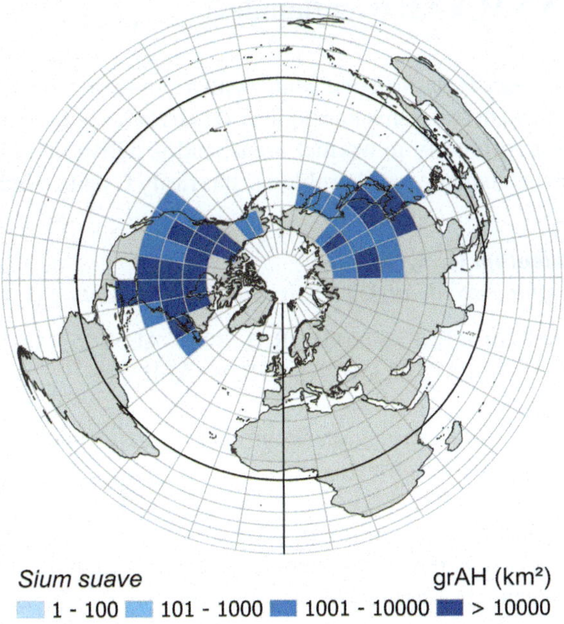

Sium suave Walter
Image copyright: A. Efremov, with permission

3.4 APOCYNACEAE (Gentianales)

The family Apocynaceae has just one genus with a single macrophyte species occurring in inland brackish or freshwater bodies: *Oxystelma esculentum.* It has a broad world distribution, primarily in warmer areas, and is native in AFR, PAL, OR and AUS. It never causes aquatic weed problems, has rarity C and endangered status LC.

***Oxystelma esculentum* (L. f.) Sm.**
Synonyms: *Periploca esculenta* L.f., *Sarcostemma esculentum* (L.f.) R.Holm

Range area (grAH km^2): PAL: 37000; OR: 118200; AUS: 25300; AFR: 115000
Ecozone origin-status: PAL: N; OR: N; AUS: N; AFR: N
World rarity status: C
World ecozone-endemic macroregion occurrence: not endemic to any ecozone
Life form(s): EM (climbing shrub)
Ploidy state: NA
Speciation age (Ma): NA
Populations also occur in terrestrial, wetland and/ or riparian habitats: YES
Long distance movement vectors: NA
Nuisance status and uses: no nuisance issues; used in traditional medicine, edible fruits
Endangered status (IUCN): LC

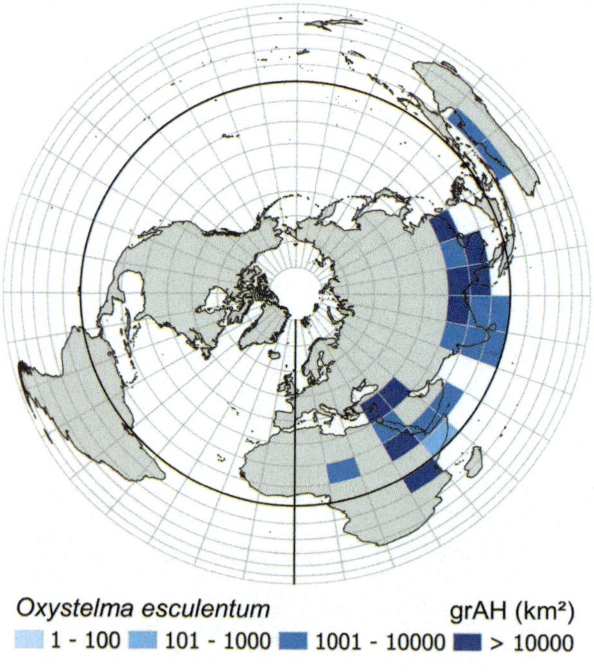

Oxystelma esculentum (L. f.) Sm.
Image copyright: iNaturalist CC-BY @meghaa

3.5 ARALIACEAE (Asterales)

The family Araliaceae has only one genus with species occurring in inland brackish or freshwater bodies. *Hydrocotyle* contains eight macrophyte species, which are widely distributed across the world, occurring in all world ecozones except ANT (seven species recorded in AUS and PAL, six in OR, five each in NEA and NEO, and four in each of AFR and OCE). Macrophyte species of the Araliaceae show low endemicity, with only one species having E or E(I) origin status, in each of PAL and AUS. Anthropochory vectors play a major role in influencing the world distribution of the Araliaceae, with I-origin (introduced or invasive) records for four species in each of OR and AUS, three in OCE, and one apiece in each of AFR, NEA and NEO. and one in AFR. Three species are aquatic weeds in at least one world ecozone, with one of these, *Hydrocotyle ranunculoides*, being a major invasive aquatic weed species in PAL and OR. The Araliaceae are dominated by common species with six species C, and two R. Three species have no endangered status data (NA/DD) and five are LC.

Hydrocotyle bonariensis **Comm. ex Lam.**
Synonyms:

Range area (grAH km²): PAL: 26200; NEA: 34000; NEO: 306700; AUS: 19100; AFR: 210400
Ecozone origin-status: PAL: I; NEA: N; NEO: N; AUS: I; AFR: N
World rarity status: C
Ecozone-endemic macroregion occurrence: not endemic to any world ecozone
Life form(s): EM (subshrub)
Ploidy state: P
Speciation age (Ma): NA
Populations also occur in terrestrial, wetland and/ or riparian habitats: YES
Long distance movement vectors: ANTHROPOCHORY
Nuisance status and uses: considered a minor aquatic weed in parts of Australia; folk medicine uses
Endangered status (IUCN): LC

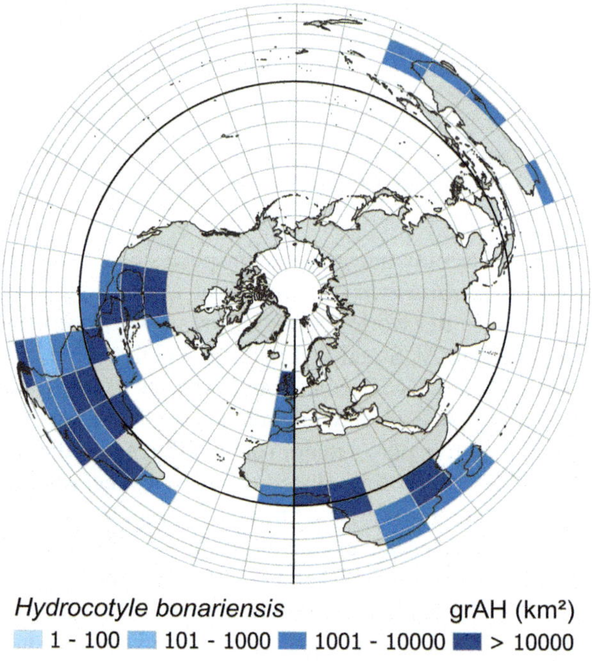

Hydrocotyle bonariensis Comm. ex Lam.
Image copyright: iNaturalist CC-BY @scottward (Scott Ward)

***Hydrocotyle javanica* Thunb.**
Synonyms:

Range area (grAH km^2): PAL: 30500; OR: 132100; AUS: 54000; present in OCE
Ecozone origin-status: PAL: N; OR: N; AUS: N; OCE: N
World rarity status: C
Ecozone-endemic macroregion occurrence: not endemic to any world ecozone
Life form(s): EM
Ploidy state: P
Speciation age (Ma): NA
Populations also occur in terrestrial, wetland and/ or riparian habitats: YES
Long distance movement vectors: NA
Nuisance status and uses: low to no nuisance issues; some traditional medicine uses, especially in India
Endangered status (IUCN): LC

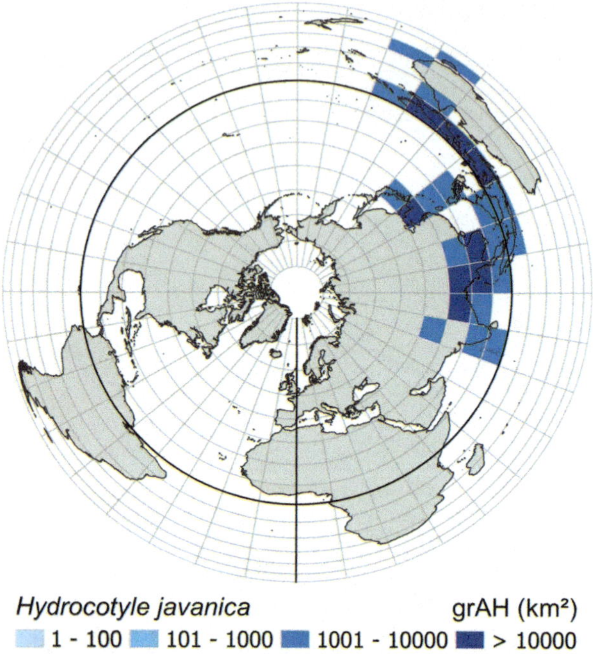

Hydrocotyle javanica Thunb.
Image copyright: iNaturalist CC-BY @srinivasan2

Hydrocotyle ranunculoides **L.f.**
Synonyms:

Range area (grAH km2): PAL: 73000; NEA: 77400; NEO: 296700; OR: 17900; AUS: 1500; AFR: 156800
Ecozone origin-status: PAL: I; NEA: N; NEO: N; OR: I; AUS: I; AFR: I
World rarity status: C
Ecozone-endemic macroregion occurrence: not endemic to any world ecozone
Life form(s): EM
Ploidy state: DP
Speciation age (Ma): NA
Populations also occur in terrestrial, wetland and/ or riparian habitats: YES
Long distance movement vectors: ANTHROPOCHORY
Nuisance status and uses: major aquatic weed in at least one world ecozone outside home range [54, 80]; folk medicine and phytoremediation uses
Endangered status (IUCN): NA

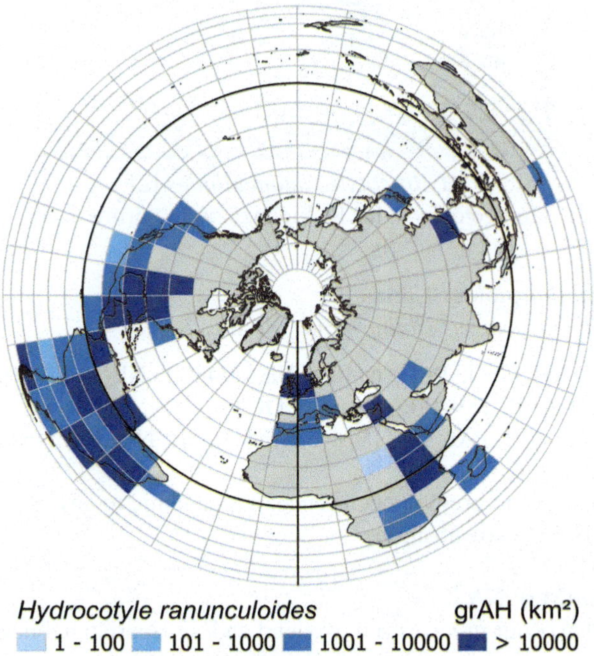

Hydrocotyle ranunculoides L.f.
Image copyright: K. Fidanza, with permission

Hydrocotyle sibthorpioides Lam.
Synonyms: *Chondrocarpus sibthorpioides* (Lam.) Sweet

Range area (grAH km^2): PAL: 61100; NEA:94800; NEO: 32700; OR: 126300; AUS: 61500; AFR: 189400; present in OCE
Ecozone origin-status: PAL: N; NEA: I; NEO: I; OR: N; AUS: N; AFR: N; OCE: I
World rarity status: C
Ecozone-endemic macroregion occurrence: not endemic to any world ecozone
Life form(s): EM (hydrosubshrub)
Ploidy state: P
Speciation age (Ma): NA
Populations also occur in terrestrial, wetland and/ or riparian habitats: YES
Long distance movement vectors: ANTHROPOCHORY
Nuisance status and uses: considered an aquatic weed in lawns, boggy areas and on the banks of small rivers and rivulets in some regions; water garden plant, medicinal use
Endangered status (IUCN): LC

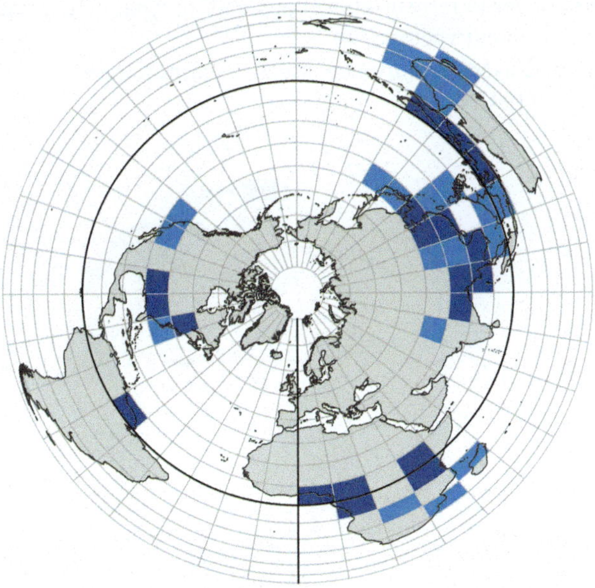

Hydrocotyle sibthorpioides Lam.

Hydrocotyle tripartita R.Br. ex A.Rich.
Synonyms:

Range area (grAH km2): AUS: 43000
Ecozone origin-status: AUS: E
World rarity status: R
Ecozone-endemic macroregion occurrence: AUS: NA, ESA, NZ
Life form(s): EM, SU
Ploidy state: P
Speciation age (Ma): NA

Populations also occur in terrestrial, wetland and/ or riparian habitats: YES
Long distance movement vectors: NA
Nuisance status and uses: no nuisance issues; aquarium trade
Endangered status (IUCN): NA

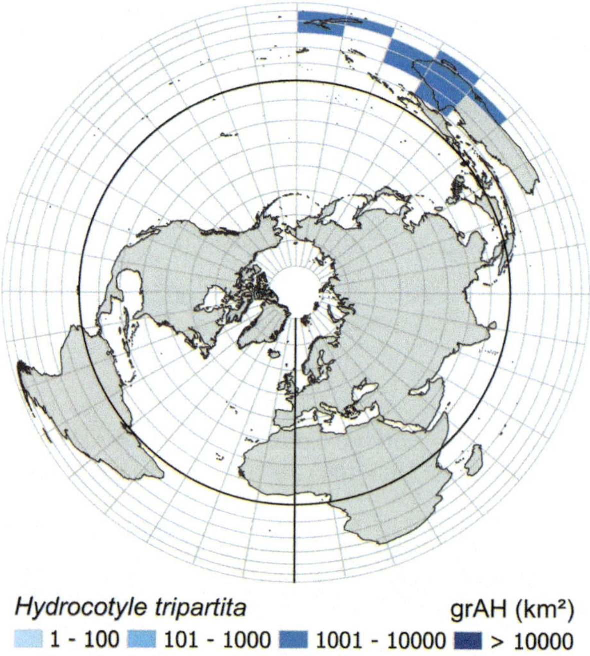

Hydrocotyle tripartita R.Br. ex A.Rich.

***Hydrocotyle umbellata* L.**

Synonyms:

Range area (grAH km2): PAL: 5000; NEA: 110300; NEO: 251600; OR: 6700; AUS: 5400; AFR: 35800

Ecozone origin-status: PAL: I; NEA: N; NEO: N; OR: I; AUS: I; AFR: I

World rarity status: C

Ecozone-endemic macroregion occurrence: not endemic to any world ecozone

Life form(s): EM

Ploidy state: P

Speciation age (Ma): NA

Populations also occur in terrestrial, wetland and/ or riparian habitats: YES

Long distance movement vectors: ANTHROPOCHORY

Nuisance status and uses: no nuisance issues; some use as water garden plant, medicinal and phytoremediation uses are also reported

Endangered status (IUCN): LC

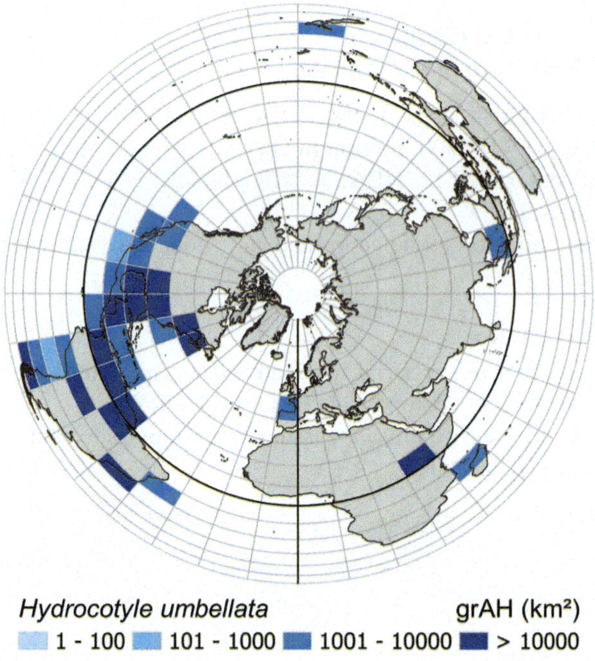

Hydrocotyle umbellata L.
Image copyright: G. Ervin, with permission

Hydrocotyle verticillata **Thunb.**
Synonyms: *Centella verticillata* (Thunb.) Fourc.

Range area (grAH km2): PAL: 35400; NEA: 334000; NEO: 270900; OR: 29400; AUS: 17700 ; AFR: 62700; present in OCE
Ecozone origin-status: PAL: I; NEA: N; NEO: N; OR: I; AUS: I; AFR: N; OCE: I
World rarity status: C
Ecozone-endemic macroregion occurrence: not endemic to any world ecozone
Life form(s): EM
Ploidy state: P
Speciation age (Ma): NA
Populations also occur in terrestrial, wetland and/ or riparian habitats: YES
Long distance movement vectors: ANTHROPOCHORY
Nuisance status and uses: no nuisance issues; aquarium trade, water garden plant
Endangered status (IUCN): LC

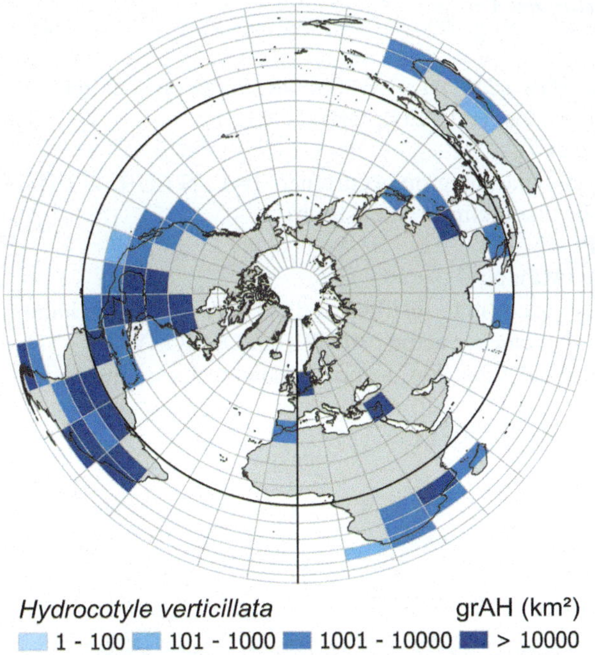

Hydrocotyle verticillata Thunb.
Image copyright: A. Efremov, with permission

***Hydrocotyle vulgaris* L.**
Synonyms:

Range area (grAH km2): PAL: 151800; OR: 9600; present in OCE
Ecozone origin-status: PAL: E(I); OR: I; OCE: I
World rarity status: C
Ecozone-endemic macroregion occurrence: PAL: N EUR, W EUR, E EUR, MED
Life form(s): EM (sometimes a subshrub)
Ploidy state: P
Speciation age (Ma): NA
Populations also occur in terrestrial, wetland and/ or riparian habitats: YES
Long distance movement vectors: ANTHROPOCHORY
Nuisance status and uses: low to no nuisance issues; aquarium trade, water garden plant
Endangered status (IUCN): LC

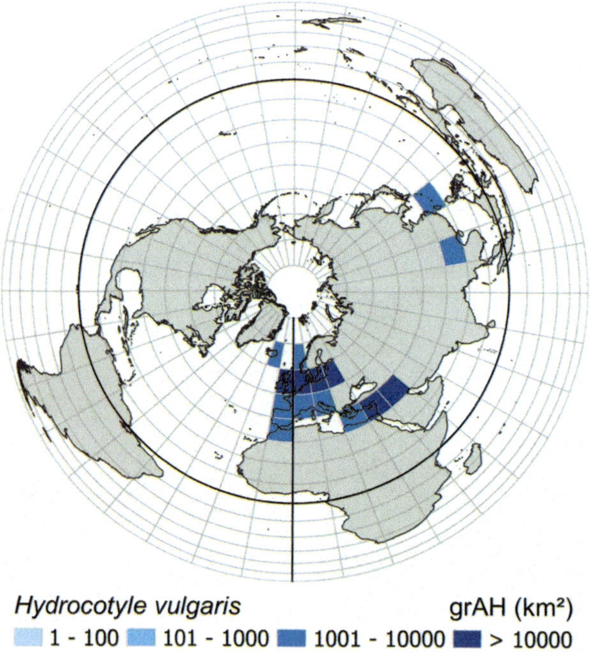

Hydrocotyle vulgaris L.
Image copyright: J.C. Schou, with permission

3.6 ASTERACEAE (Apiales)

The Asteraceae are a large family with 26 genera containing 60 macrophyte species covering a widespread world distribution from the Arctic to the Tropics. There are 33 species in NEO, 23 in OR, 18 in AFR, 17 in PAL, 13 in NEA, 11 in AUS and five in OCE. Ecozone-endemic macrophytes (E or E(I) origin status) of the Asteraceae show particularly high occurrences in NEO (21 species), with a further six in AFR, five in OR, four in PAL, and three in NEA. There are no ecozone-endemic macrophyte species in either AUS or OCE. Anthropochory vectors have played a relatively limited role in influencing the world distribution of macrophyte species of Asteraceae. However, six of the 11 species occurring in AUS are introduced, with a further six in OR and five in AFR. Four of the five species found in OCE are introduced, with the same number of I-origin species present in the Asteraceae macrophyte flora of PAL and NEO. NEA has only one introduced species. *Cotula coronopifolia* and *Eclipta prostrata* cause localised aquatic weed issues in different parts of the world. *Gymnocoronis spilanthoides* is considered a major invasive aquatic weed in AUS. Otherwise, the Asteraceae macrophyte species rarely or never cause problems. The family is dominated by world-rare species with VR 35%, R 27% and C 38%. In terms of endangered status, the Asteraceae have 63% NA/DD, 33% LC and 4% in the more endangered categories (2% of Asteraceae macrophyte species have CR status).

Quick Links to Genera

1. *Acmella*
2. *Bidens*
3. *Bothriocline*
4. *Cadiscus*
5. *Caesulia*
6. *Centipeda*
7. *Cotula*
8. *Cyathocline*
9. *Eclipta*
10. *Enydra*
11. *Epaltes*
12. *Erigeron*
13. *Grangea*
14. *Gymnocoronis*
15. *Hydropectis*
16. *Jaegeria*
17. *Pacourina*
18. *Petasites*
19. *Sclerolepis*
20. *Senecio*
21. *Shinnersia*
22. *Sphaeranthus*
23. *Stenops*
24. *Symphyllocarpus*
25. *Tagetes*
26. *Tephroseris*

***Acmella ciliata* (Kunth) Cass.**

Synonyms: *Spilanthes ciliata* Kunth, *Ceratocephalus ciliatus* (Kunth) Kuntze

Range area (grAH km^2): NEO: 221900; OR: 28700
Ecozone origin-status: NEO: E(I); OR: I
World rarity status: C
Ecozone-endemic macroregion occurrence: NEO: MES AM, NWSA, AMZ, AND, CHA, PAR
Life form(s): EM
Ploidy state: NA
Speciation age (Ma): NA
Populations also occur in terrestrial, wetland and/ or riparian habitats: YES
Long distance movement vectors: ANTHROPOCHORY
Nuisance status and uses: no nuisance issues; folk medicine
Endangered status (IUCN): NA

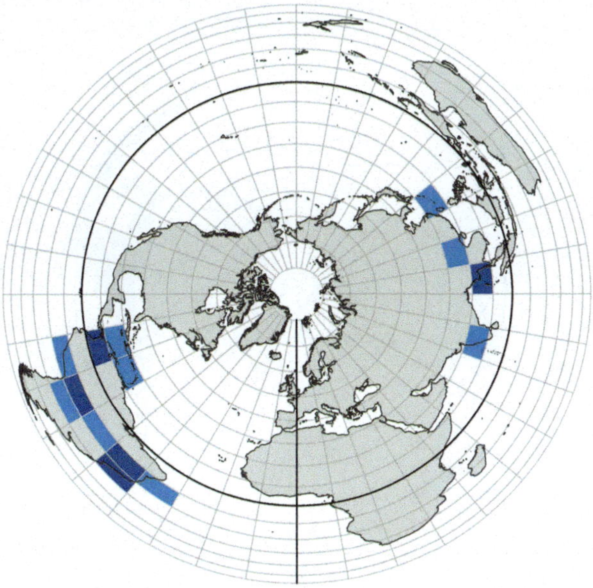

***Acmella paniculata* (Wall. ex DC.) R.K.Jansen**
Synonyms: *Spilanthes paniculata* Wall. ex DC.

Range area (grAH km^2): NEO: 74100; OR: 130900; AUS: 38500; present in OCE
Ecozone origin-status: NEO: I; OR: N; AUS: N; OCE: I
World rarity status: C
Ecozone-endemic macroregion occurrence: not endemic to any world ecozone
Life form(s): EM
Ploidy state: NA
Speciation age (Ma): NA
Populations also occur in terrestrial, wetland and/ or riparian habitats: YES
Long distance movement vectors: ANTHROPOCHORY
Nuisance status and uses: no nuisance issues; folk medicine, used to treat toothache, and throat and gum infections
Endangered status (IUCN): LC

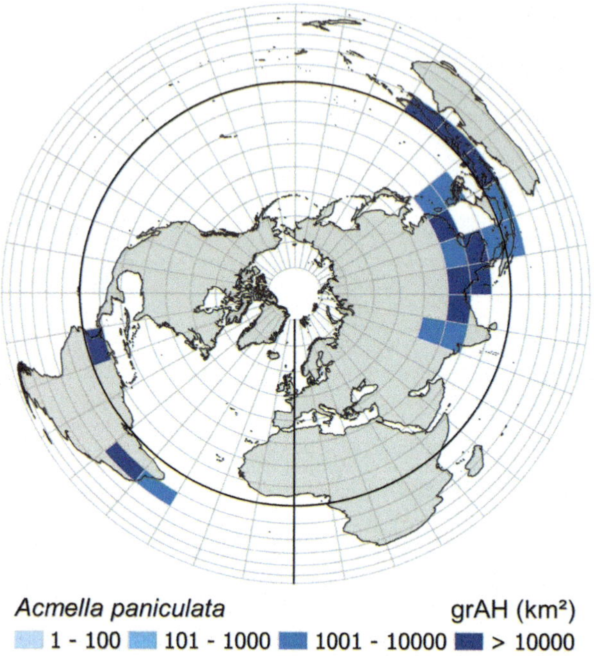

Acmella paniculata (Wall. ex DC.) R.K.Jansen
Image copyright: S.Roy, with permission

***Acmella uliginosa* (Sw.) Cass.**
Synonyms: *Spilanthes uliginosa* Sw., *Jaegeria uliginosa* (Sw.) Spreng.

Range area (grAH km^2): NEO: 164600; OR: 122300; AUS: 331002; AFR: 175700; present in OCE
Ecozone origin-status: NEO: E(I); OR: I; AUS: I; AFR: I; OCE: I
World rarity status: C

Ecozone-endemic macroregion occurrence: NEO: MES AM, NWSA, AMZ, CHA, PAR
Life form(s): EM
Ploidy state: NA
Speciation age (Ma): NA
Populations also occur in terrestrial, wetland and/ or riparian habitats: YES
Long distance movement vectors: ANTHROPOCHORY
Nuisance status and uses: low to no nuisance issues; food, folk medicine
Endangered status (IUCN): LC

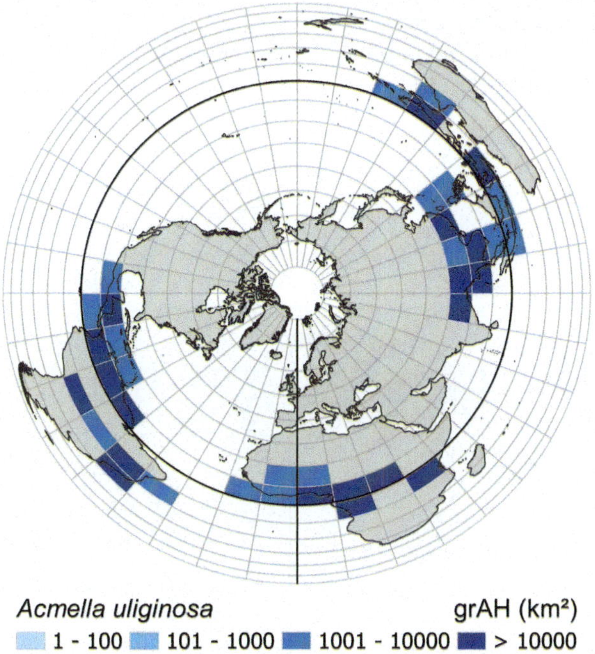

Acmella uliginosa (Sw.) Cass.
Image copyright: A.T. Dung-Le, with permission

Bidens beckii **(Sw.) Torr. ex Spreng.**
Synonyms: *Megalodonta beckii* (Torr. ex Spreng.) Greene

Range area (grAH km²): NEA: 532900
Ecozone origin-status: NEA: E
World rarity status: C
Ecozone-endemic macroregion occurrence: NEA: ARC CAN, W NAM, E NAM
Life form(s): EM, SU
Ploidy state: D
Speciation age (Ma): NA
Populations also occur in terrestrial, wetland and/ or riparian habitats: YES
Long distance movement vectors: NA
Nuisance status and uses: no nuisance issues; no known uses
Endangered status (IUCN): LC

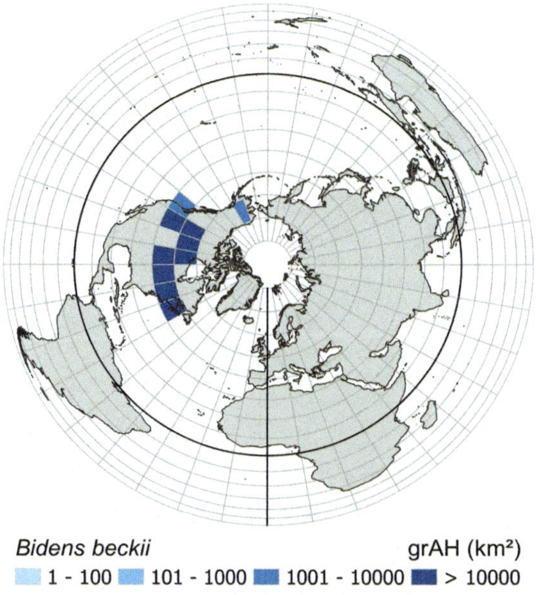

Bidens cernua **L.**
Synonyms:

Range area (grAH km²): PAL: 580500; NEA: 606900; NEO: 500; OR: 36100
Ecozone origin-status: PAL: N; NEA: N; NEO: N; OR: N: N
World rarity status: C
Ecozone-endemic macroregion occurrence: not endemic to any world ecozone
Life form(s): EM
Ploidy state: DP
Speciation age (Ma): NA
Populations also occur in terrestrial, wetland and/ or riparian habitats: YES
Long distance movement vectors: BIRD: *Anas platyrhynchos* (Northern Mallard)
Nuisance status and uses: no nuisance issues; medicinal uses
Endangered status (IUCN): LC

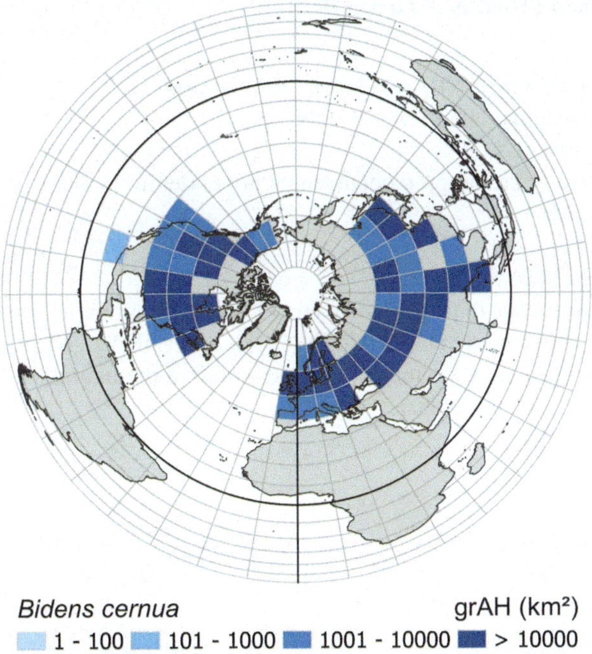

Bidens cernua L.
Image copyright: J. C. Schou, with permission

Bidens discoidea (Torr. & A.Gray) Britton
Synonyms:

Range area (grAH km²): NEA: 355500; NEO: 10300
Ecozone origin-status: NEA: N; NEO: N
World rarity status: C
Ecozone-endemic macroregion occurrence: not endemic to any world ecozone
Life form(s): EM
Ploidy state: D
Speciation age (Ma): NA
Populations also occur in terrestrial, wetland and/ or riparian habitats: YES
Long distance movement vectors: HYDROCHORY (NEA to NEO)
Nuisance status and uses: no nuisance issues; limited medicinal use
Endangered status (IUCN): NA

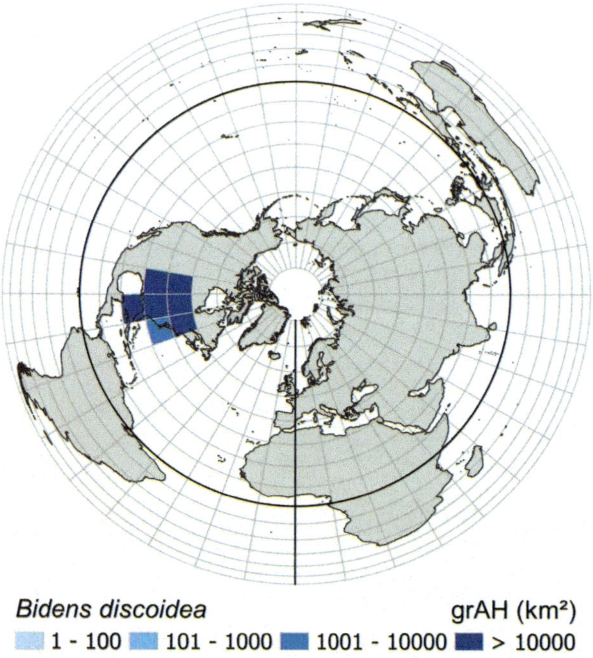

Bidens eatonii Fernald
Synonyms:

Range area (grAH km²): NEA: 71800
Ecozone origin-status: NEA: E
World rarity status: R
Ecozone-endemic macroregion occurrence: NEA: E NAM
Life form(s): EM

Ploidy state: P
Speciation age (Ma): NA
Populations also occur in terrestrial, wetland and/ or riparian habitats: YES
Long distance movement vectors: NA
Nuisance status and uses: no nuisance issues; no known uses
Endangered status (IUCN): VU

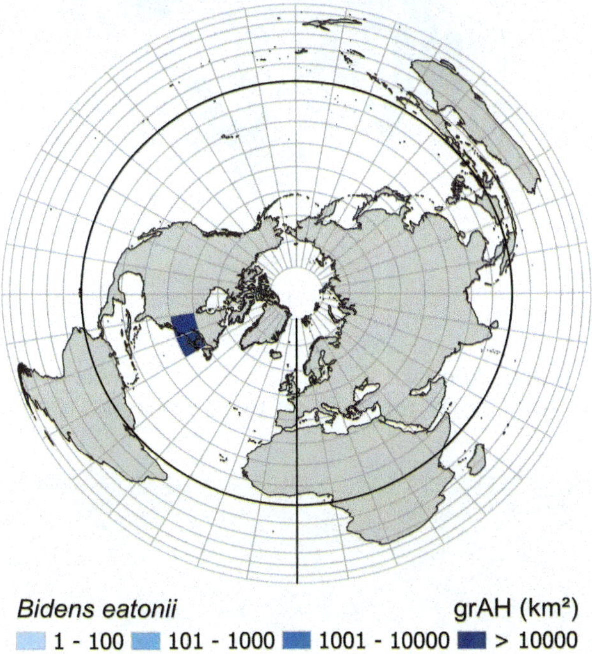

Bidens frondosa **L.**
Synonyms:

Range area (grAH km^2): PAL: 26000; NEA: 409300; NEO: 24100; OR: 17900; AUS: 12800; AFR: 8700
Ecozone origin-status: PAL: I; NEA: N; NEO: N; OR: I; AUS: I; AFR: I
World rarity status: C
Ecozone-endemic macroregion occurrence: not endemic to any world ecozone
Life form(s): EM
Ploidy state: DP
Speciation age (Ma): NA
Populations also occur in terrestrial, wetland and/ or riparian habitats: YES
Long distance movement vectors: HYDROCHORY (NEA to NEO); BIRD: *Anas platyrhynchos* (Northern Mallard)
Nuisance status and uses: no nuisance issues; medicinal use for herbal remedies
Endangered status (IUCN): LC

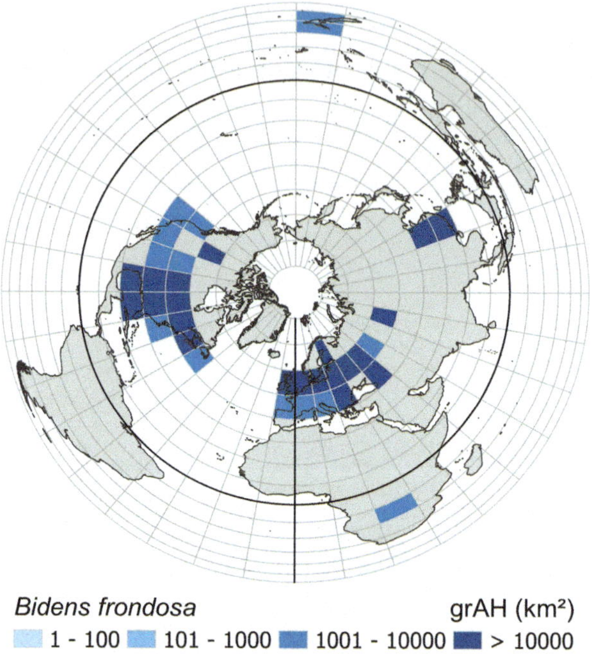

Bidens frondosa L.
Image copyright: A. Efremov, with permission

Bidens kamtschatica **Vassilcz.**
Synonyms:

Range area (grAH km^2): PAL: 6500
Ecozone origin-status: PAL: E
World rarity status: VR
Ecozone-endemic macroregion occurrence: PAL: FER/J
Life form(s): EM
Ploidy state: NA
Speciation age (Ma): NA
Populations also occur in terrestrial, wetland and/ or riparian habitats: YES
Long distance movement vectors: NA
Nuisance status and uses: no nuisance issues; no known uses
Endangered status (IUCN): NA

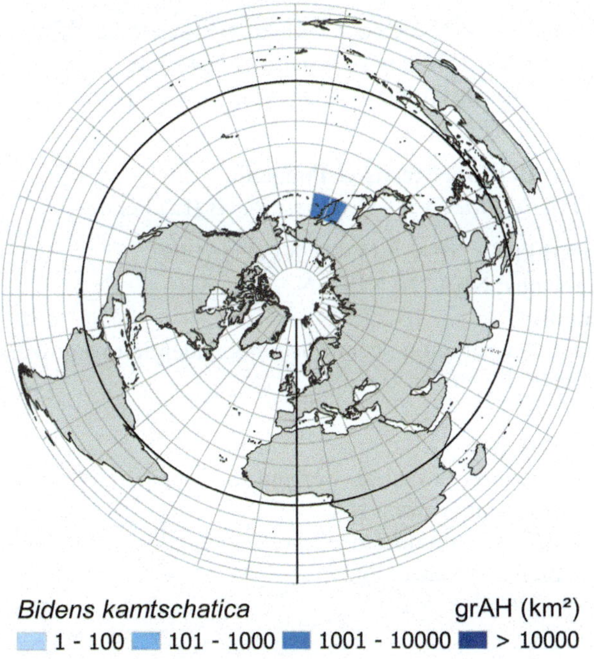

Bidens kamtschatica Vassilcz.
Image copyright: O. Chernyagina, with permission

***Bidens laevis* (L.) Britton, Sterns & Poggenb.**
Synonyms: *Helianthus laevis* L., *Heliopsis laevis* (L.) Pers.

Range area (grAH km^2): NEA: 145300; NEO: 161500
Ecozone origin-status: NEA: N; NEO: N
World rarity status: C
Ecozone-endemic macroregion occurrence: not endemic to any world ecozone
Life form(s): EM
Ploidy state: DP
Speciation age (Ma): NA
Populations also occur in terrestrial, wetland and/ or riparian habitats: YES
Long distance movement vectors: NA
Nuisance status and uses: no nuisance issues; young leaves are edible, some use as water garden plant, folk medicine
Endangered status (IUCN): NA

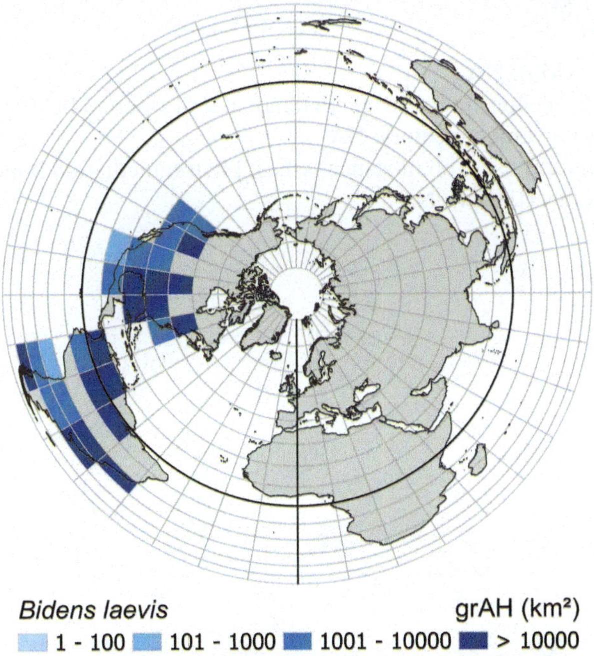

Bidens laevis (L.) Britton, Sterns & Poggenb.
Image copyright: S. Mailhos, with permission

***Bidens maximowicziana* Oett.**
Synonyms:

Range area (grAH km^2): PAL: 144600
Ecozone origin-status: PAL: E
World rarity status: R
Ecozone-endemic macroregion occurrence: PAL: FER/J, E SIB, CH, C ASIA

Life form(s): EM
Ploidy state: P
Speciation age (Ma): NA
Populations also occur in terrestrial, wetland and/ or riparian habitats: YES
Long distance movement vectors: NA
Nuisance status and uses: no nuisance issues; phytoremediation potential
Endangered status (IUCN): NA

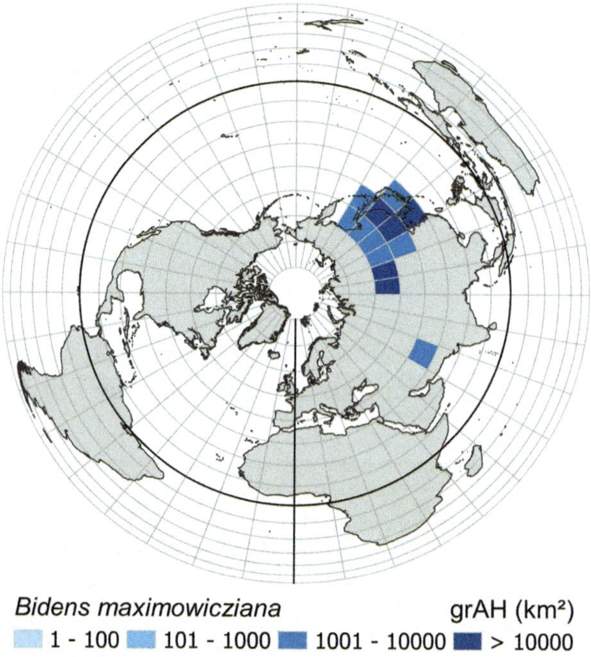

***Bidens parviflora* Willd.**
Synonyms:

Range area (grAH km^2): PAL: 188900; OR: 7000
Ecozone origin-status: PAL: N; OR: N
World rarity status: C
Ecozone-endemic macroregion occurrence: not endemic to any world ecozone
Life form(s): EM
Ploidy state: DP
Speciation age (Ma): NA
Populations also occur in terrestrial, wetland and/ or riparian habitats: YES
Long distance movement vectors: HYDROCHORY (PAL to OR)
Nuisance status and uses: no nuisance issues; harvested from the wild for local use as a medicine and probably also as a food
Endangered status (IUCN): NA

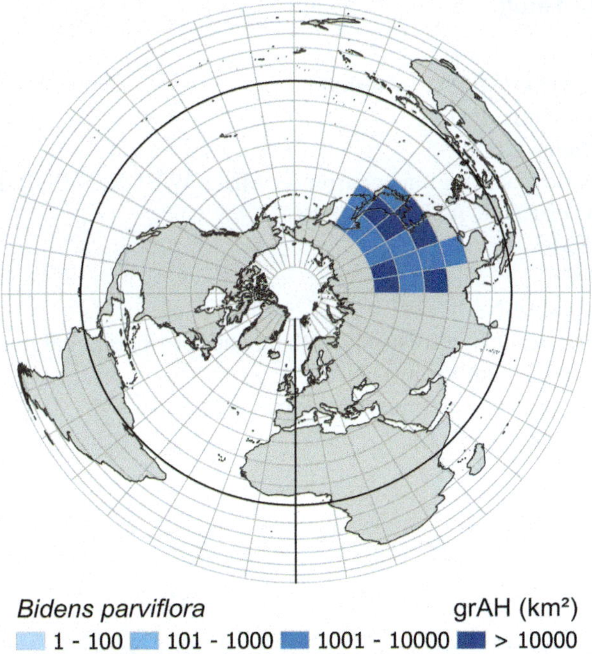

Bidens parviflora Willd.
Image copyright: iNaturalist CC0 @alzov (Alexey Zyryanov)

***Bidens radiata* Thuill.**
Synonyms:

Range area (grAH km²): PAL: 746000
Ecozone origin-status: PAL: E
World rarity status: C
Ecozone-endemic macroregion occurrence: MED, W EUR, N EUR, E EUR, UR/ W SIB, E SIB, FER/J, C ASIA, CH, ARC
Life form(s): EM
Ploidy state: DP
Speciation age (Ma): NA
Populations also occur in terrestrial, wetland and/ or riparian habitats: YES
Long distance movement vectors: NA
Nuisance status and uses: no nuisance issues; no known uses
Endangered status (IUCN): LC

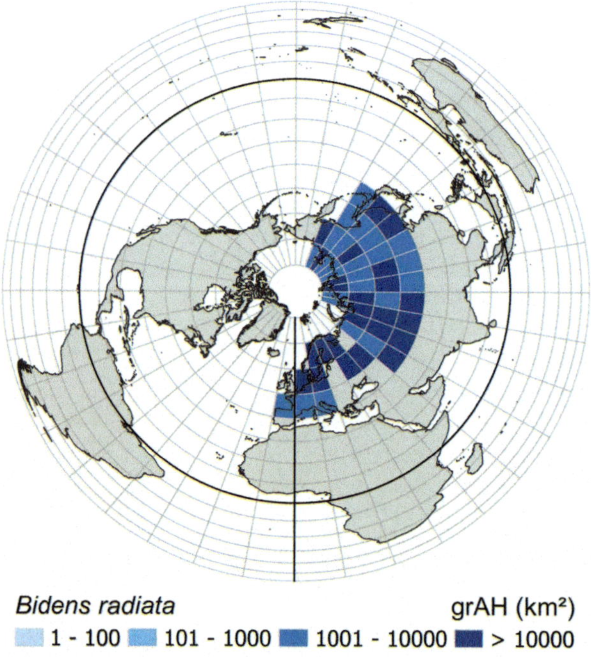

Bidens radiata Thuill.
Image copyright: A. Efremov, with permission

Bidens tripartita **L.**
Synonyms:

Range area (grAH km^2): PAL: 788300; NEA: 373600; NEO: 1400; OR: 74900; AUS: 26000
Ecozone origin-status: PAL: N; NEA: N; NEO: I; OR: N; AUS: I
World rarity status: C
Ecozone-endemic macroregion occurrence: not endemic to any world ecozone
Life form(s): EM
Ploidy state: DP
Speciation age (Ma): NA
Populations also occur in terrestrial, wetland and/ or riparian habitats: YES
Long distance movement vectors: ANTHROPOCHORY; BIRD: *Anas platyrhynchos* (Northern Mallard); *Anas crecca* (Eurasian Teal)
Nuisance status and uses: no nuisance issues; young leaves are edible when cooked, medicinal uses: formerly considered an important remedy for conditions of bleeding or haemorrhage, kidney problems and ulcers
Endangered status (IUCN): LC

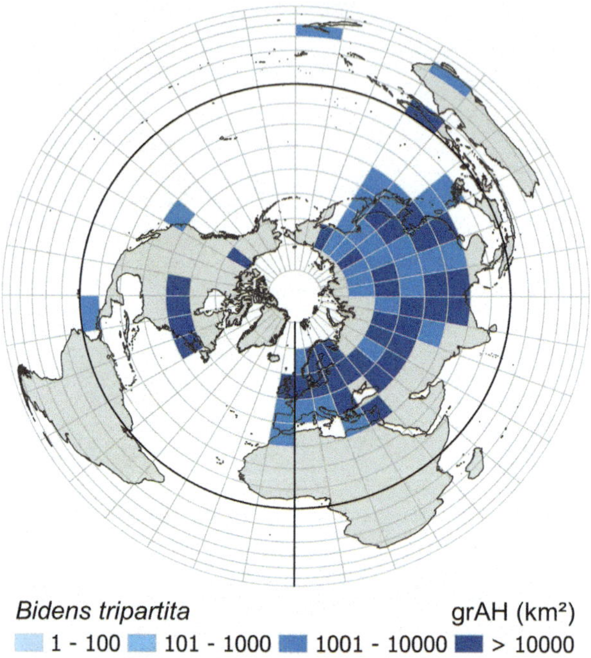

Bidens tripartita L.
Image copyright: J. C. Schou, with permission

Bothriocline monocephala (Hiern) Wild & G.V.Pope
Synonyms: *Ethulia monocephala* Hiern

Range area (grAH km^2): AFR: 11700
Ecozone origin-status: AFR: E
World rarity status: VR
Ecozone-endemic macroregion occurrence: AFR: ZAM
Life form(s): EM
Ploidy state: NA
Speciation age (Ma): NA
Populations also occur in terrestrial, wetland and/ or riparian habitats: YES
Long distance movement vectors: NA
Nuisance status and uses: no nuisance issues; no known uses
Endangered status (IUCN): NA

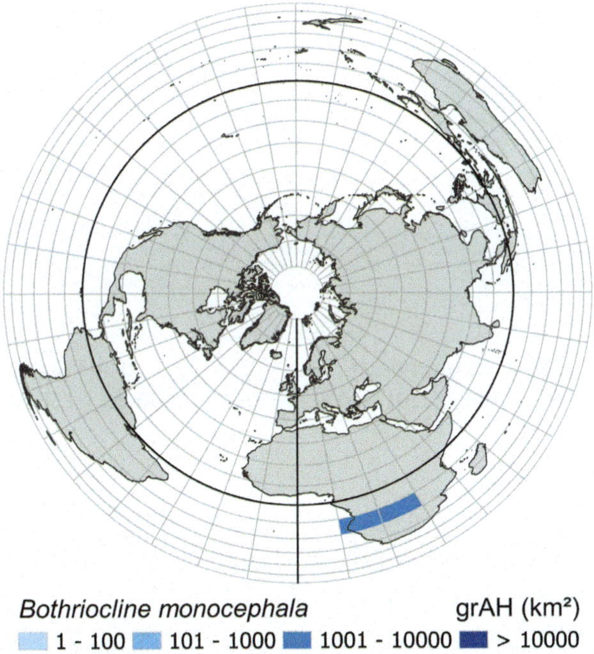

Cadiscus aquaticus E.Mey. ex DC.
Synonyms: *Senecio cadiscus* B.Nord. & Pelser

Range area (grAH km^2): AFR: 1000
Ecozone origin-status: AFR: E
World rarity status: VR
Ecozone-endemic macroregion occurrence: AFR: CAPE
Life form(s): EM

Ploidy state: NA
Speciation age (Ma): NA
Populations also occur in terrestrial, wetland and/ or riparian habitats: NO
Long distance movement vectors: NA
Nuisance status and uses: no nuisance issues; no known uses
Endangered status (IUCN): CR

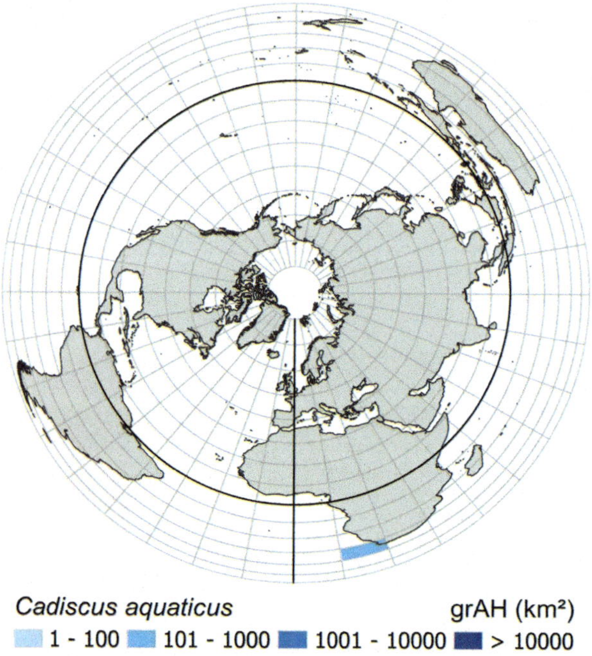

Caesulia axillaris **Roxb.**
Synonyms:

Range area (grAH km^2): PAL: 8600; OR: 28500
Ecozone origin-status: PAL: N; OR: N
World rarity status: R
Ecozone-endemic macroregion occurrence: not endemic to any world ecozone
Life form(s): EM
Ploidy state: NA
Speciation age (Ma): NA
Populations also occur in terrestrial, wetland and/ or riparian habitats: YES
Long distance movement vectors: NA
Nuisance status and uses: no nuisance issues; local uses for food and medicinal purposes
Endangered status (IUCN): LC

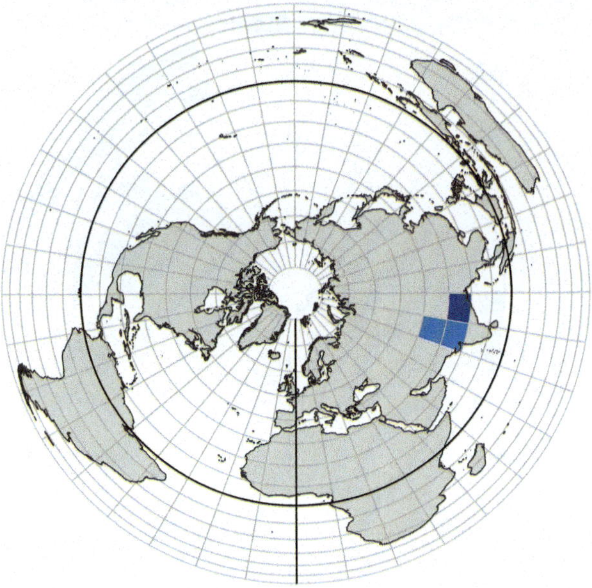

***Centipeda minima* (L.) A.Braun & Asch.**
Synonyms: *Artemisia minima* L., *Cotula minima* (L.) Willd., *Dichrocephala minima* (L.) Bojer, *Grangea minima* (L.) Dum.Cours.

Range area (grAH km^2): PAL: 36400; NEO: 35400; OR: 76800; AUS: 59100; AFR: 5400; present in OCE
Ecozone origin-status: PAL: N; NEO: I; OR: N; AUS: N; AFR: I; OCE: N
World rarity status: C
Ecozone-endemic macroregion occurrence: not endemic to any world ecozone
Life form(s): EM
Ploidy state: P
Speciation age (Ma): NA
Populations also occur in terrestrial, wetland and/ or riparian habitats: YES
Long distance movement vectors: ANTHROPOCHORY
Nuisance status and uses: no nuisance issues; medicial use, folk medicine
Endangered status (IUCN): LC

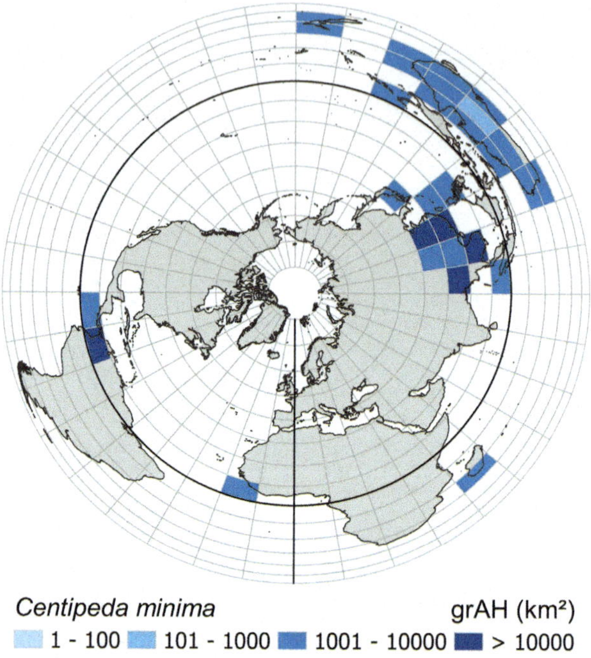

***Cotula coronopifolia* L.**
Synonyms: *Lancisia coronopifolia* (L.) Rydb.

Range area (grAH km^2): PAL: 159000; NEA: 94900; NEO: 112500; OR: 26800; AUS: 47300; AFR: 15600
Ecozone origin-status: PAL: I; NEA: I; NEO: I; OR: I; AUS: I; AFR: E(I)
World rarity status: C
Ecozone-endemic macroregion occurrence: AFR: ZAM, SDES, CAPE
Life form(s): EM
Ploidy state: D
Speciation age (Ma): 12.0
Populations also occur in terrestrial, wetland and/ or riparian habitats: YES
Long distance movement vectors: ANTHROPOCHORY; ?AV
Nuisance status and uses: low nuisance issues: in California, known to be invasive, but its spread appears to be relatively slow; water garden plant, medicinal potential
Endangered status (IUCN): NA

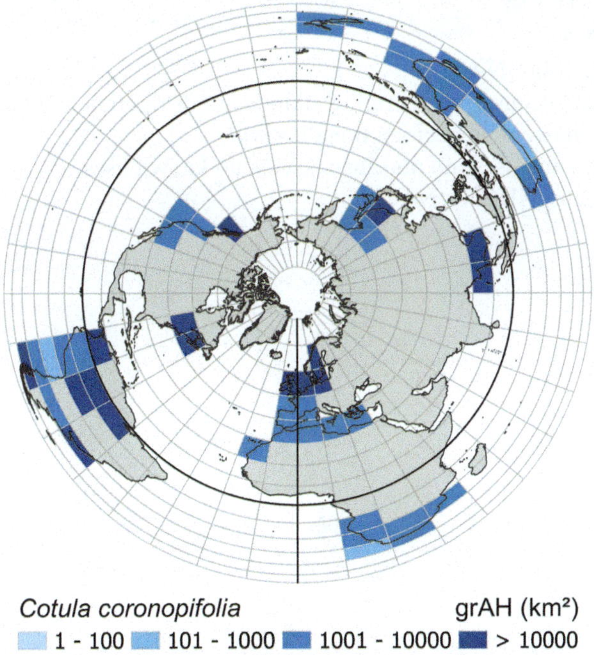

Cotula coronopifolia L.
Image copyright: J. C.Schou, with permission

***Cotula myriophylloides* Harv.**
Synonyms:

Range area (grAH km^2): AFR: 1000
Ecozone origin-status: AFR: E
World rarity status: VR
Ecozone-endemic macroregion occurrence: AFR: CAPE

Life form(s): SU, EM
Ploidy state: NA
Speciation age (Ma): NA
Populations also occur in terrestrial, wetland and/ or riparian habitats: NO
Long distance movement vectors: none known
Nuisance status and uses: no nuisance issues; no known uses
Endangered status (IUCN): NA

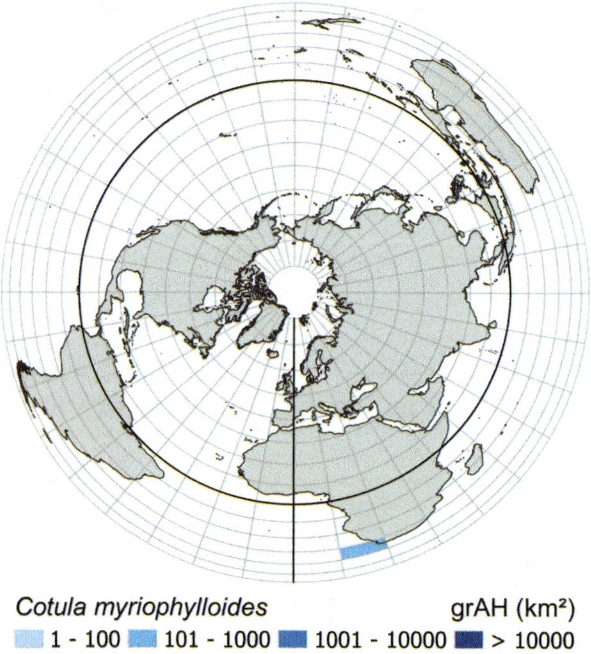

Cyathocline lutea **Law ex Wight**
Synonyms: *Blumea lutea* (J.S.Law ex Wight) K.C.Mohan

Range area (grAH km^2): OR: 10300
Ecozone origin-status: OR: E
World rarity status: VR
Ecozone-endemic macroregion occurrence: OR: IND
Life form(s): EM
Ploidy state: NA
Speciation age (Ma): NA
Populations also occur in terrestrial, wetland and/ or riparian habitats: YES
Long distance movement vectors: NA
Nuisance status and uses: no nuisance issues; bioactive
Endangered status (IUCN): LC

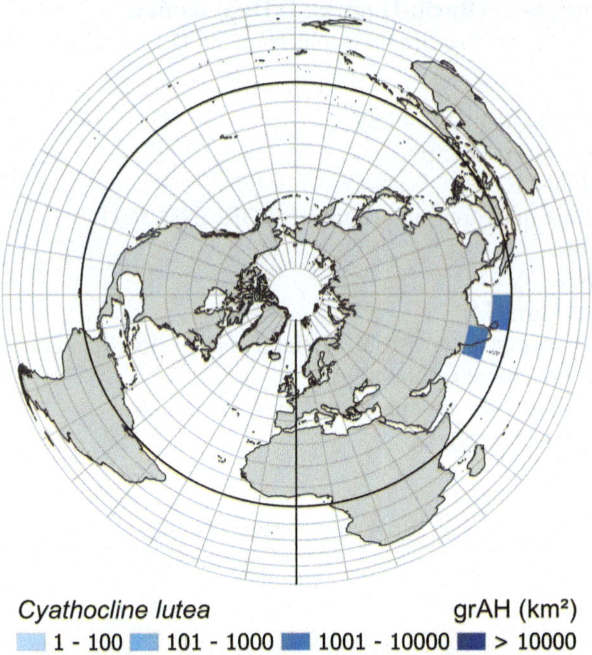

Cyathocline lutea Law ex Wight

***Cyathocline purpurea* (Buch.-Ham. ex D.Don) Kuntze**

Synonyms: *Tanacetum purpureum* Buch.-Ham. ex D.Don, *Cyathocline jacquemontii* Gagnep.

Range area (grAH km^2): OR: 49500
Ecozone origin-status: OR: E
World rarity status: R
Ecozone-endemic macroregion occurrence: OR: IND, ICH
Life form(s): EM
Ploidy state: NA
Speciation age (Ma): NA
Populations also occur in terrestrial, wetland and/ or riparian habitats: YES
Long distance movement vectors: NA
Nuisance status and uses: no nuisance issues; medicinal
Endangered status (IUCN): LC

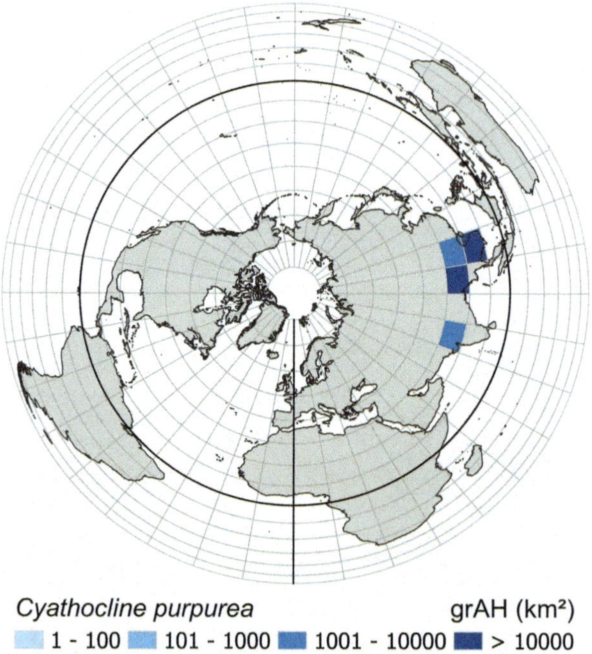

Cyathocline purpurea (Buch.-Ham. ex D.Don) Kuntze
Image copyright: iNaturalist CC0 @radhaveach

***Eclipta megapotamica* (Spreng.) Sch.Bip. ex S.F.Blake**
Synonyms:

Range area (grAH km^2): NEO: 36500
Ecozone origin-status: NEO: E
World rarity status: R
Ecozone-endemic macroregion occurrence: NEO: CHA, PAR
Life form(s): EM
Ploidy state: NA
Speciation age (Ma): NA
Populations also occur in terrestrial, wetland and/ or riparian habitats: YES
Long distance movement vectors: NA
Nuisance status and uses: no nuisance issues; medicinal uses
Endangered status (IUCN): NA

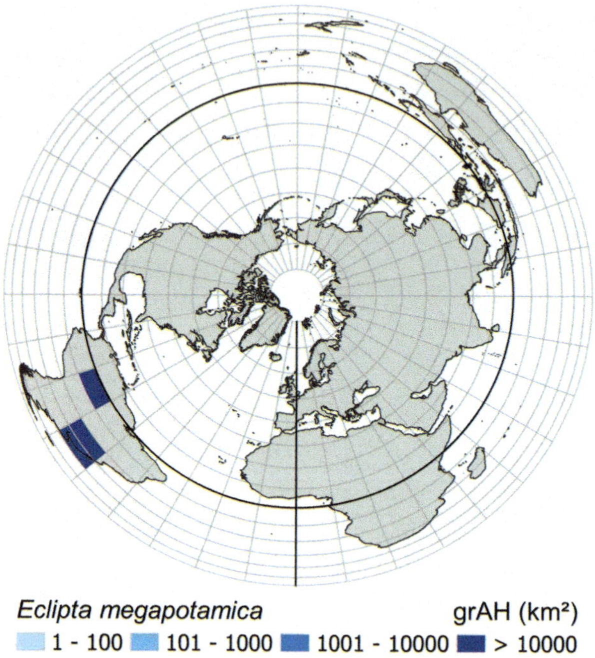

***Eclipta prostrata* (L.) L.**

Synonyms: *Eclipta alba* (L.) Hassk., *Cotula prostrata* (L.) L., *Eclipta erecta* var. *prostrata* (L.) Baker, *Eleutheranthera prostrata* (L.) Sch.Bip., *Verbesina prostrata* L.

Range area (grAH km^2): PAL: 221400; NEA: 396200; NEO: 368600; OR: 154000; AUS: 82100; AFR: 352900; present in OCE
Ecozone origin-status: PAL: I; NEA: N; NEO: N; OR: I; AUS: I; AFR: I; OCE: I
World rarity status: C
Ecozone-endemic macroregion occurrence: not endemic to any world ecozone
Life form(s): EM
Ploidy state: D
Speciation age (Ma): NA
Populations also occur in terrestrial, wetland and/ or riparian habitats: YES
Long distance movement vectors: ANTHROPOCHORY
Nuisance status and uses: can cause low to moderate nuisance issues (considered a weed in parts of the USA, and also a common weed of rainfed lowland rice in the Philippines, Indonesia, and India; medicinal (human and veterinary) and phytoremediation uses
Endangered status (IUCN): LC

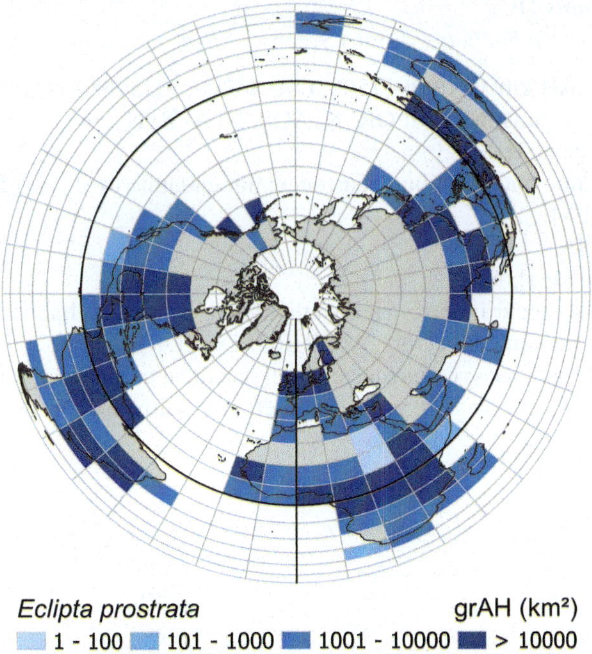

Eclipta prostrata (L.) L.
Image copyright: A.T. Dung-Le, with permission

***Enydra fluctuans* DC.**

Synonyms: *Meyera fluctuans* (Lour.) Spreng., *Enydra anagallis* Gardner

Range area (grAH km^2): NEO: 17500; OR: 83200; AUS: 8800; AFR: 13900
Ecozone origin-status: NEO: I; OR: N; AUS: N; AFR: N
World rarity status: R
Ecozone-endemic macroregion occurrence: not endemic to any world ecozone
Life form(s): EM
Ploidy state: NA
Speciation age (Ma): NA
Populations also occur in terrestrial, wetland and/ or riparian habitats: YES
Long distance movement vectors: ANTHROPOCHORY
Nuisance status and uses: no nuisance issues; medical uses, harvested from the wild for its edible leaves, and is also sometimes cultivated in Java, Malaya and Cambodia
Endangered status (IUCN): LC

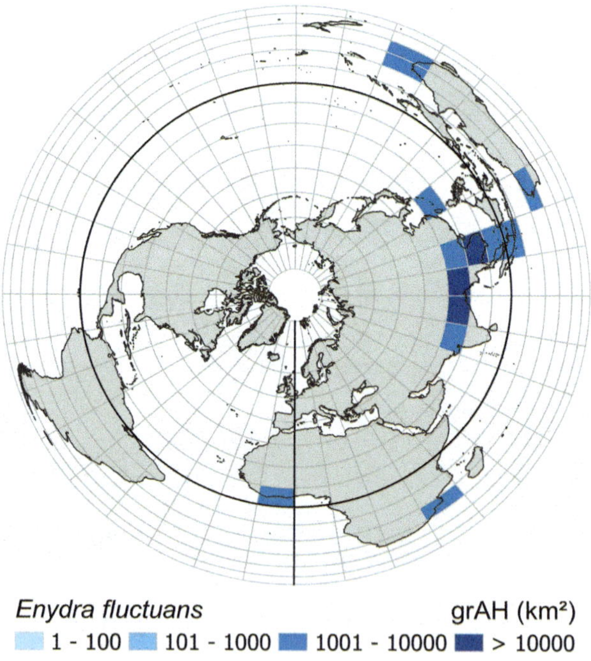

Enydra fluctuans DC.
Image copyright: iNaturalist CC0 @hhulsberg (Hugo Hulsberg)

Enydra radicans (Willd.) Lack
Synonyms: *Caesulia radicans* Willd., *Meyera guineensis* Spreng.

Range area (grAH km^2): NEO: 85100; AFR: 21300
Ecozone origin-status: NEO: N; AFR: N
World rarity status: R
Ecozone-endemic macroregion occurrence: not endemic to any world ecozone
Life form(s): EM
Ploidy state: NA
Speciation age (Ma): NA
Populations also occur in terrestrial, wetland and/ or riparian habitats: YES
Long distance movement vectors: NA
Nuisance status and uses: no nuisance issues; no known uses
Endangered status (IUCN): NA

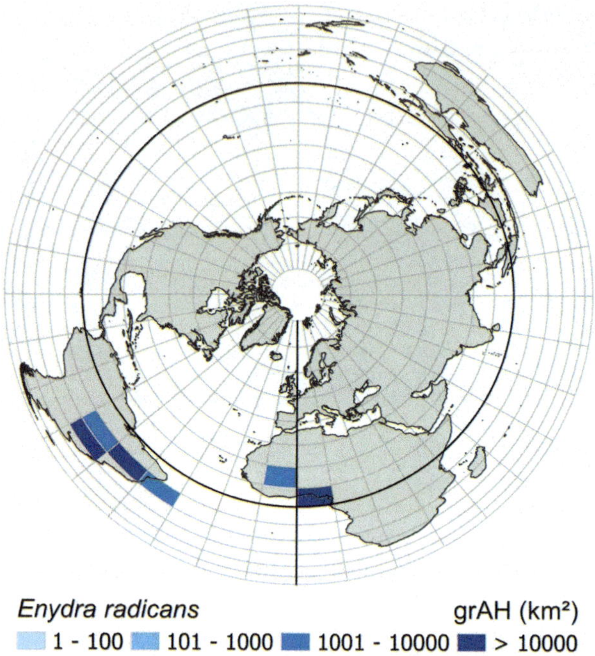

Enydra sessilifolia (Ruiz & Pav.) Cabrera
Synonyms: *Sobreyra sessilifolia* Ruiz & Pav., *Enydra rivularis* Gardner; *Enydra oblonga* (Ruiz & Pav.) DC., *Enydra maritima* (Kunth) DC.

Range area (grAH km^2): NEO: 83000; AFR: 34700
Ecozone origin-status: NEO: N; AFR: N
World rarity status: R
Ecozone-endemic macroregion occurrence: not endemic to any world ecozone
Life form(s): EM
Ploidy state: NA
Speciation age (Ma): NA
Populations also occur in terrestrial, wetland and/ or riparian habitats: YES
Long distance movement vectors: NA
Nuisance status and uses: no nuisance issues; no known uses
Endangered status (IUCN): NA

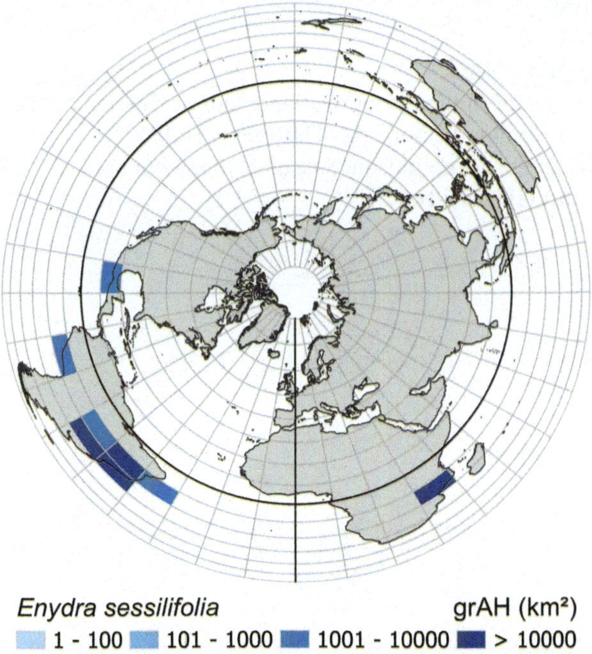

Enydra sessilifolia (Ruiz & Pav.) Cabrera

***Enydra sessilis* (Sw.) DC.**
Synonyms:

Range area (grAH km^2): NEO: 46800; AFR: 8000
Ecozone origin-status: NEO: E(I); AFR: I
World rarity status: R
Ecozone-endemic macroregion occurrence: NEO: MG MX, MES AM, PAR, CHA

Life form(s): EM
Ploidy state: NA
Speciation age (Ma): NA
Populations also occur in terrestrial, wetland and/ or riparian habitats: YES
Long distance movement vectors: ANTHROPOCHORY
Nuisance status and uses: no nuisance issues; no known uses
Endangered status (IUCN): NA

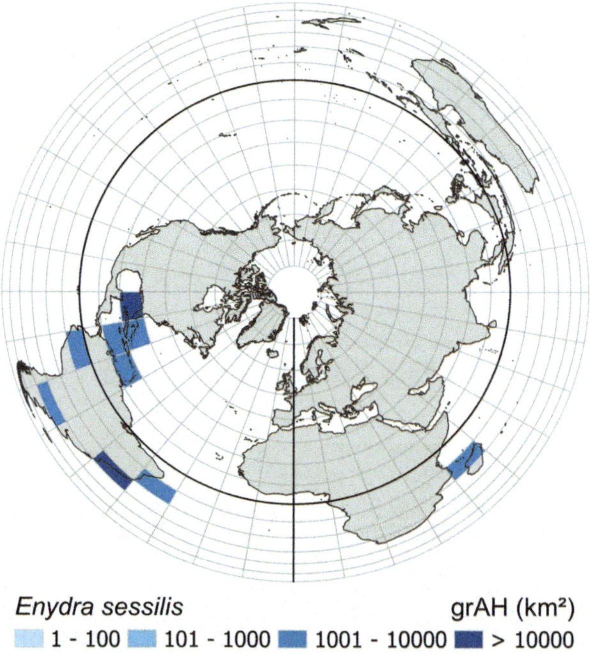

Enydra sessilis (Sw.) DC.

***Epaltes divaricata* Cass.**
Synonyms:

Range area (grAH km^2): OR: 37300
Ecozone origin-status: OR: E
World rarity status: R
Ecozone-endemic macroregion occurrence: OR: IND, ICH
Life form(s): EM
Ploidy state: NA
Speciation age (Ma): NA
Populations also occur in terrestrial, wetland and/ or riparian habitats: YES
Long distance movement vectors: NA
Nuisance status and uses: no nuisance issues; used in traditional medicine in India, some insecticidal properties
Endangered status (IUCN): NA

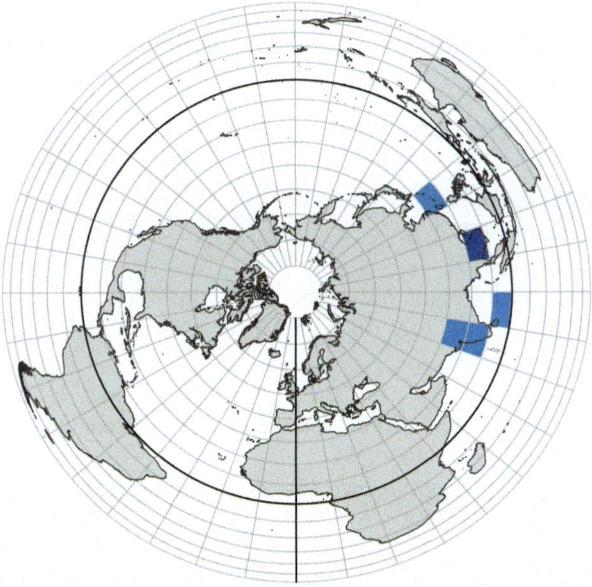

Epaltes divaricata Cass.

Erigeron heteromorphus **B.L.Rob.**
Synonyms:

Range area (grAH km^2): NEO: 13800
Ecozone origin-status: NEO: E
World rarity status: VR
Ecozone-endemic macroregion occurrence: NEO: MG MX
Life form(s): EM
Ploidy state: D
Speciation age (Ma): NA
Populations also occur in terrestrial, wetland and/ or riparian habitats: YES
Long distance movement vectors: NA
Nuisance status and uses: no nuisance issues; no known uses
Endangered status (IUCN): NA

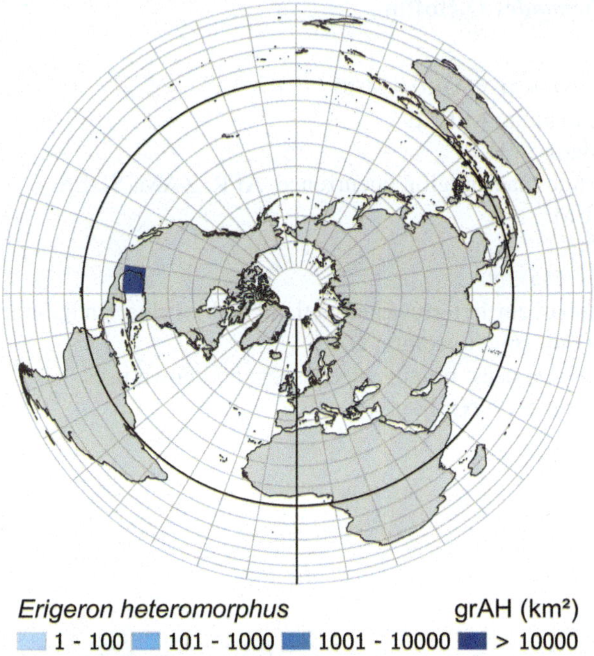

Erigeron heteromorphus B.L.Rob.
Image copyright: iNaturalist CC0 @aztekium (Carlos G Velazco-Macias)

***Grangea anthemoides* O.Hoffm.**
Synonyms:

Range area (grAH km^2): AFR: 13100
Ecozone origin-status: AFR: E
World rarity status: VR
Ecozone-endemic macroregion occurrence: AFR: ZAM, SDES
Life form(s): EM
Ploidy state: NA
Speciation age (Ma): NA
Populations also occur in terrestrial, wetland and/ or riparian habitats: YES
Long distance movement vectors: NA
Nuisance status and uses: no nuisance issues; no known uses
Endangered status (IUCN): NA

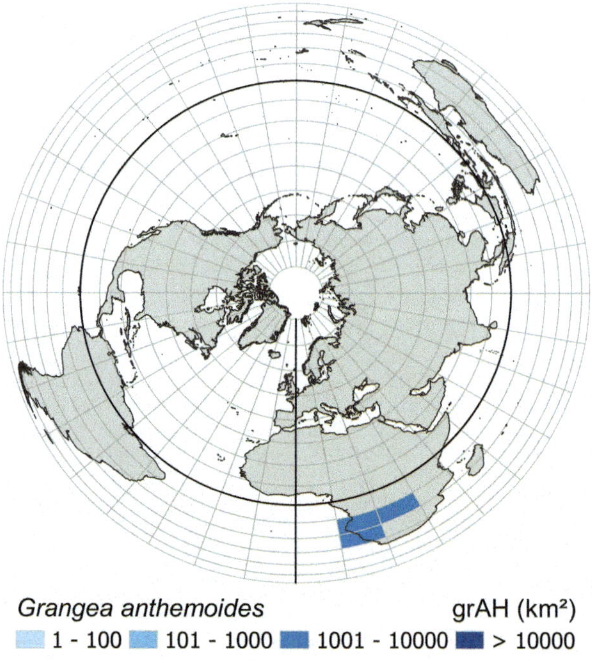

***Grangea maderaspatana* (L.) Poir.**

Synonyms: *Artemisia maderaspatana* L., *Cotula maderaspatana* (L.) Willd.

Range area (grAH km^2): OR: 100500; AFR: 260800
Ecozone origin-status: OR: N; AFR: N
World rarity status: C
Ecozone-endemic macroregion occurrence: not endemic to any world ecozone
Life form(s): EM
Ploidy state: D
Speciation age (Ma): NA
Populations also occur in terrestrial, wetland and/ or riparian habitats: YES
Long distance movement vectors: NA
Nuisance status and uses: no nuisance issues; leaves are harvested from the wild for local use as a food and medicine
Endangered status (IUCN): LC

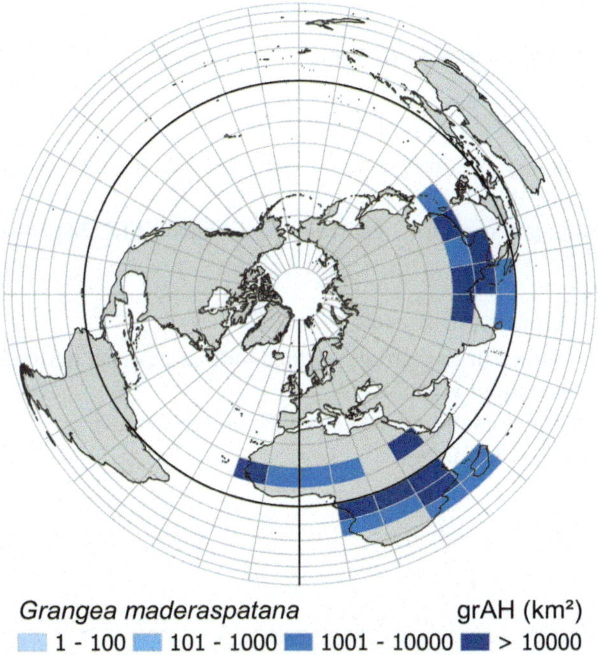

Grangea maderaspatana (L.) Poir.
Image copyright: A.T. Dung-Le, with permission

Gymnocoronis latifolia **Hook. & Arn.**
Synonyms:

Range area (grAH km^2): NEO: 41300
Ecozone origin-status: NEO: E
World rarity status: R
Ecozone-endemic macroregion occurrence: NEO: MG MX, MES AM
Life form(s): EM
Ploidy state: NA
Speciation age (Ma): NA
Populations also occur in terrestrial, wetland and/ or riparian habitats: YES
Long distance movement vectors: NA
Nuisance status and uses: no nuisance issues; no known uses
Endangered status (IUCN): NA

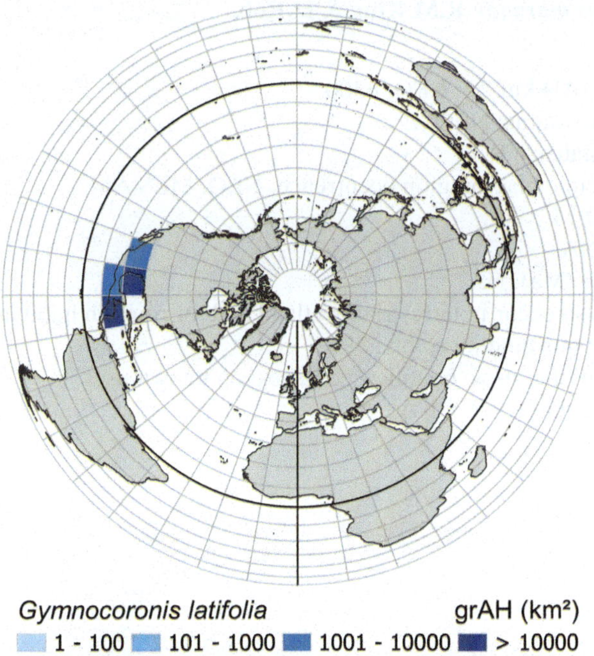

Gymnocoronis latifolia Hook. & Arn.
Image copyright: iNaturalist CC0 @annemirdl (Anne-Marie Veith)

Gymnocoronis matudae **R.M.King & H.Rob.**
Synonyms:

Range area (grAH km 2): NEO: 6300
Ecozone origin-status: NEO: E
World rarity status: VR
Ecozone-endemic macroregion occurrence: NEO: MES AM
Life form(s): EM
Ploidy state: NA
Speciation age (Ma): NA
Populations also occur in terrestrial, wetland and/ or riparian habitats: YES
Long distance movement vectors: NA
Nuisance status and uses: no nuisance issues; no known uses
Endangered status (IUCN): NA

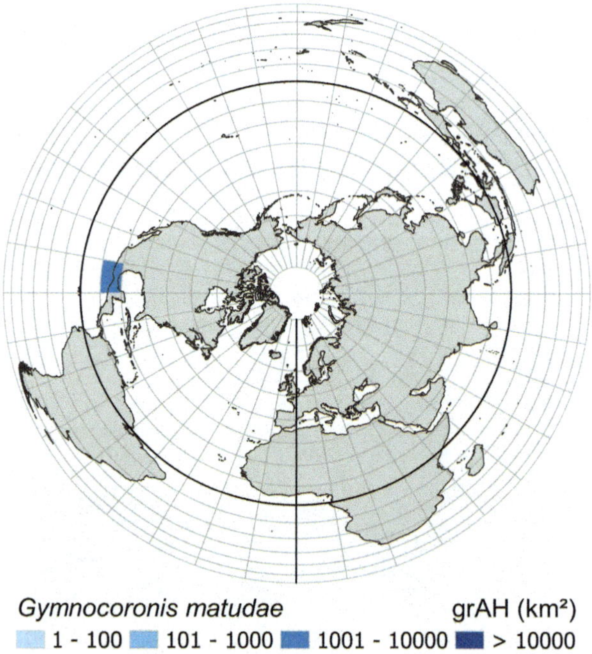

Gymnocoronis sessilis **S.F.Blake**
Synonyms:

Range area (grAH km^2): NEO: 16000
Ecozone origin-status: NEO: E
World rarity status: VR
Ecozone-endemic macroregion occurrence: NEO: MES AM
Life form(s): EM

Ploidy state: NA
Speciation age (Ma): NA
Populations also occur in terrestrial, wetland and/ or riparian habitats: YES
Long distance movement vectors: NA
Nuisance status and uses: no nuisance issues; no known uses
Endangered status (IUCN): NA

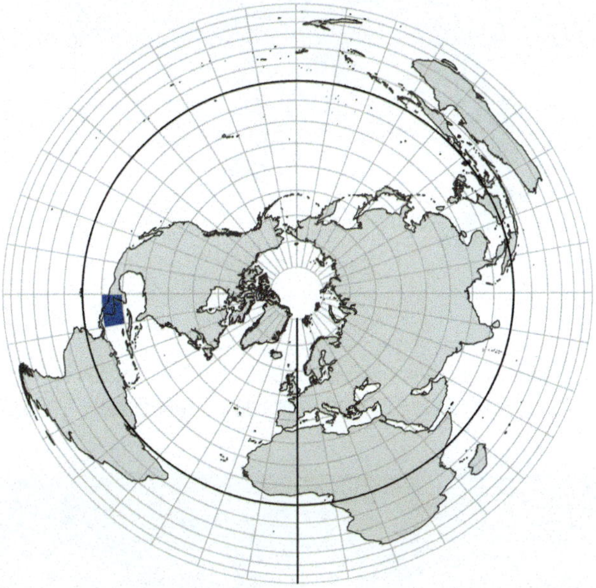

Gymnocoronis spilanthoides **(D.Don ex Hook. & Arn.) DC.**
Synonyms:

Range area (grAH km^2): PAL: 28300; NEO: 143600; OR: 2600; AUS: 28900; present in OCE
Ecozone origin-status: PAL: I; NEO: E(I); OR: I; AUS: I; OCE: I
World rarity status: C
Ecozone-endemic macroregion occurrence: NEO: AMZ, CHA, PAR
Life form(s): EM, SU
Ploidy state: D
Speciation age (Ma): NA
Populations also occur in terrestrial, wetland and/ or riparian habitats: YES
Long distance movement vectors: ANTHROPOCHORY
Nuisance status and uses: major invasive aquatic weed in AUS [112]; aquarium trade and water garden plant, medicinal
Endangered status (IUCN): NA

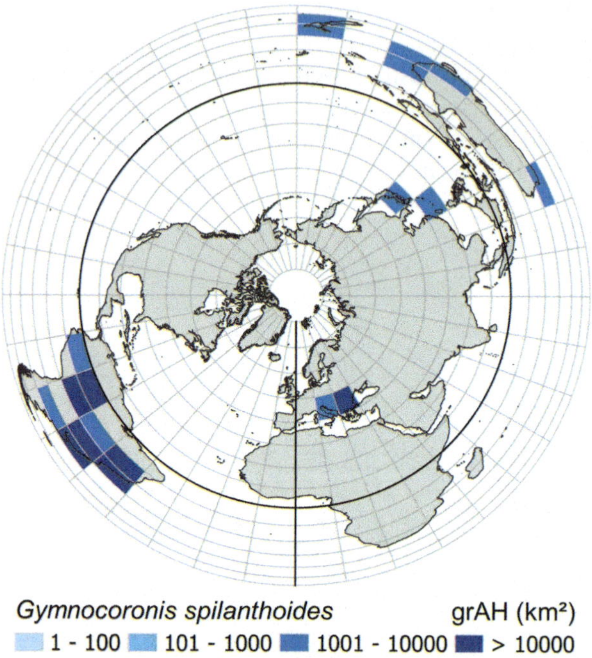

Gymnocoronis spilanthoides (D.Don ex Hook. & Arn.) DC.
Image copyright: iNaturalist CC-BY @yunglunlin (Yung-Lun Lin)

Hydropectis aquatica **(S.Watson) Rydb.**
Synonyms:

Range area (grAH km²): NEO: 5200
Ecozone origin-status: NEO: E
World rarity status: VR
Ecozone-endemic macroregion occurrence: NEO: MG MX
Life form(s): EM
Ploidy state: NA
Speciation age (Ma): NA
Populations also occur in terrestrial, wetland and/ or riparian habitats: YES
Long distance movement vectors: NA
Nuisance status and uses: no nuisance issues; no known uses
Endangered status (IUCN): NA

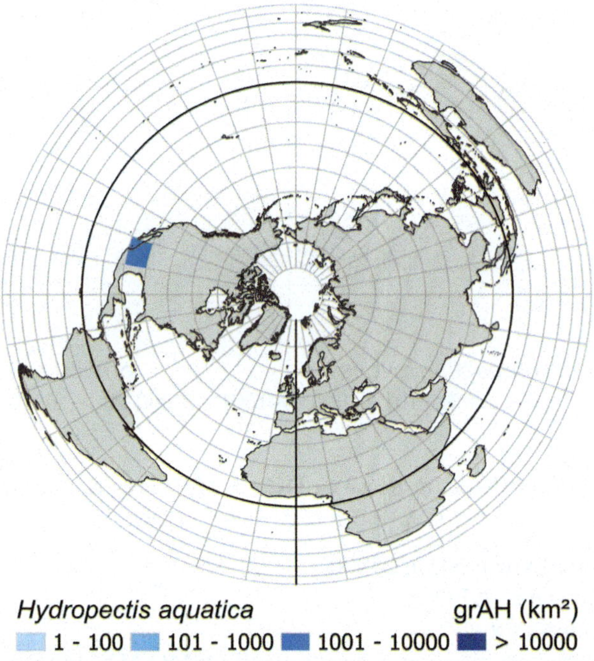

Hydropectis stevensii **McVaugh**
Synonyms: *Hydrodyssodia stevensii* (McVaugh) B.L.Turner

Range area (grAH km²): NEO: 5200
Ecozone origin-status: NEO: E
World rarity status: VR
Ecozone-endemic macroregion occurrence: NEO: MG MX
Life form(s): EM
Ploidy state: NA

Speciation age (Ma): NA
Populations also occur in terrestrial, wetland and/ or riparian habitats: YES
Long distance movement vectors: NA
Nuisance status and uses: no nuisance issues; no known uses
Endangered status (IUCN): NA

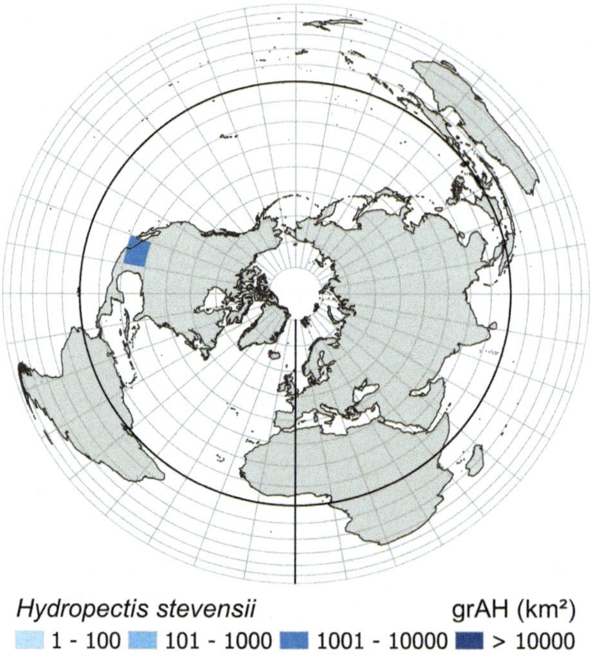

***Jaegeria bellidiflora* (Moc. & Sessé ex DC.) A.M.Torres & Beaman**
Synonyms: *Aganippea bellidiflora* Moc. & Sessé ex DC.

Range area (grAH km^2): NEO: 25900
Ecozone origin-status: NEO: E
World rarity status: VR
Ecozone-endemic macroregion occurrence: NEO: MG MX, MES AM
Life form(s): EM
Ploidy state: NA
Speciation age (Ma): NA
Populations also occur in terrestrial, wetland and/ or riparian habitats: YES
Long distance movement vectors: NA
Nuisance status and uses: no nuisance issues; no known uses
Endangered status (IUCN): NA

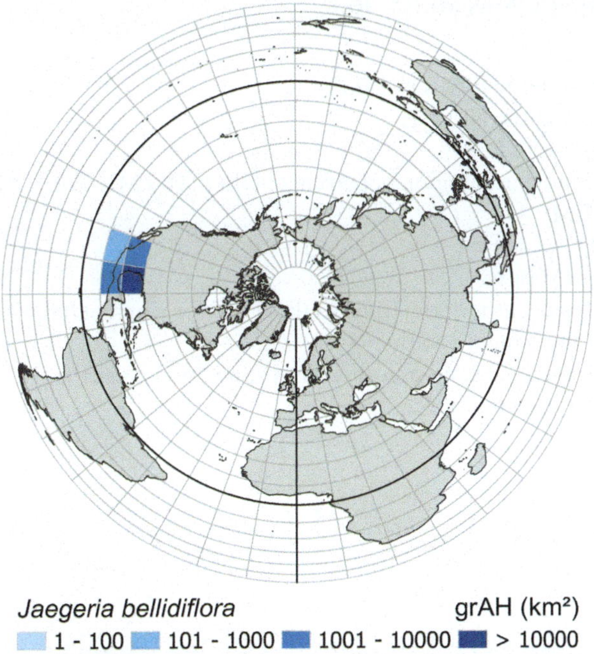

Jaegeria bellidiflora (Moc. & Sessé ex DC.) A.M.Torres & Beaman

Jaegeria glabra **(S.Watson) B.L.Rob.**
Synonyms:

Range area (grAH km^2): NEO: 12000
Ecozone origin-status: NEO: E
World rarity status: VR
Ecozone-endemic macroregion occurrence: NEO: MG MX, MES AM
Life form(s): EM
Ploidy state: D
Speciation age (Ma): NA
Populations also occur in terrestrial, wetland and/ or riparian habitats: YES
Long distance movement vectors: NA
Nuisance status and uses: no nuisance issues; no known uses
Endangered status (IUCN): NA

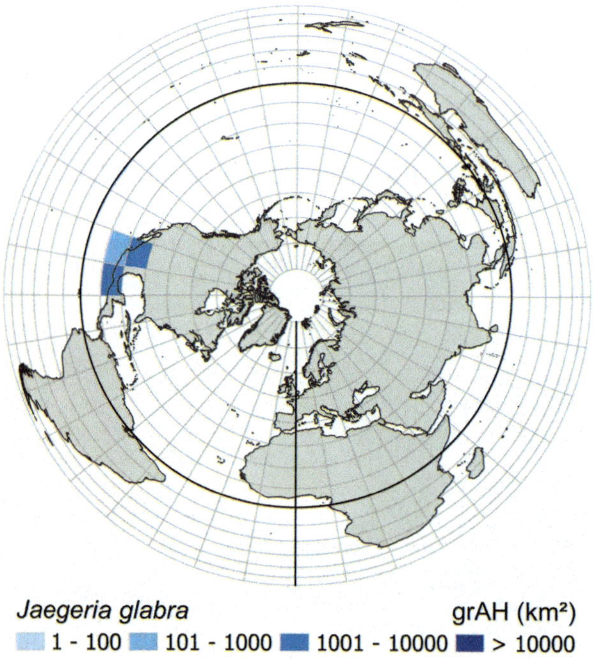

Jaegeria glabra (S.Watson) B.L.Rob.
Image copyright: T. Lobato-de Magalhães

***Jaegeria hirta* (Lag.) Less.**
Synonyms: *Acmella hirta* Lag., *Macella hirta* (Lag.) K.Koch, *Spilanthes hirta* (Lag.) Steud.

Range area (grAH km^2): NEO: 240300
Ecozone origin-status: NEO: E
World rarity status: C
Ecozone-endemic macroregion occurrence: NEO: MG MX, MES AM, NWSA, AMZ, CHA, PAR
Life form(s): EM
Ploidy state: P
Speciation age (Ma): NA
Populations also occur in terrestrial, wetland and/ or riparian habitats: YES
Long distance movement vectors: NA
Nuisance status and uses: no nuisance issues; some insecticidal value
Endangered status (IUCN): NA

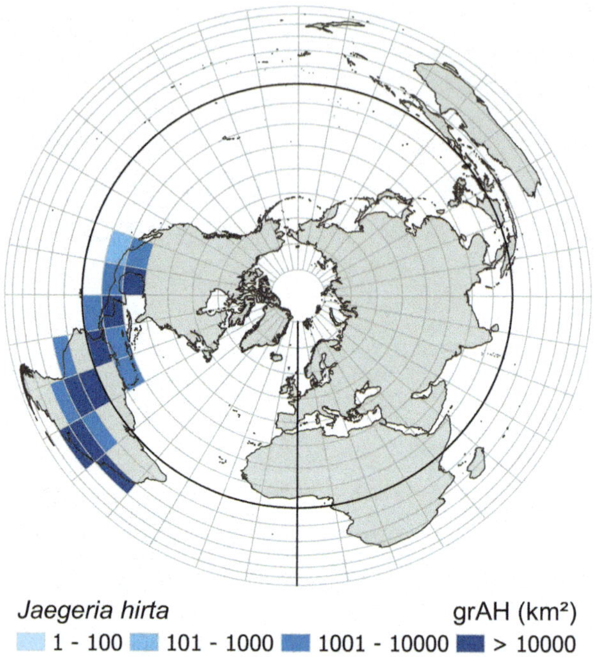

Jaegeria hirta (Lag.) Less.

***Jaegeria pedunculata* Hook. & Arn.**
Synonyms:

Range area (grAH km^2): NEO: 9100
Ecozone origin-status: NEO: E
World rarity status: VR
Ecozone-endemic macroregion occurrence: NEO: MG MX, MES AM
Life form(s): EM
Ploidy state: DP
Speciation age (Ma): NA
Populations also occur in terrestrial, wetland and/ or riparian habitats: YES
Long distance movement vectors: NA
Nuisance status and uses: no nuisance issues; no known uses
Endangered status (IUCN): NA

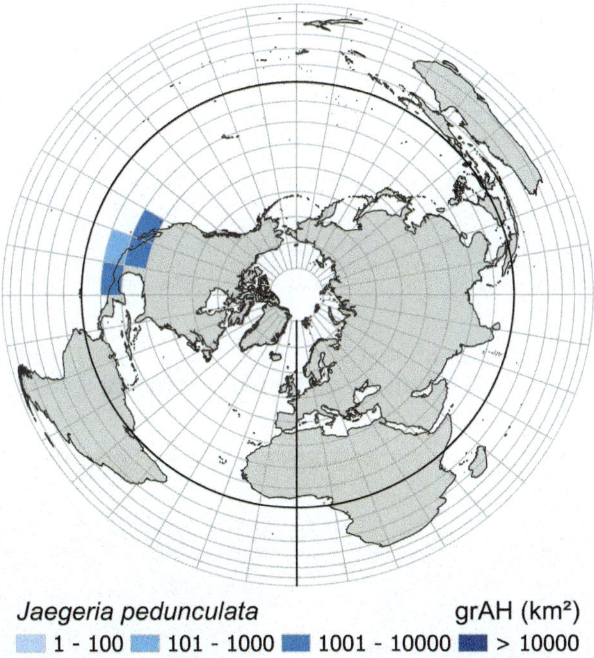

***Jaegeria purpurascens* B.L. Rob**
Synonyms:

Range area (grAH km^2): NEO: 15900
Ecozone origin-status: NEO: E
World rarity status: VR
Ecozone-endemic macroregion occurrence: NEO: MG MX
Life form(s): EM
Ploidy state: NA

Speciation age (Ma): NA
Populations also occur in terrestrial, wetland and/ or riparian habitats: YES
Long distance movement vectors: NA
Nuisance status and uses: no nuisance issues; no known uses
Endangered status (IUCN): NA

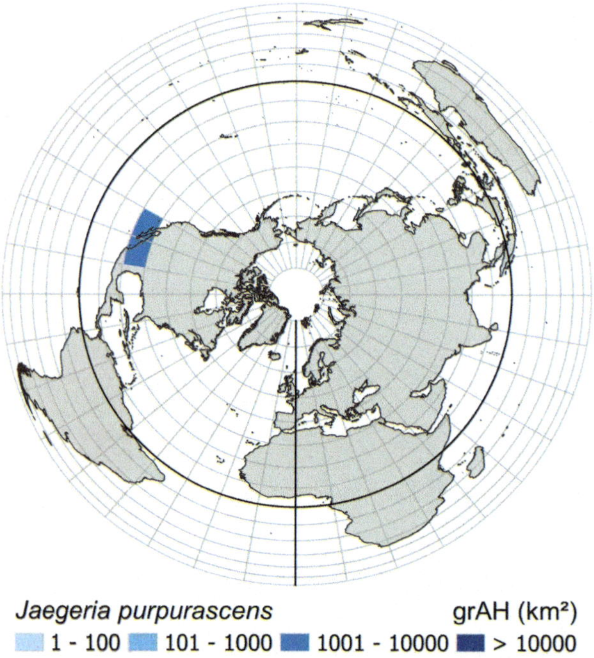

Jaegeria purpurascens B.L. Rob
Images copyright: T. Lobato-de Magalhães

Pacourina edulis **Aubl.**

Synonyms: *Calea sessiliflora* Stokes, nom. superfl., *Haynea edulis* (Aubl.) Willd., *Vernonia edulis* (Aubl.) Steud.

Range area (grAH km^2): NEO: 206000
Ecozone origin-status: NEO: E
World rarity status: C
Ecozone-endemic macroregion occurrence: NEO: MES AM, NWSA, AMZ, AND, CHA, PAR
Life form(s): EM
Ploidy state: NA
Speciation age (Ma): NA
Populations also occur in terrestrial, wetland and/ or riparian habitats: YES
Long distance movement vectors: NA
Nuisance status and uses: no nuisance issues; food and medical uses
Endangered status (IUCN): NA

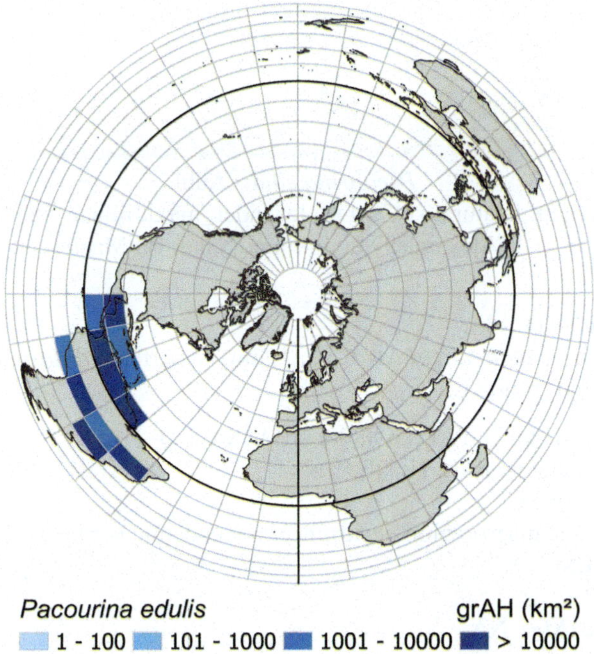

Pacourina edulis Aubl.
Image copyright: iNaturalist CC0 @alexander_maximiliano_sandoval_maciel (Alexander Maximiliano Sandoval Maciel)

Petasites frigidus **(L.) Fr.**
Synonyms: *Nardosmia angulosa* Cass., nom. superfl., *Nardosmia frigida* (L.) Hook., *Tussilago frigida* L.

Range area (grAH km^2): PAL: 598800; NEA: 895200
Ecozone origin-status: PAL: N; NEA: N
World rarity status: C
Ecozone-endemic macroregion occurrence: not endemic to any world ecozone
Life form(s): EM
Ploidy state: P
Speciation age (Ma): NA
Populations also occur in terrestrial, wetland and/ or riparian habitats: YES
Long distance movement vectors: NA
Nuisance status and uses: no nuisance issues; Native Americans of Alaska and northern Canada made extensive use of the plant: to make a cough medicine, and a salad of young leaves, stems and flowers.The mature leaves were used to create a makeshift funnel, a temporary container for berries, and to cover kegs of berries and rhubarb to discourage mould growth during winter storage, occasional use as a water garden plant
Endangered status (IUCN): LC

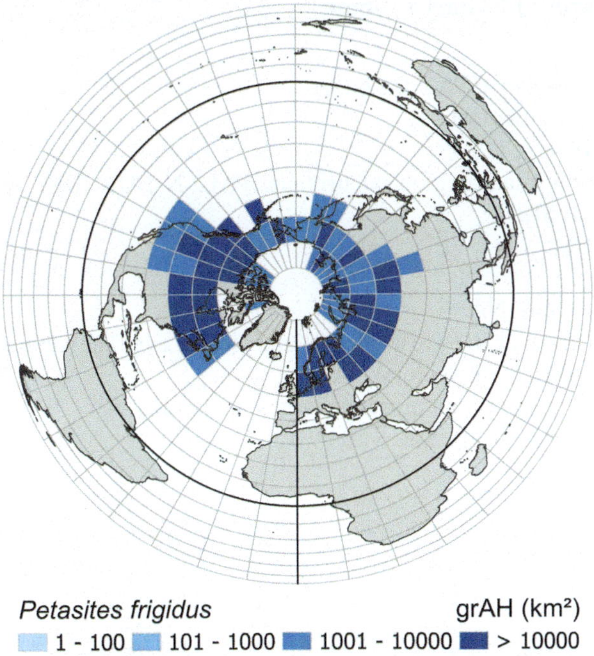

Petasites frigidus (L.) Fr.

***Petasites radiatus* (J.F.Gmel.) Toman**

Synonyms:

Range area (grAH km^2): PAL: 327400

Ecozone origin-status: PAL: E

World rarity status: C

Ecozone-endemic macroregion occurrence: PAL: N EUR, UR/ W SIB, ARC, C ASIA, CH, E SIB

Life form(s): EM

Ploidy state: P

Speciation age (Ma): NA

Populations also occur in terrestrial, wetland and/ or riparian habitats: YES

Long distance movement vectors: NA

Nuisance status and uses: no nuisance issues; no known uses

Endangered status (IUCN): NA

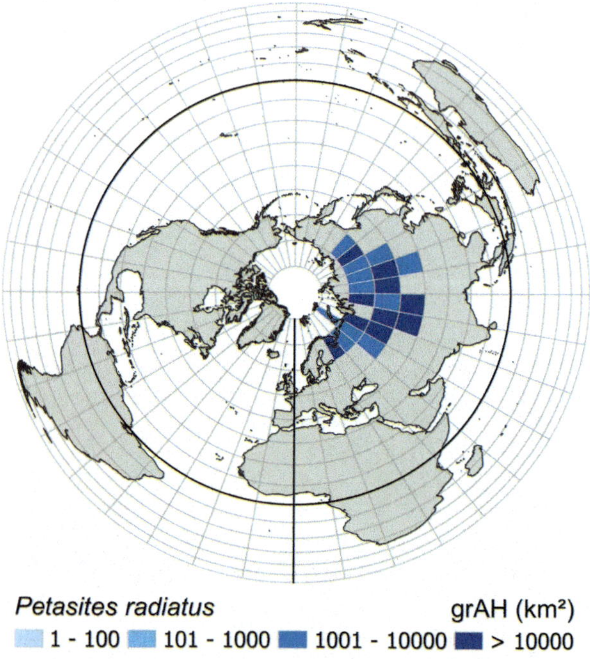

Petasites radiatus (J.F.Gmel.) Toman
Image copyright: iNaturalist CC-BY @windof (Marina Potapova)

Sclerolepis uniflora **(Walter) Britton, Sterns & Poggenb.**
Synonyms: *Ethulia uniflora* Walter

Range area (grAH km^2): NEA: 77500
Ecozone origin-status: NEA: E
World rarity status: R
Ecozone-endemic macroregion occurrence: NEA: E NAM
Life form(s): EM
Ploidy state: NA
Speciation age (Ma): NA
Populations also occur in terrestrial, wetland and/ or riparian habitats: YES
Long distance movement vectors: NA
Nuisance status and uses: no nuisance issues; no known uses
Endangered status (IUCN): NA

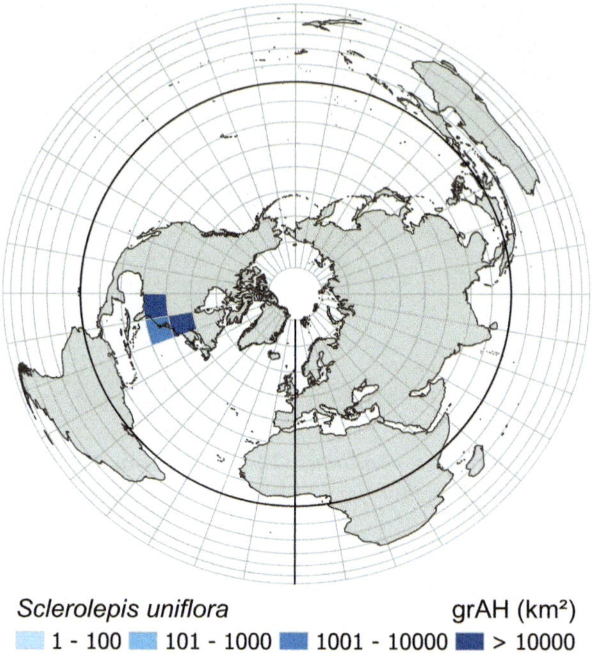

Sclerolepis uniflora (Walter) Britton, Sterns & Poggenb.

***Senecio bonariensis* Hook. & Arn.**

Synonyms:

Range area (grAH km^2): NEO: 82000
Ecozone origin-status: NEO: E
World rarity status: R
Ecozone-endemic macroregion occurrence: NEO: AMZ, CHA, PAR
Life form(s): EM
Ploidy state: NA
Speciation age (Ma): NA
Populations also occur in terrestrial, wetland and/ or riparian habitats: YES
Long distance movement vectors: NA
Nuisance status and uses: no nuisance issues; some phytoremediation and medicinal potential
Endangered status (IUCN): NA

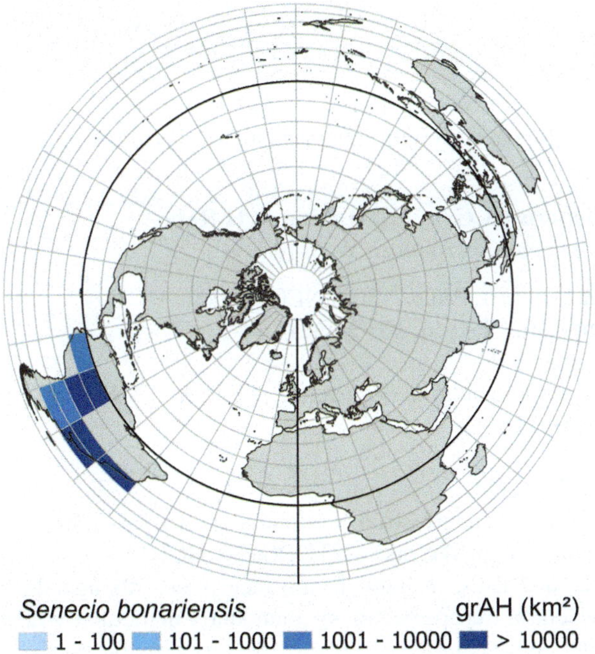

Senecio bonariensis Hook. & Arn.
Image copyright: iNaturalist CC0 @christiangilli

Shinnersia rivularis (A.Gray) R.M.King & H.Rob.
Synonyms: *Trichocoronis rivularis* A.Gray

Range area (grAH km^2): PAL: 21300; NEO: 5200
Ecozone origin-status: PAL: I; NEO: E(I)
World rarity status: VR
Ecozone-endemic macroregion occurrence: NEO: MG MX
Life form(s): EM, SU
Ploidy state: P
Speciation age (Ma): NA
Populations also occur in terrestrial, wetland and/ or riparian habitats: YES
Long distance movement vectors: ANTHROPOCHORY
Nuisance status and uses: no known nuisance issues, though the plant has been reported in Central Europe occurring abundantly in thermally-heated springs and rivers, giving some cause for concern over its possible further spread; aquarium trade
Endangered status (IUCN): NA

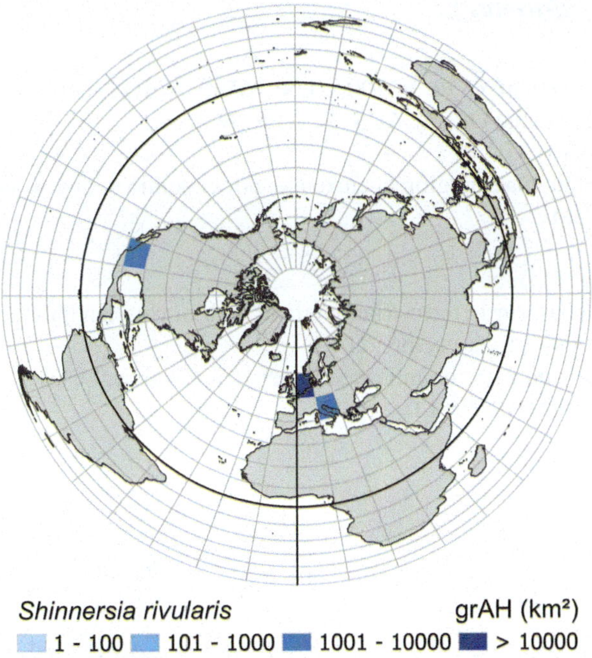

Shinnersia rivularis (A.Gray) R.M.King & H.Rob.
Image copyright: iNaturalist CC-BY @jorgetrevino

***Sphaeranthus africanus* L.**
Synonyms:

Range area (grAH km^2): OR: 82400; AUS: 16500; AFR: 136400
Ecozone origin-status: OR: N; AUS: N; AFR: N
World rarity status: C
Ecozone-endemic macroregion occurrence: not endemic to any world ecozone
Life form(s): EM
Ploidy state: D
Speciation age (Ma): NA
Populations also occur in terrestrial, wetland and/ or riparian habitats: YES
Long distance movement vectors: NA
Nuisance status and uses: no nuisance issues; some medical uses
Endangered status (IUCN): LC

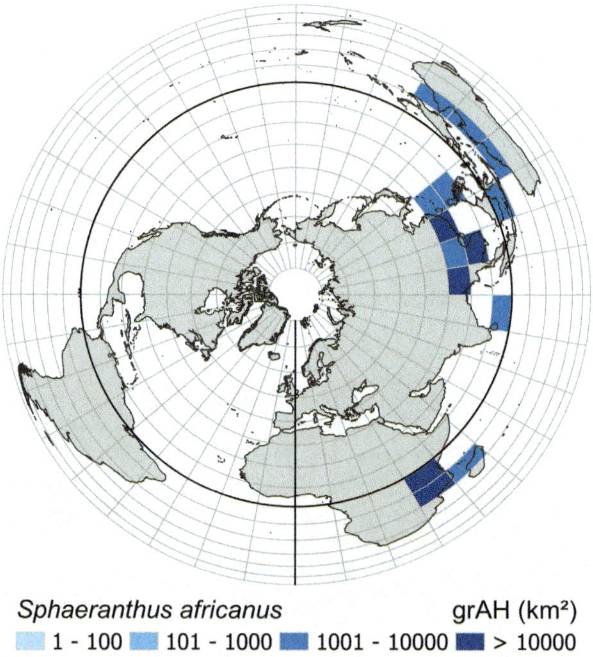

Sphaeranthus africanus L.

Sphaeranthus amaranthoides Burm.f.
Synonyms: *Oligolepis amaranthoides* (Burm.f.) Wight

Range area (grAH km^2): OR: 10800; AFR: 7800
Ecozone origin-status: OR: E(I); AFR: I
World rarity status: VR
Ecozone-endemic macroregion occurrence: OR: IND
Life form(s): EM
Ploidy state: NA
Speciation age (Ma): NA
Populations also occur in terrestrial, wetland and/ or riparian habitats: YES
Long distance movement vectors: ANTHROPOCHORY
Nuisance status and uses: no nuisance issues; some medicinal potential
Endangered status (IUCN): LC

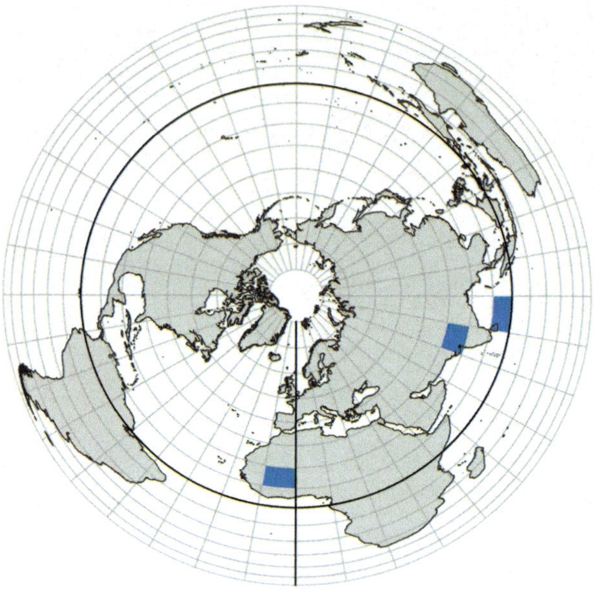

***Sphaeranthus indicus* L.**
Synonyms: *Oligolepis indicus* (L.) Cass. ex Steud.

Range area (grAH km^2): OR: 28500; AUS: 30100
Ecozone origin-status: OR: N; AUS: N
World rarity status: R
Ecozone-endemic macroregion occurrence: not endemic to any world ecozone
Life form(s): EM
Ploidy state: D
Speciation age (Ma): NA
Populations also occur in terrestrial, wetland and/ or riparian habitats: YES
Long distance movement vectors: NA
Nuisance status and uses: no nuisance issues; used in traditional medicine in India
Endangered status (IUCN): LC

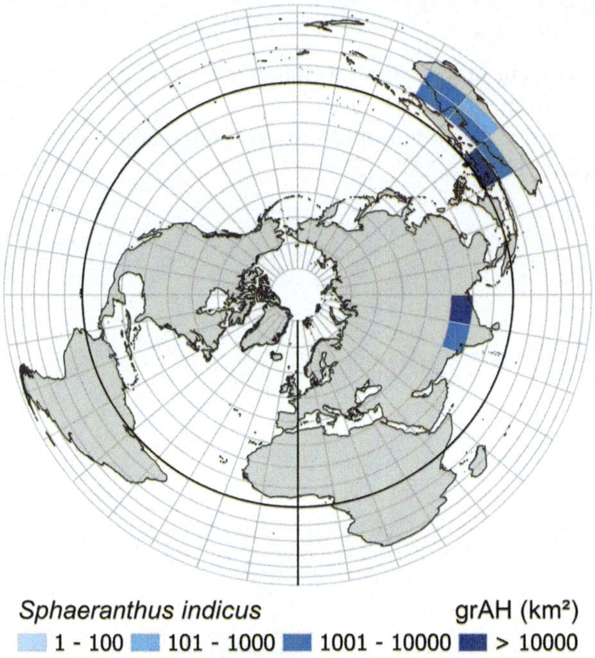

Sphaeranthus indicus L.
Image copyright: A.T. Dung-Le, with permission

***Sphaeranthus peguensis* Kurz ex C.B.Clarke**
Synonyms:

Range area (grAH km^2): OR: 11000
Ecozone origin-status: OR: E
World rarity status: VR
Ecozone-endemic macroregion occurrence: OR: ICH
Life form(s): EM

Ploidy state: NA
Speciation age (Ma): NA
Populations also occur in terrestrial, wetland and/ or riparian habitats: YES
Long distance movement vectors: NA
Nuisance status and uses: no nuisance issues; food, medical uses
Endangered status (IUCN): LC

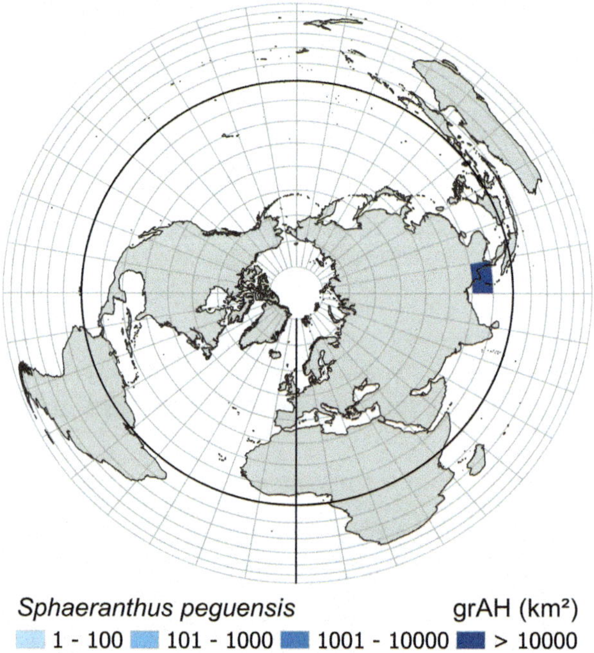

***Sphaeranthus senegalensis* DC.**
Synonyms:

Range area (grAH km^2): OR: 36100; AFR: 138100
Ecozone origin-status: OR: N; AFR: N
World rarity status: C
Ecozone-endemic macroregion occurrence: not endemic to any world ecozone
Life form(s): EM
Ploidy state: NA
Speciation age (Ma): NA
Populations also occur in terrestrial, wetland and/ or riparian habitats: YES
Long distance movement vectors: NA
Nuisance status and uses: no nuisance issues; some medicinal uses
Endangered status (IUCN): LC

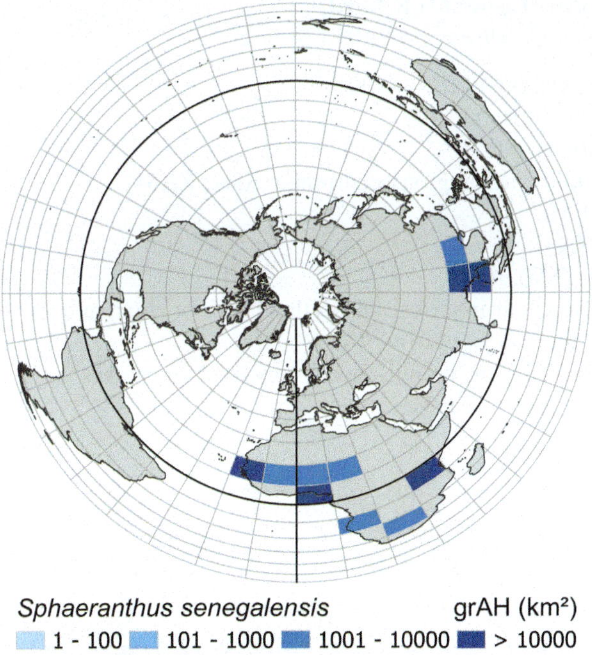

Sphaeranthus senegalensis DC.
Image copyright: iNaturalist CC0

***Stenops zairensis* (Lisowski) B.Nord.**
Synonyms: *Pseudocadiscus zairensis* Lisowski

Range area (grAH km^2): AFR: 8700
Ecozone origin-status: AFR: E
World rarity status: VR
Ecozone-endemic macroregion occurrence: AFR: ZAM
Life form(s): EM
Ploidy state: NA
Speciation age (Ma): NA
Populations also occur in terrestrial, wetland and/ or riparian habitats: YES
Long distance movement vectors: NA
Nuisance status and uses: no nuisance issues; no known uses
Endangered status (IUCN): NA

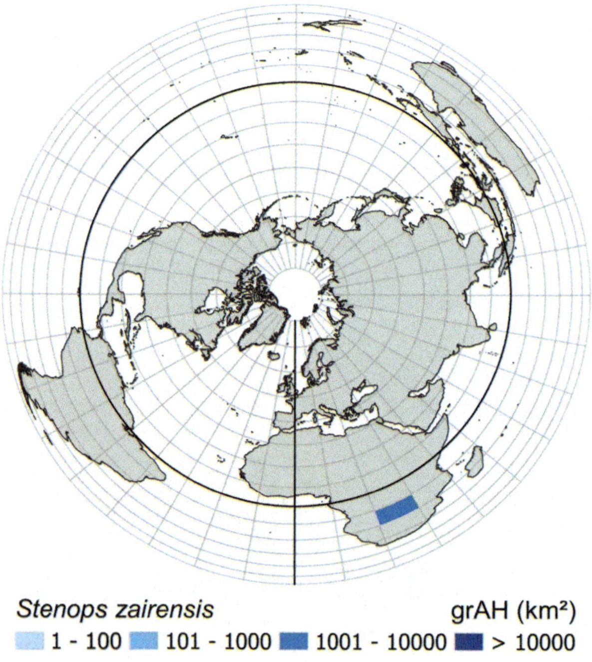

***Symphyllocarpus exilis* Maxim.**
Synonyms:

Range area (grAH km^2): PAL: 45700; OR: 7000
Ecozone origin-status: PAL: N; OR: N
World rarity status: R
Ecozone-endemic macroregion occurrence: not endemic to any world ecozone
Life form(s): EM

Ploidy state: NA
Speciation age (Ma): NA
Populations also occur in terrestrial, wetland and/ or riparian habitats: YES
Long distance movement vectors: NA
Nuisance status and uses: no nuisance issues; no known uses
Endangered status (IUCN): NA

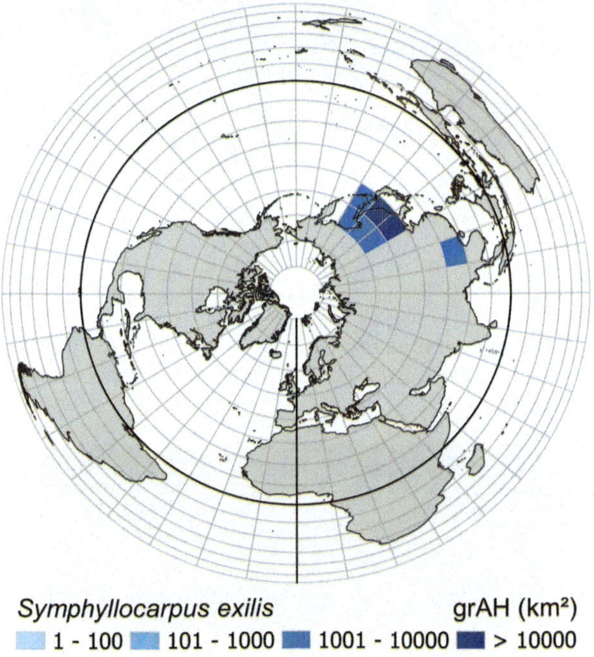

***Tagetes epapposa* B.L.Turner**
Synonyms:

Range area (grAH km^2): NEO: 5200
Ecozone origin-status: NEO: E
World rarity status: VR
Ecozone-endemic macroregion occurrence: NEO: MG MX
Life form(s): EM
Ploidy state: NA
Speciation age (Ma): NA
Populations also occur in terrestrial, wetland and/ or riparian habitats: YES
Long distance movement vectors: NA
Nuisance status and uses: no nuisance issues; no known uses
Endangered status (IUCN): NA

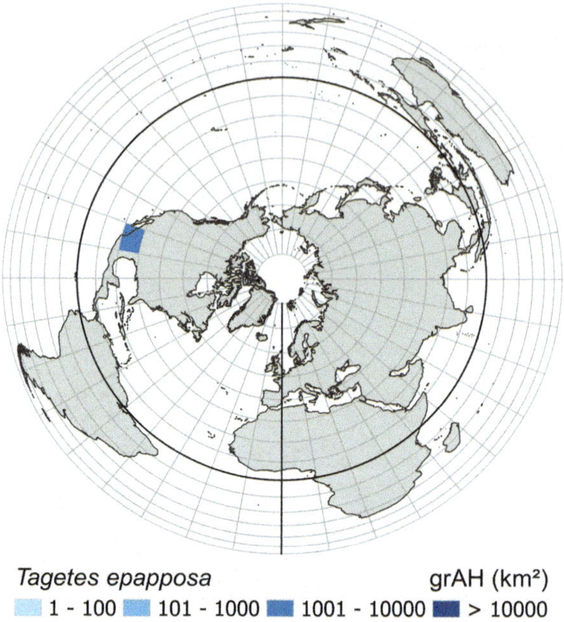

Tagetes epapposa B.L.Turner
Image copyright: iNaturalist CC-BY @arturocastro (Arturo Castro Castro)

***Tagetes filifolia* Lag.**

Synonyms:

Range area (grAH km^2): NEA: 8200; NEO: 114600
Ecozone origin-status: NEA: N; NEO: N
World rarity status: R
Ecozone-endemic macroregion occurrence: not endemic to any world ecozone
Life form(s): EM
Ploidy state: D
Speciation age (Ma): NA
Populations also occur in terrestrial, wetland and/ or riparian habitats: YES
Long distance movement vectors: NA
Nuisance status and uses: no nuisance issues; the plant is harvested from the wild and used locally as a food flavouring, tea and medicinal herb, insect repellent, garden plant
Endangered status (IUCN): NA

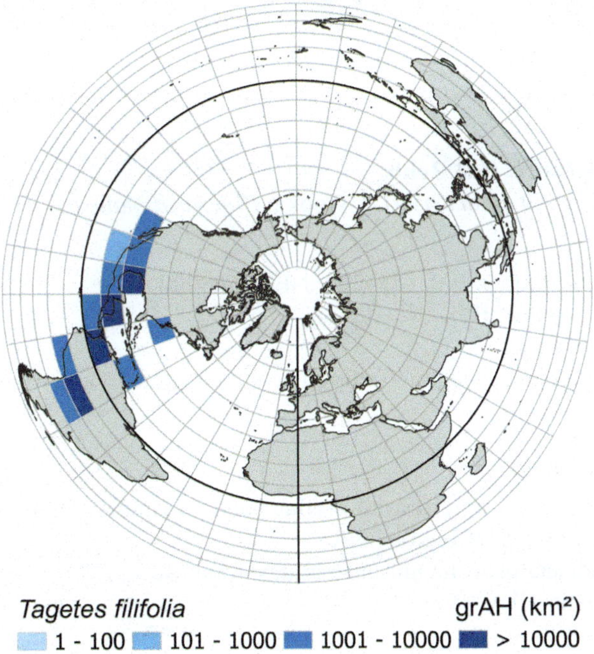

Tagetes filifolia Lag.
Image copyright: E.R. Félix, with permission

Tephroseris palustris (L.) Rchb.
Synonyms: *Senecio palustris* (L.) Hook., *Cineraria palustris* (L.) L., *Othonna palustris* L.

Range area (grAH km^2): PAL: 454200; NEA: 568000
Ecozone origin-status: PAL: N; NEA: N
World rarity status: C
Ecozone-endemic macroregion occurrence: not endemic to any world ecozone
Life form(s): EM
Ploidy state: DP
Speciation age (Ma): NA
Populations also occur in terrestrial, wetland and/ or riparian habitats: YES
Long distance movement vectors: NA
Nuisance status and uses: no nuisance issues; food use
Endangered status (IUCN): NA

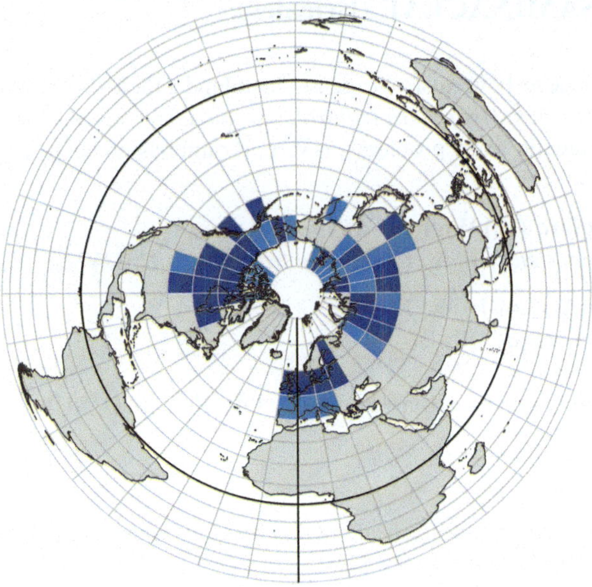

Tephroseris palustris (L.) Rchb.
Images copyright: J.C. Schou, with permission

3.7 BALSAMINACEAE (Ericales)

The Balsaminaceae have just one genus in inland brackish or freshwater bodies, with a single species, *Hydrocera trifloral*, a tropical species, endemic to OR, which never causes aquatic weed problems, is rare, R, and has endangered status LC.

Hydrocera triflora **(L.) Wight & Arn.**
Synonyms: *Impatiens triflora* L., *Tytonia triflora* (L.) C.E.Wood

Range area (grAH km^2): OR: 59800
Ecozone origin-status: OR: E
World rarity status: R
Ecozone-endemic macroregion occurrence: OR: IND, ICH, SUP
Life form(s): EM
Ploidy state: NA
Speciation age (Ma): NA
Populations also occur in terrestrial, wetland and/ or riparian habitats: YES
Long distance movement vectors: NA
Nuisance status and uses: no nuisance issues; harvested from the wild for local use as a dye
Endangered status (IUCN): LC

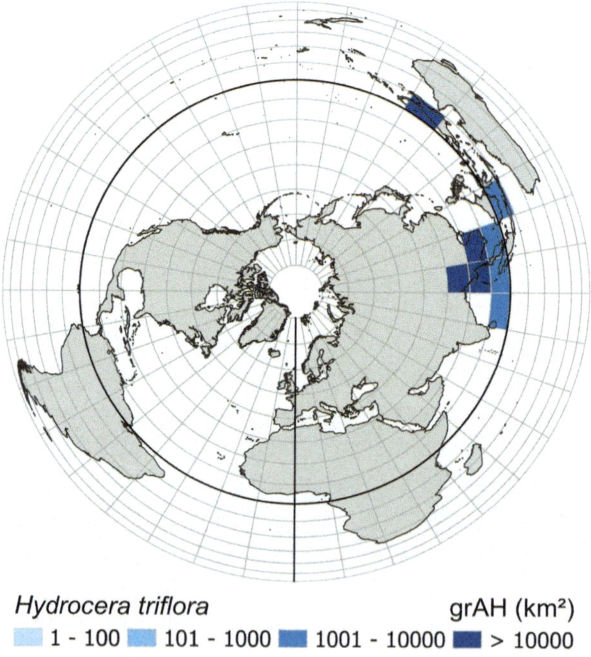

Hydrocera triflora (L.) Wight & Arn.
Image copyright: Z. Li, with permission

3.8 BORAGINACEAE (Boraginales)

The Boraginaceae contain four genera with 12 freshwater macrophyte species widely distributed in all world ecozones, except OCE and ANT. Nine species are present in PAL, six in OR, six in AFR, five species recorded in AUS, three in NEO, two in NEA, and none in either ANT or OCE. They show low ecozone-endemicity, except for PAL, which has six E or E(I) species, all being *Myosotis* spp., whilst *Heliotropium keralense* is limited in its world distribution to the Indian macroregion (IND) of OR. Anthropochory vectors play a limited role in the world distribution of Boraginaceae macrophyte species, though four of the five such species present in AUS have I-origin status, with a further two in NEO and two in AFR. *Coldenia procumbens* occasionally causes aquatic weed problems, mostly in rice paddies. The Boraginaceae have seven VR species and six C. The Boraginaceae have six species with endangered status NA/DD and five with LC. *Myosotis rehsteineri* is at higher risk: EN.

Quick Links to Genera

1. *Coldenia*
2. *Heliotropium*
3. *Myosotis*
4. *Rotula*

***Coldenia procumbens* L.**

Synonyms:

Range area (grAH km^2): OR: 29200; AUS: 42800; AFR: 208800
Ecozone origin-status: OR: N; AUS: N; AFR: N
World rarity status: C
Ecozone-endemic macroregion occurrence: not endemic to any world ecozone
Life form(s): EM
Ploidy state: NA
Speciation age (Ma): NA
Populations also occur in terrestrial, wetland and/ or riparian habitats: YES
Long distance movement vectors: NA
Nuisance status and uses: some nuisance issues – locally considered to be a weed, often growing in rice fields within its native range; the plant is often harvested from the wild and traded on the Internet, mainly for local use as a medicine
Endangered status (IUCN): NA

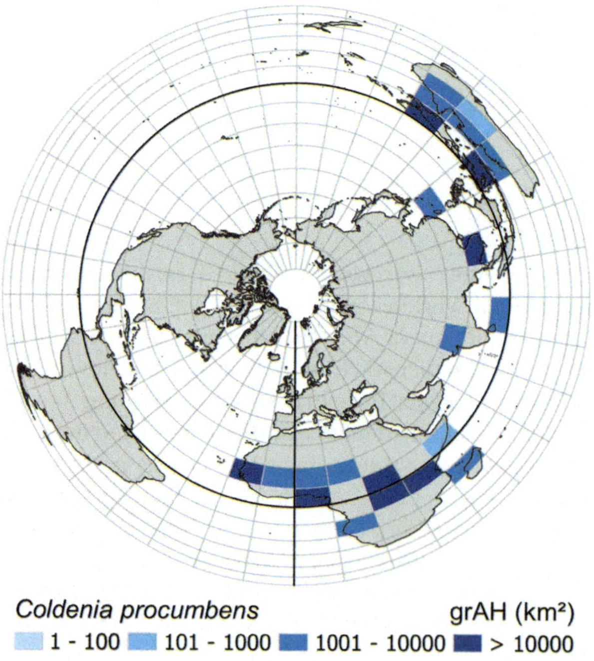

Coldenia procumbens L.
Image copyright: iNaturalist CC0 @dineshvalke (F164Dinesh Valke)

Heliotropium keralense **Sivarajan & Manilal**
Synonyms:

Range area (grAH km^2): OR: 8100
Ecozone origin-status: OR: E
World rarity status: VR
Ecozone-endemic macroregion occurrence: OR: IND
Life form(s): EM
Ploidy state: NA
Speciation age (Ma): NA
Populations also occur in terrestrial, wetland and/ or riparian habitats: YES
Long distance movement vectors: NA
Nuisance status and uses: no nuisance issues; some medicinal uses
Endangered status (IUCN): NA

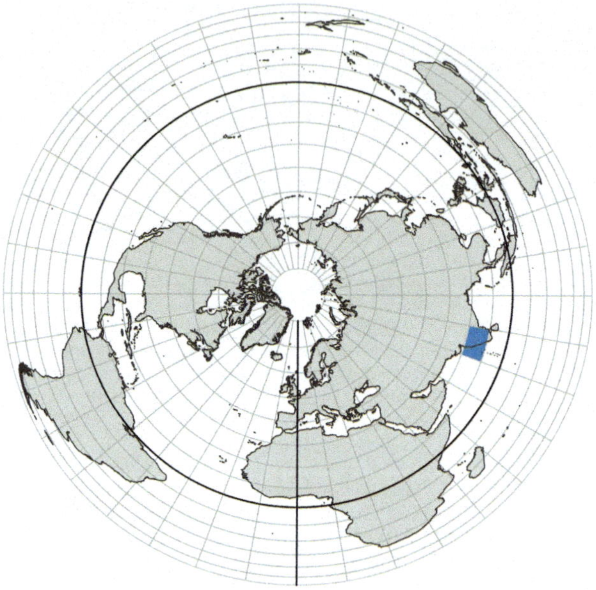

Heliotropium ovalifolium **Forssk.**
Synonyms: *Euploca ovalifolia* (Forssk.) Diane & Hilger

Range area (grAH km^2): PAL: 6400; OR: 11200; AUS: 25100; AFR: 321300
Ecozone origin-status: PAL: N; OR: N; AUS: N; AFR: N
World rarity status: C
Ecozone-endemic macroregion occurrence: not endemic to any world ecozone
Life form(s): EM (sometimes a subshrub)
Ploidy state: P
Speciation age (Ma): NA
Populations also occur in terrestrial, wetland and/ or riparian habitats: YES
Long distance movement vectors: NA
Nuisance status and uses: no nuisance issues; traditionally used to treat wounds and to treat inflammations or other skin diseases, in various African countries is used against syphilis and as an analgesic
Endangered status (IUCN): LC

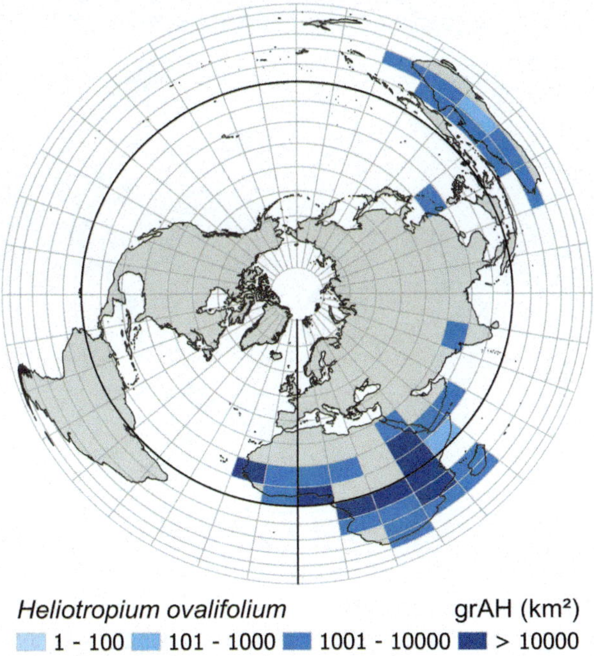

***Heliotropium supinum* L.**
Synonyms: *Piptoclaina supina* (L.) G.Don

Range area (grAH km^2): PAL: 91100; OR: 28500; AUS: 13400; AFR: 201900
Ecozone origin-status: PAL: N; OR: N; AUS: I; AFR: N
World rarity status: C
Ecozone-endemic macroregion occurrence: not endemic to any world ecozone
Life form(s): EM
Ploidy state: D
Speciation age (Ma): NA
Populations also occur in terrestrial, wetland and/ or riparian habitats: YES
Long distance movement vectors: ANTHROPOCHORY
Nuisance status and uses: no nuisance issues; harvested from the wild for local use as a medicine
Endangered status (IUCN): NA

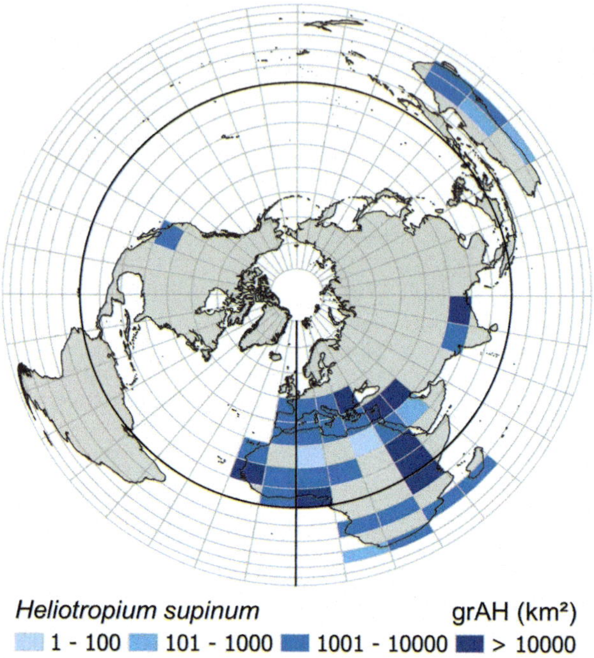

Myosotis debilis **Pomel**
Synonyms:

Range area (grAH km^2): PAL: 10200
Ecozone origin-status: PAL: E
World rarity status: VR
Ecozone-endemic macroregion occurrence: PAL: MED, W EUR
Life form(s): EM
Ploidy state: NA
Speciation age (Ma): NA
Populations also occur in terrestrial, wetland and/ or riparian habitats: YES
Long distance movement vectors: NA
Nuisance status and uses: no nuisance issues; no known uses
Endangered status (IUCN): LC

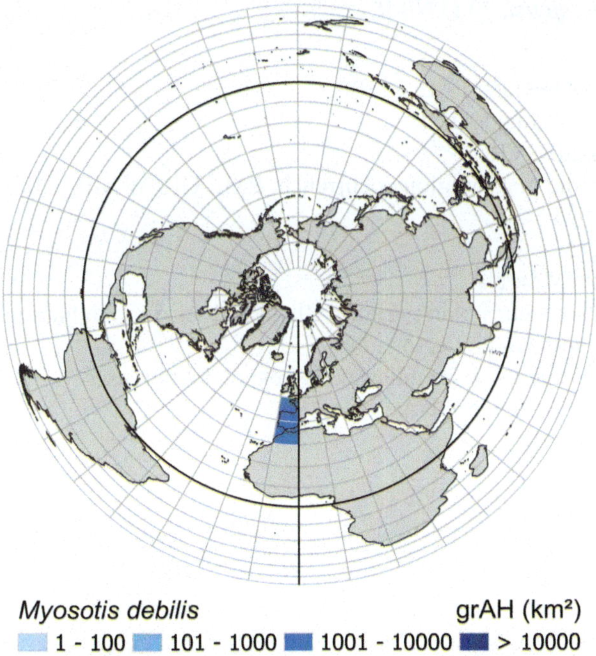

Myosotis debilis Pomel
Image copyright: J.T.R Tavares, with permission

Myosotis jenissejensis **O.D.Nikif.**
Synonyms:

Range area (grAH km²): PAL: 22800
Ecozone origin-status: PAL: E
World rarity status: VR
Ecozone-endemic macroregion occurrence: PAL: UR/ W SIB
Life form(s): EM
Ploidy state: NA
Speciation age (Ma): NA
Populations also occur in terrestrial, wetland and/ or riparian habitats: YES
Long distance movement vectors: NA
Nuisance status and uses: no nuisance issues; no known uses
Endangered status (IUCN): NA

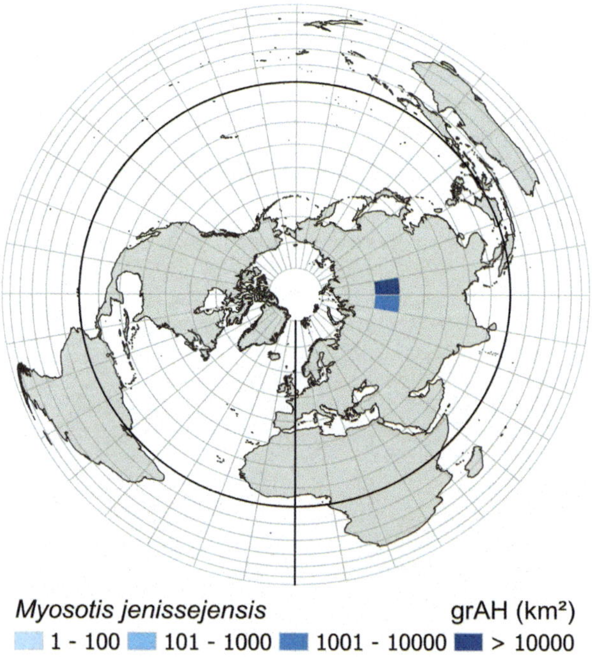

Myosotis laxa **Lehm.**
Synonyms:

Range area (grAH km²): PAL: 869100; NEA: 386100; NEO: 48200; OR: 7000; AUS: 35300; AFR: 3900
Ecozone origin-status: PAL: N; NEA: N; NEO: I; OR: N; AUS: I; AFR: N
World rarity status: C
Ecozone-endemic macroregion occurrence: not endemic to any world ecozone
Life form(s): EM
Ploidy state: P
Speciation age (Ma): 7.0
Populations also occur in terrestrial, wetland and/ or riparian habitats: YES
Long distance movement vectors: ANTHROPOCHORY; ?AV
Nuisance status and uses: no nuisance issues; occasional use as a garden plant, folk medicine use
Endangered status (IUCN): LC

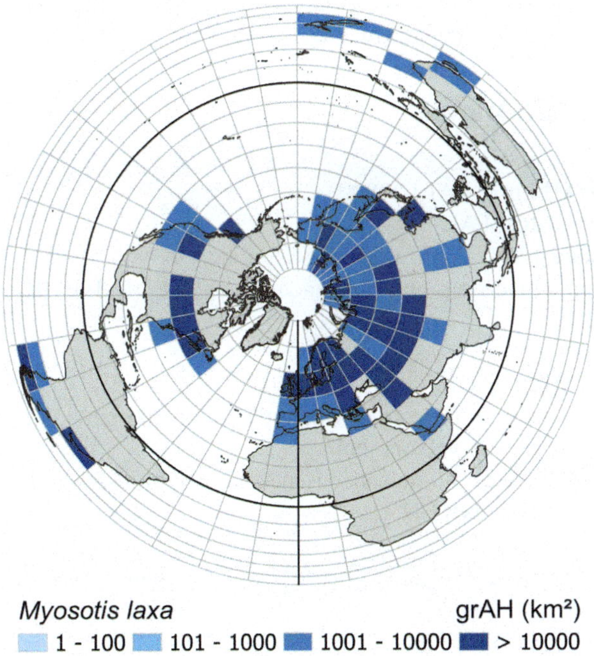

Myosotis laxa Lehm.
Image copyright: J.C. Schou, with permission

***Myosotis rehsteineri* Wartm.**
Synonyms:

Range area (grAH km^2): PAL: 29800
Ecozone origin-status: PAL: E
World rarity status: VR
Ecozone-endemic macroregion occurrence: PAL: W EUR, E EUR
Life form(s): EM
Ploidy state: NA
Speciation age (Ma): NA
Populations also occur in terrestrial, wetland and/ or riparian habitats: YES
Long distance movement vectors: NA
Nuisance status and uses: no nuisance issues; no known uses
Endangered status (IUCN): EN

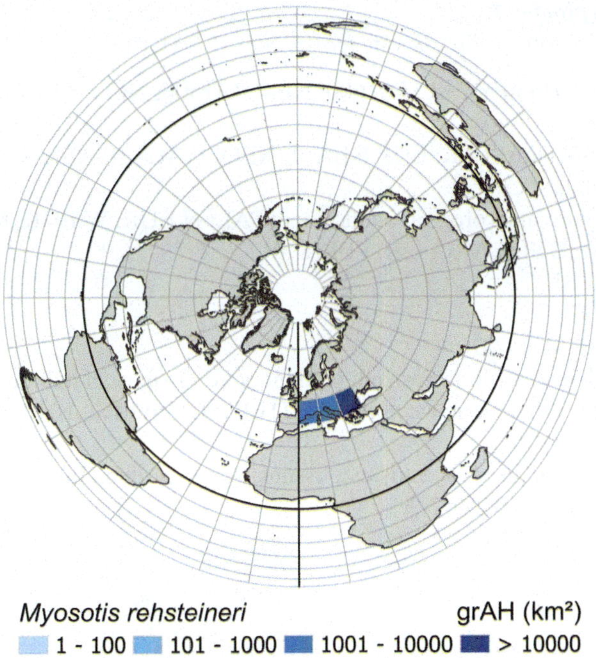

Myosotis rehsteineri Wartm.

Myosotis scorpioides **L.**

Synonyms: *Myosotis palustris* var. *scorpioides* (L.) Hartm.

Range area (grAH km^2): PAL: 701600; NEA: 431500; NEO: 121900; AUS: 28600; AFR: 2100

Ecozone origin-status: PAL: E(I); NEA: I; NEO: I; AUS: I; AFR: I

World rarity status: C

Ecozone-endemic macroregion occurrence: PAL: N EUR, W EUR, MED, C ASIA, UR/ W SIB; E SIB, ARC, CH, FER/J

Life form(s): EM

Ploidy state: DP

Speciation age (Ma): 7.0

Populations also occur in terrestrial, wetland and/ or riparian habitats: YES

Long distance movement vectors: ANTHROPOCHORY; BIRD: *Anas crecca* (Eurasian Teal); ?AV

Nuisance status and uses: no nuisance issues; water garden plant, folk medicine

Endangered status (IUCN): NA

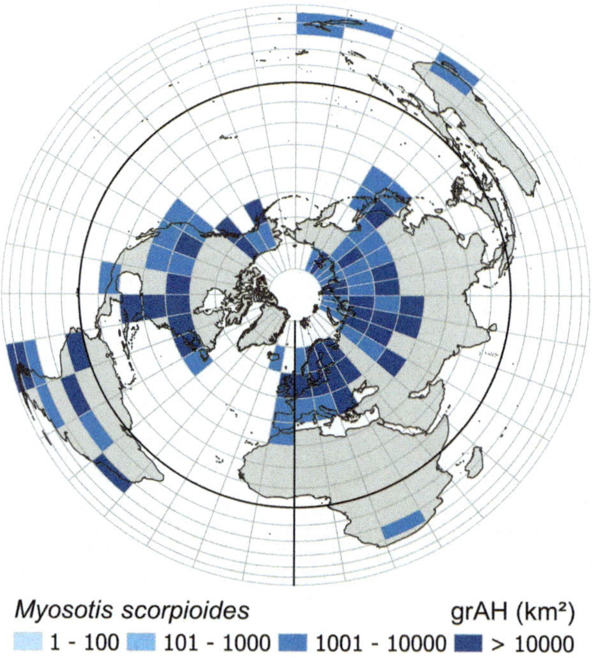

Myosotis scorpioides L.
Image copyright: J.C. Schou, with permission

Myosotis soleirolii **Godr.**
Synonyms: *Myosotis pyrenaica* subsp. *soleirolii* (Godr.) Nyman, *Myosotis sylvatica* subsp. *soleirolii* (Godr.) P.Fourn.

Range area (grAH km^2): PAL: 10300
Ecozone origin-status: PAL: E
World rarity status: VR
Ecozone-endemic macroregion occurrence: PAL: MED
Life form(s): EM
Ploidy state: NA
Speciation age (Ma): NA
Populations also occur in terrestrial, wetland and/ or riparian habitats: YES
Long distance movement vectors: NA
Nuisance status and uses: no nuisance issues; no known uses
Endangered status (IUCN): LC

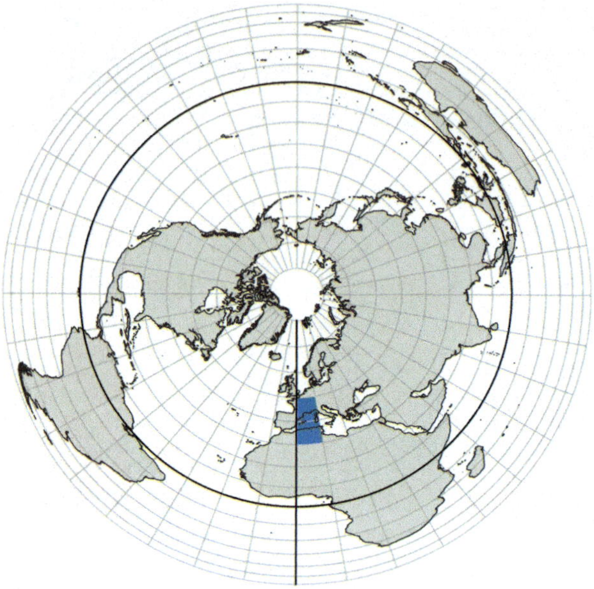

***Myosotis welwitschii* Boiss. & Reut.**
Synonyms:

Range area (grAH km^2): PAL: 10200
Ecozone origin-status: PAL: E
World rarity status: VR
Ecozone-endemic macroregion occurrence: PAL: MED, W EUR
Life form(s): EM
Ploidy state: NA
Speciation age (Ma): NA
Populations also occur in terrestrial, wetland and/ or riparian habitats: YES
Long distance movement vectors: NA
Nuisance status and uses: no nuisance issues; no known uses
Endangered status (IUCN): LC

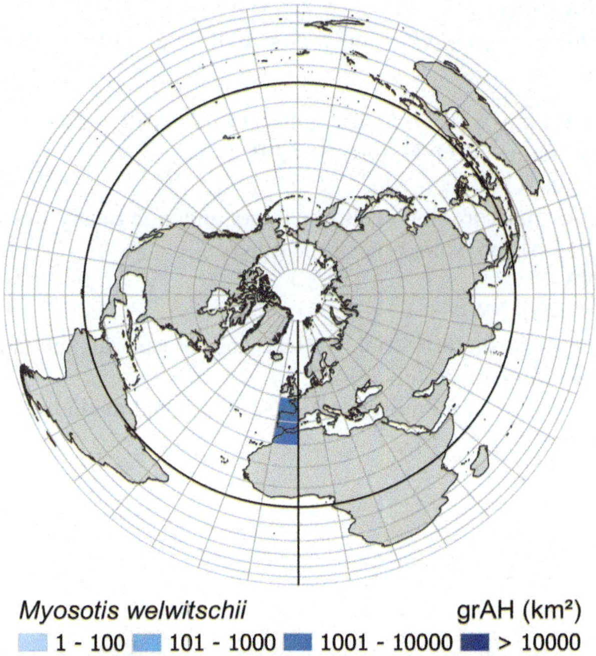

Myosotis welwitschii Boiss. & Reut.
Image copyright: iNaturalist CC-BY @cminda

***Rotula aquatica* Lour.**

Synonyms: *Ehretia aquatica* (Lour.) Gottschling & Hilger, *Rhabdia aquatica* (Lour.) Kuntze

Range area (grAH km^2): NEO: 51500; OR: 115700; AFR: 42800
Ecozone origin-status: NEO: N; OR: N; AFR: N
World rarity status: C
Ecozone-endemic macroregion occurrence: not endemic to any world ecozone
Life form(s): EM (shrub, sometimes rhoeophytic)
Ploidy state: NA
Speciation age (Ma): NA
Populations also occur in terrestrial, wetland and/ or riparian habitats: YES
Long distance movement vectors: NA
Nuisance status and uses: no nuisance issues; traditional medicine uses
Endangered status (IUCN): LC

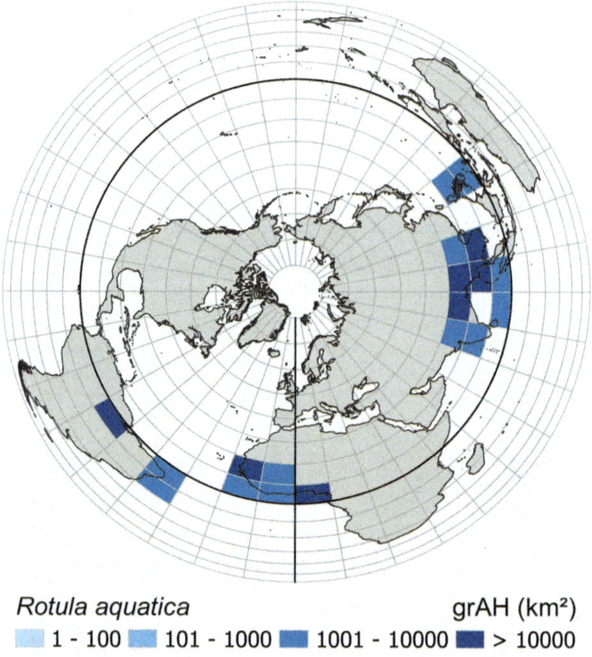

Rotula aquatica Lour.
Image copyright: iNaturalist CC-BY @renju

Rotula aquatica Lour.
Image copyright: iNaturalist CC-BY @renju

3.9 BRASSICACEAE (Brassicales)

The Brassicaceae have 29 macrophyte species within six genera, covering a widespread though mainly temperate to sub-tropical world distribution. There are 19 species occurring in NEA, 13 in PAL, 12 in NEO, four apiece in OR, AFR and AUS, two in OCE and one in ANT. Ecozone-endemic macrophytes (E or E(I) origin status) of Brassicaceae are most prominent in NEA, which has nine such species, with a further five in PAL, four in NEO and one in AFR. Anthropochory vectors are of limited importance to their world distribution, with four introduced species in NEA, three in AUS and three in NEO, two in OCE and two in PAL, and one, *Rorippa islandica*, in ANT. *Nasturtium officinale* is reported as causing major invasive aquatic weed problems in NEA, and *Rorippa palustris* causes weed problems in parts of AUS. Otherwise, the macrophyte species of the Brassicaceae rarely cause nuisance problems. *Nasturtium officinale* (watercress), is widely cultivated in several parts of the world as a crop. The Brassicaceae have VR 28%, R 21% and C 52%, and endangered status 62% with NA/DD, 31% LC, while two species, 7%, are classed as VU.

Quick Links to Genera

1. *Armoracia*
2. *Cardamine*
3. *Nasturtium*
4. *Neobeckia*
5. *Rorippa*
6. *Subularia*

***Armoracia macrocarpa* (Waldst. & Kit.) Kit. ex Baumg.**
Synonyms: *Cochlearia macrocarpa* Waldst. & Kit.

Range area (grAH km^2): PAL: 16700
Ecozone origin-status: PAL: E
World rarity status: VR
Ecozone-endemic macroregion occurrence: PAL: E EUR, C ASIA
Life form(s): EM
Ploidy state: NA
Speciation age (Ma): NA
Populations also occur in terrestrial, wetland and/ or riparian habitats: YES
Long distance movement vectors: NA
Nuisance status and uses: no nuisance issues; some medicinal potential, food: young leaves can be eaten raw or cooked like spinach or cabbage
Endangered status (IUCN): DD

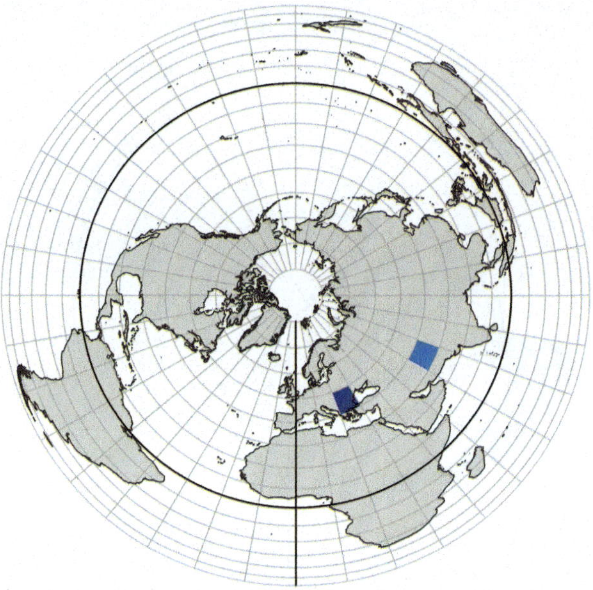

Cardamine bonariensis **Juss. ex. Pers.**
Synonyms:

Range area (grAH km^2): NEO: 215400
Ecozone origin-status: NEO: E
World rarity status: C
Ecozone-endemic macroregion occurrence: NEO: MG MX, MES AM, NWSA, AND, AMZ, CHA, PAR
Life form(s): EM
Ploidy state: D
Speciation age (Ma): NA
Populations also occur in terrestrial, wetland and/ or riparian habitats: YES
Long distance movement vectors: NA
Nuisance status and uses: no nuisance issues; harvested from the wild for local use as a food, medicinal uses
Endangered status (IUCN): NA

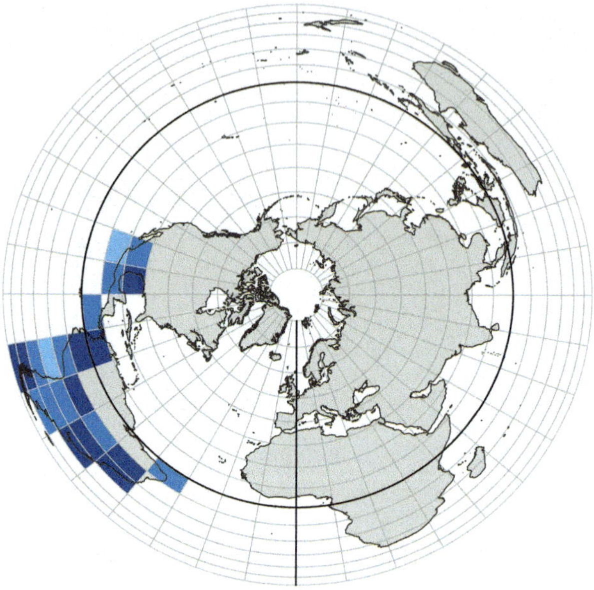

***Cardamine geraniifolia* (Poir.) DC.**
Synonyms: *Sisymbrium geraniifolium* Poir.

Range area (grAH km^2): NEO: 21500
Ecozone origin-status: NEO: E
World rarity status: VR
Ecozone-endemic macroregion occurrence: NEO: AND
Life form(s): EM
Ploidy state: P
Speciation age (Ma): NA
Populations also occur in terrestrial, wetland and/ or riparian habitats: YES
Long distance movement vectors: NA
Nuisance status and uses: no nuisance issues; no known uses
Endangered status (IUCN): NA

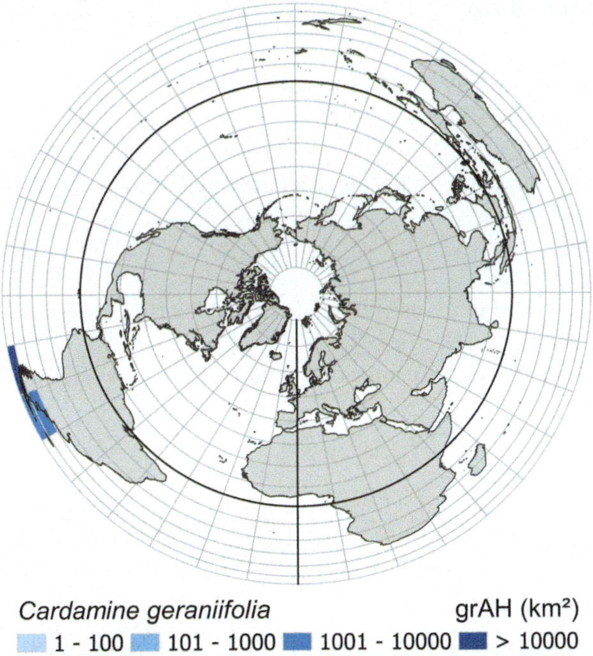

Cardamine geraniifolia (Poir.) DC.
Image copyright: iNaturalist CC0 @arneholgersson (Arne Holgersson)

***Cardamine lyrata* Bunge**
Synonyms:

Range area (grAH km^2): PAL: 173300; OR: 27500
Ecozone origin-status: PAL: N; OR: N
World rarity status: C
Ecozone-endemic macroregion occurrence: not endemic to any world ecozone
Life form(s): EM
Ploidy state: NA
Speciation age (Ma): NA
Populations also occur in terrestrial, wetland and/ or riparian habitats: YES
Long distance movement vectors: NA
Nuisance status and uses: no nuisance issues; aquarium trade and water garden plant, edible
Endangered status (IUCN): NA

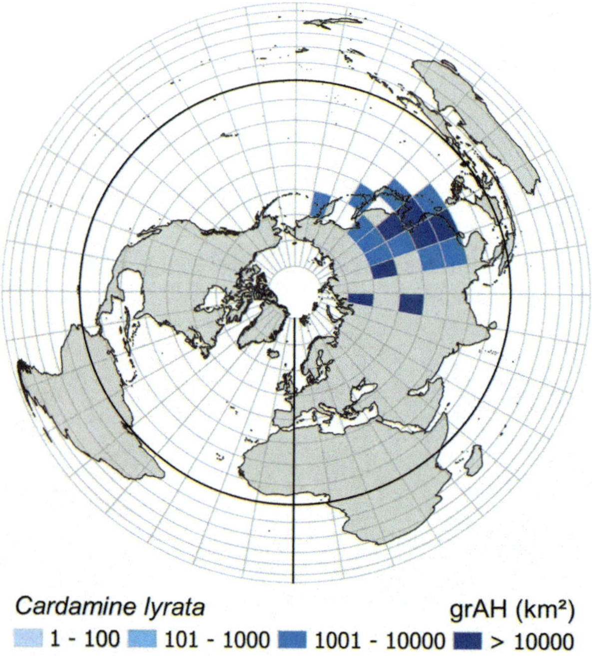

Cardamine lyrata Bunge
Image copyright: iNaturalist CC0 @gbno

***Cardamine pensylvanica* Muhl. ex Willd.**
Synonyms: *Dracamine pensylvanica* (Muhl. ex Willd.) Nieuwl.

Range area (grAH km^2): NEA: 70400; NEO: 24100
Ecozone origin-status: NEA: N; NEO: N
World rarity status: C
Ecozone-endemic macroregion occurrence: not endemic to any world ecozone
Life form(s): EM
Ploidy state: P
Speciation age (Ma): NA
Populations also occur in terrestrial, wetland and/ or riparian habitats: YES
Long distance movement vectors: HYDROCHORY (NEA to NEO)
Nuisance status and uses: no nuisance issues; young stems and leaves are edible
Endangered status (IUCN): LC

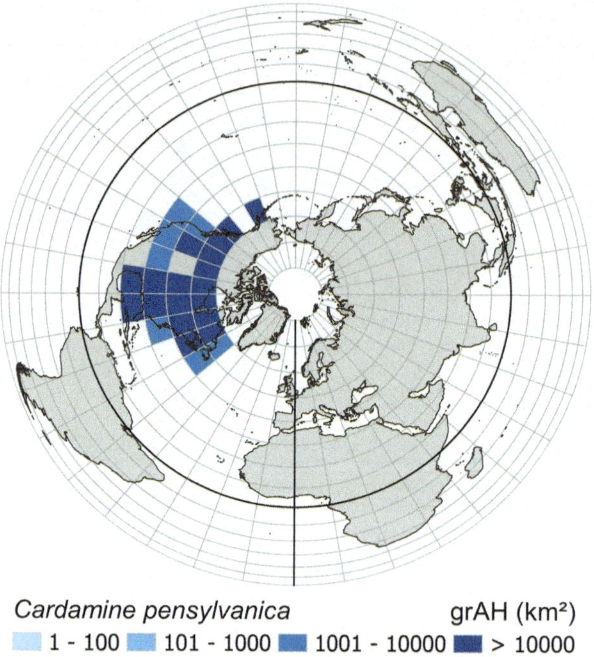

Cardamine pensylvanica Muhl. ex Willd.
Image copyright: iNaturalist CC0 @peakaytea

Cardamine prorepens **Fisch. ex DC.**
Synonyms:

Range area (grAH km²): PAL: 279200
Ecozone origin-status: PAL: E
World rarity status: C
Ecozone-endemic macroregion occurrence: PAL: UR/ W SIB, E SIB, ARC, CH, FER/J
Life form(s): EM
Ploidy state: D
Speciation age (Ma): NA
Populations also occur in terrestrial, wetland and/ or riparian habitats: YES
Long distance movement vectors: NA
Nuisance status and uses: no nuisance issues; leaves and young plant edible if cooked
Endangered status (IUCN): NA

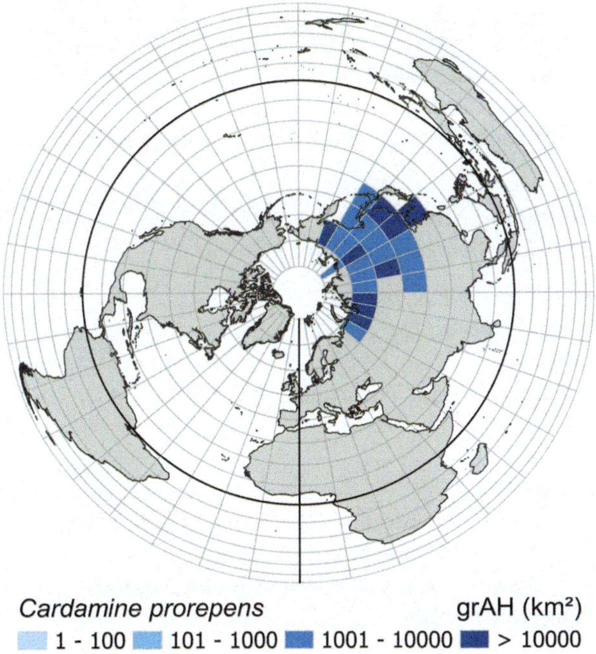

Cardamine prorepens Fisch. ex DC.
Image copyright: A. Efremov, with permission

Cardamine rotundifolia **Michx.**
Synonyms: *Dentaria rotundifolia* (Michx.) Greene

Range area (grAH km^2): NEA: 90300
Ecozone origin-status: NEA: E
World rarity status: R
Ecozone-endemic macroregion occurrence: NEA: E NAM
Life form(s): EM
Ploidy state: NA
Speciation age (Ma): NA
Populations also occur in terrestrial, wetland and/ or riparian habitats: YES
Long distance movement vectors: NA
Nuisance status and uses: no nuisance issues; no known uses
Endangered status (IUCN): NA

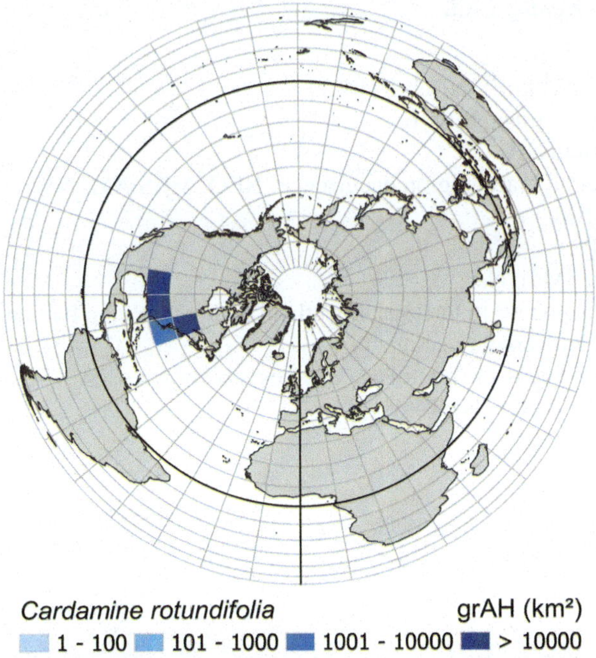

Cardamine rotundifolia Michx.
Image copyright: iNaturalist CC0 @shaunpogacnik95 (Shaun Pogacnik)

***Cardamine variabilis* Phil.**
Synonyms:

Range area (grAH km^2): NEO: 59400
Ecozone origin-status: NEO: E
World rarity status: R
Ecozone-endemic macroregion occurrence: NEO: MES AM, NWSA, CHA, AND
Life form(s): EM
Ploidy state: P
Speciation age (Ma): NA
Populations also occur in terrestrial, wetland and/ or riparian habitats: YES
Long distance movement vectors: NA
Nuisance status and uses: no nuisance issues; no known uses
Endangered status (IUCN): NA

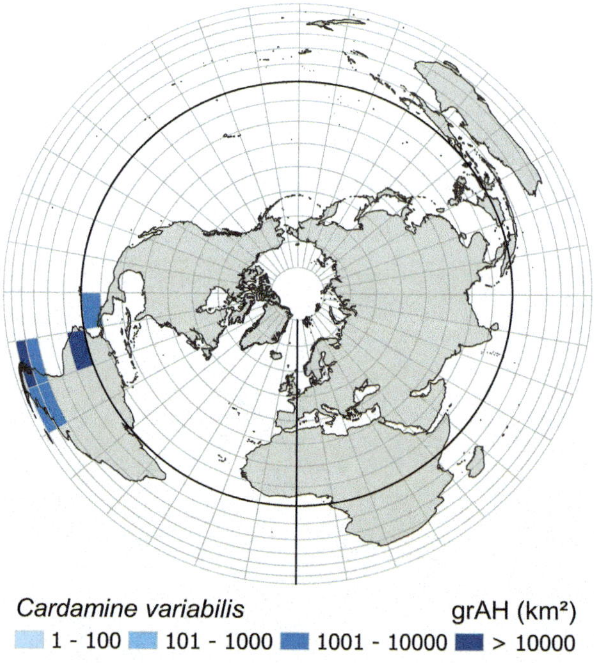

***Nasturtium gambelii* (S. Watson) O.E. Schulz**
Synonyms: *Cardamine gambelii* S.Watson, *Rorippa gambelii* (S.Watson) Rollins & Al-Shehbaz

Range area (grAH km^2): NEA: 3500; NEO: 12000
Ecozone origin-status: NEA: N; NEO: N
World rarity status: VR
Ecozone-endemic macroregion occurrence: not endemic to any world ecozone
Life form(s): EM

Ploidy state: NA
Speciation age (Ma): NA
Populations also occur in terrestrial, wetland and/ or riparian habitats: YES
Long distance movement vectors: NA
Nuisance status and uses: no nuisance issues; no known uses
Endangered status (IUCN): NA

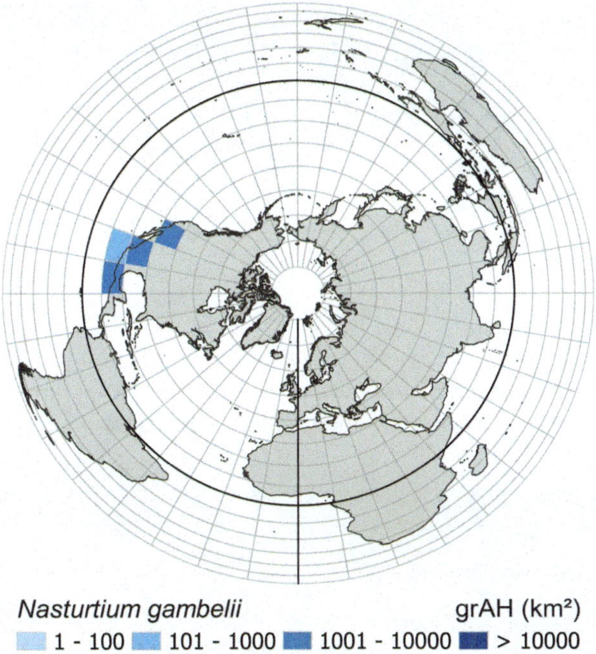

***Nasturtium microphyllum* (Boenn. ex Rchb.) Rchb.**
Synonyms: *Rorippa microphylla* (Boenn. ex Rchb.) Hyl.

Range area (grAH km^2): PAL: 84500; NEA: 89800; NEO: 25700; AUS: 25000; AFR: 10500; present in OCE
Ecozone origin-status: PAL: N; NEA: I; NEO: I; AUS: I; AFR: N; OCE: I
World rarity status: C
Ecozone-endemic macroregion occurrence: not endemic to any world ecozone
Life form(s): EM
Ploidy state: P
Speciation age (Ma): NA
Populations also occur in terrestrial, wetland and/ or riparian habitats: YES
Long distance movement vectors: ANTHROPOCHORY; BIRD: *Anas crecca* (Eurasian Teal)
Nuisance status and uses: no nuisance issues; medicinal, phytoremediation uses
Endangered status (IUCN): LC

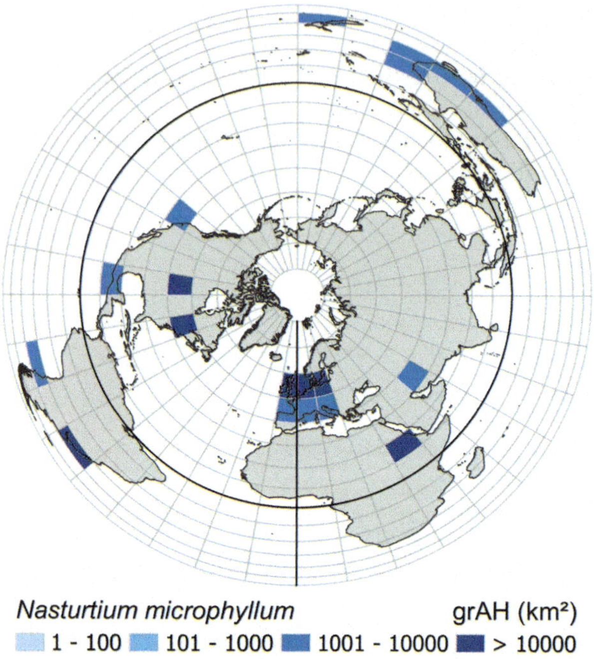

Nasturtium microphyllum (Boenn. ex Rchb.) Rchb
Image copyright: J. C.Schou, with permission

***Nasturtium officinale* R.Br.**

Synonyms: *Rorippa nasturtium-aquaticum* (L.) Hayek, *Radicula officinalis* (W.T.Aiton) J.Groves, *Rorippa officinalis* (W.T.Aiton) P.Royen

Range area (grAH km^2): PAL: 353800; NEA: 458000; NEO: 259800; OR: 90600; AUS: 57600; AFR: 69000; present in OCE

Ecozone origin-status: PAL: N; NEA: I; NEO: I; OR: N; AUS: I; AFR: N; OCE: I

World rarity status: C

Ecozone-endemic macroregion occurrence: not endemic to any world ecozone

Life form(s): EM

Ploidy state: DP

Speciation age (Ma): NA

Populations also occur in terrestrial, wetland and/ or riparian habitats: YES

Long distance movement vectors: ANTHROPOCHORY; BIRD: *Anas platyrhynchos* (Northern Mallard); *Charadrius vociferus* (Killdeer)

Nuisance status and uses: invasive aquatic weed in NEA, causes few or no problems elsewhere, both within and outside its native range; water garden plant, medicinal, phytoremediaton and food uses (water cress)

Endangered status (IUCN): LC

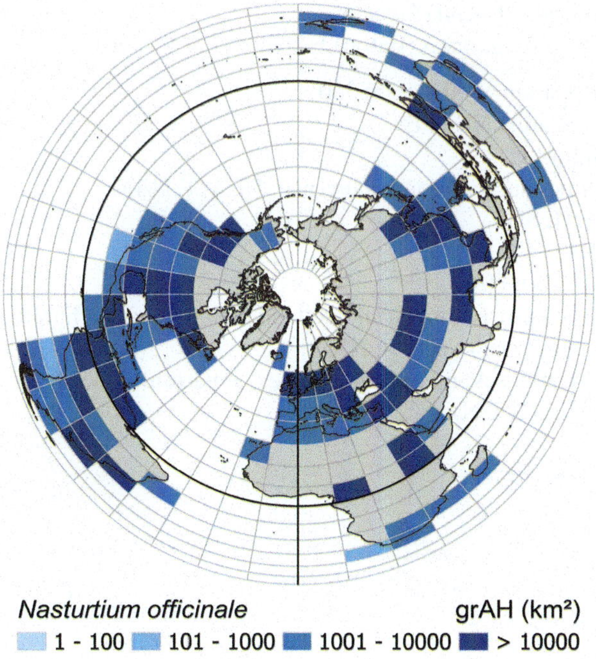

Nasturtium officinale R.Br.
Image copyright: J. C.Schou, with permission

***Neobeckia aquatica* (Eaton) Greene**
Synonyms: *Rorippa aquatica* (Eaton) E.J.Palmer & Steyerm.

Range area (grAH km^2): NEA: 355500
Ecozone origin-status: NEA: E
World rarity status: C
Ecozone-endemic macroregion occurrence: NEA: E NAM
Life form(s): EM, SU
Ploidy state: NA
Speciation age (Ma): NA
Populations also occur in terrestrial, wetland and/ or riparian habitats: YES
Long distance movement vectors: NA
Nuisance status and uses: no nuisance issues; aquarium trade
Endangered status (IUCN): LC

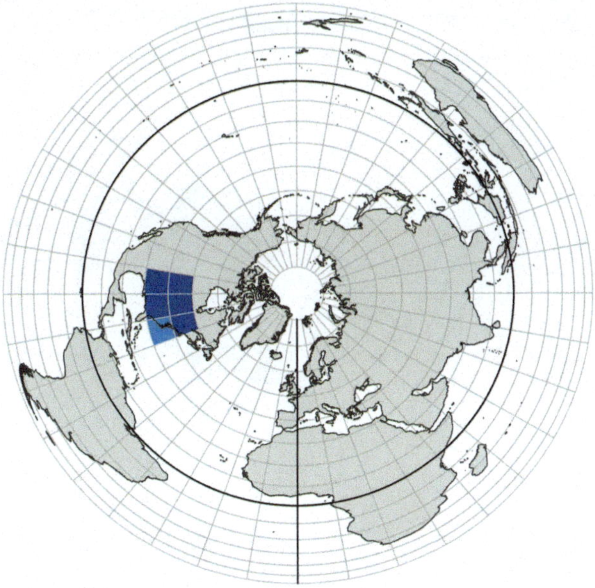

***Rorippa amphibia* (L.) Besser**

Synonyms: *Armoracia amphibia* (L.) Peterm., *Brachiolobos amphibius* (L.) All., *Crucifera amphibia* (L.) E.H.L.Krause, *Myagrum amphibium* (L.) Loisel., *Nasturtium amphibium* (L.) W.T.Aiton, *Sisymbrianthus amphibius* (L.) Chevall. *Sisymbrium amphibium* L.,

Range area (grAH km^2): PAL: 543400; NEA: 89400; AUS: 5400
Ecozone origin-status: PAL: E(I); NEA: I; AUS: I
World rarity status: C
Ecozone-endemic macroregion occurrence: PAL: N EUR, W EUR, MED, E EUR, UR/ W SIB, C ASIA, E SIB, FER/J, CH
Life form(s): EM
Ploidy state: DP
Speciation age (Ma): NA
Populations also occur in terrestrial, wetland and/ or riparian habitats: YES
Long distance movement vectors: ANTHROPOCHORY; BIRD: *Anas platyrhynchos* (Northern Mallard); *Anas crecca* (Eurasian Teal)
Nuisance status and uses: no nuisance issues; edible, aquarium trade
Endangered status (IUCN): LC

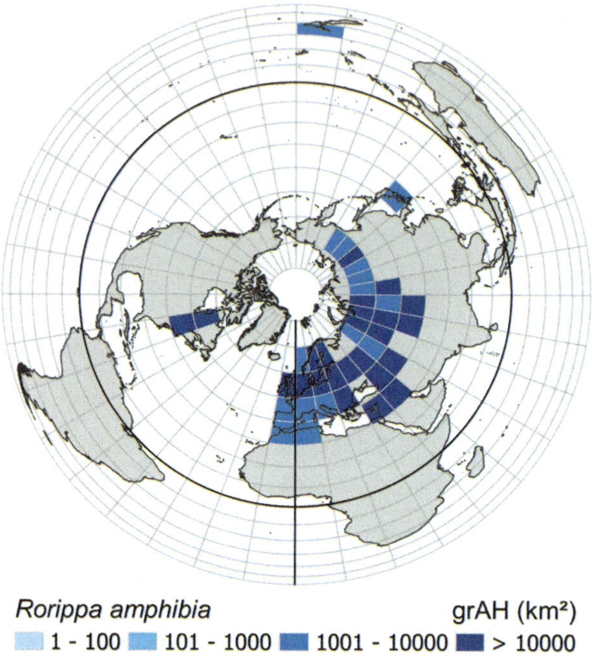

Rorippa amphibia (L.) Besser
Image copyright: J. C. Schou, with permission

***Rorippa aurea* (Boiss. & Heldr.) Hub.-Mor.**
Synonyms: *Nasturtium aureum* Boiss. & Heldr.

Range area (grAH km^2): PAL: 11200
Ecozone origin-status: PAL: E
World rarity status: VR
Ecozone-endemic macroregion occurrence: PAL: MED
Life form(s): EM
Ploidy state: NA
Speciation age (Ma): NA
Populations also occur in terrestrial, wetland and/ or riparian habitats: YES
Long distance movement vectors: NA
Nuisance status and uses: no nuisance issues; no known uses
Endangered status (IUCN): LC

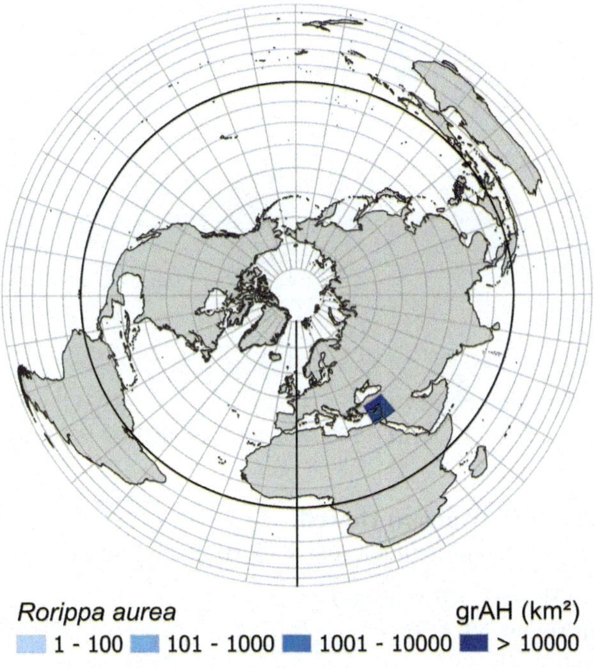

***Rorippa calycina* (Engelm. ex Hayden) Rydb.**
Synonyms:

Range area (grAH km^2): NEA: 88400
Ecozone origin-status: NEA: E
World rarity status: R
Ecozone-endemic macroregion occurrence: NEA: ARC CAN, W NAM, E NAM
Life form(s): EM

Ploidy state: D
Speciation age (Ma): NA
Populations also occur in terrestrial, wetland and/ or riparian habitats: YES
Long distance movement vectors: NA
Nuisance status and uses: no nuisance issues; no known uses
Endangered status (IUCN): NA (but its current status may be VU, and its range over-estimated in the US West, possibly now restricted to W NAM)

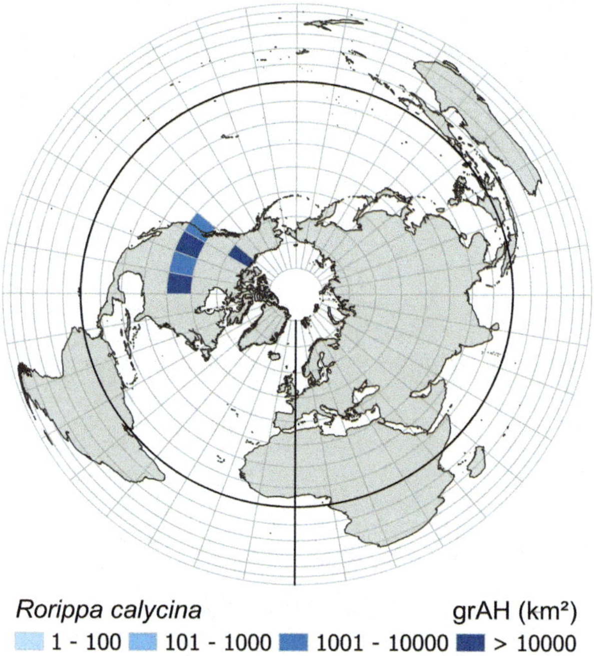

Rorippa columbiae **(S. Watson) Howell**
Synonyms: *Nasturtium columbiae* (Suksd. ex B.L.Rob.) Suksd.

Range area (grAH km^2): NEA: 16600
Ecozone origin-status: NEA: E
World rarity status: VR
Ecozone-endemic macroregion occurrence: NEA: W NAM
Life form(s): EM
Ploidy state: NA
Speciation age (Ma): NA
Populations also occur in terrestrial, wetland and/ or riparian habitats: YES
Long distance movement vectors: NA

Nuisance status and uses: no nuisance issues; no known uses
Endangered status (IUCN): NA

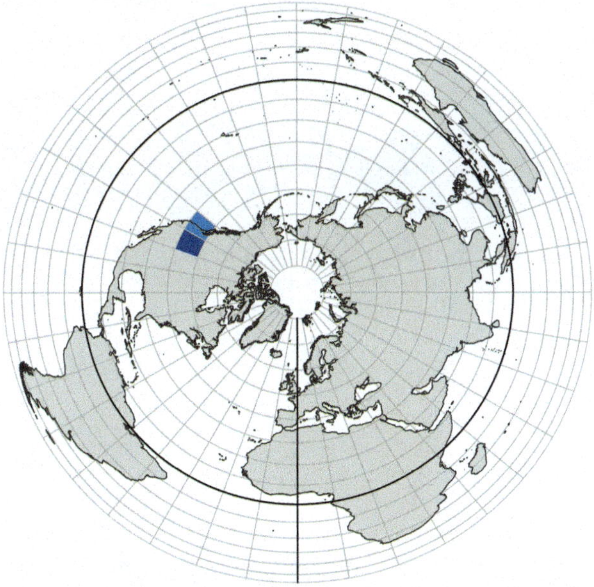

Rorippa curvipes **Greene**
Synonyms: *Radicula curvipes* (Greene) Greene

Range area (grAH km^2): NEA: 486900
Ecozone origin-status: NEA: E
World rarity status: C
Ecozone-endemic macroregion occurrence: NEA: ARC CAN, W NAM, SW NAM, E NAM
Life form(s): EM
Ploidy state: D
Speciation age (Ma): NA
Populations also occur in terrestrial, wetland and/ or riparian habitats: YES
Long distance movement vectors: NA
Nuisance status and uses: no nuisance issues; no known uses
Endangered status (IUCN): NA

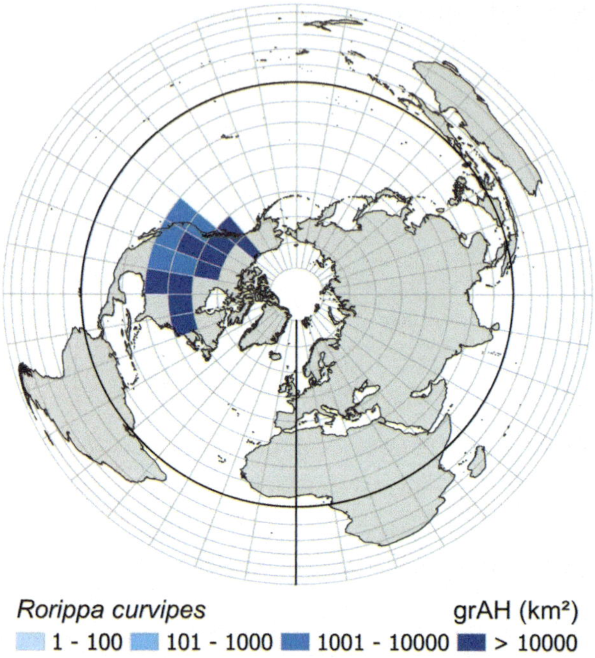

Rorippa curvipes Greene
Images copyright: iNaturalist CC-BY @matsonburger (Steve Matson)

***Rorippa curvisiliqua* (Hook.) Bessey ex Britton**

Synonyms: *Cardamine indica* subvar. *curvisiliqua* (Hook.) Kuntze, *Nasturtium curvisiliquum* (Hook.) Nutt., *Sisymbrium curvisiliquum* Hook.

Range area (grAH km^2): PAL: 5000; NEA: 119500
Ecozone origin-status: PAL: I; NEA: E(I)
World rarity status: R
Ecozone-endemic macroregion occurrence: NEA: ARC CAN, W NAM, SW NAM, E NAM
Life form(s): EM
Ploidy state: D
Speciation age (Ma): NA
Populations also occur in terrestrial, wetland and/ or riparian habitats: YES
Long distance movement vectors: ANTHROPOCHORY
Nuisance status and uses: no nuisance issues; no known uses
Endangered status (IUCN): LC

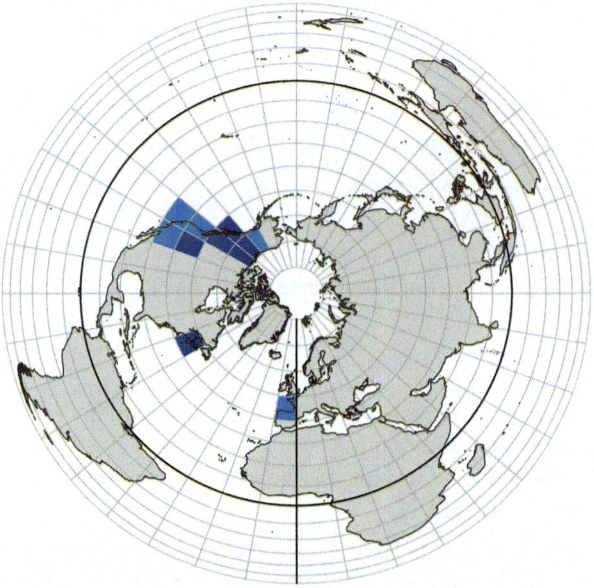

Rorippa curvisiliqua grAH (km²)
1 - 100 101 - 1000 1001 - 10000 > 10000

Rorippa curvisiliqua (Hook.) Bessey ex Britton

Rorippa hayanica **Maire**
Synonyms: *Nasturtium hayanicum* (Maire) O.E.Schulz

Range area (grAH km^2): PAL: 5200
Ecozone origin-status: PAL: E
World rarity status: VR
Ecozone-endemic macroregion occurrence: PAL: MED
Life form(s): EM
Ploidy state: NA
Speciation age (Ma): NA
Populations also occur in terrestrial, wetland and/ or riparian habitats: YES
Long distance movement vectors: NA
Nuisance status and uses: no nuisance issues; no known uses
Endangered status (IUCN): VU

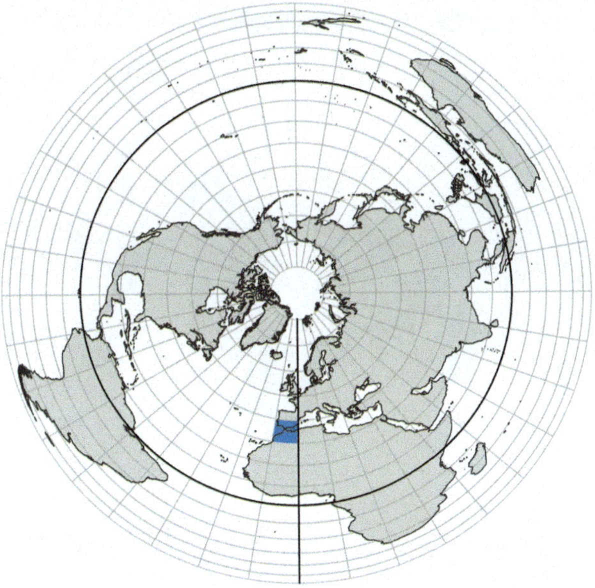

***Rorippa islandica* (Oeder) Borbás**
Synonyms: *Nasturtium islandicum* (Oeder) Sprague, *Sisymbrium islandicum* Oeder

Range area (grAH km^2): PAL: 552200; NEA: 660200; NEO: 18100; OR: 25100; present in ANT
Ecozone origin-status: PAL: E(I); NEA: I; NEO: I; OR: I; ANT: I
World rarity status: C
Ecozone-endemic macroregion occurrence: PAL: N EUR, W EUR, E EUR, UR/ W SIB, E SIB, FER/J, CH, ARC
Life form(s): EM
Ploidy state: D
Speciation age (Ma): NA
Populations also occur in terrestrial, wetland and/ or riparian habitats: YES
Long distance movement vectors: ANTHROPOCHORY
Nuisance status and uses: no nuisance issues; edible; Eskimo and Inuit used this wild mustard as a condiment in fish soup
Endangered status (IUCN): LC

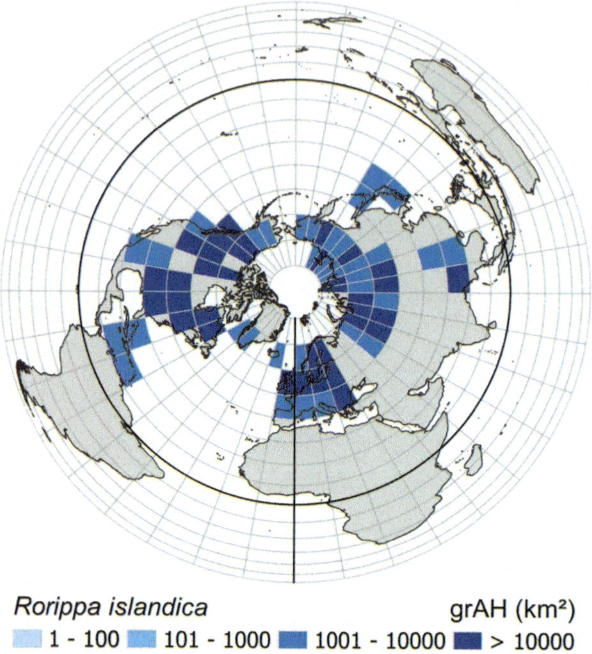

Rorippa islandica (Oeder) Borbás

***Rorippa mexicana* (Michx.) Stuckey**
Synonyms: *Cardamine palustris* var. *mexicana* (Moc., Sessé & Cerv. ex DC.) Kuntze, *Nasturtium mexicanum* Moc., Sessé & Cerv. ex DC., *Radicula mexicana* (Moc., Sessé & Cerv. ex DC.) Standl.

Range area (grAH km^2): NEO: 59800
Ecozone origin-status: NEO: E
World rarity status: R
Ecozone-endemic macroregion occurrence: NEO: MG MX, MES AM, NWSA
Life form(s): EM
Ploidy state: NA
Speciation age (Ma): NA
Populations also occur in terrestrial, wetland and/ or riparian habitats: YES
Long distance movement vectors: NA
Nuisance status and uses: no nuisance issues; no known uses
Endangered status (IUCN): NA

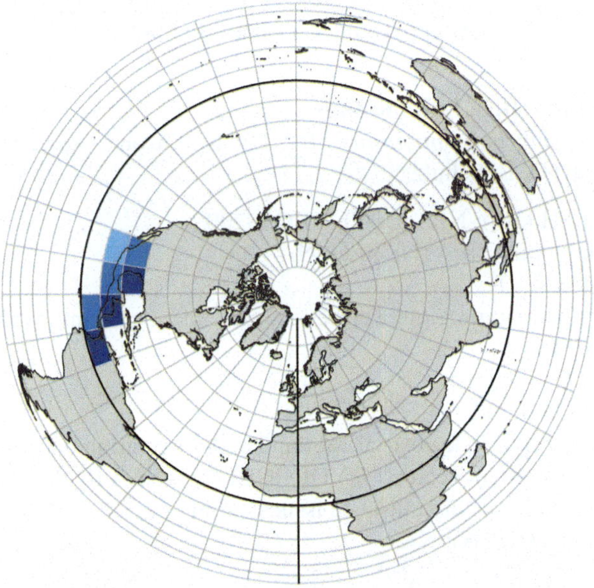

Rorippa mexicana grAH (km²)
1 - 100 101 - 1000 1001 - 10000 > 10000

Rorippa mexicana (Michx.) Stuckey
Image copyright: T. Lobato-de Magalhães

Rorippa microtitis (B.L. Rob.) Rollins
Synonyms: *Sisymbrium microtitis* B. L. Robinson, *Nasturtium microtitis* (B. L. Robinson) O. E. Schulz

Range area (grAH km^2): NEA: 5000; NEO: 5200
Ecozone origin-status: NEA: N; NEO: N
World rarity status: VR
Ecozone-endemic macroregion occurrence: not endemic to any world ecozone
Life form(s): EM
Ploidy state: NA
Speciation age (Ma): NA
Populations also occur in terrestrial, wetland and/ or riparian habitats: YES
Long distance movement vectors: HYDROCHORY (NEA to NEO)
Nuisance status and uses: no nuisance issues; edible
Endangered status (IUCN): NA

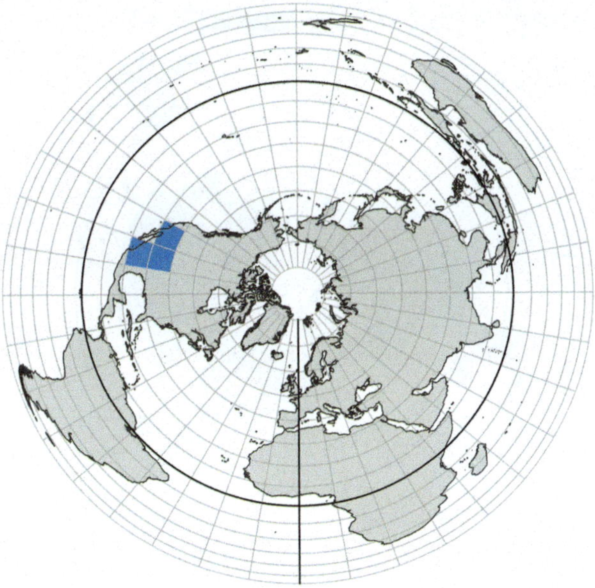

Rorippa palustris **(L.) Besser**

Synonyms: *Brachiolobos palustris* (L.) Clairv., *Cardamine palustris* (L.) Kuntze, nom. illeg., *Crucifera palustris* (L.) E.H.L.Krause, *Myagrum palustre* (L.) Lam., *Nasturtium palustre* (L.) Crantz, *Radicula palustris* (L.) Moench, *Sisymbrium palustre* (L.) Leyss.

Range area (grAH km^2): PAL: 911200; NEA: 954000; NEO: 123500; OR: 65500; AUS: 55800; AFR: 39200

Ecozone origin-status: PAL: N; NEA: N; NEO: N; OR: N; AUS: N; AFR: N

World rarity status: C

Ecozone-endemic macroregion occurrence: not endemic to any world ecozone

Life form(s): EM

Ploidy state: P

Speciation age (Ma): 2.8

Populations also occur in terrestrial, wetland and/ or riparian habitats: YES

Long distance movement vectors: NA

Nuisance status and uses: mainly low nuisance issues, but regarded as an aquatic weed in AUS (Victoria, Queensland and Tasmania); some medicinal and phytoremediation uses

Endangered status (IUCN): NA

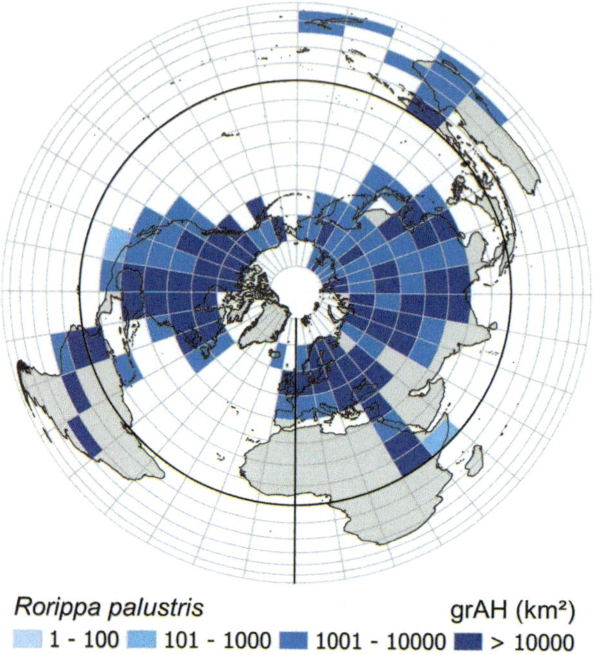

Rorippa palustris (L.) Besser
Image copyright: J.C. Schou, with permission

***Rorippa sphaerocarpa* (A. Gray) Britton**

Synonyms: *Nasturtium sphaerocarpum* A.Gray, *Radicula sphaerocarpa* (A.Gray) Greene

Range area (grAH km^2): NEA: 285200
Ecozone origin-status: NEA: E
World rarity status: C
Ecozone-endemic macroregion occurrence: NEA: W NAM, SW NAM, E NAM
Life form(s): EM
Ploidy state: NA
Speciation age (Ma): NA
Populations also occur in terrestrial, wetland and/ or riparian habitats: YES
Long distance movement vectors: NA
Nuisance status and uses: no nuisance issues; no known uses
Endangered status (IUCN): NA

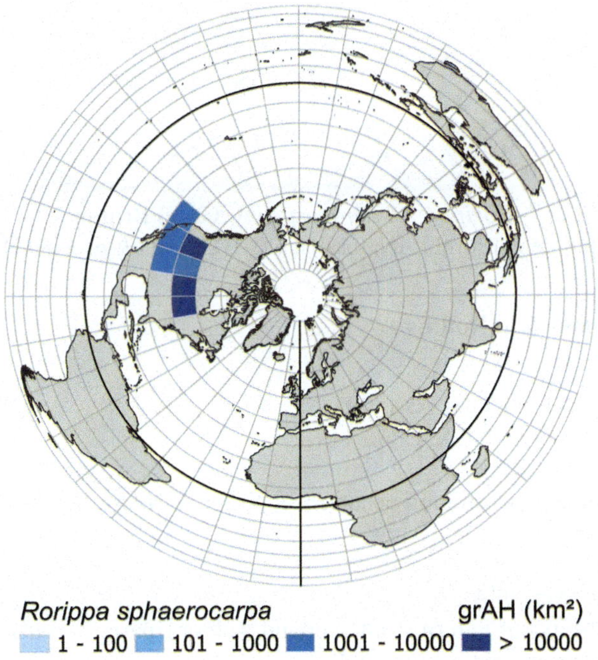

Rorippa sphaerocarpa (A. Gray) Britton

***Rorippa subumbellata* Rollins**
Synonyms:

Range area (grAH km^2): NEA: 4500
Ecozone origin-status: NEA: E
World rarity status: VR
Ecozone-endemic macroregion occurrence: NEA: W NAM, SW NAM
Life form(s): EM
Ploidy state: D
Speciation age (Ma): NA
Populations also occur in terrestrial, wetland and/ or riparian habitats: YES
Long distance movement vectors: NA
Nuisance status and uses: no nuisance issues; no known uses
Endangered status (IUCN): NA

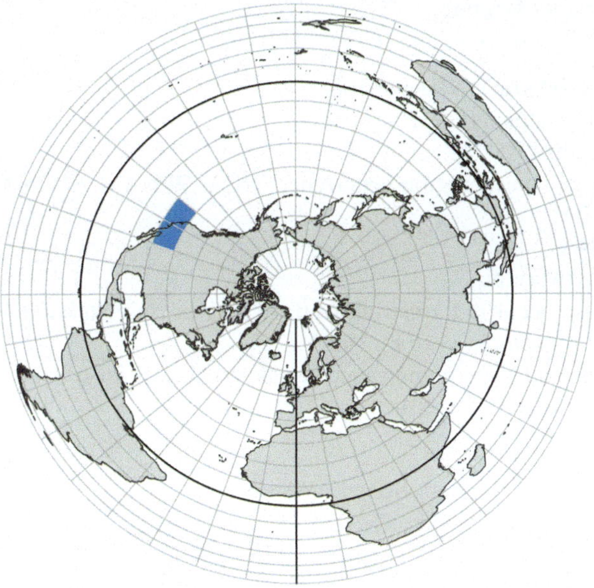

Rorippa subumbellata Rollins

Rorippa tenerrima **Greene**
Synonyms: *Radicula tenerrima* (Greene) Greene

Range area (grAH km^2): NEA: 66100
Ecozone origin-status: NEA: E
World rarity status: R
Ecozone-endemic macroregion occurrence: NEA: W NAM, SW NAM, E NAM
Life form(s): EM
Ploidy state: NA
Speciation age (Ma): NA
Populations also occur in terrestrial, wetland and/ or riparian habitats: YES
Long distance movement vectors: NA
Nuisance status and uses: no nuisance issues; no known uses
Endangered status (IUCN): NA

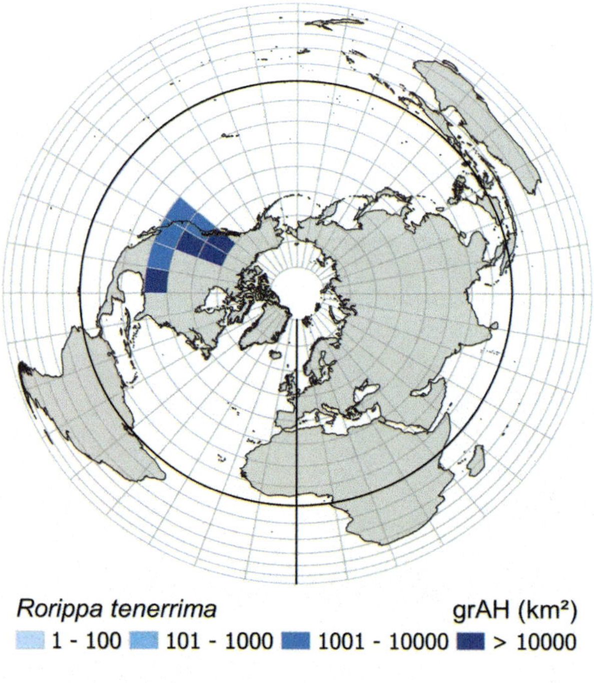

Rorippa tenerrima Greene

Rorippa teres (Michx.) Stuckey
Synonyms: *Cardamine teres* Michx.

Range area (grAH km^2): PAL: 7500; NEA: 285000; NEO: 56000
Ecozone origin-status: PAL: I; NEA: N; NEO: N
World rarity status: C
Ecozone-endemic macroregion occurrence: not endemic to any world ecozone
Life form(s): EM
Ploidy state: P
Speciation age (Ma): NA
Populations also occur in terrestrial, wetland and/ or riparian habitats: YES
Long distance movement vectors: ANTHROPOCHORY
Nuisance status and uses: no nuisance issues; no known uses
Endangered status (IUCN): NA

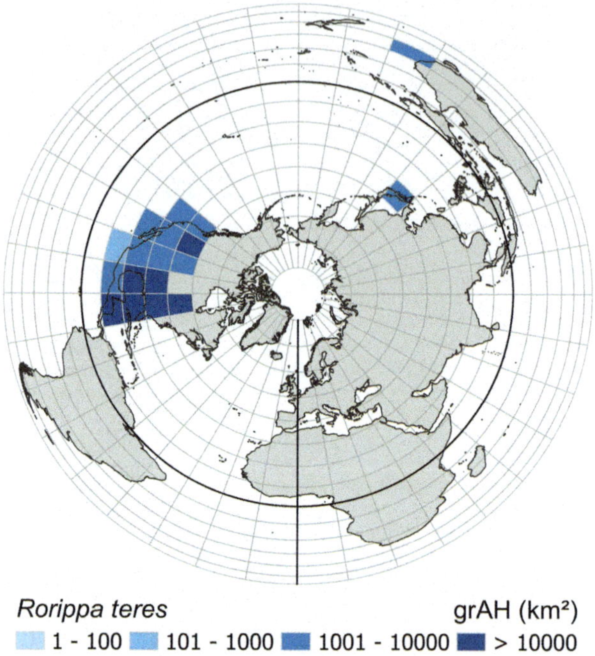

***Subularia aquatica* L.**

Synonyms: *Crucifera subularia* (Lam.) E.H.L.Krause, *Draba subularia* Lam.

Range area (grAH km^2): PAL: 664500; NEA: 798800
Ecozone origin-status: PAL: N; NEA: N
World rarity status: C
Ecozone-endemic macroregion occurrence: not endemic to any world ecozone
Life form(s): SU
Ploidy state: P
Speciation age (Ma): NA
Populations also occur in terrestrial, wetland and/ or riparian habitats: NO
Long distance movement vectors: NA
Nuisance status and uses: no nuisance issues; no known uses
Endangered status (IUCN): LC

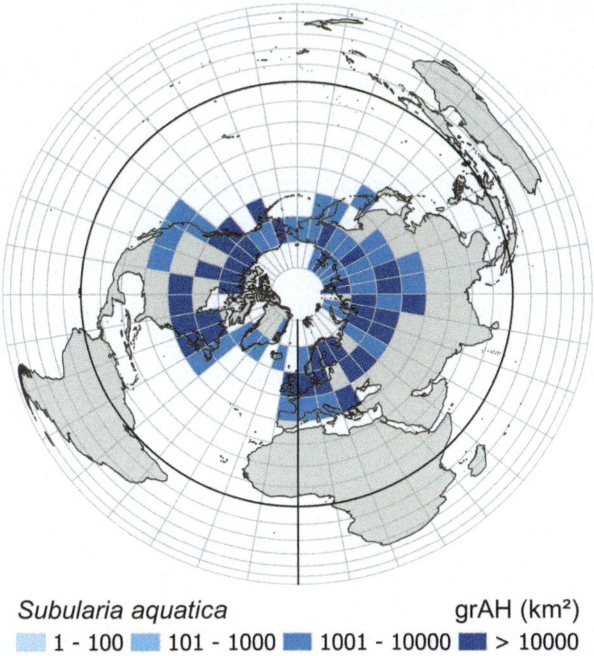

Subularia aquatica L.
Image copyright: J. C. Schou, with permission

***Subularia monticola* A. Braun ex Schweinf.**
Synonyms:

Range area (grAH km^2): AFR: 195900
Ecozone origin-status: AFR: E
World rarity status: C
Ecozone-endemic macroregion occurrence: AFR: SAH/ AR, FOR, EA
Life form(s): EM, SU
Ploidy state: P
Speciation age (Ma): NA
Populations also occur in terrestrial, wetland and/ or riparian habitats: YES
Long distance movement vectors: NA
Nuisance status and uses: no nuisance issues; no known uses
Endangered status (IUCN): NA

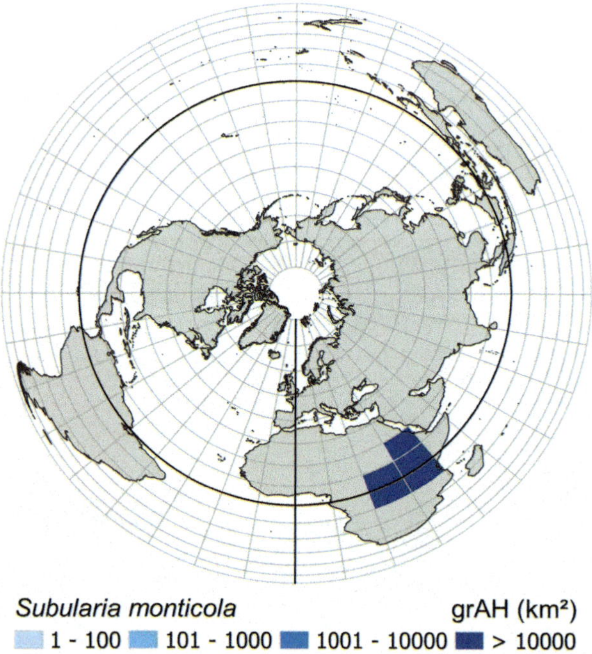

Subularia monticola A. Braun ex Schweinf.
Image copyright: E. Bidault, with permission

3.10 BURMANNIACEAE (Dioscoreales)

The Burmanniaceae contain five freshwater macrophyte species within a single genus, all with a largely tropical to sub-tropical distribution. Three species are present in OR (one endemic), two in AFR (one with E(I) status) and two in AUS, one in NEO (endemic) and one in OCE. No nuisance problems are reported for these five species. One species is VR at world, three R and one C. One species, *Burmannia bicolor*, is introduced in AFR. Four species are classed as endangered status LC.

***Burmannia bicolor* Mart.**
Synonyms: *Tripterella bicolor* (Mart.) Schult.

Range area (grAH km^2): NEO: 251100; AFR: 34700
Ecozone origin-status:NEO: E(I); AFR: I
World rarity status: C

Ecozone-endemic macroregion occurrence: NEO: MG MX, NWSA, AMZ, CHA, PAR
Life form(s): EM
Ploidy state: NA
Speciation age (Ma): NA
Populations also occur in terrestrial, wetland and/ or riparian habitats: YES
Long distance movement vectors: NA
Nuisance status and uses: no nuisance issues; no known uses
Endangered status (IUCN): NA

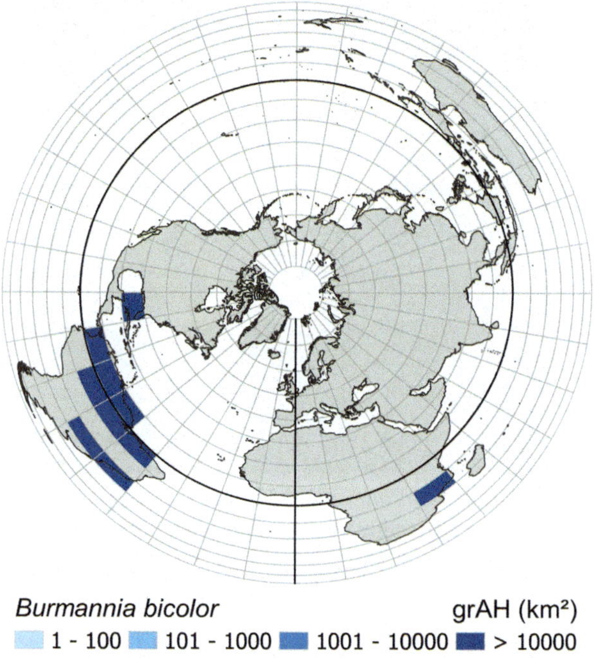

Burmannia coelestis **D.Don**
Synonyms:

Range area (grAH km^2): OR: 100400; AUS: 22100; present in OCE
Ecozone origin-status:OR: N; AUS: N; OCE: N
World rarity status: R
Ecozone-endemic macroregion occurrence: not endemic to any world ecozone
Life form(s): EM
Ploidy state: DP
Speciation age (Ma): NA
Populations also occur in terrestrial, wetland and/ or riparian habitats: YES

Long distance movement vectors: NA
Nuisance status and uses: no nuisance issues; no known uses
Endangered status (IUCN): LC

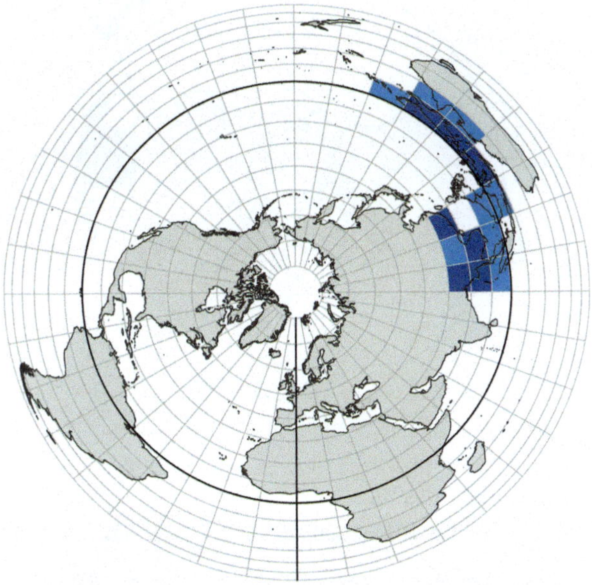

Burmannia disticha **L.**
Synonyms:

Range area (grAH km^2): OR: 75400; AUS: 44600
Ecozone origin-status:OR: N; AUS: N
World rarity status: R
Ecozone-endemic macroregion occurrence: not endemic to any world ecozone
Life form(s): EM
Ploidy state: P
Speciation age (Ma): NA
Populations also occur in terrestrial, wetland and/ or riparian habitats: YES
Long distance movement vectors: NA
Nuisance status and uses: no nuisance issues; no known uses
Endangered status (IUCN): LC

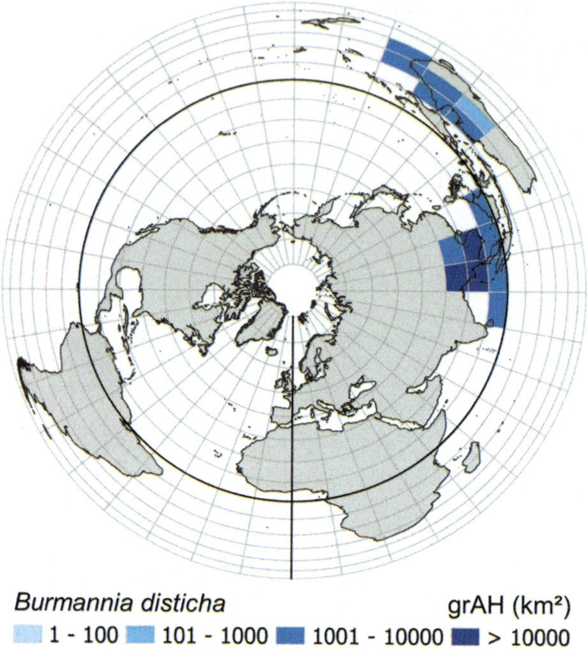

***Burmannia madagascariensis* L.**
Synonyms:

Range area (grAH km^2): AFR: 93300
Ecozone origin-status:AFR: E
World rarity status: R
Ecozone-endemic macroregion occurrence: AFR: FOR, ZAM, SDES, MAD
Life form(s): EM
Ploidy state: NA
Speciation age (Ma): NA
Populations also occur in terrestrial, wetland and/ or riparian habitats: YES
Long distance movement vectors: NA
Nuisance status and uses: no nuisance issues; no known uses
Endangered status (IUCN): LC

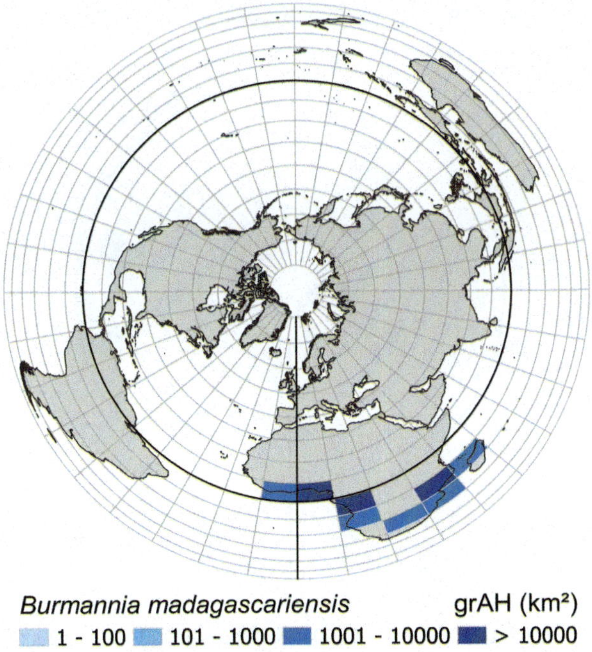

***Burmannia pusilla* (Miers) Thwaites**

Synonyms: *Burmannia coelestis* var. *pusilla* (Miers) Trimen; *Cyananthus pusillus* (Miers) Wall.; *Gonianthes pusilla* Miers

Range area (grAH km^2): OR: 26100
Ecozone origin-status:OR: E
World rarity status: VR
Ecozone-endemic macroregion occurrence: OR: IND, ICH
Life form(s): EM
Ploidy state: NA
Speciation age (Ma): NA
Populations also occur in terrestrial, wetland and/ or riparian habitats: YES
Long distance movement vectors: NA
Nuisance status and uses: no nuisance issues; no known uses
Endangered status (IUCN): LC

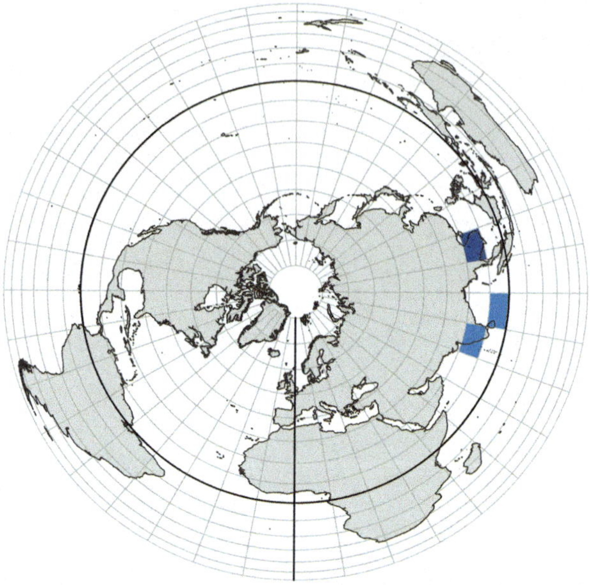

3.11 CABOMBACEAE (Nymphaeales)

Six species within two genera in the Cabombaceae are freshwater macrophyte species, with a widespread, mainly sub-tropical to tropical world distribution. All six occur in NEO, with four of these being endemic to that ecozone. No macrophyte species of Cabombaceae are endemic to any other world ecozone, and anthropochory is important in explaining the distribution of these plants outside NEO, with two of the three species present in PAL and both of those found in OR being introduced, while *Cabomba caroliniana* is a major invasive aquatic weed in AUS. *Brasenia schreberi* is also reported as causing aquatic weed issues within its native range in NEA, in Alaska, but very rarely, or never, elsewhere in the world. The family Cabombaceae has no VR species, two have R and four C status. For five of the species endangered status has not been assessed (NA), and the sixth is LC.

Quick Links to Genera

1. *Brasenia*
2. *Cabomba*

***Brasenia schreberi* J.F.Gmel.**

Synonyms:

Range area (grAH km^2): PAL: 128600; NEA: 508800; NEO: 84000; OR: 45600; AUS: 17500; AFR: 90700

Ecozone origin-status: PAL: N; NEA: N; NEO: N; OR: N; AUS: N; AFR: N

World rarity status: C

Ecozone-endemic macroregion occurrence: not endemic to any world ecozone

Life form(s): FR (hydrosubshrub)

Ploidy state: D

Speciation age (Ma): NA

Populations also occur in terrestrial, wetland and/ or riparian habitats: NO

Long distance movement vectors: NA

Nuisance status and uses: considered a native invasive weed species in Alaska, but usually causing only low or no nuisance issues elsewhere; water garden plant, medicinal, human food and forage uses

Endangered status (IUCN): LC

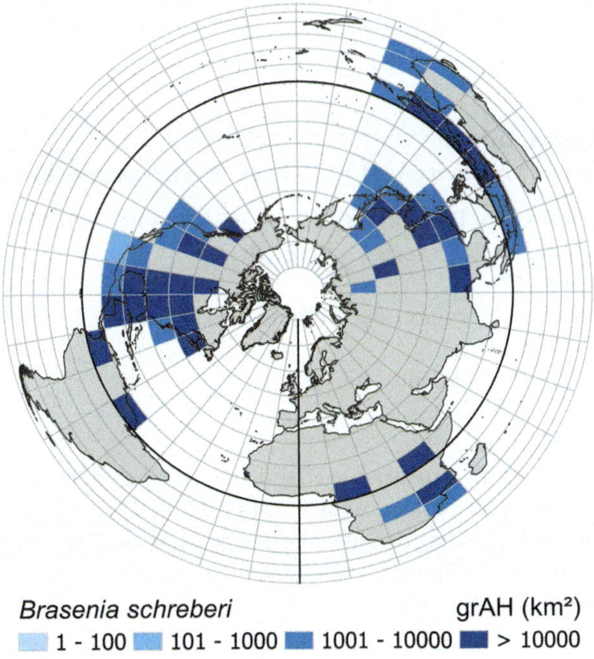

Brasenia schreberi J.F.Gmel.
Image copyright: A. Efremov, with permission

Cabomba aquatica **Aubl.**
Synonyms: *Nectris aquatica* (Aubl.) J.F.Gmel., *Nectris peltata* Pursh

Range area (grAH km^2): PAL: 11400; NEO: 203600; OR: 24700
Ecozone origin-status: PAL: I; NEO: E(I); OR: I
World rarity status: C
Ecozone-endemic macroregion occurrence: NEO: MG MX, MES AM, NWSA, AMZ, CHA, PAR
Life form(s): SU, FR (hydrosubshrub)
Ploidy state: D
Speciation age (Ma): NA
Populations also occur in terrestrial, wetland and/ or riparian habitats: NO
Long distance movement vectors: ANTHROPOCHORY
Nuisance status and uses: no or only rare nuisance issues; aquarium trade, water garden plant in warm climates, some medicinal properties
Endangered status (IUCN): NA

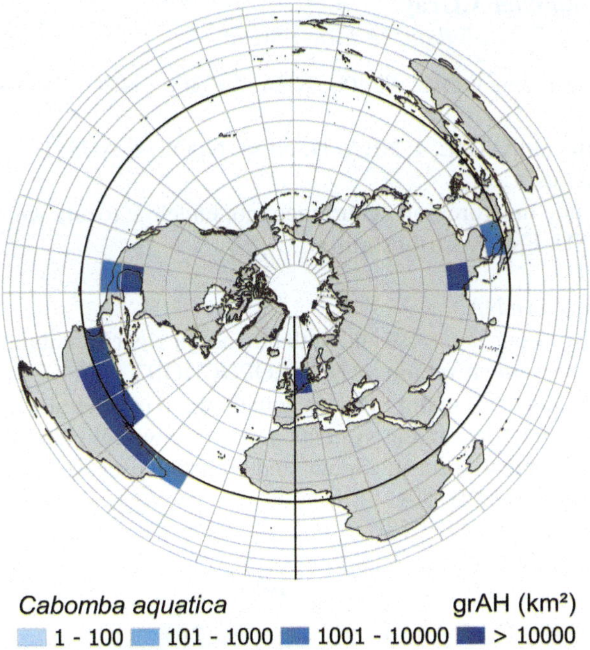

Cabomba aquatica Aubl.

***Cabomba caroliniana* A.Gray**

Synonyms: *Nectris caroliniana* (A.Gray) Steud.

Range area (grAH km^2): PAL: 92400; NEA: 113300 ; NEO: 235400 ; OR: 29400; AUS: 42000

Ecozone origin-status: PAL: I; NEA: N; NEO: N; OR: I; AUS: I

World rarity status: C

Ecozone-endemic macroregion occurrence: not endemic to any world ecozone

Life form(s): SU, FR (hydrosubshrub)

Ploidy state: D

Speciation age (Ma): NA

Populations also occur in terrestrial, wetland and/ or riparian habitats: NO

Long distance movement vectors: ANTHROPOCHORY

Nuisance status and uses: major invasive aquatic weed in Australia, and poses a serious invasive threat to aquatic systems across the world, particularly in warmer regions, some potential for medicinal and bioactive uses, aquarium trade

Endangered status (IUCN): NA

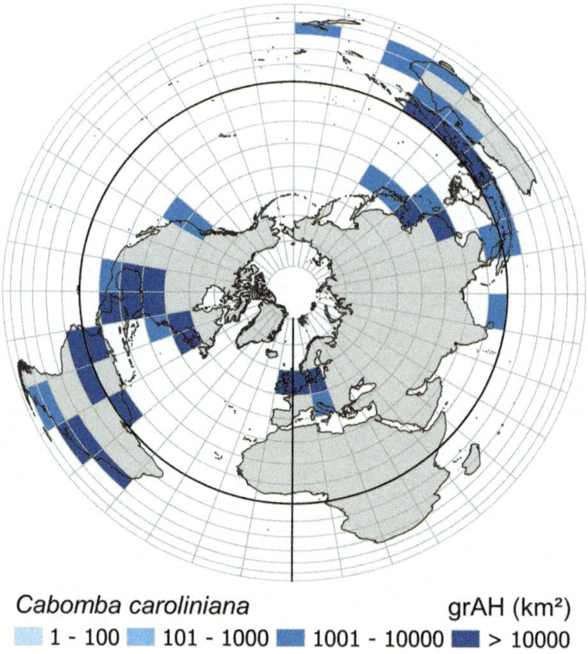

Cabomba caroliniana A.Gray
Image copyright: iNaturalist CC-BY @romigaleota (Romi Galeota Lencina)

Cabomba furcata Schult. & Schult.f.
Synonyms: *Nectris furcata* (Schult. & Schult.f.) Leandro ex Steud.

Range area (grAH km^2): NEO: 305600
Ecozone origin-status: NEO: E
World rarity status: C
Ecozone-endemic macroregion occurrence: NEO: MG MX, MES AM, NWSA, AMZ, CHA, PAR
Life form(s): SU, FR (hydrosubshrub)
Ploidy state: D
Speciation age (Ma): NA
Populations also occur in terrestrial, wetland and/ or riparian habitats: NO
Long distance movement vectors: NA
Nuisance status and uses: no confirmed instances of nuisance growth though there is a recent (unquantified) report of aquatic weed problems caused by this species in Sri Lanka; aquarium trade
Endangered status (IUCN): NA

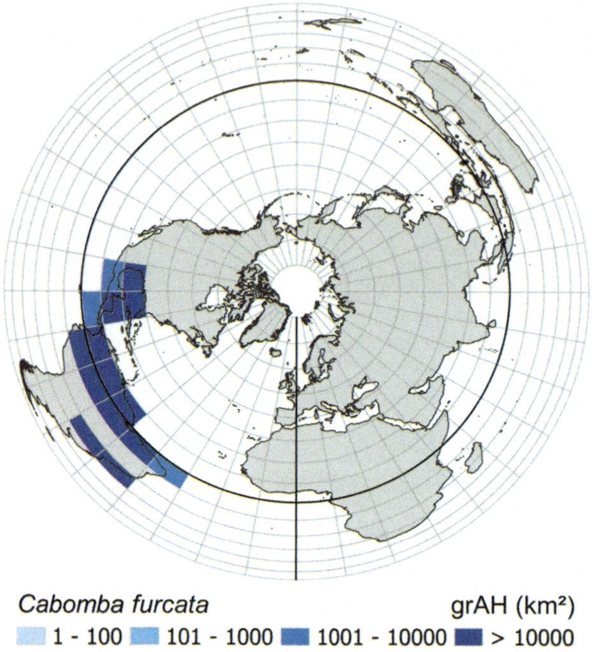

Cabomba furcata Schult. & Schult.f.
Image copyright: iNaturalist CC-BY @bernhard_hiller

Cabomba haynesii **Wiersema**
Synonyms:

Range area (grAH km^2): NEO: 105300
Ecozone origin-status: NEO: E
World rarity status: R
Ecozone-endemic macroregion occurrence: NEO: MES AM, NWSA, AMZ, CHA, PAR
Life form(s): SU (hydrosubshrub)
Ploidy state: D
Speciation age (Ma): NA
Populations also occur in terrestrial, wetland and/ or riparian habitats: NO
Long distance movement vectors: NA
Nuisance status and uses: no nuisance issues; aquarium trade, fodder
Endangered status (IUCN): NA

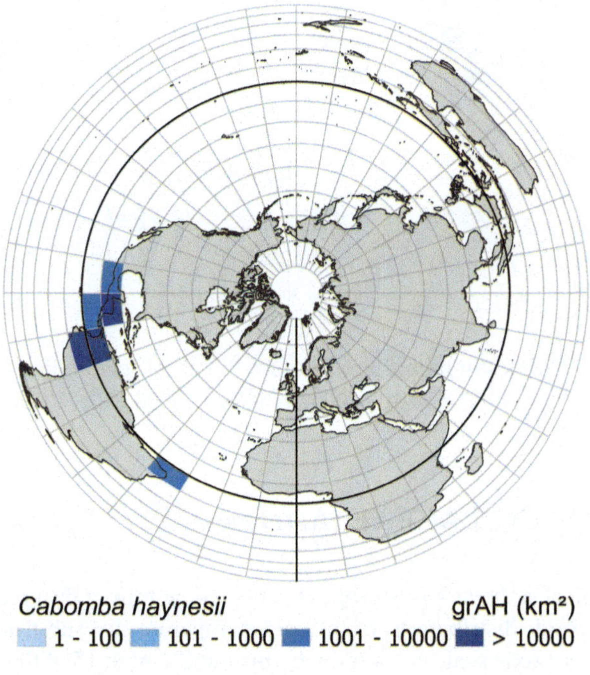

Cabomba palaeformis **Fassett**
Synonyms:

Range area (grAH km^2): NEO: 42700
Ecozone origin-status: NEO: E
World rarity status: R
Ecozone-endemic macroregion occurrence: NEO: MG MX, MES AM
Life form(s): SU (hydrosubshrub)

Ploidy state: D
Speciation age (Ma): NA
Populations also occur in terrestrial, wetland and/ or riparian habitats: NO
Long distance movement vectors: NA
Nuisance status and uses: no nuisance issues; aquarium trade
Endangered status (IUCN): NA

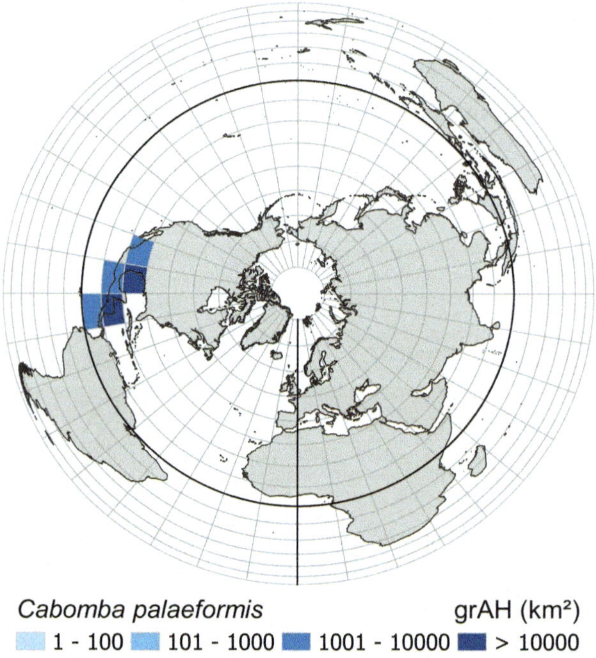

3.12 CAMPANULACEAE (Asterales)

The Campanulaceae have 44 macrophyte species, within eight genera, covering a widespread world distribution from cool temperate to sub-tropical latitudes. Endemicity at ecozone scale is common in this family, with 15 of the 18 macrophyte species of Campanulaceae present in NEO being endemic to that ecozone. In addition, eight of 13 species present in NEA, seven of the nine occurring in AFR, and four of the nine species found in AFR are endemic to their respective ecozones. Campanulaceae has no endemic macrophyte species present in the four species found in each of PAL and OR, or the single species occurring in OCE (there are none in ANT). Anthropochorous vectors are rather uncommon in this group, with only two of the Campanulaceae macrophyte species found in AUS and PAL being introduced, together with one apiece in AFR, NEA and OCE. Only one species is considered to cause nuisance issues, *Isotoma fluviatilis*, which is an aggressive alien in parts of NEA. The Campanulaceae are dominated by world-rare macrophyte species, with VR 52%, R32% and C16%, and endangered status of 60% NA/DD and 40% LC.

Quick Links to Genera

1. *Downingia*
2. *Grammatotheca*
3. *Howellia*
4. *Isotoma*
5. *Legenere*
6. *Lobelia*
7. *Porterella*
8. *Unigenes*

Downingia bacigalupii **Weiler**
Synonyms:

Range area (grAH km^2): NEA: 21100
Ecozone origin-status: NEA: E
World rarity status: VR
Ecozone-endemic macroregion occurrence: NEA: W NAM, SW NAM
Life form(s): EM
Ploidy state: DP
Speciation age (Ma): 5.5
Populations also occur in terrestrial, wetland and/ or riparian habitats: YES
Long distance movement vectors: ?AV
Nuisance status and uses: no nuisance issues; no known uses
Endangered status (IUCN): NA

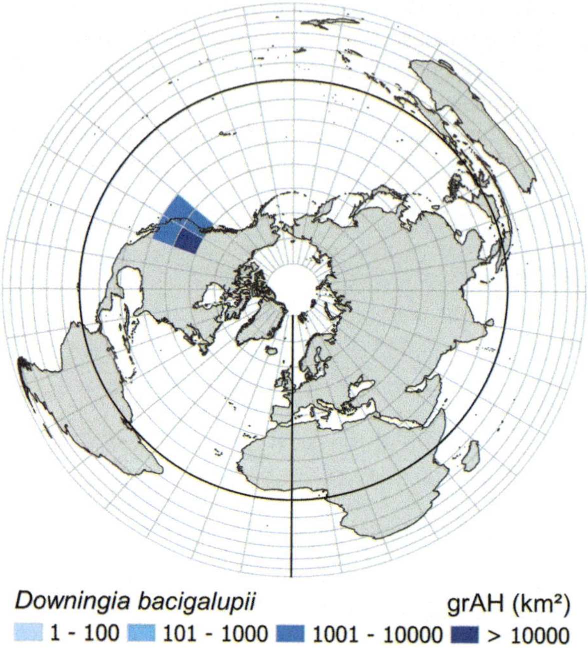

Downingia bacigalupii Weiler
Image copyright: iNaturalist CC-BY @matsonburger (Steve Matson)

Downingia bella **Hoover**
Synonyms:

Range area (grAH km^2): NEA: 11000
Ecozone origin-status: NEA: E
World rarity status: VR
Ecozone-endemic macroregion occurrence: NEA: W NAM, SW NAM
Life form(s): EM
Ploidy state: DP
Speciation age (Ma): 8.0
Populations also occur in terrestrial, wetland and/ or riparian habitats: YES
Long distance movement vectors: ?AV
Nuisance status and uses: no nuisance issues; no known uses
Endangered status (IUCN): NA

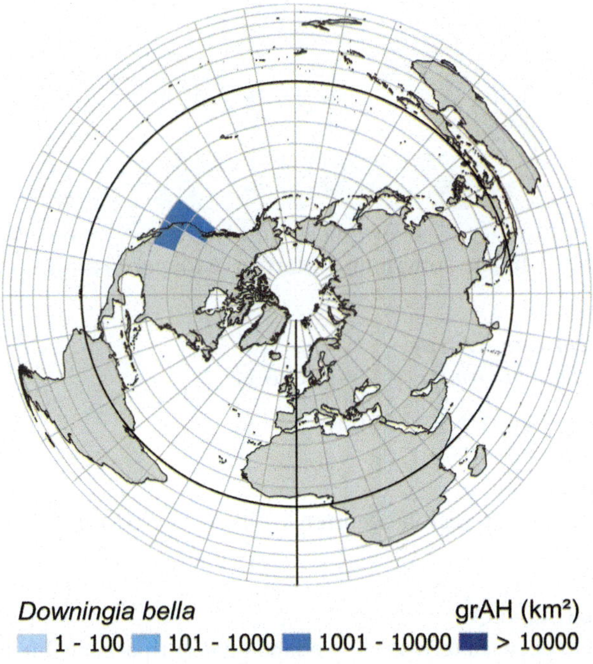

Downingia bella Hoover
Image copyright: J Marfori, with permission

***Downingia bicornuta* A.Gray**
Synonyms: *Bolelia bicornuta* (A.Gray) Greene

Range area (grAH km^2): NEA: 21100
Ecozone origin-status: NEA: E
World rarity status: VR
Ecozone-endemic macroregion occurrence: NEA: W NAM, SW NAM
Life form(s): EM
Ploidy state: DP
Speciation age (Ma): NA
Populations also occur in terrestrial, wetland and/ or riparian habitats: YES
Long distance movement vectors: NA
Nuisance status and uses: no nuisance issues; no known uses
Endangered status (IUCN): NA

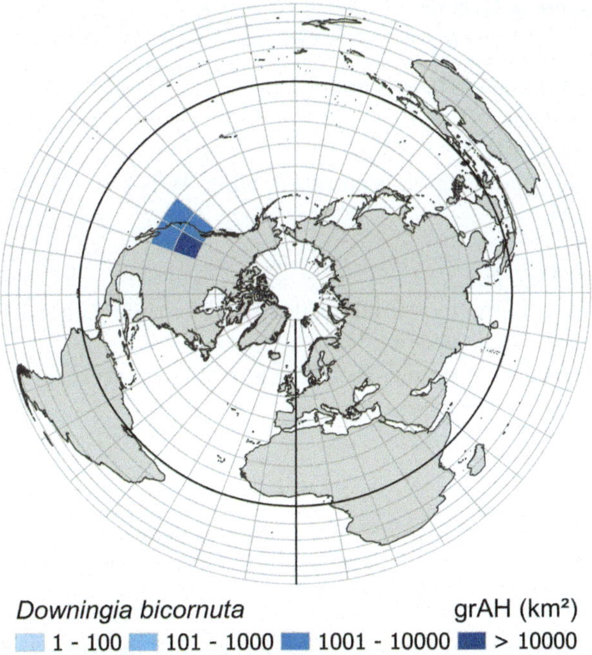

Downingia bicornuta A.Gray
Image copyright: iNaturalist CC0 @benjamintryon (Benjamin Tryon)

Downingia concolor **Greene**
Synonyms: *Bolelia concolor* (Greene) Greene

Range area (grAH km^2): NEA: 11000
Ecozone origin-status: NEA: E
World rarity status: VR
Ecozone-endemic macroregion occurrence: NEA: W NAM, SW NAM
Life form(s): EM
Ploidy state: D
Speciation age (Ma): NA
Populations also occur in terrestrial, wetland and/ or riparian habitats: YES
Long distance movement vectors: NA
Nuisance status and uses: no nuisance issues; no known uses
Endangered status (IUCN): NA

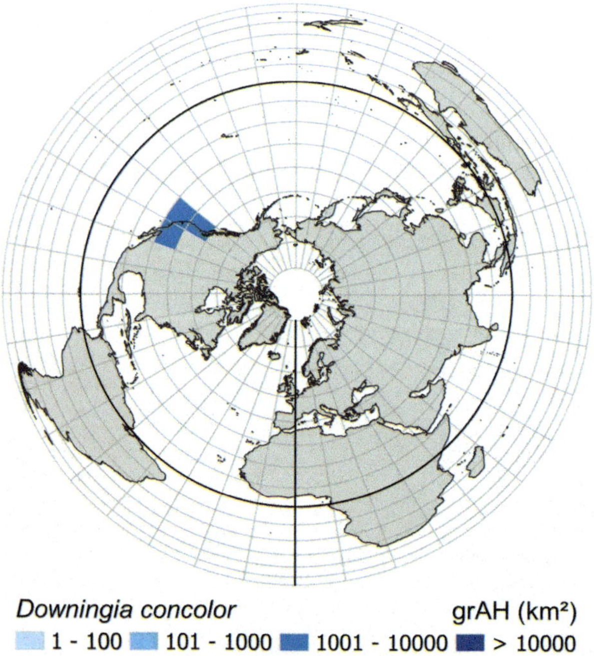

Downingia concolor Greene

***Downingia cuspidata* (Greene) Rattan**
Synonyms: *Bolelia cuspidata* Greene

Range area (grAH km^2): NEA: 11000
Ecozone origin-status: NEA: E
World rarity status: VR
Ecozone-endemic macroregion occurrence: NEA: W NAM, SW NAM
Life form(s): EM
Ploidy state: DP
Speciation age (Ma): NA
Populations also occur in terrestrial, wetland and/ or riparian habitats: YES
Long distance movement vectors: NA
Nuisance status and uses: no nuisance issues; no known uses
Endangered status (IUCN): NA

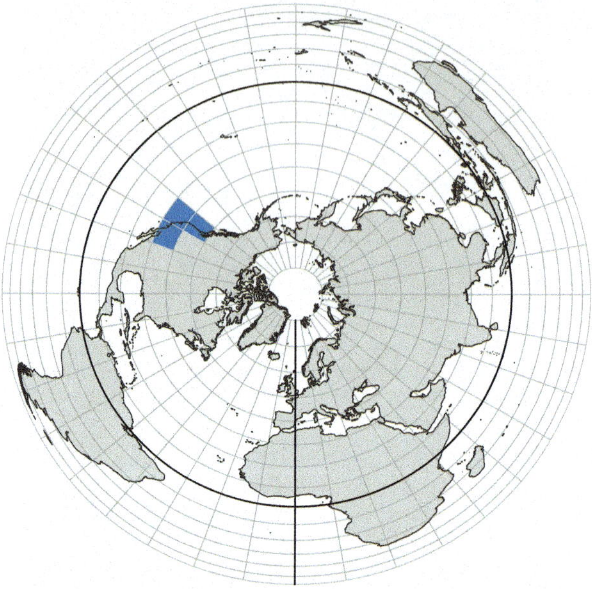

***Downingia elegans* (Douglas ex Lindl.) Torr.**
Synonyms: *Bolelia elegans* (Torr.) Greene, *Clintonia elegans* Douglas ex Lindl., *Lobelia clintonia* F.Dietr.

Range area (grAH km^2): PAL: 27300; NEA: 21100
Ecozone origin-status: PAL: I; NEA: E(I)
World rarity status: R
Ecozone-endemic macroregion occurrence: NEA: W NAM, SW NAM
Life form(s): EM
Ploidy state: DP
Speciation age (Ma): 5.5
Populations also occur in terrestrial, wetland and/ or riparian habitats: YES
Long distance movement vectors: ?AV; ANTHROPOCHORY
Nuisance status and uses: no nuisance issues; water garden plant
Endangered status (IUCN): LC

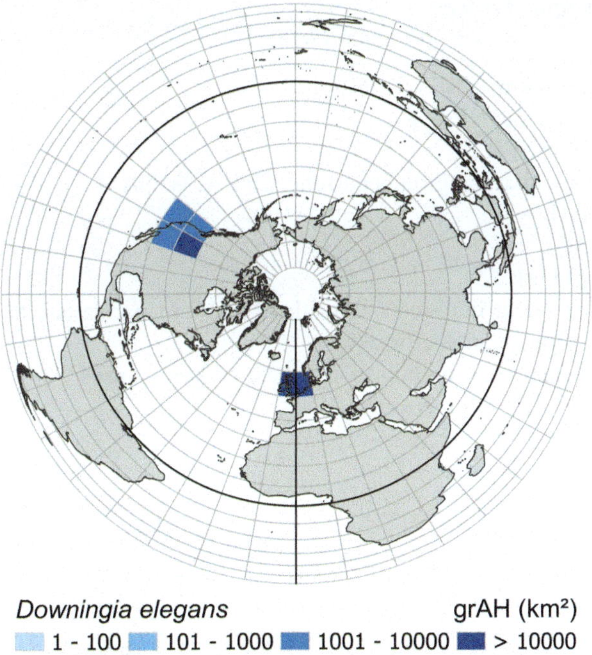

Downingia elegans (Douglas ex Lindl.) Torr.

Downingia insignis **Greene**
Synonyms: *Bolelia insignis* (Greene) Greene

Range area (grAH km^2): NEA: 17600
Ecozone origin-status: NEA: E
World rarity status: VR
Ecozone-endemic macroregion occurrence: NEA: W NAM
Life form(s): EM
Ploidy state: DP
Speciation age (Ma): 2.9
Populations also occur in terrestrial, wetland and/ or riparian habitats: YES
Long distance movement vectors: NA
Nuisance status and uses: no nuisance issues; no known uses
Endangered status (IUCN): NA

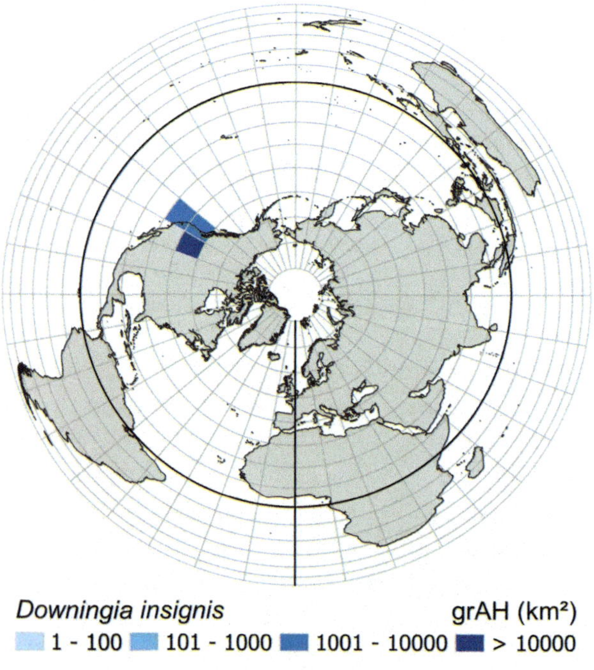

Downingia insignis Greene

Downingia laeta **(Greene) Greene**
Synonyms: *Bolelia laeta* Greene

Range area (grAH km^2): NEA: 88400
Ecozone origin-status: NEA: E
World rarity status: R
Ecozone-endemic macroregion occurrence: NEA: ARC CAN, W NAM, SW NAM
Life form(s): EM
Ploidy state: DP
Speciation age (Ma): 2.9
Populations also occur in terrestrial, wetland and/ or riparian habitats: YES
Long distance movement vectors: NA
Nuisance status and uses: no nuisance issues; no known uses
Endangered status (IUCN): LC

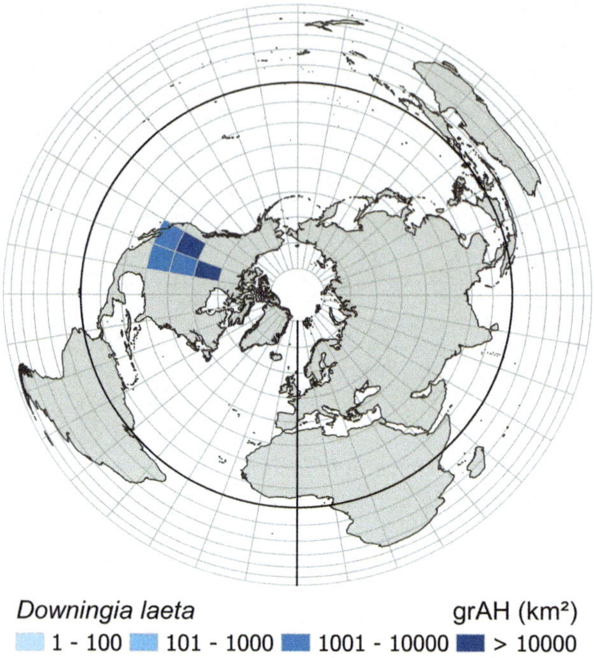

Downingia montana **Greene**
Synonyms: *Bolelia montana* (Greene) Greene

Range area (grAH km^2): NEA: 11000
Ecozone origin-status: NEA: E
World rarity status: VR
Ecozone-endemic macroregion occurrence: NEA: W NAM, SW NAM
Life form(s): EM
Ploidy state: DP
Speciation age (Ma): NA
Populations also occur in terrestrial, wetland and/ or riparian habitats: YES
Long distance movement vectors: NA
Nuisance status and uses: no nuisance issues; no known uses
Endangered status (IUCN): NA

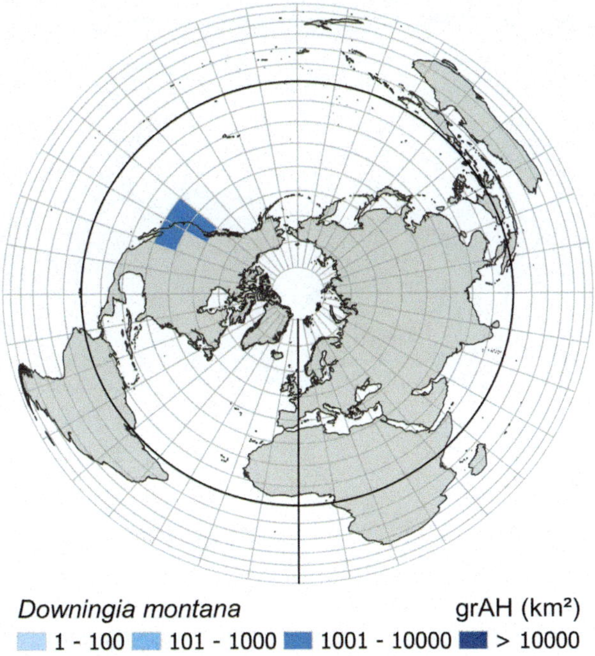

Downingia ornatissima **Greene**
Synonyms: *Bolelia ornatissima* (Greene) Greene

Range area (grAH km^2): NEA: 11000
Ecozone origin-status: NEA: E
World rarity status: VR
Ecozone-endemic macroregion occurrence: NEA: W NAM, SW NAM
Life form(s): EM
Ploidy state: DP
Speciation age (Ma): NA
Populations also occur in terrestrial, wetland and/ or riparian habitats: YES
Long distance movement vectors: NA
Nuisance status and uses: no nuisance issues; no known uses
Endangered status (IUCN): NA

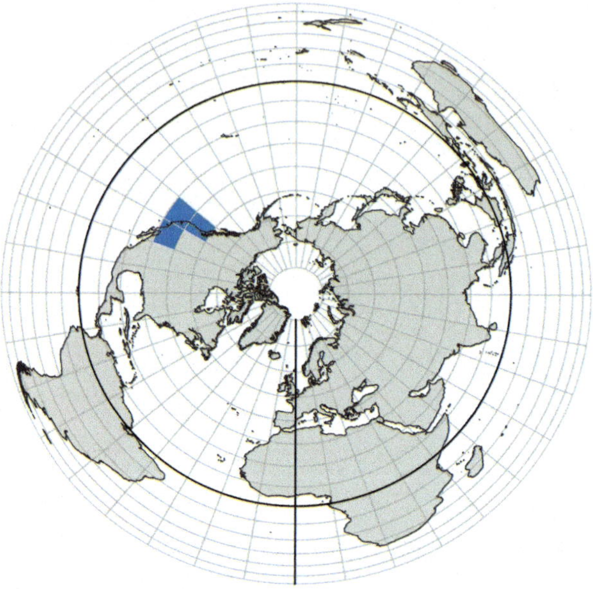

***Downingia pulchella* (Lindl.) Torr.**
Synonyms: *Bolelia pulchella* (Lindl.) Greene, *Clintonia pulchella* Lindl.

Range area (grAH km^2): NEA: 4500
Ecozone origin-status: NEA: E
World rarity status: VR
Ecozone-endemic macroregion occurrence: NEA: W NAM, SW NAM
Life form(s): EM
Ploidy state: DP
Speciation age (Ma): NA
Populations also occur in terrestrial, wetland and/ or riparian habitats: YES
Long distance movement vectors: NA
Nuisance status and uses: no nuisance issues; no known uses
Endangered status (IUCN): NA

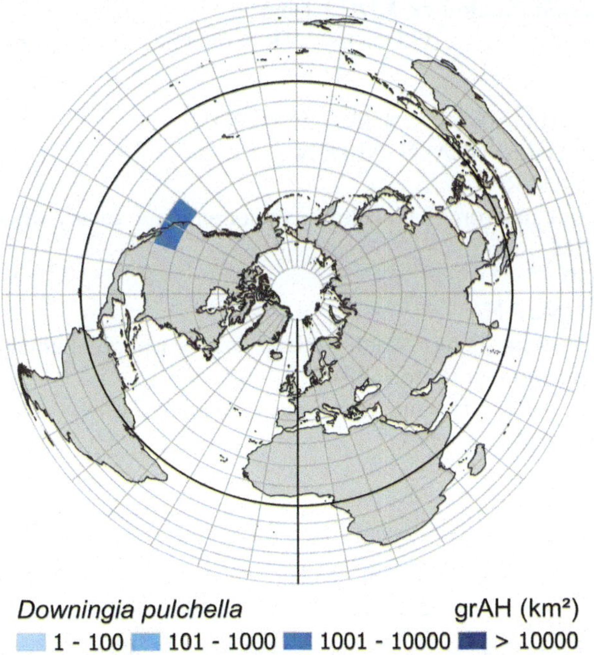

Downingia pulchella (Lindl.) Torr.

Downingia pusilla **(G.Don ex A.DC.) Torr.**
Synonyms: *Bolelia pusilla* (G.Don ex A.DC.) Greene, *Clintonia pusilla* G.Don ex A.DC.

Range area (grAH km^2): NEA: 4500; NEO: 32000
Ecozone origin-status: NEA: N; NEO: N
World rarity status: R
Ecozone-endemic macroregion occurrence: not endemic to any world ecozone
Life form(s): EM
Ploidy state: DP
Speciation age (Ma): NA
Populations also occur in terrestrial, wetland and/ or riparian habitats: YES
Long distance movement vectors: NA
Nuisance status and uses: no nuisance issues; no known uses
Endangered status (IUCN): NA

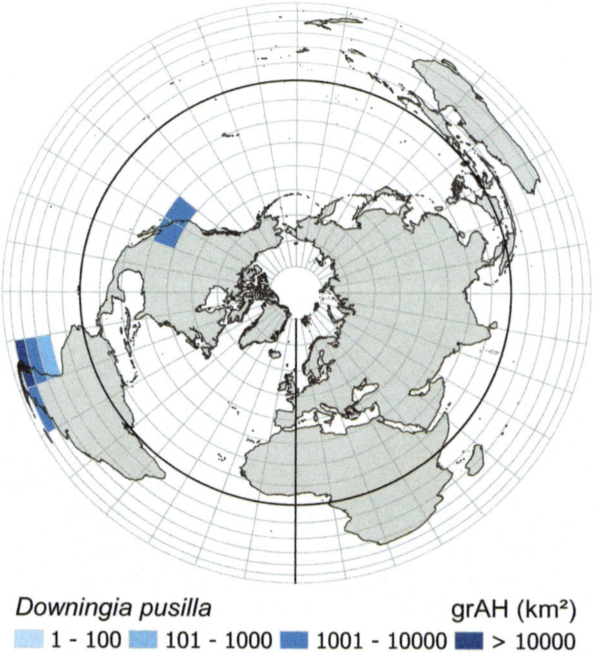

Downingia yina **Applegate**
Synonyms: *Downingia willamettensis* M.Peck, *Downingia pulcherrima* M.Peck

Range area (grAH km^2): NEA: 17600
Ecozone origin-status: NEA: E
World rarity status: VR
Ecozone-endemic macroregion occurrence: NEA: W NAM

Life form(s): EM
Ploidy state: D
Speciation age (Ma): NA
Populations also occur in terrestrial, wetland and/ or riparian habitats: YES
Long distance movement vectors: NA
Nuisance status and uses: no nuisance issues; no known uses
Endangered status (IUCN): NA

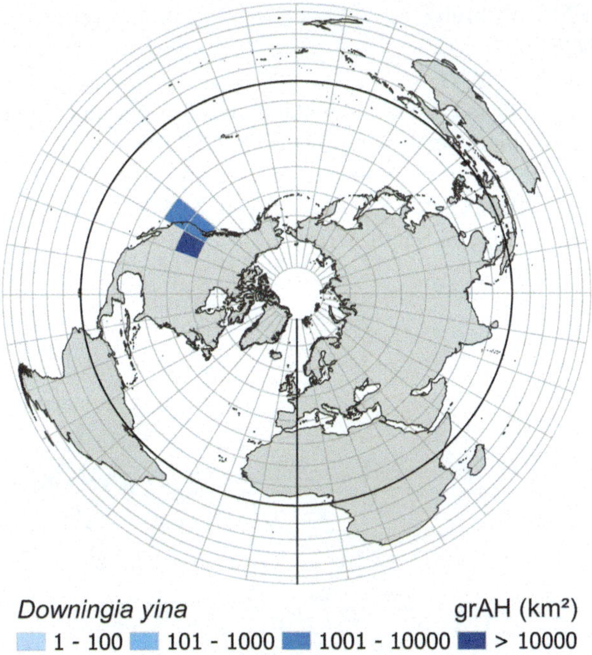

Downingia yina Applegate
Image copyright: iNaturalist CC-BY @jonaleef (M226JN)

Grammatotheca bergiana **(Cham.) C.Presl.**
Synonyms: *Clintonia bergiana* (Cham.) G.Don, *Dortmanna bergiana* (Cham.) Kuntze., *Lobelia bergiana* Cham.

Range area (grAH km^2): AUS: 2200; AFR: 17800
Ecozone origin-status: AUS: I; AFR: E(I)
World rarity status: VR
Ecozone-endemic macroregion occurrence: AFR: ZAM, CAPE
Life form(s): EM (subshrub)
Ploidy state: NA
Speciation age (Ma): NA
Populations also occur in terrestrial, wetland and/ or riparian habitats: YES
Long distance movement vectors: ANTHROPOCHORY
Nuisance status and uses: no nuisance issues; no known uses
Endangered status (IUCN): NA (LC according to SANBI Red List of South African Plants, in native range in South Africa)

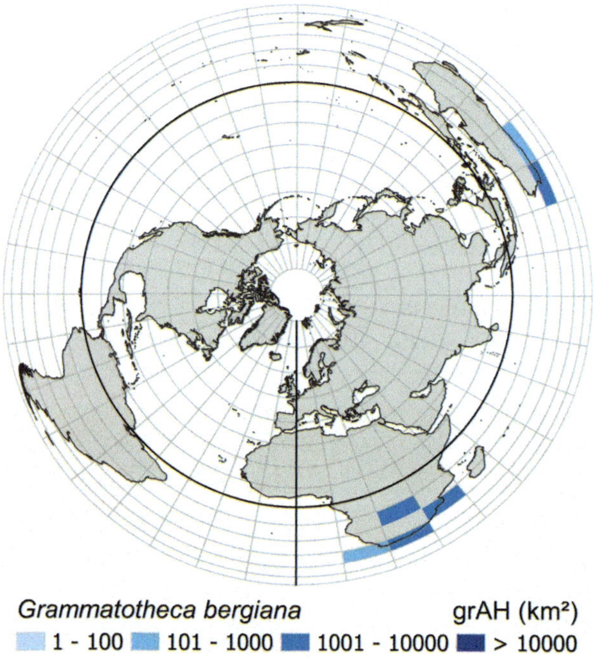

Grammatotheca bergiana (Cham.) C.Presl.

***Howellia aquatilis* A.Gray**
Synonyms:

Range area (grAH km^2): NEA: 20100
Ecozone origin-status: NEA: E
World rarity status: VR
Ecozone-endemic macroregion occurrence: NEA: W NAM, SW NAM
Life form(s): SU, FR, EM
Ploidy state: D
Speciation age (Ma): NA
Populations also occur in terrestrial, wetland and/ or riparian habitats: YES
Long distance movement vectors: NA
Nuisance status and uses: no nuisance issues; no known uses
Endangered status (IUCN): NA

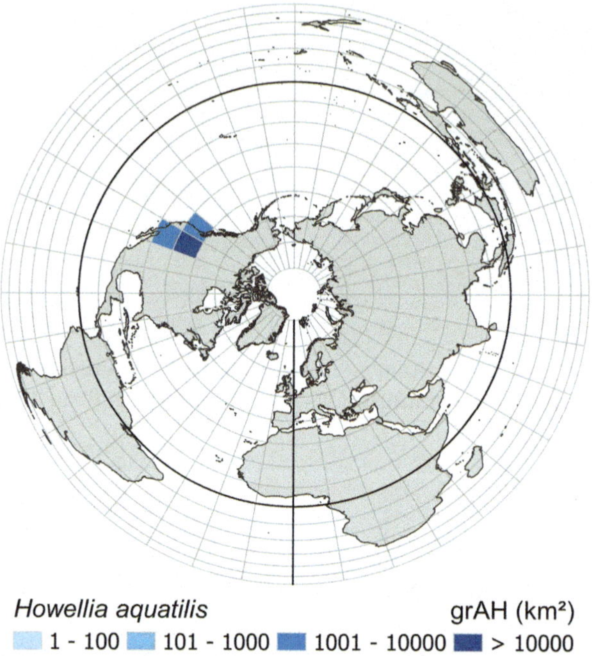

Howellia aquatilis A.Gray

***Isotoma fluviatilis* (R.Br.) F.Muell. ex Benth.**
Synonyms: *Hypsela sessiliflora* E.Wimm.; *Laurentia fluviatilis* (R.Br.) E.Wimm.

Range area (grAH km^2): NEA: 11000; AUS: 54500
Ecozone origin-status: NEA: I; AUS: E(I)
World rarity status: R
Ecozone-endemic macroregion occurrence: AUS: NG, ESA, NA, WA
Life form(s): EM
Ploidy state: DP
Speciation age (Ma): NA
Populations also occur in terrestrial, wetland and/ or riparian habitats: YES
Long distance movement vectors: ANTHROPOCHORY
Nuisance status and uses: considered an aggressive alien in the USA but few major nuisance issues have been reported; some use as a garden plant
Endangered status (IUCN): NA

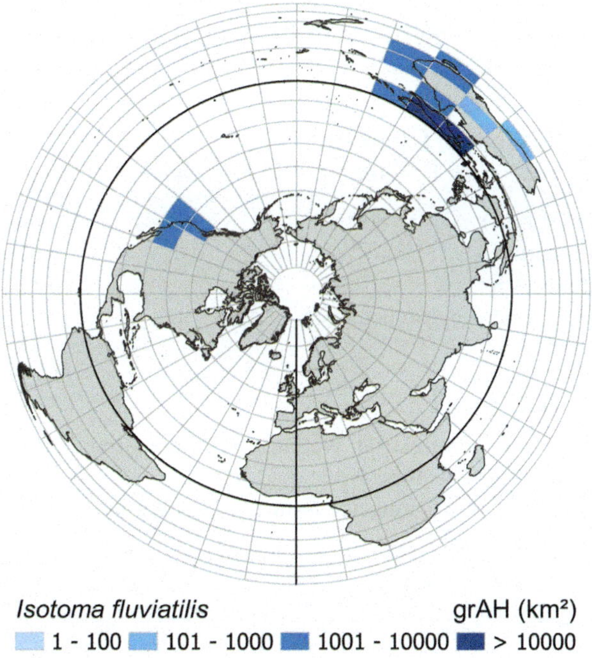

Isotoma fluviatilis (R.Br.) F.Muell. ex Benth.

***Isotoma rivalis* (E.Wimm.) Lammers**
Synonyms: *Hypsela rivalis* E.Wimm. *Lobelia ionantha* Heenan

Range area (grAH km^2): AUS: 7800
Ecozone origin-status: AUS: E
World rarity status: VR
Ecozone-endemic macroregion occurrence: AUS: NZ
Life form(s): EM
Ploidy state: NA
Speciation age (Ma): NA
Populations also occur in terrestrial, wetland and/ or riparian habitats: YES
Long distance movement vectors: NA
Nuisance status and uses: no nuisance issues; no known uses
Endangered status (IUCN): NA

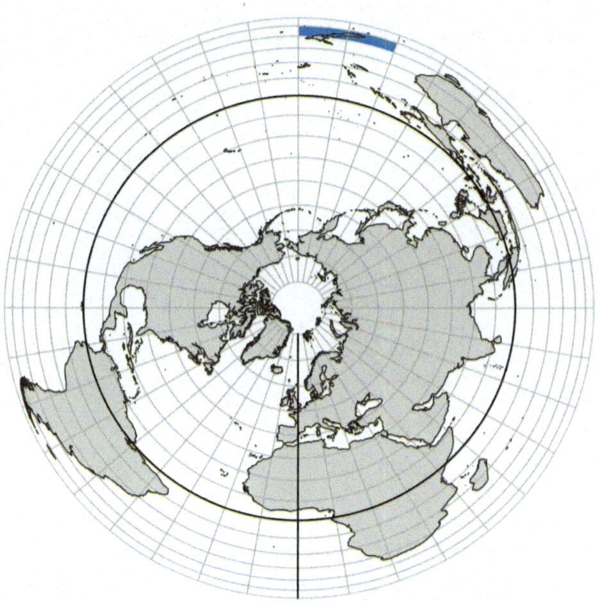

***Isotoma tridens* (E.Wimm.) Lammers**
Synonyms: *Hypsela tridens* E.Wimm.

Range area (grAH km^2): AUS: 11500
Ecozone origin-status: AUS: E
World rarity status: VR
Ecozone-endemic macroregion occurrence: AUS: ESA
Life form(s): EM
Ploidy state: NA
Speciation age (Ma): NA
Populations also occur in terrestrial, wetland and/ or riparian habitats: YES
Long distance movement vectors: NA
Nuisance status and uses: no nuisance issues; no known uses
Endangered status (IUCN): NA

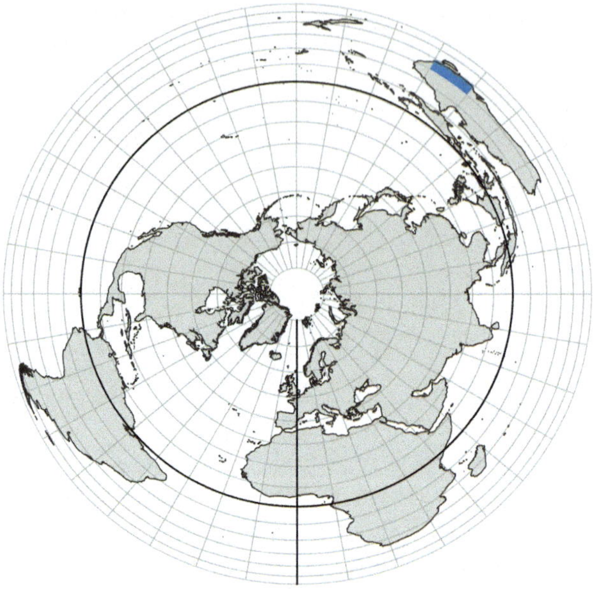

***Legenere valdiviana* (Phil.) E.Wimm.**
Synonyms: *Legenere limosa* (Greene) McVaugh; *Mezleria valdiviana* Phil., *Dortmanna valdiviana* (Phil.) Kuntze, *Howellia valdiviana* (Phil.) E.Wimm.

Range area (grAH km^2): NEA: 7500; NEO: 28800
Ecozone origin-status: NEA: N; NEO: N
World rarity status: R
Ecozone-endemic macroregion occurrence: not endemic to any world ecozone
Life form(s): EM
Ploidy state: NA
Speciation age (Ma): NA
Populations also occur in terrestrial, wetland and/ or riparian habitats: YES
Long distance movement vectors: NA
Nuisance status and uses: no nuisance issues; no known uses
Endangered status (IUCN): NA

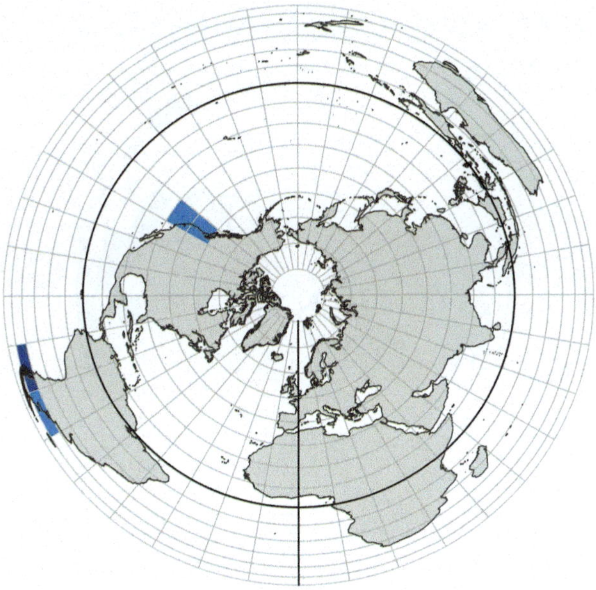

Lobelia alsinoides **Lam.**
Synonyms: *Dortmanna alsinodes* (Lam.) Kuntze, *Rapuntium alsinoides* (Lam.) C.Presl

Range area (grAH km^2): OR: 76400; AUS: 12800
Ecozone origin-status: OR: N; AUS: N
World rarity status: R
Ecozone-endemic macroregion occurrence: not endemic to any world ecozone
Life form(s): EM
Ploidy state: D
Speciation age (Ma): NA
Populations also occur in terrestrial, wetland and/ or riparian habitats: YES
Long distance movement vectors: NA
Nuisance status and uses: no nuisance issues; folk medicine
Endangered status (IUCN): LC

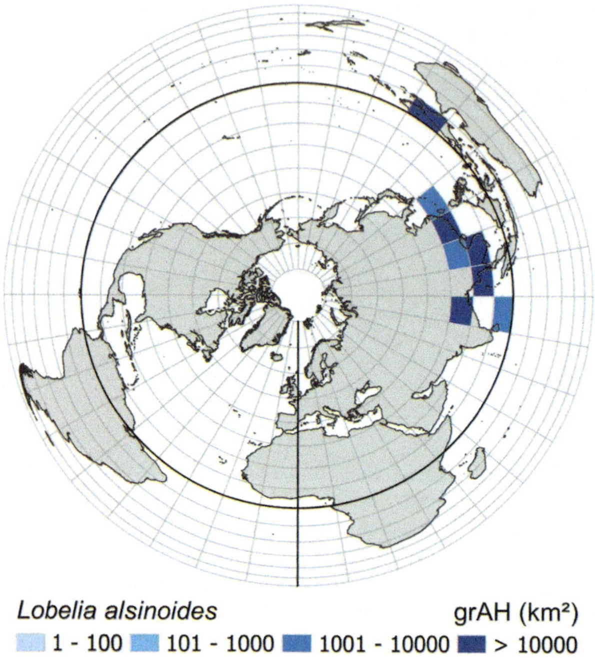

Lobelia alsinoides Lam.
Image copyright: iNaturalist CC-BY @renju

Lobelia aquatica **Cham.**
Synonyms: *Dortmanna aquatica* (Cham.), *Rapuntium aquaticum* (Cham.) C.Presl

Range area (grAH km^2): NEO: 262500
Ecozone origin-status: NEO: E
World rarity status: C
Ecozone-endemic macroregion occurrence: NEO: MES AM, NWSA, AMZ, CHA, PAR
Life form(s): EM
Ploidy state: NA
Speciation age (Ma): 6.0
Populations also occur in terrestrial, wetland and/ or riparian habitats: YES
Long distance movement vectors: ?AV
Nuisance status and uses: no nuisance issues; used in folk medicine
Endangered status (IUCN): NA

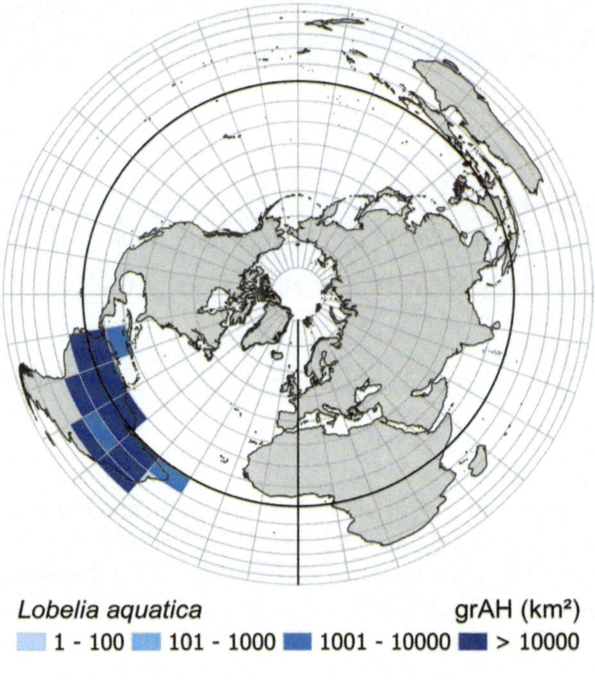

Lobelia aquatica Cham.
Image copyright: iNaturalist CC0 @rvgalli (Hervé Galliffet)

***Lobelia cardinalis* L.**
Synonyms: *Dortmanna cardinalis* (L.) Kuntze, *Rapuntium cardinale* (L.) Mill.

Range area (grAH km^2): NEA: 387400; NEO: 53500; AUS: 10900
Ecozone origin-status: NEA: N; NEO: N; AUS: I
World rarity status: C
Ecozone-endemic macroregion occurrence: not endemic to any world ecozone
Life form(s): EM
Ploidy state: D
Speciation age (Ma): 6.0
Populations also occur in terrestrial, wetland and/ or riparian habitats: YES
Long distance movement vectors: ?AV; ANTHROPOCHORY
Nuisance status and uses: no nuisance issues; medicinal and garden plant uses
Endangered status (IUCN): LC

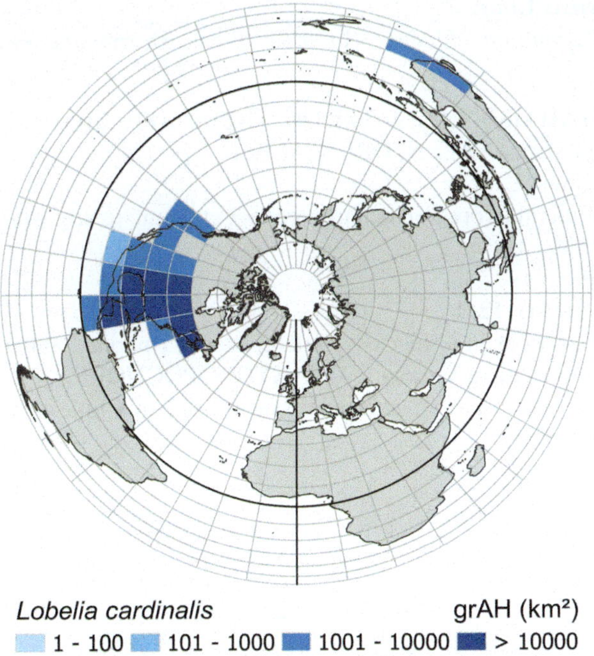

Lobelia cardinalis L.
Image copyright: G. Ervin, with permission

Lobelia chinensis **Lour.**

Synonyms: *Rapuntium chinense* (Lour.) C.Presl, *Dortmanna chinensis* (Lour.) Kuntze

Range area (grAH km^2): PAL: 46300; OR: 110500; AFR: 2100
Ecozone origin-status: PAL: N; OR: N; AFR: I
World rarity status: C
Ecozone-endemic macroregion occurrence: not endemic to any world ecozone
Life form(s): EM
Ploidy state: P
Speciation age (Ma): 12.0
Populations also occur in terrestrial, wetland and/ or riparian habitats: YES
Long distance movement vectors: ?AV; ANTHROPOCHORY
Nuisance status and uses: no nuisance issues; used in Chinese herbal medicine, water garden plant
Endangered status (IUCN): LC

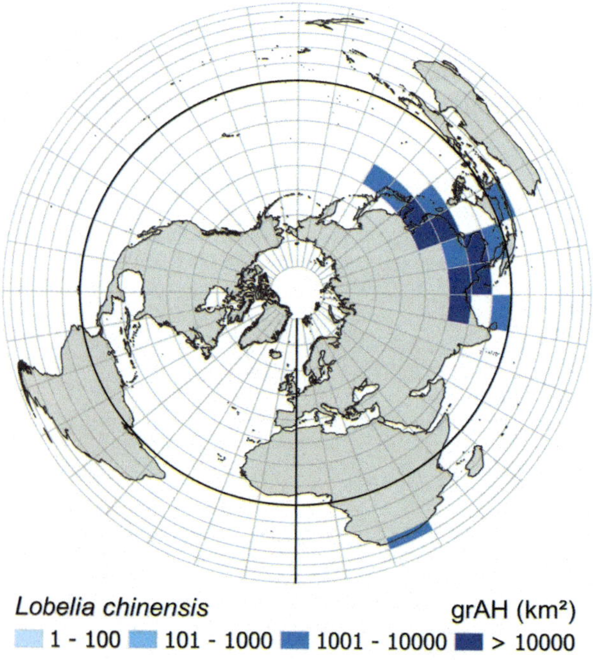

Lobelia chinensis Lour.
Image copyright: Z. Li, with permission

Lobelia concolor R.Br.
Synonyms: *Isolobus concolor* (R.Br.) A.DC., *Pratia concolor* (R.Br.) Druce

Range area (grAH km^2): AUS: 22400
Ecozone origin-status: AUS: E
World rarity status: VR
Ecozone-endemic macroregion occurrence: AUS: NA, ESA
Life form(s): EM (subshrub)
Ploidy state: P
Speciation age (Ma): NA
Populations also occur in terrestrial, wetland and/ or riparian habitats: YES
Long distance movement vectors: NA
Nuisance status and uses: no nuisance issues; toxic to livestook
Endangered status (IUCN): NA

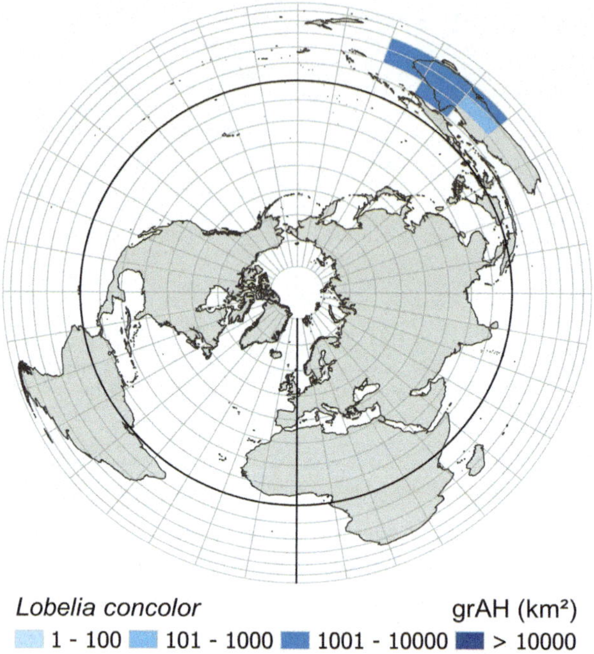

Lobelia concolor R.Br.
Image copyright: K. Ward, with permission

***Lobelia dortmanna* L.**

Synonyms: *Dortmanna lacustris* (Salisb.) G.Don, *Rapuntium dortmanna* (L.) Chaz.

Range area (grAH km^2): PAL: 171800; NEA: 518100
Ecozone origin-status: PAL: N; NEA: N
World rarity status: C
Ecozone-endemic macroregion occurrence: not endemic to any world ecozone
Life form(s): SU
Ploidy state: DP
Speciation age (Ma): 7.0
Populations also occur in terrestrial, wetland and/ or riparian habitats: NO
Long distance movement vectors: ?AV
Nuisance status and uses: no nuisance issues; folk medicine
Endangered status (IUCN): LC

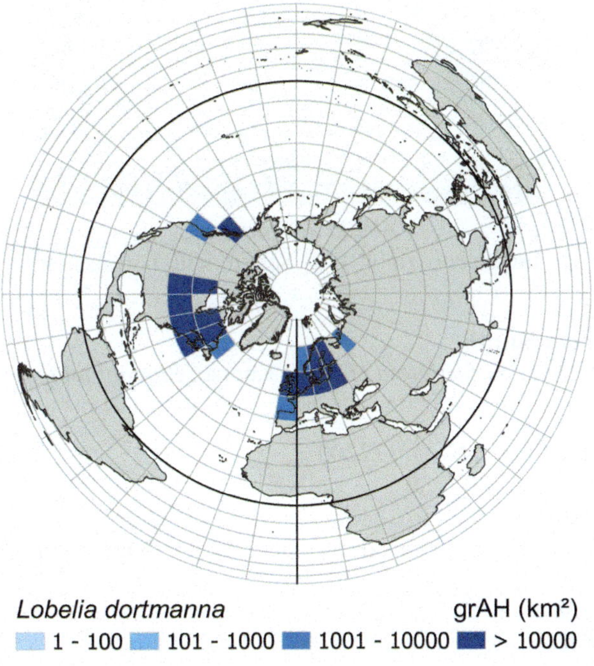

Lobelia dortmanna L.
Image copyright: J. C. Schou, with permission

Lobelia fastigiata **Kunth**

Synonyms: *Dortmanna fastigiata* (Kunth) Kuntze, *Rapuntium fastigiatum* (Kunth) C.Presl

Range area (grAH km^2): NEO: 112200
Ecozone origin-status: NEO: E
World rarity status: R
Ecozone-endemic macroregion occurrence: NEO: MES AM, NWSA, AMZ, CHA, PAR
Life form(s): EM
Ploidy state: NA
Speciation age (Ma): NA
Populations also occur in terrestrial, wetland and/ or riparian habitats: YES
Long distance movement vectors: NA
Nuisance status and uses: no nuisance issues; no known uses
Endangered status (IUCN): NA

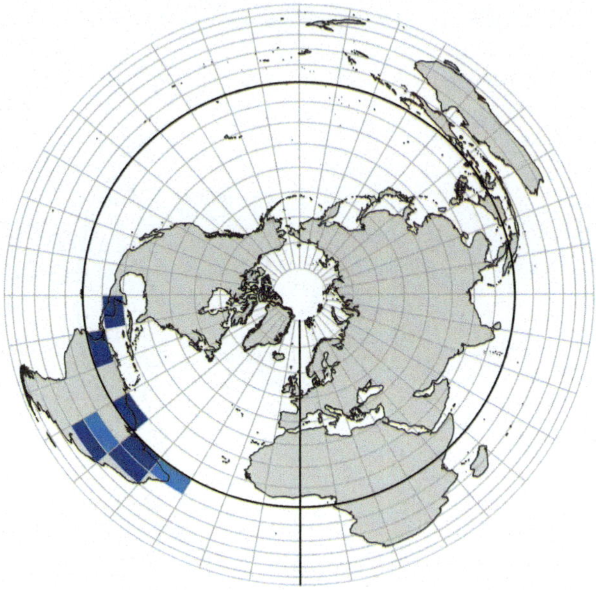

Lobelia filicaulis (C.Presl) Schönland
Synonyms: *Mezleria filicaulis* C.Presl

Range area (grAH km^2): AFR: 3000
Ecozone origin-status: AFR: E
World rarity status: VR
Ecozone-endemic macroregion occurrence: AFR: CAPE
Life form(s): ?EM
Ploidy state: NA
Speciation age (Ma): NA
Populations also occur in terrestrial, wetland and/ or riparian habitats: ?YES
Long distance movement vectors: NA
Nuisance status and uses: no nuisance issues; no known uses
Endangered status (IUCN): NA

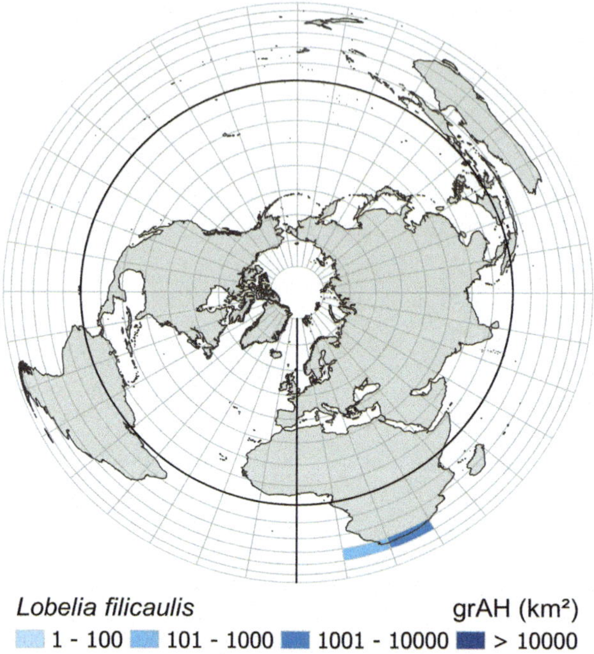

Lobelia filicaulis (C.Presl) Schönland
Image copyright: N. Helme, with permission

Lobelia glandulosa Walter
Synonyms: *Dortmanna glandulosa* (Walter) Kuntze, *Rapuntium glandulosum* (Walter) C.Presl

Range area (grAH km^2): NEA: 21200; NEO: 24100
Ecozone origin-status: NEA: N; NEO: N
World rarity status: R

Ecozone-endemic macroregion occurrence: not endemic to any world ecozone
Life form(s): EM
Ploidy state: P
Speciation age (Ma): NA
Populations also occur in terrestrial, wetland and/ or riparian habitats: YES
Long distance movement vectors: HYDROCHORY (NEA to NEO)
Nuisance status and uses: no nuisance issues; no known uses
Endangered status (IUCN): NA

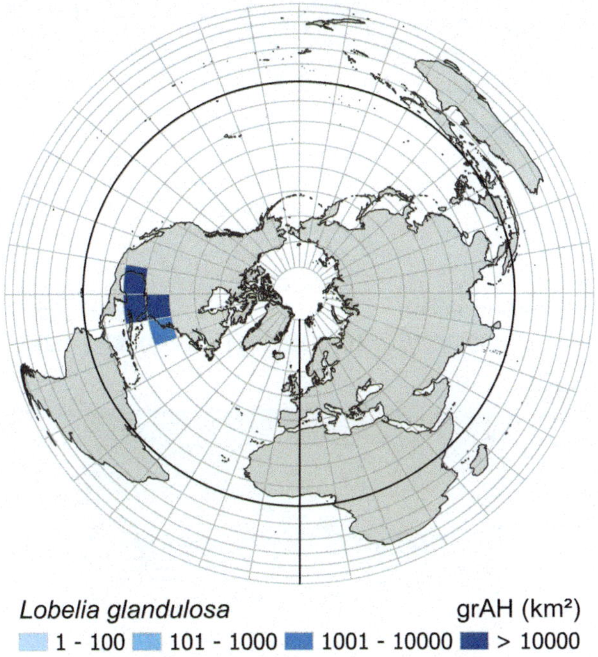

Lobelia hassleri **Zahlbr.**
Synonyms:

Range area (grAH km^2): NEO: 39000
Ecozone origin-status: NEO: E
World rarity status: R
Ecozone-endemic macroregion occurrence: NEO: AMZ, PAR
Life form(s): EM (sometimes a subshrub)
Ploidy state: P
Speciation age (Ma): NA
Populations also occur in terrestrial, wetland and/ or riparian habitats: YES
Long distance movement vectors: NA
Nuisance status and uses: no nuisance issues; some medicinal uses
Endangered status (IUCN): NA

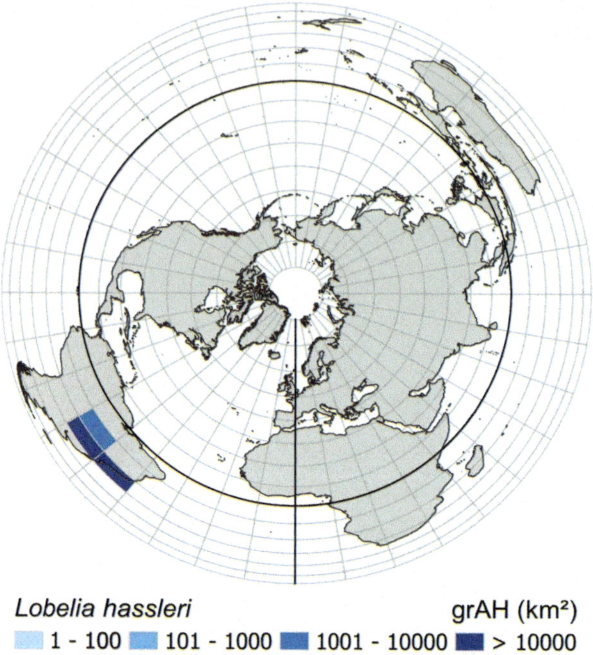

Lobelia hassleri Zahlbr.
Image copyright: N. Helme, with permission

***Lobelia heyneana* Schult.**
Synonyms: *Lobelia decurrens* Roth

Range area (grAH km^2): OR: 90500; AUS: 24600; AFR: 177600
Ecozone origin-status: OR: N; AUS: N; AFR: N
World rarity status: C
Ecozone-endemic macroregion occurrence: not endemic to any world ecozone
Life form(s): EM
Ploidy state: NA
Speciation age (Ma): NA
Populations also occur in terrestrial, wetland and/ or riparian habitats: YES
Long distance movement vectors: NA
Nuisance status and uses: no nuisance issues; no known uses
Endangered status (IUCN): LC

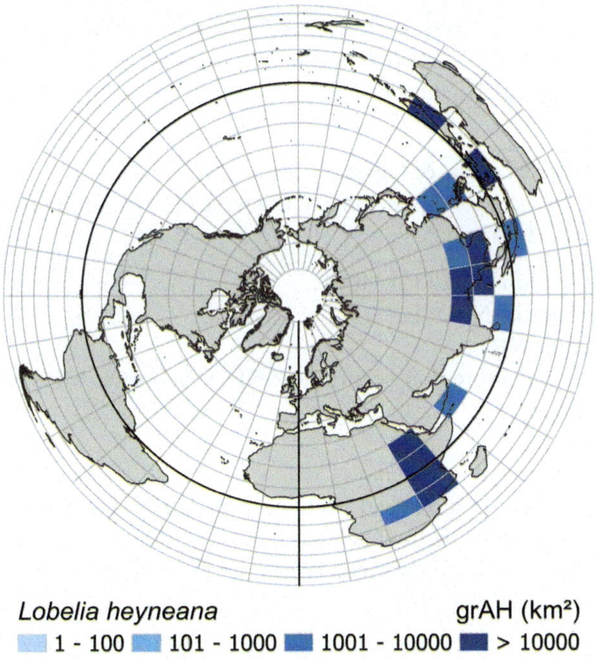

Lobelia heyneana Schult.

***Lobelia irasuensis* Planch. & Oerst.**
Synonyms: *Diastatea irasuensis* (Planch. & Oerst.) E.Wimm., *Laurentia irasuensis* (Planch. & Oerst.) E.Wimm.

Range area (grAH km^2): NEO: 13400
Ecozone origin-status: NEO: E
World rarity status: VR
Ecozone-endemic macroregion occurrence: NEO: MG MX, MES AM
Life form(s): EM
Ploidy state: NA
Speciation age (Ma): NA
Populations also occur in terrestrial, wetland and/ or riparian habitats: YES
Long distance movement vectors: NA
Nuisance status and uses: no nuisance issues; no known uses
Endangered status (IUCN): NA

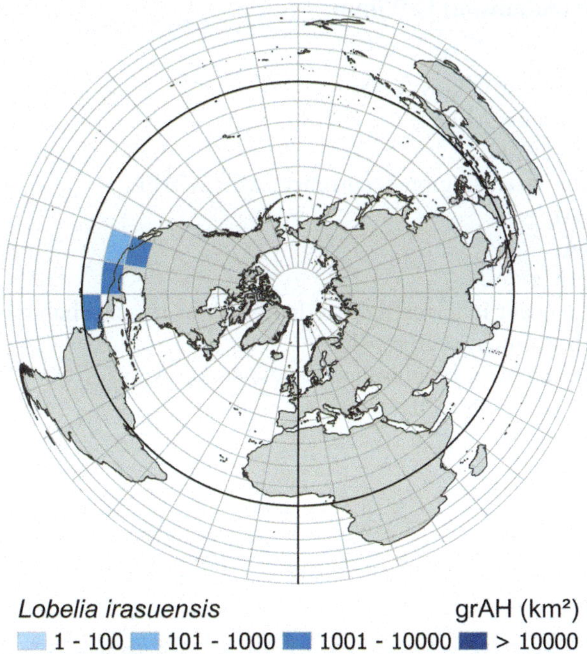

Lobelia irasuensis Planch. & Oerst.
Image copyright: iNaturalist CC-BY @chloe_and_trevor (Chloe and Trevor Van Loon)

***Lobelia limosa* (Adamson) E.Wimm.**
Synonyms: *Mezleria limosa* Adamson

Range area (grAH km^2): AFR: 1000
Ecozone origin-status: AFR: E
World rarity status: VR
Ecozone-endemic macroregion occurrence: AFR: CAPE
Life form(s): EM
Ploidy state: NA
Speciation age (Ma): NA
Populations also occur in terrestrial, wetland and/ or riparian habitats: YES
Long distance movement vectors: NA
Nuisance status and uses: no nuisance issues; no known uses
Endangered status (IUCN): DD

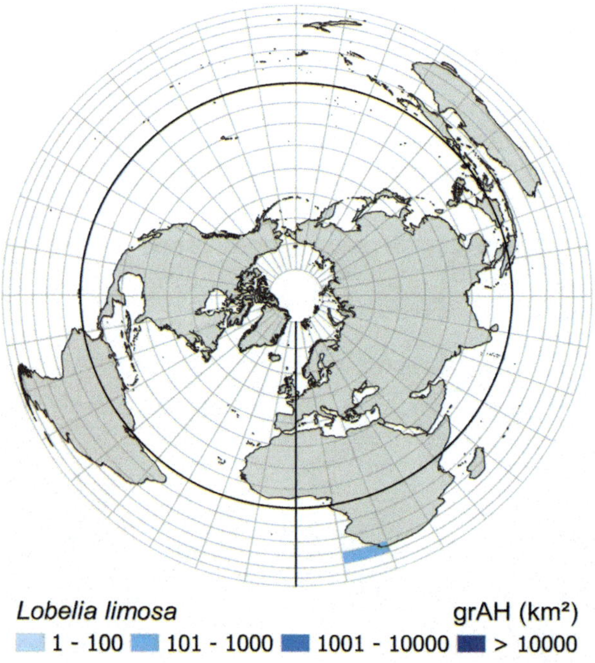

***Lobelia nummularioides* Cham**
Synonyms: *Rapuntium nummularioides* (Cham.) C.Presl, *Dortmanna nummulari-odes* (Cham.) Kuntze

Range area (grAH km^2): NEO: 72200;
Ecozone origin-status: NEO: E
World rarity status: R
Ecozone-endemic macroregion occurrence: NEO: AMZ, CHA, PAR

Life form(s): EM (sometimes a subshrub)
Ploidy state: NA
Speciation age (Ma): NA
Populations also occur in terrestrial, wetland and/ or riparian habitats: YES
Long distance movement vectors: NA
Nuisance status and uses: no nuisance issues; no known uses
Endangered status (IUCN): NA

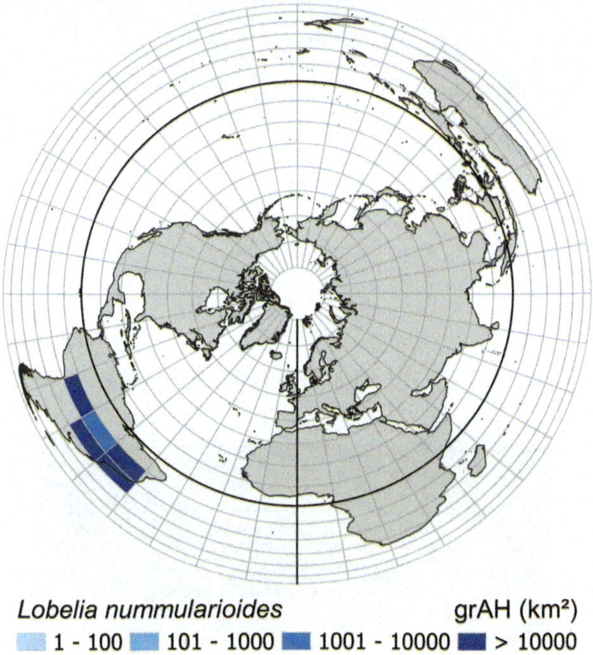

Lobelia oligophylla **(Wedd.) Lammers**
Synonyms: *Hypsela oligophylla* (Wedd.) Benth. & Hook.f. ex Zahlbr., Pratia oligophylla Wedd.

Range area (grAH km^2): NEO: 104900
Ecozone origin-status: NEO: E
World rarity status: R
Ecozone-endemic macroregion occurrence: NEO: NWSA, AND, CHA, PAR
Life form(s): EM (subshrub)
Ploidy state: NA
Speciation age (Ma): NA
Populations also occur in terrestrial, wetland and/ or riparian habitats: YES
Long distance movement vectors: NA
Nuisance status and uses: no nuisance issues; water garden plant and forage uses
Endangered status (IUCN): NA

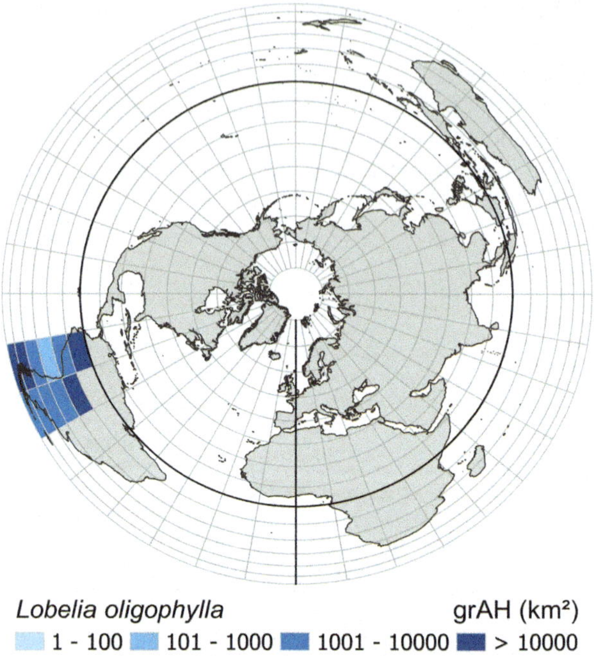

Lobelia oligophylla (Wedd.) Lammers
Image copyright: iNaturalist CC-BY @nodora (Nodora L. Moyano)

Lobelia paludosa **Nutt.**
Synonyms: *Dortmanna paludosa* (Nutt.) G.Don, *Rapuntium paludosum* (Nutt.) C.Presl

Range area (grAH km^2): NEA: 21200; NEO: 10300
Ecozone origin-status: NEA: N; NEO: N
World rarity status: R
Ecozone-endemic macroregion occurrence: not endemic to any world ecozone
Life form(s): EM
Ploidy state: P
Speciation age (Ma): NA
Populations also occur in terrestrial, wetland and/ or riparian habitats: YES
Long distance movement vectors: HYDROCHORY (NEA to NEO)
Nuisance status and uses: no nuisance issues; water garden plant
Endangered status (IUCN): NA

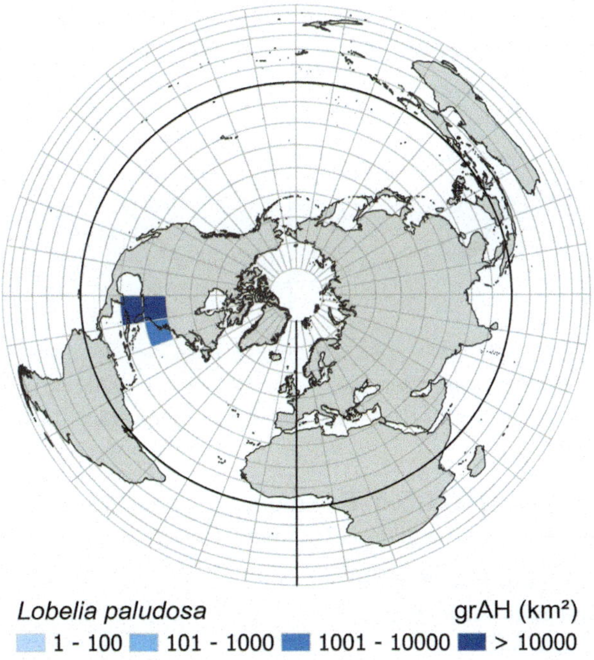

Lobelia paranaensis **R.Braga**
Synonyms:

Range area (grAH km^2): NEO: 31400
Ecozone origin-status: NEO: E
World rarity status: R
Ecozone-endemic macroregion occurrence: NEO: PAR

Life form(s): EM
Ploidy state: NA
Speciation age (Ma): NA
Populations also occur in terrestrial, wetland and/ or riparian habitats: YES
Long distance movement vectors: NA
Nuisance status and uses: no nuisance issues; no known uses
Endangered status (IUCN): NA

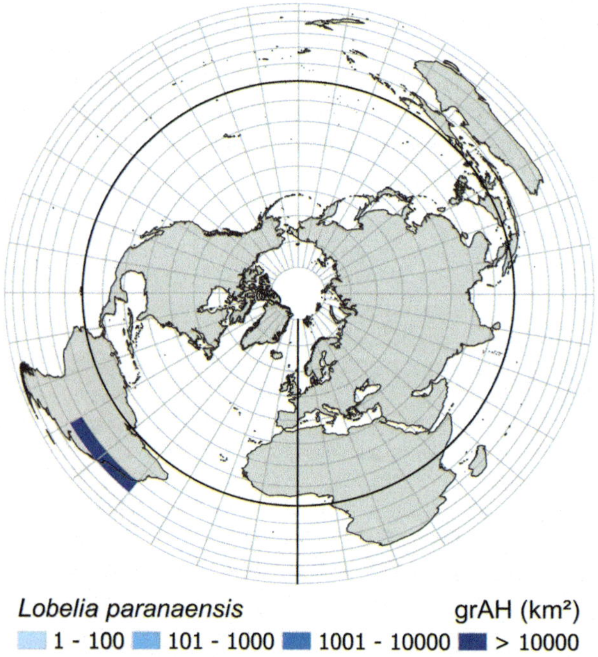

***Lobelia quadrisepala* (R.D.Good) E.Wimm.**
Synonyms: *Mezleria quadrisepala* R.D.Good

Range area (grAH km^2): AFR: 1000
Ecozone origin-status: AFR: E
World rarity status: VR
Ecozone-endemic macroregion occurrence: AFR: CAPE
Life form(s): EM
Ploidy state: NA
Speciation age (Ma): NA
Populations also occur in terrestrial, wetland and/ or riparian habitats: YES
Long distance movement vectors: NA
Nuisance status and uses: no nuisance issues; no known uses
Endangered status (IUCN): DD

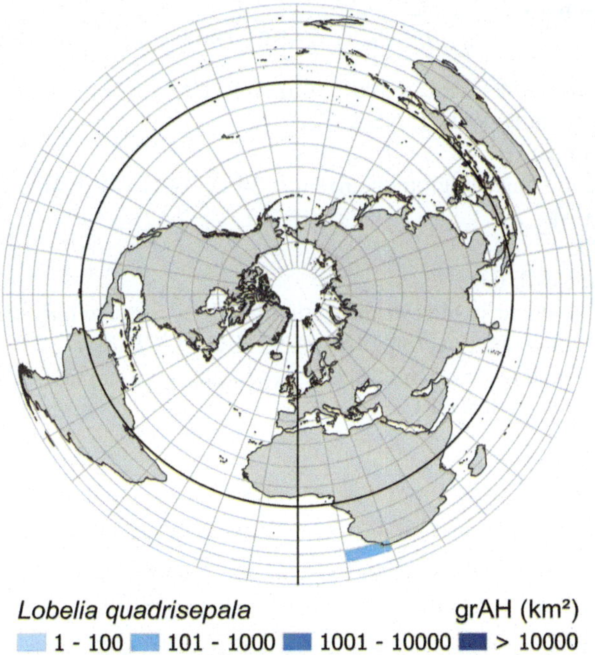

Lobelia quadrisepala (R.D.Good) E.Wimm.
Image copyright: N. Helme, with permission

***Lobelia siphilitica* L.**

Synonyms: *Rapuntium siphiliticum* (L.) Mill., *Dortmanna siphilitica* (L.) Kuntze

Range area (grAH km^2): PAL: 64000; NEA: 391700
Ecozone origin-status: PAL: I; NEA: E(I)
World rarity status: C
Ecozone-endemic macroregion occurrence: NEA: W NAM, E NAM
Life form(s): EM
Ploidy state: D
Speciation age (Ma): 6.0
Populations also occur in terrestrial, wetland and/ or riparian habitats: YES
Long distance movement vectors: ?AV; ANTHROPOCHORY
Nuisance status and uses: no nuisance issues; water garden plant, medicinal uses
Endangered status (IUCN): LC

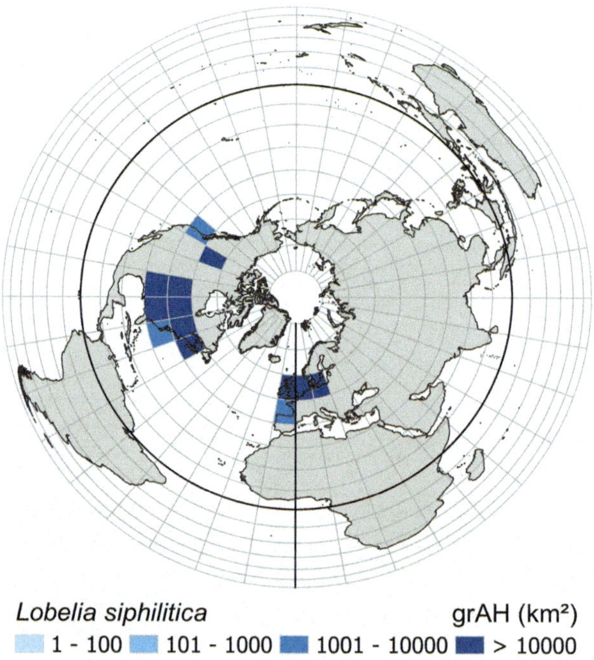

Lobelia siphilitica L.
Image copyright: J. C.Schou, with permission

***Lobelia sonderiana* (Kuntze) Lammers**
Synonyms: *Lobelia angolensis* Engl. & Diels, *Dortmanna sonderiana* Kuntze, *Mezleria dregeana* Sond.

Range area (grAH km^2): AFR: 59000
Ecozone origin-status: AFR: E
World rarity status: R
Ecozone-endemic macroregion occurrence: AFR: ZAM, SDES, CAPE
Life form(s): EM (sometimes a subshrub)
Ploidy state: NA
Speciation age (Ma): 52.0
Populations also occur in terrestrial, wetland and/ or riparian habitats: YES
Long distance movement vectors: ?AV
Nuisance status and uses: no nuisance issues; no known uses
Endangered status (IUCN): NA

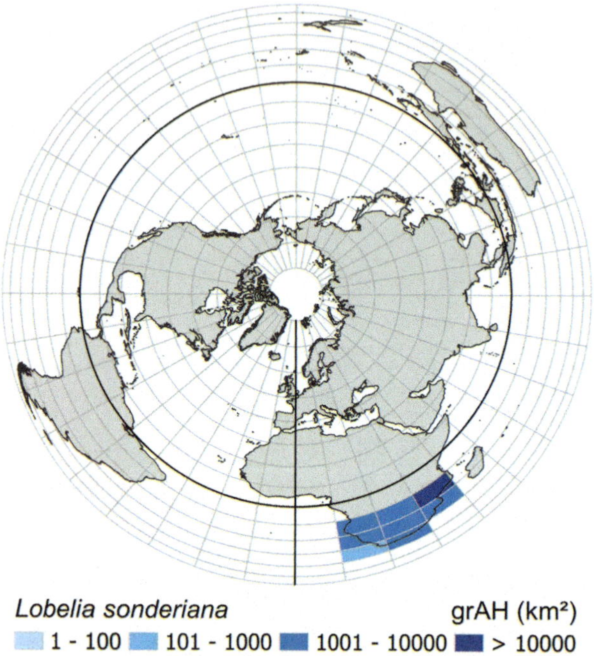

Lobelia sonderiana (Kuntze) Lammers
Image copyright: iNaturalist CC0 @botswanabugs (Tony Benn)

Lobelia stenosiphon **(Adamson) E.Wimm.**
Synonyms: *Mezleria stenosiphon* Adamson

Range area (grAH km^2): AFR: 1000
Ecozone origin-status: AFR: E
World rarity status: VR
Ecozone-endemic macroregion occurrence: AFR: CAPE
Life form(s): EM
Ploidy state: NA
Speciation age (Ma): NA
Populations also occur in terrestrial, wetland and/ or riparian habitats: YES
Long distance movement vectors: NA
Nuisance status and uses: no nuisance issues; no known uses
Endangered status (IUCN): NA (LC according to SANBI Red List of South African Plants)

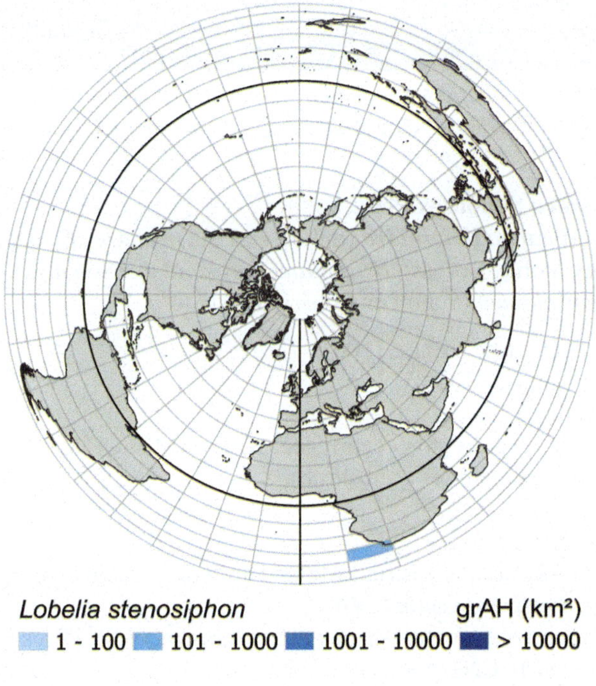

Lobelia stenosiphon (Adamson) E.Wimm.
Image copyright: iNaturalist CC-BY @botaneek (Nick Helme)

Lobelia thapsoidea **Schott ex Pohl**
Synonyms: *Dortmanna thapsoidea* (Schott ex Pohl) Kuntze, *Euhaynaldia thapsoidea* (Schott ex Pohl) Borbás, *Haynaldia thapsioidea* (Schott ex Pohl) Kanitz, *Rapuntium thapsoideum* (Schott ex Pohl) C.Presl

Range area (grAH km^2): NEO: 31100
Ecozone origin-status: NEO: E
World rarity status: VR
Ecozone-endemic macroregion occurrence: NEO: AMZ, CHA, PAR
Life form(s): EM (sometimes a subshrub)
Ploidy state: P
Speciation age (Ma): NA
Populations also occur in terrestrial, wetland and/ or riparian habitats: YES
Long distance movement vectors: NA
Nuisance status and uses: no nuisance issues; some medicinal uses
Endangered status (IUCN): NA

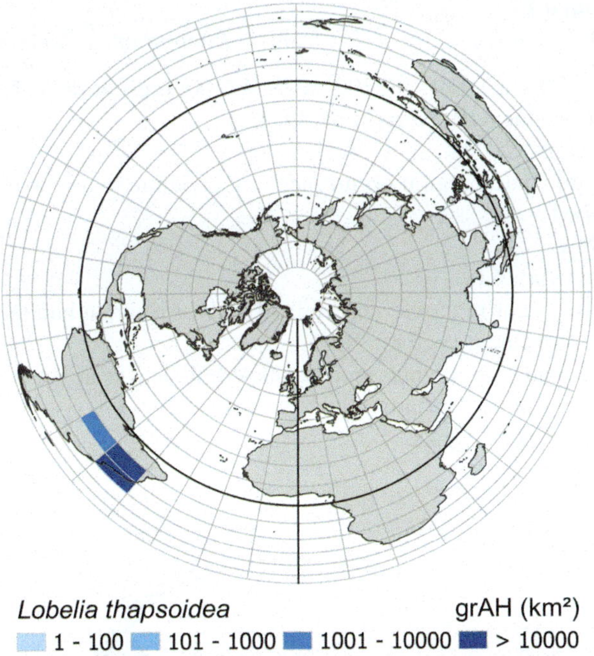

Lobelia thapsoidea Schott ex Pohl
Image copyright: P.G. Carvalho, with permission

***Lobelia zeylanica* L.**

Synonyms: *Dortmanna zeylanica* (L.) Kuntze, *Rapuntium zeylanicum* (L.) Chaz.

Range area (grAH km^2): OR: 132300; AUS: 44200; present in OCE
Ecozone origin-status: OR: N; AUS: N; OCE: I
World rarity status: C
Ecozone-endemic macroregion occurrence: not endemic to any world ecozone
Life form(s): EM
Ploidy state: NA
Speciation age (Ma): 6.0
Populations also occur in terrestrial, wetland and/ or riparian habitats: YES
Long distance movement vectors: ?AV
Nuisance status and uses: no nuisance issues; edible leaves are sometimes gathered from the wild and used locally
Endangered status (IUCN): LC

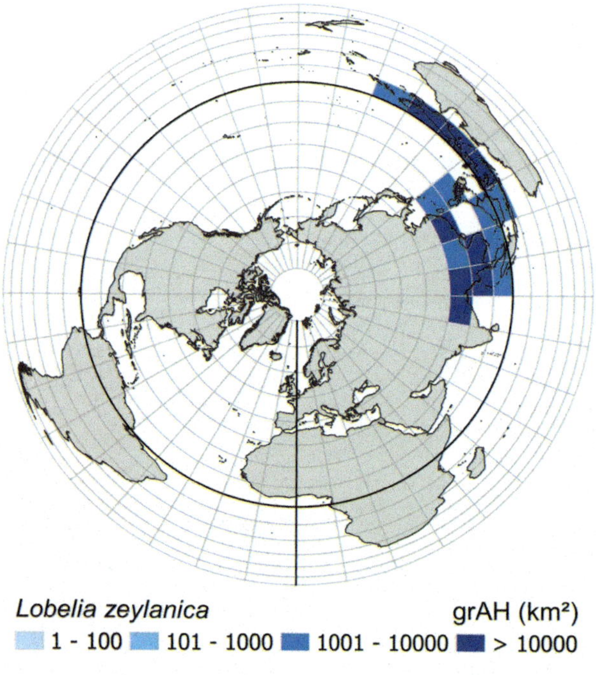

***Porterella carnosula* (Hook. & Arn.) Torr.**

Synonyms: *Porterella eximia* A.Nelson, *Laurentia carnosula* (Hook. & Arn.) Benth. ex A.Gray, *Lobelia carnosula* Hook. & Arn.

Range area (grAH km^2): NEA: 21100
Ecozone origin-status: NEA: E
World rarity status: VR
Ecozone-endemic macroregion occurrence: NEA: W NAM, SW NAM
Life form(s): EM
Ploidy state: D

Speciation age (Ma): NA
Populations also occur in terrestrial, wetland and/ or riparian habitats: YES
Long distance movement vectors: NA
Nuisance status and uses: no nuisance issues; no known uses
Endangered status (IUCN): NA

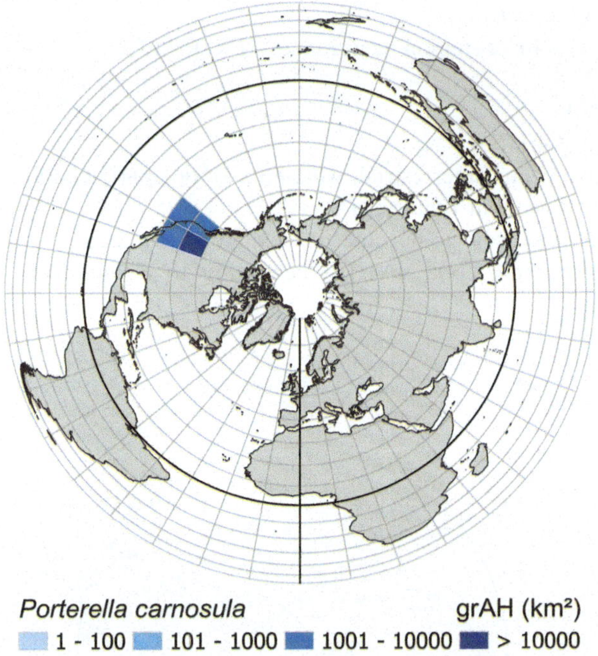

Porterella carnosula (Hook. & Arn.) Torr.

***Unigenes humifusa* (A.DC.) E.Wimm.**

Synonyms: *Mezleria humifusa* A.DC.; *Lobelia disperma* E.Wimm., *Dortmanna humifusa* (A.DC.) Kuntze

Range area (grAH km^2): AFR: 3000
Ecozone origin-status: AFR: E
World rarity status: VR
Ecozone-endemic macroregion occurrence: AFR: CAPE
Life form(s): EM
Ploidy state: NA
Speciation age (Ma): NA
Populations also occur in terrestrial, wetland and/ or riparian habitats: YES
Long distance movement vectors: NA
Nuisance status and uses: no nuisance issues; no known uses
Endangered status (IUCN): NA (LC according to SANBI Red List of South African Plants)

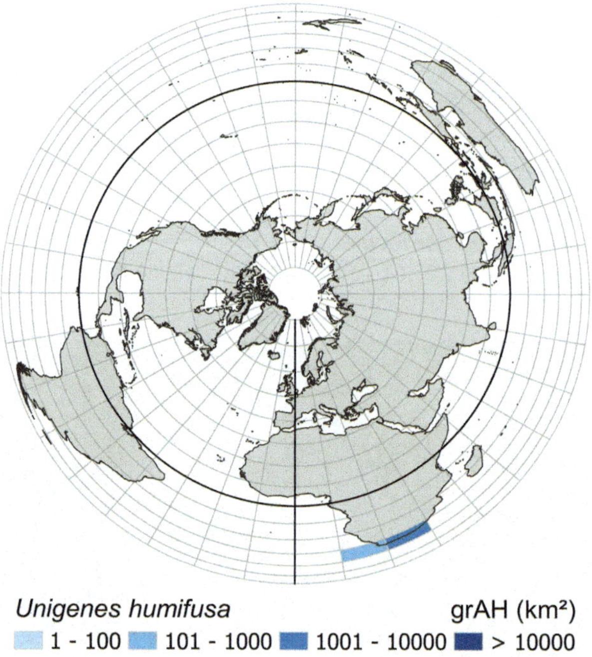

Unigenes humifusa (A.DC.) E.Wimm.

3.13 CARYOPHYLLACEAE (Caryophyllales)

The Caryophyllaceae have two genera occurring in inland brackish or freshwater bodies, each with a single species. Both species are native in the Old World, *Illecebrum verticillatum* is rare at world scale, R, but of Least Concern, LC, endemic to PAL and introduced in NEA. *Stellaria aquatica* is common at world scale, C, native in PAL and OR, also introduced in NEA. Neither species causes aquatic weed problems.

Quick Links to Genera

1. *Illecebrum*
2. *Stellaria*

***Illecebrum verticillatum* L.**
Synonyms:

Range area (g
rAH km^2): PAL: 97500; NEA: 6500; AUS: 6900
Ecozone origin-status: PAL: E(I); NEA: I; AUS: I
World rarity status: R
Ecozone-endemic macroregion occurrence: PAL: W EUR, MED
Life form(s): EM
Ploidy state: D
Speciation age (Ma): NA
Populations also occur in terrestrial, wetland and/ or riparian habitats: YES
Long distance movement vectors: ANTHROPOCHORY
Nuisance status and uses: no nuisance issues; folk medicine
Endangered status (IUCN): LC

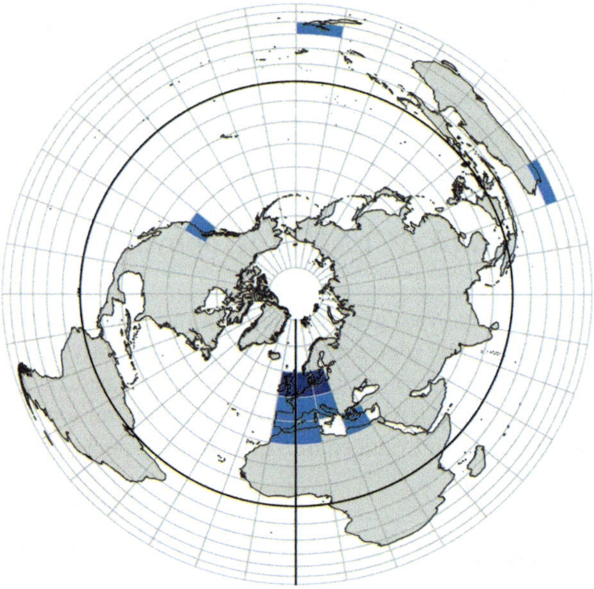

Illecebrum verticillatum L.
Image copyright: J. C.Schou, with permission

***Stellaria aquatica* (L.) Scop.**
Synonyms: *Myosoton aquaticum* (L.) Moench, *Alsine aquatica* (L.) Britton, *Cerastium aquaticum* L., *Larbrea aquatica* (L.) Ser., *Malachium aquaticum* (L.) Fr. ex Rchb., *Myosanthus aquaticus* (L.) Desv.

Range area (grAH km^2): PAL: 616800; NEA: 353600; OR: 61300
Ecozone origin-status: PAL: N; NEA: I; OR: N
World rarity status: C
Ecozone-endemic macroregion occurrence: not endemic to any world ecozone
Life form(s): EM
Ploidy state: D
Speciation age (Ma): NA
Populations also occur in terrestrial, wetland and/ or riparian habitats: YES
Long distance movement vectors: ANTHROPOCHORY
Nuisance status and uses: no nuisance issues; some medicinal uses
Endangered status (IUCN): NA

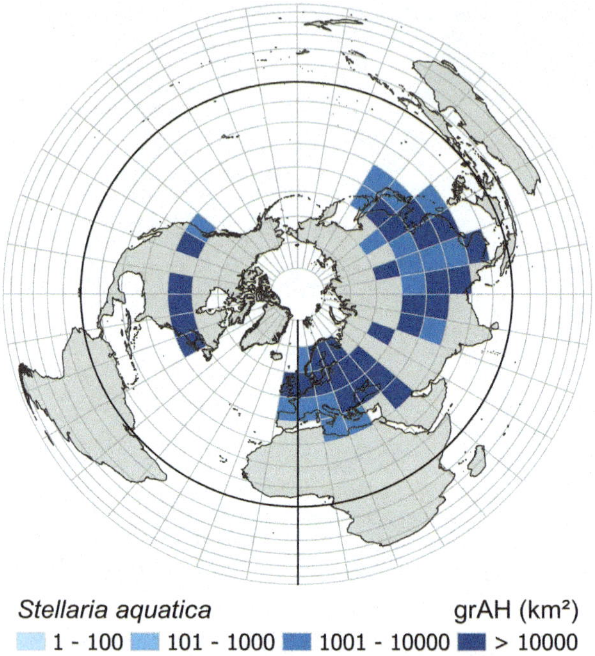

Stellaria aquatica (L.) Scop.
Image copyright: J. C.Schou, with permission

3.14 CERATOPHYLLACEAE (Ceratophyllales)

The Ceratophyllaceae contain a single genus with four macrophyte species. *Ceratophyllum demersum* is one of the commonest freshwater macrophyte species on Earth, native and widespread in every world ecozone except ANT, though some sources consider it introduced in New Zealand. Between two and four *Ceratophyllum* species occur in each of the world ecozones where the family is represented, and one species is introduced in NEO, OCE and AUS. Other than localised issues in parts of New Zealand, there are no reports of aquatic weed problems caused by members of this family. All four species are common at world scale, C. In terms of endangered status, *Ceratophyllum platyacanthum* is unassessed and the other three are LC.

***Ceratophyllum demersum* L.**
Synonyms: *Dichotophyllum demersum* (L.) Moench

Range area (grAH km^2): PAL: 879600; NEA: 507600; NEO: 339700; OR: 120400; AUS: 82600; AFR: 302700; present in OCE
Ecozone origin-status: PAL: N; NEA: N; NEO: N; OR: N; AUS: N; AFR: N; OCE: N
World rarity status: C
Ecozone-endemic macroregion occurrence: not endemic to any world ecozone
Life form(s): SU (sometimes hydrosubshrub)
Ploidy state: DP
Speciation age (Ma): 1.0
Populations also occur in terrestrial, wetland and/ or riparian habitats: NO
Long distance movement vectors: BIRD: *Mareca strepera* (Gadwall); *Anas platyrhynchos* (Northern Mallard); *Anas crecca* (Eurasian Teal); ANTHROPOCHORY (within AUS)
Nuisance status and uses: considered an introduced aquatic weed in New Zealand, low or no nuisance issues elsewhere; medicinal, water garden plant, phytoremediation, aquarium trade uses
Endangered status (IUCN): LC

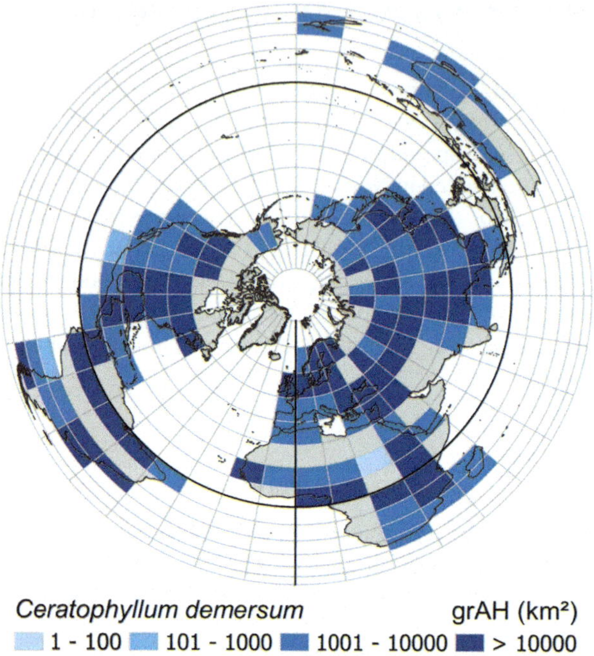

Ceratophyllum demersum L.
Image copyright: J. C.Schou, with permission

Ceratophyllum muricatum **Cham.**
Synonyms: *Ceratophyllum tanaiticum* Sapjegin; *Ceratophyllum australe* Griseb.

Range area (grAH km^2): PAL: 182400; NEA: 347100; NEO: 143000; OR: 51000; AUS: 30100; AFR: 97300
Ecozone origin-status: PAL: N; NEA: N; NEO: N; OR: N; AUS: N; AFR: N
World rarity status: C
Ecozone-endemic macroregion occurrence: not endemic to any world ecozone
Life form(s): SU (sometimes hydrosubshrub)
Ploidy state: D
Speciation age (Ma): 10.5
Populations also occur in terrestrial, wetland and/ or riparian habitats: NO
Long distance movement vectors: ?AV
Nuisance status and uses: no nuisance issues; no known uses
Endangered status (IUCN): LC

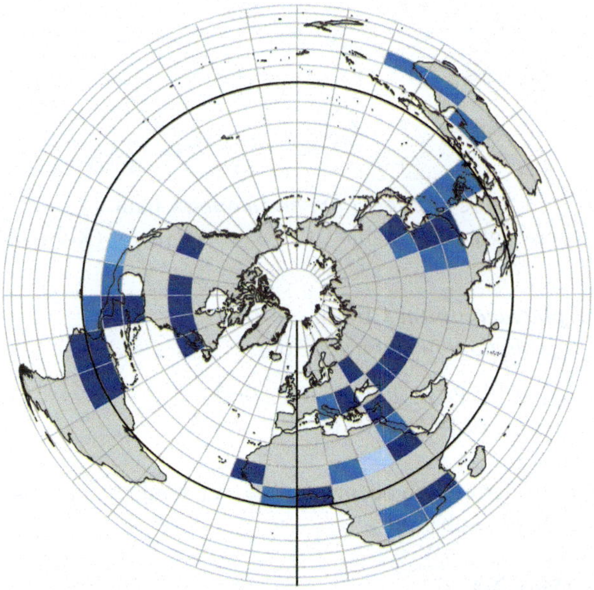

Ceratophyllum muricatum grAH (km²)
1 - 100 101 - 1000 1001 - 10000 > 10000

Ceratophyllum muricatum Cham.
Image copyright: iNaturalist CC0 @robert_taylor (Robert Taylor)

Ceratophyllum platyacanthum Cham.
Synonyms: *Ceratophyllum oryzetorum* Kom.

Range area (grAH km^2): PAL: 288900; OR: 9600
Ecozone origin-status: PAL: N; OR: N
World rarity status: C
Ecozone-endemic macroregion occurrence: not endemic to any world ecozone
Life form(s): SU (hydrosubshrub)
Ploidy state: NA
Speciation age (Ma): NA
Populations also occur in terrestrial, wetland and/ or riparian habitats: NO
Long distance movement vectors: NA
Nuisance status and uses: no nuisance issues; no known uses
Endangered status (IUCN): NA

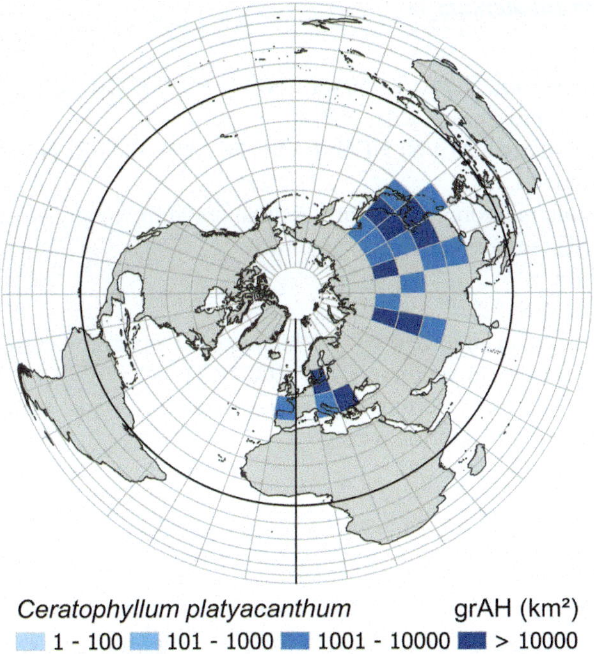

Ceratophyllum platyacanthum Cham.
Image copyright: A. Efremov, with permission

Ceratophyllum submersum **L.**
Synonyms:

Range area (grAH km^2): PAL: 359400; NEO: 63200; OR: 40500; AUS: 15700; AFR: 116000; present in OCE
Ecozone origin-status: PAL: N; NEO: I; OR: N; AUS: I; AFR: N; OCE: I
World rarity status: C
Ecozone-endemic macroregion occurrence: not endemic to any world ecozone
Life form(s): SU (sometimes hydrosubshrub)
Ploidy state: D
Speciation age (Ma): 5.5
Populations also occur in terrestrial, wetland and/ or riparian habitats: NO
Long distance movement vectors: ANTHROPOCHORY; ?AV
Nuisance status and uses: no nuisance issues; aquarium trade, water garden plant, medicinal, phytoremediation uses
Endangered status (IUCN): LC

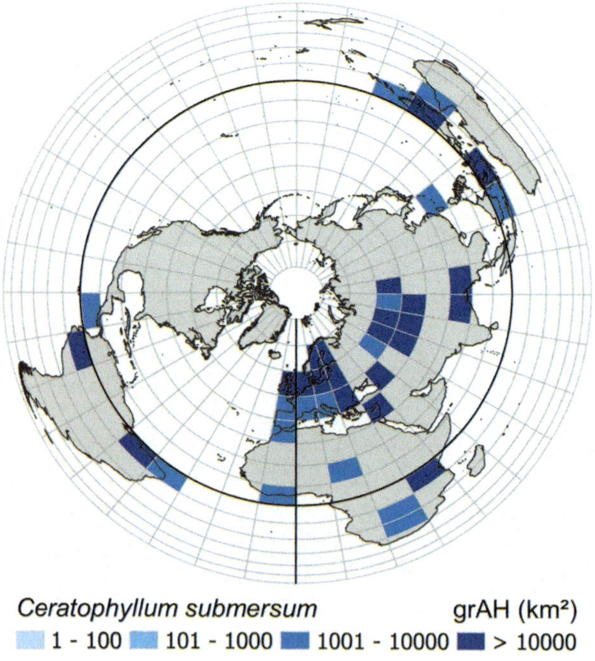

Ceratophyllum submersum L.
Image copyright: J. C. Schou, with permission

3.15 CONVOLVULACEAE (Solanales)

Four freshwater macrophytes within the genus *Ipomoea* are in the Convolvulaceae with a widespread world distribution, though many of the records for these species are from wetland, riparian or terrestrial habitats. All four species occur in OR and NEO, and two species in every ecozone except ANT. NEO is the only ecozone in which two *Ipomoea* species are endemic, though with E(I) origin status. Anthropochorous dispersal has played a major role in the broad distribution of these plants. One or two species in every world ecozone are introduced, while in NEA *Ipomoea aquatica* and *Ipomoea carnea* are considered serious invasive weeds. All four species are common, C, at world scale, so C 100%, two are unassessed for endangered status and two are LC.

***Ipomoea alba* L.**
Synonyms: *Calonyction album* (L.) Hous, *Calonyction roxburghii* G.Don, *Convolvulus bona-nox* Spreng., *Ipomoea grandiflora* Lam.

Range area (grAH km^2): PAL: 9800; NEA: 356900; NEO: 339700; OR: 74800; AUS: 32300; AFR: 228100; present in OCE
Ecozone origin-status: PAL: I; NEA: I; NEO: E(I); OR: I; AUS: I; AFR: I; OCE: I
World rarity status: C
Ecozone-endemic macroregion occurrence: NEO: MG MX, MES AM, NWSA, AND, AMZ, CHA, PAR
Life form(s): EM (climber)

Ploidy state: D
Speciation age (Ma): NA
Populations also occur in terrestrial, wetland and/ or riparian habitats: YES
Long distance movement vectors: ANTHROPOCHORY
Nuisance status and uses: garden plant, medicinal uses
Endangered status (IUCN): LC

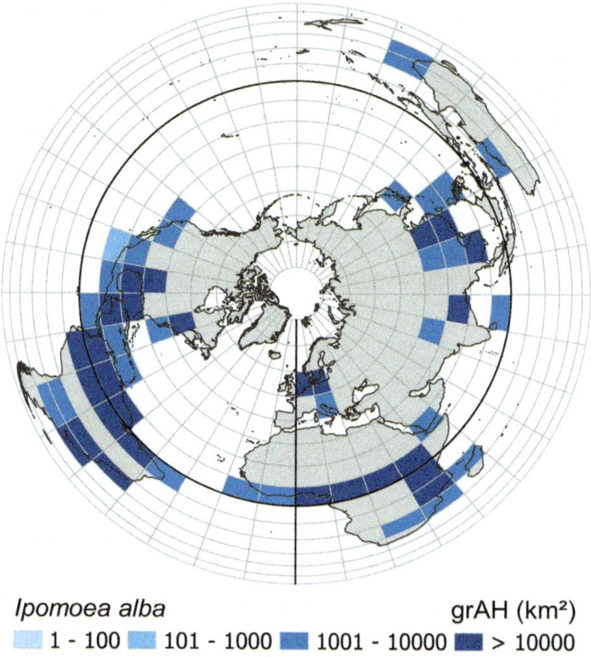

Ipomoea alba L.
Image copyright: iNaturalist CC-BY @maryury

***Ipomoea aquatica* Forssk.**

Synonyms:

Range area (grAH km^2): PAL: 72200; NEA: 45000; NEO: 133600; OR: 147700; AUS: 72500; AFR: 309000; present in OCE

Ecozone origin-status: PAL: I; NEA: I; NEO: I; OR: N; AUS: N; AFR: N; OCE: I

World rarity status: C

Ecozone-endemic macroregion occurrence: not endemic to any world ecozone

Life form(s): EM, FR

Ploidy state: D

Speciation age (Ma): NA

Populations also occur in terrestrial, wetland and/ or riparian habitats: YES

Long distance movement vectors: ANTHROPOCHORY

Nuisance status and uses: major invasive aquatic weed in North America, also causes nuisance problems in parts of its native range, e.g. the Philippines; edible and widely cultivated in warmer regions as water spinach, medicinal and phytoremediation uses

Endangered status (IUCN): LC

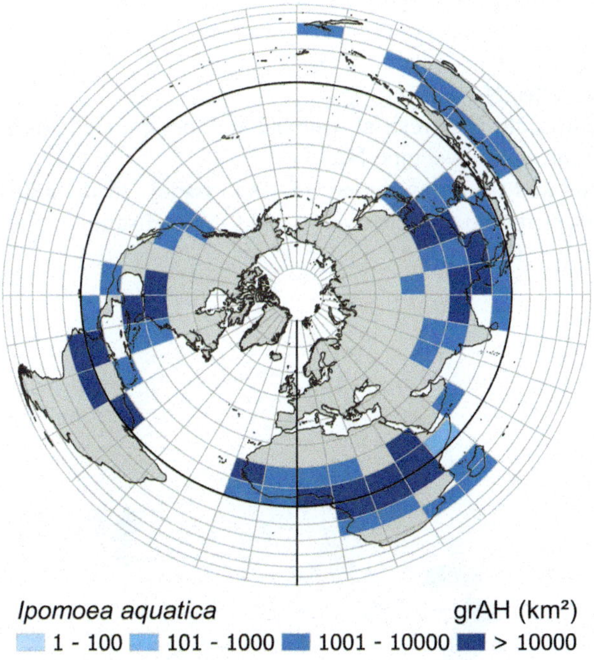

Ipomoea aquatica Forssk.
Image copyright: A. Efremov, with permission

***Ipomoea asarifolia* (Desr.) Roem. & Schult.**
Synonyms: *Amphione asarifolia* (Desr.) Raf., *Convolvulus asarifolius* Desr.

Range area (grAH km^2): NEO: 299900; OR: 26800; AFR: 82200
Ecozone origin-status: NEO: N; OR: N ; AFR: N
World rarity status: C
Ecozone-endemic macroregion occurrence: not endemic to any world ecozone
Life form(s): EM (scrambler or climber; subshrub)
Ploidy state: D
Speciation age (Ma): NA
Populations also occur in terrestrial, wetland and/ or riparian habitats: YES
Long distance movement vectors: NA
Nuisance status and uses: low or no nuisance issues; medicinal uses, mainly in African folk medicine, used in dyeing and for stabilisation of sand dunes
Endangered status (IUCN): NA

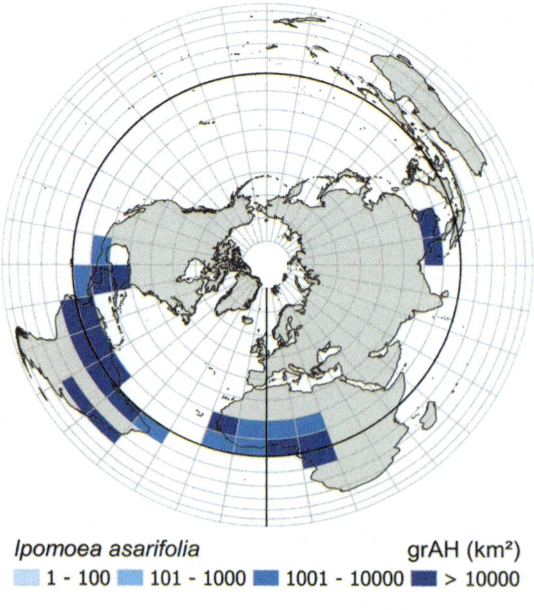

Ipomoea asarifolia (Desr.) Roem. & Schult.

***Ipomoea carnea* Jacq.**
Synonyms: *Convolvulus carneus* (Jacq.) Spreng.

Range area (grAH km^2): NEO: 356700; OR: 84300; AUS: 73400; AFR: 297600; present in OCE
Ecozone origin-status: NEO: E(I); OR: I; AUS: I; AFR: I; OCE: I
World rarity status: C
Ecozone-endemic macroregion occurrence: NEO: MG MX, MES AM, NWSA, AND, AMZ, CHA, PAR
Life form(s): EM (scrambling shrub or small tree)
Ploidy state: D
Speciation age (Ma): NA
Populations also occur in terrestrial, wetland and/ or riparian habitats: YES
Long distance movement vectors: ANTHROPOCHORY
Nuisance status and uses: naturalized or invasive mostly in disturbed sites, riparian areas and wetlands, and behaves as a weed in rice fields, throughout much of its introduced range; harvested from the wild for local use as food and for medicinal uses, and sometimes grown as an ornamental
Endangered status (IUCN): NA

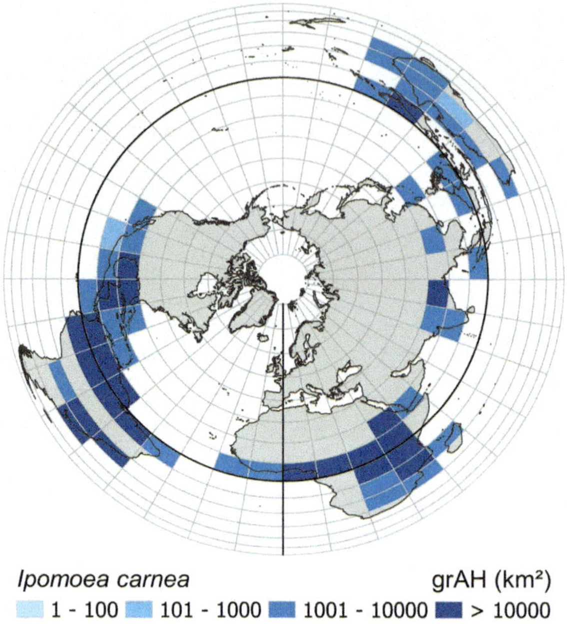

Ipomoea carnea Jacq.
Image copyright: K. Fidanza, with permission

3.16 CRASSULACEAE (Saxifragales)

There are nine macrophyte species within the Crassulaceae, all members of the genus *Crassula*. Between one and three species occur in every ecozone except OCE and ANT, with two endemic species present in each of AFR and AUS. *Crassula helmsii* is introduced in PAL and considered invasive, causing severe weed problems, in Europe. *Crassula natans* is introduced in AUS and *Crassula aquatica* in NEO. *Crassula sinclairii* is classed as VR, four species R and four species C. In terms of endangered status, six species are NA and three LC.

***Crassula aquatica* (L.) Schönland**
Synonyms: *Tillaea aquatica* L., *Hydrophila aquatica* (L.) House, *Tillaeastrum aquaticum* (L.) Britton, *Bulliarda aquatica* (L.) DC.

Range area (grAH km^2): PAL: 348500; NEA: 221500; NEO: 53300; OR: 18100
Ecozone origin-status: PAL: N; NEA: N; NEO: I; OR: N
World rarity status: C
Ecozone-endemic macroregion occurrence: not endemic to any world ecozone
Life form(s): SU, EM
Ploidy state: P
Speciation age (Ma): NA
Populations also occur in terrestrial, wetland and/ or riparian habitats: YES
Long distance movement vectors: ANTHROPOCHORY
Nuisance status and uses: no nuisance issues; folk medicine
Endangered status (IUCN): NA

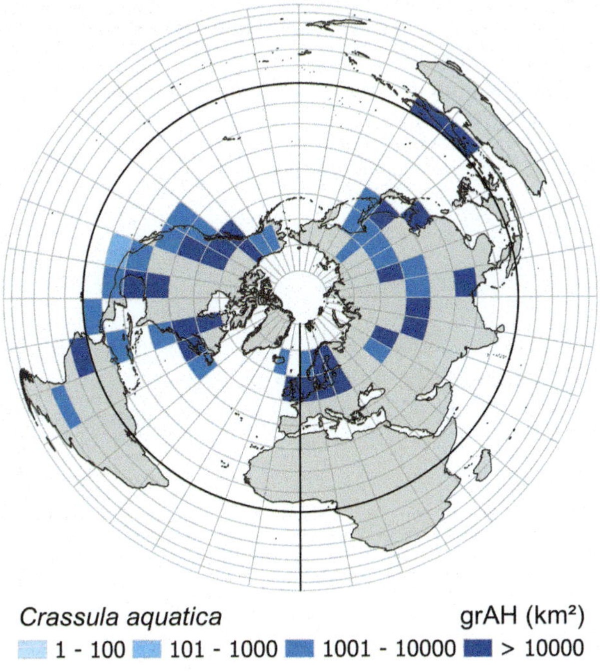

Crassula aquatica (L.) Schönland
Image copyright: J.C. Schou, with permission

Crassula granvikii **Mildbr.**
Synonyms: *Crassula hedbergii* Wickens & M.Bywater, *Tillaea granvikii* (Mildbr.) V.V.Byalt

Range area (grAH km^2): AFR: 199000
Ecozone origin-status: AFR: E
World rarity status: C
Ecozone-endemic macroregion occurrence: AFR: SAH/ AR, FOR, EA, SDES
Life form(s): EM
Ploidy state: NA
Speciation age (Ma): NA
Populations also occur in terrestrial, wetland and/ or riparian habitats: YES
Long distance movement vectors: NA
Nuisance status and uses: no nuisance issues; no known uses
Endangered status (IUCN): LC

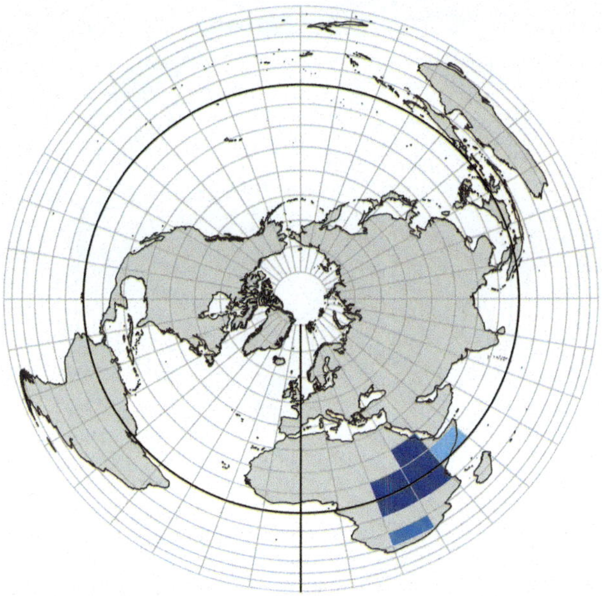

Crassula helmsii **(Kirk) Cockayne**
Synonyms: *Tillaea helmsii* Kirk

Range area (grAH km^2): PAL: 82900; AUS: 43700
Ecozone origin-status: PAL: I; AUS: E(I)
World rarity status: R
Ecozone-endemic macroregion occurrence: AUS: WA, NA, ESA, NZ
Life form(s): SU, EM
Ploidy state: D
Speciation age (Ma): NA
Populations also occur in terrestrial, wetland and/ or riparian habitats: YES
Long distance movement vectors: ANTHROPOCHORY
Nuisance status and uses: invasive aquatic weed in the UK and other parts of northern Europe; no known use other than occasional use as a water garden plant or in the aquarium trade, sale of the species has been banned since 2014 in the UK owing to the high risk of escape and further spread
Endangered status (IUCN): NA

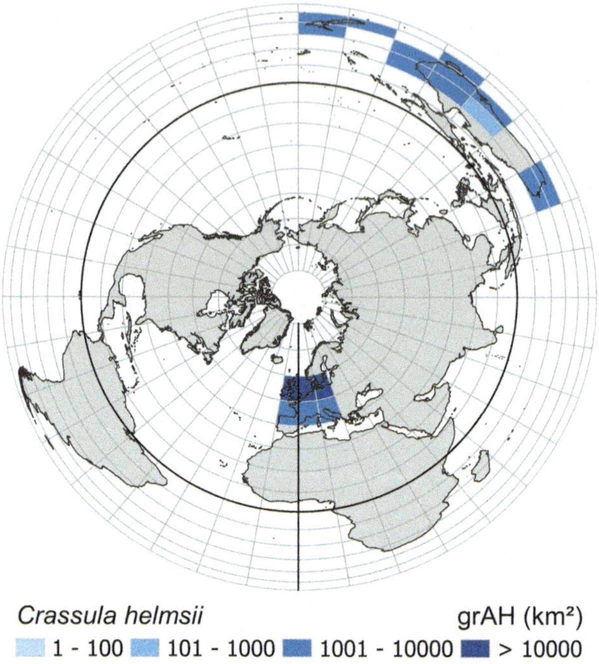

Crassula helmsii (Kirk) Cockayne
Image copyright: J.C. Schou, with permission

***Crassula natans* Thunb.**
Synonyms:

Range area (grAH km²): AUS: 25400; AFR: 12700
Ecozone origin-status: AUS: I; AFR: E(I)
World rarity status: R
Ecozone-endemic macroregion occurrence: AFR: SDES, ZAM, CAPE
Life form(s): FR, EM
Ploidy state: NA
Speciation age (Ma): NA
Populations also occur in terrestrial, wetland and/ or riparian habitats: YES
Long distance movement vectors: ANTHROPOCHORY
Nuisance status and uses: no nuisance issues; no known uses
Endangered status (IUCN): LC

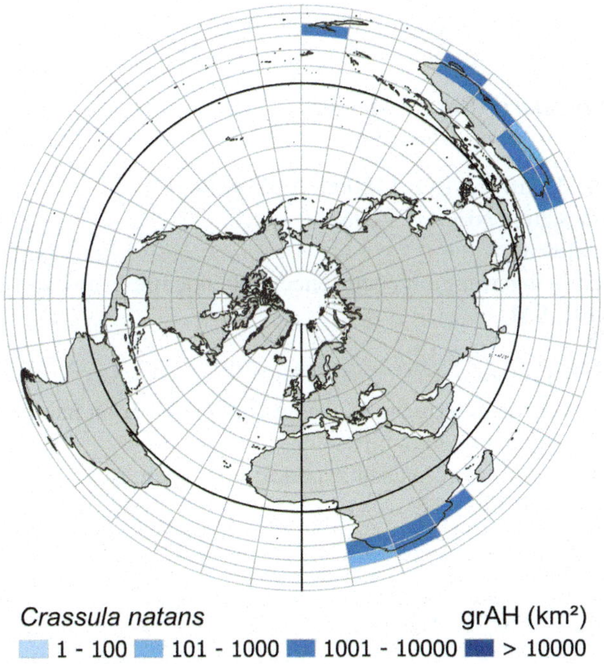

Crassula natans Thunb.
Image copyright: iNaturalist CC0 @klauswehrlin

***Crassula peduncularis* (Sm.) F.Meigen**
Synonyms: *Tillaea peduncularis* Sm.

Range area (grAH km^2): NEO: 148000; AUS 49700
Ecozone origin-status: NEO: N; AUS: N
World rarity status: C
Ecozone-endemic macroregion occurrence: not endemic to any world ecozone
Life form(s): FR, EM
Ploidy state: P
Speciation age (Ma): NA
Populations also occur in terrestrial, wetland and/ or riparian habitats: YES
Long distance movement vectors: NA
Nuisance status and uses: no nuisance issues; no known uses
Endangered status (IUCN): NA

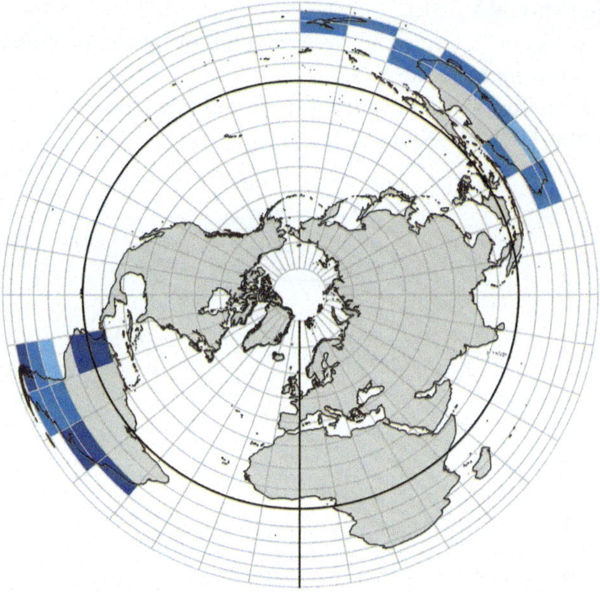

Crassula peduncularis (Sm.) F.Meigen
Image copyright: K. Ward, with permission

***Crassula sinclairii* (Hook.f.) A.P.Druce et D.R.Given**
Synonyms: *Tillaea sinclairii* Hook.f.

Range area (grAH km^2): AUS: 16900
Ecozone origin-status: AUS: E
World rarity status: VR
Ecozone-endemic macroregion occurrence: AUS: NZ
Life form(s): SU, EM
Ploidy state: DP

Speciation age (Ma): NA
Populations also occur in terrestrial, wetland and/ or riparian habitats: YES
Long distance movement vectors: NA
Nuisance status and uses: no nuisance issues; no known uses
Endangered status (IUCN): NA

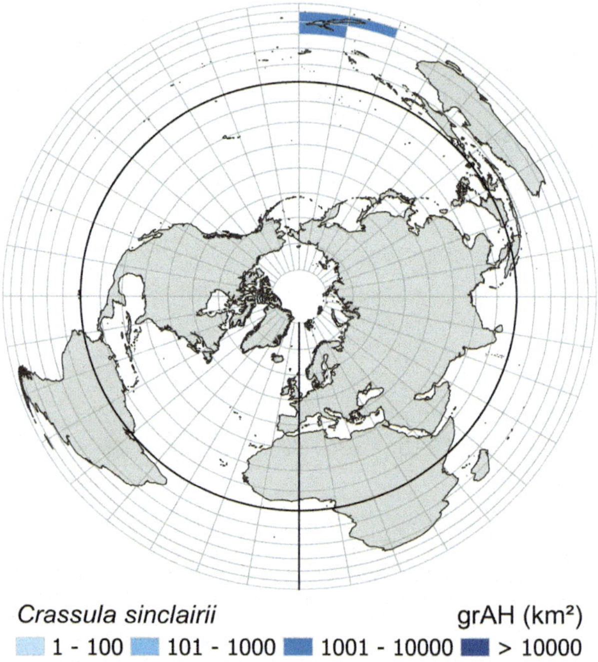

Crassula sinclairii (Hook.f.) A.P.Druce et D.R.Given

***Crassula vaillantii* (Willd.) Roth**

Synonyms: *Bulliarda vaillantii* (Willd.) DC., *Hydrophila vaillantii* (Willd.) House, *Tillaea vaillantii* Willd., *Tillaeastrum vaillantii* (Willd.) Britton

Range area (grAH km^2): PAL: 146700; AFR: 12700
Ecozone origin-status: PAL: N; AFR: N
World rarity status: C
Ecozone-endemic macroregion occurrence: not endemic to any world ecozone
Life form(s): SU, EM
Ploidy state: P
Speciation age (Ma): NA
Populations also occur in terrestrial, wetland and/ or riparian habitats: YES
Long distance movement vectors: NA
Nuisance status and uses: no nuisance issues; no known uses
Endangered status (IUCN): LC (Mediterranean)

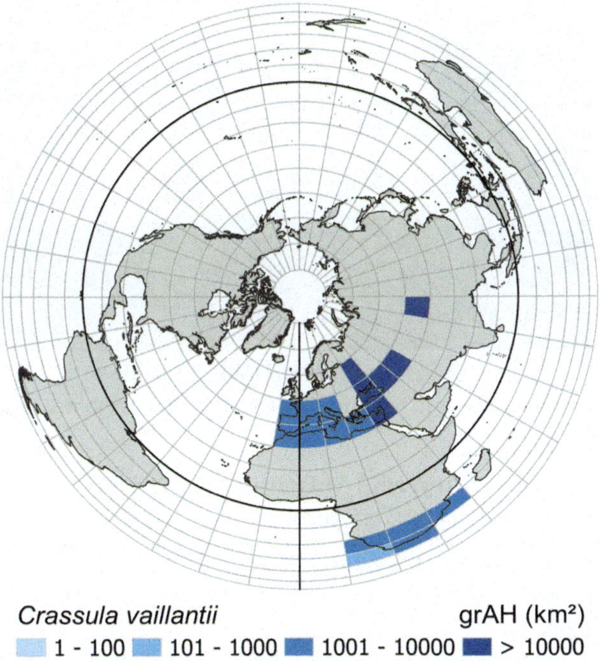

Crassula vaillantii (Willd.) Roth

***Crassula venezuelensis* (Steyerm.) M.Bywater & Wickens**
Synonyms: *Tillaea venezuelensis* Steyerm.

Range area (grAH km^2): NEO: 89800
Ecozone origin-status: NEO: E
World rarity status: R
Ecozone-endemic macroregion occurrence: NEO: NWSA, AND, CHA
Life form(s): SU, EM
Ploidy state: NA
Speciation age (Ma): NA
Populations also occur in terrestrial, wetland and/ or riparian habitats: YES
Long distance movement vectors: NA
Nuisance status and uses: no nuisance issues; no known uses
Endangered status (IUCN): NA

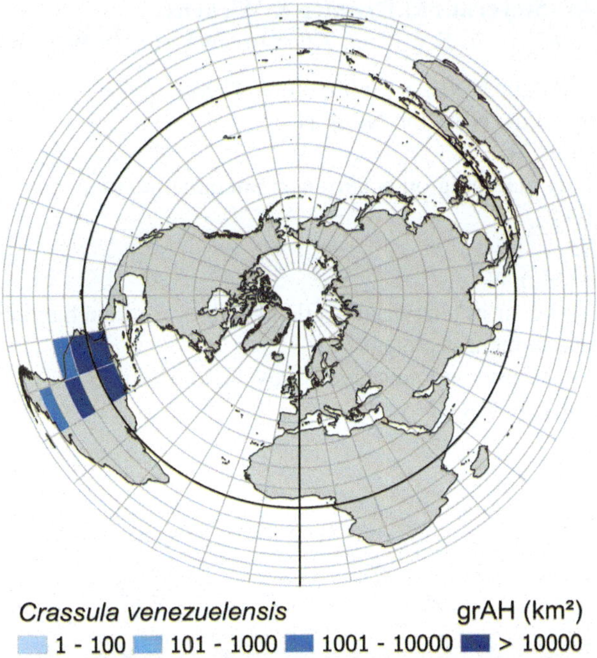

Crassula venezuelensis (Steyerm.) M.Bywater & Wickens

***Crassula viridis* (Steyerm.) M.Bywater & Wickens**
Synonyms: *Tillaea viridis* S.Watson, *Tillaeastrum viride* (S.Watson) Britton

Range area (grAH km^2): NEA: 13600; NEO: 41900
Ecozone origin-status: NEA: N; NEO: N
World rarity status: R
Ecozone-endemic macroregion occurrence: not endemic to any world ecozone
Life form(s): EM
Ploidy state: NA
Speciation age (Ma): NA
Populations also occur in terrestrial, wetland and/ or riparian habitats: YES
Long distance movement vectors: NA
Nuisance status and uses: no nuisance issues; no known uses
Endangered status (IUCN): NA

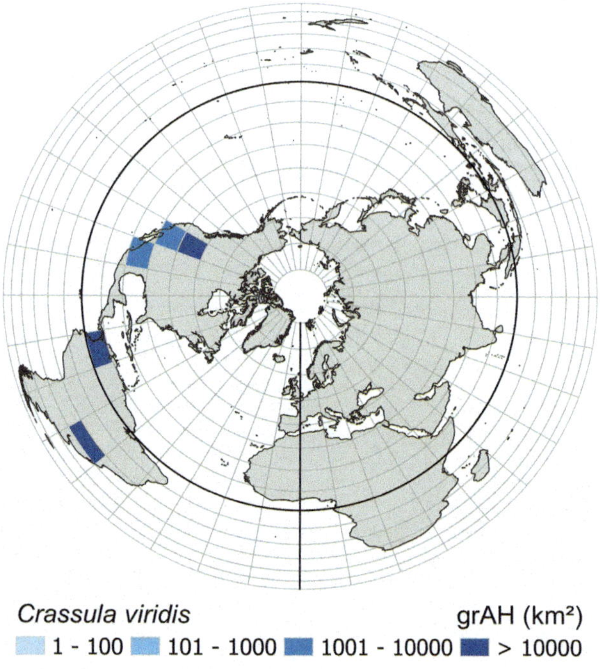

3.17 DROSERACEAE (Caryophyllales)

There is one macrophyte species in the Droseraceae, *Aldrovanda vesiculosa*, an Old World species native in PAL, OR, AUS and AFR, and introduced in NEA, endangered (EN) at the global scale.

***Aldrovanda vesiculosa* L.**
Synonyms: *Drosera aldrovanda* F.Muell.

Range area (grAH km^2): PAL: 308200; NEA: 64500; OR: 18000; AUS: 23800; AFR: 90700
Ecozone origin-status: PAL: N; NEA: I; OR: N (rediscovered after 60 years in India, in 2019: [107]); AUS: N; AFR: N
World rarity status: C
Ecozone-endemic macroregion occurrence: not endemic to any world ecozone
Life form(s): SU
Ploidy state: DP
Speciation age (Ma): 1.6
Populations also occur in terrestrial, wetland and/ or riparian habitats: NO
Long distance movement vectors: ANTHROPOCHORY; BIRD: Waterfowl (species not named)
Nuisance status and uses: no nuisance issues; limited specialist use in aquarium trade
Endangered status (IUCN): EN

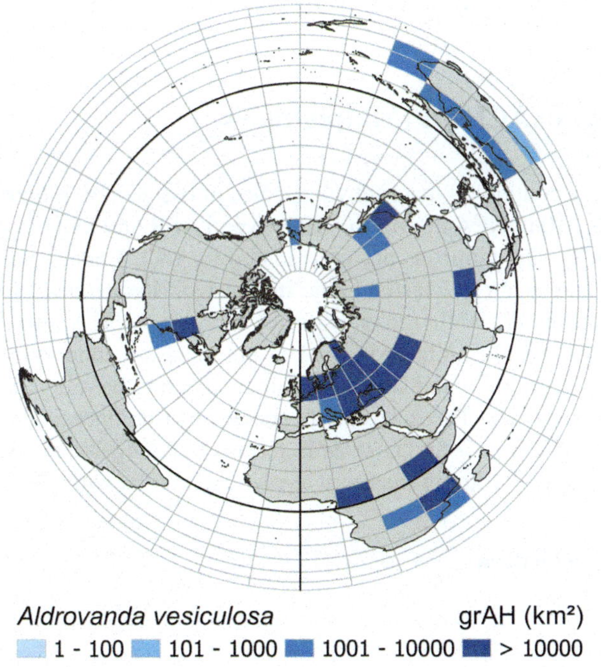

Aldrovanda vesiculosa L.
Image copyright: A. Efremov, with permission

3.18 ELATINACEAE (Malpighiales)

The Elatinaceae contain 29 macrophyte species in two genera. There are 12 species present in PAL, 10 in NEA and 10 in NEO, five in AFR and five in OR, four in AUS and one in OCE. The family is not represented in ANT. Endemism is strong, with between two and seven species endemic in five world ecozones. No endemic species are present in OR, while the one species found in OCE is introduced. None of the 29 species is introduced in PAL or AUS, but there are either one or two I-origin species in AFR, OR, NEA and NEO. No aquatic weed issues are reported for any of the macrophyte species. The macrophyte species of Elatinaceae are mainly world-rare, with VR 24%, R 48% and C 28%. In terms of endangered status, 30% are NA/DD, 52% LC, and 17% are in the more at-risk categories, with two species CR.

Quick Links to Genera
1. *Bergia*
2. *Elatine*

Bergia ammannioides **Roxb. ex Roth**
Synonyms: *Lechea verticillata* Willd.

Range area (grAH km^2): PAL: 25600; OR: 89800; AUS: 29300; AFR: 223000
Ecozone origin-status: PAL: N; OR: N; AUS: N; AFR: N
World rarity status: C

Ecozone-endemic macroregion occurrence: not endemic to any world ecozone
Life form(s): EM (sometimes a subshrub)
Ploidy state: DP
Speciation age (Ma): NA
Populations also occur in terrestrial, wetland and/ or riparian habitats: YES
Long distance movement vectors: NA
Nuisance status and uses: no nuisance issues; some medicinal uses
Endangered status (IUCN): LC

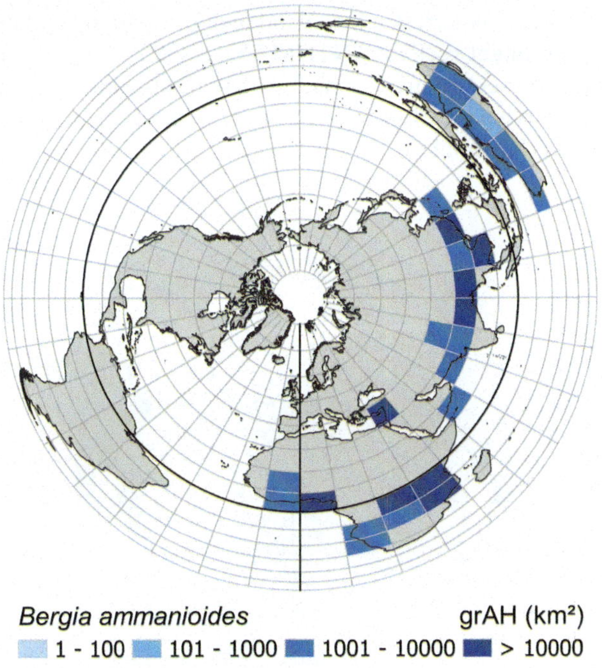

Bergia ammannioides Roxb. ex Roth
Image copyright: iNaturalist CC0 @radhaveach (Radha Veach)

***Bergia capensis* L.**
Synonyms:

Range area (grAH km²): PAL: 5200; NEO: 7600; OR: 65700; AFR: 41800
Ecozone origin-status: PAL: N; NEO: I; OR: N; AFR: N
World rarity status: C
Ecozone-endemic macroregion occurrence: not endemic to any world ecozone
Life form(s): EM
Ploidy state: D
Speciation age (Ma): NA
Populations also occur in terrestrial, wetland and/ or riparian habitats: YES
Long distance movement vectors: ANTHROPOCHORY
Nuisance status and uses: often associated with rice cultivation in both its native and introduced ranges but only low or no nuisance issues are reported; medicinal uses
Endangered status (IUCN): LC

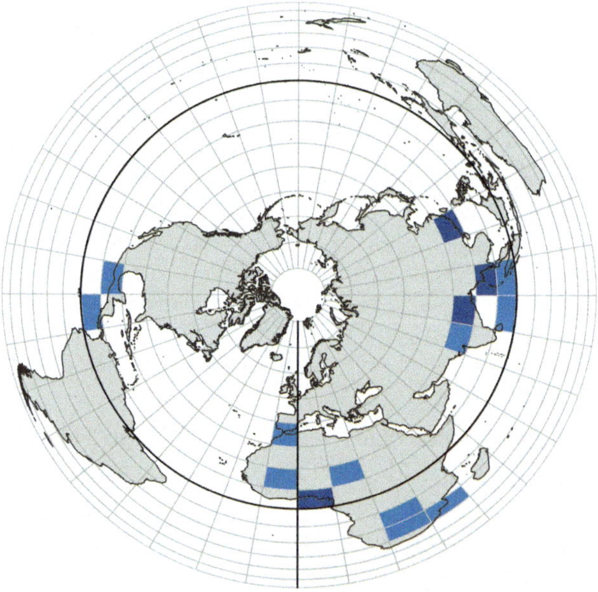

***Bergia texana* (Hook.) Seub.**
Synonyms: *Elatine texana* (Hook.) Torr. & A.Gray, *Merimea texana* Hook.

Range area (grAH km²): NEA: 40600; NEO: 22900
Ecozone origin-status: NEA: N; NEO: N
World rarity status: C
Ecozone-endemic macroregion occurrence: not endemic to any world ecozone
Life form(s): EM (sometimes a subshrub)

Ploidy state: NA
Speciation age (Ma): NA
Populations also occur in terrestrial, wetland and/ or riparian habitats: YES
Long distance movement vectors: HYDROCHORY (NEA to NEO)
Nuisance status and uses: no nuisance issues; no known uses
Endangered status (IUCN): NA

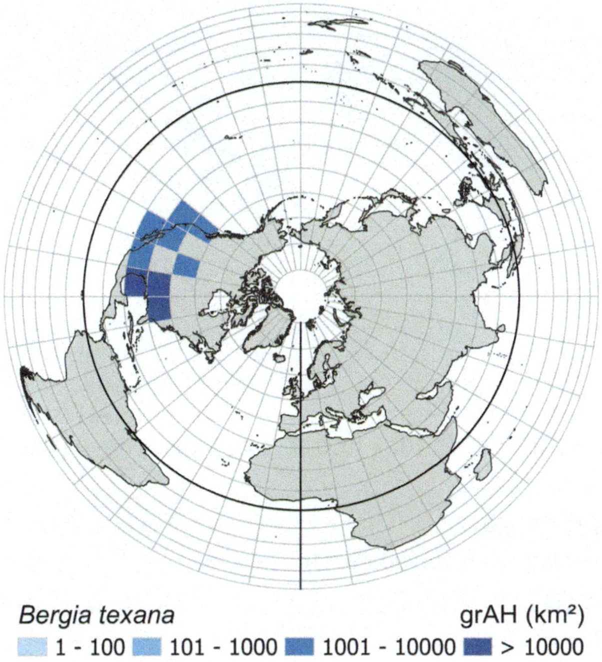

***Elatine alsinastrum* L.**

Synonyms: *Alsine alsinastrum* (L.) Crantz, *Elatinella alsinastrum* (L.) Opiz, *Potamopitys alsinastrum* (L.) Fourr.

Range area (grAH km^2): PAL: 288300
Ecozone origin-status: PAL: E
World rarity status: C
Ecozone-endemic macroregion occurrence: PAL: N EUR, W EUR, E EUR, MED, UR/ W SIB, C ASIA
Life form(s): EM, SU
Ploidy state: P
Speciation age (Ma): NA
Populations also occur in terrestrial, wetland and/ or riparian habitats: YES
Long distance movement vectors: NA
Nuisance status and uses: no nuisance issues; aquarium trade
Endangered status (IUCN): NT

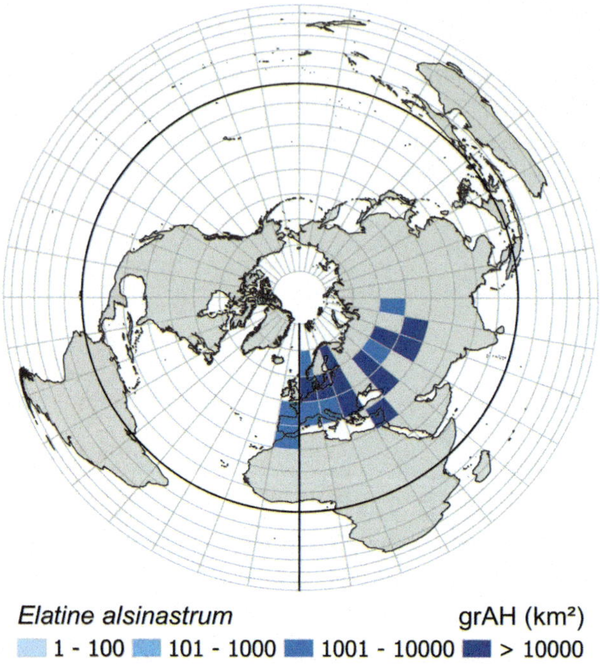

Elatine ambigua **Wight**
Synonyms: *Potamopitys ambigua* (Wight) Kuntze

Range area (grAH km^2): PAL: 46900; NEA: 7500; OR: 81600; AFR: 20500; present in OCE
Ecozone origin-status: PAL: N; NEA: I; OR: N; AFR: I; OCE: I
World rarity status: R
Ecozone-endemic macroregion occurrence: not endemic to any world ecozone
Life form(s): EM, SU
Ploidy state: P
Speciation age (Ma): NA
Populations also occur in terrestrial, wetland and/ or riparian habitats: YES
Long distance movement vectors: ANTHROPOCHORY
Nuisance status and uses: no nuisance issues; aquarium trade
Endangered status (IUCN): LC

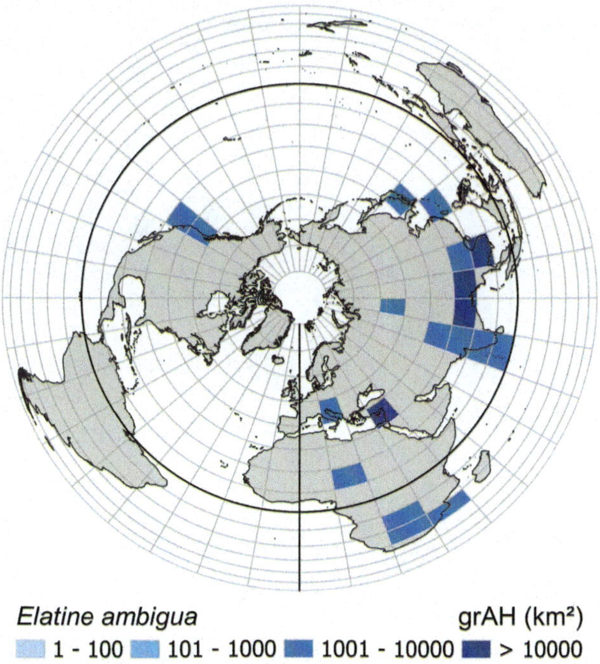

Elatine americana (Pursh) Arn.
Synonyms: *Alsinastrum americanum* (Pursh) Torr. & A.Gray ex Rupr., *Peplis americana* Pursh

Range area (grAH km^2): PAL: 17322; NEA: 392400; OR: 17900
Ecozone origin-status: PAL: N; NEA: N; OR: I
World rarity status: C
Ecozone-endemic macroregion occurrence: not endemic to any world ecozone
Life form(s): EM, SU
Ploidy state: DP
Speciation age (Ma): NA
Populations also occur in terrestrial, wetland and/ or riparian habitats: YES
Long distance movement vectors: ANTHROPOCHORY
Nuisance status and uses: no nuisance issues; no known uses
Endangered status (IUCN): LC

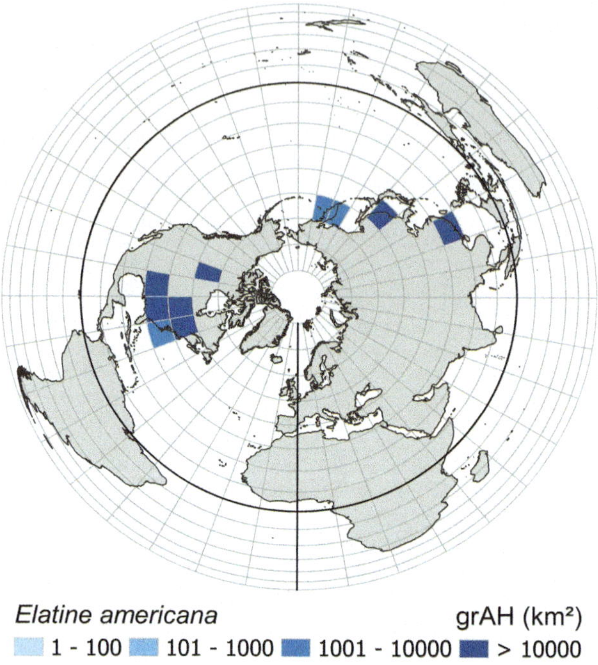

Elatine americana (Pursh) Arn.
Image copyright: iNaturalist CC0 @allan7 (Allan Harris)

***Elatine brachysperma* A.Gray**

Synonyms: *Potamopitys brachysperma* (A.Gray) Kuntze, *Alsinastrum brachyspermum* (A.Gray) Greene

Range area (grAH km^2): NEA: 38300 ; NEO: 11500
Ecozone origin-status: NEA: N; NEO: N
World rarity status: R
Ecozone-endemic macroregion occurrence: not endemic to any world ecozone
Life form(s): EM, SU
Ploidy state: NA
Speciation age (Ma): NA
Populations also occur in terrestrial, wetland and/ or riparian habitats: YES
Long distance movement vectors: NA
Nuisance status and uses: no nuisance issues; no known uses
Endangered status (IUCN): LC

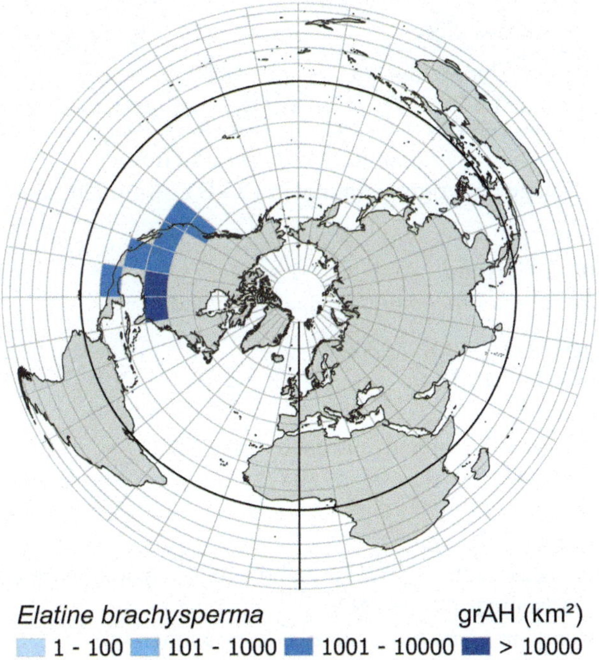

Elatine brachysperma A.Gray

Elatine brochonii Clavaud
Synonyms: *Elatinopsis brochonii* (Clavaud) Clavaud

Range area (grAH km^2): PAL: 19300
Ecozone origin-status: PAL: E
World rarity status: VR
Ecozone-endemic macroregion occurrence: PAL: MED
Life form(s): EM, SU
Ploidy state: NA
Speciation age (Ma): NA
Populations also occur in terrestrial, wetland and/ or riparian habitats: YES
Long distance movement vectors: NA
Nuisance status and uses: no nuisance issues; no known uses
Endangered status (IUCN): VU

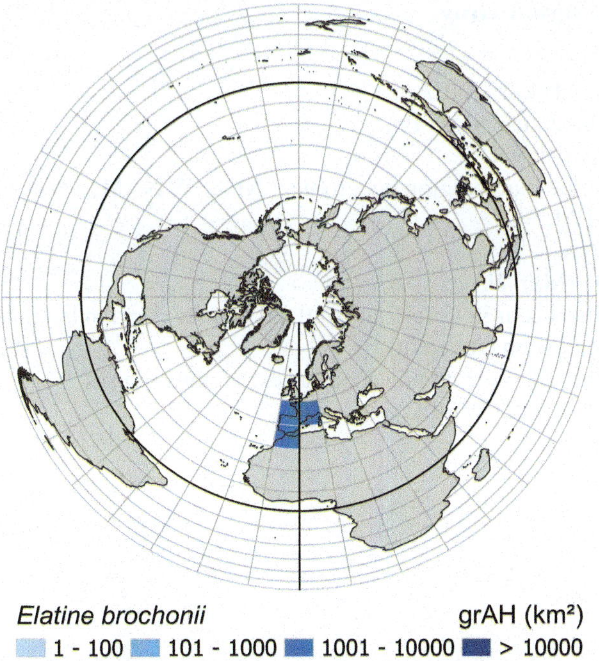

Elatine brochonii Clavaud

***Elatine californica* A.Gray**

Synonyms: *Potamopitys californica* (A.Gray) Kuntze

Range area (grAH km^2): NEA: 21100
Ecozone origin-status: NEA: E
World rarity status: VR
Ecozone-endemic macroregion occurrence: NEA: W NAM, SW NAM
Life form(s): EM, SU
Ploidy state: NA
Speciation age (Ma): NA
Populations also occur in terrestrial, wetland and/ or riparian habitats: YES
Long distance movement vectors: NA
Nuisance status and uses: no nuisance issues; no known uses
Endangered status (IUCN): LC

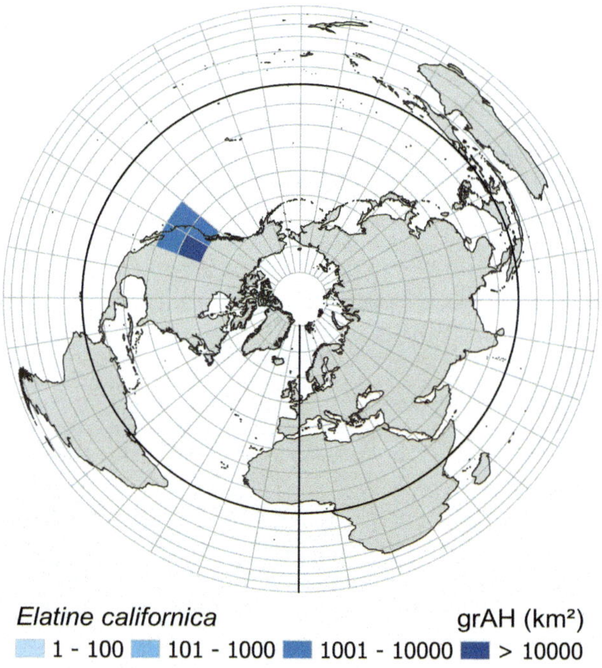

Elatine californica A.Gray
Image copyright: J. Bailey, with permission

***Elatine ecuadoriensis* Malau**
Synonyms:

Range area (grAH km^2): NEO: 34100
Ecozone origin-status: NEO: E
World rarity status: R
Ecozone-endemic macroregion occurrence: NEO: NWSA
Life form(s): EM, SU
Ploidy state: NA
Speciation age (Ma): NA
Populations also occur in terrestrial, wetland and/ or riparian habitats: YES
Long distance movement vectors: NA
Nuisance status and uses: no nuisance issues; no known uses
Endangered status (IUCN): LC

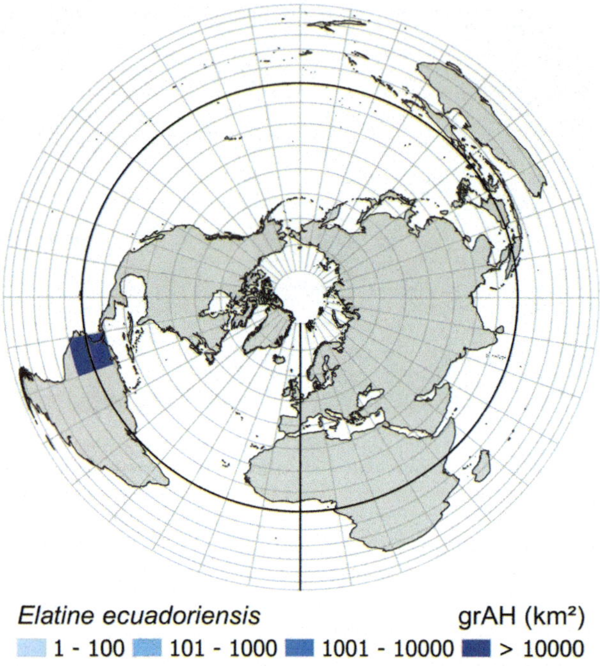

Elatine fassettiana **Steyerm.**
Synonyms:

Range area (grAH km^2): NEO: 16600
Ecozone origin-status: NEO: E
World rarity status: VR
Ecozone-endemic macroregion occurrence: NEO: NWSA
Life form(s): EM, SU
Ploidy state: NA
Speciation age (Ma): NA
Populations also occur in terrestrial, wetland and/ or riparian habitats: YES
Long distance movement vectors: NA
Nuisance status and uses: no nuisance issues; no known uses
Endangered status (IUCN): LC

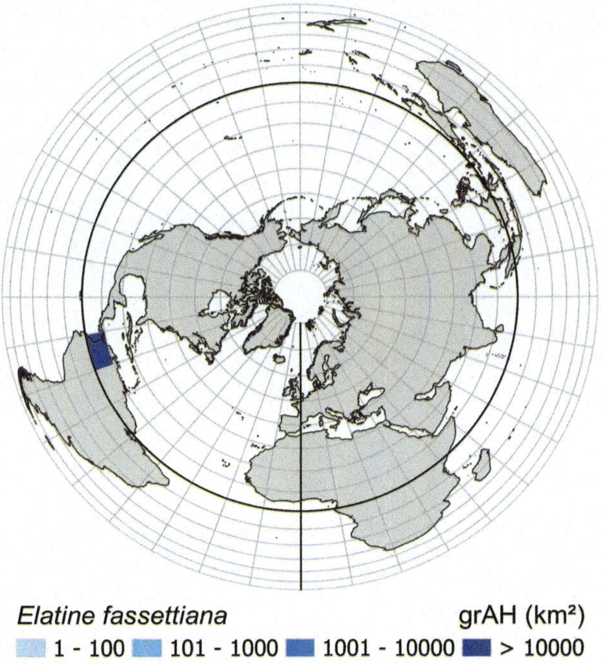

Elatine fauquei **Monod**
Synonyms:

Range area (grAH km^2): AFR: 7800
Ecozone origin-status: AFR: E
World rarity status: VR
Ecozone-endemic macroregion occurrence: AFR: SAH/ AR
Life form(s): EM, SU
Ploidy state: NA
Speciation age (Ma): NA
Populations also occur in terrestrial, wetland and/ or riparian habitats: YES
Long distance movement vectors: NA
Nuisance status and uses: no nuisance issues; no known uses
Endangered status (IUCN): CR

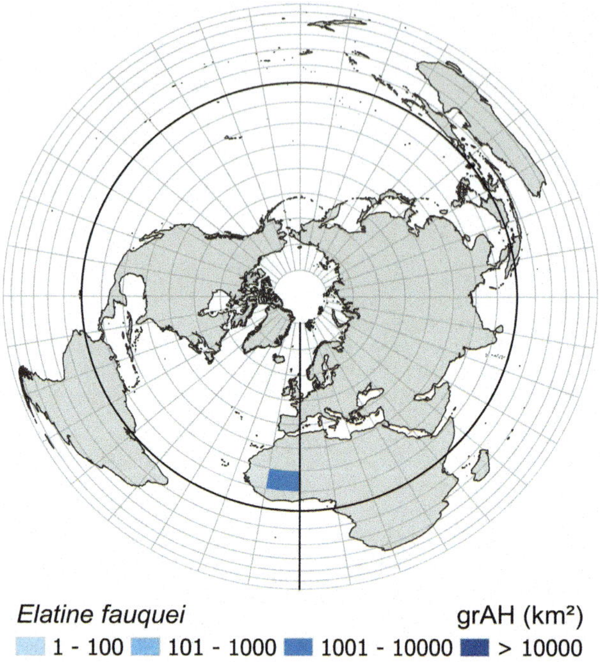

Elatine gratioloides **A.Cunn.**
Synonyms:

Range area (grAH km^2): AUS: 67600
Ecozone origin-status: AUS: E
World rarity status: R
Ecozone-endemic macroregion occurrence: AUS: NG, NA, WA, ESA, NZ
Life form(s): EM, SU
Ploidy state: D
Speciation age (Ma): NA
Populations also occur in terrestrial, wetland and/ or riparian habitats: YES
Long distance movement vectors: NA
Nuisance status and uses: no nuisance issues; aquarium trade
Endangered status (IUCN): NA

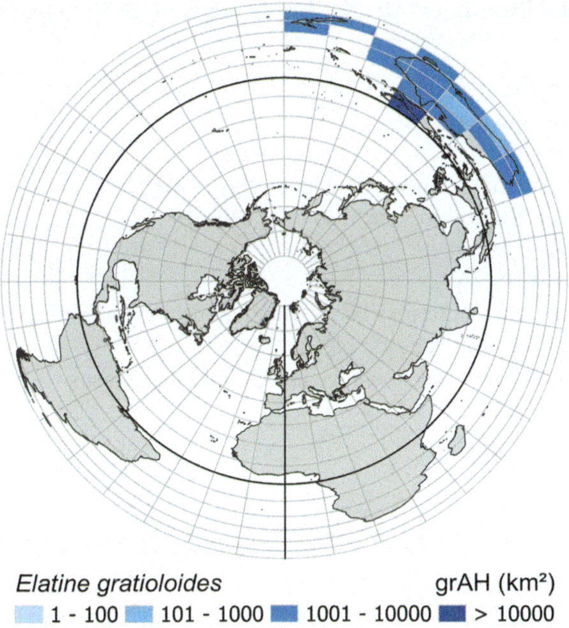

Elatine gratioloides A.Cunn.

***Elatine gussonei* (Sommier) Brullo, Lanfr., Pavone & Ronsisv.**
Synonyms: *Elatine hydropiper* var. *gussonei* Sommier

Range area (grAH km^2): PAL: 33700
Ecozone origin-status: PAL: E
World rarity status: R
Ecozone-endemic macroregion occurrence: PAL: MED, W EUR
Life form(s): EM, SU (hydrosubshrub or subshrub)
Ploidy state: NA
Speciation age (Ma): NA
Populations also occur in terrestrial, wetland and/ or riparian habitats: YES
Long distance movement vectors: NA
Nuisance status and uses: no nuisance issues; no known uses
Endangered status (IUCN): LC

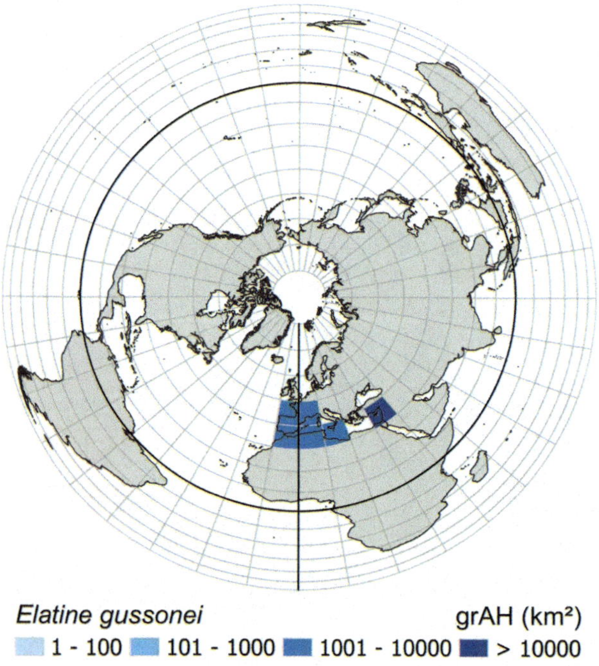

Elatine gussonei (Sommier) Brullo, Lanfr., Pavone & Ronsisv.
Image copyright: Ron Frumkin, with permission

Elatine heterandra **H.Mason**
Synonyms:

Range area (grAH km^2): NEA: 4500
Ecozone origin-status: NEA: E
World rarity status: VR
Ecozone-endemic macroregion occurrence: NEA: W NAM, SW NAM
Life form(s): EM, SU
Ploidy state: NA
Speciation age (Ma): NA
Populations also occur in terrestrial, wetland and/ or riparian habitats: YES
Long distance movement vectors: NA
Nuisance status and uses: no nuisance issues; no known uses
Endangered status (IUCN): NA

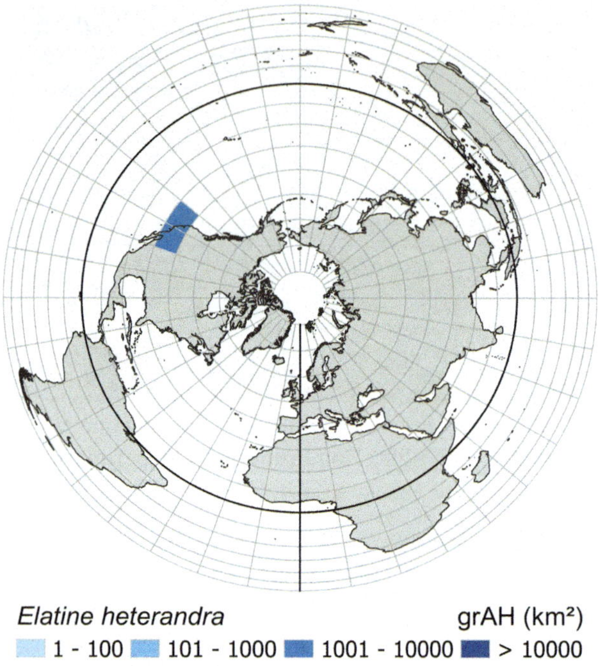

Elatine hexandra **H.Mason**
Synonyms: *Tillaea hexandra* Lapierre, *Elatine hydropiper* var. *hexandra* (DC.) Gaudin

Range area (grAH km^2): PAL: 128000
Ecozone origin-status: PAL: E
World rarity status: R
Ecozone-endemic macroregion occurrence: PAL: MED, W EUR, N EUR, E EUR
Life form(s): EM, SU
Ploidy state: P
Speciation age (Ma): 36.0
Populations also occur in terrestrial, wetland and/ or riparian habitats: YES
Long distance movement vectors: ?AV
Nuisance status and uses: no nuisance issues; aquarium trade
Endangered status (IUCN): LC

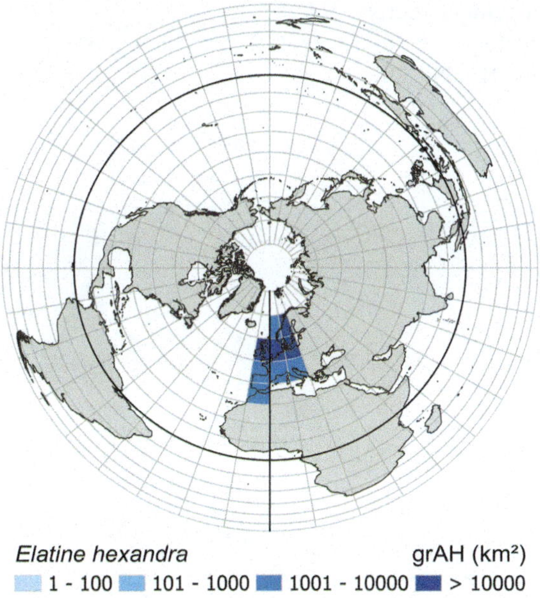

Elatine hexandra H.Mason
Image copyright: J.C. Schou, with permission

***Elatine hungarica* Moesz**
Synonyms:

Range area (grAH km^2): PAL: 83400
Ecozone origin-status: PAL: E
World rarity status: R
Ecozone-endemic macroregion occurrence: PAL: W EUR, E EUR, UR/ W SIB, C ASIA
Life form(s): EM, SU
Ploidy state: NA
Speciation age (Ma): NA
Populations also occur in terrestrial, wetland and/ or riparian habitats: YES

Long distance movement vectors: NA
Nuisance status and uses: no nuisance issues; no known uses
Endangered status (IUCN): DD

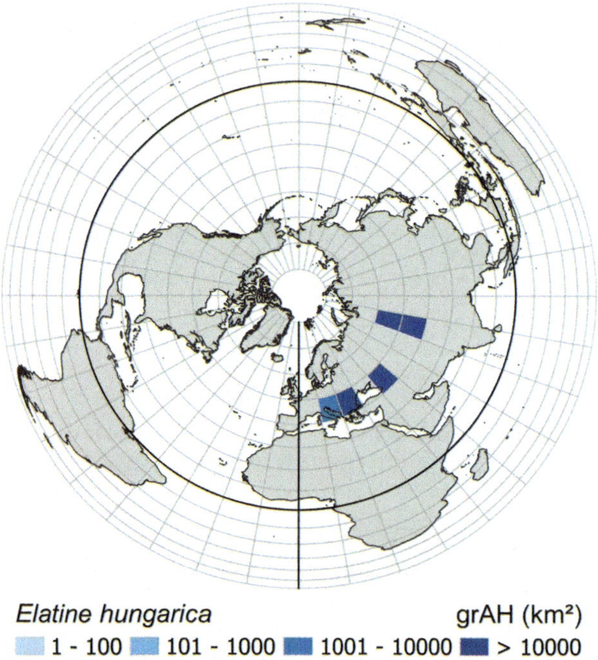

Elatine hungarica Moesz
Image copyright: A. Efremov, with permission

***Elatine hydropiper* L.**
Synonyms: *Elatinella hydropiper* (L.) Opiz, *Potamopitys hydropiper* (L.) Kuntze

Range area (grAH km^2): PAL: 551800
Ecozone origin-status: PAL: E

World rarity status: C
Ecozone-endemic macroregion occurrence: PAL: N EUR, W EUR, E EUR, MED, UR/ W SIB, C ASIA, E SIB, FER/J
Life form(s): EM, SU
Ploidy state: P
Speciation age (Ma): 36.0
Populations also occur in terrestrial, wetland and/ or riparian habitats: YES
Long distance movement vectors: BIRD: *Anas crecca* (Eurasian Teal); ?AV
Nuisance status and uses: no nuisance issues; aquarium trade
Endangered status (IUCN): LC

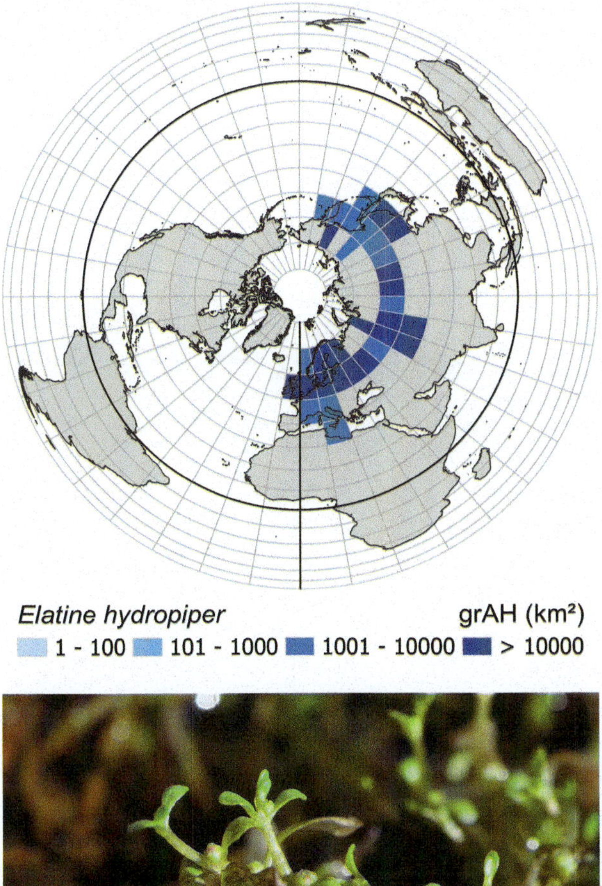

Elatine hydropiper L.
Image copyright: J.C. Schou, with permission

Elatine lindbergii **Rohrb.**
Synonyms:

Range area (grAH km^2): NEO: 31400
Ecozone origin-status: NEO: E
World rarity status: R
Ecozone-endemic macroregion occurrence: NEO: PAR
Life form(s): EM, SU
Ploidy state: NA
Speciation age (Ma): NA
Populations also occur in terrestrial, wetland and/ or riparian habitats: YES
Long distance movement vectors: NA
Nuisance status and uses: no nuisance issues; no known uses
Endangered status (IUCN): NA

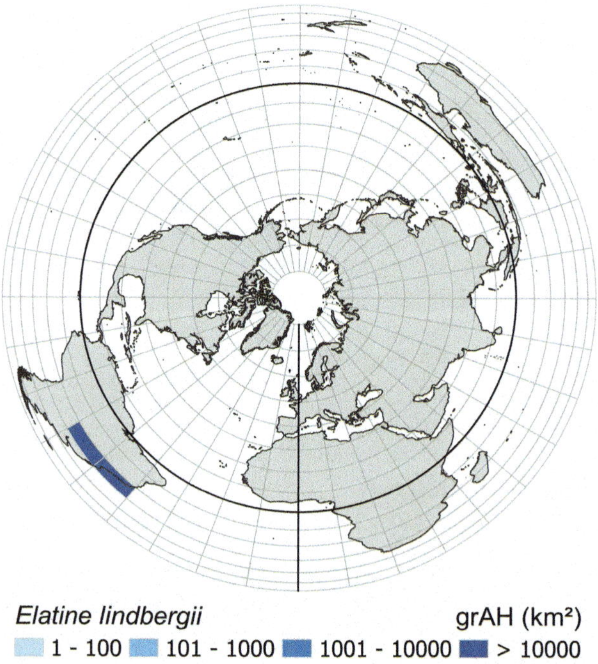

Elatine lorentziana **Hunz.**
Synonyms:

Range area (grAH km^2): NEO: 45400
Ecozone origin-status: NEO: E
World rarity status: R
Ecozone-endemic macroregion occurrence: NEO: CHA, PAR
Life form(s): EM, SU

Ploidy state: NA
Speciation age (Ma): NA
Populations also occur in terrestrial, wetland and/ or riparian habitats: YES
Long distance movement vectors: NA
Nuisance status and uses: no nuisance issues; no known uses
Endangered status (IUCN): NA

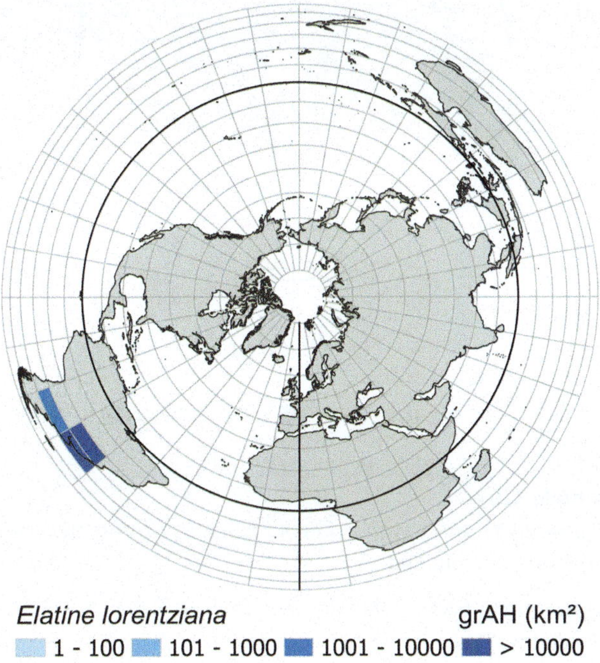

***Elatine macrocalyx* Albr.**
Synonyms:

Range area (grAH km^2): AUS: 7100
Ecozone origin-status: AUS: E
World rarity status: VR
Ecozone-endemic macroregion occurrence: AUS: WA, NA
Life form(s): EM, SU
Ploidy state: NA
Speciation age (Ma): NA
Populations also occur in terrestrial, wetland and/ or riparian habitats: YES
Long distance movement vectors: NA
Nuisance status and uses: no nuisance issues; no known uses
Endangered status (IUCN): NA

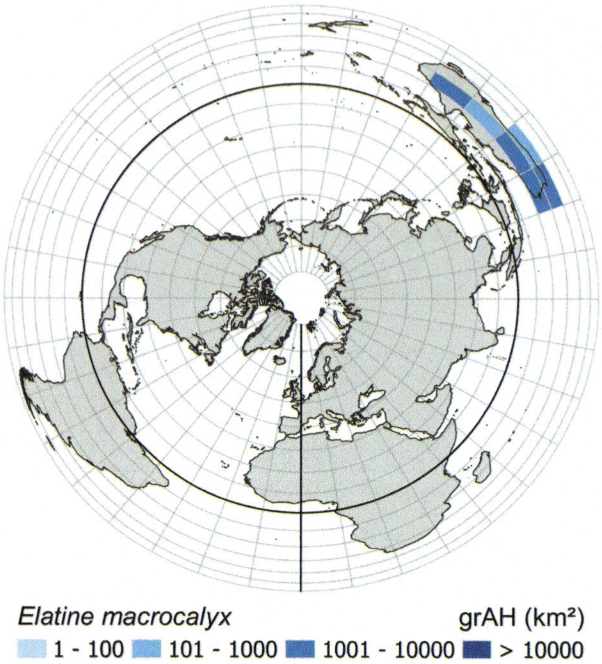

***Elatine macropoda* Guss.**
Synonyms: *Potamopitys macropoda* (Guss.) Kuntze, *Elatine hexandra* subsp. *macropoda* (Guss.) Rouy & Foucaud

Range area (grAH km^2): PAL: 31800
Ecozone origin-status: PAL: E
World rarity status: R
Ecozone-endemic macroregion occurrence: PAL: MED, W EUR
Life form(s): EM, SU
Ploidy state: D
Speciation age (Ma): NA
Populations also occur in terrestrial, wetland and/ or riparian habitats: YES
Long distance movement vectors: NA
Nuisance status and uses: no nuisance issues; no known uses
Endangered status (IUCN): LC

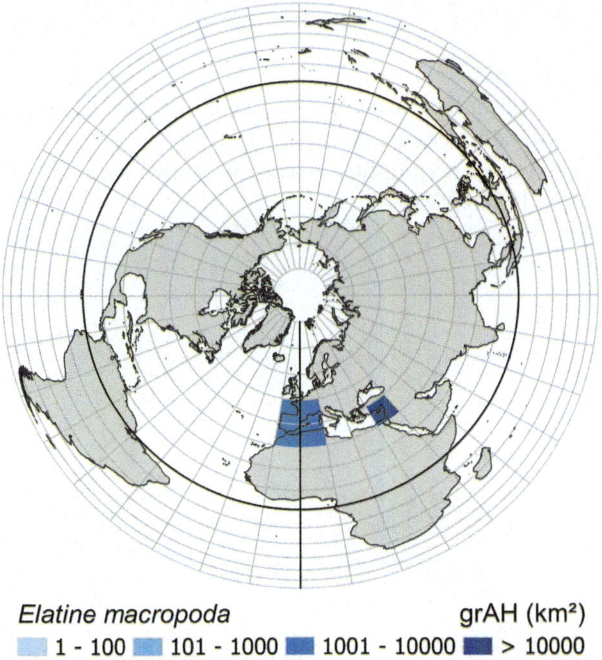

Elatine macropoda Guss.
Image copyright: Y Ur, with permission

***Elatine madagascariensis* H.Perrier**
Synonyms:

Range area (grAH km^2): AFR: 2000
Ecozone origin-status: AFR: E
World rarity status: VR
Ecozone-endemic macroregion occurrence: AFR: MAD

Life form(s): EM, SU
Ploidy state: NA
Speciation age (Ma): NA
Populations also occur in terrestrial, wetland and/ or riparian habitats: YES
Long distance movement vectors: NA
Nuisance status and uses: no nuisance issues; no known uses
Endangered status (IUCN): VU

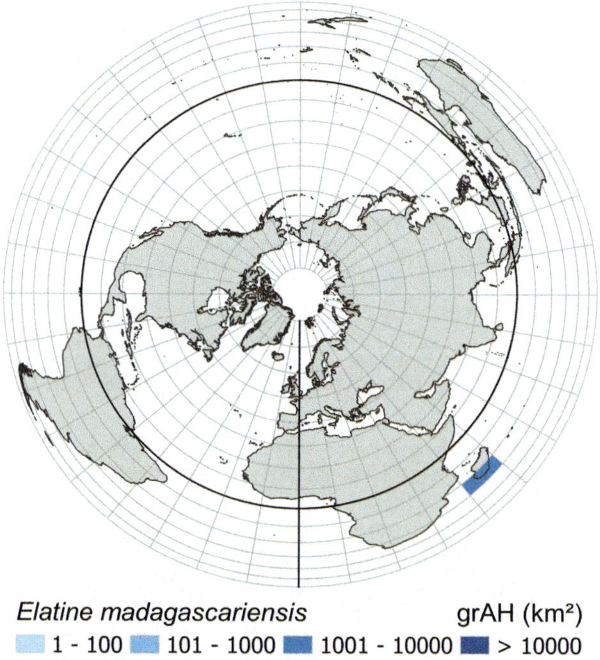

***Elatine minima* (Nutt.) Fisch. & C.A.Mey.**
Synonyms: *Crypta minima* Nutt.

Range area (grAH km^2): NEA: 378300
Ecozone origin-status: NEA: E
World rarity status: C
Ecozone-endemic macroregion occurrence: NEA: ARC CAN, E NAM
Life form(s): SU
Ploidy state: NA
Speciation age (Ma): 10.0
Populations also occur in terrestrial, wetland and/ or riparian habitats: NO
Long distance movement vectors: ?AV
Nuisance status and uses: no nuisance issues; no known uses
Endangered status (IUCN): LC

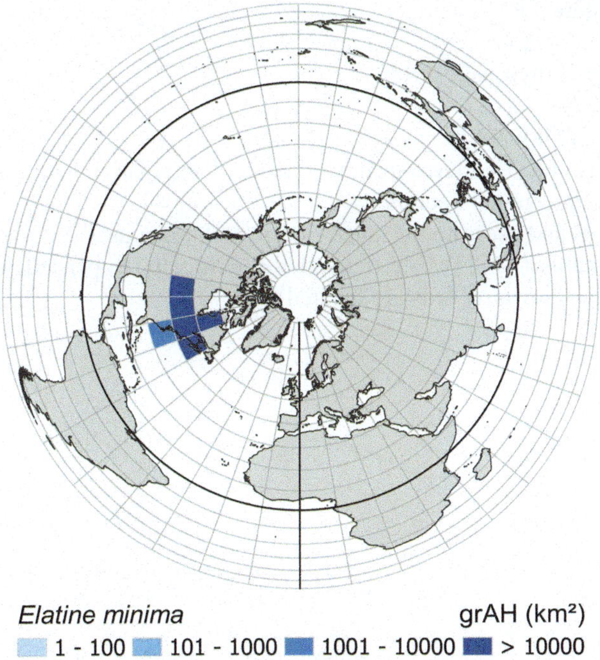

Elatine minima (Nutt.) Fisch. & C.A.Mey.
Image copyright: iNaturalist CC-BY @rob21 (Rob Foster)

Elatine ojibwayensis **Garneau**
Synonyms:

Range area (grAH km^2): NEA: 33100
Ecozone origin-status: NEA: E

World rarity status: R
Ecozone-endemic macroregion occurrence: NEA: ARC CAN
Life form(s): SU (hydrosubshrub)
Ploidy state: NA
Speciation age (Ma): NA
Populations also occur in terrestrial, wetland and/ or riparian habitats: NO
Long distance movement vectors: NA
Nuisance status and uses: no nuisance issues; no known uses
Endangered status (IUCN): NT (considered Critically Imperiled within Québec, where all its known records are located)

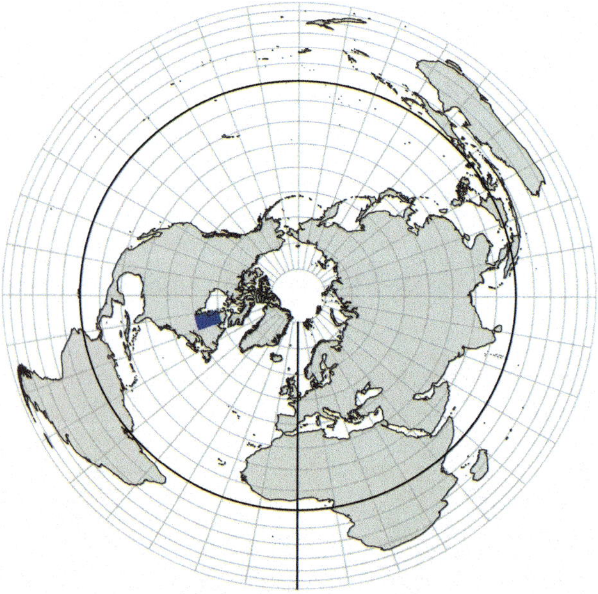

Elatine ojibwayensis grAH (km²)
1 - 100 101 - 1000 1001 - 10000 > 10000

Elatine paramoana **Schmidt-M. & Bernal**
Synonyms:

Range area (grAH km^2): NEO: 37700
Ecozone origin-status: NEO: E
World rarity status: R
Ecozone-endemic macroregion occurrence: NEO: NWSA, AND
Life form(s): SU, EM
Ploidy state: NA
Speciation age (Ma): NA
Populations also occur in terrestrial, wetland and/ or riparian habitats: YES
Long distance movement vectors: NA

Nuisance status and uses: no nuisance issues; no known uses
Endangered status (IUCN): NA

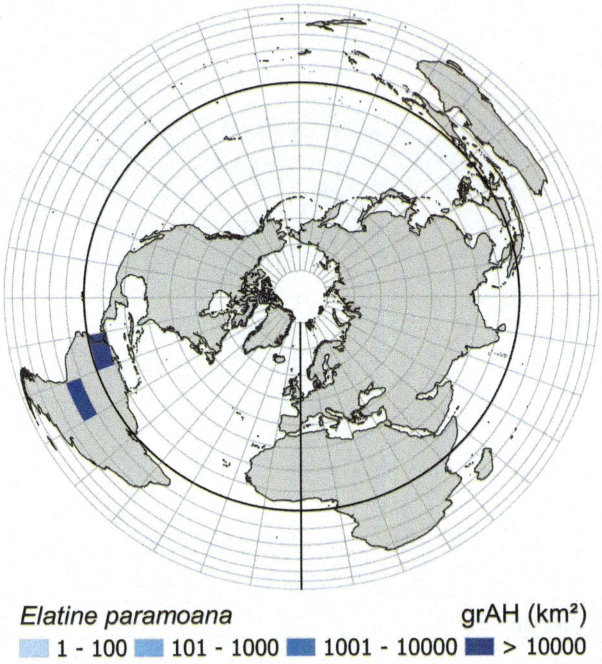

Elatine peruviana **Baehni & J.F.Macbr.**
Synonyms:

Range area (grAH km^2): NEO: 38600
Ecozone origin-status: NEO: E
World rarity status: R
Ecozone-endemic macroregion occurrence: NEO: NWSA, AND
Life form(s): SU, EM
Ploidy state: NA
Speciation age (Ma): NA
Populations also occur in terrestrial, wetland and/ or riparian habitats: YES
Long distance movement vectors: NA
Nuisance status and uses: no nuisance issues; no known uses
Endangered status (IUCN): NA

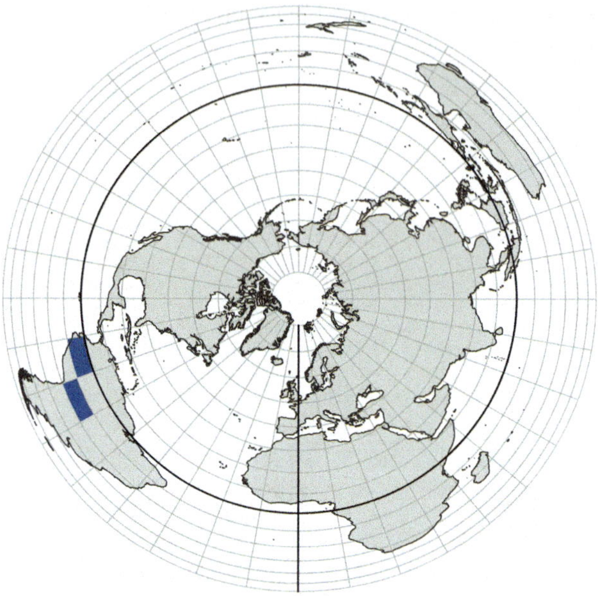

Elatine rubella **Rydb.**
Synonyms:

Range area (grAH km^2): NEA: 48200
Ecozone origin-status: NEA: E
World rarity status: R
Ecozone-endemic macroregion occurrence: NEA: ARC CAN, W NAM, SW NAM, E NAM
Life form(s): SU, EM
Ploidy state: NA
Speciation age (Ma): NA
Populations also occur in terrestrial, wetland and/ or riparian habitats: YES
Long distance movement vectors: NA
Nuisance status and uses: occasionally occurs in rice fields but only low or no nuisance issues reported; no known uses
Endangered status (IUCN): LC

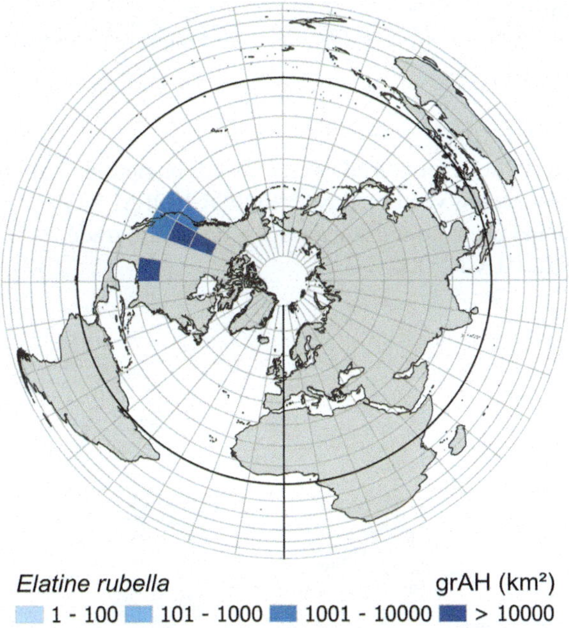

Elatine rubella Rydb.

***Elatine triandra* Schkuhr**
Synonyms: *Ilyphilos triandrus* (Schkuhr) Lunell, *Potamopitys triandra* (Schkuhr) Kuntze, *Alsinastrum triandrum* (Schkuhr) Rupr.

Range area (grAH km^2): PAL: 471600; NEA: 222200; NEO: 127400; OR: 49800; AUS: 13900; AFR: 74300
Ecozone origin-status: PAL: N; NEA: I; NEO: I; OR: N; AUS: N; AFR: N

World rarity status: C
Ecozone-endemic macroregion occurrence: not endemic to any world ecozone
Life form(s): SU, EM
Ploidy state: P
Speciation age (Ma): 10.0
Populations also occur in terrestrial, wetland and/ or riparian habitats: YES
Long distance movement vectors: ANTHROPOCHORY; ?AV
Nuisance status and uses: no nuisance issues; aquarium trade
Endangered status (IUCN): LC

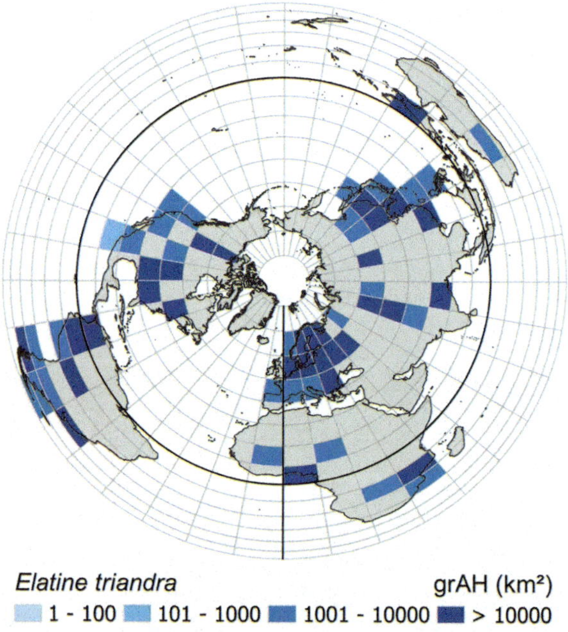

Elatine triandra Schkuhr
Image copyright: Z. Li, with permission

3.19 EUPHORBIACEAE (Malpighiales)

There are three macrophyte species in the Euphorbiaceae, in the genus *Caperonia*, endemic to NEO. *Caperonia palustris* has also been introduced to warm-water habitats in AFR, NEA and OCE. None of the species cause aquatic weed problems. One species is rare, two are common. None of the species has been assessed for endangered status.

***Caperonia castaneifolia* (L.) A.St.-Hil.**

Synonyms: *Argythamnia castaneifolia* (L.) Kuntze., *Croton castaneifolius* L., *Ditaxis castaneifolia* (L.) Baill., *Meterana castaneifolia* (L.) Raf., *Tournesol castaneifolia* (L.) M.Gómez

Range area (grAH km^2): NEO: 344800
Ecozone origin-status: NEO: E
World rarity status: C
Ecozone-endemic macroregion occurrence: NEO: MG MX, MES AM, NWSA, AMZ, CHA, PAR
Life form(s): EM
Ploidy state: NA
Speciation age (Ma): NA
Populations also occur in terrestrial, wetland and/ or riparian habitats: YES
Long distance movement vectors: NA
Nuisance status and uses: no nuisance issues; medicinal uses
Endangered status (IUCN): NA

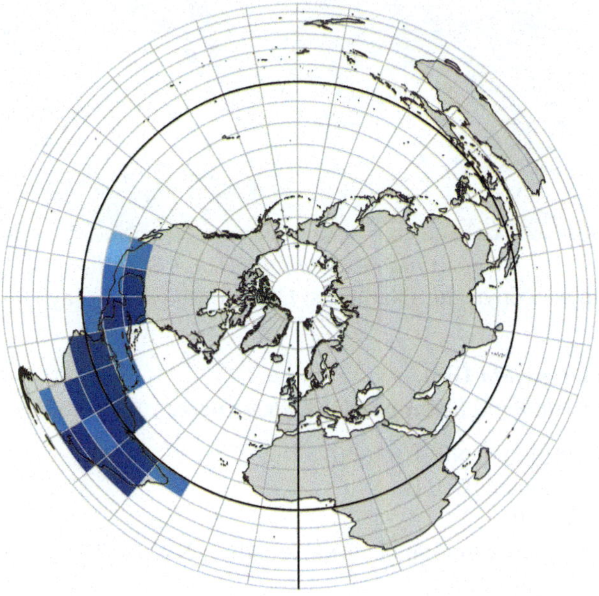

Caperonia castaneifolia (L.) A.St.-Hil.

Caperonia heteropetala **Didr.**
Synonyms: *Argythamnia heteropetala* (Didr.) Kuntze

Range area (grAH km^2): NEO: 93600
Ecozone origin-status: NEO: E
World rarity status: R
Ecozone-endemic macroregion occurrence: NEO: NWSA, CHA, AMZ, PAR
Life form(s): EM (subshrub)
Ploidy state: NA
Speciation age (Ma): NA

Populations also occur in terrestrial, wetland and/ or riparian habitats: YES
Long distance movement vectors: NA
Nuisance status and uses: no nuisance issues; no known uses
Endangered status (IUCN): NA

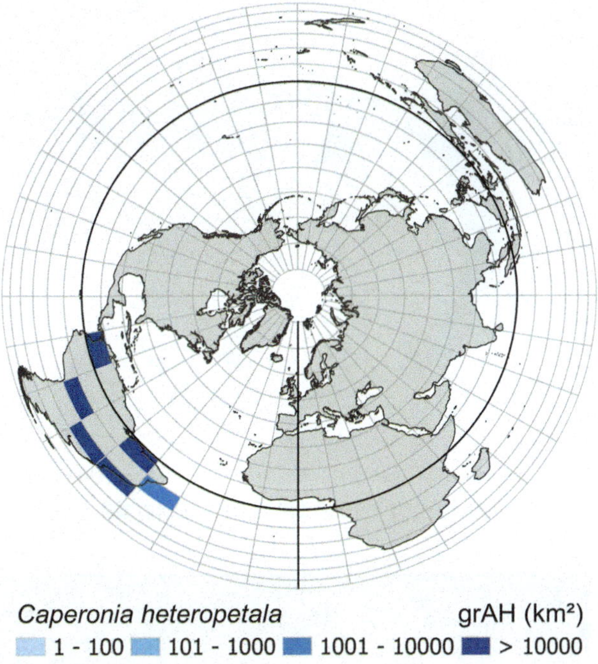

***Caperonia palustris* (L.) A.St.-Hil.**
Synonyms: *Argythamnia palustris* (L.) Kuntze, *Croton palustris* L., *Meterana palustris* (L.) Raf.

Range area (grAH km^2): NEA: 12800; NEO: 340600; AFR: 51800; present in OCE
Ecozone origin-status: NEA: I; NEO: E(I); AFR: I; OCE: I
World rarity status: C
Ecozone-endemic macroregion occurrence: NEO: MG MX, MES AM, NWSA, AMZ, AND, CHA, PAR
Life form(s): EM
Ploidy state: P
Speciation age (Ma): NA
Populations also occur in terrestrial, wetland and/ or riparian habitats: YES
Long distance movement vectors: ANTHROPOCHORY
Nuisance status and uses: no nuisance issues; folk medicine, edible
Endangered status (IUCN): NA

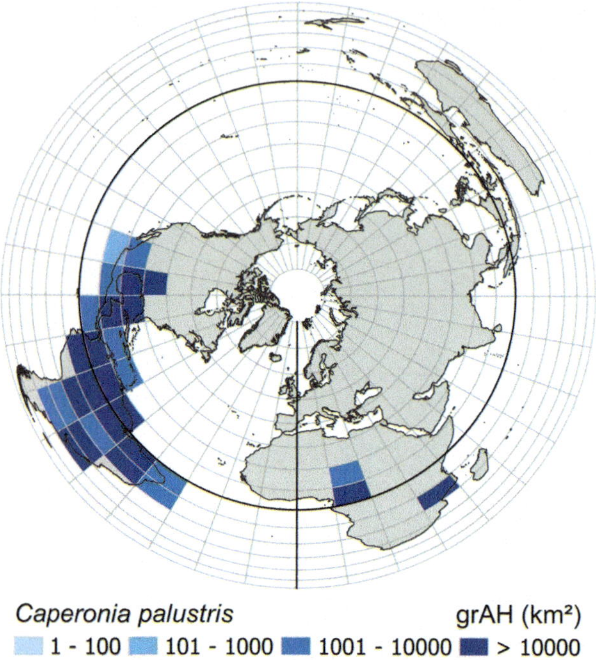

Caperonia palustris (L.) A.St.-Hil.
Image copyright: iNaturalist CC-BY @huracan (Neptalí Ramírez Marcial)

3.20 FABACEAE (Fabales)

The Fabaceae contain 24 macrophyte species, within five genera, mainly having warm-temperate to tropical distributions, and especially present in NEO, where 19 species are found, of which 13 are endemic. *Aeschynomene fluitans* occurs as an E(I) species, and *A. schimperi* as an E species in AFR. *A fluitans* is also introduced in NEO. Outside NEO, there are seven species in AFR and six in OR (two of them introduced in AFR and three in OR), five, of which three are introduced, in AUS and five, of which two are introduced, in NEA, two in PAL (one introduced) and two in OCE (both introduced). There are no Fabaceae \species in ANT. Of the species in this family, 35% are R and 65% C. About 71% of the Fabaceae have not been assessed for endangered status, the remaining 29% are LC.

Quick Links to Genera

1. *Aeschynomene*
2. *Discolobium*
3. *Neptunia*
4. *Sesbania*
5. *Vigna*

Aeschynomene americana **L.**
Synonyms:

Range area (grAH km^2): NEA: 13000; NEO: 352100; OR: 108000; AUS: 66000; AFR: 108800
Ecozone origin-status: NEA: N; NEO: N; OR: I; AUS: I; AFR: I
World rarity status: C
Ecozone-endemic macroregion occurrence: not endemic to any world ecoregion
Life form(s): EM (sometimes a small shrub)
Ploidy state: D
Speciation age (Ma): 36.0
Populations also occur in terrestrial, wetland and/ or riparian habitats: YES
Long distance movement vectors: HYDROCHORY (NEO to NEA); ANTHROPOCHORY; ?AV
Nuisance status and uses: no or only low nuisance issues; widely naturalised across its introduced range in the tropics and subtropical regions, because of its value as a tropical forage and green manure crop; some medicinal use
Endangered status (IUCN): NA

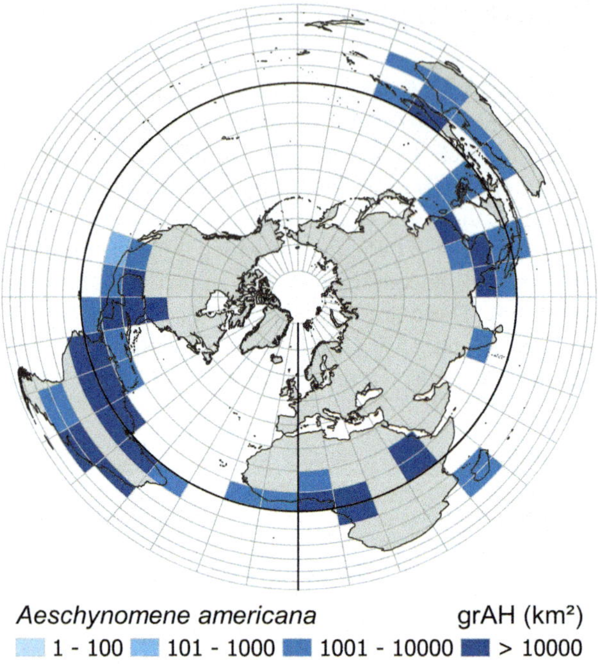
Aeschynomene americana
grAH (km²)
1 - 100
101 - 1000
1001 - 10000
> 10000

Aeschynomene americana L.
Image copyright: iNaturalist CC0 @stephanie_naturalist (Stephanie C)

***Aeschynomene aspera* L.**
Synonyms: *Aeschynomene indica* var. *aspera* (L.) Hassk. ex Miq.

Range area (grAH km^2): OR: 106000; AUS: 12900; AFR: 19000
Ecozone origin-status: OR: N; AUS: N; AFR: I
World rarity status: R
Ecozone-endemic macroregion occurrence: not endemic to any world ecoregion
Life form(s): EM (shrub)

Ploidy state: P
Speciation age (Ma): NA
Populations also occur in terrestrial, wetland and/ or riparian habitats: YES
Long distance movement vectors: ANTHROPOCHORY
Nuisance status and uses: no nuisance issues; one of the main sources of ambatch wood: the pithy wood from the stems is very light and is used to make art paper products, artificial flowers, and as a cork substitute
Endangered status (IUCN): LC

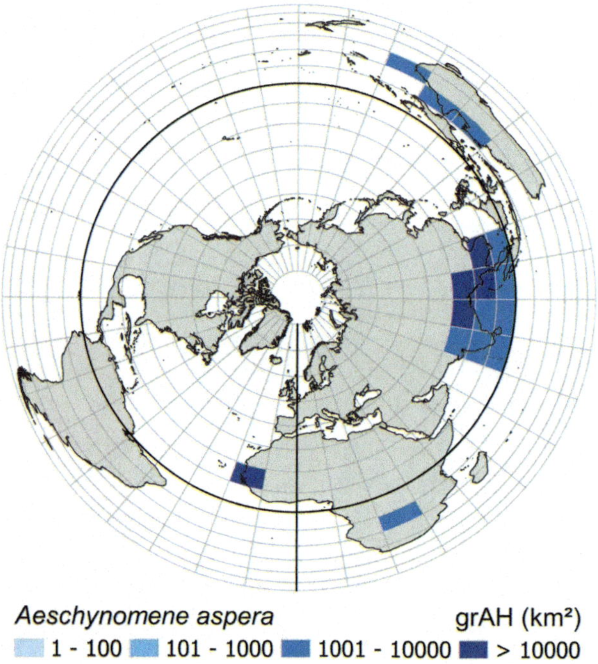

Aeschynomene ciliata **Vogel**
Synonyms:

Range area (grAH km^2): NEO: 203600
Ecozone origin-status: NEO: E
World rarity status: C
Ecozone-endemic macroregion occurrence: NEO: MES AM, NWSA, AMZ, CHA, PAR
Life form(s): EM (sometimes shrubby)
Ploidy state: NA
Speciation age (Ma): NA

Populations also occur in terrestrial, wetland and/ or riparian habitats: YES
Long distance movement vectors: NA
Nuisance status and uses: no nuisance issues; soil recuperation
Endangered status (IUCN): LC

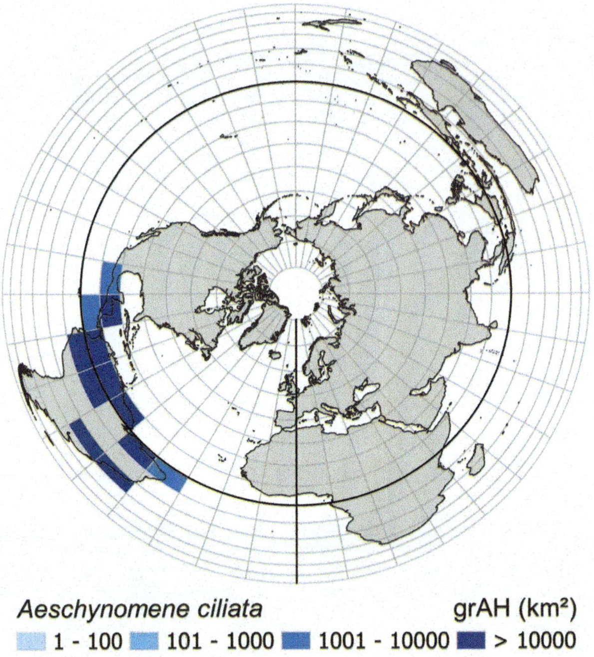

Aeschynomene denticulata **Rudd**
Synonyms:

Range area (grAH km^2): NEO: 205000
Ecozone origin-status: NEO: E
World rarity status: C
Ecozone-endemic macroregion occurrence: NEO: AMZ, CHA, PAR
Life form(s): EM
Ploidy state: D
Speciation age (Ma): NA
Populations also occur in terrestrial, wetland and/ or riparian habitats: YES
Long distance movement vectors: NA
Nuisance status and uses: no nuisance issues; medicinal uses
Endangered status (IUCN): NA

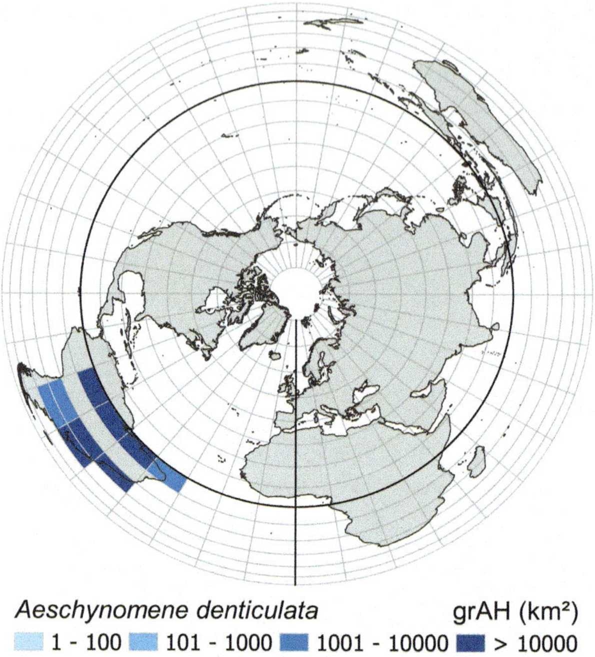

Aeschynomene denticulata Rudd
Image copyright: iNaturalist CC0 @andrespautasso (Andrés Pautasso)

***Aeschynomene elegans* Cham. & Schltdl.**
Synonyms: *Ctenodon elegans* (Schltdl. & Cham.) D.B.O.S.Cardoso & A.Delgado

Range area (grAH km^2): NEO: 181800
Ecozone origin-status: NEO: E
World rarity status: C
Ecozone-endemic macroregion occurrence: NEO: MG MX, MES AM, NWSA, AMZ, CHA, PAR
Life form(s): EM (sometimes shrubby)

Ploidy state: D
Speciation age (Ma): NA
Populations also occur in terrestrial, wetland and/ or riparian habitats: YES
Long distance movement vectors: NA
Nuisance status and uses: no nuisance issues; no known uses
Endangered status (IUCN): NA

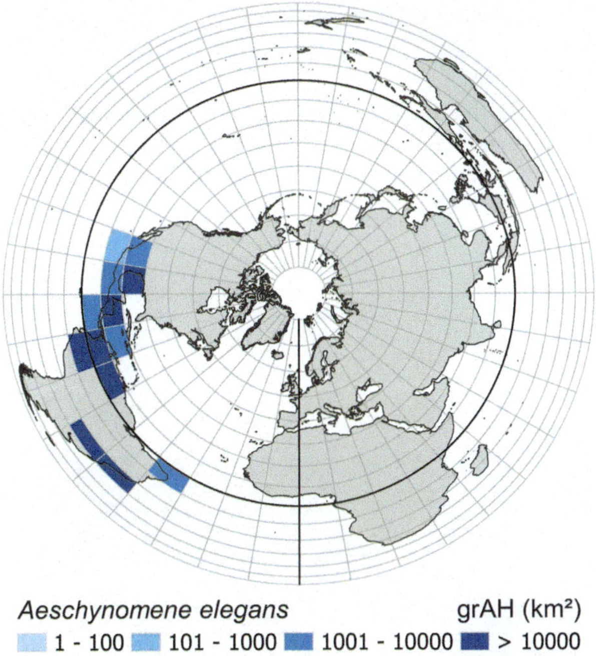

Aeschynomene elegans Cham. & Schltdl.
Image copyright: iNaturalist CC-BY @mauricio_mercadante (Mauricio Mercadante)

Aeschynomene evenia **C.Wright**
Synonyms:

Range area (grAH km^2): NEO: 228500
Ecozone origin-status: NEO: E
World rarity status: C
Ecozone-endemic macroregion occurrence: NEO: MES AM, NWSA, AMZ, CHA, PAR
Life form(s): EM (sometimes shrubby)
Ploidy state: NA
Speciation age (Ma): NA
Populations also occur in terrestrial, wetland and/ or riparian habitats: YES
Long distance movement vectors: NA
Nuisance status and uses: no nuisance issues; medicinal potential, forage
Endangered status (IUCN): NA

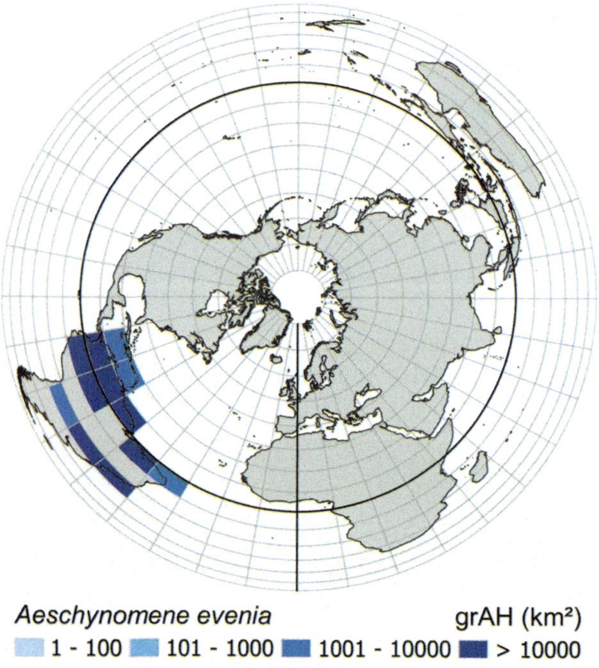

Aeschynomene filosa **Mart. ex Benth.**
Synonyms:

Range area (grAH km^2): NEO: 184600
Ecozone origin-status: NEO: E
World rarity status: C
Ecozone-endemic macroregion occurrence: NEO: MES AM, NWSA, AMZ, CHA, PAR

Life form(s): EM (shrub or subshrub)
Ploidy state: NA
Speciation age (Ma): NA
Populations also occur in terrestrial, wetland and/ or riparian habitats: YES
Long distance movement vectors: NA
Nuisance status and uses: no nuisance issues; medicinal potential
Endangered status (IUCN): LC

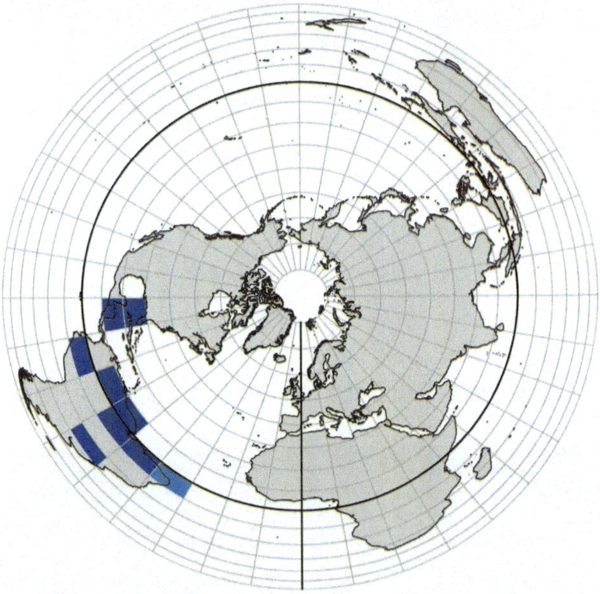

Aeschynomene fluitans **Peter**
Synonyms:

Range area (grAH km^2): NEO: 10300; AFR: 105400
Ecozone origin-status: NEO: I; AFR: E(I)
World rarity status: R
Ecozone-endemic macroregion occurrence: AFR: EA, ZAM
Life form(s): FR
Ploidy state: NA
Speciation age (Ma): NA
Populations also occur in terrestrial, wetland and/ or riparian habitats: NO
Long distance movement vectors: ANTHROPOCHORY
Nuisance status and uses: some nuisance issues reported from the USA, mainly in Florida; aquarium trade and water garden plant, medicinal uses
Endangered status (IUCN): NA

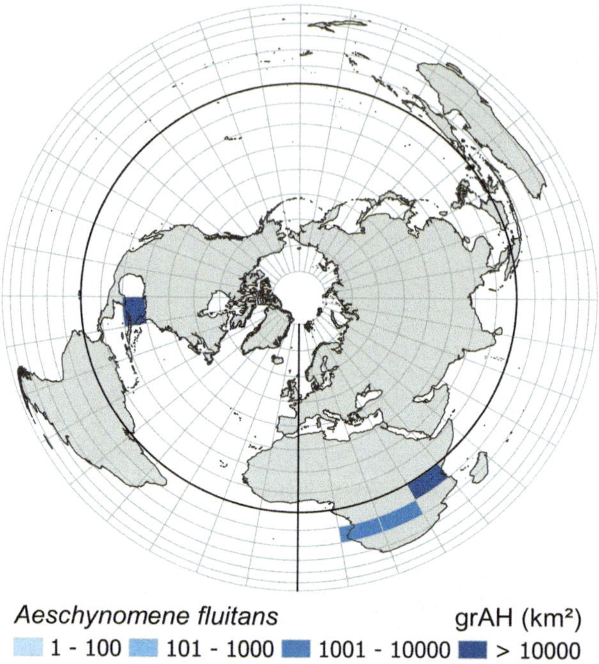

Aeschynomene fluitans Peter

Aeschynomene fluminensis **Vell**
Synonyms:

Range area (grAH km^2): NEO: 228700
Ecozone origin-status: NEO: E
World rarity status: C
Ecozone-endemic macroregion occurrence: NEO: NWSA, AMZ, CHA, PAR
Life form(s): EM (subshrub or shrub)

Ploidy state: NA
Speciation age (Ma): NA
Populations also occur in terrestrial, wetland and/ or riparian habitats: YES
Long distance movement vectors: NA
Nuisance status and uses: no nuisance issues; medicinal potential uses
Endangered status (IUCN): NA

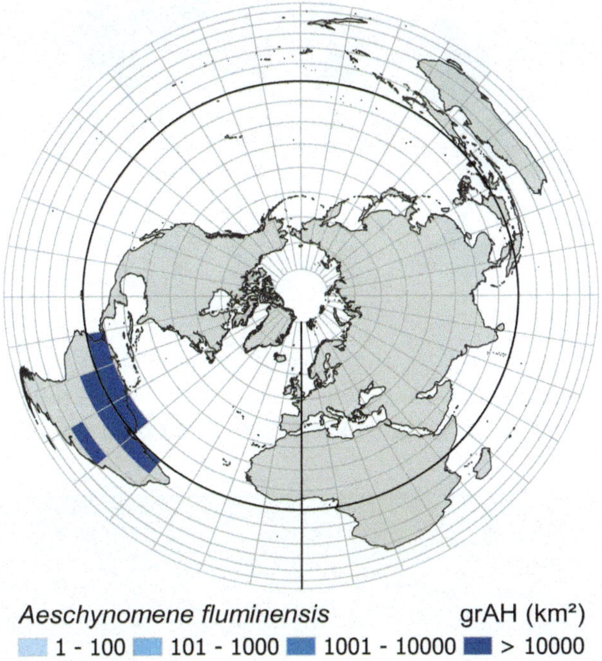

***Aeschynomene indica* L.**
Synonyms: *Hedysarum neli-tali* Roxb.

Range area (grAH km^2): PAL: 121800; NEA: 25800; NEO: 112400; OR: 157200; AUS: 83800; AFR: 260600; present in OCE
Ecozone origin-status: PAL: N; NEA: N; NEO: N; OR: N; AUS: N; AFR: N; OCE: I
World rarity status: C
Ecozone-endemic macroregion occurrence: not endemic to any world ecozone
Life form(s): EM (sometimes shrubby)
Ploidy state: P
Speciation age (Ma): 10.0
Populations also occur in terrestrial, wetland and/ or riparian habitats: YES
Long distance movement vectors: ?AV; ANTHROPOCHORY
Nuisance status and uses: no nuisance issues; medicinal, phytoremediation potential, also gathered from the wild for local use as a food, medicine and source of material for craftwork, and sometimes grown as a green manure
Endangered status (IUCN): LC

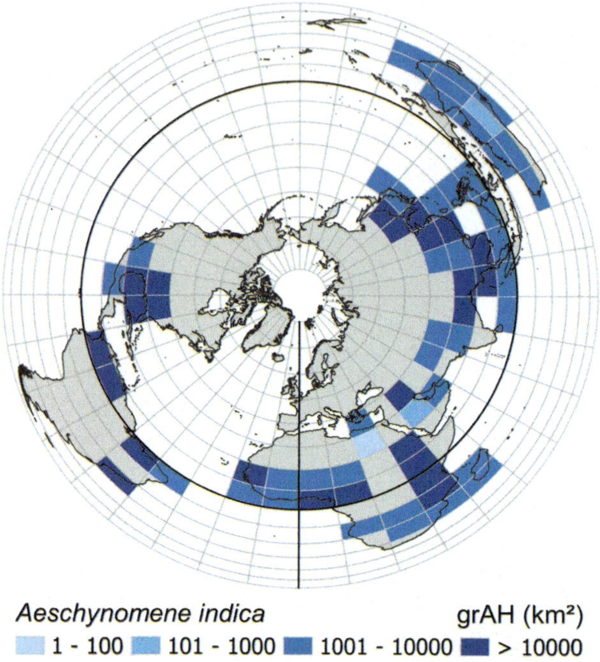

Aeschynomene indica L.

***Aeschynomene montevidensis* Vogel**
Synonyms:

Range area (grAH km²): NEO: 113900
Ecozone origin-status: NEO: E
World rarity status: R
Ecozone-endemic macroregion occurrence: NEO: AMZ, CHA, PAR
Life form(s): EM (shrub)
Ploidy state: D
Speciation age (Ma): NA
Populations also occur in terrestrial, wetland and/ or riparian habitats: YES
Long distance movement vectors: NA
Nuisance status and uses: no nuisance issues; no known uses
Endangered status (IUCN): NA

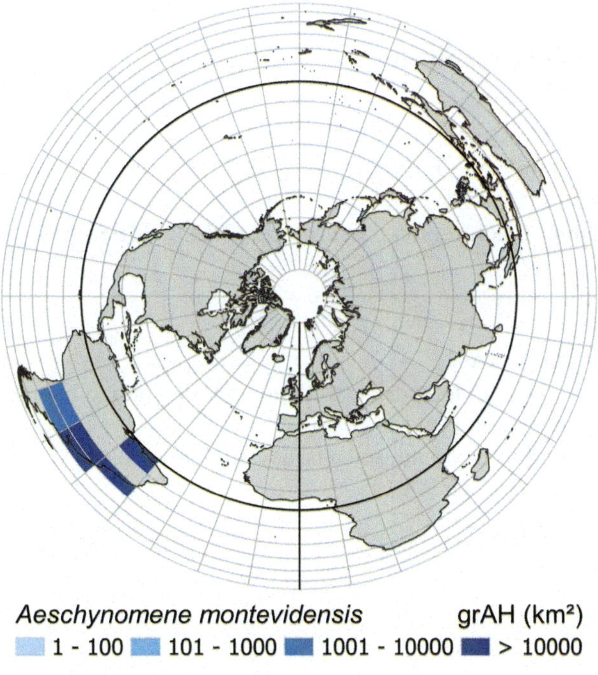

Aeschynomene montevidensis Vogel
Image copyright: iNaturalist CC-BY @romigaleota Romi Galeota Lencina)

Aeschynomene pratensis **Small**
Synonyms:

Range area (grAH km²): NEO: 217800
Ecozone origin-status: NEO: E
World rarity status: C
Ecozone-endemic macroregion occurrence: NEO: MES AM, NWSA, AMZ, CHA, PAR
Life form(s): EM (shrub)
Ploidy state: DP
Speciation age (Ma): NA
Populations also occur in terrestrial, wetland and/ or riparian habitats: YES
Long distance movement vectors: NA
Nuisance status and uses: no nuisance issues; no known uses
Endangered status (IUCN): NA

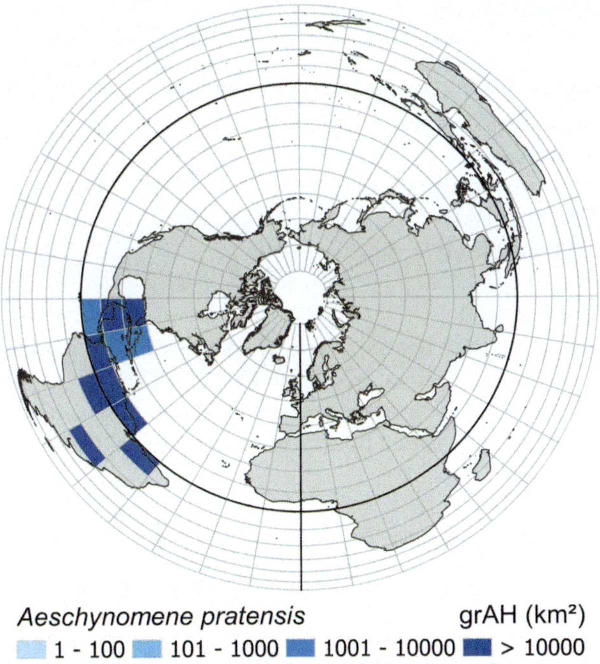

Aeschynomene pratensis Small
Image copyright: iNaturalist CC-BY @blazeclaw (Jared Shorma)

***Aeschynomene rudis* Benth.**
Synonyms:

Range area (grAH km^2): NEA: 13800; NEO: 309800
Ecozone origin-status: NEA: I; NEO: E(I)
World rarity status: C
Ecozone-endemic macroregion occurrence: NEO: MG MX, MES AM, NWSA, AMZ, AND, CHA, PAR
Life form(s): EM (subshrub or shrub)
Ploidy state: P
Speciation age (Ma): 10.0
Populations also occur in terrestrial, wetland and/ or riparian habitats: YES
Long distance movement vectors: ANTHROPOCHORY; ?AV
Nuisance status and uses: in North America considered to be a noxious weed, especially of wet areas such as rice fields; limited medicinal uses
Endangered status (IUCN): NA

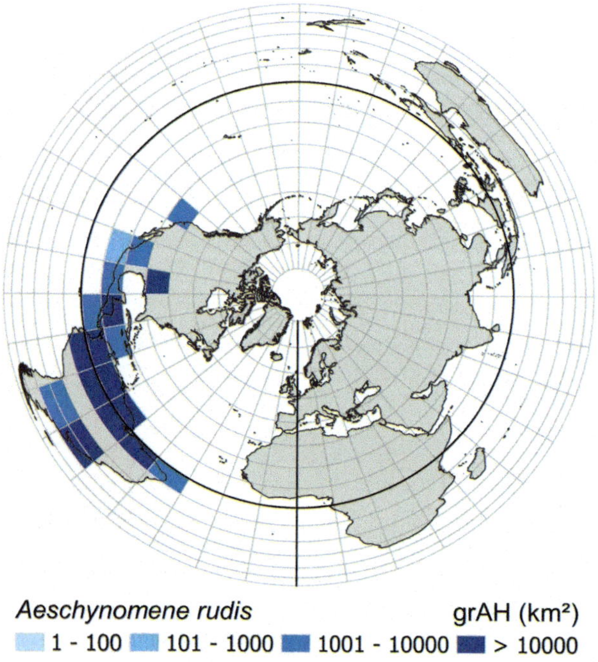

Aeschynomene rudis Benth.
Image copyright: iNaturalist CC-BY @jim_keesling

Aeschynomene schimperi Hochst. ex A.Rich.
Synonyms:

Range area (grAH km^2): AFR: 317400
Ecozone origin-status: AFR: E
World rarity status: C
Ecozone-endemic macroregion occurrence: AFR: SAH/ AR, FOR, EA, ZAM, MAD
Life form(s): EM (subshrub or shrub)
Ploidy state: NA
Speciation age (Ma): NA
Populations also occur in terrestrial, wetland and/ or riparian habitats: YES
Long distance movement vectors: NA
Nuisance status and uses: no nuisance issues; limited medicinal use
Endangered status (IUCN): LC

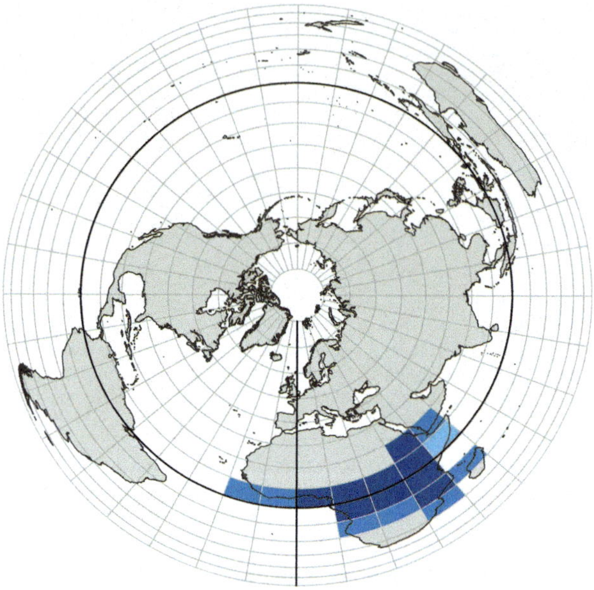

Aeschynomene selloi Vogel.
Synonyms:

Range area (grAH km 2): NEO: 45800
Ecozone origin-status:; NEO: E
World rarity status: R
Ecozone-endemic macroregion occurrence: NEO: AMZ, CHA, PAR
Life form(s): EM (subshrub or shrub)
Ploidy state: NA
Speciation age (Ma): NA
Populations also occur in terrestrial, wetland and/ or riparian habitats: YES
Long distance movement vectors: NA
Nuisance status and uses: no nuisance issues; no known uses
Endangered status (IUCN): NA

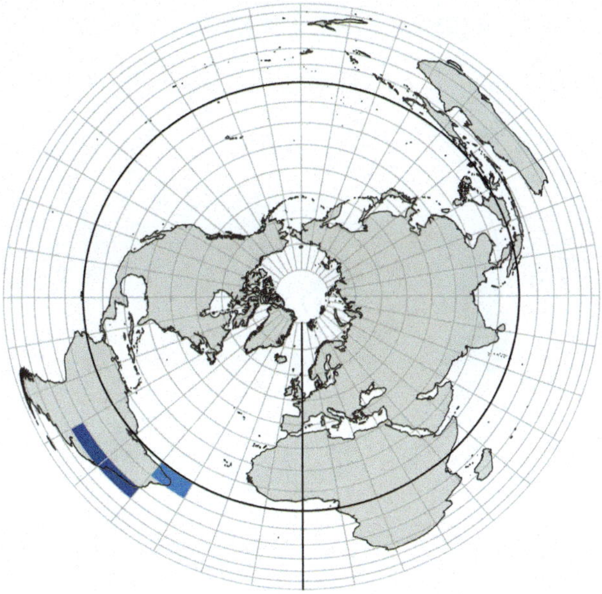

Aeschynomene selloi grAH (km²)
1 - 100 101 - 1000 1001 - 10000 > 10000

***Aeschynomene sensitiva* Sw.**
Synonyms:

Range area (grAH km²): NEO: 297700; OR: 2600; AFR: 191900
Ecozone origin-status: NEO: N; OR: I; AFR: N
World rarity status: C
Ecozone-endemic macroregion occurrence: not endemic to any world ecozone
Life form(s): EM (subshrub or shrub)
Ploidy state: D
Speciation age (Ma): NA
Populations also occur in terrestrial, wetland and/ or riparian habitats: YES
Long distance movement vectors: ANTHROPOCHORY
Nuisance status and uses: no nuisance issues; medicinal uses
Endangered status (IUCN): NA

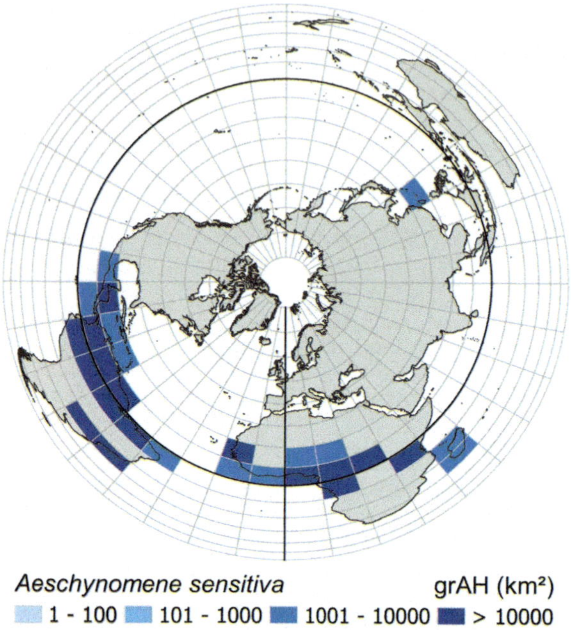

Aeschynomene sensitiva Sw.
Image copyright: S.M. Silvestrini, with permission

***Aeschynomene virginica* (L.) Britton, Sterns & Poggenb.**
Synonyms: *Hedysarum virginicum* L.

Range area (grAH km^2): PAL: 7500; NEA: 8200; NEO: 39100
Ecozone origin-status: PAL: I; NEA: N; NEO: N
World rarity status: R
Ecozone-endemic macroregion occurrence: not endemic to any world ecozone
Life form(s): EM
Ploidy state: P
Speciation age (Ma): NA
Populations also occur in terrestrial, wetland and/ or riparian habitats: YES
Long distance movement vectors: ANTHROPOCHORY
Nuisance status and uses: no nuisance issues; no known uses
Endangered status (IUCN): NA

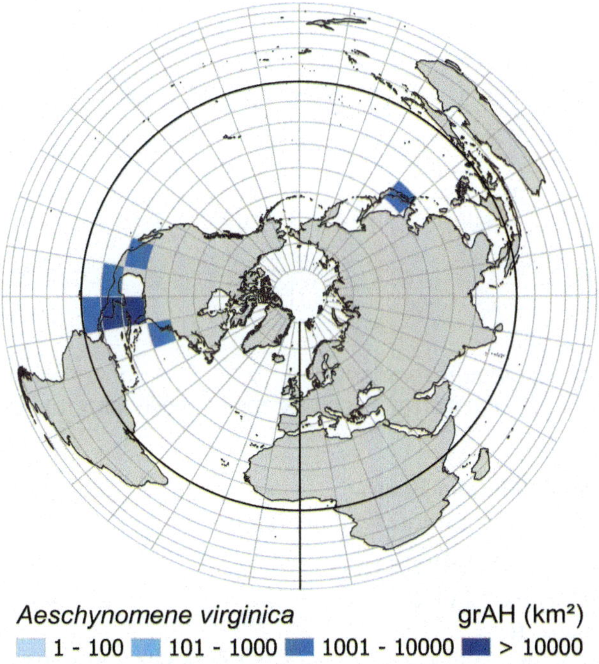

***Discolobium leptophyllum* Benth.**
Synonyms:

Range area (grAH km^2): NEO: 51100
Ecozone origin-status: NEO: E
World rarity status: R
Ecozone-endemic macroregion occurrence: NEO: AMZ, CHA, PAR
Life form(s): EM (subshrub or shrub)
Ploidy state: NA

Speciation age (Ma): NA
Populations also occur in terrestrial, wetland and/ or riparian habitats: YES
Long distance movement vectors: NA
Nuisance status and uses: no nuisance issues; folk medicine
Endangered status (IUCN): NA

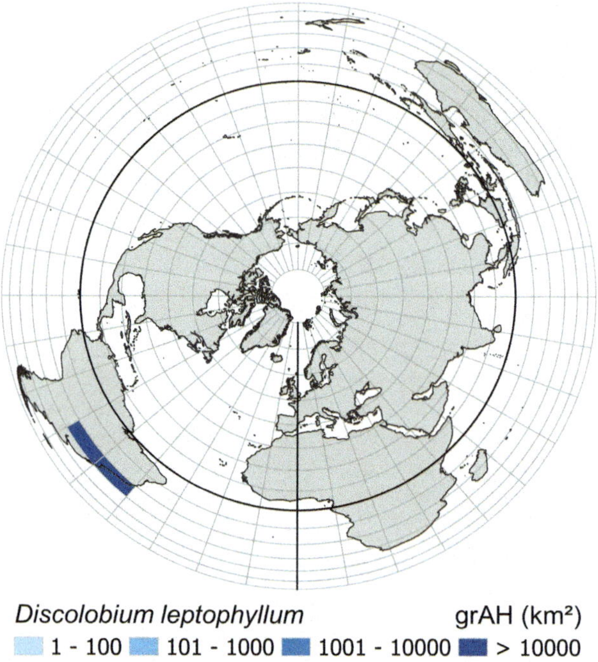

Discolobium psoraliifolium **Benth.**
Synonyms: *Discolobium psoraleaefolium* Benth.

Range area (grAH km^2): NEO: 44100
Ecozone origin-status: NEO: E
World rarity status: R
Ecozone-endemic macroregion occurrence: NEO: AMZ, CHA, PAR
Life form(s): EM (subshrub)
Ploidy state: NA
Speciation age (Ma): NA
Populations also occur in terrestrial, wetland and/ or riparian habitats: YES
Long distance movement vectors: NA
Nuisance status and uses: no nuisance issues; no known uses
Endangered status (IUCN): NA

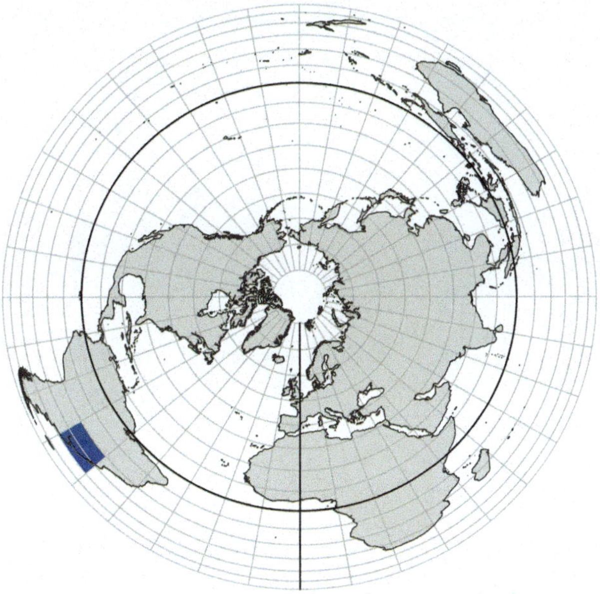

Discolobium pulchellum **Benth.**
Synonyms:

Range area (grAH km^2): NEO: 48700
Ecozone origin-status: NEO: E
World rarity status: R
Ecozone-endemic macroregion occurrence: NEO: AMZ, PAR
Life form(s): EM (shrub)
Ploidy state: NA
Speciation age (Ma): NA
Populations also occur in terrestrial, wetland and/ or riparian habitats: YES
Long distance movement vectors: NA
Nuisance status and uses: no nuisance issues; medicinal potential
Endangered status (IUCN): NA

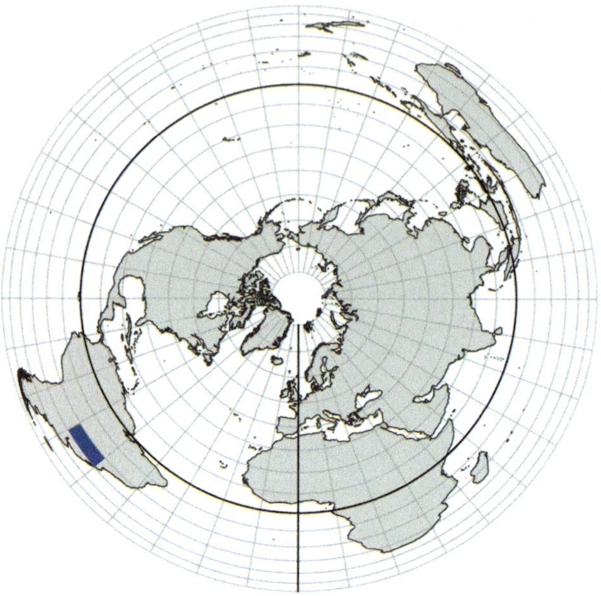

***Neptunia oleracea* Lour.**

Synonyms: *Neptunia natans* W.Theob., *Desmanthus natans* Willd.

Range area (grAH km^2): NEA: 12800; NEO: 297700; OR: 111600; AUS: 3600; AFR: 319000; present in OCE
Ecozone origin-status: NEA: I; NEO: N; OR: N; AUS: I; AFR: N; OCE: I
World rarity status: C
Ecozone-endemic macroregion occurrence: not endemic to any world ecozone
Life form(s): FR, EM (hydrosubshrub)
Ploidy state: P
Speciation age (Ma): NA
Populations also occur in terrestrial, wetland and/ or riparian habitats: YES
Long distance movement vectors: ANTHROPOCHORY
Nuisance status and uses: considered a troublesome aquatic weed in many parts of its tropical to subtropical range; medicinal, forage, ornamental uses, cultivated in some tropical areas of Asia for its edible leaves and young shoots
Endangered status (IUCN): LC

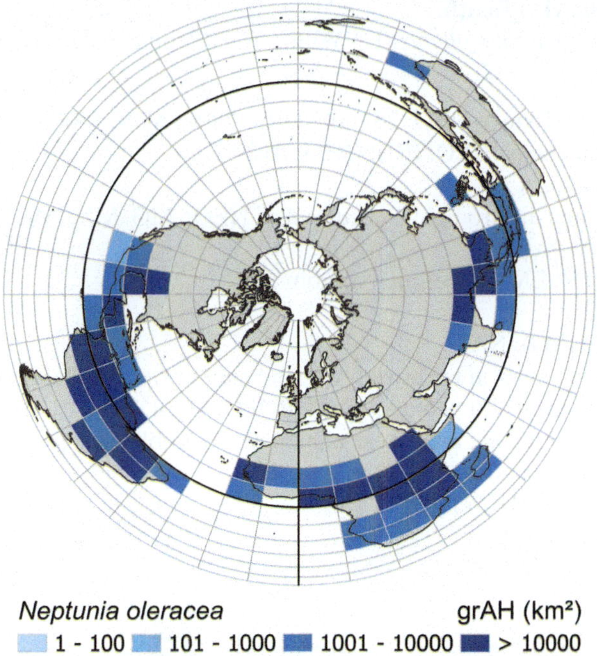

Neptunia oleracea Lour.
Image copyright: iNaturalist CC-BY @robert_taylor (Robert Taylor)

Neptunia plena **(L.) Benth.**
Synonyms: *Desmanthus plenus* (L.) Willd., *Hemidesmas plenus* (L.) Raf., *Mimosa plena* L.

Range area (grAH km²): NEO: 233600; OR: 37800; AUS: 16000
Ecozone origin-status: NEO: E(I); OR: I; AUS: I
World rarity status: C
Ecozone-endemic macroregion occurrence: NEO: MG MX, MES AM, NWSA, AMZ, CHA, PAR
Life form(s): FR, EM (sometimes shrubby)
Ploidy state: P
Speciation age (Ma): NA
Populations also occur in terrestrial, wetland and/ or riparian habitats: YES
Long distance movement vectors: ANTHROPOCHORY
Nuisance status and uses: may cause aquatic weed problems, e.g. in Australia; water garden plant, bioactive
Endangered status (IUCN): LC

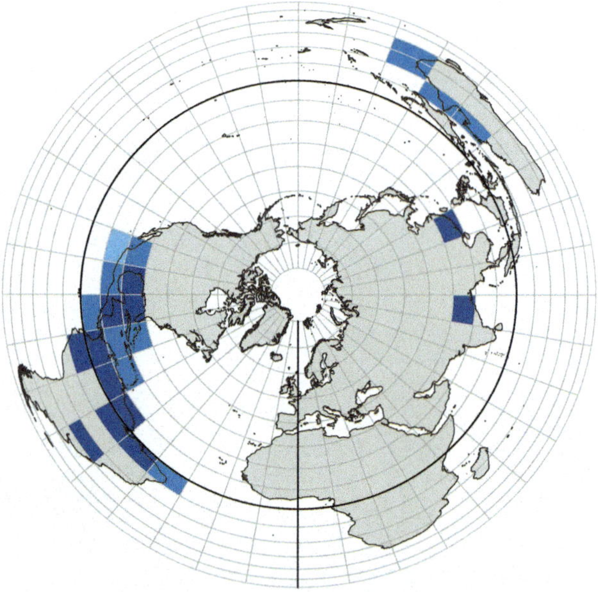

Neptunia plena (L.) Benth.

***Sesbania exasperata* Kunth**
Synonyms: Emerus exasperatus (Kunth) Kuntze

Range area (grAH km 2): NEO: 322800
Ecozone origin-status: NEO: E
World rarity status: C
Ecozone-endemic macroregion occurrence: NEO: MG MX, MES AM, NWSA, AMZ, CHA, PAR
Life form(s): EM (subshrub or shrub)
Ploidy state: D
Speciation age (Ma): NA
Populations also occur in terrestrial, wetland and/ or riparian habitats: YES
Long distance movement vectors: NA
Nuisance status and uses: no nuisance issues; phytoremediation potential
Endangered status (IUCN): NA

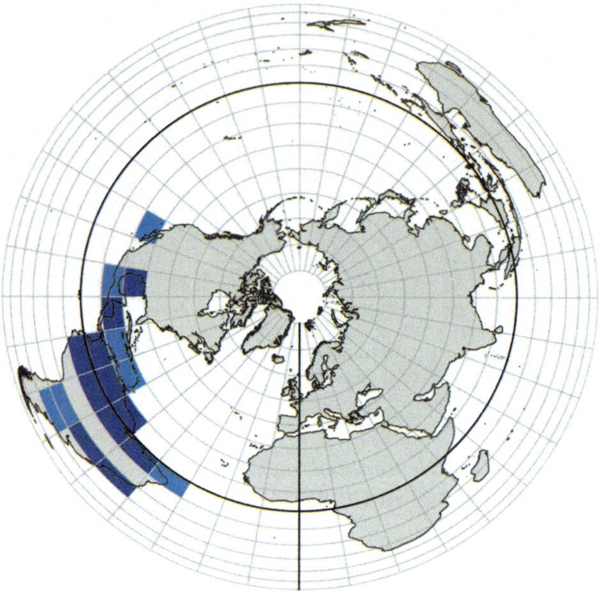

Vigna lasiocarpa **(Benth.) Verdc.**
Synonyms: Phaseolus lasiocarpus Mart. ex Benth.

Range area (grAH km 2): NEO: 296500
Ecozone origin-status: NEO: E
World rarity status: C
Ecozone-endemic macroregion occurrence: NEO: MES AM, NWSA, AMZ, CHA, PAR
Life form(s): EM (climber)
Ploidy state: D
Speciation age (Ma): 1.2
Populations also occur in terrestrial, wetland and/ or riparian habitats: YES
Long distance movement vectors: NA
Nuisance status and uses: no nuisance issues; no known uses
Endangered status (IUCN): NA

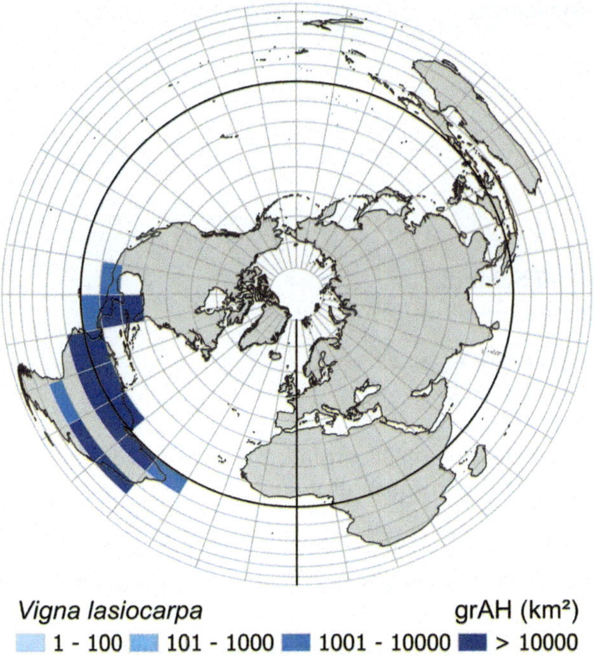

3.21 HALORAGACEAE (Saxifragales)

The Haloragaceae are a large family of macrophytes, with 69 species in six genera, widespread in every ecozone worldwide, but especially in AUS, where 44 species are present, 35 endemic to the ecozone. There are also substantial numbers of species in OR (18 species, five being ecozone-endemics), NEA (17, with one endemic species), and NEO (15 species, including three endemics). Elsewhere the diversity of Haloragaceae macrophyte species is lower, with 11 species present in PAL (none endemic), eight in AFR (four endemic), two in OCE (both introduced) and one in ANT. With one notable exception, I-origin is relatively uncommon in these plants, though there are four introduced species in NEA, four in PAL, three in OR, two in AUS, two in OCE, and one in AFR. The exception is *Myriophyllum aquaticum*, which is an E(I) species in NEO that has been introduced to every other world ecozone except ANT and is considered a highly invasive and aggressive alien species in many warm-temperate to tropical areas of PAL, NEA, OR and AUS. The Haloragaceae have VR 57%, R 17% and C 26%, with 65% unassessed for Endangered Status, 27% LC, and 8% in the higher risk categories, with two species (3%) classed as CR.

Quick Links to Genera

1. *Haloragis*
2. *Laurembergia*
3. *Meziella*
4. *Myriophyllum*
5. *Proserpinaca*
6. *Vinkia*

Haloragis brownii **Schindl.**
Synonyms: *Meionectes brownii* Hook.f.

Range area (grAH km²): AUS: 19300
Ecozone origin-status: AUS: E
World rarity status: VR
Ecozone-endemic macroregion occurrence: AUS: WA, ESA
Life form(s): SU, EM (hydrosubshrub)
Ploidy state: NA
Speciation age (Ma): NA
Populations also occur in terrestrial, wetland and/ or riparian habitats: YES
Long distance movement vectors: NA
Nuisance status and uses: no nuisance issues; no known uses
Endangered status (IUCN): NA

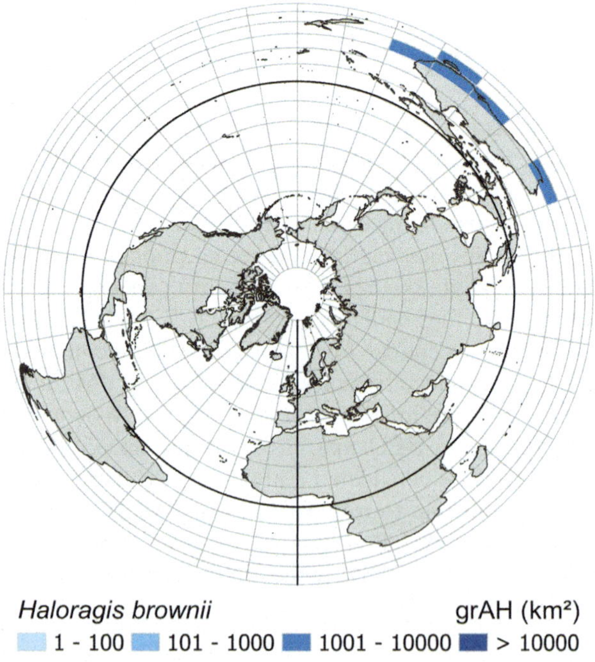

***Haloragis halconensis* Merr.**
Synonyms: *Gonocarpus halconensis* (Merr.) Orchard

Range area (grAH km^2): OR: 18900; AUS: 44200
Ecozone origin-status: OR: N; AUS: N
World rarity status: R
Ecozone-endemic macroregion occurrence: not endemic to any world ecoregion
Life form(s): EM (scrambling subshrub)
Ploidy state: NA
Speciation age (Ma): NA
Populations also occur in terrestrial, wetland and/ or riparian habitats: YES
Long distance movement vectors: NA
Nuisance status and uses: no nuisance issues; no known uses
Endangered status (IUCN): NA

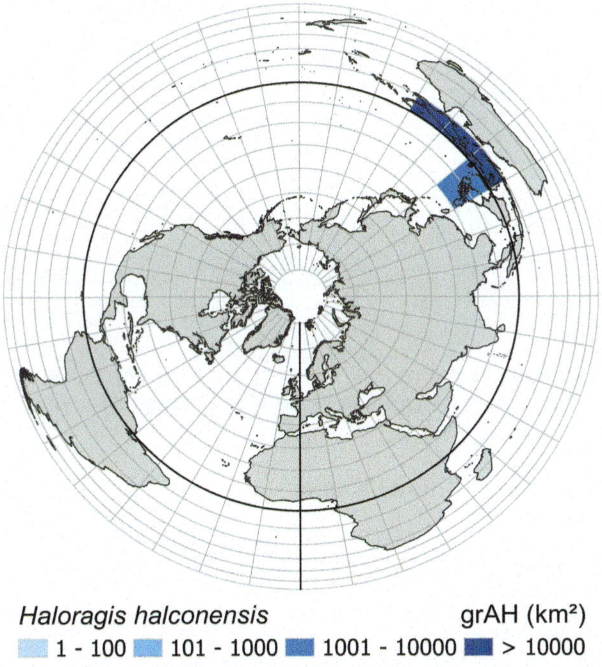

***Haloragis micrantha* (Thunb.) R.Br. ex Sieb. & Zucc.**
Synonyms: *Gonocarpus micranthus* Thunb.

Range area (grAH km^2): PAL: 58800; OR: 77400; AUS: 47900
Ecozone origin-status: PAL: N; OR: N; AUS: N
World rarity status: C
Ecozone-endemic macroregion occurrence: not endemic to any world ecoregion
Life form(s): EM

Ploidy state: NA
Speciation age (Ma): NA
Populations also occur in terrestrial, wetland and/ or riparian habitats: YES
Long distance movement vectors: NA
Nuisance status and uses: no nuisance issues; no known uses
Endangered status (IUCN): NA

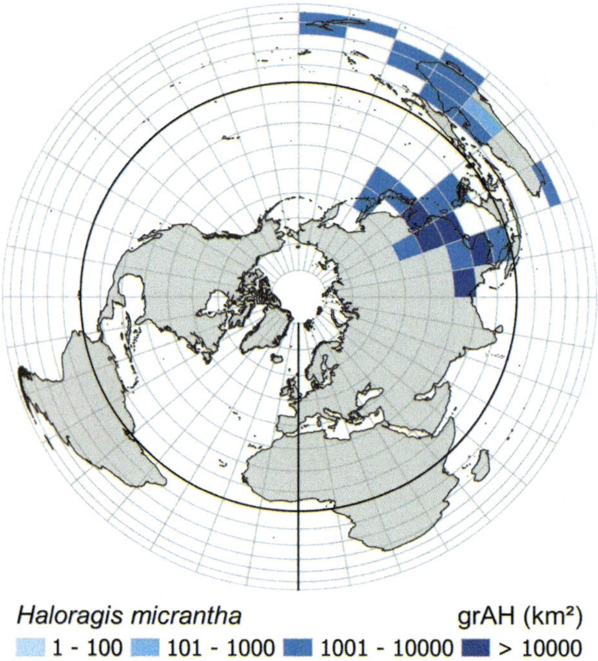

***Laurembergia coccinea* (Blume) Kanitz**
Synonyms:

Range area (grAH km^2): OR: 23500
Ecozone origin-status: OR: E
World rarity status: VR
Ecozone-endemic macroregion occurrence: OR: IND, SUP
Life form(s): EM
Ploidy state: NA
Speciation age (Ma): NA
Populations also occur in terrestrial, wetland and/ or riparian habitats: YES
Long distance movement vectors: NA
Nuisance status and uses: no nuisance issues; no known uses
Endangered status (IUCN): NA

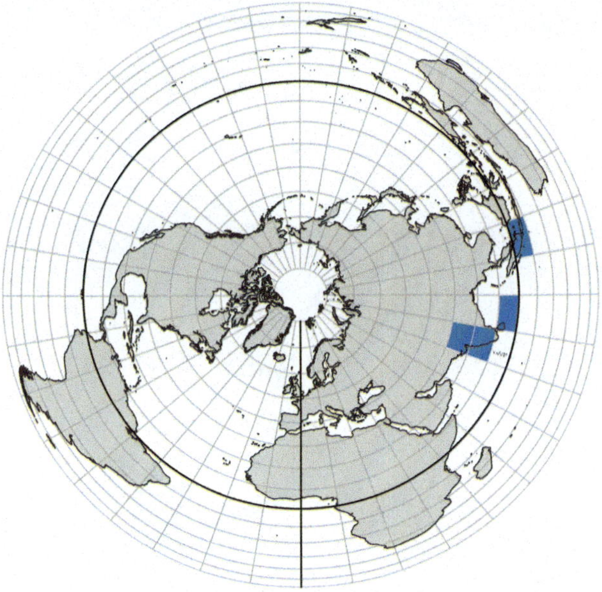

Laurembergia repens **(L.) P.J.Bergius**
Synonyms: *Serpicula repens* (P.J.Bergius) L.

Range area (grAH km^2): AFR: 282100
Ecozone origin-status: AFR: E
World rarity status: C
Ecozone-endemic macroregion occurrence: AFR: FOR, EA, ZAM, SDES, CAPE, MAD
Life form(s): EM (subshrub)
Ploidy state: NA
Speciation age (Ma): NA
Populations also occur in terrestrial, wetland and/ or riparian habitats: YES
Long distance movement vectors: NA
Nuisance status and uses: no nuisance issues; no known uses
Endangered status (IUCN): LC

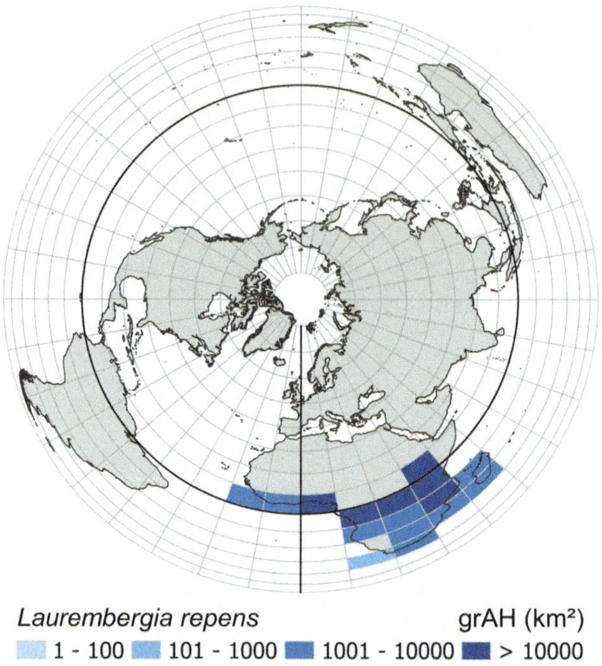

Laurembergia repens (L.) P.J.Bergius
Image copyright: iNaturalist CC-BY @jeremygilmore (Jeremy Gilmore)

***Laurembergia tetrandra* (Schott) Kanitz**
Synonyms: *Haloragis tetrandra* Schott

Range area (grAH km^2): NEO: 131800; AFR: 158300
Ecozone origin-status: NEO: N; AFR: N
World rarity status: C
Ecozone-endemic macroregion occurrence: not endemic to any world ecozone
Life form(s): EM (subshrub)
Ploidy state: NA
Speciation age (Ma): NA
Populations also occur in terrestrial, wetland and/ or riparian habitats: YES
Long distance movement vectors: NA
Nuisance status and uses: no nuisance issues; no known uses
Endangered status (IUCN): LC

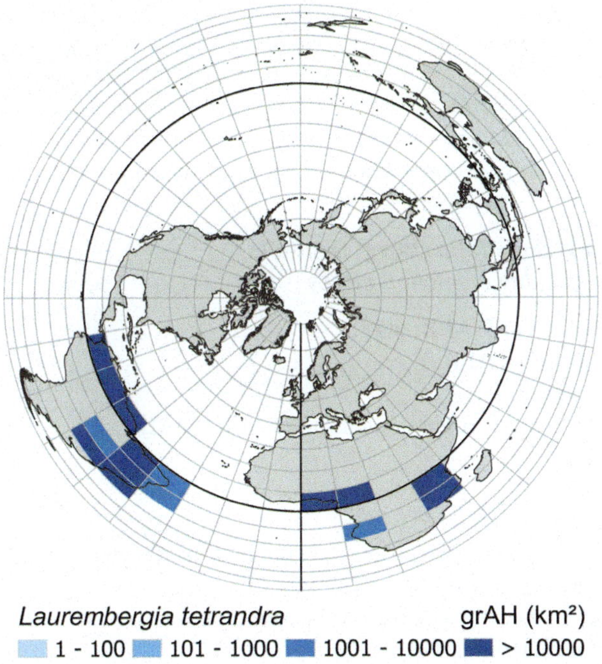

Laurembergia tetrandra (Schott) Kanitz
Image copyright: N. Helme, with permission

***Laurembergia veronicifolia* (Bory) Schindl.**
Synonyms: *Serpicula veronicifolia* Bory

Range area (grAH km^2): AFR: 8000
Ecozone origin-status: AFR: E
World rarity status: VR
Ecozone-endemic macroregion occurrence: AFR: MAD
Life form(s): ?EM
Ploidy state: NA
Speciation age (Ma): NA
Populations also occur in terrestrial, wetland and/ or riparian habitats: YES
Long distance movement vectors: NA
Nuisance status and uses: no nuisance issues; no known uses
Endangered status (IUCN): NA

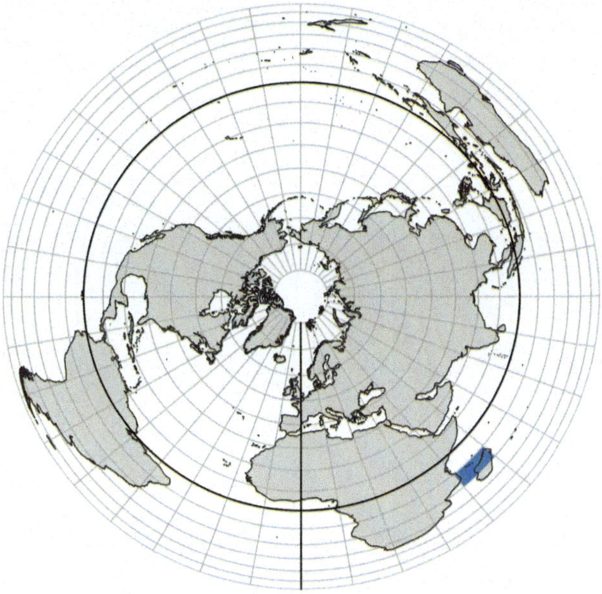

***Meziella trifida* (Nees) Schindl.**
Synonyms: *Myriophyllum trifidum* (Nees) M.L.Moody & Les

Range area (grAH km^2): AUS: 2200
Ecozone origin-status: AUS: E
World rarity status: VR
Ecozone-endemic macroregion occurrence: AUS: WA
Life form(s): SU, EM (subshrub)
Ploidy state: NA
Speciation age (Ma): NA
Populations also occur in terrestrial, wetland and/ or riparian habitats: YES
Long distance movement vectors: NA
Nuisance status and uses: no nuisance issues; no known uses
Endangered status (IUCN): NA

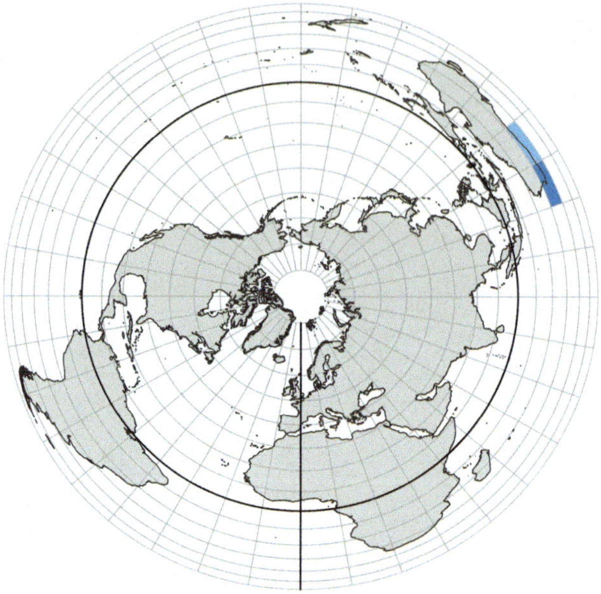

Myriophyllum alpinum **Orchard**
Synonyms:

Range area (grAH km^2): AUS: 14600
Ecozone origin-status: AUS: E
World rarity status: VR
Ecozone-endemic macroregion occurrence: AUS: ESA
Life form(s): SU, EM
Ploidy state: NA
Speciation age (Ma): 7.0
Populations also occur in terrestrial, wetland and/ or riparian habitats: NO
Long distance movement vectors: ?AV
Nuisance status and uses: no nuisance issues; no known uses
Endangered status (IUCN): NA

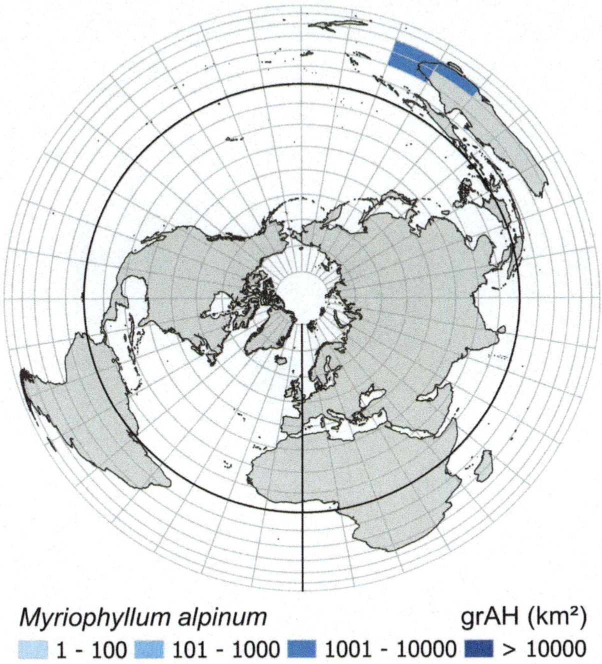

***Myriophyllum alterniflorum* DC.**
Synonyms:

Range area (grAH km²): PAL: 289500; NEA: 615700
Ecozone origin-status: PAL: N; NEA: N
Life form(s): SU
World rarity status: C
Ecozone-endemic macroregion occurrence: not endemic to any world ecozone
Ploidy state: D
Speciation age (Ma): 6.0
Populations also occur in terrestrial, wetland and/ or riparian habitats: NO
Long distance movement vectors: ?AV
Nuisance status and uses: no nuisance issues; remediation
Endangered status (IUCN): LC

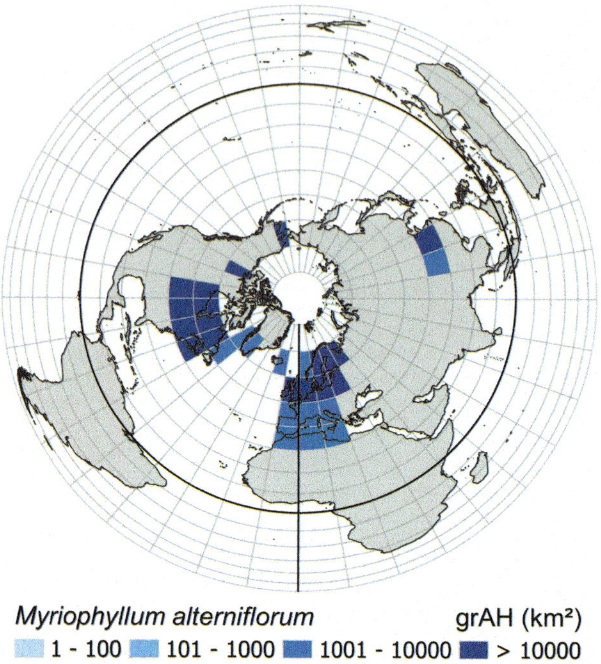

***Myriophyllum amphibium* Labill.**
Synonyms:

Range area (grAH km^2): AUS: 17800
Ecozone origin-status: AUS: E
Life form(s): EM
World rarity status: VR
Ecozone-endemic macroregion occurrence: AUS: ESA, WA
Ploidy state: NA
Speciation age (Ma): 3.0
Populations also occur in terrestrial, wetland and/ or riparian habitats: YES
Long distance movement vectors: NA
Nuisance status and uses: no nuisance issues; water garden plant in Australia
Endangered status (IUCN): NA

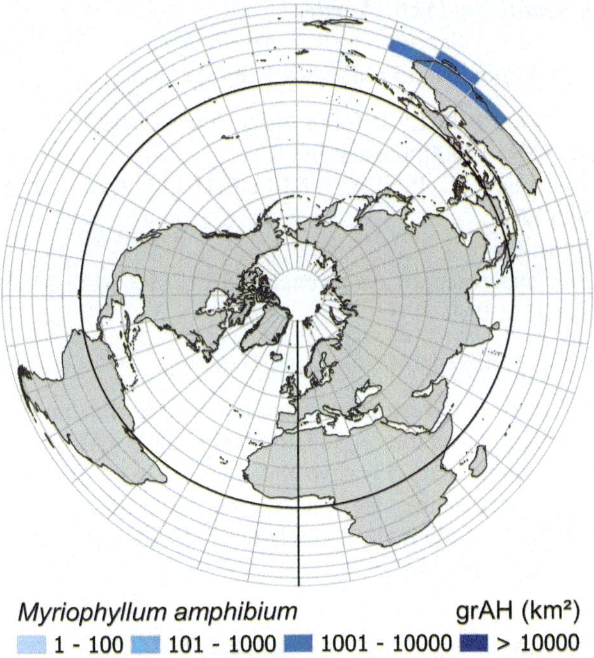

Myriophyllum amphibium Labill.

***Myriophyllum aquaticum* (Vell.) Verdc.**

Synonyms: *Enydria aquatica* Vell., *Myriophyllum brasiliense* Cambess.; *Myriophyllum proserpinacoides* Gillies ex Hook. & Arn.

Range area (grAH km^2): PAL: 261600; NEA: 94600; NEO: 228500; OR: 77000; AUS: 50900; AFR: 19900; present in OCE

Ecozone origin-status: PAL: I; NEA: I; NEO: E(I); OR: I; AUS: I; AFR: I; OCE: I

Life form(s): SU, EM

World rarity status: C

Ecozone-endemic macroregion occurrence: NEO: MG MX, MES AM, NWSA, AND, AMZ, CHA, PAR

Ploidy state: DP

Speciation age (Ma): 17.0

Populations also occur in terrestrial, wetland and/ or riparian habitats: YES

Long distance movement vectors: ANTHROPOCHORY; ?AV

Nuisance status and uses: major invasive aquatic weed in all world ecozones outside its native Neotropical range; uses include aquarium trade (banned from sale in the UK since 2014), medicinal, food, phytoremediation, water garden plant

Endangered status (IUCN): NA

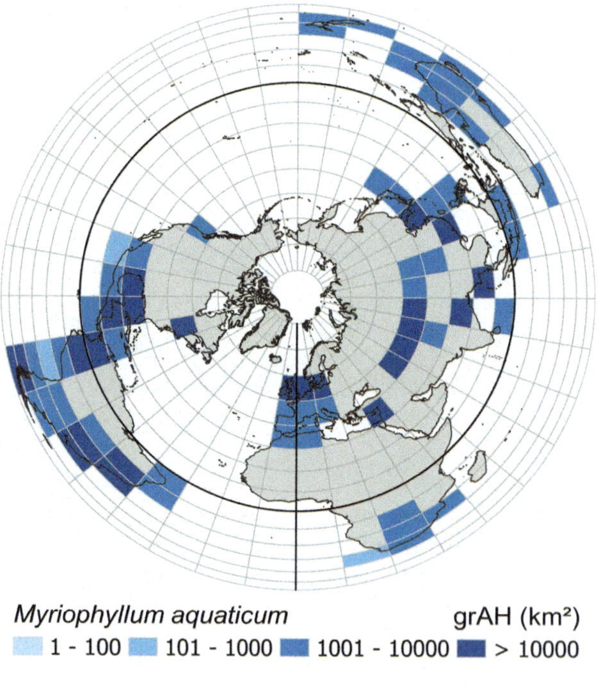

Myriophyllum aquaticum (Vell.) Verdc.
Image copyright: A. Efremov, with permission

Myriophyllum artesium **Halford & Fensham**
Synonyms:

Range area (grAH km^2): AUS: 5300
Ecozone origin-status: AUS: E
Life form(s): SU, EM
World rarity status: VR
Ecozone-endemic macroregion occurrence: AUS: NA
Ploidy state: NA
Speciation age (Ma): NA
Populations also occur in terrestrial, wetland and/ or riparian habitats: YES
Long distance movement vectors: NA
Nuisance status and uses: no nuisance issues; no known uses
Endangered status (IUCN): NA

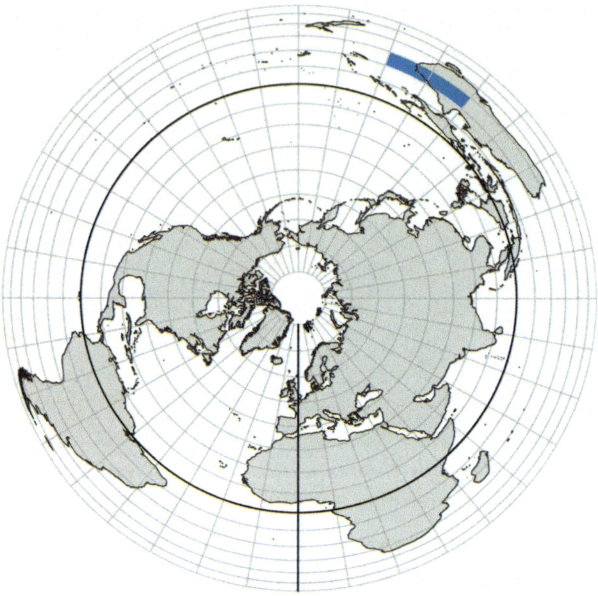

Myriophyllum austropygmaeum **Orchard**
Synonyms:

Range area (grAH km^2): AUS: 3800
Ecozone origin-status: AUS: E
Life form(s): SU, EM
World rarity status: VR
Ecozone-endemic macroregion occurrence: AUS: ESA
Ploidy state: NA
Speciation age (Ma): NA
Populations also occur in terrestrial, wetland and/ or riparian habitats: YES
Long distance movement vectors: NA
Nuisance status and uses: no nuisance issues; no known uses
Endangered status (IUCN): NA

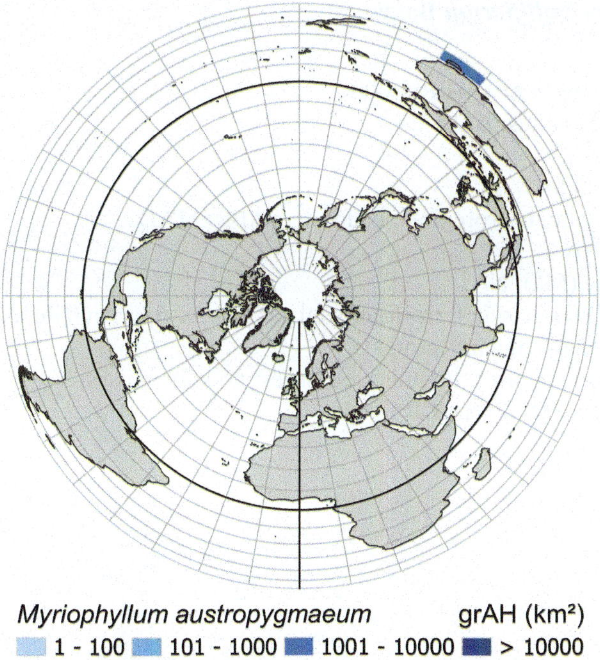

Myriophyllum austropygmaeum Orchard
Image copyright: T. Rudman, with permission

Myriophyllum axilliflorum **Baker**
Synonyms:

Range area (grAH km²): AFR: 8000
Ecozone origin-status: AFR: E
Life form(s): SU
World rarity status: VR
Ecozone-endemic macroregion occurrence: AFR: MAD
Ploidy state: NA
Speciation age (Ma): NA
Populations also occur in terrestrial, wetland and/ or riparian habitats: NO
Long distance movement vectors: NA
Nuisance status and uses: no nuisance issues; no known uses
Endangered status (IUCN): CR

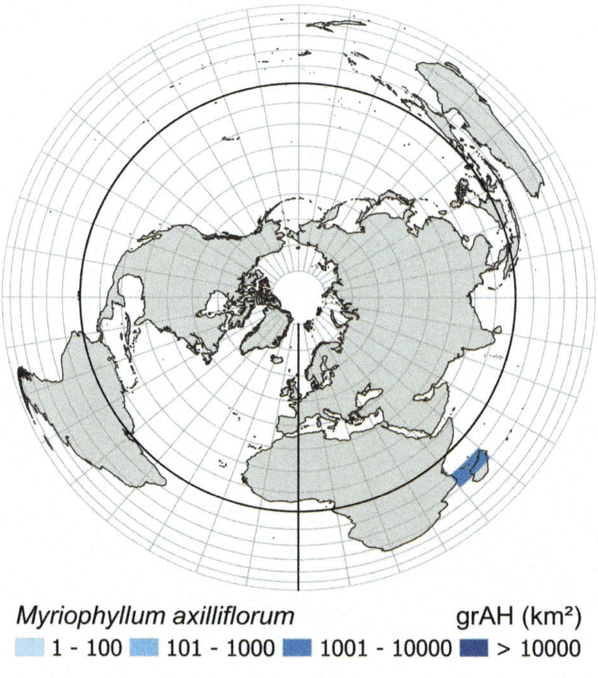

Myriophyllum balladoniense **Orchard**
Synonyms:

Range area (grAH km²): AUS: 2700
Ecozone origin-status: AUS: E
Life form(s): EM
World rarity status: VR
Ecozone-endemic macroregion occurrence: AUS: WA

Ploidy state: NA
Speciation age (Ma): 6.5
Populations also occur in terrestrial, wetland and/ or riparian habitats: YES
Long distance movement vectors: ?AV
Nuisance status and uses: no nuisance issues; no known uses
Endangered status (IUCN): NA

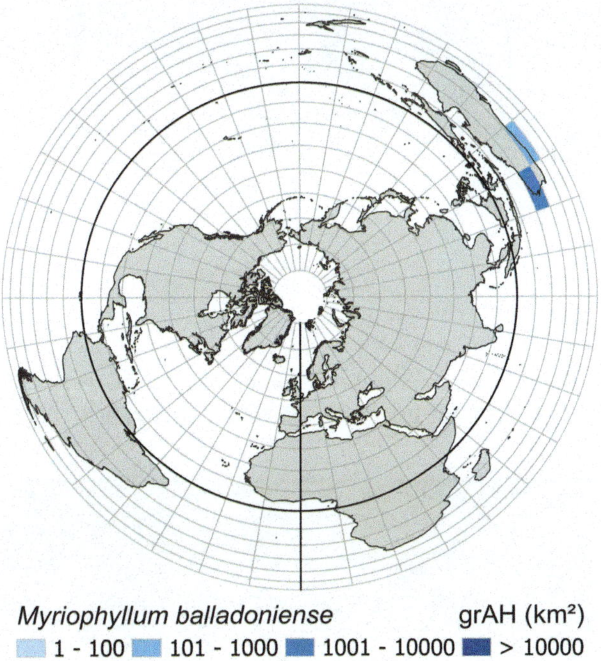

***Myriophyllum caput-medusae* Orchard**
Synonyms:

Range area (grAH km²): AUS: 14000
Ecozone origin-status: AUS: E
Life form(s): SU, EM
World rarity status: VR
Ecozone-endemic macroregion occurrence: AUS: WA, ESA
Ploidy state: NA
Speciation age (Ma): 5.0
Populations also occur in terrestrial, wetland and/ or riparian habitats: NO
Long distance movement vectors: NA
Nuisance status and uses: no nuisance issues; no known uses
Endangered status (IUCN): NA

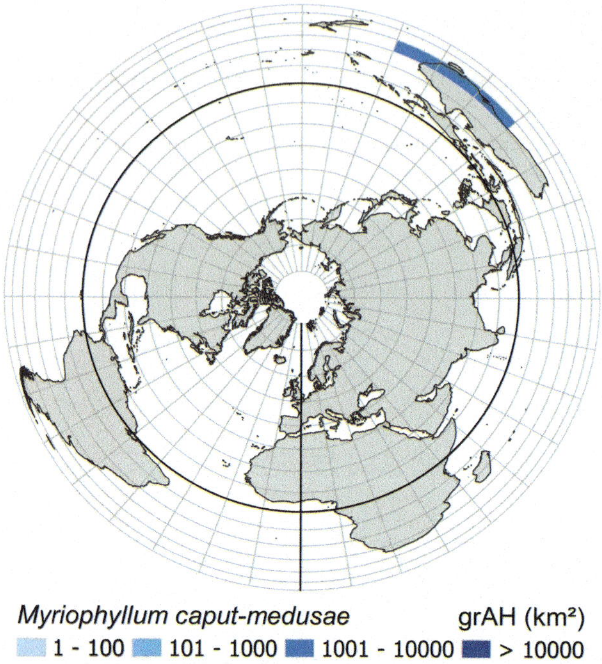

Myriophyllum caput-medusae Orchard
Image copyright: J. Ling, with permission

Myriophyllum coronatum **Orchard**
Synonyms:

Range area (grAH km^2): AUS: 22100
Ecozone origin-status: AUS: E
Life form(s): EM

World rarity status: VR
Ecozone-endemic macroregion occurrence: AUS: NA, NG
Ploidy state: NA
Speciation age (Ma): 9.0
Populations also occur in terrestrial, wetland and/ or riparian habitats: YES
Long distance movement vectors: ?AV
Nuisance status and uses: no nuisance issues; no known uses
Endangered status (IUCN): NA

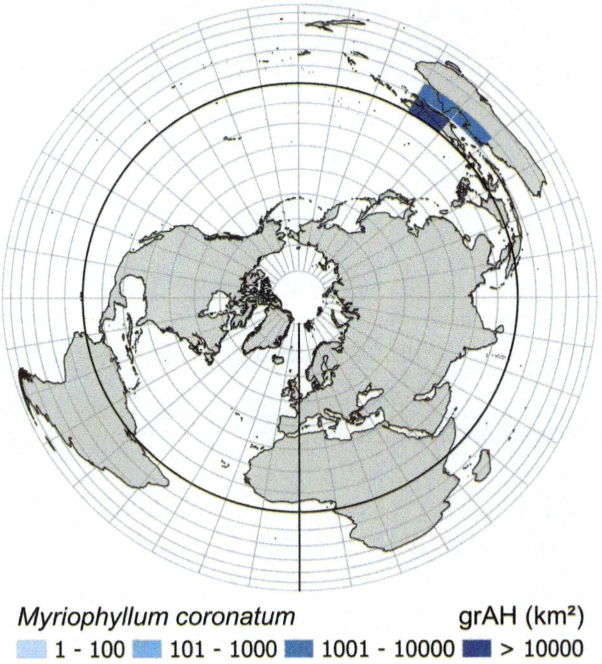

***Myriophyllum costatum* Orchard**
Synonyms:

Range area (grAH km^2): AUS: 7200
Ecozone origin-status: AUS: E
Life form(s): SU, EM
World rarity status: VR
Ecozone-endemic macroregion occurrence: AUS: NA
Ploidy state: NA
Speciation age (Ma): 9.0
Populations also occur in terrestrial, wetland and/ or riparian habitats: NO
Long distance movement vectors: ?AV
Nuisance status and uses: no nuisance issues; no known uses
Endangered status (IUCN): NA

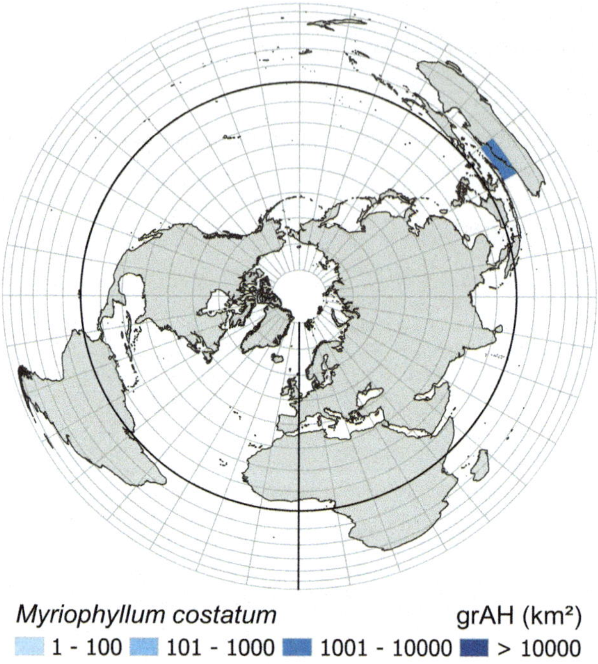

Myriophyllum crispatum **Orchard**
Synonyms:

Range area (grAH km^2): AUS: 34800
Ecozone origin-status: AUS: E
Life form(s): SU, EM
World rarity status: R
Ecozone-endemic macroregion occurrence: AUS: NA, WA, ESA
Ploidy state: NA
Speciation age (Ma): 2.0
Populations also occur in terrestrial, wetland and/ or riparian habitats: NO
Long distance movement vectors: BIRD: *Cygnus atratus* (Black Swan); *Fulica atra* (Eurasian Coot); *Pelecanus conspicillatus* (Australian Pelican); *Anas gracilis* (Grey Teal)
Nuisance status and uses: no nuisance issues; aquarium trade, water garden plant uses
Endangered status (IUCN): NA

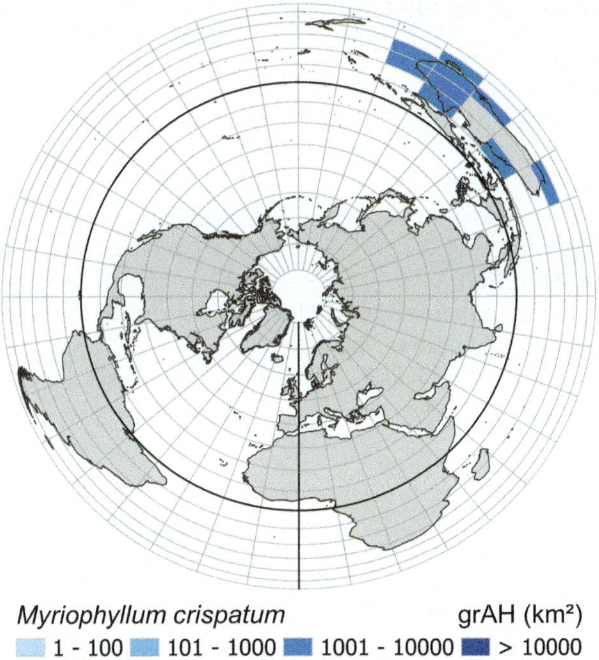

Myriophyllum crispatum Orchard

Myriophyllum decussatum **Orchard**
Synonyms:

Range area (grAH km²): AUS: 4700
Ecozone origin-status: AUS: E
Life form(s): FR, EM
World rarity status: VR
Ecozone-endemic macroregion occurrence: AUS: WA
Ploidy state: NA
Speciation age (Ma): 14.0
Populations also occur in terrestrial, wetland and/ or riparian habitats: YES
Long distance movement vectors: ?AV
Nuisance status and uses: no nuisance issues; no known uses
Endangered status (IUCN): NA

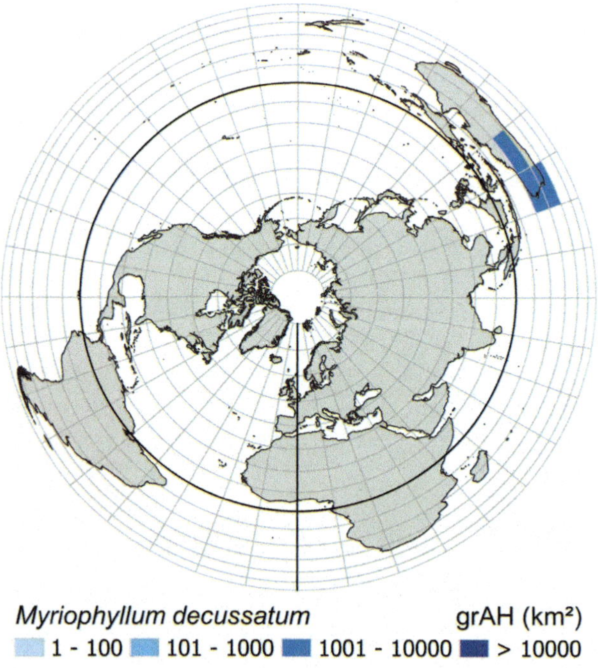

Myriophyllum dicoccum **F. Muell.**
Synonyms:

Range area (grAH km²): OR: 52700; AUS: 29300
Ecozone origin-status: OR: N; AUS: N
Life form(s): SU
World rarity status: R
Ecozone-endemic macroregion occurrence: not endemic to any world ecozone

Ploidy state: NA
Speciation age (Ma): 13.0
Populations also occur in terrestrial, wetland and/ or riparian habitats: YES
Long distance movement vectors: ?AV
Nuisance status and uses: no nuisance issues; aquarium trade
Endangered status (IUCN): NA

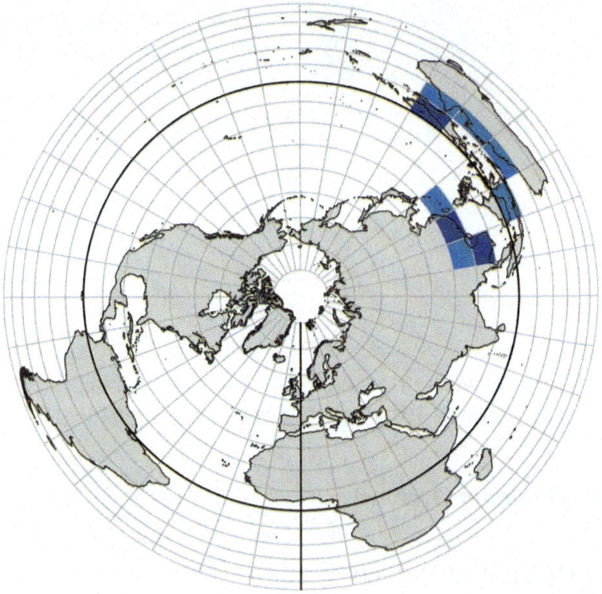

***Myriophyllum drummondii* Benth.**
Synonyms:

Range area (grAH km²): AUS: 3600
Ecozone origin-status: AUS: E
Life form(s): SU, EM
World rarity status: VR
Ecozone-endemic macroregion occurrence: AUS: WA
Ploidy state: NA
Speciation age (Ma): 0.5
Populations also occur in terrestrial, wetland and/ or riparian habitats: YES
Long distance movement vectors: NA
Nuisance status and uses: no nuisance issues; no known uses
Endangered status (IUCN): NA

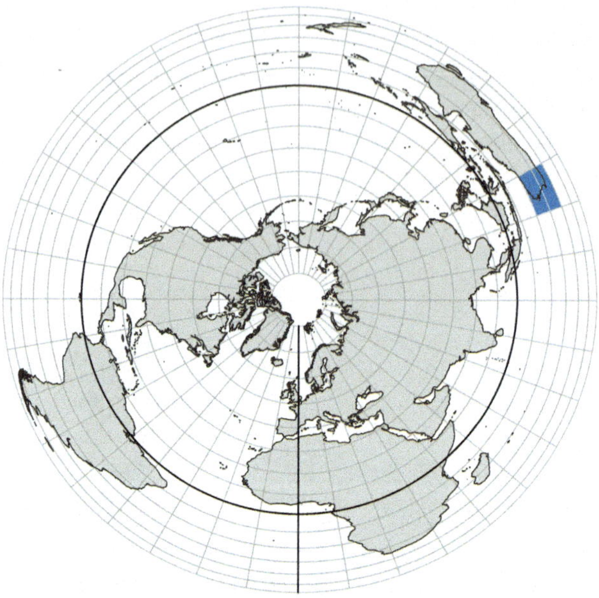

Myriophyllum echinatum **Orchard**
Synonyms:

Range area (grAH km^2): AUS: 1500
Ecozone origin-status: AUS: E
Life form(s): SU, EM
World rarity status: VR
Ecozone-endemic macroregion occurrence: AUS: WA
Ploidy state: NA
Speciation age (Ma): 1.5
Populations also occur in terrestrial, wetland and/ or riparian habitats: YES
Long distance movement vectors: NA
Nuisance status and uses: no nuisance issues; no known uses
Endangered status (IUCN): NA

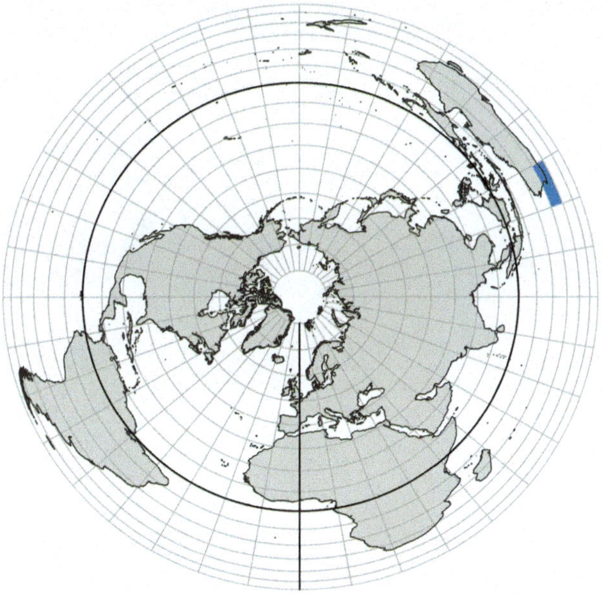

***Myriophyllum exasperatum* D. Wang, D.Yu & Z.Yu Li**
Synonyms:

Range area (grAH km^2): OR: 24900
Ecozone origin-status: OR: E
Life form(s): SU, EM
World rarity status: VR
Ecozone-endemic macroregion occurrence: OR: ICH
Ploidy state: NA
Speciation age (Ma): NA
Populations also occur in terrestrial, wetland and/ or riparian habitats: YES
Long distance movement vectors: NA
Nuisance status and uses: no nuisance issues; no known uses
Endangered status (IUCN): NA

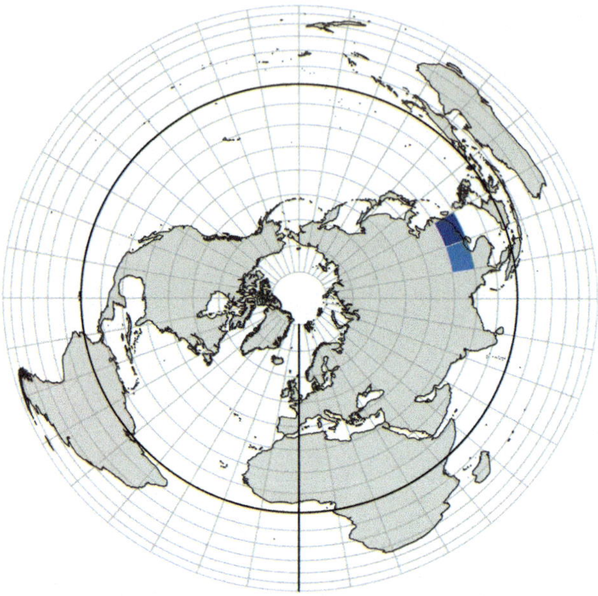

Myriophyllum farwellii **Morong**
Synonyms:

Range area (grAH km^2): NEA: 477800; NEO: 6300
Ecozone origin-status: NEA: N; NEO: N
Life form(s): SU
World rarity status: C
Ecozone-endemic macroregion occurrence: not endemic to any world ecozone
Ploidy state: NA
Speciation age (Ma): 2.5
Populations also occur in terrestrial, wetland and/ or riparian habitats: NO
Long distance movement vectors: NA
Nuisance status and uses: no nuisance issues; no known uses
Endangered status (IUCN): LC

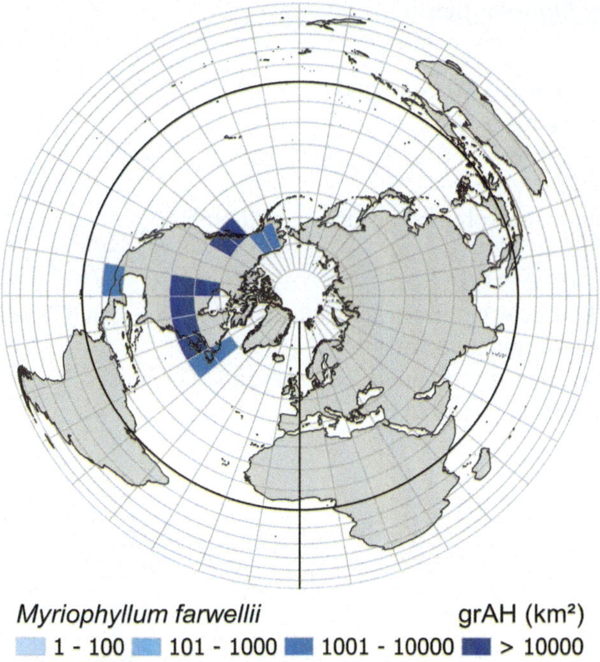

Myriophyllum farwellii Morong
Image copyright: iNaturalist CC-0 @allan7 (Allan Harris)

***Myriophyllum filiforme* Benth**
Synonyms:

Range area (grAH km^2): AUS: 18200
Ecozone origin-status: AUS: E
Life form(s): SU, EM
World rarity status: VR
Ecozone-endemic macroregion occurrence: AUS: NA
Ploidy state: NA
Speciation age (Ma): 11.0
Populations also occur in terrestrial, wetland and/ or riparian habitats: YES
Long distance movement vectors: ?AV
Nuisance status and uses: no nuisance issues; no known uses
Endangered status (IUCN): NA

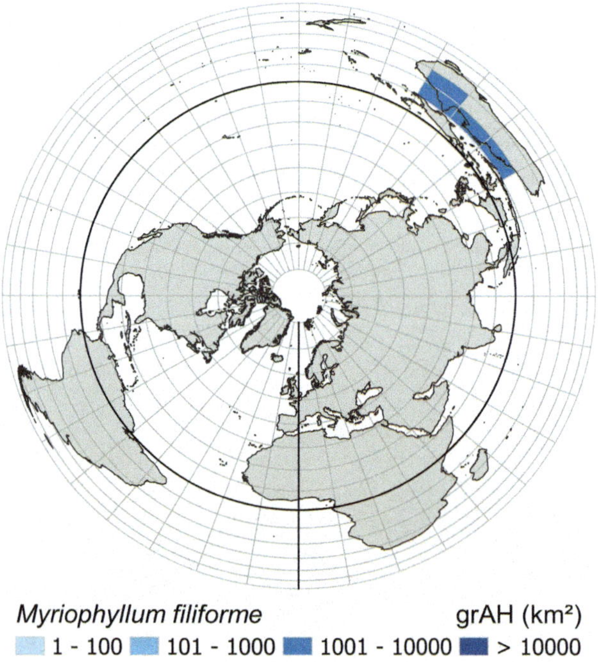

***Myriophyllum glomeratum* Schindl.**
Synonyms:

Range area (grAH km^2): AUS: 17800
Ecozone origin-status: AUS: E
Life form(s): SU, EM
World rarity status: VR
Ecozone-endemic macroregion occurrence: AUS: ESA

Ploidy state: NA
Speciation age (Ma): NA
Populations also occur in terrestrial, wetland and/ or riparian habitats: YES
Long distance movement vectors: NA
Nuisance status and uses: no nuisance issues; no known uses
Endangered status (IUCN): NA

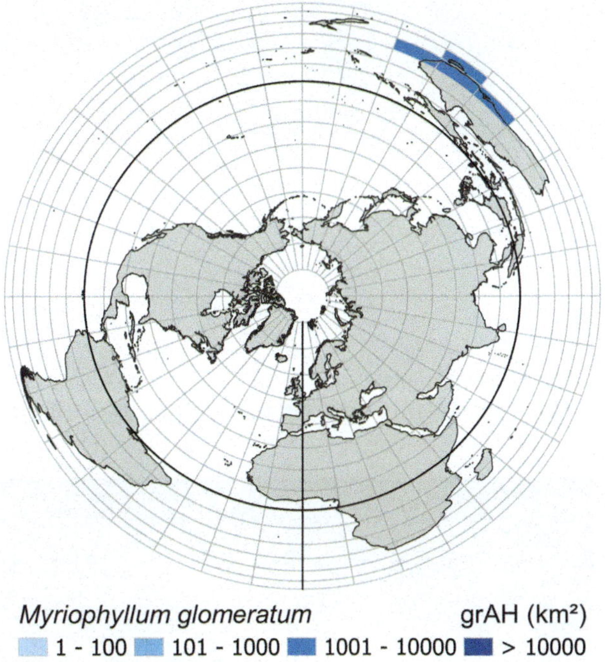

Myriophyllum gracile **Benth.**
Synonyms:

Range area (grAH km^2): AUS: 19200
Ecozone origin-status: AUS: E
Life form(s): SU, EM
World rarity status: VR
Ecozone-endemic macroregion occurrence: AUS: NA, ESA
Ploidy state: NA
Speciation age (Ma): NA
Populations also occur in terrestrial, wetland and/ or riparian habitats: YES
Long distance movement vectors: NA
Nuisance status and uses: no nuisance issues; aquarium trade
Endangered status (IUCN): NA

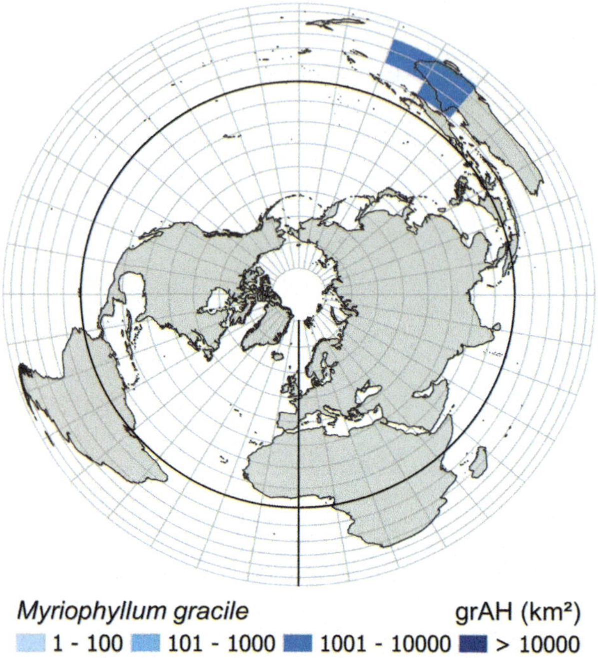

Myriophyllum gracile Benth.
Image copyright: D. Davey, with permission

***Myriophyllum heterophyllum* Michx.**
Synonyms:

Range area (grAH km^2): PAL: 78300; NEA: 395000; NEO: 36100; OR: 17900
Ecozone origin-status: PAL: I; NEA: N; NEO: N; OR: I
Life form(s): SU, EM
World rarity status: C
Ecozone-endemic macroregion occurrence: not endemic to any world ecozone
Ploidy state: NA
Speciation age (Ma): 13.0
Populations also occur in terrestrial, wetland and/ or riparian habitats: YES
Long distance movement vectors: ANTHROPOCHORY; ?AV
Nuisance status and uses: considered an aquatic weed in parts of Europe, and also in parts of its Nearctic home range (New England, north western USA: banned from sale in some New England states and Washington state); some phytoremediation potential, aquarium trade and water garden plant uses
Endangered status (IUCN): LC

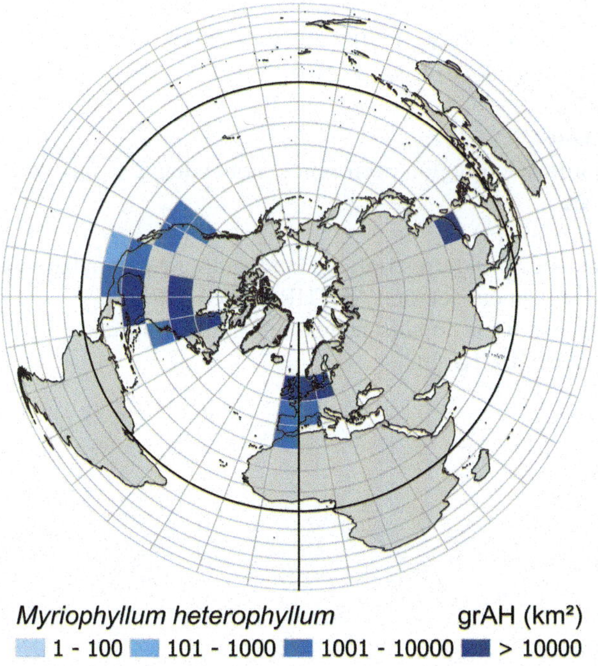

Myriophyllum heterophyllum Michx.
Image copyright: iNaturalist CC-BY @a_schmidy_botanist (Samuel A. Schmid)

***Myriophyllum hippuroides* Nutt.**
Synonyms:

Range area (grAH km^2): PAL: 31700; NEA: 296800; NEO: 25900
Ecozone origin-status: PAL: I; NEA: N; NEO: N
Life form(s): SU
World rarity status: C
Ecozone-endemic macroregion occurrence: not endemic to any world ecozone
Ploidy state: NA
Speciation age (Ma): 1.0
Populations also occur in terrestrial, wetland and/ or riparian habitats: NO
Long distance movement vectors: ANTHROPOCHORY
Nuisance status and uses: no nuisance issues; aquarium trade
Endangered status (IUCN): LC

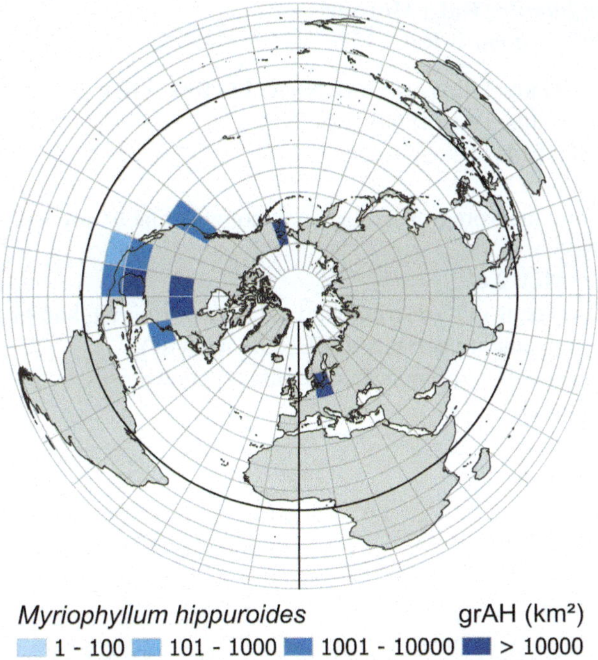

Myriophyllum hippuroides Nutt.
Image copyright: iNaturalist CC-BY @fern_friend (Nolan Exe)

***Myriophyllum humile* (Raf.) Morong**
Synonyms:*Burshia humilis* Raf.

Range area (grAH km^2): PAL: 45400; NEA: 118100; NEO: 20800; OR: 18300
Ecozone origin-status: PAL: I; NEA: N; NEO: N; OR: N
Life form(s): SU, EM
World rarity status: C
Ecozone-endemic macroregion occurrence: not endemic to any world ecozone
Ploidy state: NA
Speciation age (Ma): 2.6
Populations also occur in terrestrial, wetland and/ or riparian habitats: YES
Long distance movement vectors: ANTHROPOCHORY
Nuisance status and uses: no nuisance issues; no known uses
Endangered status (IUCN): LC

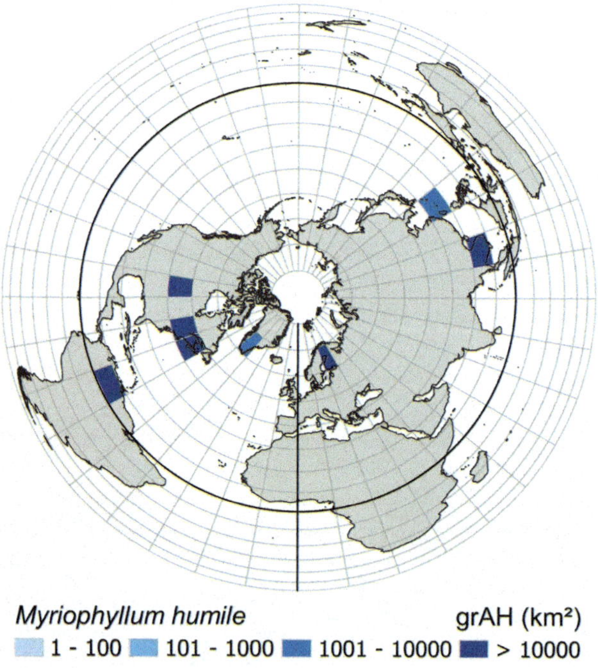

Myriophyllum humile (Raf.) Morong
Image copyright: iNaturalist CC0 @hagladit (Haley Gladitsch)

***Myriophyllum implicatum* Orchard**
Synonyms:

Range area (grAH km^2): AUS: 25500
Ecozone origin-status: AUS: E
Life form(s): EM
World rarity status: VR
Ecozone-endemic macroregion occurrence: AUS: NA, ESA
Ploidy state: NA
Speciation age (Ma): NA
Populations also occur in terrestrial, wetland and/ or riparian habitats: YES
Long distance movement vectors: NA
Nuisance status and uses: no nuisance issues; no known uses
Endangered status (IUCN): NA

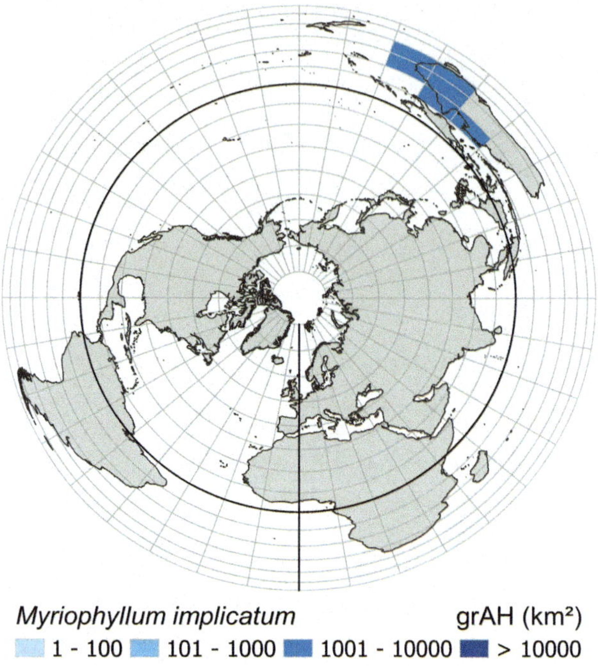

***Myriophyllum indicum* Willd.**
Synonyms:

Range area (grAH km^2): OR: 53300
Ecozone origin-status: OR: E
Life form(s): SU, EM
World rarity status: R
Ecozone-endemic macroregion occurrence: OR: IND
Ploidy state: D
Speciation age (Ma): NA
Populations also occur in terrestrial, wetland and/ or riparian habitats: NO
Long distance movement vectors: NA
Nuisance status and uses: no nuisance issues; no known uses
Endangered status (IUCN): LC

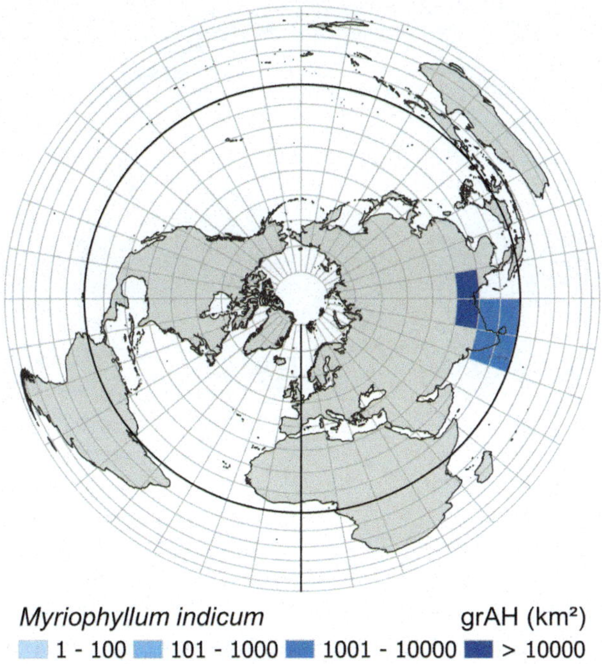

***Myriophyllum latifolium* F.Muell.**
Synonyms:

Range area (grAH km^2): AUS: 14500
Ecozone origin-status: AUS: E
Life form(s): SU, EM
World rarity status: VR
Ecozone-endemic macroregion occurrence: AUS: ESA
Ploidy state: NA
Speciation age (Ma): 12.0
Populations also occur in terrestrial, wetland and/ or riparian habitats: NO
Long distance movement vectors: ?AV
Nuisance status and uses: no nuisance issues; no known uses
Endangered status (IUCN): NA

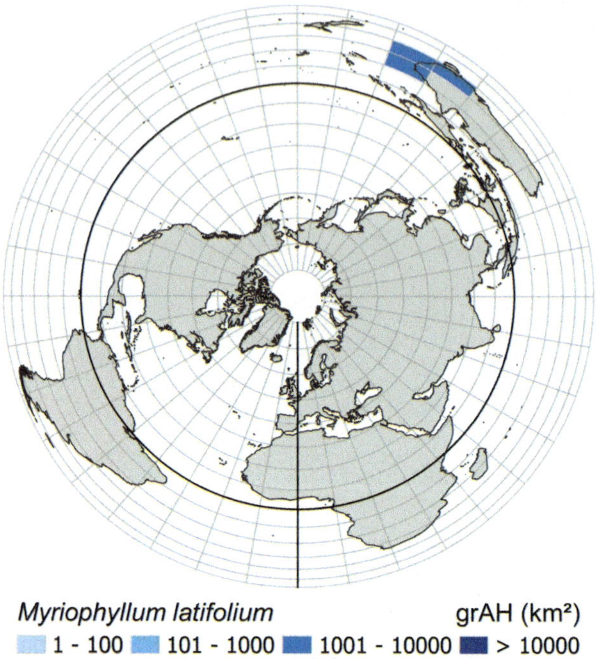

Myriophyllum laxum **Shuttlew. ex Chapm.**
Synonyms:

Range area (grAH km^2): NEA: 34000; NEO: 10300
Ecozone origin-status: NEA: N; NEO: N
Life form(s): SU
World rarity status: R
Ecozone-endemic macroregion occurrence: not endemic to any world ecozone
Ploidy state: NA
Speciation age (Ma): 2.5
Populations also occur in terrestrial, wetland and/ or riparian habitats: NO
Long distance movement vectors: HYDROCHORY (NEA to NEO)
Nuisance status and uses: no nuisance issues; no known uses
Endangered status (IUCN): NA (listed by the U.S. Fish and Wildlife Service as Potentially Endangered)

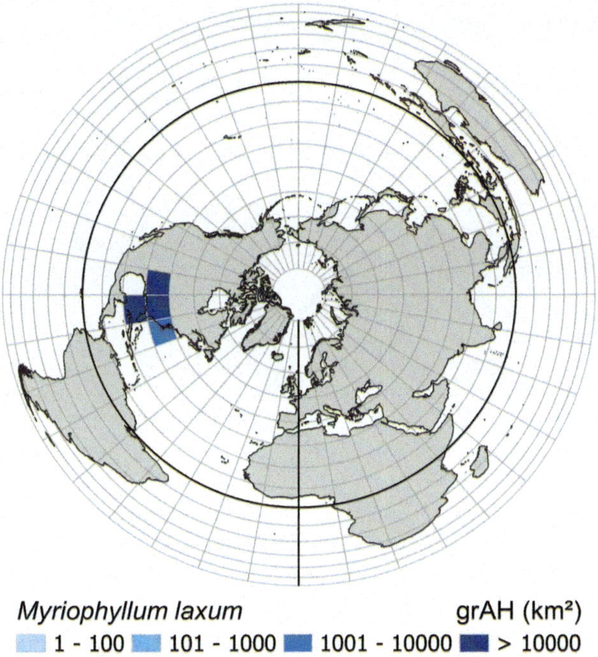

***Myriophyllum limnophilum* Orchard**
Synonyms:

Range area (grAH km^2): AUS: 1500
Ecozone origin-status: AUS: E
Life form(s): EM
World rarity status: VR
Ecozone-endemic macroregion occurrence: AUS: WA
Ploidy state: NA
Speciation age (Ma): 0.5
Populations also occur in terrestrial, wetland and/ or riparian habitats: YES
Long distance movement vectors: NA
Nuisance status and uses: no nuisance issues; no known uses
Endangered status (IUCN): NA

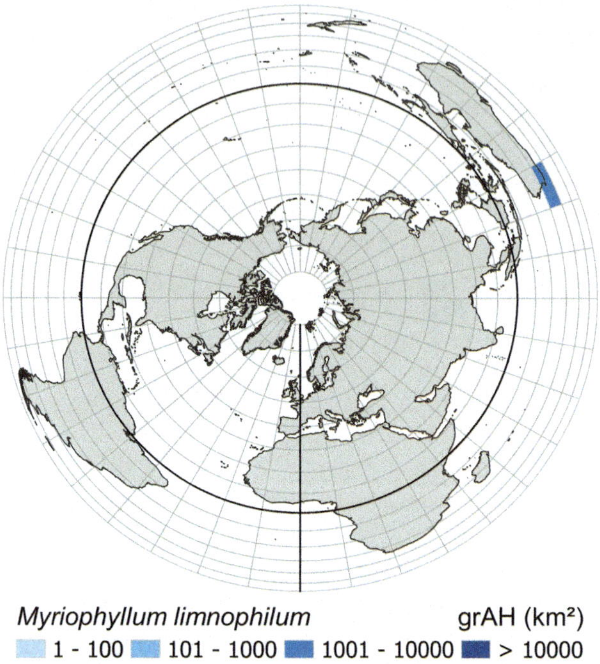

Myriophyllum lophatum **Orchard**
Synonyms:

Range area (grAH km^2): AUS: 16100
Ecozone origin-status: AUS: E
Life form(s): EM
World rarity status: VR
Ecozone-endemic macroregion occurrence: AUS: ESA, WA
Ploidy state: NA
Speciation age (Ma): 7.0
Populations also occur in terrestrial, wetland and/ or riparian habitats: YES
Long distance movement vectors: ?AV
Nuisance status and uses: no nuisance issues; no known uses
Endangered status (IUCN): NA

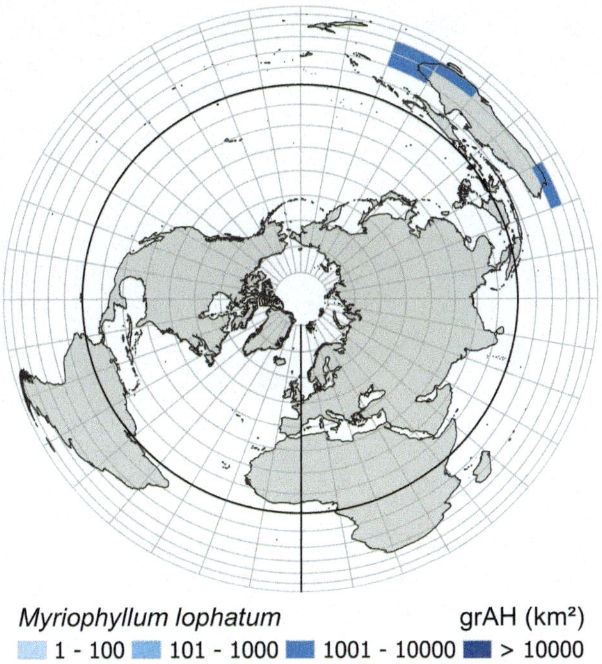

Myriophyllum mattogrossense **Hoehne**
Synonyms:

Range area (grAH km^2): NEO: 46100
Ecozone origin-status: NEO: E
Life form(s): SU, EM (hydrosubshrub)
World rarity status: R
Ecozone-endemic macroregion occurrence: NEO: AMZ
Ploidy state: NA
Speciation age (Ma): 17.0
Populations also occur in terrestrial, wetland and/ or riparian habitats: YES
Long distance movement vectors: ?AV
Nuisance status and uses: no nuisance issues, though considered a potential aquatic weed risk in the USA; aquarium trade
Endangered status (IUCN): NA

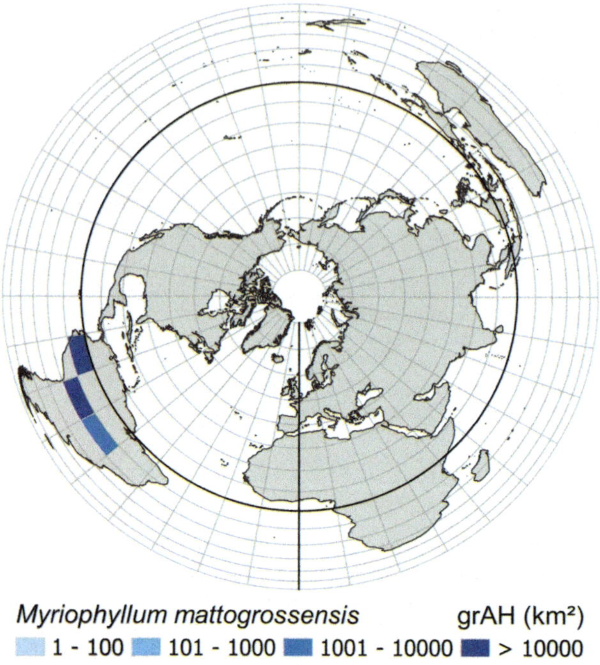

Myriophyllum mezianum **Schindl.**
Synonyms:

Range area (grAH km^2): AFR: 10100
Ecozone origin-status: AFR: E
Life form(s): SU, EM
World rarity status: VR
Ecozone-endemic macroregion occurrence: AFR: MAD
Ploidy state: NA
Speciation age (Ma): NA
Populations also occur in terrestrial, wetland and/ or riparian habitats: YES
Long distance movement vectors: NA
Nuisance status and uses: no nuisance issues; aquarium trade
Endangered status (IUCN): EN

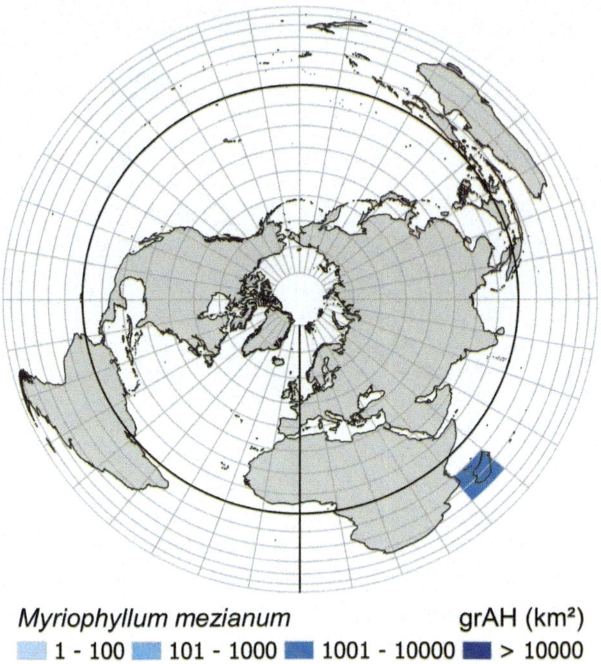

***Myriophyllum muelleri* Sond.**
Synonyms:

Range area (grAH km^2): AUS: 18000
Ecozone origin-status: AUS: E
Life form(s): SU, EM
World rarity status: VR
Ecozone-endemic macroregion occurrence: AUS: WA, ESA
Ploidy state: NA
Speciation age (Ma): NA
Populations also occur in terrestrial, wetland and/ or riparian habitats: YES
Long distance movement vectors: NA
Nuisance status and uses: no nuisance issues; no known uses
Endangered status (IUCN): NA

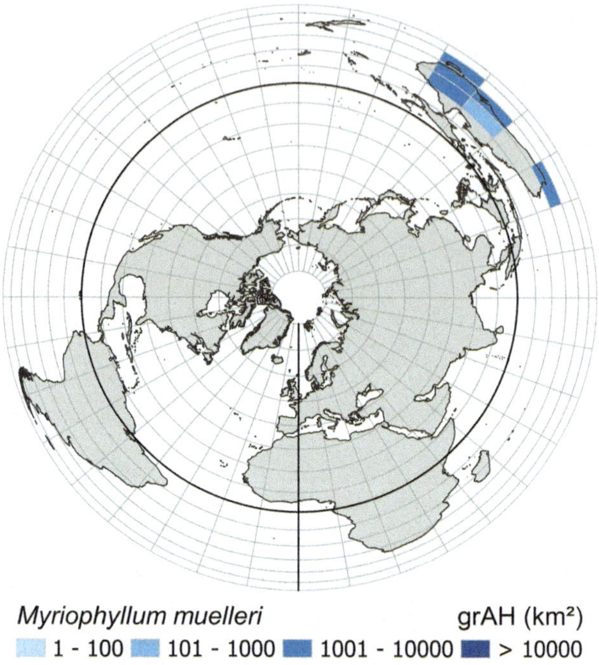

Myriophyllum muelleri Sond.
Image copyright: iNaturalist CC0 @mattintas (Matthew Baker)

Myriophyllum muricatum **Orchard**
Synonyms:

Range area (grAH km^2): OR: 25800; AUS: 14600
Ecozone origin-status: OR: N; AUS: N
Life form(s): SU
World rarity status: R
Ecozone-endemic macroregion occurrence: not endemic to any world ecozone
Ploidy state: NA
Speciation age (Ma): 13.0
Populations also occur in terrestrial, wetland and/ or riparian habitats: NO
Long distance movement vectors: ?AV
Nuisance status and uses: no nuisance issues; aquarium trade
Endangered status (IUCN): NA

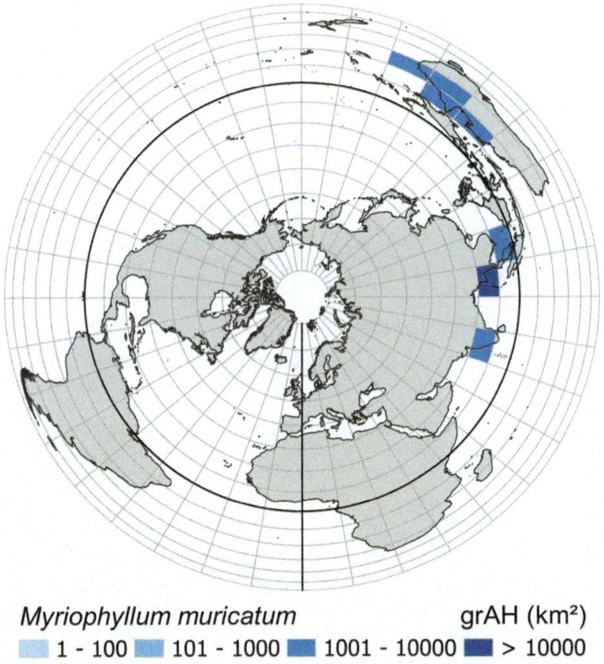

Myriophyllum oguraense **Miki**
Synonyms:

Range area (grAH km^2): PAL: 55300; OR: 20500
Ecozone origin-status: PAL: N; OR: N
Life form(s): SU, EM
World rarity status: R
Ecozone-endemic macroregion occurrence: not endemic to any world ecozone
Ploidy state: NA

Speciation age (Ma): 5.0
Populations also occur in terrestrial, wetland and/ or riparian habitats: NO
Long distance movement vectors: NA
Nuisance status and uses: no nuisance issues; no known uses
Endangered status (IUCN): NA

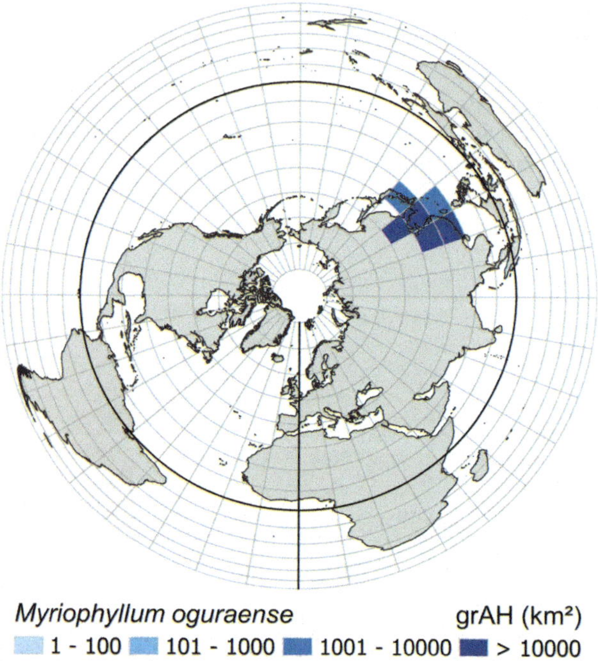

Myriophyllum oguraense Miki
Image copyright: Y. Cho, with permission

Myriophyllum oliganthum (Wight & Arn.) F.Muell.
Synonyms: *Haloragis oligantha* Wight & Arn.

Range area (grAH km^2): OR: 10300
Ecozone origin-status: OR: E
Life form(s): SU, FR, EM
World rarity status: VR
Ecozone-endemic macroregion occurrence: OR: IND
Ploidy state: NA
Speciation age (Ma): NA
Populations also occur in terrestrial, wetland and/ or riparian habitats: NO
Long distance movement vectors: NA
Nuisance status and uses: no nuisance issues; no known uses
Endangered status (IUCN): LC

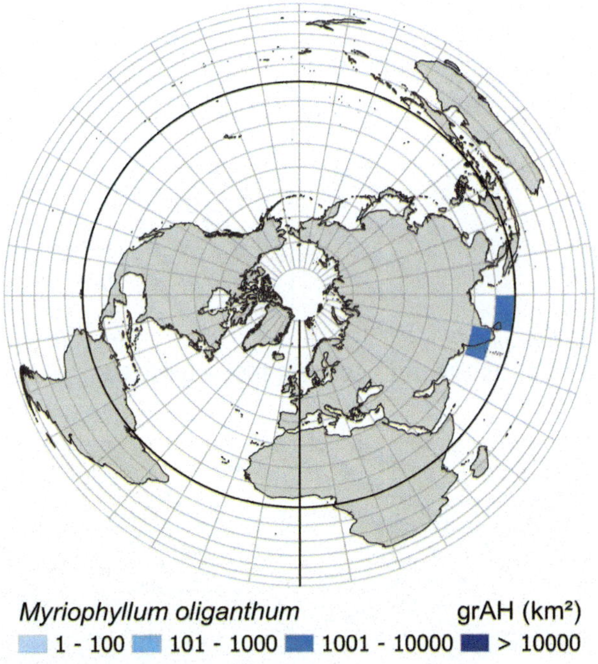

Myriophyllum papillosum Orchard
Synonyms:

Range area (grAH km^2): AUS: 19300
Ecozone origin-status: AUS: E
Life form(s): SU, EM
World rarity status: VR
Ecozone-endemic macroregion occurrence: AUS: ESA

Ploidy state: NA
Speciation age (Ma): 11.0
Populations also occur in terrestrial, wetland and/ or riparian habitats: YES
Long distance movement vectors: ?AV
Nuisance status and uses: no nuisance issues; aquarium trade
Endangered status (IUCN): NA

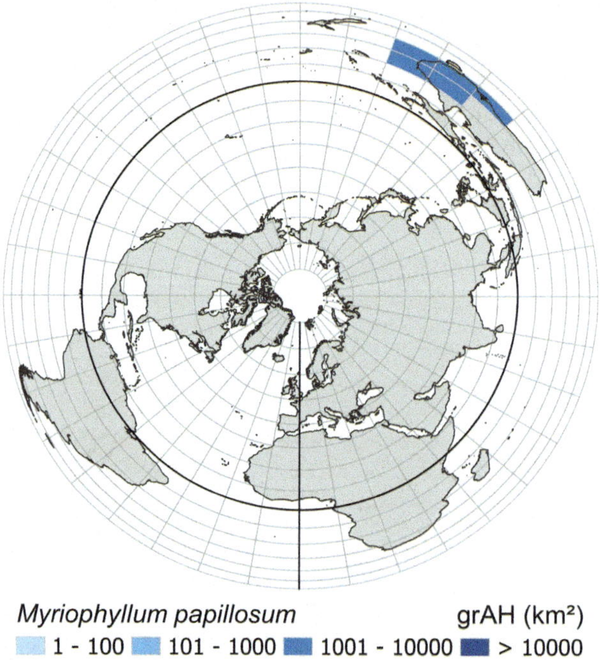

Myriophyllum papillosum Orchard
Image copyright: G. Holland, with permission

Myriophyllum pedunculatum **Hook.f.**
Synonyms:

Range area (grAH km^2): AUS: 48200
Ecozone origin-status: AUS: E
Life form(s): SU, EM (hydrosubshrub or subshrub)
World rarity status: VR
Ecozone-endemic macroregion occurrence: AUS: NG, NA, ESA, NZ
Ploidy state: NA
Speciation age (Ma): 21.0
Populations also occur in terrestrial, wetland and/ or riparian habitats: YES
Long distance movement vectors: ?AV
Nuisance status and uses: no nuisance issues; water garden plant
Endangered status (IUCN): NA

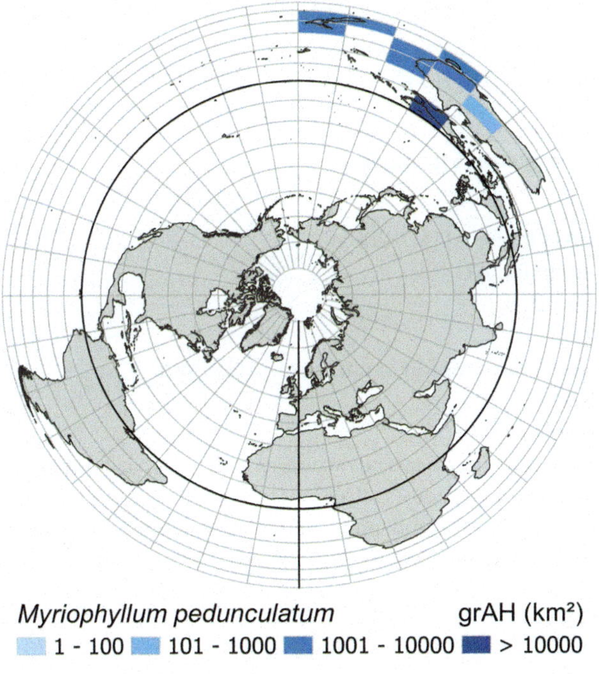

Myriophyllum pedunculatum Hook.f.
Image copyright: iNaturalist CC0 @pjd1 (Peter de Lange)

Myriophyllum petraeum Orchard
Synonyms:

Range area (grAH km^2): AUS: 2200
Ecozone origin-status: AUS: E
Life form(s): SU, EM
World rarity status: VR
Ecozone-endemic macroregion occurrence: AUS: WA
Ploidy state: NA
Speciation age (Ma): 6.0
Populations also occur in terrestrial, wetland and/ or riparian habitats: NO
Long distance movement vectors: ?AV
Nuisance status and uses: no nuisance issues; no known uses
Endangered status (IUCN): NA

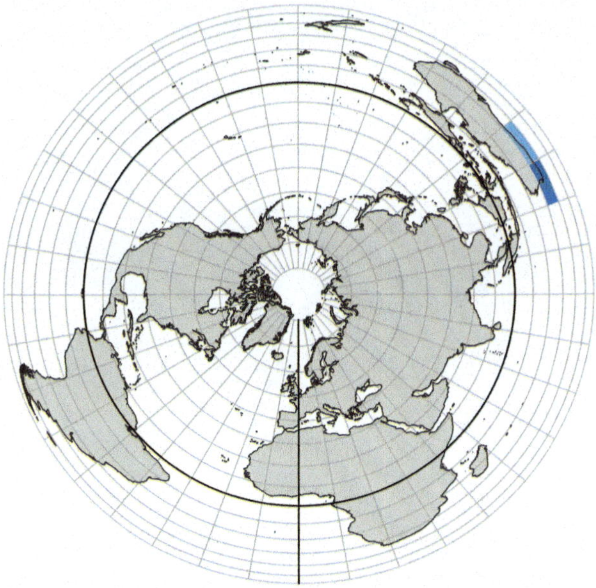

***Myriophyllum pinnatum* (Walter) Britton, Sterns & Poggenb.**
Synonyms: *Myriophyllum scabratum* Michx., *Potamogeton pinnatum* Walter

Range area (grAH km^2): NEA: 362000; NEO: 30400
Ecozone origin-status: NEA: N; NEO: N
Life form(s): SU
World rarity status: C
Ecozone-endemic macroregion occurrence: not endemic to any world ecozone
Ploidy state: NA
Speciation age (Ma): 2.0
Populations also occur in terrestrial, wetland and/ or riparian habitats: NO
Long distance movement vectors: NA
Nuisance status and uses: no nuisance issues; aquarium trade, water garden plant
Endangered status (IUCN): LC

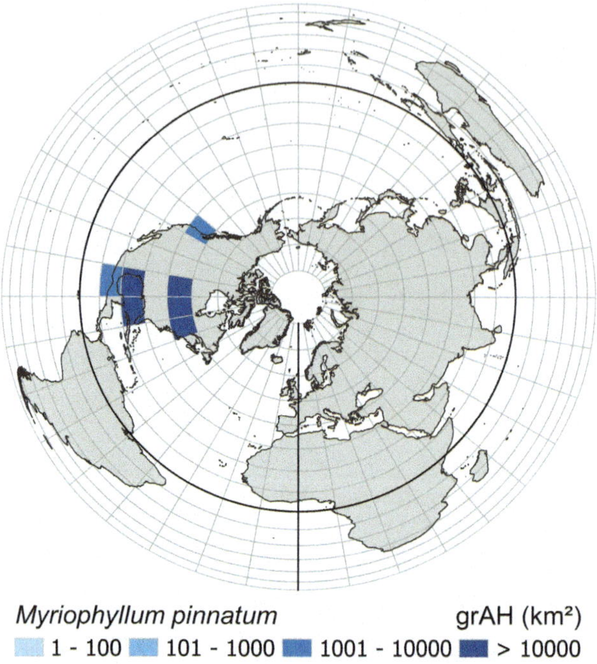

Myriophyllum pinnatum (Walter) Britton, Sterns & Poggenb.
Image copyright: iNaturalist CC-BY @pufferchung

Myriophyllum porcatum **Orchard**
Synonyms:

Range area (grAH km^2): AUS: 7800
Ecozone origin-status: AUS: E
Life form(s): SU, EM
World rarity status: VR

Ecozone-endemic macroregion occurrence: AUS: ESA
Ploidy state: NA
Speciation age (Ma): NA
Populations also occur in terrestrial, wetland and/ or riparian habitats: YES
Long distance movement vectors: NA
Nuisance status and uses: no nuisance issues; no known uses
Endangered status (IUCN): NA (considered Critically Endangered in Victoria, where all its records are located)

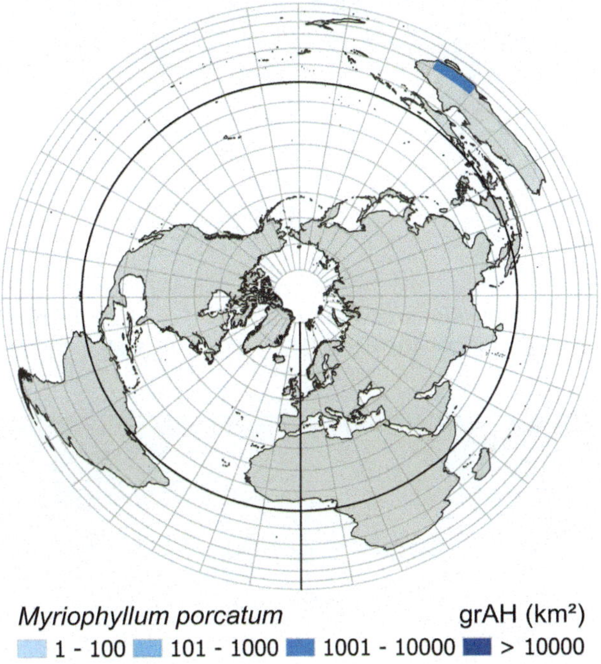

***Myriophyllum propinquum* A. Cunn.**
Synonyms:

Range area (grAH km^2): OR: 2600; AUS: 41500
Ecozone origin-status: OR: I; AUS: E(I)
Life form(s): SU, EM (hydrosubshrub)
World rarity status: R
Ecozone-endemic macroregion occurrence: AUS: NA, ESA, NZ
Ploidy state: D
Speciation age (Ma): NA
Populations also occur in terrestrial, wetland and/ or riparian habitats: YES
Long distance movement vectors: ANTHROPOCHORY
Nuisance status and uses: no nuisance issues; aquarium trade, water garden plant
Endangered status (IUCN): NA

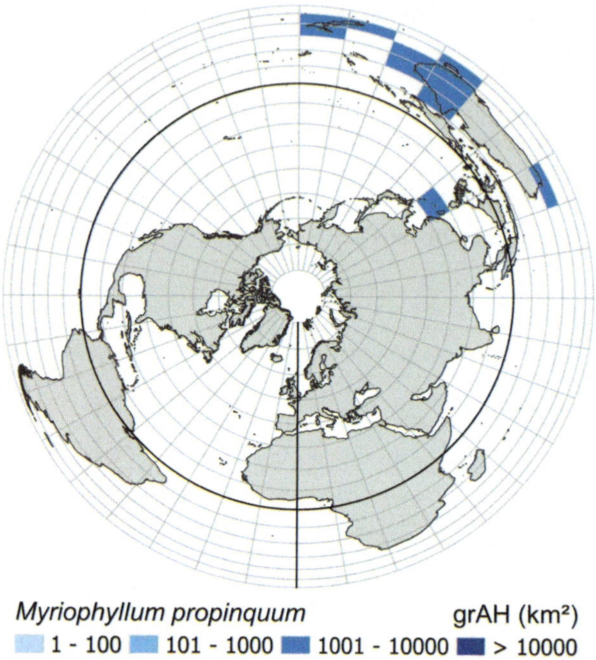

Myriophyllum propinquum A. Cunn.

Myriophyllum quitense **Kunth**
Synonyms:

Range area (grAH km^2): NEA: 38600; NEO: 144300
Ecozone origin-status: NEA: N; NEO: N; Records from AUS are questionable [110]
Life form(s): SU, EM
World rarity status: C
Ecozone-endemic macroregion occurrence: not endemic to any world ecozone
Ploidy state: DP
Speciation age (Ma): 2.5
Populations also occur in terrestrial, wetland and/ or riparian habitats: YES
Long distance movement vectors: NA
Nuisance status and uses: no nuisance issues; uncertain reports of use in the aquarium trade, or as a water garden plant, biomonitoring and phytoremediation uses
Endangered status (IUCN): LC

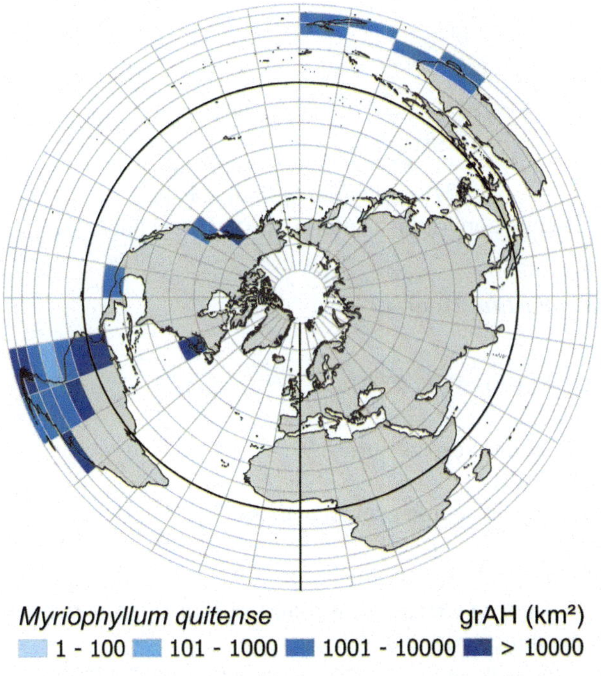

Myriophyllum quitense Kunth
Image copyright: iNaturalist CC-BY @pablo_jusim

Myriophyllum robustum **Hook.f.**
Synonyms:

Range area (grAH km^2): AUS: 12800
Ecozone origin-status: AUS: E
Life form(s): SU, EM (hydrosubshrub)
World rarity status: VR
Ecozone-endemic macroregion occurrence: AUS: NZ
Ploidy state: NA
Speciation age (Ma): NA
Populations also occur in terrestrial, wetland and/ or riparian habitats: YES
Long distance movement vectors: NA
Nuisance status and uses: no nuisance issues; aquarium trade
Endangered status (IUCN): NA (considered At Risk – Declining in its native range of New Zealand)

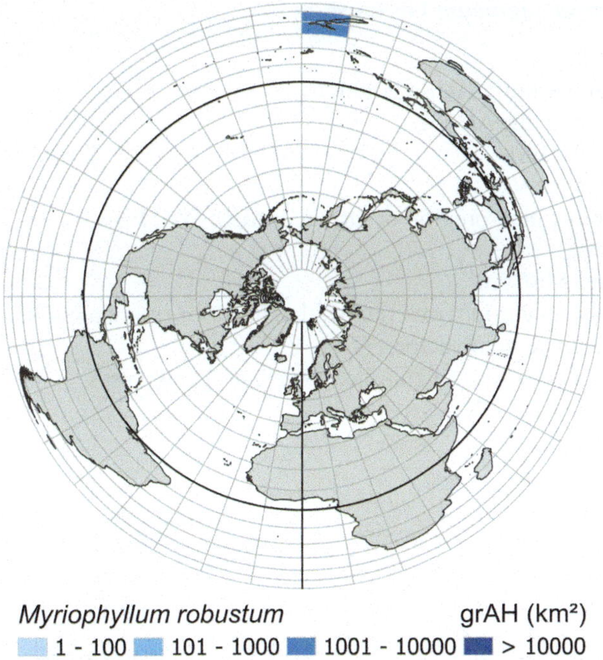

Myriophyllum robustum Hook.f.

***Myriophyllum salsugineum* Orchard**
Synonyms:

Range area (grAH km^2): AUS: 21500
Ecozone origin-status: AUS: E
Life form(s): SU, EM
World rarity status: VR
Ecozone-endemic macroregion occurrence: AUS: WA, ESA
Ploidy state: NA
Speciation age (Ma): 5.0
Populations also occur in terrestrial, wetland and/ or riparian habitats: NO
Long distance movement vectors: NA
Nuisance status and uses: no nuisance issues; no known uses
Endangered status (IUCN): NA

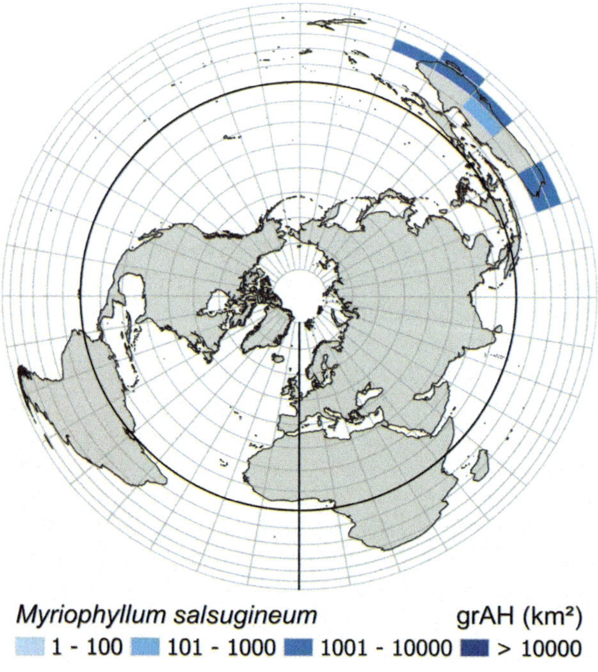

Myriophyllum salsugineum Orchard
Image copyright: K. Harris, with permission

***Myriophyllum sibiricum* Komarov**
Synonyms: *Myriophyllum exalbescens* Fernald

Range area (grAH km^2): PAL: 830000; NEA: 980500; OR: 36100
Ecozone origin-status: PAL: N; NEA: N; OR: N
Life form(s): SU
World rarity status: C
Ecozone-endemic macroregion occurrence: not endemic to any world ecozone
Ploidy state: P
Speciation age (Ma): 3.0
Populations also occur in terrestrial, wetland and/ or riparian habitats: NO
Long distance movement vectors: NA
Nuisance status and uses: no nuisance issues; folk medicine
Endangered status (IUCN): LC

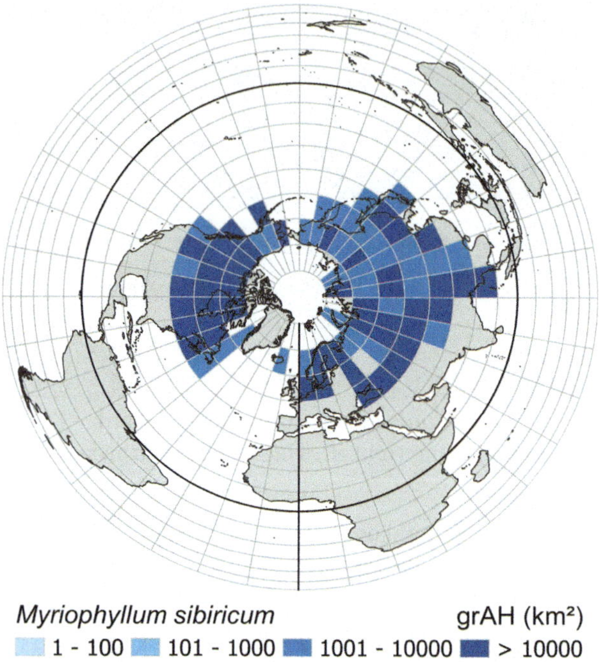

Myriophyllum sibiricum Komarov
Image copyright: A. Efremov, with permission

***Myriophyllum simulans* Orchard**
Synonyms:

Range area (grAH km^2): AUS: 31500
Ecozone origin-status: AUS: E
Life form(s): SU, EM
World rarity status: R
Ecozone-endemic macroregion occurrence: AUS: NA, ESA, NZ
Ploidy state: NA
Speciation age (Ma): NA
Populations also occur in terrestrial, wetland and/ or riparian habitats: YES
Long distance movement vectors: NA
Nuisance status and uses: no nuisance issues; aquarium trade
Endangered status (IUCN): NA

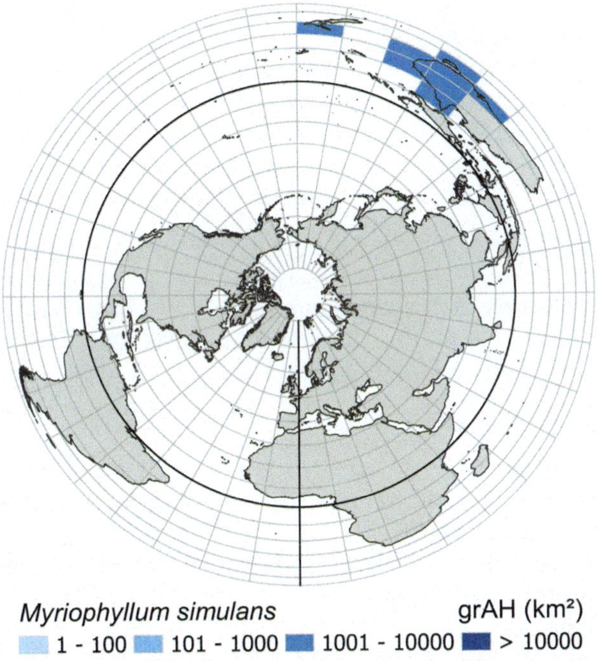

***Myriophyllum sparsiflorum* C.Wright**
Synonyms:

Range area (grAH km^2): NEO: 4800
Ecozone origin-status: NEO: E
Life form(s): SU, EM
World rarity status: VR
Ecozone-endemic macroregion occurrence: NEO: MES AM
Ploidy state: NA

Speciation age (Ma): NA
Populations also occur in terrestrial, wetland and/ or riparian habitats: YES
Long distance movement vectors: NA
Nuisance status and uses: no nuisance issues; no known uses
Endangered status (IUCN): NA (possibly Extinct; but GBIF has a 2011 record)

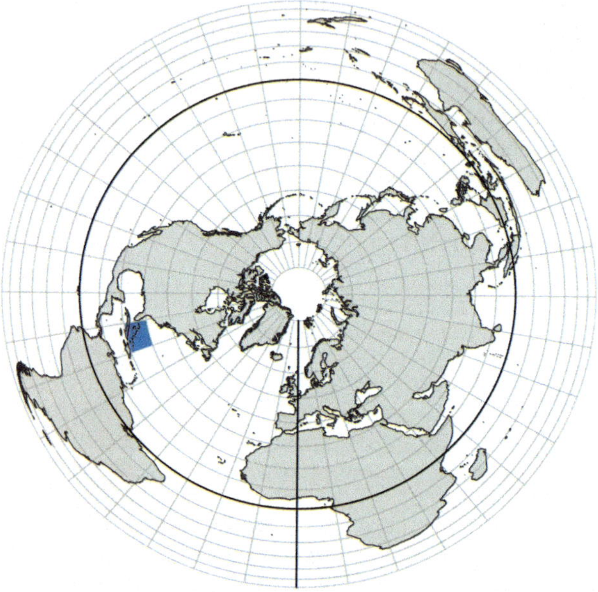

Myriophyllum spicatum **L.**
Synonyms:

Range area (grAH km^2): PAL: 821100; NEA: 463700; NEO: 45900; OR: 114400; AFR: 33500; present in OCE
Ecozone origin-status: PAL: N; NEA: I; NEO: I; OR: N; AFR: N; OCE: I
Life form(s): SU (hydrosubshrub)
World rarity status: C
Ecozone-endemic macroregion occurrence: not endemic to any world ecozone
Ploidy state: P
Speciation age (Ma): 3.0
Populations also occur in terrestrial, wetland and/ or riparian habitats: NO
Long distance movement vectors: ANTHROPOCHORY; BIRD: *Mareca strepera* (Gadwall); *Anas platyrhynchos* (Northern Mallard); *Anas acuta* (Northern Pintail); *Spatula clypeata* (Northern Shoveler)
Nuisance status and uses: major invasive aquatic weed in Nearctic (USA and Canada); aquarium trade, water garden plant, medicinal and phytoremediation uses
Endangered status (IUCN): LC

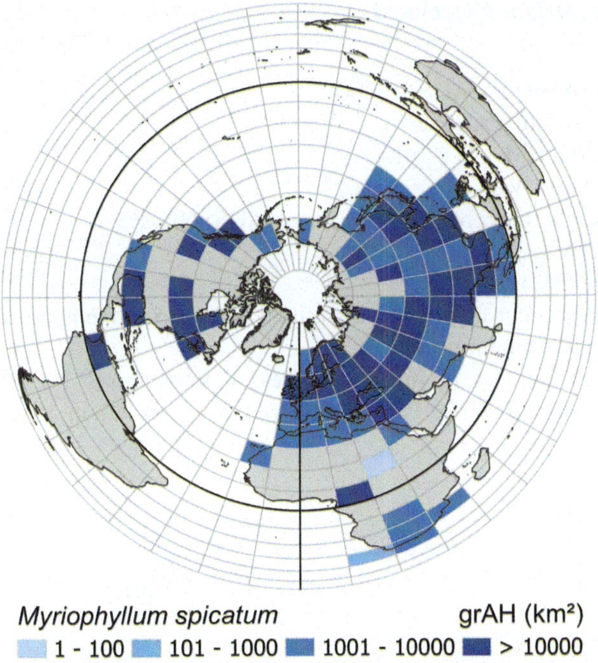

Myriophyllum spicatum L.
Image copyright: A. Efremov, with permission

Myriophyllum striatum **Orchard**
Synonyms:

Range area (grAH km²): AUS: 19200
Ecozone origin-status: AUS: E-
Life form(s): SU, EM
World rarity status: VR
Ecozone-endemic macroregion occurrence: AUS: NA, ESA
Ploidy state: NA
Speciation age (Ma): NA
Populations also occur in terrestrial, wetland and/ or riparian habitats: YES
Long distance movement vectors: NA
Nuisance status and uses: no nuisance issues; no known uses
Endangered status (IUCN): NA (considered Endangered in its native range in Australia)

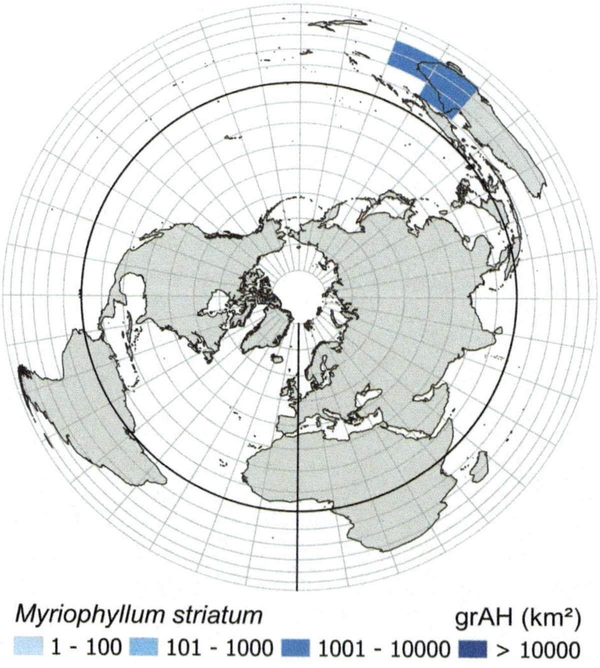

Myriophyllum striatum Orchard

Myriophyllum tenellum **Bigelow**
Synonyms:

Range area (grAH km^2): NEA: 416800
Ecozone origin-status: NEA: E
Life form(s): SU, EM
World rarity status: C
Ecozone-endemic macroregion occurrence: NEA: ARC CAN, E NAM
Ploidy state: D
Speciation age (Ma): 2.5
Populations also occur in terrestrial, wetland and/ or riparian habitats: YES
Long distance movement vectors: NA
Nuisance status and uses: no nuisance issues; aquarium trade
Endangered status (IUCN): LC

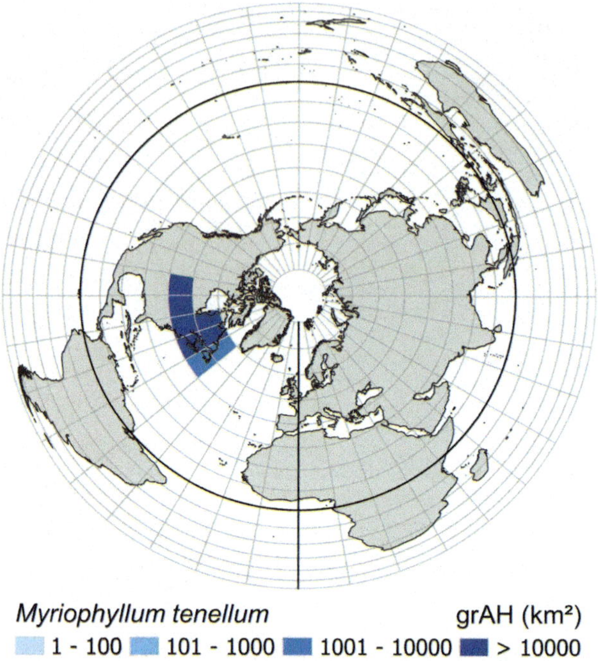

Myriophyllum tenellum Bigelow

Myriophyllum tetrandrum **Roxb. ex Andrews**
Synonyms:

Range area (grAH km^2): OR: 48400
Ecozone origin-status: OR: E
Life form(s): SU, EM
World rarity status: R
Ecozone-endemic macroregion occurrence: OR: IND, ICH, SUP
Ploidy state: NA
Speciation age (Ma): NA
Populations also occur in terrestrial, wetland and/ or riparian habitats: NO
Long distance movement vectors: NA
Nuisance status and uses: no nuisance issues; aquarium trade
Endangered status (IUCN): NA

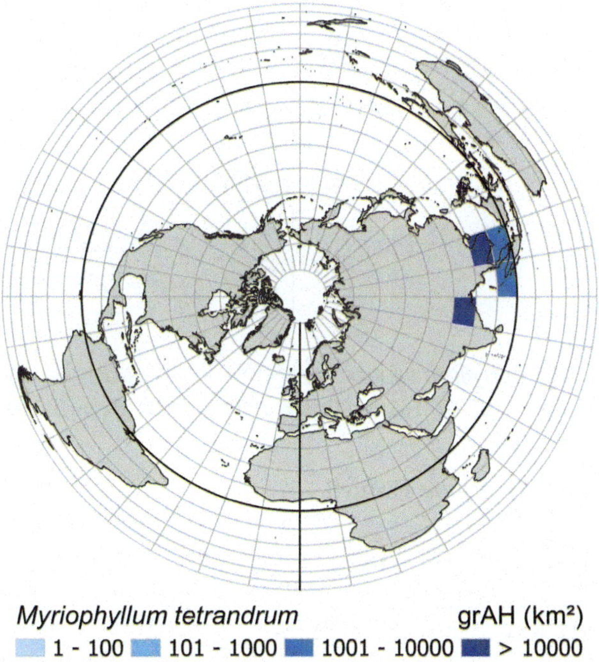

Myriophyllum tillaeoides **Diels**
Synonyms:

Range area (grAH km^2): AUS: 2300
Ecozone origin-status: AUS: E
Life form(s): SU, EM
World rarity status: VR
Ecozone-endemic macroregion occurrence: AUS: NA, WA

Ploidy state: NA
Speciation age (Ma): NA
Populations also occur in terrestrial, wetland and/ or riparian habitats: YES
Long distance movement vectors: NA
Nuisance status and uses: no nuisance issues; no known uses
Endangered status (IUCN): NA

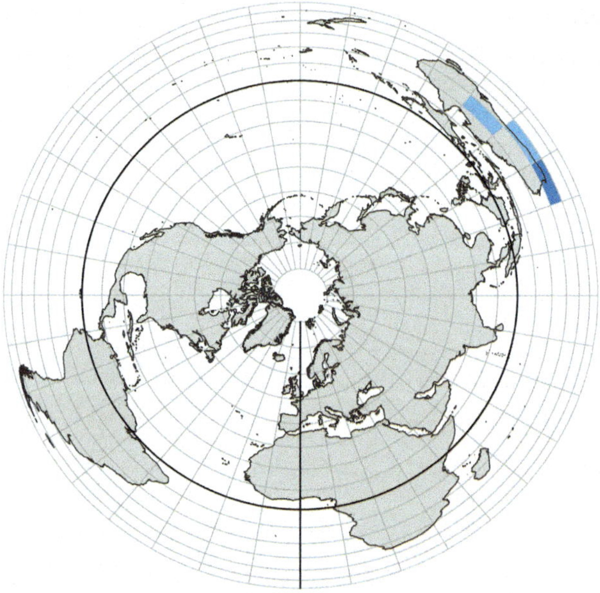

***Myriophyllum trachycarpum* F.Muell.**
Synonyms:

Range area (grAH km^2): AUS: 13500
Ecozone origin-status: AUS: E
Life form(s): SU, EM
World rarity status: VR
Ecozone-endemic macroregion occurrence: AUS: NA, ESA
Ploidy state: NA
Speciation age (Ma): NA
Populations also occur in terrestrial, wetland and/ or riparian habitats: YES
Long distance movement vectors: NA
Nuisance status and uses: no nuisance issues; no known uses
Endangered status (IUCN): NA

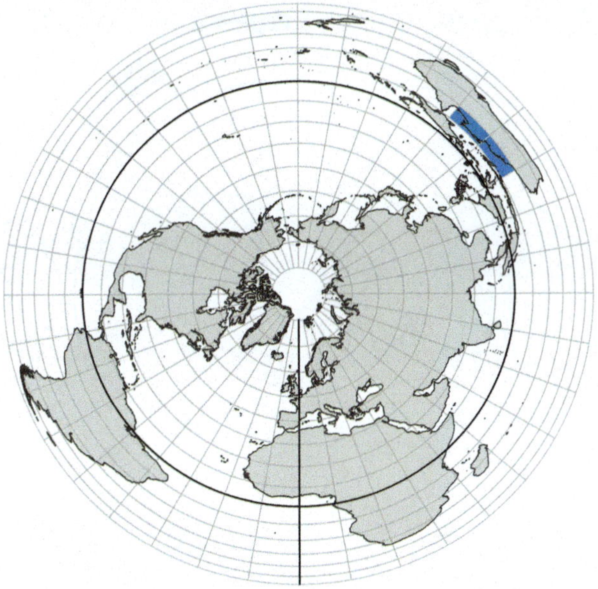

***Myriophyllum triphyllum* Orchard**
Synonyms:

Range area (grAH km^2): NEA: 69100; AUS: 26500; present in ANT
Ecozone origin-status: NEA: I; AUS: N; ANT: N
Life form(s): SU, EM
World rarity status: VR
Ecozone-endemic macroregion occurrence: not endemic to any world ecozone
Ploidy state: P
Speciation age (Ma): 2.5
Populations also occur in terrestrial, wetland and/ or riparian habitats: YES
Long distance movement vectors: ANTHROPOCHORY
Nuisance status and uses: no nuisance issues; aquarium trade
Endangered status (IUCN): NA

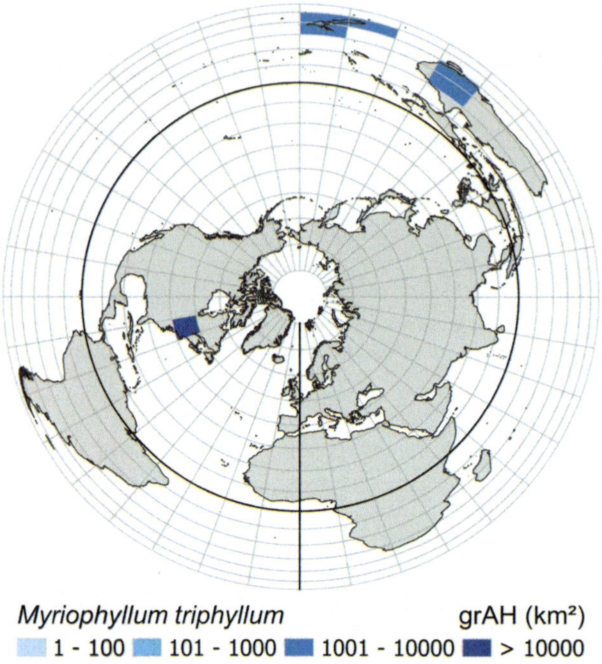

Myriophyllum triphyllum Orchard

***Myriophyllum ussuriense* (Regel) Maxim.**
Synonyms:

Range area (grAH km^2): PAL: 173400; NEA: 33000; OR: 27500
Ecozone origin-status: PAL:N; NEA: I; OR: N
Life form(s): SU, EM
World rarity status: C
Ecozone-endemic macroregion occurrence: not endemic to any world ecozone
Ploidy state: DP
Speciation age (Ma): 2.0
Populations also occur in terrestrial, wetland and/ or riparian habitats: YES
Long distance movement vectors: ANTHROPOCHORY
Nuisance status and uses: no nuisance issues; no known uses
Endangered status (IUCN): LC

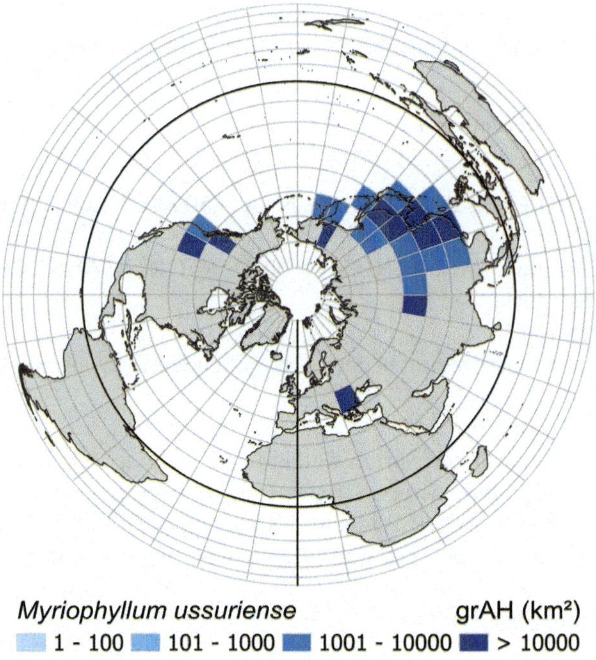

Myriophyllum ussuriense (Regel) Maxim.
Image copyright: A. Efremov, with permission

***Myriophyllum verticillatum* L.**
Synonyms:

Range area (grAH km^2): PAL: 895700; NEA: 850900; NEO: 33800; OR: 92900; AUS: 14000
Ecozone origin-status: PAL: N; NEA: N; NEO: N; OR: N; AUS: N
Life form(s): SU, EM
World rarity status: C
Ecozone-endemic macroregion occurrence: not endemic to any world ecozone
Ploidy state: DP
Speciation age (Ma): 5.0
Populations also occur in terrestrial, wetland and/ or riparian habitats: YES
Long distance movement vectors: BIRD: *Anas platyrhynchos* (Northern Mallard); *Mareca penelope* (Eurasian Widgeon); *Anas crecca* (Eurasian Teal)
Nuisance status and uses: no nuisance issues; water garden plant, biocidal potential
Endangered status (IUCN): LC

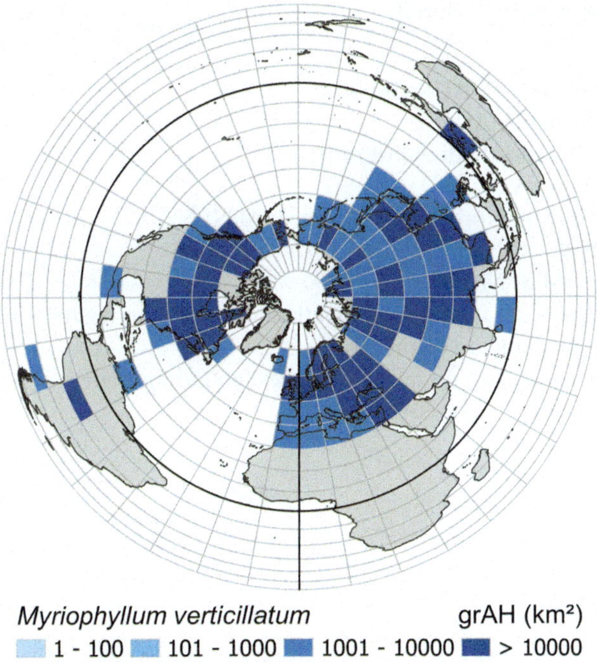

Myriophyllum verticillatum L.
Image copyright: A. Efremov, with permission

Myriophyllum votschii **Schindl.**
Synonyms:

Range area (grAH km^2): AUS: 16900
Ecozone origin-status: AUS: E
Life form(s): SU, EM (hydrosubshrub)
World rarity status: VR
Ecozone-endemic macroregion occurrence: AUS: NZ
Ploidy state: NA
Speciation age (Ma): NA
Populations also occur in terrestrial, wetland and/ or riparian habitats: YES
Long distance movement vectors: NA
Nuisance status and uses: no nuisance issues; aquarium trade
Endangered status (IUCN): NA

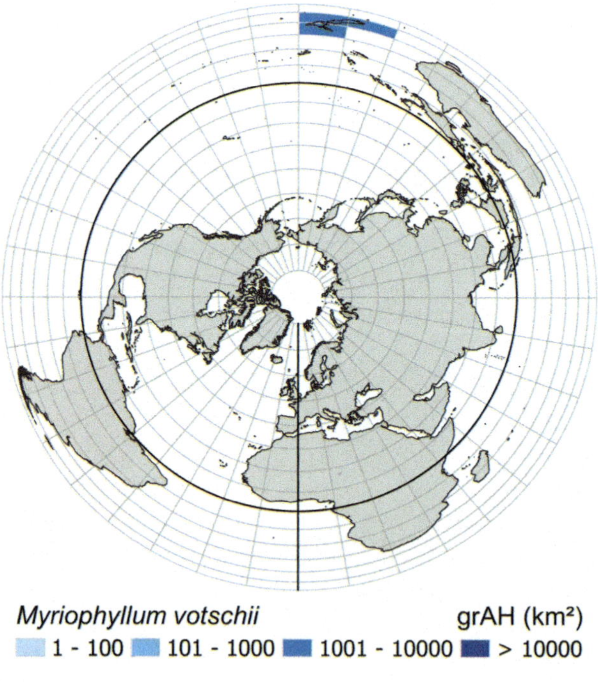

Myriophyllum votschii Schindl.

***Proserpinaca palustris* L.**
Synonyms:

Range area (grAH km^2): NEA: 371000; NEO: 79400
Ecozone origin-status: NEA: N; NEO: N
Life form(s): SU, EM (hydrosubshrub)
World rarity status: C
Ecozone-endemic macroregion occurrence: not endemic to any world ecozone
Ploidy state: D
Speciation age (Ma): NA
Populations also occur in terrestrial, wetland and/ or riparian habitats: YES
Long distance movement vectors: NA
Nuisance status and uses: no nuisance issues; aquarium trade, medicinal and fish-keeping uses
Endangered status (IUCN): LC

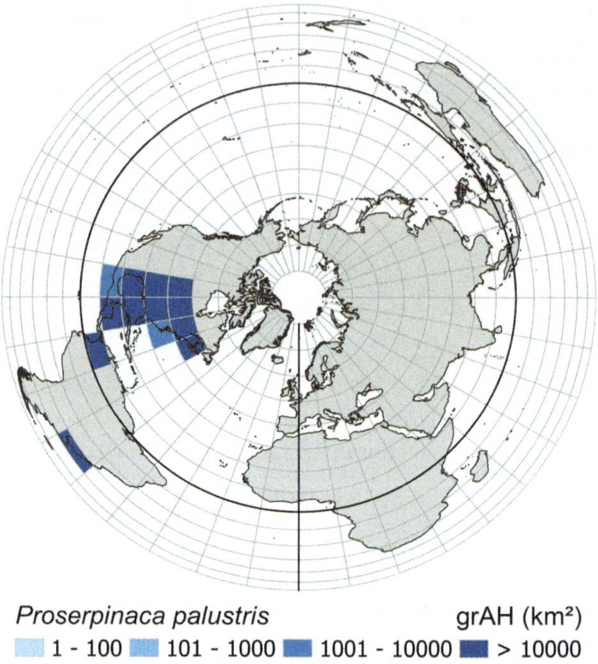

Proserpinaca palustris L.
Image copyright: G. Ervin, with permission

Proserpinaca pectinata **Lam.**
Synonyms:

Range area (grAH km^2): NEA: 138300; NEO: 30400; AUS: 6300
Ecozone origin-status: NEA: N; NEO: N; AUS: I
Life form(s): EM
World rarity status: C
Ecozone-endemic macroregion occurrence: not endemic to any world ecozone
Ploidy state: D
Speciation age (Ma): NA
Populations also occur in terrestrial, wetland and/ or riparian habitats: YES
Long distance movement vectors: ANTHROPOCHORY
Nuisance status and uses: no nuisance issues; fish keeping and phytoremediation uses
Endangered status (IUCN): LC

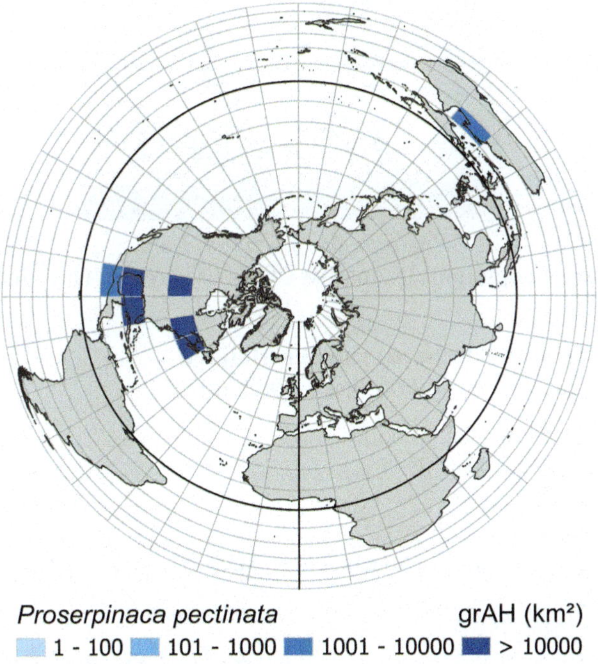

Proserpinaca pectinata Lam.
Image copyright: iNaturalist CC-BY @jayhorn (Jay Horn)

Vinkia callitrichoides (Orchard) van der Meijden
Synonyms: *Myriophyllum callitrichoides* Orchard

Range area (grAH km^2): AUS: 6300
Ecozone origin-status: AUS: E
Life form(s): SU, FR
World rarity status: VR
Ecozone-endemic macroregion occurrence: AUS: NA
Ploidy state: NA
Speciation age (Ma): NA
Populations also occur in terrestrial, wetland and/ or riparian habitats: NO
Long distance movement vectors: NA
Nuisance status and uses: no nuisance issues; no known uses
Endangered status (IUCN): NA

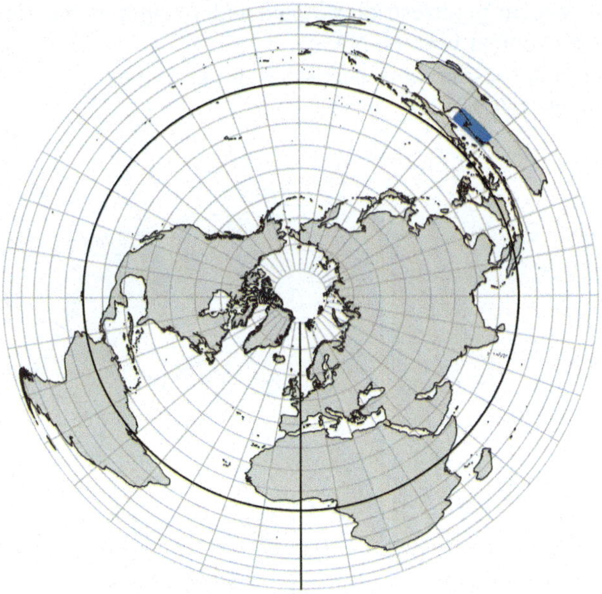

3.22 HYDATELLACEAE (Nympheales)

All 13 macrophyte species of *Trithuria* in the Hydatellaceae have VR status, with one species being an OR endemic, and the rest endemic to AUS. None ever cause nuisance issues. This family was formerly placed among the monocots, either in an order of its own or within Poales, but *Trithuria* spp. are now known to be early-divergent angiosperms, within Nympheales [145]. Only one of the 13 Hydatellaceae species has been assessed by IUCN: *Trithuria lanterna* is classed as LC.

Trithuria austinensis **D.D.Sokoloff, Remizowa, T.D.Macfarl. & Rudall**
Synonyms:

Range area (grAH km^2): AUS: 1500
Ecozone origin-status: AUS: E
World rarity status: VR
Ecozone-endemic macroregion occurrence: AUS: WA
Life form(s): SU, EM
Ploidy state: NA
Speciation age (Ma): NA

Populations also occur in terrestrial, wetland and/ or riparian habitats: YES
Long distance movement vectors: NA
Nuisance status and uses: no nuisance issues; no known uses
Endangered status (IUCN): NA

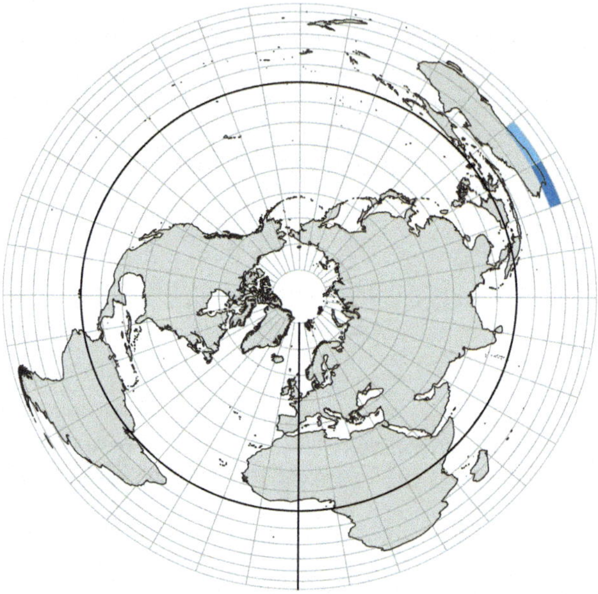

Trithuria australis **(Diels) D.D.Sokoloff, Remizowa, T.D.Macfarl. & Rudall**
Synonyms: *Hydatella australis* Diels

Range area (grAH km^2): AUS: 1500
Ecozone origin-status: AUS: E
World rarity status: VR
Ecozone-endemic macroregion occurrence: AUS: WA
Life form(s): SU, EM
Ploidy state: DP [146]
Speciation age (Ma): NA
Populations also occur in terrestrial, wetland and/ or riparian habitats: YES
Long distance movement vectors: NA
Nuisance status and uses: no nuisance issues; no known uses
Endangered status (IUCN): NA

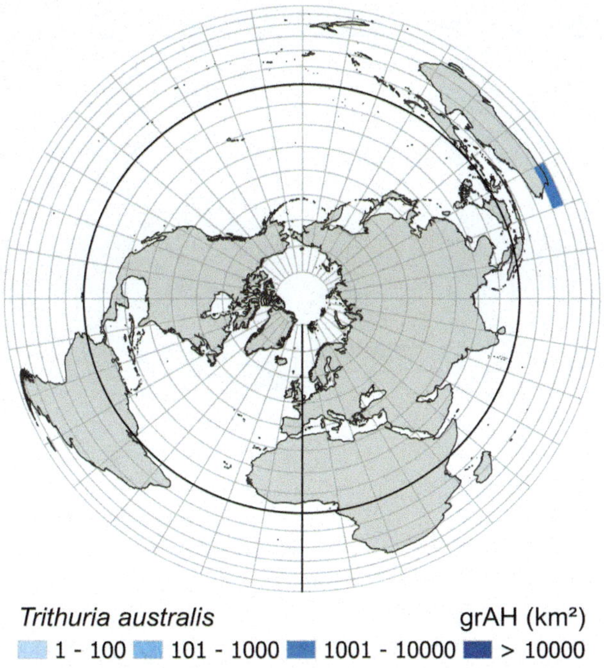

***Trithuria bibracteata* Stapf ex D.A.Cooke**
Synonyms:

Range area (grAH km^2): AUS: 2200
Ecozone origin-status: AUS: E
World rarity status: VR
Ecozone-endemic macroregion occurrence: AUS: WA
Life form(s): SU, EM
Ploidy state: NA
Speciation age (Ma): NA
Populations also occur in terrestrial, wetland and/ or riparian habitats: YES
Long distance movement vectors: NA
Nuisance status and uses: no nuisance issues; no known uses
Endangered status (IUCN): NA

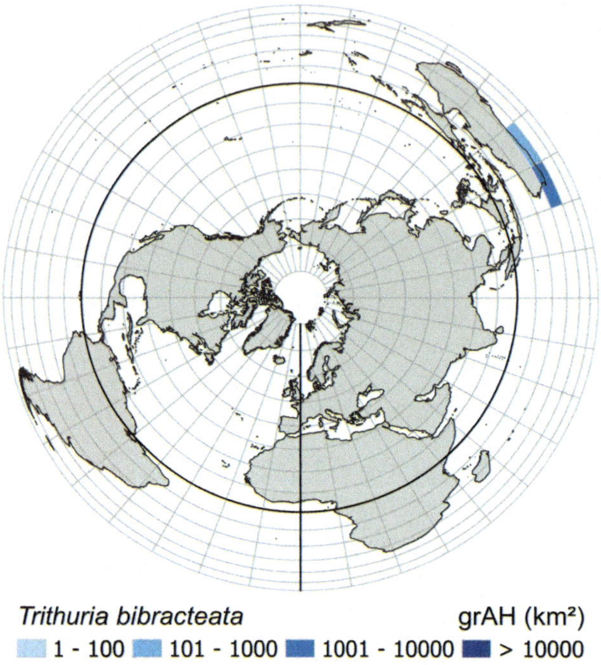

Trithuria cookeana **D.D.Sokoloff, Remizowa, T.D.Macfarl. & Rudall**
Synonyms:

Range area (grAH km^2): AUS: 6300
Ecozone origin-status: AUS: E
World rarity status: VR
Ecozone-endemic macroregion occurrence: AUS: NA
Life form(s): SU, EM
Ploidy state: NA
Speciation age (Ma): NA
Populations also occur in terrestrial, wetland and/ or riparian habitats: YES
Long distance movement vectors: NA
Nuisance status and uses: no nuisance issues; no known uses
Endangered status (IUCN): NA

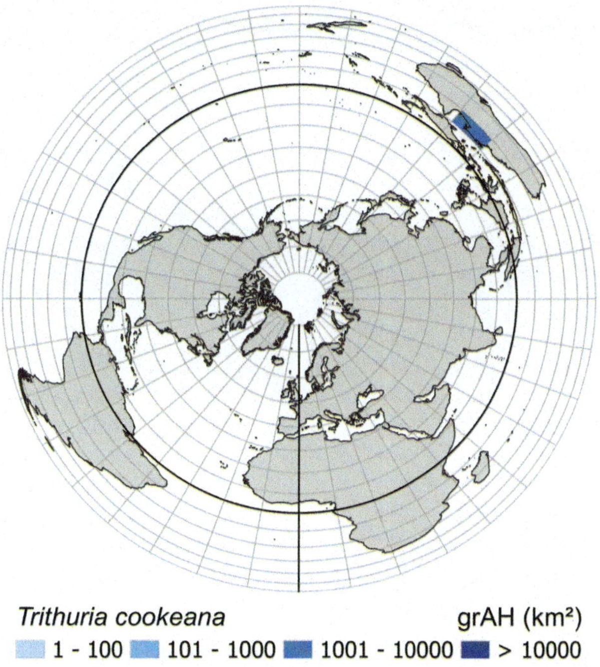

***Trithuria cowieana* D.D.Sokoloff, Remizowa, T.D.Macfarl. & Rudall**
Synonyms:

Range area (grAH km^2): AUS: 6300
Ecozone origin-status: AUS: E
World rarity status: VR
Ecozone-endemic macroregion occurrence: AUS: NA
Life form(s): SU, EM
Ploidy state: NA
Speciation age (Ma): NA
Populations also occur in terrestrial, wetland and/ or riparian habitats: YES
Long distance movement vectors: NA
Nuisance status and uses: no nuisance issues; no known uses
Endangered status (IUCN): NA

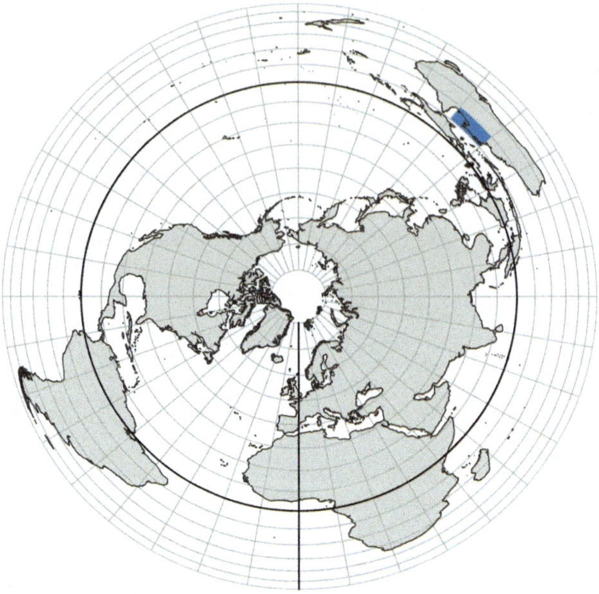

***Trithuria filamentosa* Rodway**
Synonyms: *Hydatella filamentosa* (Rodway) W.M.Curtis

Range area (grAH km^2): AUS: 3800
Ecozone origin-status: AUS: E
World rarity status: VR
Ecozone-endemic macroregion occurrence: AUS: ESA
Life form(s): SU, EM
Ploidy state: NA
Speciation age (Ma): NA
Populations also occur in terrestrial, wetland and/ or riparian habitats: YES
Long distance movement vectors: NA
Nuisance status and uses: no nuisance issues; no known uses
Endangered status (IUCN): NA

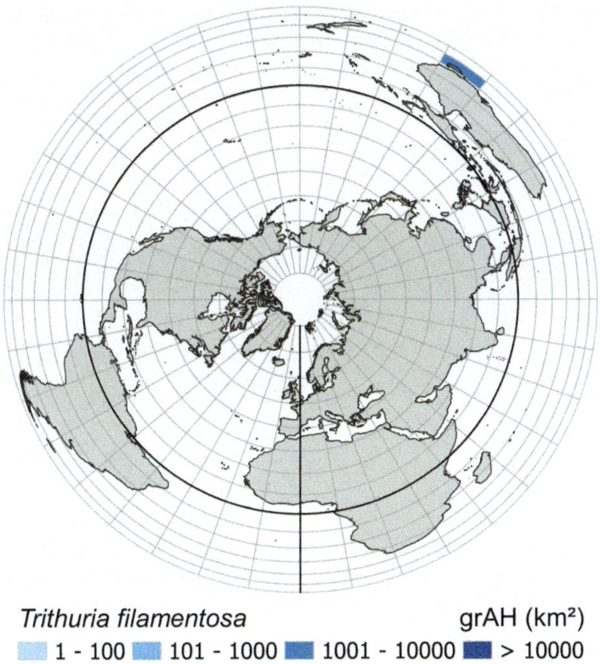

***Trithuria fitzgeraldii* D.D.Sokoloff, I.Marques, T.D.Macfarl., Rudall & S.W.Graham**
Synonyms:

Range area (grAH km^2): AUS: 1500
Ecozone origin-status: AUS: E
World rarity status: VR
Ecozone-endemic macroregion occurrence: AUS: WA
Life form(s): SU, EM
Ploidy state: NA
Speciation age (Ma): NA
Populations also occur in terrestrial, wetland and/ or riparian habitats: YES
Long distance movement vectors: NA
Nuisance status and uses: no nuisance issues; no known uses
Endangered status (IUCN): NA

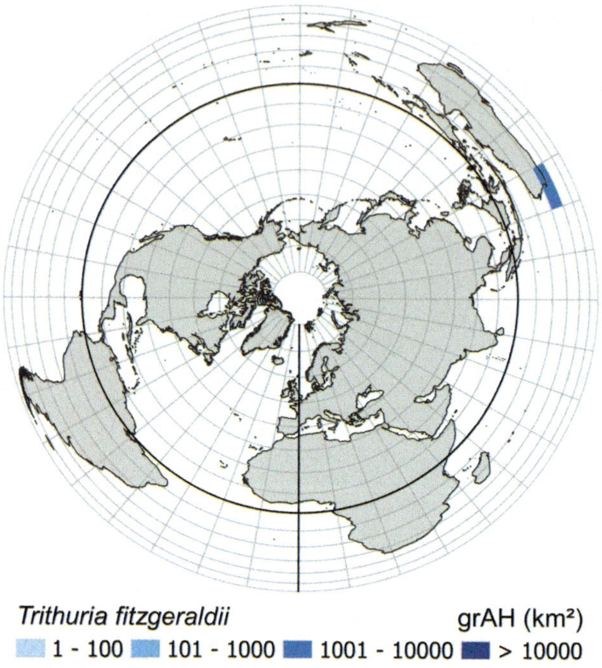

Trithuria inconspicua **Cheeseman**
Synonyms: *Hydatella inconspicua* (Cheeseman) Cheeseman

Range area (grAH km^2): AUS: 3800
Ecozone origin-status: AUS: E
World rarity status: VR
Ecozone-endemic macroregion occurrence: AUS: NZ
Life form(s): SU, EM
Ploidy state: NA
Speciation age (Ma): NA
Populations also occur in terrestrial, wetland and/ or riparian habitats: YES
Long distance movement vectors: NA
Nuisance status and uses: no nuisance issues; no known uses
Endangered status (IUCN): NA

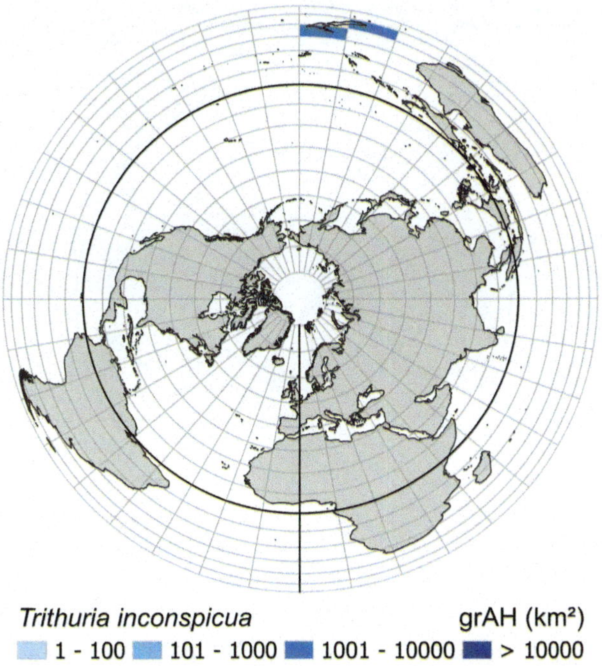

***Trithuria konkanensis* S.R.Yadav & Janarth.**
Synonyms:

Range area (grAH km^2): OR: 8100
Ecozone origin-status: OR: E
World rarity status: VR
Ecozone-endemic macroregion occurrence: OR: IND
Life form(s): SU, EM
Ploidy state: NA
Speciation age (Ma): NA
Populations also occur in terrestrial, wetland and/ or riparian habitats: YES
Long distance movement vectors: NA
Nuisance status and uses: no nuisance issues; no known uses
Endangered status (IUCN): NA

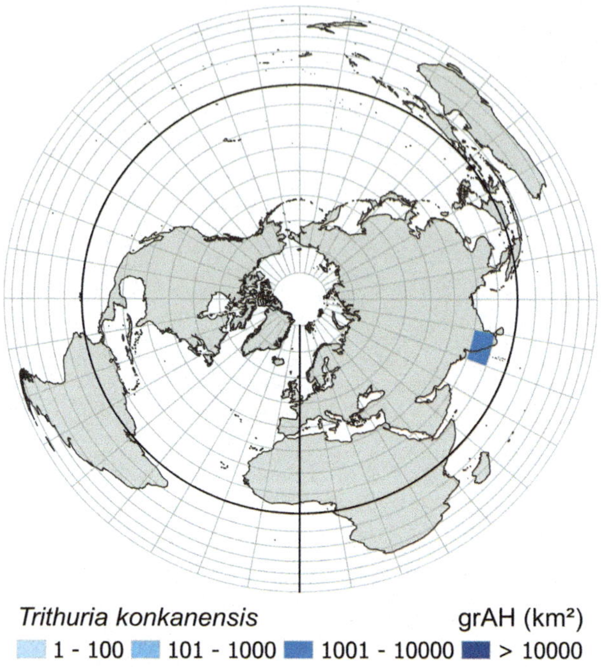

Trithuria lanterna **D.A.Cooke**
Synonyms:

Range area (grAH km^2): AUS: 18200
Ecozone origin-status: AUS: E
World rarity status: VR
Ecozone-endemic macroregion occurrence: AUS: NA
Life form(s): SU, EM
Ploidy state: NA
Speciation age (Ma): NA
Populations also occur in terrestrial, wetland and/ or riparian habitats: YES
Long distance movement vectors: NA
Nuisance status and uses: no nuisance issues; no known uses
Endangered status (IUCN): LC

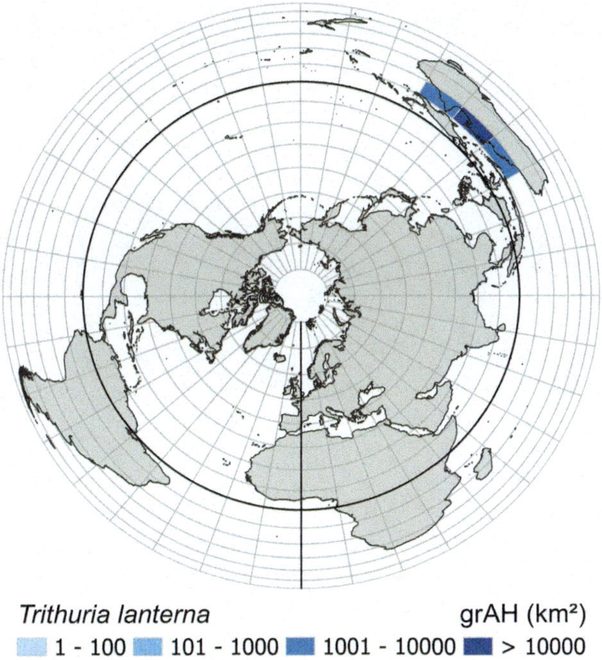

Trithuria occidentalis **Benth.**
Synonyms: *Juncella occidentalis (Benth.) Hieron.*

Range area (grAH km^2): AUS: 1500
Ecozone origin-status: AUS: E
World rarity status: VR
Ecozone-endemic macroregion occurrence: AUS: WA
Life form(s): SU, EM
Ploidy state: NA
Speciation age (Ma): NA
Populations also occur in terrestrial, wetland and/ or riparian habitats: YES
Long distance movement vectors: NA
Nuisance status and uses: no nuisance issues; no known uses
Endangered status (IUCN): NA

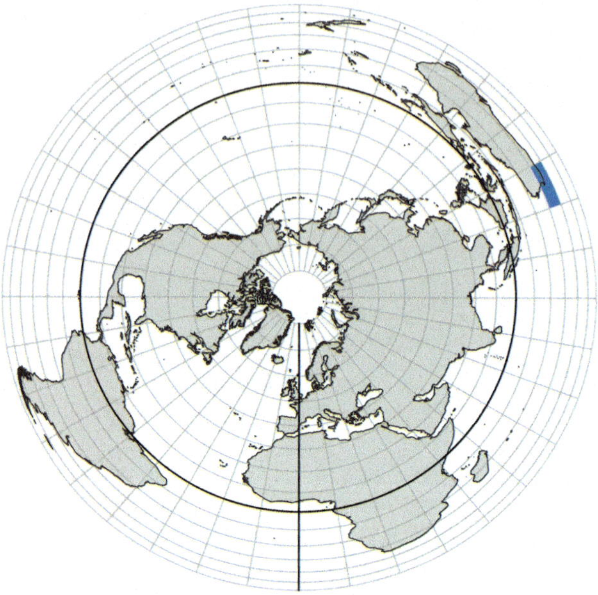

Trithuria polybracteata **D.A.Cooke ex D.D.Sokoloff, Remizowa, T.D.Macfarl. & Rudall**
Synonyms:

Range area (grAH km^2): AUS: 1500
Ecozone origin-status: AUS: E
World rarity status: VR
Ecozone-endemic macroregion occurrence: AUS: NA
Life form(s): SU, EM
Ploidy state: NA
Speciation age (Ma): NA
Populations also occur in terrestrial, wetland and/ or riparian habitats: YES
Long distance movement vectors: NA
Nuisance status and uses: no nuisance issues; no known uses
Endangered status (IUCN): NA

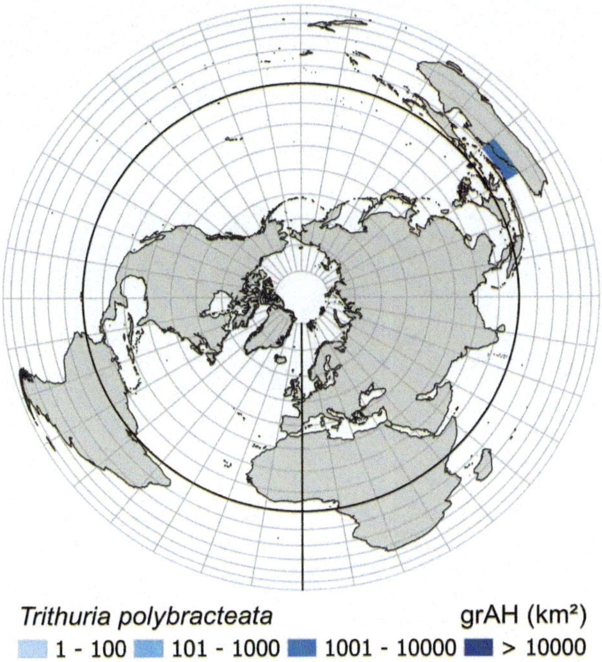

***Trithuria submersa* Hook.f.**
Synonyms: *Juncella submersa* (Hook.f.) Hieron.

Range area (grAH km^2): AUS: 16200
Ecozone origin-status: AUS: E
World rarity status: VR
Ecozone-endemic macroregion occurrence: AUS: WA, ESA
Life form(s): SU, EM
Ploidy state: P [146]
Speciation age (Ma): NA
Populations also occur in terrestrial, wetland and/ or riparian habitats: YES
Long distance movement vectors: NA
Nuisance status and uses: no nuisance issues; no known uses
Endangered status (IUCN): NA

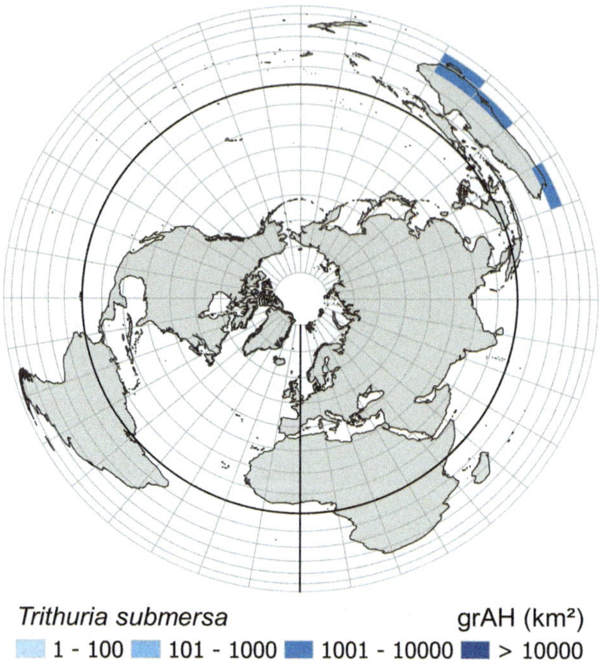

3.23 HYDROLEACEAE (Solanales)

Four freshwater macrophyte species are in the Hydroleaceae, all in the genus *Hydrolea* and largely restricted to warmer parts of the world. Three species occur in NEO (two endemic), two in OR and one each in AFR, PAL, NEA and AUS. *Hydrolea spinosa* is introduced in OR, and *Hydrolea zeylanica,* in AFR and PAL. *H. zeylanica* is the only species of this family that occasionally causes aquatic weed problems, usually in rice. Of the four macrophyte species, none are very rare at world scale, with values of R 25%, and C 75%. None of the four species has had endangered status assessed.

***Hydrolea corymbosa* Elliott**
Synonyms: *Nama corymbosa* (Elliott) Kuntze

Range area (grAH km^2): NEA: 21200; NEO: 10300
Ecozone origin-status: NEA: N; NEO: N
World rarity status: R
Ecozone-endemic macroregion occurrence: not endemic to any world ecozone
Life form(s): EM
Ploidy state: NA
Speciation age (Ma): NA

Populations also occur in terrestrial, wetland and/ or riparian habitats: YES
Long distance movement vectors: NA
Nuisance status and uses: no nuisance issues; water garden plant
Endangered status (IUCN): NA

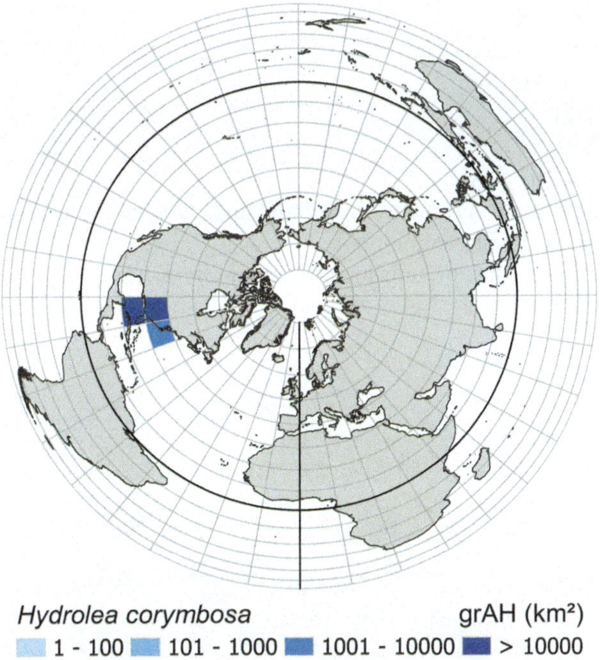

Hydrolea corymbosa Elliott

***Hydrolea elatior* Schott**
Synonyms:

Range area (grAH km^2): NEO: 250500
Ecozone origin-status: NEO: E
World rarity status: C
Ecozone-endemic macroregion occurrence: NEO: MG MX, MES AM, NWSA, AMZ, CHA, PAR
Life form(s): EM
Ploidy state: NA
Speciation age (Ma): NA
Populations also occur in terrestrial, wetland and/ or riparian habitats: YES
Long distance movement vectors: NA
Nuisance status and uses: no nuisance issues; no known uses
Endangered status (IUCN): NA

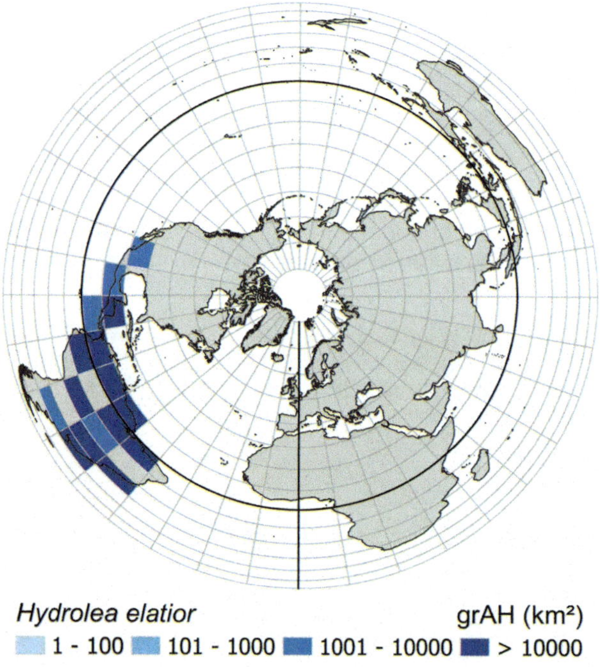

Hydrolea elatior Schott

***Hydrolea spinosa* L.**
Synonyms: *Nama spinosa* (L.) Kuntze

Range area (grAH km^2): NEO: 306000; OR: 16200;
Ecozone origin-status: NEO: E(I); OR: I
World rarity status: C
Ecozone-endemic macroregion occurrence: NEO: MG MX, MES AM, NWSA, AMZ, CHA, PAR, AND
Life form(s): EM (shrub)
Ploidy state: P
Speciation age (Ma): NA
Populations also occur in terrestrial, wetland and/ or riparian habitats: YES
Long distance movement vectors: ANTHROPOCHORY
Nuisance status and uses: no nuisance issues; medicinal, forage uses
Endangered status (IUCN): NA

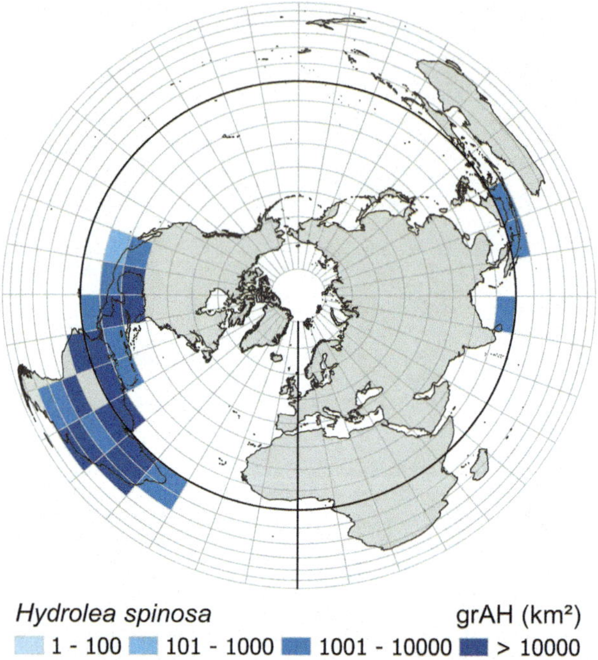

Hydrolea spinosa L.

***Hydrolea zeylanica* L.**
Synonyms: *Nama zeylanica* L.

Range area (grAH km^2): PAL: 400; OR: 117000; AUS: 16500; AFR: 112300
Ecozone origin-status: PAL: I; OR: N; AUS: N; AFR: I
World rarity status: C
Ecozone-endemic macroregion occurrence: not endemic to any world ecozone

Life form(s): EM
Ploidy state: D
Speciation age (Ma): NA
Populations also occur in terrestrial, wetland and/ or riparian habitats: YES
Long distance movement vectors: ANTHROPOCHORY
Nuisance status and uses: occasionally an aquatic weed in rice; leaves are edible, folk medicine uses
Endangered status (IUCN): NA

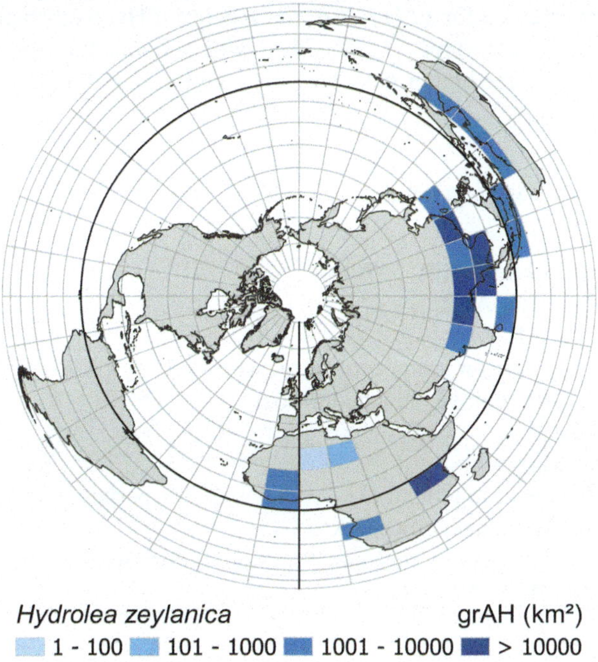

Hydrolea zeylanica L.
Image copyright: Z. Li, with permission

3.24 HYDROSTACHYACEAE (Cornales)

The Hydrostachyaceae is a single-genus family totalling 22 species, all of them freshwater macrophytes, usually found as rheophytes in fast-flowing rivers, and all endemic to AFR, with individual distributions usually very restricted within that ecozone, particularly in the Madagascar macroregion of AFR. None of the species cause aquatic weed problems. The 22 species are almost all world-rare, with values of VR 77%, R 18% and just a single species being common, C, 6%. Endangered status has been assessed for 17 (77%) of the species of Hydrostachyaceae, with one species being DD, and three LC, together with a rather high proportion (59%) assessed as being in the higher risk categories, with four NT, one VU, five EN and two in the CR class.

***Hydrostachys angustisecta* Engl.**
Synonyms:

Range area (grAH km^2): AFR: 93700
Ecozone origin-status: AFR: E
World rarity status: R
Ecozone-endemic macroregion occurrence: AFR: EA
Life form(s): SU (rheophyte, attached to rocks)
Ploidy state: NA
Speciation age (Ma): NA
Populations also occur in terrestrial, wetland and/ or riparian habitats: NO
Long distance movement vectors: NA
Nuisance status and uses: no nuisance issues; no known uses
Endangered status (IUCN): EN

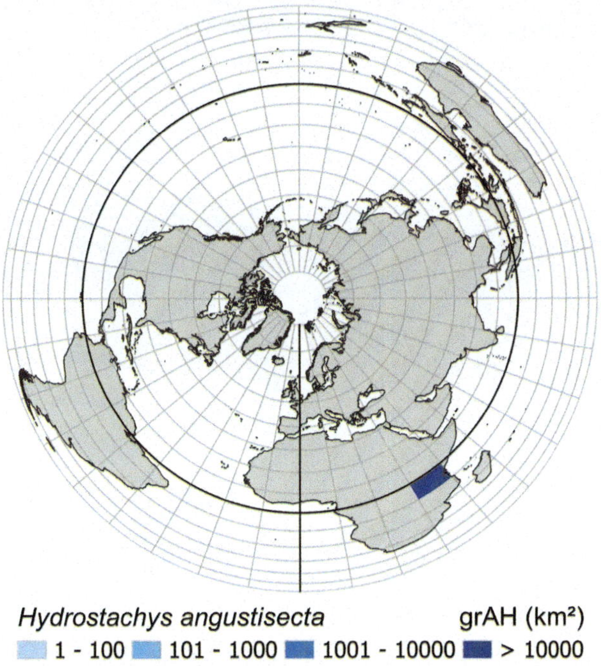

***Hydrostachys bismarckii* Engl.**
Synonyms:

Range area (grAH km^2): AFR: 50200
Ecozone origin-status: AFR: E
World rarity status: R
Ecozone-endemic macroregion occurrence: AFR: FOR
Life form(s): SU (rheophyte, attached to rocks)
Ploidy state: NA
Speciation age (Ma): NA
Populations also occur in terrestrial, wetland and/ or riparian habitats: NO
Long distance movement vectors: NA
Nuisance status and uses: no nuisance issues; no known uses
Endangered status (IUCN): NA

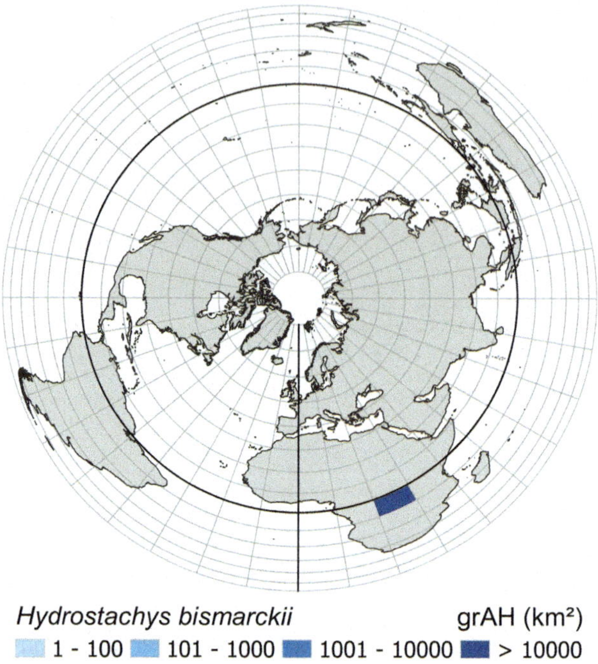

***Hydrostachys decaryi* H. Perrier**
Synonyms:

Range area (grAH km^2): AFR: 2000
Ecozone origin-status: AFR: E
World rarity status: VR
Ecozone-endemic macroregion occurrence: AFR: MAD
Life form(s): SU (rheophyte, attached to rocks)
Ploidy state: NA
Speciation age (Ma): NA
Populations also occur in terrestrial, wetland and/ or riparian habitats: NO
Long distance movement vectors: NA
Nuisance status and uses: no nuisance issues; no known uses
Endangered status (IUCN): EN

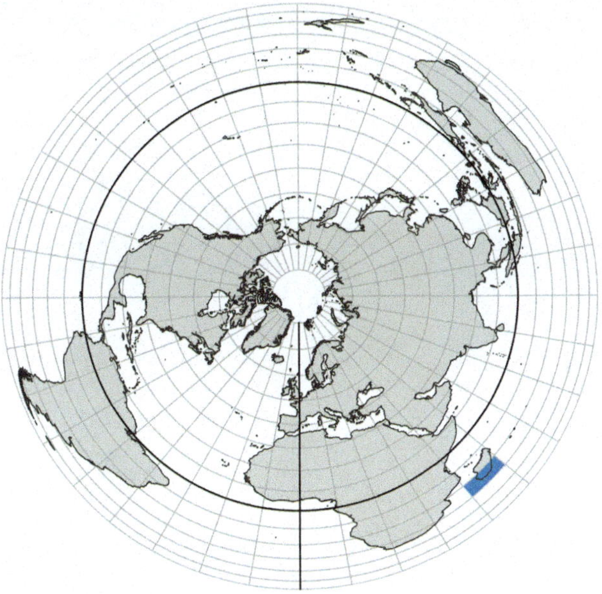

Hydrostachys distichophylla **A.Juss.**
Synonyms:

Range area (grAH km^2): AFR: 10100
Ecozone origin-status: AFR: E
World rarity status: VR
Ecozone-endemic macroregion occurrence: AFR: MAD
Life form(s): SU (rheophyte, attached to rocks)
Ploidy state: NA
Speciation age (Ma): NA
Populations also occur in terrestrial, wetland and/ or riparian habitats: NO
Long distance movement vectors: NA
Nuisance status and uses: no nuisance issues; no known uses
Endangered status (IUCN): NT

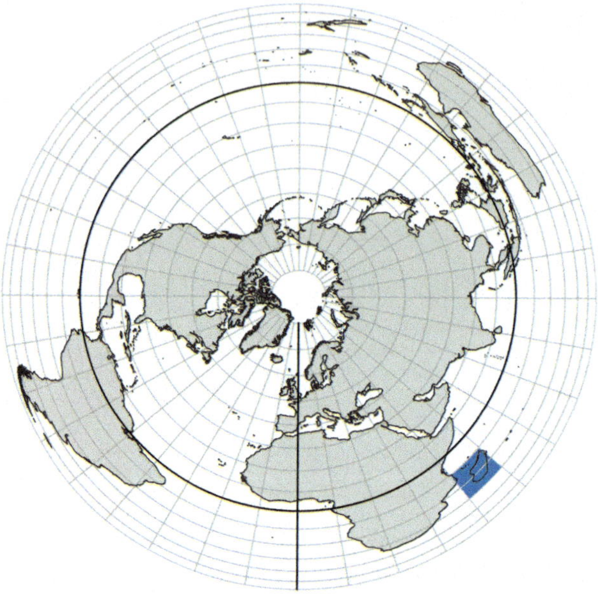

Hydrostachys fimbriata C.Cusset
Synonyms:

Range area (grAH km^2): AFR: 10100
Ecozone origin-status: AFR: E
World rarity status: VR
Ecozone-endemic macroregion occurrence: AFR: MAD
Life form(s): SU (rheophyte, attached to rocks)
Ploidy state: NA
Speciation age (Ma): NA
Populations also occur in terrestrial, wetland and/ or riparian habitats: NO
Long distance movement vectors: NA
Nuisance status and uses: no nuisance issues; no known uses
Endangered status (IUCN): EN

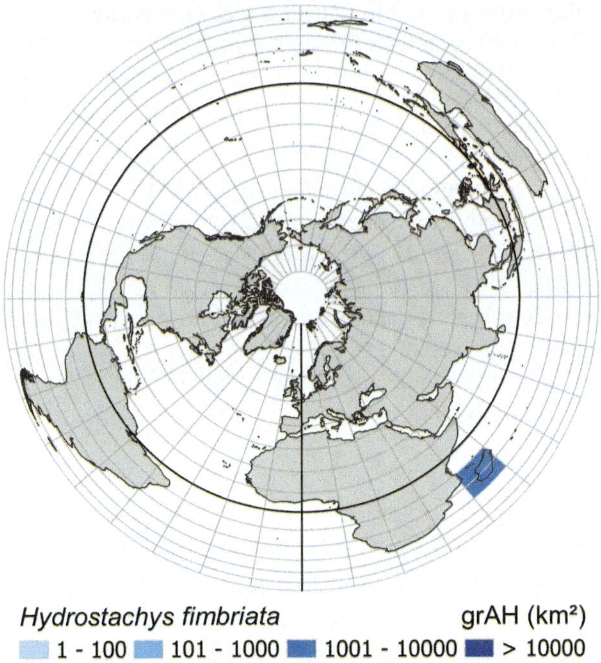

Hydrostachys fimbriata C.Cusset

***Hydrostachys flabellifera* G.W.Hu, Zhun Xu & Q.F.Wang**
Synonyms: (species first described in 2020: [166])

Range area (grAH km^2): AFR: 2000
Ecozone origin-status: AFR: E
World rarity status: VR
Ecozone-endemic macroregion occurrence: AFR: MAD
Life form(s): SU (rheophyte, attached to rocks)
Ploidy state: NA
Speciation age (Ma): NA
Populations also occur in terrestrial, wetland and/ or riparian habitats: NO
Long distance movement vectors: NA
Nuisance status and uses: no nuisance issues; no known uses
Endangered status (IUCN): NA

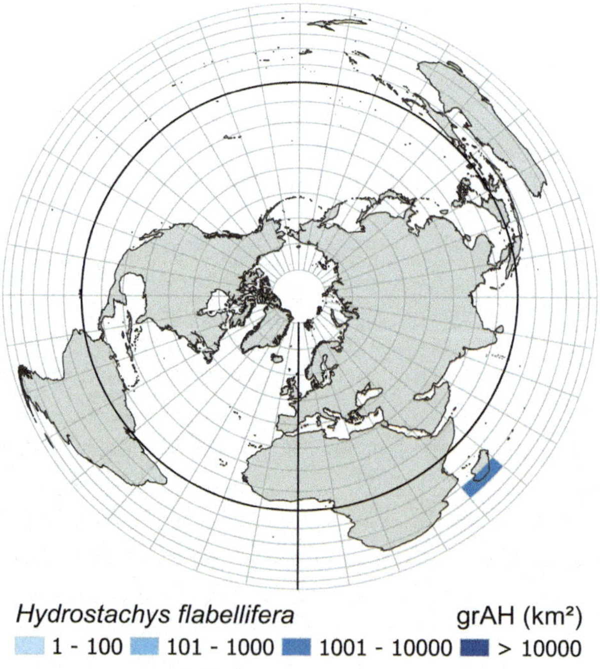

***Hydrostachys imbricata* A.Juss.**
Synonyms:

Range area (grAH km^2): AFR: 10100
Ecozone origin-status: AFR: E
World rarity status: VR
Ecozone-endemic macroregion occurrence: AFR: MAD

Life form(s): SU (rheophyte, attached to rocks)
Ploidy state: D
Speciation age (Ma): NA
Populations also occur in terrestrial, wetland and/ or riparian habitats: NO
Long distance movement vectors: NA
Nuisance status and uses: no nuisance issues; no known uses
Endangered status (IUCN): LC

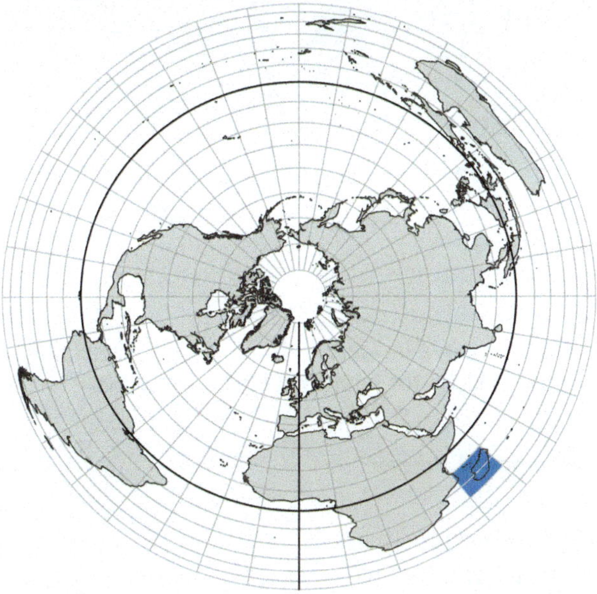

Hydrostachys imbricata grAH (km²)
1 - 100 101 - 1000 1001 - 10000 > 10000

***Hydrostachys insignis* A.Juss.**
Synonyms:

Range area (grAH km²): AFR: 93700
Ecozone origin-status: AFR: E
World rarity status: R
Ecozone-endemic macroregion occurrence: AFR: EA
Life form(s): SU (rheophyte, attached to rocks)
Ploidy state: NA
Speciation age (Ma): NA
Populations also occur in terrestrial, wetland and/ or riparian habitats: NO
Long distance movement vectors: NA
Nuisance status and uses: no nuisance issues; no known uses
Endangered status (IUCN): DD

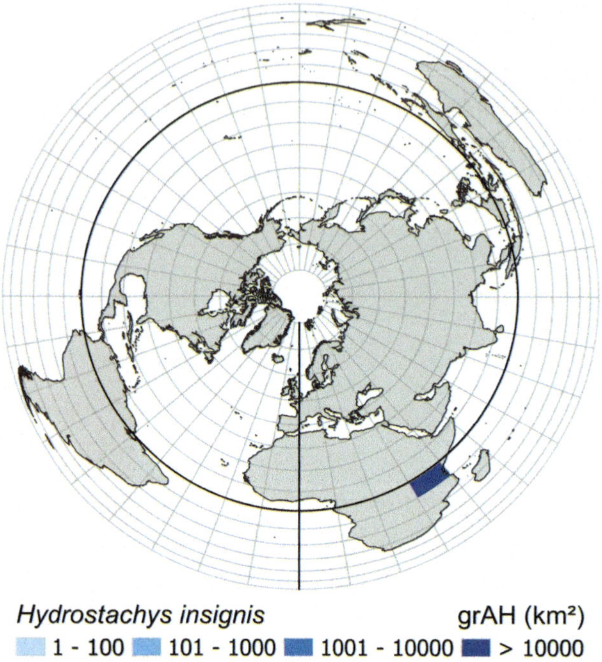

Hydrostachys longifida **H.Perrier**
Synonyms:

Range area (grAH km^2): AFR: 10100
Ecozone origin-status: AFR: E
World rarity status: VR
Ecozone-endemic macroregion occurrence: AFR: MAD
Life form(s): SU (rheophyte, attached to rocks)
Ploidy state: NA
Speciation age (Ma): NA
Populations also occur in terrestrial, wetland and/ or riparian habitats: NO
Long distance movement vectors: NA
Nuisance status and uses: no nuisance issues; no known uses
Endangered status (IUCN): NT

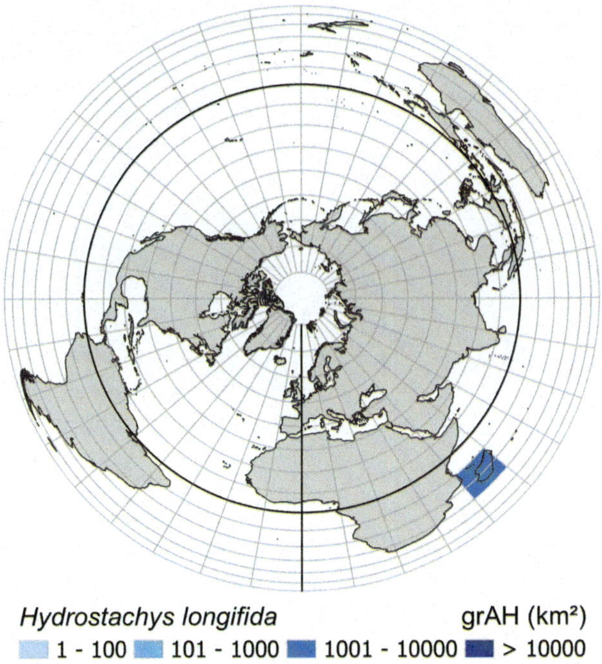

***Hydrostachys lufirensis* C.Cusset**
Synonyms:

Range area (grAH km²): AFR: 50200
Ecozone origin-status: AFR: E
World rarity status: R
Ecozone-endemic macroregion occurrence: AFR: FOR
Life form(s): SU (rheophyte, attached to rocks)
Ploidy state: NA
Speciation age (Ma): NA
Populations also occur in terrestrial, wetland and/ or riparian habitats: NO
Long distance movement vectors: NA
Nuisance status and uses: no nuisance issues; no known uses
Endangered status (IUCN): NA

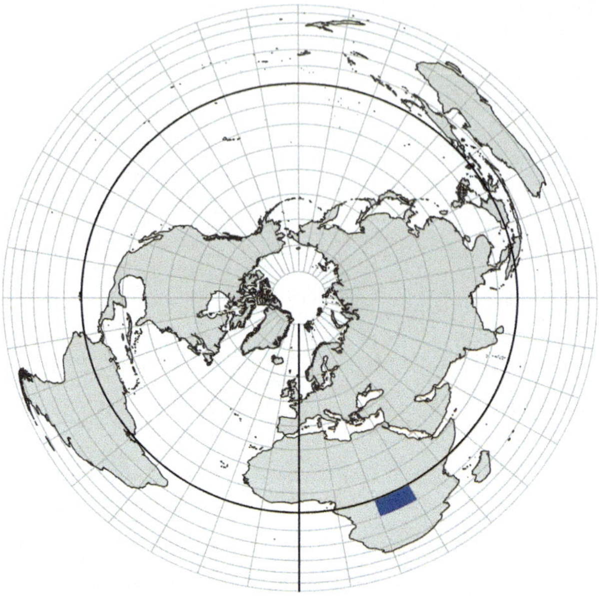

***Hydrostachys lukungensis* (Hauman) C.Cusset**
Synonyms:

Range area (grAH km^2): AFR: 18200
Ecozone origin-status: AFR: E
World rarity status: VR
Ecozone-endemic macroregion occurrence: AFR: FOR
Life form(s): SU (rheophyte, attached to rocks)
Ploidy state: NA
Speciation age (Ma): NA
Populations also occur in terrestrial, wetland and/ or riparian habitats: NO
Long distance movement vectors: NA
Nuisance status and uses: no nuisance issues; no known uses
Endangered status (IUCN): NA

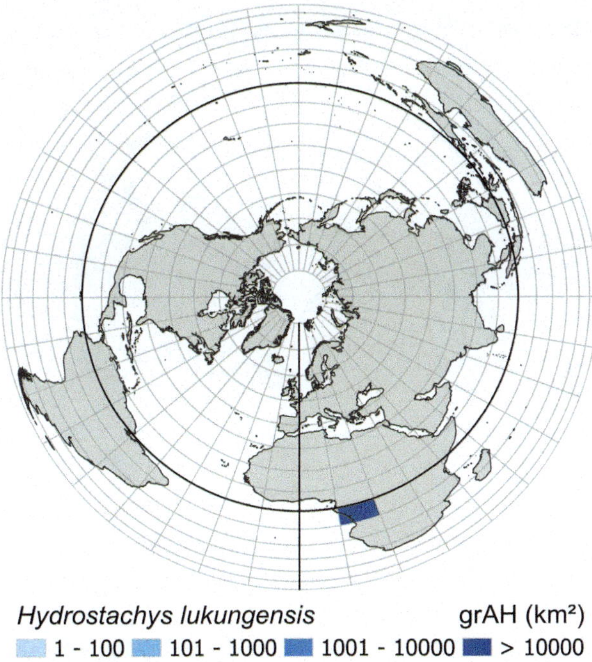

Hydrostachys maxima **H.Perrier**
Synonyms:

Range area (grAH km^2): AFR: 8000
Ecozone origin-status: AFR: E
World rarity status: VR
Ecozone-endemic macroregion occurrence: AFR: MAD
Life form(s): SU (rheophyte, attached to rocks)
Ploidy state: NA
Speciation age (Ma): NA
Populations also occur in terrestrial, wetland and/ or riparian habitats: NO
Long distance movement vectors: NA
Nuisance status and uses: no nuisance issues; no known uses
Endangered status (IUCN): VU

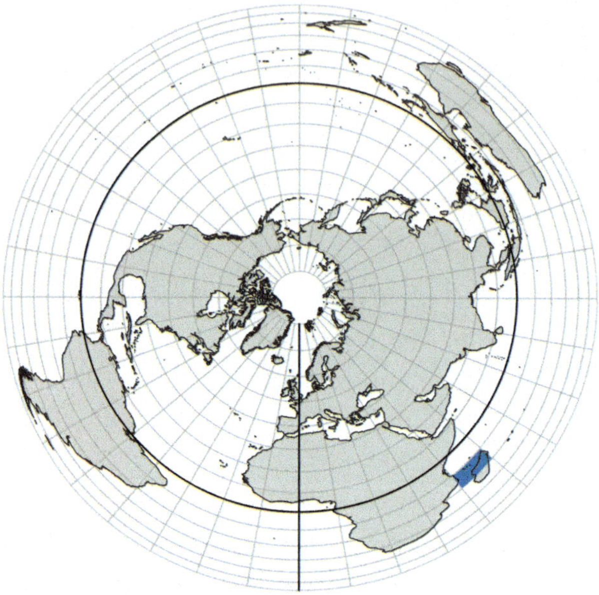

***Hydrostachys monoicus* H.Perrier**
Synonyms:

Range area (grAH km^2): AFR: 8000
Ecozone origin-status: AFR: E
World rarity status: VR
Ecozone-endemic macroregion occurrence: AFR: MAD
Life form(s): SU (rheophyte, attached to rocks)
Ploidy state: NA
Speciation age (Ma): NA
Populations also occur in terrestrial, wetland and/ or riparian habitats: NO
Long distance movement vectors: NA
Nuisance status and uses: no nuisance issues; no known uses
Endangered status (IUCN): CR

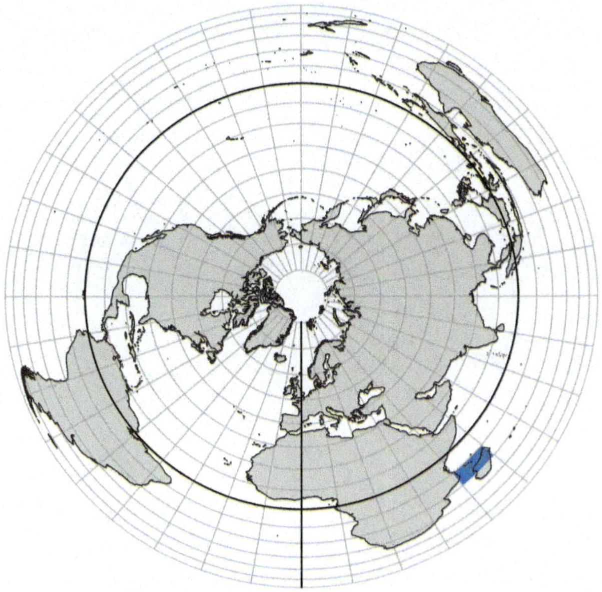

***Hydrostachys multifida* A.Juss.**
Synonyms:

Range area (grAH km^2): AFR: 10100
Ecozone origin-status: AFR: E
World rarity status: VR
Ecozone-endemic macroregion occurrence: AFR: MAD
Life form(s): SU (rheophyte, attached to rocks)
Ploidy state: NA
Speciation age (Ma): NA
Populations also occur in terrestrial, wetland and/ or riparian habitats: NO
Long distance movement vectors: NA
Nuisance status and uses: no nuisance issues; no known uses
Endangered status (IUCN): NT

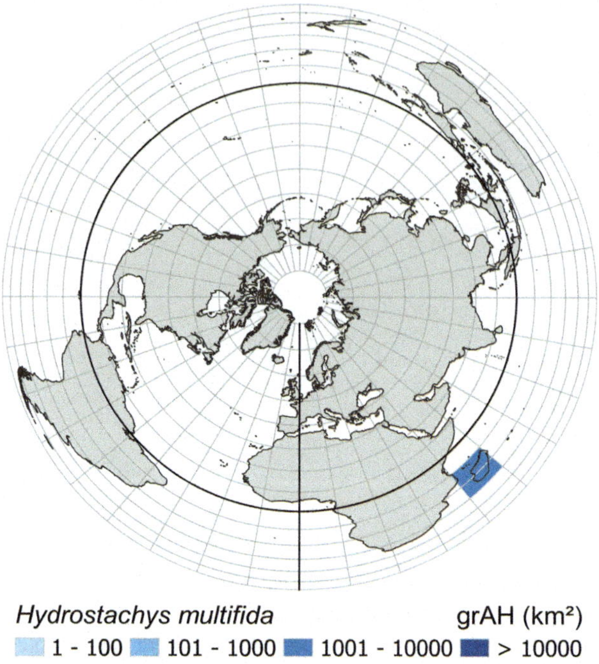

Hydrostachys myriophylla **Hauman**.
Synonyms:

Range area (grAH km^2): AFR: 10100
Ecozone origin-status: AFR: E
World rarity status: VR
Ecozone-endemic macroregion occurrence: AFR: ZAM
Life form(s): SU (rheophyte, attached to rocks)
Ploidy state: NA
Speciation age (Ma): NA
Populations also occur in terrestrial, wetland and/ or riparian habitats: NO
Long distance movement vectors: NA
Nuisance status and uses: no nuisance issues; no known uses
Endangered status (IUCN): NA

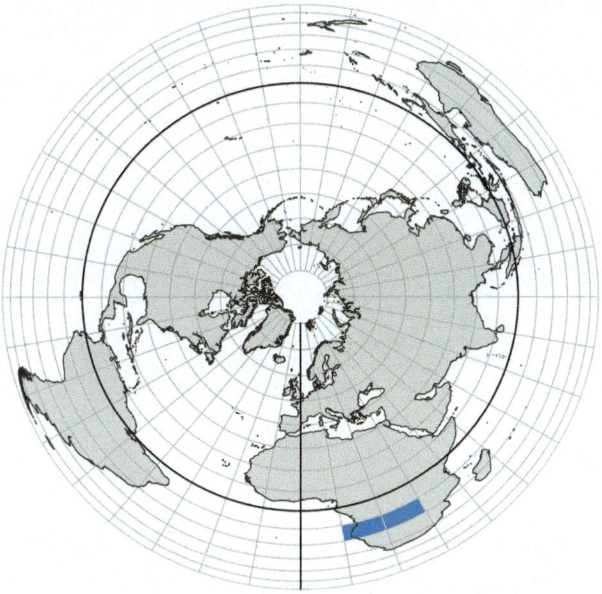

Hydrostachys perrieri **C.Cusset**
Synonyms:

Range area (grAH km^2): AFR: 8000
Ecozone origin-status: AFR: E
World rarity status: VR
Ecozone-endemic macroregion occurrence: AFR: MAD
Life form(s): SU (rheophyte, attached to rocks)
Ploidy state: NA
Speciation age (Ma): NA
Populations also occur in terrestrial, wetland and/ or riparian habitats: NO
Long distance movement vectors: NA
Nuisance status and uses: no nuisance issues; no known uses
Endangered status (IUCN): NA

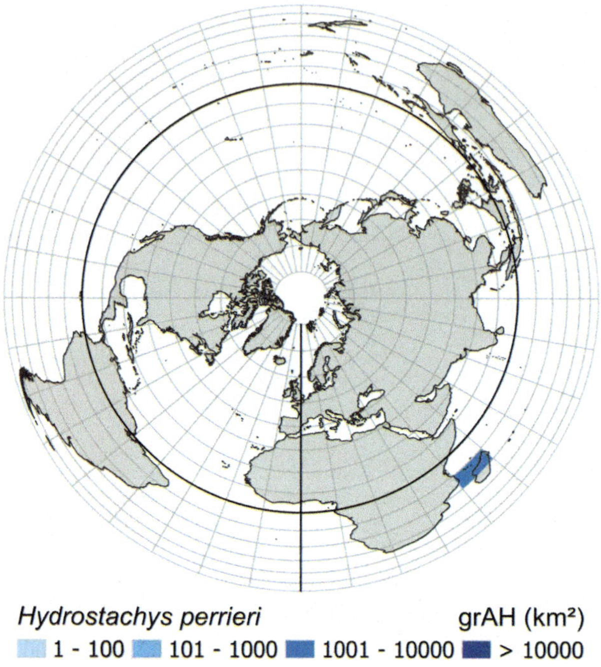

***Hydrostachys plumosa* A. Juss. ex Tul.**
Synonyms: *Hydrostachys pinnatifolia* Engl.

Range area (grAH km^2): AFR: 10100
Ecozone origin-status: AFR: E
World rarity status: VR
Ecozone-endemic macroregion occurrence: AFR: MAD
Life form(s): SU (rheophyte, attached to rocks)
Ploidy state: NA
Speciation age (Ma): NA
Populations also occur in terrestrial, wetland and/ or riparian habitats: NO
Long distance movement vectors: NA
Nuisance status and uses: no nuisance issues; folk medicine
Endangered status (IUCN): LC

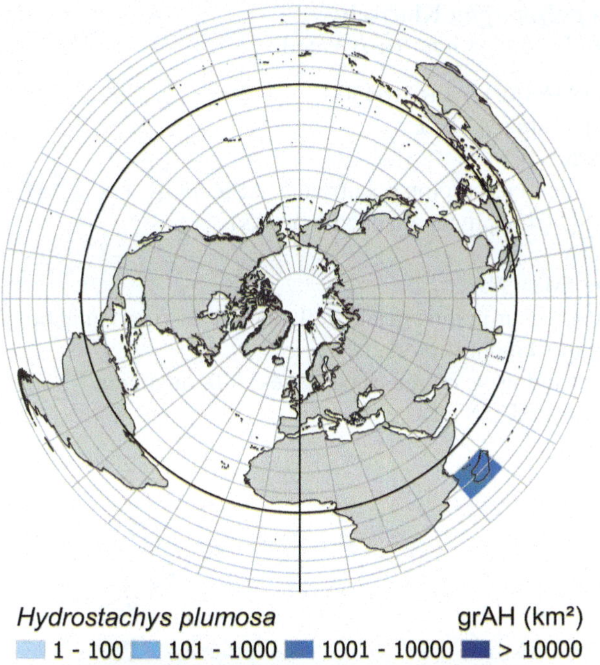

Hydrostachys plumosa A. Juss. ex Tul.
Image copyright: iNaturalist CC0 @romer (Romer Rabarijaona)

Hydrostachys polymorpha **Klotzsch**
Synonyms: *Hydrostachys natalensis* Wedd.

Range area (grAH km²): AFR: 164300
Ecozone origin-status: AFR: E
World rarity status: C
Ecozone-endemic macroregion occurrence: AFR: FOR, EA, ZAM
Life form(s): SU (rheophyte, attached to rocks)
Ploidy state: NA
Speciation age (Ma): NA
Populations also occur in terrestrial, wetland and/ or riparian habitats: NO
Long distance movement vectors: NA
Nuisance status and uses: no nuisance issues; folk medicine
Endangered status (IUCN): LC

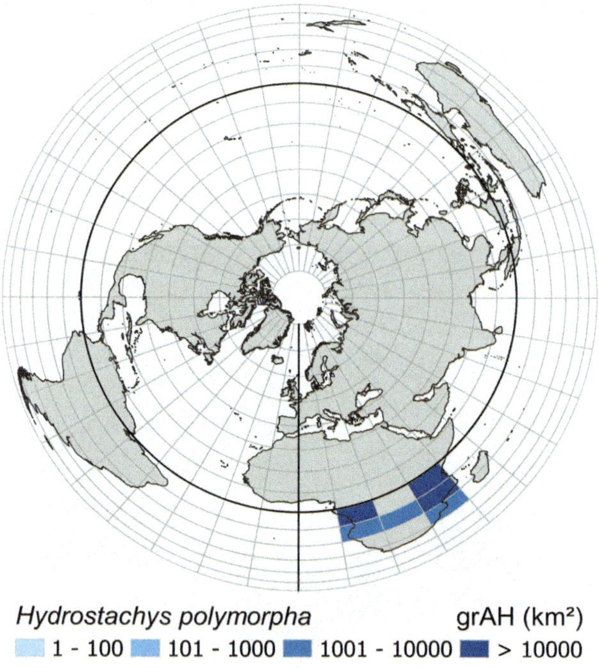

Hydrostachys polymorpha Klotzsch
Image copyright: iNaturalist CC-BY @rob_palmer (Rob Palmer)

***Hydrostachys stolonifera* Baker**
Synonyms:

Range area (grAH km^2): AFR: 10100
Ecozone origin-status: AFR: E
World rarity status: VR
Ecozone-endemic macroregion occurrence: AFR: MAD
Life form(s): SU (rheophyte, attached to rocks)
Ploidy state: NA
Speciation age (Ma): NA
Populations also occur in terrestrial, wetland and/ or riparian habitats: NO
Long distance movement vectors: NA
Nuisance status and uses: no nuisance issues; no known uses
Endangered status (IUCN): NT

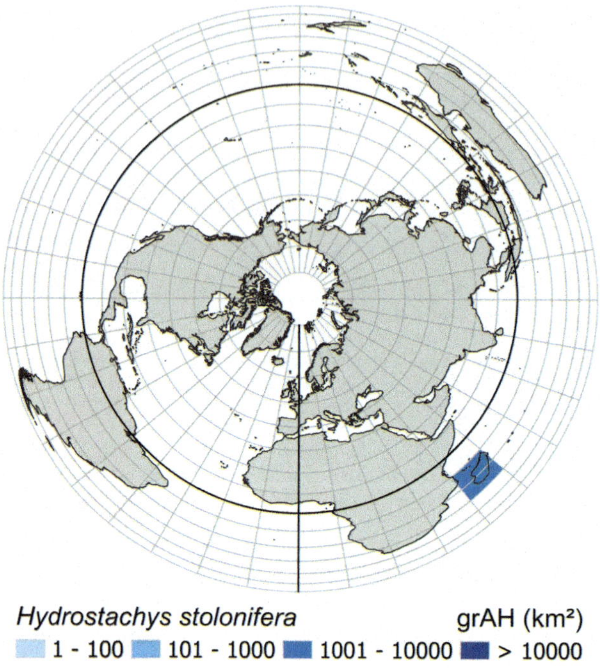

Hydrostachys triaxialis **Engl. & Gilg**
Synonyms:

Range area (grAH km^2): AFR: 21100
Ecozone origin-status: AFR: E
World rarity status: VR
Ecozone-endemic macroregion occurrence: AFR: FOR, ZAM
Life form(s): SU (rheophyte, attached to rocks)
Ploidy state: NA
Speciation age (Ma): NA
Populations also occur in terrestrial, wetland and/ or riparian habitats: NO
Long distance movement vectors: NA
Nuisance status and uses: no nuisance issues; no known uses
Endangered status (IUCN): NA

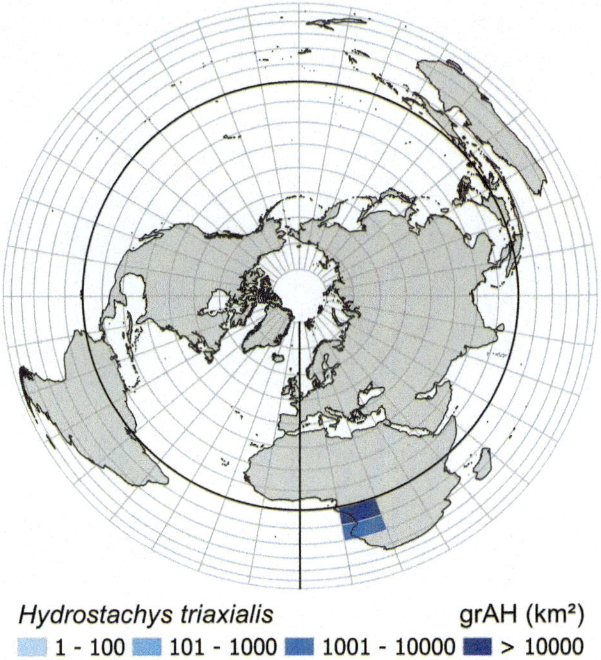

Hydrostachys trifaria **H. Perrier**
Synonyms:

Range area (grAH km^2): AFR: 10100
Ecozone origin-status: AFR: E
World rarity status: VR
Ecozone-endemic macroregion occurrence: AFR: MAD
Life form(s): SU (rheophyte, attached to rocks)
Ploidy state: NA
Speciation age (Ma): NA
Populations also occur in terrestrial, wetland and/ or riparian habitats: NO
Long distance movement vectors: NA
Nuisance status and uses: no nuisance issues; no known uses
Endangered status (IUCN): EN

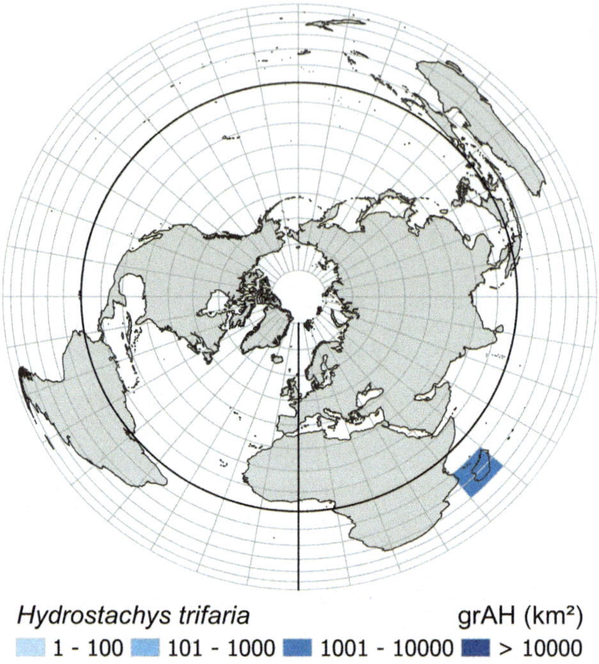

***Hydrostachys verruculosa* A.Juss**

Synonyms: *Hydrostachys laciniata* Warm.

Range area (grAH km^2): AFR: 10100
Ecozone origin-status: AFR: E
World rarity status: VR
Ecozone-endemic macroregion occurrence: AFR: MAD
Life form(s): SU (rheophyte, attached to rocks)
Ploidy state: NA
Speciation age (Ma): NA
Populations also occur in terrestrial, wetland and/ or riparian habitats: NO
Long distance movement vectors: NA
Nuisance status and uses: no nuisance issues; no known uses
Endangered status (IUCN): EN

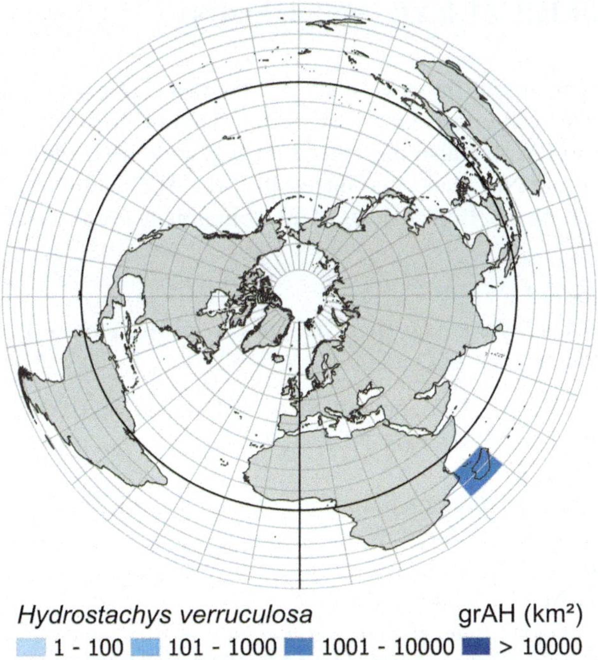

Hydrostachys verruculosa A.Juss.
Image copyright: iNaturalist CC0 @romer (Romer Rabarijaona)

3.25 HYPERICACEAE (Malpighiales)

Within the Hypericaceae, only two species in the genus *Hypericum* occur as macrophytes in inland fresh or brackish waterbodies. One of these is an Old World species, endemic to PAL but also introduced in AUS and NEA, while the other one is native in NEA and NEO, and introduced in PAL, AUS and OCE. Neither causes aquatic weed problems. Both species of are common at world scale.

***Hypericum elodes* L.**
Synonyms: *Spachelodes elodes (*L.) Y.Kimura*, Tripentas elodes* (L.) Asch., *Elodes palustris* Spach

Range area (grAH km^2): PAL: 56400; NEA: 304500; AUS: 3100
Ecozone origin-status: PAL: E(I); NEA: I; AUS: I
World rarity status: C
Ecozone-endemic macroregion occurrence: PAL: W EUR, MED
Life form(s): EM
Ploidy state: D
Speciation age (Ma): 25.0
Populations also occur in terrestrial, wetland and/ or riparian habitats: YES
Long distance movement vectors: ?AV; ANTHROPOCHORY
Nuisance status and uses: no nuisance issues; water garden plant, medicinal uses
Endangered status (IUCN): LC

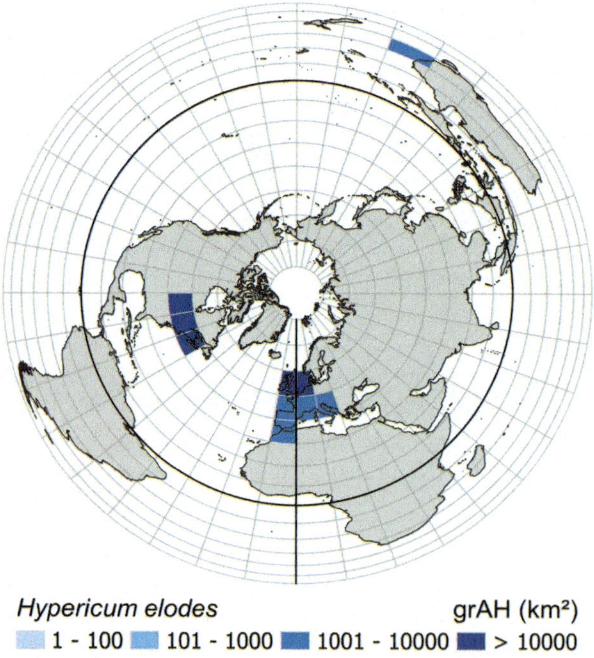

***Hypericum mutilum* L.**
Synonyms: *Sarothra mutila* (L.) Y.Kimura

Range area (grAH km^2): PAL: 48400; NEA: 392100; NEO: 125700; AUS: 5400; present in OCE
Ecozone origin-status: PAL: I; NEA: N; NEO: N; AUS: I; OCE: I
World rarity status: C
Ecozone-endemic macroregion occurrence: not endemic to any world ecozone
Life form(s): EM
Ploidy state: D
Speciation age (Ma): 1.0
Populations also occur in terrestrial, wetland and/ or riparian habitats: YES
Long distance movement vectors: ANTHROPOCHORY
Nuisance status and uses: no nuisance issues; medicinal potential
Endangered status (IUCN): NA

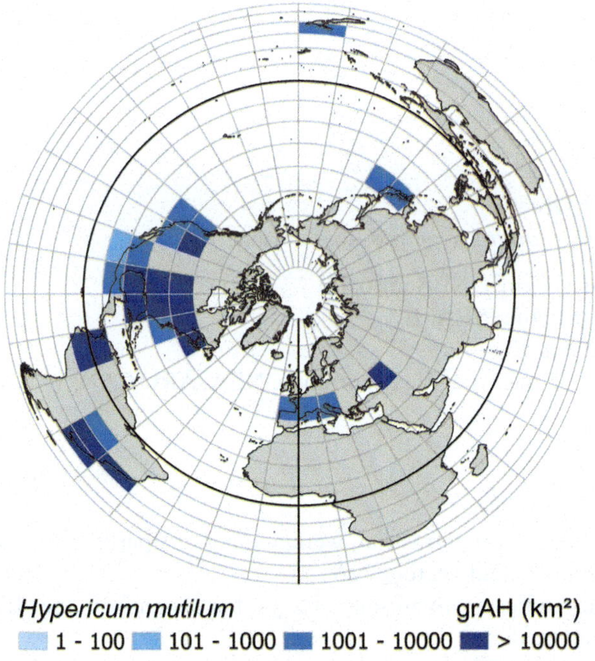

3.26 LAMIACEAE (Lamiales)

The Lamiaceae contain 23 macrophyte species within six genera occurring in all eight world ecozones, mainly at high latitudes. There are 14 species present in NEA (eight endemic), 11 in PAL (four endemic), eight each in AUS and OR (one of these endemic to OR), five each in AFR (two endemic) and NEO (no endemics), four in OCE (all introduced), and one species introduced in ANT. Other introductions include six species in NEA, five of the nine species found in AUS, all five of those occurring in NEO, and one species in AFR. None of the macrophyte species of this family causes aquatic weed problems. There are more world-rare than common macrophyte species in this family, with VR 39%, R 22%, and C 39%. One species is of CR endangered status, 44% NA/DD and 52% LC.

Quick Links to Genera

1. *Lycopus*
2. *Mentha*
3. *Platostoma*
4. *Pogogyne*
5. *Pogostemon*
6. *Teucrium*

***Lycopus asper* Greene**
Synonyms: *Phytosalpinx aspera* (Greene) Lunell

Range area (grAH km^2): NEA: 570700
Ecozone origin-status: NEA: E
World rarity status: C
Ecozone-endemic macroregion occurrence: NEA: ARC CAN, W NAM, SW NAM, E NAM
Life form(s): EM
Ploidy state: D
Speciation age (Ma): NA
Populations also occur in terrestrial, wetland and/ or riparian habitats: YES
Long distance movement vectors: NA
Nuisance status and uses: no nuisance issues; harvested from the wild for local use as a food and medicine
Endangered status (IUCN): LC

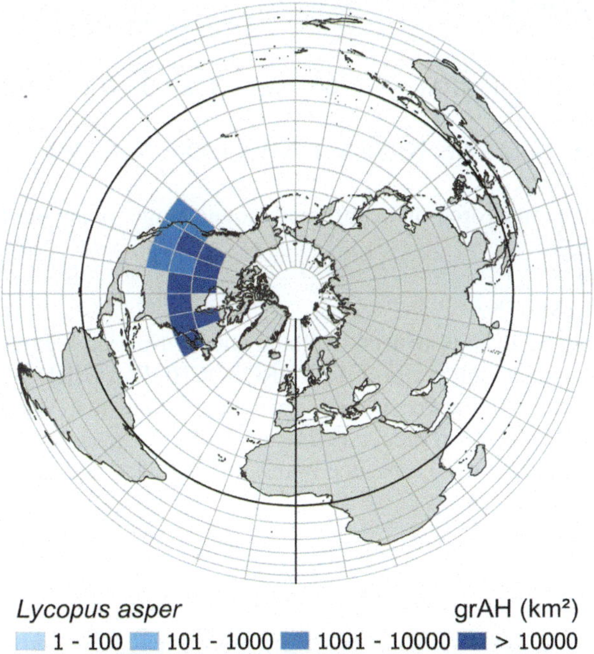

Lycopus asper Greene

***Lycopus europaeus* L.**
Synonyms:

Range area (grAH km^2): PAL: 697700; NEA: 321100; NEO: 23800; OR: 24900; AUS: 20600
Ecozone origin-status: PAL: N; NEA: I; NEO: I; OR: N; AUS: I
World rarity status: C
Ecozone-endemic macroregion occurrence: not endemic to any world ecozone
Life form(s): EM
Ploidy state: D
Speciation age (Ma): NA
Populations also occur in terrestrial, wetland and/ or riparian habitats: YES
Long distance movement vectors: ANTHROPOCHORY; BIRD: *Anas platyrhynchos* (Northern Mallard); *Mareca penelope* (Eurasian Widgeon); *Mareca strepera* (Gadwall); *Anas crecca* (Eurasian Teal)
Nuisance status and uses: no nuisance issues; medicinal, water garden plant uses
Endangered status (IUCN): LC

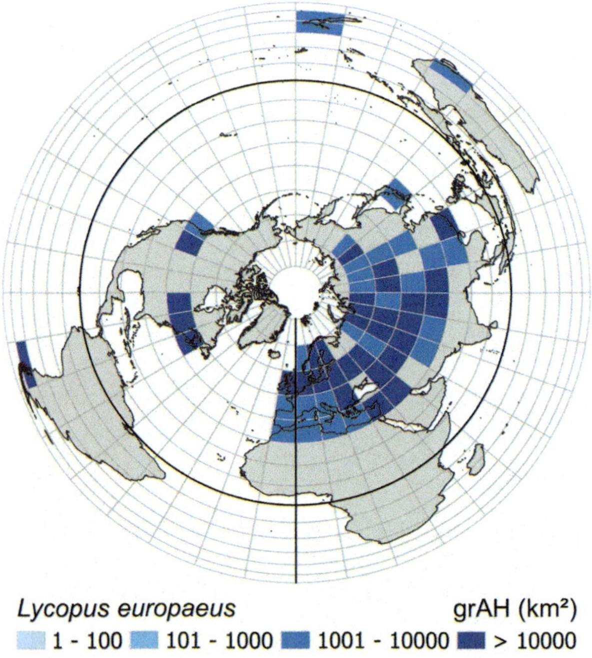

Lycopus europaeus L.
Image copyright: A. Efremov, with permission

Lycopus exaltatus **L.f.**
Synonyms:

Range area (grAH km^2): PAL: 234900
Ecozone origin-status: PAL: E
World rarity status: C
Ecozone-endemic macroregion occurrence: PAL: W EUR, E EUR, N EUR, UR/W SIB, E SIB
Life form(s): EM
Ploidy state: D
Speciation age (Ma): NA
Populations also occur in terrestrial, wetland and/ or riparian habitats: YES
Long distance movement vectors: NA
Nuisance status and uses: no nuisance issues; medicinal
Endangered status (IUCN): LC

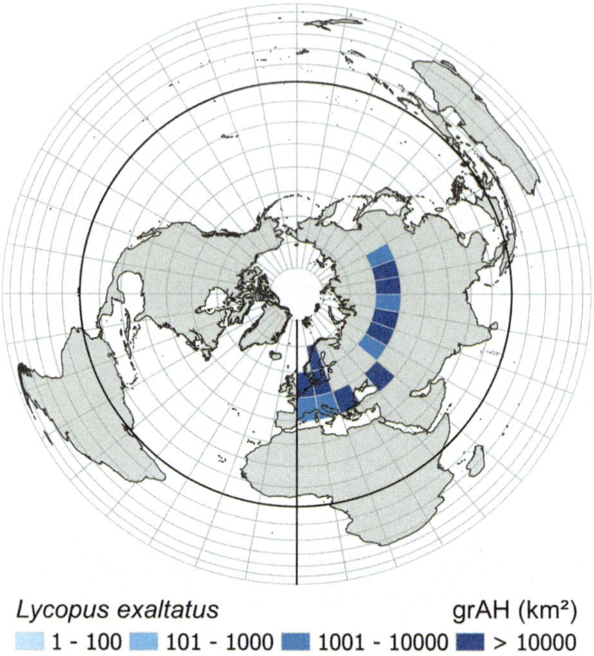

Lycopus exaltatus L.f.
Image copyright: A. Efremov, with permission

***Lycopus hirtellus* Kom.**
Synonyms:

Range area (grAH km²): PAL: 23700
Ecozone origin-status: PAL: E
World rarity status: VR
Ecozone-endemic macroregion occurrence: PAL: E SIB, FER/J
Life form(s): EM
Ploidy state: D
Speciation age (Ma): NA
Populations also occur in terrestrial, wetland and/ or riparian habitats: YES
Long distance movement vectors: NA
Nuisance status and uses: no nuisance issues; no known uses
Endangered status (IUCN): NA

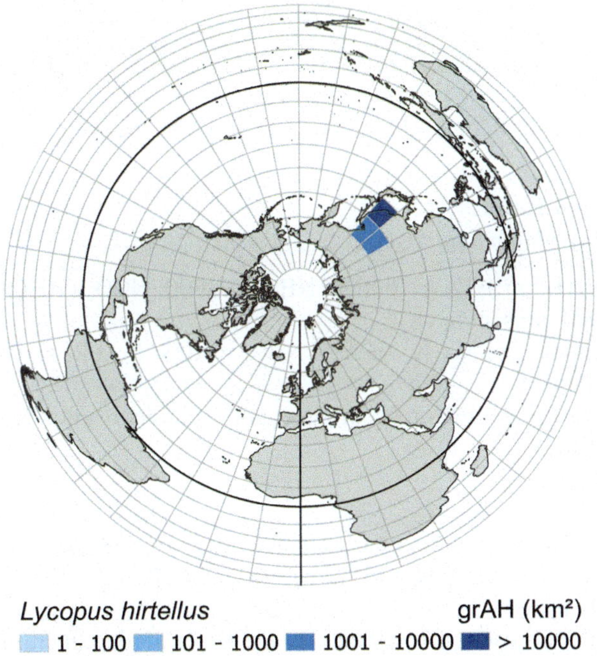

Lycopus hirtellus Kom.
Image copyright: A. Banyshenko, with permission

***Mentha aquatica* Kom.**
Synonyms: *Marrubium aquaticum* (L.) Uspenski

Range area (grAH km^2): PAL: 369600; NEA: 103800; NEO: 110300; AUS: 28300; AFR: 177000; present in OCE
Ecozone origin-status: PAL: N; NEA: I; NEO: I; AUS: I; AFR: N; OCE: I
World rarity status: C
Ecozone-endemic macroregion occurrence: not endemic to any world ecozone
Life form(s): EM
Ploidy state: P
Speciation age (Ma): 5.3
Populations also occur in terrestrial, wetland and/ or riparian habitats: YES
Long distance movement vectors: ANTHROPOCHORY; BIRD: *Anas crecca* (Eurasian Teal); ?AV
Nuisance status and uses: no nuisance issues; water garden plant, medicinal, cosmetic, phytoremediation uses
Endangered status (IUCN): LC

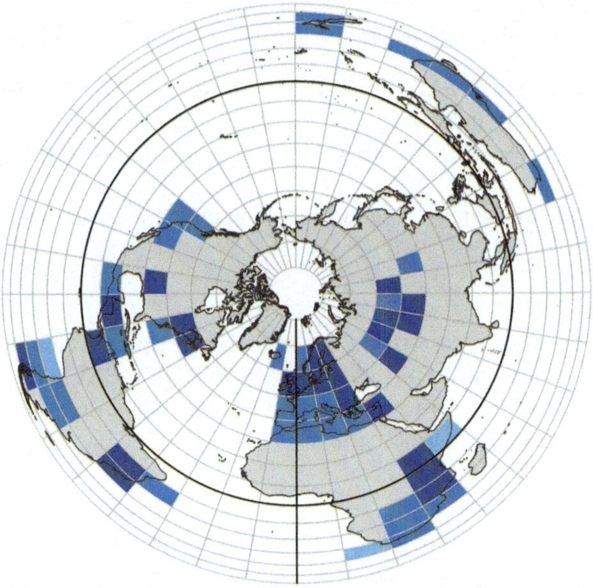

Mentha aquatica Kom.
Image copyright: J.C. Schou, with permission

***Mentha arvensis* L.**

Synonyms: *Calamintha arvensis* (L.) Garsault

Range area (grAH km²): PAL: 703500; NEA: 768000; NEO: 111600; OR: 2600; AUS: 7800; present in OCE

Ecozone origin-status: PAL: N; NEA: I; NEO: I; OR: N; AUS: I; OCE: I

World rarity status: C

Ecozone-endemic macroregion occurrence: not endemic to any world ecozone

Life form(s): EM

Ploidy state: P

Speciation age (Ma): 5.3

Populations also occur in terrestrial, wetland and/ or riparian habitats: YES

Long distance movement vectors: ANTHROPOCHORY; ?AV

Nuisance status and uses: no nuisance issues; occasional use as a culinary herb, essential oil production, folk medicine

Endangered status (IUCN): LC

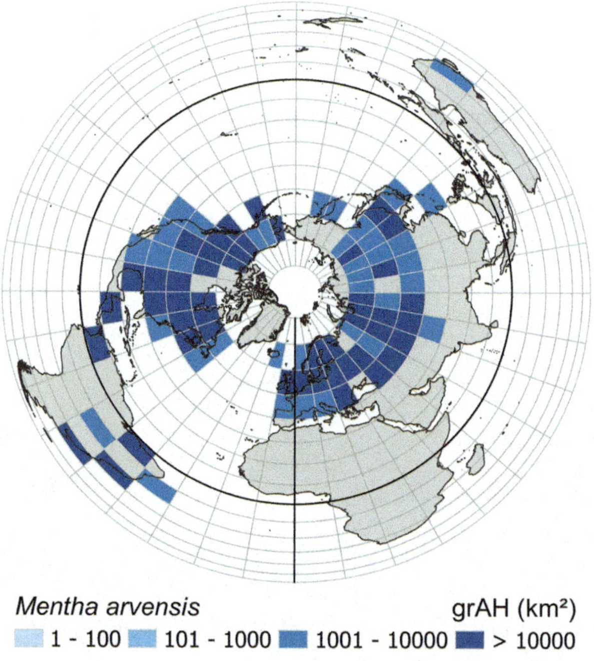

Mentha arvensis L.
Image copyright: J.C. Schou, with permission

***Mentha longifolia* (L.) L.**
Synonyms:

Range area (grAH km^2): PAL: 446800; NEA: 68000; NEO: 12100; OR: 38000; AUS: 19400; AFR: 72700
Ecozone origin-status: PAL: N; NEA: I; NEO: I; OR: N; AUS: I; AFR: N
World rarity status: C
Ecozone-endemic macroregion occurrence: not endemic to any world ecozone
Life form(s): EM
Ploidy state: DP
Speciation age (Ma): 5.3
Populations also occur in terrestrial, wetland and/ or riparian habitats: YES
Long distance movement vectors: ANTHROPOCHORY; ?AV
Nuisance status and uses: no nuisance issues; medicinal, cosmetic uses, also a garden plant
Endangered status (IUCN): LC

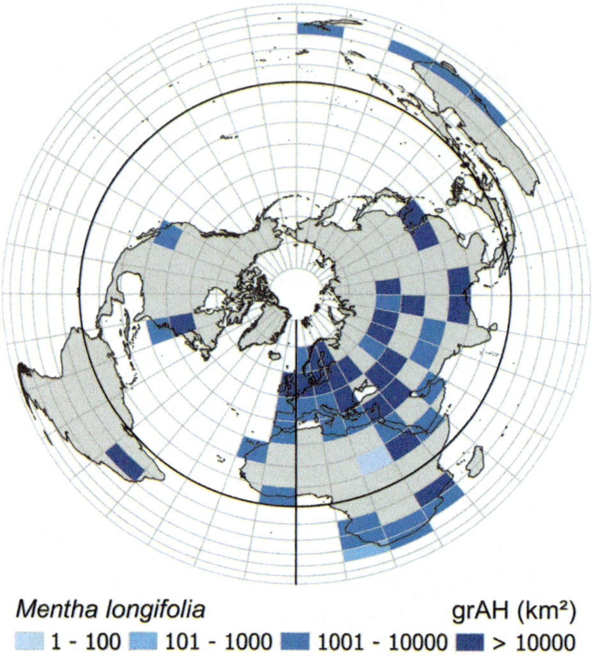

Mentha longifolia (L.) L.
Image copyright: A. Efremov, with permission

***Mentha pulegium* L.**

Synonyms: *Melissa pulegium* (L.) Griseb., *Mentha aromatica* Salisb., *Minthe pulegia* (L.) St.-Lag., *Pulegium vulgare* Mill.

Range area (grAH km^2): PAL: 252800; NEA: 243700; NEO: 246900; AUS: 37500; present in ANT, OCE
Ecozone origin-status: PAL: E(I); NEA: I; NEO: I; AUS: I; OCE: I; ANT: I
World rarity status: C
Ecozone-endemic macroregion occurrence: PAL: N EUR, W EUR, E EUR, MED, N AFR/ ARAB, E SIB, FER/J
Life form(s): EM
Ploidy state: D
Speciation age (Ma): 5.3
Populations also occur in terrestrial, wetland and/ or riparian habitats: YES
Long distance movement vectors: ANTHROPOCHORY; ?AV
Nuisance status and uses: no nuisance issues; folk medicine, cosmetic, garden plant uses
Endangered status (IUCN): LC

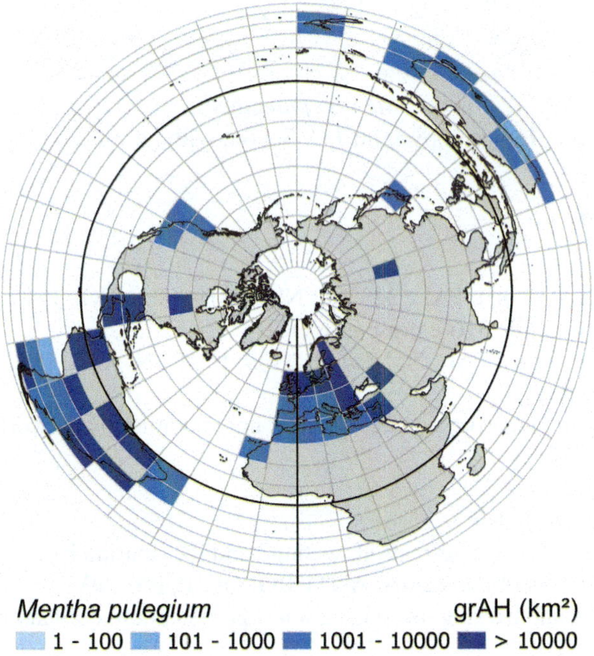

Mentha pulegium L.
Image copyright: J.C. Schou, with permission

***Mentha spicata* L.**
Synonyms:

Range area (grAH km^2): PAL: 323300; NEA: 419900; NEO: 304200; OR: 27500; AUS: 41918; AFR: 3100; present in OCE
Ecozone origin-status: PAL: N; NEA: I; NEO: I; OR: N; AUS: I; AFR: I; OCE: I
World rarity status: C
Ecozone-endemic macroregion occurrence: not endemic to any world ecozone
Life form(s): EM
Ploidy state: P
Speciation age (Ma): 5.3
Populations also occur in terrestrial, wetland and/ or riparian habitats: YES
Long distance movement vectors: ANTHROPOCHORY; ?AV
Nuisance status and uses: no nuisance issues; garden plant, culinary herb, folk medicine uses
Endangered status (IUCN): LC

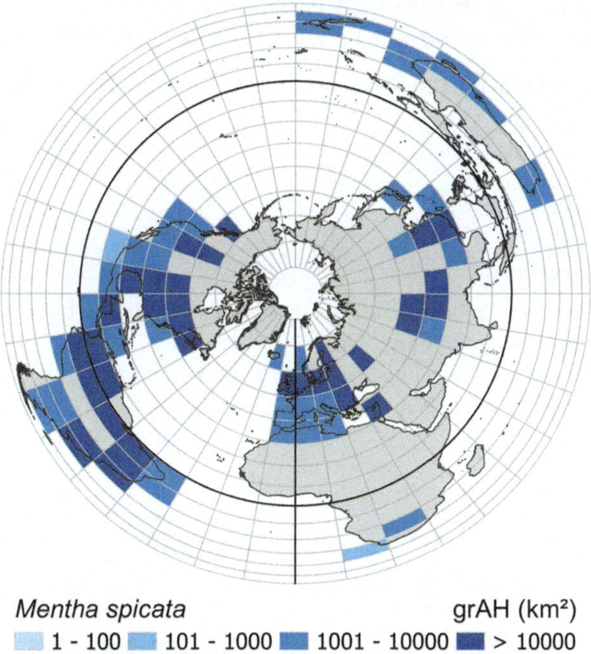

Mentha spicata L.
Image copyright: J.C. Schou, with permission

***Platostoma coeruleum* (R.E.Fr.) A.J.Paton**
Synonyms: *Limniboza coerulea* R.E.Fr.

Range area (grAH km^2): AFR: 93700
Ecozone origin-status: AFR: E
World rarity status: R
Ecozone-endemic macroregion occurrence: AFR: EA
Life form(s): EM
Ploidy state: NA
Speciation age (Ma): NA
Populations also occur in terrestrial, wetland and/ or riparian habitats: YES
Long distance movement vectors: NA
Nuisance status and uses: no nuisance issues; no known uses
Endangered status (IUCN): NA

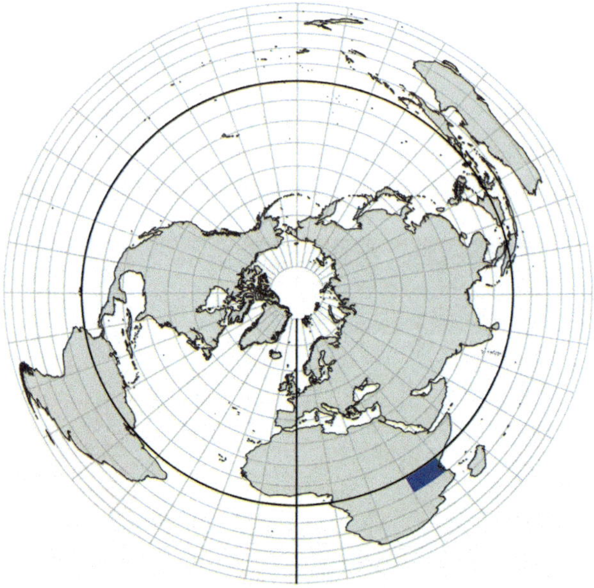

***Platostoma dilungense* (Lisowski & Mielcarek) A.J.Paton**
Synonyms: *Limniboza dilungensis* Lisowski & Mielcarek

Range area (grAH km^2): AFR: 50200-
Ecozone origin-status: AFR: E
World rarity status: R
Ecozone-endemic macroregion occurrence: AFR: FOR
Life form(s): EM (sometimes hydrosubshrub)

Ploidy state: NA
Speciation age (Ma): NA
Populations also occur in terrestrial, wetland and/ or riparian habitats: YES
Long distance movement vectors: NA
Nuisance status and uses: no nuisance issues; no known uses
Endangered status (IUCN): NA

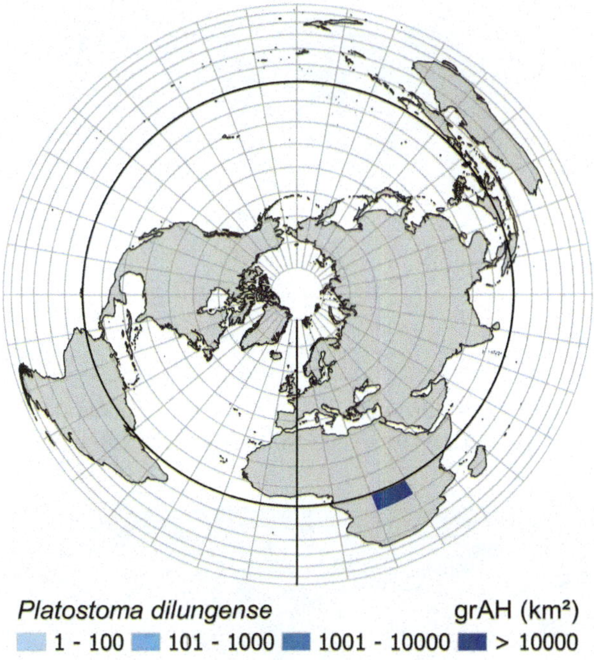

Pogogyne abramsii **J.T.Howell**
Synonyms:

Range area (grAH km^2): NEA: 3500
Ecozone origin-status: NEA: E
World rarity status: VR
Ecozone-endemic macroregion occurrence: NEA: SW NAM
Life form(s): EM
Ploidy state: NA
Speciation age (Ma): NA
Populations also occur in terrestrial, wetland and/ or riparian habitats: YES
Long distance movement vectors: NA
Nuisance status and uses: no nuisance issues; no known uses
Endangered status (IUCN): NA (considered Critically Imperiled in its native range in California)

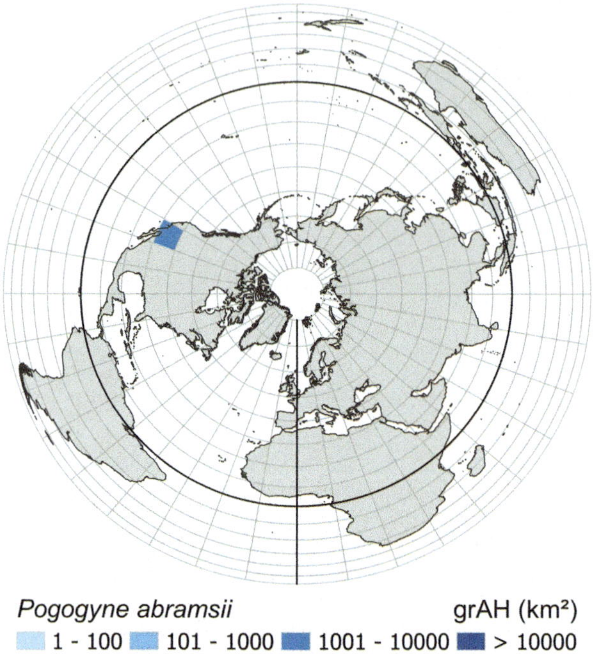

Pogogyne abramsii J.T.Howell

Pogogyne clareana **J.T.Howell**
Synonyms:

Range area (grAH km^2): NEA: 4500
Ecozone origin-status: NEA: E
World rarity status: VR
Ecozone-endemic macroregion occurrence: NEA: SW NAM
Life form(s): EM
Ploidy state: D
Speciation age (Ma): NA
Populations also occur in terrestrial, wetland and/ or riparian habitats: YES
Long distance movement vectors: NA
Nuisance status and uses: no nuisance issues; no known uses
Endangered status (IUCN): NA

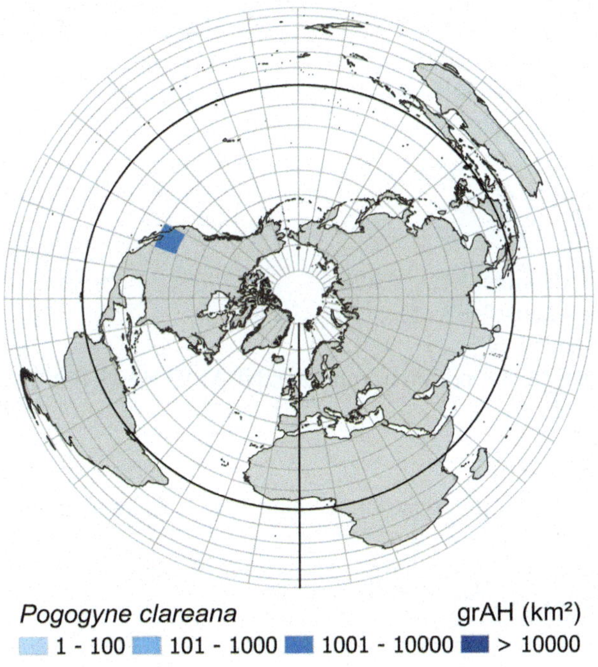

Pogogyne clareana J.T.Howell
Image copyright: iNaturalist CC-BY @chloe_and_trevor (Chloe and Trevor Van Loon)

Pogogyne douglasii **Benth.**
Synonyms:

Range area (grAH km^2): NEA: 4500
Ecozone origin-status: NEA: E
World rarity status: VR
Ecozone-endemic macroregion occurrence: NEA: W NAM, SW NAM
Life form(s): EM
Ploidy state: NA
Speciation age (Ma): NA
Populations also occur in terrestrial, wetland and/ or riparian habitats: NO
Long distance movement vectors: NA
Nuisance status and uses: no nuisance issues; medicinal
Endangered status (IUCN): NA

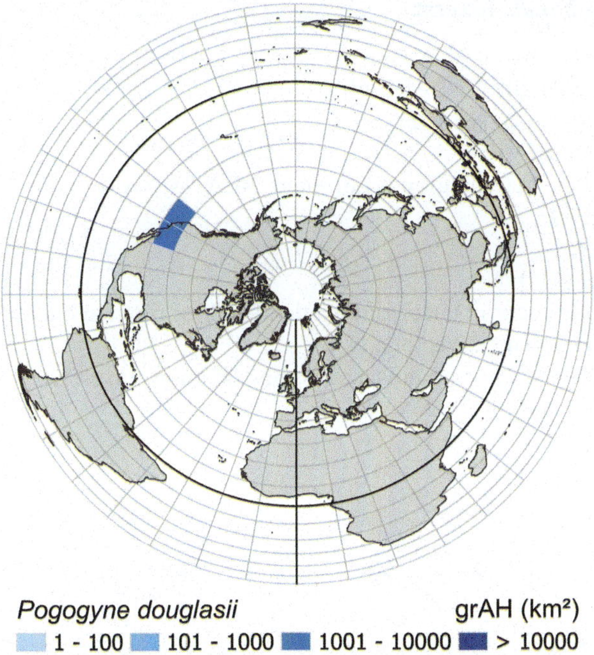

Pogogyne douglasii Benth.

Pogogyne floribunda **Jokerst**
Synonyms:

Range area (grAH km^2): NEA: 17600
Ecozone origin-status: NEA: E
World rarity status: VR
Ecozone-endemic macroregion occurrence: NEA: W NAM
Life form(s): EM
Ploidy state: NA
Speciation age (Ma): NA
Populations also occur in terrestrial, wetland and/ or riparian habitats: NO
Long distance movement vectors: NA
Nuisance status and uses: no nuisance issues; no known uses
Endangered status (IUCN): NA

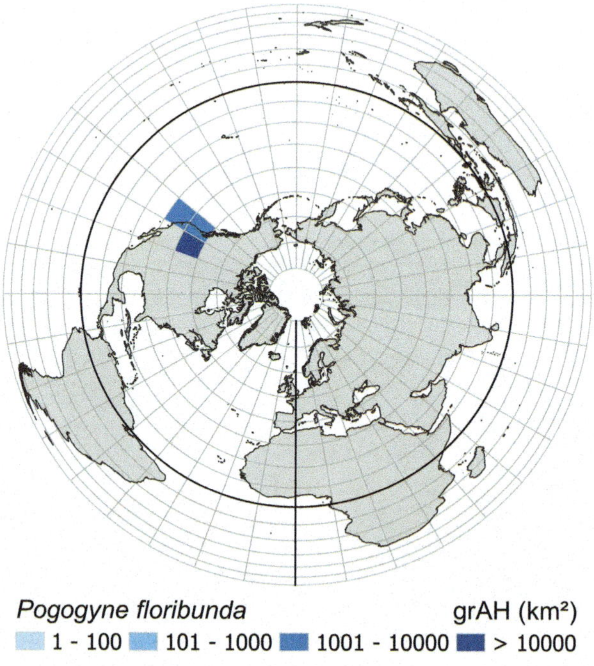

Pogogyne nudiuscula **A.Gray**
Synonyms:

Range area (grAH km^2): NEA: 3500
Ecozone origin-status: NEA: E
World rarity status: VR
Ecozone-endemic macroregion occurrence: NEA: SW NAM
Life form(s): EM
Ploidy state: NA

Speciation age (Ma): NA
Populations also occur in terrestrial, wetland and/ or riparian habitats: NO
Long distance movement vectors: NA
Nuisance status and uses: no nuisance issues; no known uses
Endangered status (IUCN): NA

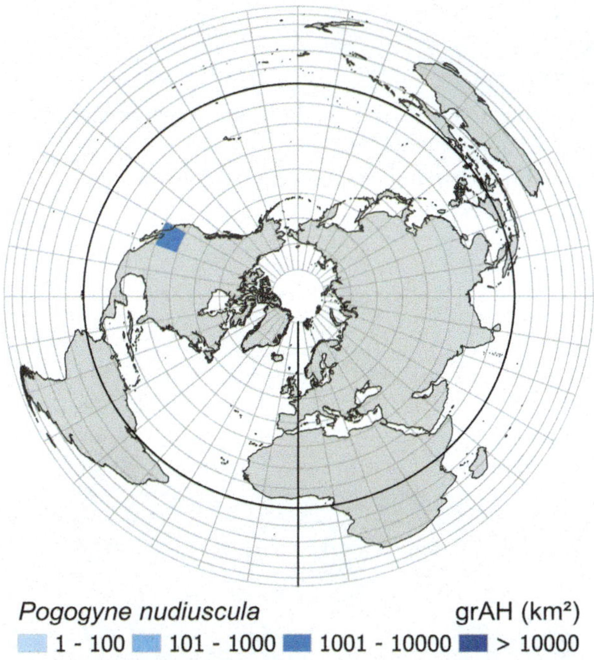

Pogogyne nudiuscula A.Gray

***Pogogyne serpylloides* (Torr.) A.Gray**
Synonyms: *Hedeomoides serpylloides* (Torr.) Briq., *Hedeoma serpylloides* Torr.

Range area (grAH km^2): NEA: 11000
Ecozone origin-status: NEA: E
World rarity status: VR
Ecozone-endemic macroregion occurrence: NEA: W NAM, SW NAM
Life form(s): EM
Ploidy state: NA
Speciation age (Ma): NA
Populations also occur in terrestrial, wetland and/ or riparian habitats: YES
Long distance movement vectors: NA
Nuisance status and uses: no nuisance issues; no known uses
Endangered status (IUCN): NA

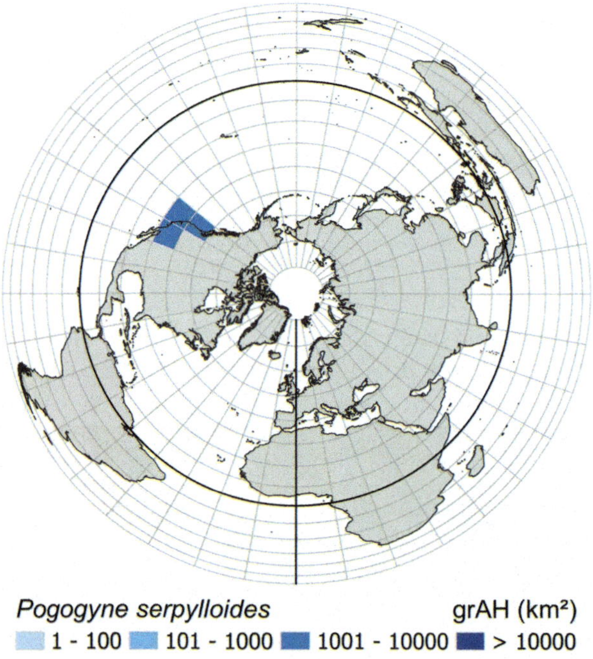

Pogogyne serpylloides (Torr.) A.Gray
Image copyright: iNaturalist CC-BY @kueda (Ken-ichi Ueda)

***Pogogyne zizyphoroides* Benth.**
Synonyms: *Hedeomoides ziziphoroides* (Benth.) Briq.

Range area (grAH km^2): NEA: 11000
Ecozone origin-status: NEA: E
World rarity status: VR
Ecozone-endemic macroregion occurrence: NEA: W NAM, SW NAM
Life form(s): EM
Ploidy state: D
Speciation age (Ma): NA
Populations also occur in terrestrial, wetland and/ or riparian habitats: YES
Long distance movement vectors: NA
Nuisance status and uses: no nuisance issues; no known uses
Endangered status (IUCN): NA

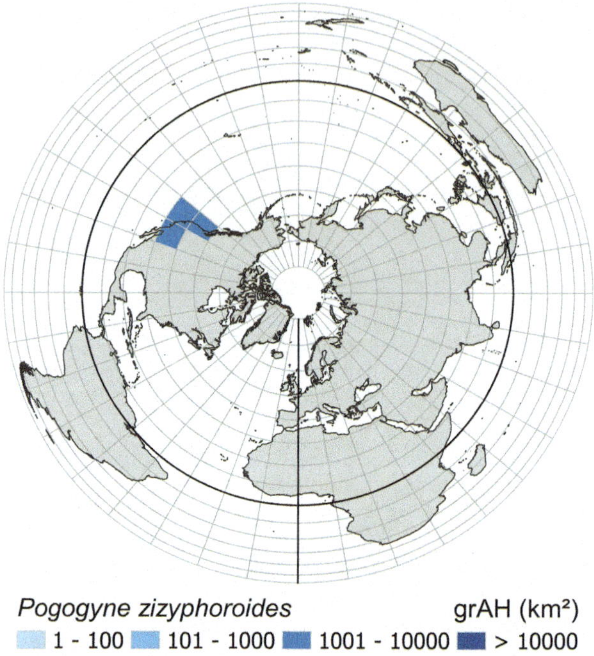

Pogogyne zizyphoroides Benth.

***Pogostemon helferi* (Hook.f.) Press**

Synonyms: *Anuragia helferi* (Hook.f.) Raizada, *Dysophylla helferi* Hook.f., *Eusteralis helferi* (Hook.f.) Panigrahi

Range area (grAH km^2): OR: 29100
Ecozone origin-status: OR: E
World rarity status: VR
Ecozone-endemic macroregion occurrence: OR: IND, ICH
Life form(s): SU, EM
Ploidy state: NA
Speciation age (Ma): NA
Populations also occur in terrestrial, wetland and/ or riparian habitats: YES
Long distance movement vectors: NA
Nuisance status and uses: no nuisance issues; aquarium trade
Endangered status (IUCN): NA

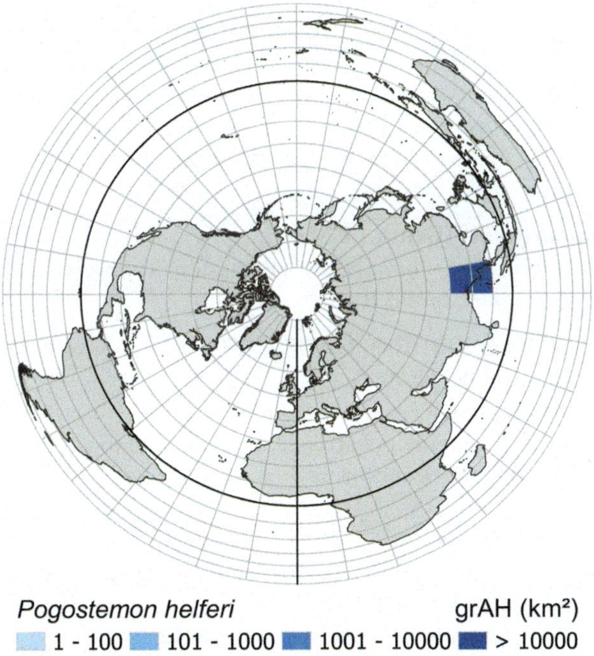

Pogostemon helferi (Hook.f.) Press

Pogostemon quadrifolius (Benth.) F.Muell.
Synonyms: *Eusteralis quadrifolia* (Benth.) Panigrahi, *Anuragia quadrifolia* (Benth.) Raizada, *Dysophylla quadrifolia* Benth.

Range area (grAH km^2): OR: 70400; AUS: 38500
Ecozone origin-status: OR: N; AUS: N
World rarity status: R
Ecozone-endemic macroregion occurrence: not endemic to any world ecozone
Life form(s): EM (sometimes subshrub)
Ploidy state: D
Speciation age (Ma): 5.4
Populations also occur in terrestrial, wetland and/ or riparian habitats: YES
Long distance movement vectors: ?AV
Nuisance status and uses: no nuisance issues; aquarium trade
Endangered status (IUCN): DD

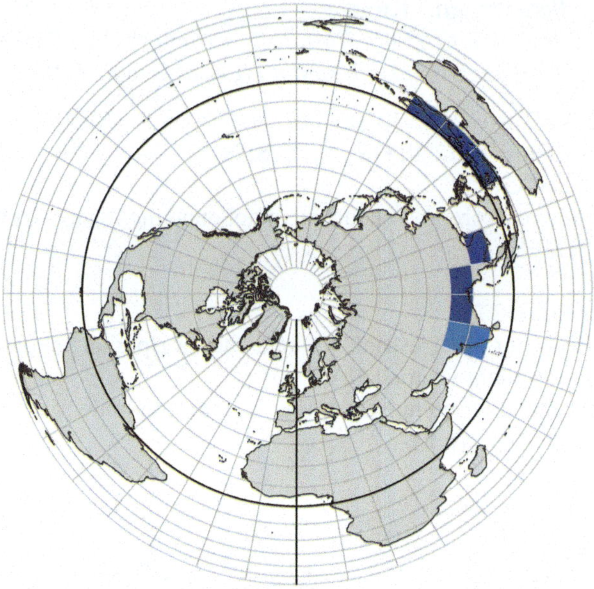

Pogostemon quadrifolius (Benth.) F.Muell.

***Pogostemon stellatus* (Lour.) Kuntze**
Synonyms: *Mentha stellata* Lour., *Eusteralis stellata* (Lour.) Panigrahi, *Anuragia stellata* (Lour.) Raizada, *Dysophylla stellata* (Lour.) Benth. ex Wall.

Range area (grAH km^2): PAL: 30200; OR: 86400; AUS: 48900
Ecozone origin-status: PAL: N; OR: N; AUS: N
World rarity status: C
Ecozone-endemic macroregion occurrence: not endemic to any world ecozone
Life form(s): SU, EM
Ploidy state: P
Speciation age (Ma): 2.2
Populations also occur in terrestrial, wetland and/ or riparian habitats: YES
Long distance movement vectors: NA
Nuisance status and uses: no nuisance issues; aquarium trade
Endangered status (IUCN): LC

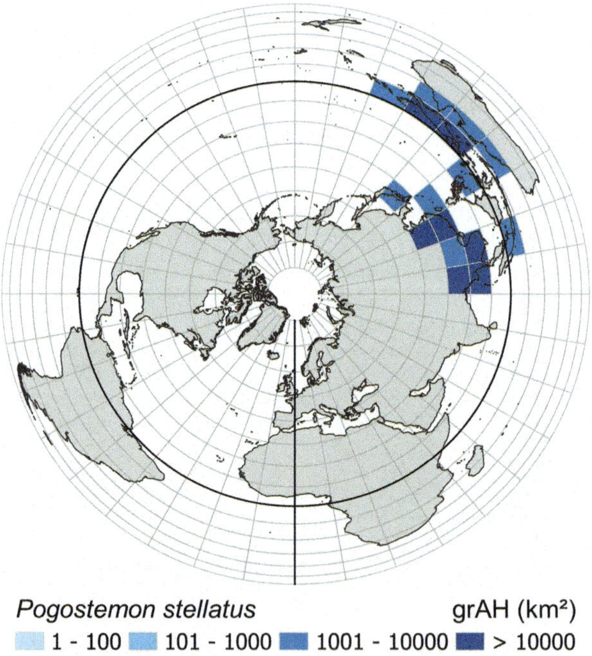

Pogostemon stellatus (Lour.) Kuntze

***Pogostemon yatabeanus* (Makino) Press**
Synonyms: *Eusteralis yatabeana* (Makino) Panigrahi, *Dysophylla yatabeana* Makino

Range area (grAH km^2): PAL: 68400; OR: 27500
Ecozone origin-status: PAL: N; OR: N
World rarity status: R
Ecozone-endemic macroregion occurrence: not endemic to any world ecozone
Life form(s): SU, EM
Ploidy state: NA
Speciation age (Ma): NA
Populations also occur in terrestrial, wetland and/ or riparian habitats: YES
Long distance movement vectors: NA
Nuisance status and uses: no nuisance issues; aquarium trade
Endangered status (IUCN): LC

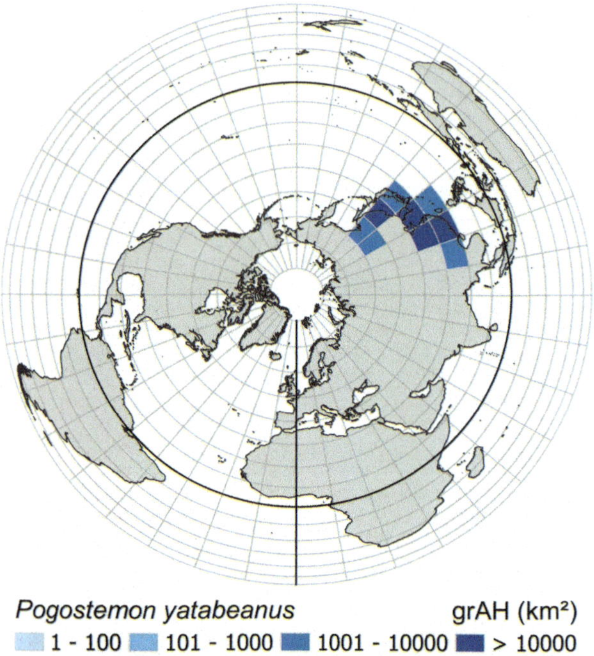

Pogostemon yatabeanus (Makino) Press

***Teucrium scordium* L.**
Synonyms:

Range area (grAH km^2): PAL: 149100
Ecozone origin-status: PAL: E
World rarity status: R
Ecozone-endemic macroregion occurrence: PAL: MED, W EUR, E EUR, N EUR, C ASIA
Life form(s): EM
Ploidy state: P
Speciation age (Ma): NA
Populations also occur in terrestrial, wetland and/ or riparian habitats: YES
Long distance movement vectors: NA
Nuisance status and uses: no nuisance issues; harvested from the wild for local use as a medicine and source of materials
Endangered status (IUCN): LC

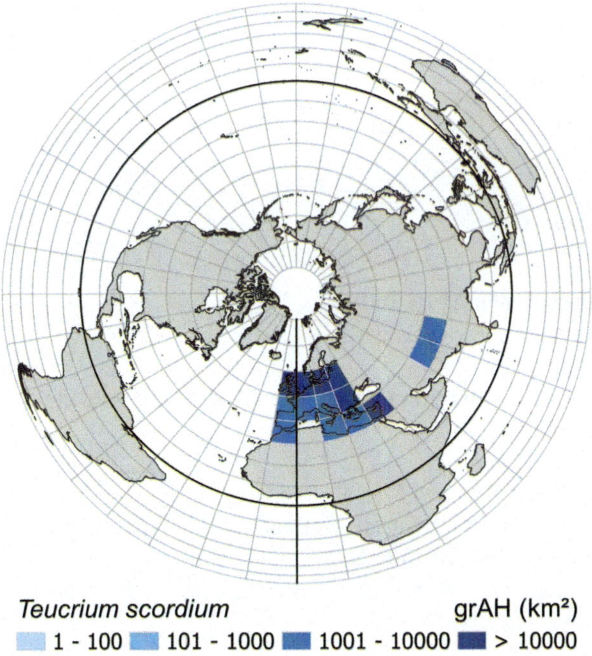

3.27 LENTIBULARIACEAE (Lamiales)

The Lentibulariaceae include many mainly carnivorous macrophyte species, 103 in total, representing one small and one large genus, and occurring widely across the planet, except in ANT. In total 50 species occur in NEO, 35 of them endemic. A hotspot for these plants is AUS, where there are 40 species, of which 28 are endemic to that ecozone. In addition, 19 species occur in AFR (nine endemic), 18 in OR (four endemic), 16 in NEA (one endemic), 11 in PAL (one endemic), and two in OCE (one native, the other introduced). Anthropochory plays a limited role in influencing the world distribution of these macrophytes. Aside from the single introduced species in OCE there are just three species with I-origin status, all in NEO. The only one with any reports of nuisance issues is *Utricularia gibba*, which occasionally causes aquatic weed problems in New Zealand, but otherwise they cause no problems. The species are predominantly world-rare, with values of VR 42%, R 27% and C 31%. In terms of endangered status, 24% of the species are NA/DD, and 73% LC. Just one species is VU and another EN, but there are no CR species in this family.

Quick Links to Genera

Genlisea africana **Oliv.**
Synonyms:

Range area (grAH km^2): AFR: 160800
Ecozone origin-status: AFR: E
World rarity status: C
Ecozone-endemic macroregion occurrence: AFR: SAH/ AR, FOR, EA, ZAM
Life form(s): EM
Ploidy state: P
Speciation age (Ma): NA
Populations also occur in terrestrial, wetland and/ or riparian habitats: YES
Long distance movement vectors: NA
Nuisance status and uses: no nuisance issues; carnivorous plants hobby trade
Endangered status (IUCN): NA

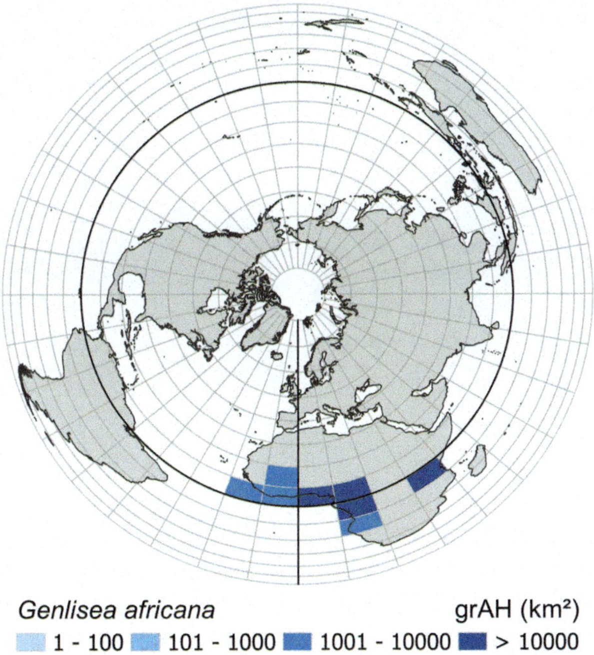

Genlisea angolensis **R.D.Good**
Synonyms:

Range area (grAH km^2): AFR: 11700
Ecozone origin-status: AFR: E
World rarity status: VR
Ecozone-endemic macroregion occurrence: AFR: ZAM
Life form(s): EM
Ploidy state: NA
Speciation age (Ma): NA
Populations also occur in terrestrial, wetland and/ or riparian habitats: YES
Long distance movement vectors: NA
Nuisance status and uses: no nuisance issues; no known uses
Endangered status (IUCN): EN

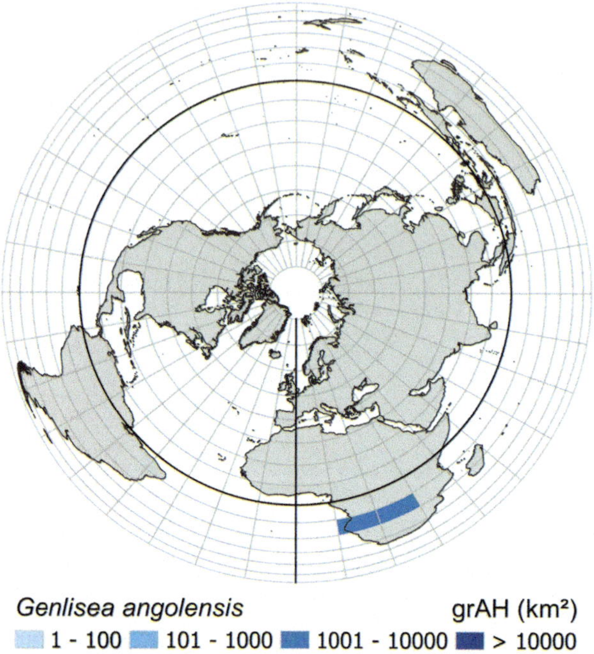

Genlisea angolensis R.D.Good

Genlisea aurea **A.St.-Hil.**
Synonyms:

Range area (grAH km^2): NEO: 54400
Ecozone origin-status: NEO: E
World rarity status: R
Ecozone-endemic macroregion occurrence: NEO: AMZ, CHA, PAR
Life form(s): SU, EM
Ploidy state: P
Speciation age (Ma): NA
Populations also occur in terrestrial, wetland and/ or riparian habitats: YES
Long distance movement vectors: NA
Nuisance status and uses: no nuisance issues; carnivorous plants hobby trade, medicinal uses
Endangered status (IUCN): NA

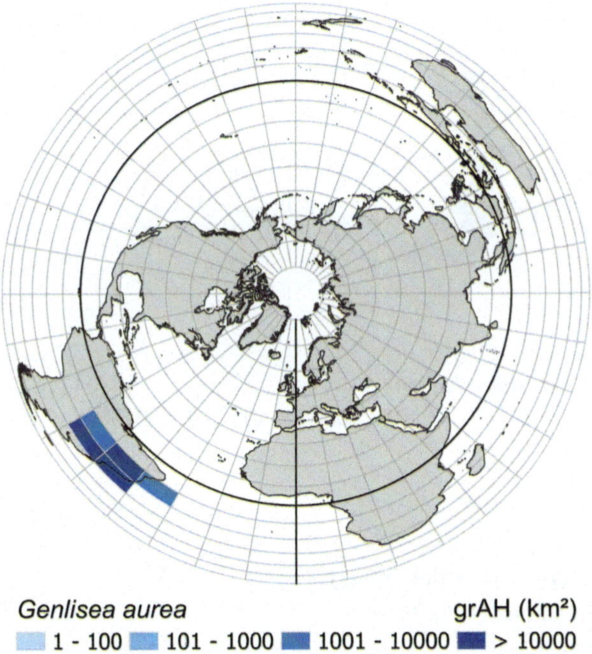

Genlisea filiformis **A.St.-Hil.**
Synonyms:

Range area (grAH km^2): NEO: 202100
Ecozone origin-status: NEO: E
World rarity status: C
Ecozone-endemic macroregion occurrence: NEO: MES AM. NWSA, AMZ, CHA, PAR

Life form(s): ?EM
Ploidy state: P
Speciation age (Ma): NA
Populations also occur in terrestrial, wetland and/ or riparian habitats: YES
Long distance movement vectors: NA
Nuisance status and uses: no nuisance issues; no known uses
Endangered status (IUCN): NA

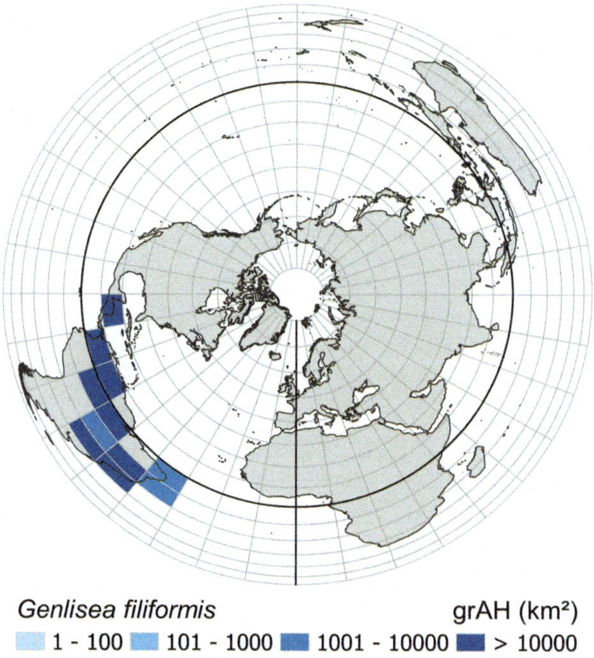

***Genlisea glabra* P.Taylor**
Synonyms:

Range area (grAH km^2): NEO: 20800
Ecozone origin-status: NEO: E
World rarity status: VR
Ecozone-endemic macroregion occurrence: NEO: NWSA
Life form(s): EM
Ploidy state: NA
Speciation age (Ma): NA
Populations also occur in terrestrial, wetland and/ or riparian habitats: YES
Long distance movement vectors: NA
Nuisance status and uses: no nuisance issues; no known uses
Endangered status (IUCN): NA

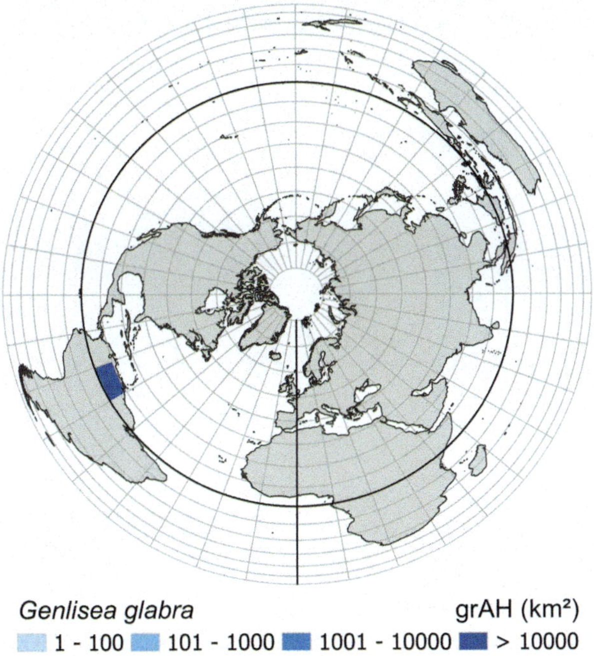

Genlisea glandulosissima **R.E.Fr.**
Synonyms:

Range area (grAH km^2): AFR: 93700
Ecozone origin-status: AFR: E
World rarity status: R
Ecozone-endemic macroregion occurrence: AFR: EA
Life form(s): ?EM
Ploidy state: P
Speciation age (Ma): NA
Populations also occur in terrestrial, wetland and/ or riparian habitats: YES
Long distance movement vectors: NA
Nuisance status and uses: no nuisance issues; ornamental
Endangered status (IUCN): DD

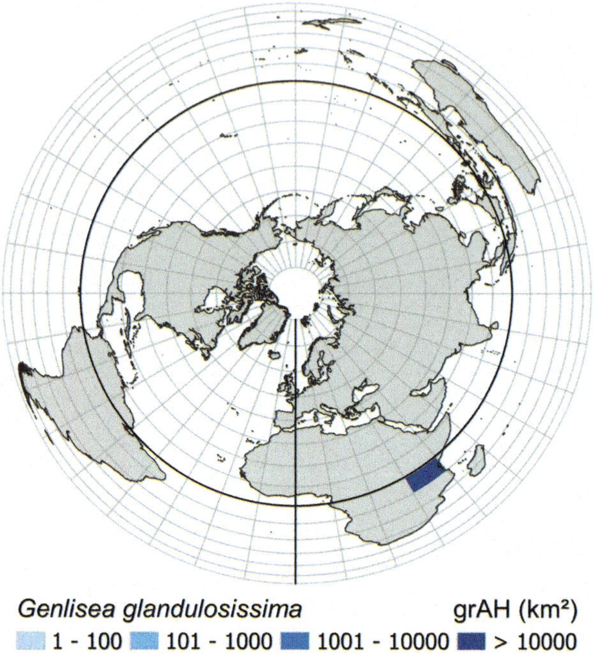

Genlisea guianensis **N.E.Br.**
Synonyms:

Range area (grAH km^2): NEO: 80200
Ecozone origin-status: NEO: E
World rarity status: R
Ecozone-endemic macroregion occurrence: NEO: NWSA, AMZ, CHA, PAR
Life form(s): EM
Ploidy state: P
Speciation age (Ma): NA
Populations also occur in terrestrial, wetland and/ or riparian habitats: YES
Long distance movement vectors: NA
Nuisance status and uses: no nuisance issues; no known uses
Endangered status (IUCN): NA

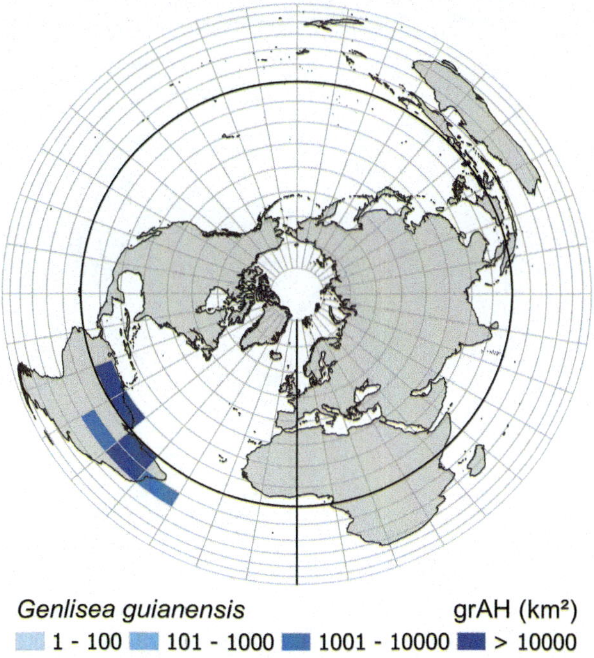

***Genlisea hispidula* Stapf**
Synonyms:

Range area (grAH km^2): AFR: 223600
Ecozone origin-status: AFR: E
World rarity status: C
Ecozone-endemic macroregion occurrence: AFR: SAH/ AR, FOR, EA, ZAM, SDES
Life form(s): EM
Ploidy state: P
Speciation age (Ma): NA
Populations also occur in terrestrial, wetland and/ or riparian habitats: YES
Long distance movement vectors: NA
Nuisance status and uses: no nuisance issues; carnivorous plants hobby trade
Endangered status (IUCN): LC

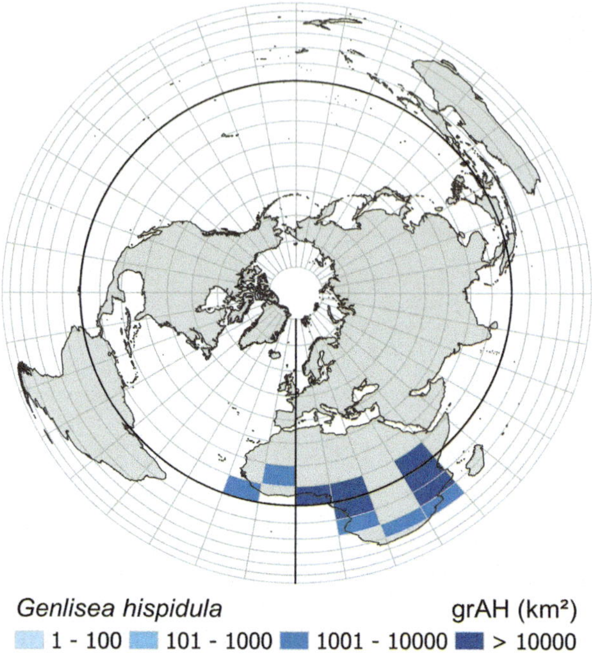

Genlisea hispidula Stapf
Image copyright: iNaturalist CC0 @jimmy_whatmore (Jimmy Whatmore)

***Genlisea lobata* Fromm**
Synonyms:

Range area (grAH km^2): World: NEO: 11400
Ecozone origin-status: NEO: E
World rarity status: VR
Ecozone-endemic macroregion occurrence: NEO: PAR
Life form(s): EM
Ploidy state: D
Speciation age (Ma): NA
Populations also occur in terrestrial, wetland and/ or riparian habitats: YES
Long distance movement vectors: NA
Nuisance status and uses: no nuisance issues; carnivorous plant hobby species
Endangered status (IUCN): NA

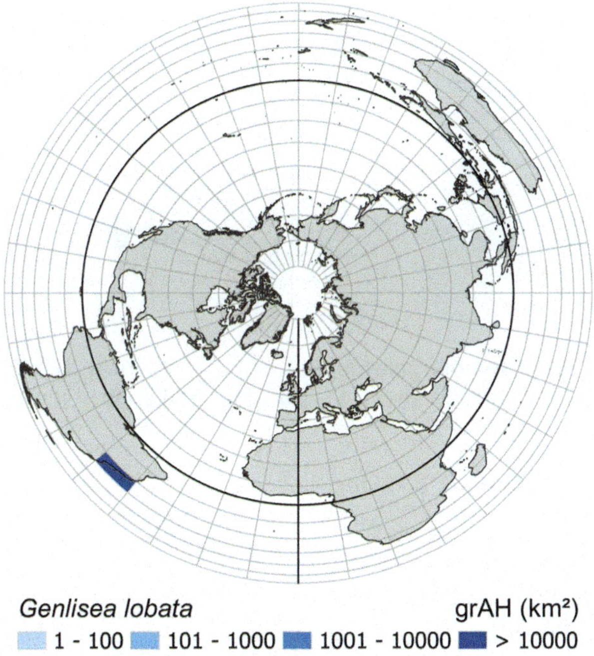

***Genlisea margaretae* Hutch.**
Synonyms:

Range area (grAH km^2): AFR: 103800
Ecozone origin-status: AFR: E
World rarity status: R
Ecozone-endemic macroregion occurrence: AFR: EA, MAD
Life form(s): EM

Ploidy state: P
Speciation age (Ma): NA
Populations also occur in terrestrial, wetland and/ or riparian habitats: YES
Long distance movement vectors: NA
Nuisance status and uses: no nuisance issues; carnivorous plant hobby species
Endangered status (IUCN): NA

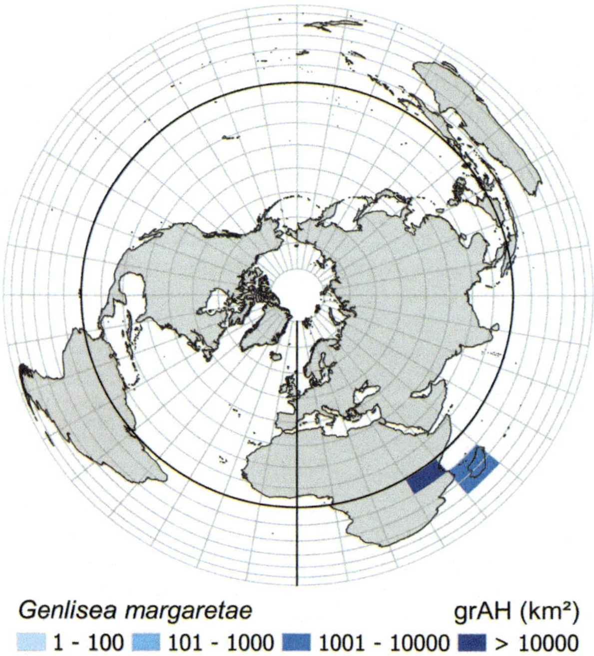

***Genlisea pallida* Fromm & P.Taylor**
Synonyms:

Range area (grAH km^2): AFR: 2900
Ecozone origin-status: AFR: E
World rarity status: VR
Ecozone-endemic macroregion occurrence: AFR: ZAM
Life form(s): EM
Ploidy state: NA
Speciation age (Ma): NA
Populations also occur in terrestrial, wetland and/ or riparian habitats: YES
Long distance movement vectors: NA
Nuisance status and uses: no nuisance issues; occasionally grown as a carnivorous plant hobby species
Endangered status (IUCN): NA

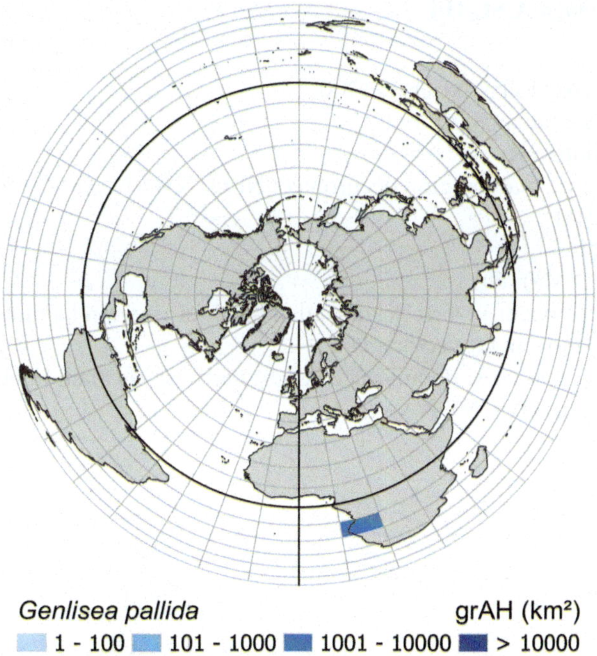

Genlisea pallida Fromm & P.Taylor

***Genlisea pygmaea* A.St.-Hil.**
Synonyms:

Range area (grAH km²): NEO: 121700
Ecozone origin-status: NEO: E
World rarity status: R
Ecozone-endemic macroregion occurrence: NEO: NWSA, AMZ, CHA
Life form(s): EM
Ploidy state: NA
Speciation age (Ma): NA
Populations also occur in terrestrial, wetland and/ or riparian habitats: YES
Long distance movement vectors: NA
Nuisance status and uses: no nuisance issues; carnivorous plant hobby species
Endangered status (IUCN): NA

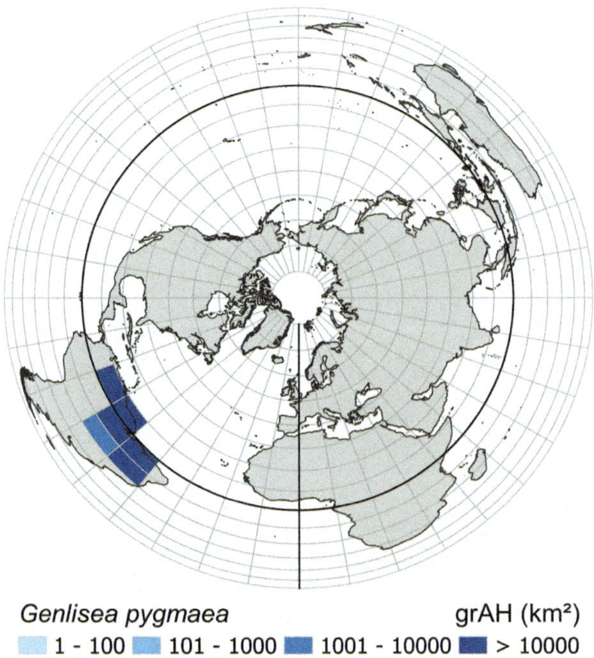

***Genlisea repens* Benj.**
Synonyms:

Range area (grAH km²): NEO: 157700
Ecozone origin-status: NEO: E
World rarity status: C
Ecozone-endemic macroregion occurrence: NEO: NWSA, AMZ, CHA, PAR
Life form(s): EM

Ploidy state: P
Speciation age (Ma): NA
Populations also occur in terrestrial, wetland and/ or riparian habitats: YES
Long distance movement vectors: NA
Nuisance status and uses: no nuisance issues; carnivorous plant hobby species
Endangered status (IUCN): NA

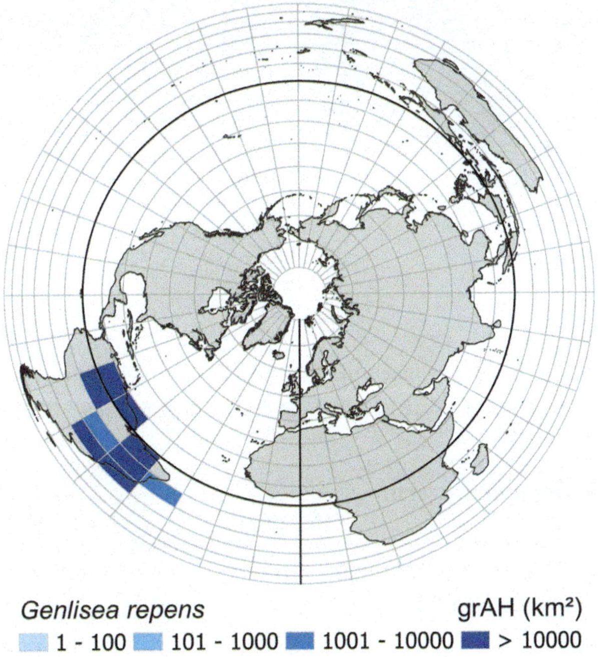

***Genlisea roraimensis* N.E.Br.**
Synonyms:

Range area (grAH km^2): NEO: 37500
Ecozone origin-status: NEO: E
World rarity status: R
Ecozone-endemic macroregion occurrence: NEO: NWSA
Life form(s): EM
Ploidy state: P
Speciation age (Ma): NA
Populations also occur in terrestrial, wetland and/ or riparian habitats: YES
Long distance movement vectors: NA
Nuisance status and uses: no nuisance issues; occasionally grown as a carnivorous plant hobby species
Endangered status (IUCN): NA

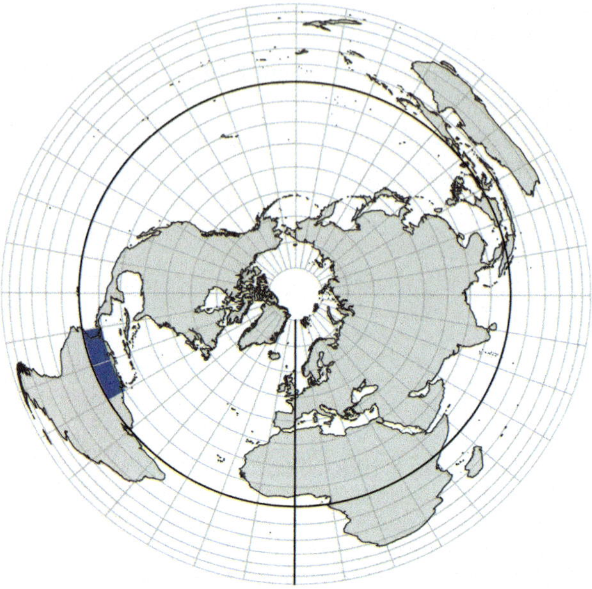

Genlisea sanariapoana **Steyerm.**
Synonyms:

Range area (grAH km^2): NEO: 20800
Ecozone origin-status: NEO: E
World rarity status: VR
Ecozone-endemic macroregion occurrence: NEO: NWSA
Life form(s): EM
Ploidy state: NA
Speciation age (Ma): NA
Populations also occur in terrestrial, wetland and/ or riparian habitats: YES
Long distance movement vectors: NA
Nuisance status and uses: no nuisance issues; occasionally grown as a carnivorous plant hobby species
Endangered status (IUCN): NA

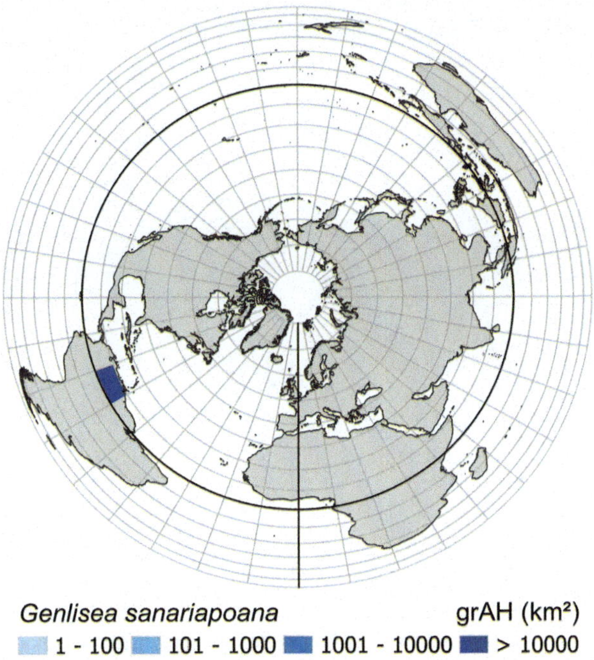

Genlisea uncinata **P.Taylor & Fromm**
Synonyms:

Range area (grAH km^2): NEO: 12100
Ecozone origin-status: NEO: E
World rarity status: VR
Ecozone-endemic macroregion occurrence: NEO: CHA
Life form(s): EM
Ploidy state: D
Speciation age (Ma): NA
Populations also occur in terrestrial, wetland and/ or riparian habitats: YES
Long distance movement vectors: NA
Nuisance status and uses: no nuisance issues; occasionally grown as a carnivorous plant hobby species
Endangered status (IUCN): NA

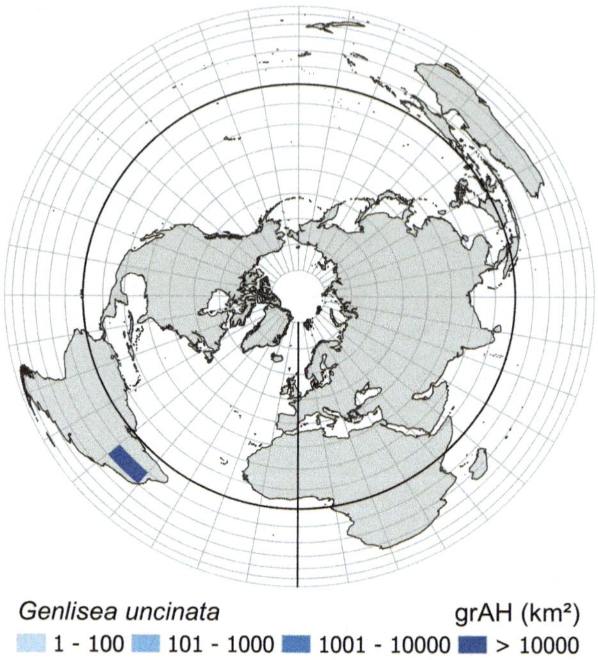

***Genlisea violacea* A.St.-Hil.**
Synonyms:

Range area (grAH km^2): NEO: 31100
Ecozone origin-status: NEO: E
World rarity status: VR
Ecozone-endemic macroregion occurrence: NEO: AMZ, CHA, PAR
Life form(s): EM
Ploidy state: D
Speciation age (Ma): NA
Populations also occur in terrestrial, wetland and/ or riparian habitats: YES
Long distance movement vectors: NA
Nuisance status and uses: no nuisance issues; carnivorous plant hobby species
Endangered status (IUCN): NA

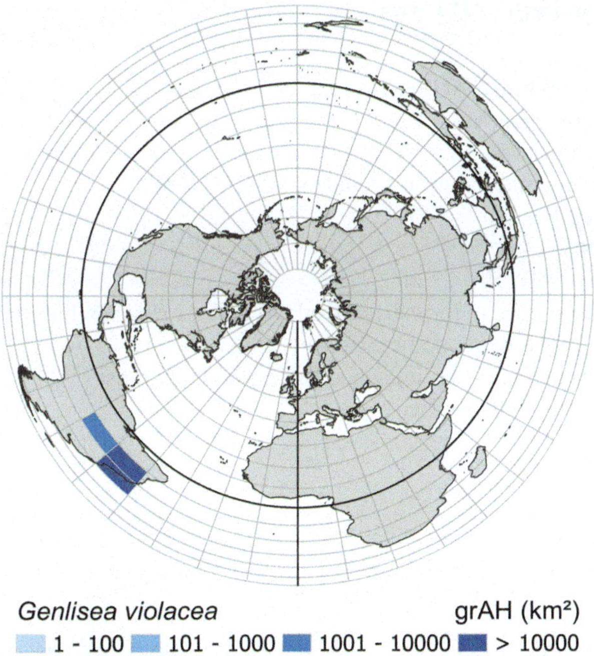

Genlisea violacea A.St.-Hil.

Utricularia antennifera **P.Taylor**
Synonyms:

Range area (grAH km^2): AUS: 7200
Ecozone origin-status: AUS: E
World rarity status: VR
Ecozone-endemic macroregion occurrence: AUS: NA
Life form(s): EM
Ploidy state: NA
Speciation age (Ma): NA
Populations also occur in terrestrial, wetland and/ or riparian habitats: YES
Long distance movement vectors: NA
Nuisance status and uses: no nuisance issues; no known uses
Endangered status (IUCN): LC

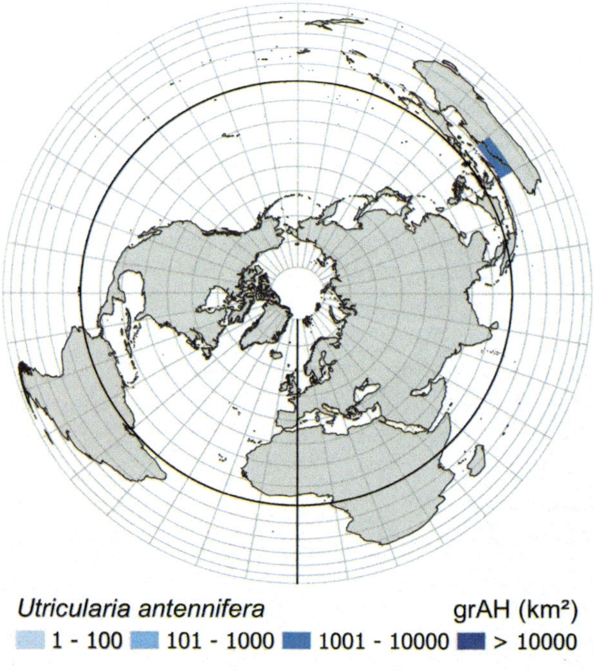

Utricularia antennifera P.Taylor
Image copyright: iNaturalist CC-BY @thilokrueger (Thilo Krueger)

***Utricularia arnhemica* P.Taylor**
Synonyms:

Range area (grAH km^2): AUS: 13500
Ecozone origin-status: AUS: E
World rarity status: VR
Ecozone-endemic macroregion occurrence: AUS: NA
Life form(s): SU, EM
Ploidy state: NA
Speciation age (Ma): NA
Populations also occur in terrestrial, wetland and/ or riparian habitats: YES
Long distance movement vectors: NA
Nuisance status and uses: no nuisance issues; grown and sold as part of carnivorous plants hobby trade
Endangered status (IUCN): LC

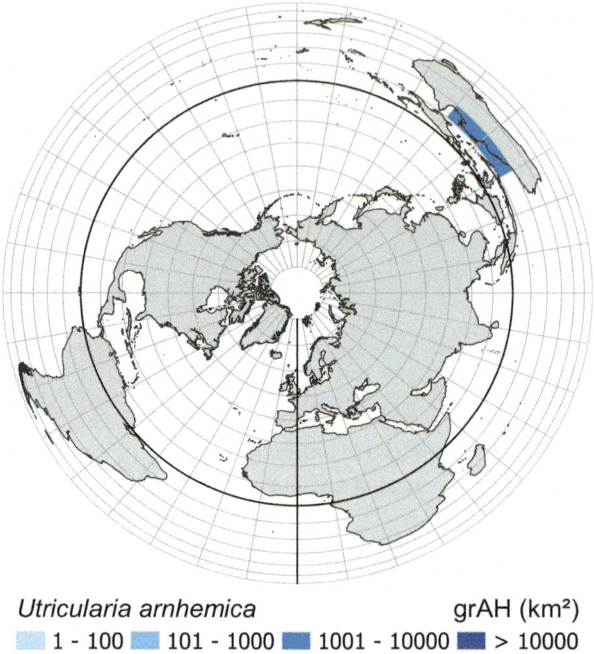

Utricularia arnhemica P.Taylor
Image copyright: iNaturalist CC-BY @thilokrueger

***Utricularia aurea* Lour.**
Synonyms:

Range area (grAH km²): PAL: 117100; NEO: 11400; OR: 126000; AUS: 47500
Ecozone origin-status: PAL: N; NEO: I; OR: N; AUS: N
World rarity status: C
Ecozone-endemic macroregion occurrence: not endemic to any world ecozone
Life form(s): SU, FF (hydrosubshrub)
Ploidy state: DP
Speciation age (Ma): NA
Populations also occur in terrestrial, wetland and/ or riparian habitats: NO
Long distance movement vectors: ANTHROPOCHORY
Nuisance status and uses: no nuisance issues; water garden plant, carnivorous plant hobby species, aquarium trade, medicinal and phytoremediation potential uses
Endangered status (IUCN): LC

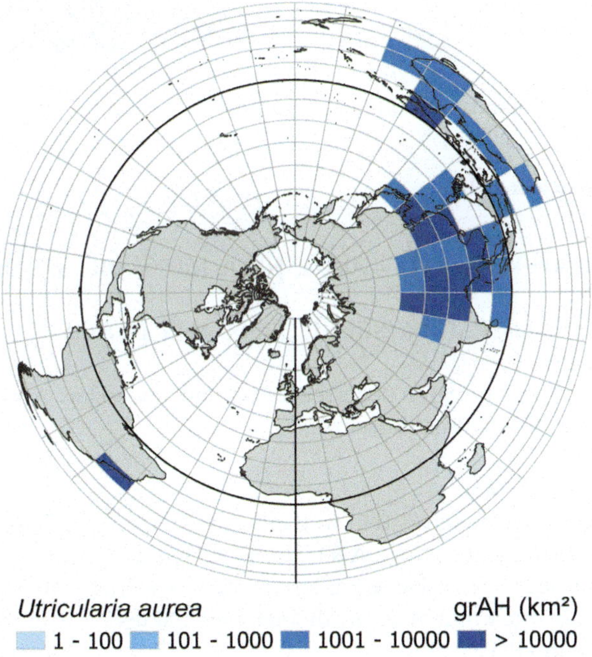

Utricularia aurea Lour.
Image copyright: A. Efremov, with permission

Utricularia australis **R.Br.**
Synonyms:

Range area (grAH km^2): PAL: 490500; NEO: 16600; OR: 85800; AUS: 66200; AFR: 24200
Ecozone origin-status: PAL: N; NEO: I; OR: N; AUS: N; AFR: N
World rarity status: C
Ecozone-endemic macroregion occurrence: not endemic to any world ecozone
Life form(s): SU, FF (hydrosubshrub)
Ploidy state: D
Speciation age (Ma): 6.5
Populations also occur in terrestrial, wetland and/ or riparian habitats: NO
Long distance movement vectors: ANTHROPOCHORY; ?AV
Nuisance status and uses: no nuisance issues; water garden plant, carnivorous plant hobby species, aquarium trade, medicinal potential uses
Endangered status (IUCN): LC

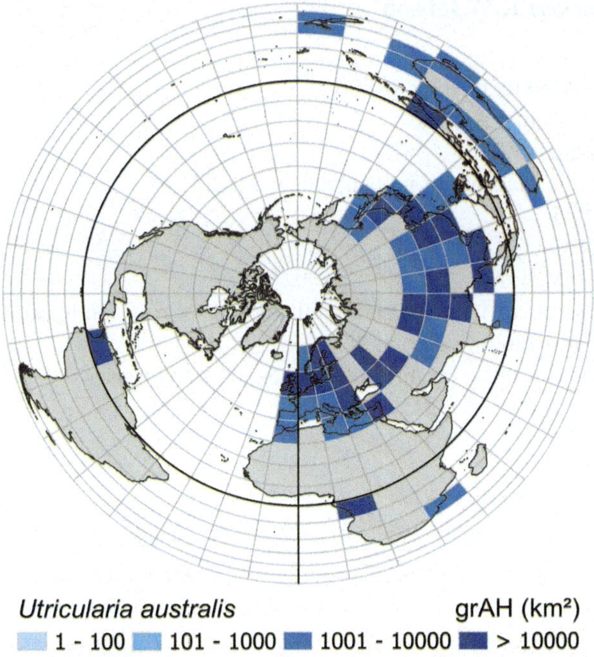

Utricularia australis R.Br.
Image copyright: J.C. Schou, with permission

***Utricularia barkeri* R.W.Jobson**
Synonyms:

Range area (grAH km²): AUS: 14700
Ecozone origin-status: AUS: E
World rarity status: VR
Ecozone-endemic macroregion occurrence: AUS: ESA
Life form(s): EM
Ploidy state: NA
Speciation age (Ma): NA
Populations also occur in terrestrial, wetland and/ or riparian habitats: YES
Long distance movement vectors: NA
Nuisance status and uses: no nuisance issues; no known uses
Endangered status (IUCN): LC

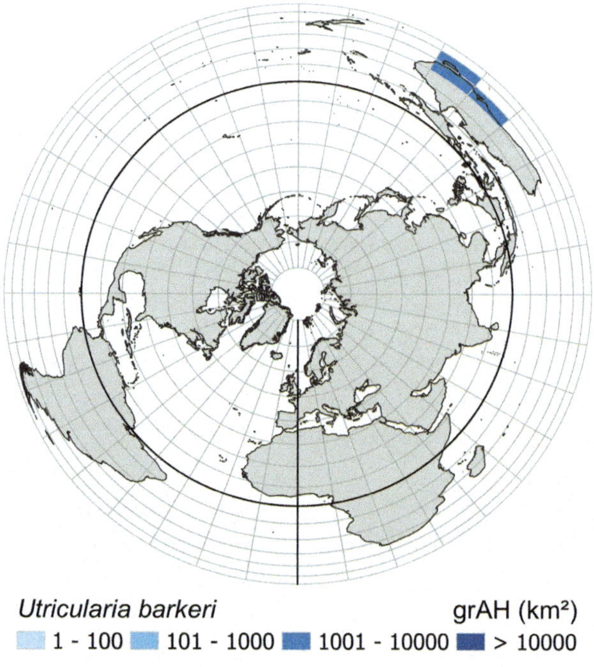

Utricularia barkeri R.W.Jobson

***Utricularia benjaminiana* Oliv.**
Synonyms: *Akentra inflata* Benj.

Range area (grAH km^2): NEO: 52100; AFR: 114800
Ecozone origin-status: NEO: N; AFR: N
World rarity status: C
Ecozone-endemic macroregion occurrence: not endemic to any world ecozone
Life form(s): SU, FF (hydrosubshrub)
Ploidy state: NA
Speciation age (Ma): NA
Populations also occur in terrestrial, wetland and/ or riparian habitats: NO
Long distance movement vectors: NA
Nuisance status and uses: no nuisance issues; water garden plant, carnivorous plant hobby species
Endangered status (IUCN): LC

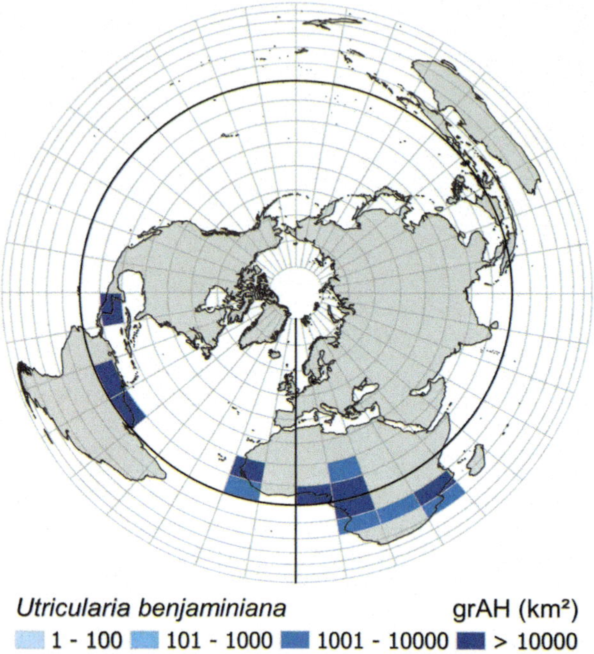

Utricularia benjaminiana Oliv.

***Utricularia biloba* R.Br.**
Synonyms: *Nelipus biloba* (R.Br.) Raf.

Range area (grAH km²): AUS: 25500
Ecozone origin-status: AUS: E
World rarity status: VR
Ecozone-endemic macroregion occurrence: AUS: ESA, NA
Life form(s): SU, EM
Ploidy state: NA
Speciation age (Ma): NA
Populations also occur in terrestrial, wetland and/ or riparian habitats: YES
Long distance movement vectors: NA
Nuisance status and uses: no nuisance issues; carnivorous plant hobby species
Endangered status (IUCN): LC

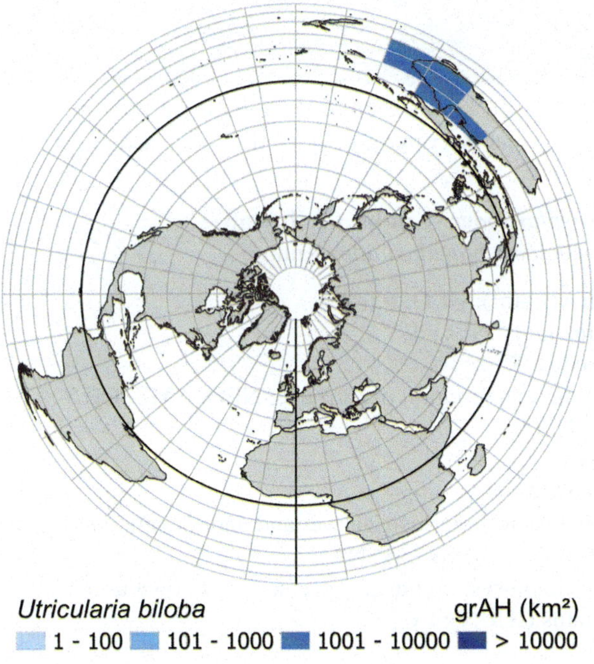

Utricularia biloba R.Br.
Image copyright: G. Neill, with permission

***Utricularia biovularioides* (Kuhlm.) P. Taylor**
Synonyms: *Saccolaria biovularioides* Kuhlm.

Range area (grAH km^2): NEO: 65700
Ecozone origin-status: NEO: E
World rarity status: R
Ecozone-endemic macroregion occurrence: NEO: AMZ, CHA
Life form(s): SU, FF
Ploidy state: NA
Speciation age (Ma): NA
Populations also occur in terrestrial, wetland and/ or riparian habitats: NO
Long distance movement vectors: NA
Nuisance status and uses: no nuisance issues; no known uses
Endangered status (IUCN): LC

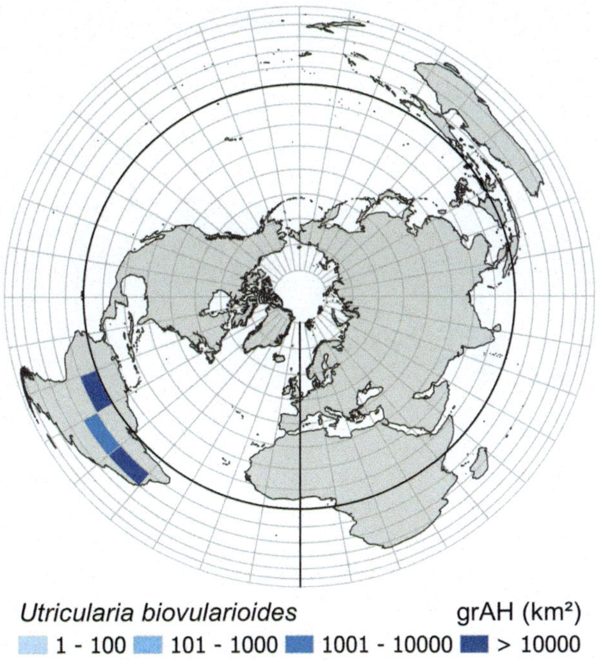

***Utricularia breviscapa* C.Wright ex Griseb.**
Synonyms:

Range area (grAH km^2): NEO: 221900
Ecozone origin-status: NEO: E
World rarity status: C
Ecozone-endemic macroregion occurrence: NEO: NWSA, AMZ, CHA
Life form(s): SU, FF
Ploidy state: NA
Speciation age (Ma): NA
Populations also occur in terrestrial, wetland and/ or riparian habitats: NO
Long distance movement vectors: NA
Nuisance status and uses: no nuisance issues; ornamental use
Endangered status (IUCN): LC

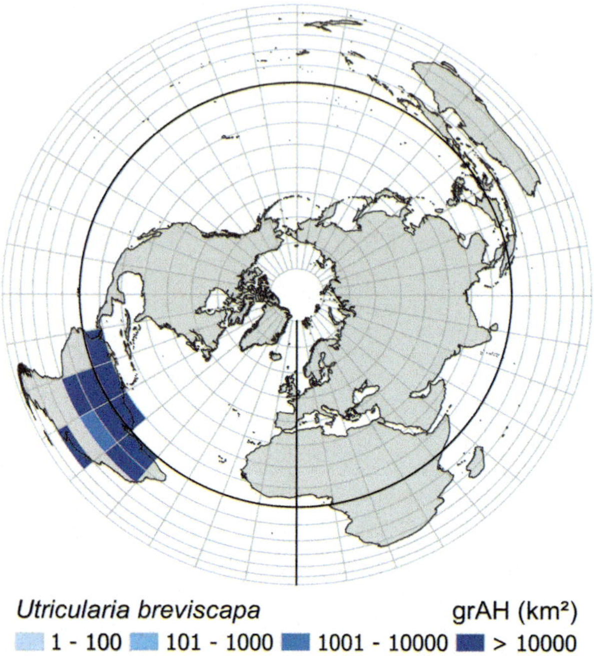

Utricularia breviscapa C.Wright ex Griseb.
Image copyright: Y. Domínguez, with permission

***Utricularia chiribiquetensis* A.Fernández**
Synonyms:

Range area (grAH km²): NEO: 54900
Ecozone origin-status: NEO: E
World rarity status: R
Ecozone-endemic macroregion occurrence: NEO: NWSA, AMZ
Life form(s): EM
Ploidy state: NA
Speciation age (Ma): NA
Populations also occur in terrestrial, wetland and/ or riparian habitats: YES
Long distance movement vectors: NA
Nuisance status and uses: no nuisance issues; no known uses
Endangered status (IUCN): LC

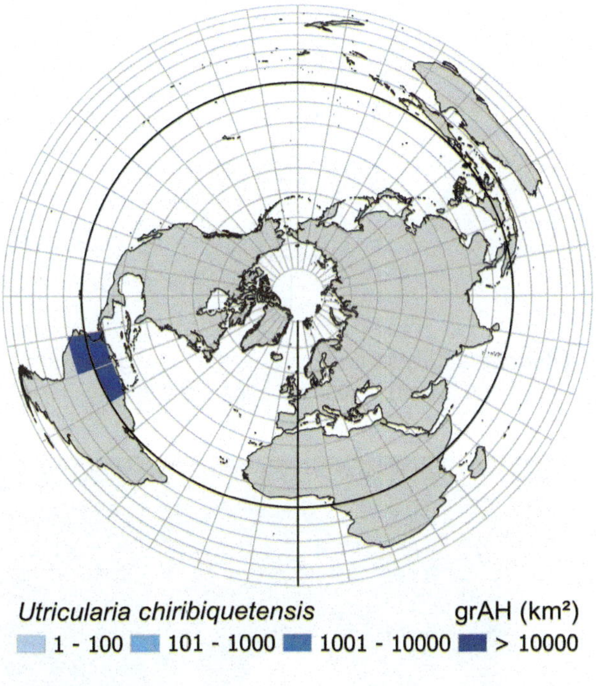

***Utricularia chrysantha* R.Br.**
Synonyms: *Enskide chrysantha* (R.Br.) Raf.

Range area (grAH km²): NEO: 21100; AUS: 31000
Ecozone origin-status: NEO: I; AUS: E(I)
World rarity status: R
Ecozone-endemic macroregion occurrence: AUS: NG, NA
Life form(s): EM

Ploidy state: NA
Speciation age (Ma): NA
Populations also occur in terrestrial, wetland and/ or riparian habitats: YES
Long distance movement vectors: ANTHROPOCHORY
Nuisance status and uses: no nuisance issues; carnivorous plant hobby species
Endangered status (IUCN): LC

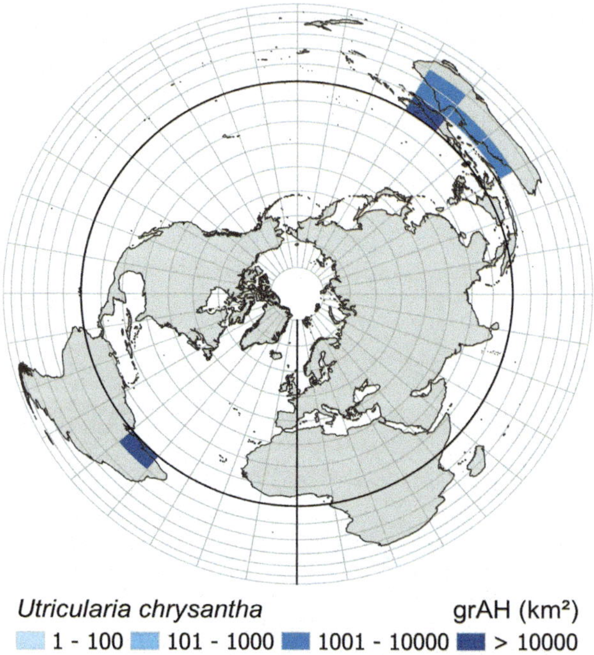

Utricularia chrysantha R.Br.

Utricularia cucullata A.St.-Hil. & Girard
Synonyms: *Vesiculina cucullata* (A.St.-Hil. & Girard) Barnhart

Range area (grAH km^2): NEO: 199200
Ecozone origin-status: NEO: E
World rarity status: C
Ecozone-endemic macroregion occurrence: NEO: NWSA, AMZ, CHA, PAR
Life form(s): SU, FF
Ploidy state: NA
Speciation age (Ma): NA
Populations also occur in terrestrial, wetland and/ or riparian habitats: NO
Long distance movement vectors: NA
Nuisance status and uses: no nuisance issues; no known uses
Endangered status (IUCN): LC

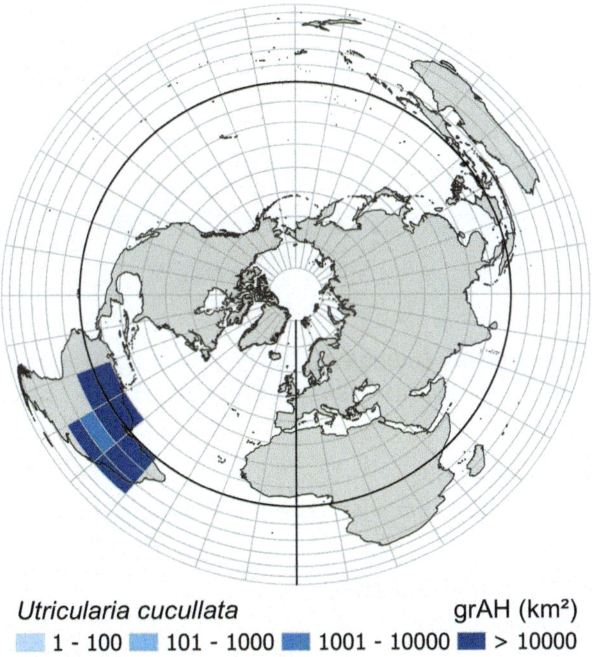

Utricularia cymbantha Oliv.
Synonyms: *Biovularia cymbantha* (Oliv.) Kuhlm.

Range area (grAH km^2): AFR: 27900
Ecozone origin-status: AFR: E
World rarity status: VR
Ecozone-endemic macroregion occurrence: AFR: ZAM, SDES, MAD
Life form(s): SU, FF

Ploidy state: NA
Speciation age (Ma): NA
Populations also occur in terrestrial, wetland and/ or riparian habitats: NO
Long distance movement vectors: NA
Nuisance status and uses: no nuisance issues; no known uses
Endangered status (IUCN): LC

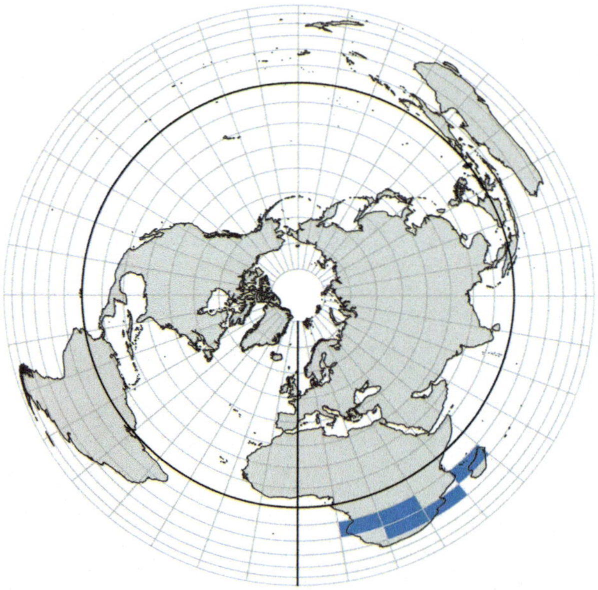

Utricularia dichotoma **Labill.**

Synonyms: *Pleiochasia dichotoma* (Labill.) Barnhart, *Utricularia billardierei* F.Muell.

Range area (grAH km^2): AUS: 47300
Ecozone origin-status: AUS: E
World rarity status: R
Ecozone-endemic macroregion occurrence: AUS: WA, ESA, NA, NZ
Life form(s): EM
Ploidy state: NA
Speciation age (Ma): NA
Populations also occur in terrestrial, wetland and/ or riparian habitats: YES
Long distance movement vectors: NA
Nuisance status and uses: no nuisance issues; grown as a carnivorous plant hobby species
Endangered status (IUCN): LC

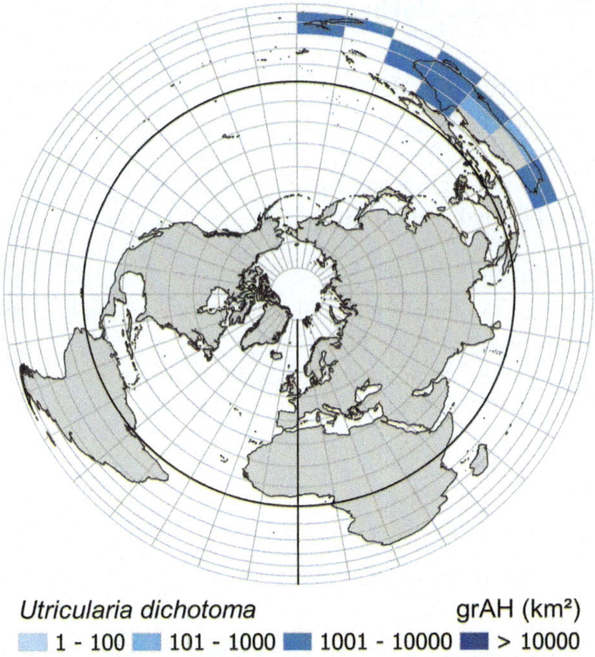

Utricularia dichotoma Labill.

Utricularia dimorphantha **Makino**
Synonyms:

Range area (grAH km^2): PAL: 9200
Ecozone origin-status: PAL: E
World rarity status: VR
Ecozone-endemic macroregion occurrence: PAL: FER/J
Life form(s): SU, FF (hydrosubshrub)
Ploidy state: D
Speciation age (Ma): NA
Populations also occur in terrestrial, wetland and/ or riparian habitats: NO
Long distance movement vectors: NA
Nuisance status and uses: no nuisance issues; no known uses
Endangered status (IUCN): NA

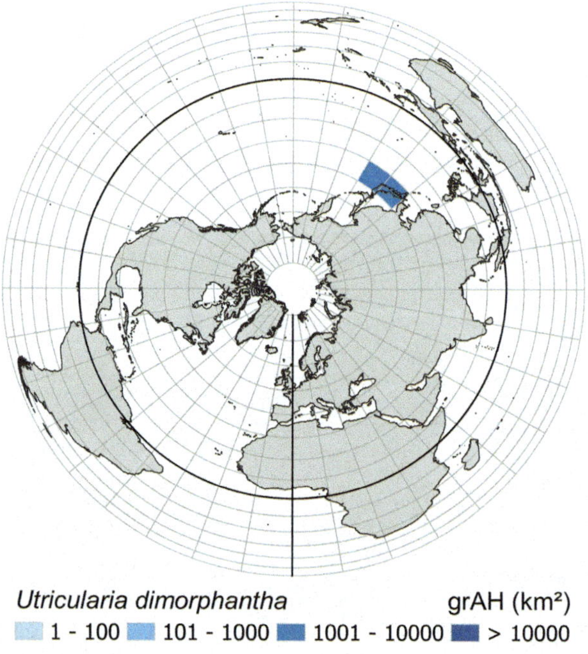

Utricularia dunlopii **P.Taylor**
Synonyms:

Range area (grAH km²): AUS: 13500
Ecozone origin-status: AUS: E
World rarity status: VR
Ecozone-endemic macroregion occurrence: AUS: NA
Life form(s): EM
Ploidy state: NA
Speciation age (Ma): NA
Populations also occur in terrestrial, wetland and/ or riparian habitats: YES
Long distance movement vectors: NA
Nuisance status and uses: no nuisance issues; no known uses
Endangered status (IUCN): LC

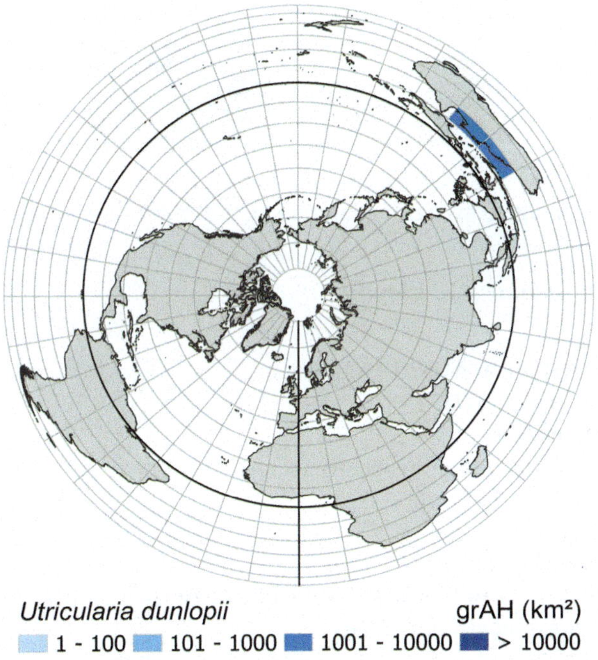

Utricularia dunlopii P.Taylor

***Utricularia dunstaniae* F.E. Lloyd**
Synonyms:

Range area (grAH km^2): AUS: 13500
Ecozone origin-status: AUS: E
World rarity status: VR
Ecozone-endemic macroregion occurrence: AUS: NA
Life form(s): EM
Ploidy state: NA
Speciation age (Ma): NA
Populations also occur in terrestrial, wetland and/ or riparian habitats: YES
Long distance movement vectors: NA
Nuisance status and uses: no nuisance issues; no known uses
Endangered status (IUCN): LC

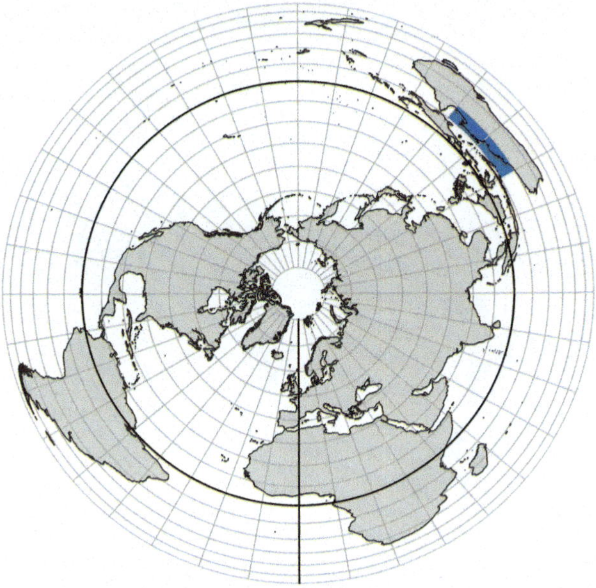

Utricularia erectiflora **A.St.-Hil. & Girard**
Synonyms:

Range area (grAH km^2): NEO: 143600
Ecozone origin-status: NEO: E
World rarity status: R
Ecozone-endemic macroregion occurrence: NEO: MG MX, MES AM, NWSA, AMZ, CHA, PAR
Life form(s): EM
Ploidy state: NA
Speciation age (Ma): NA
Populations also occur in terrestrial, wetland and/ or riparian habitats: YES
Long distance movement vectors: NA
Nuisance status and uses: no nuisance issues; no known uses
Endangered status (IUCN): LC

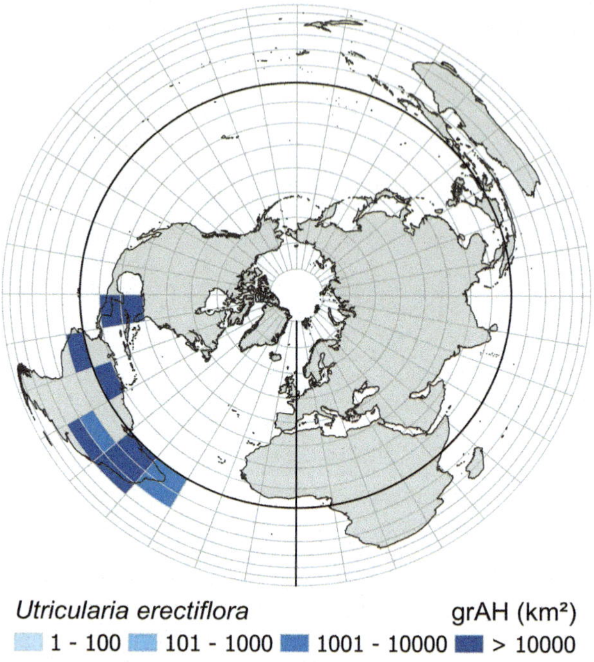

Utricularia fenshamii **R.W.Dobson**
Synonyms:

Range area (grAH km²): AUS: 15500
Ecozone origin-status: AUS: E
World rarity status: R
Ecozone-endemic macroregion occurrence: AUS: ESA, NA
Life form(s): EM
Ploidy state: NA
Speciation age (Ma): NA
Populations also occur in terrestrial, wetland and/ or riparian habitats: YES
Long distance movement vectors: NA
Nuisance status and uses: no nuisance issues; no known uses
Endangered status (IUCN): LC

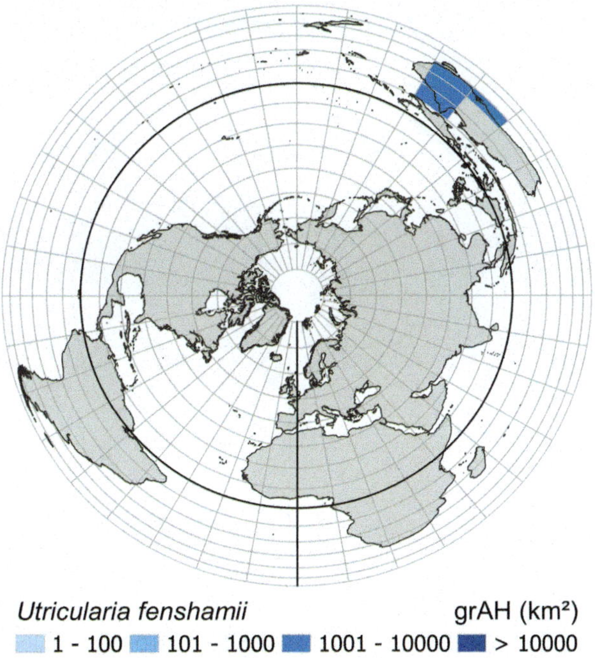

***Utricularia fistulosa* P.Taylor**
Synonyms:

Range area (grAH km^2): AUS: 13500
Ecozone origin-status: AUS: E
World rarity status: VR
Ecozone-endemic macroregion occurrence: AUS: NA
Life form(s): SU
Ploidy state: NA
Speciation age (Ma): NA
Populations also occur in terrestrial, wetland and/ or riparian habitats: NO
Long distance movement vectors: NA
Nuisance status and uses: no nuisance issues; no known uses
Endangered status (IUCN): LC

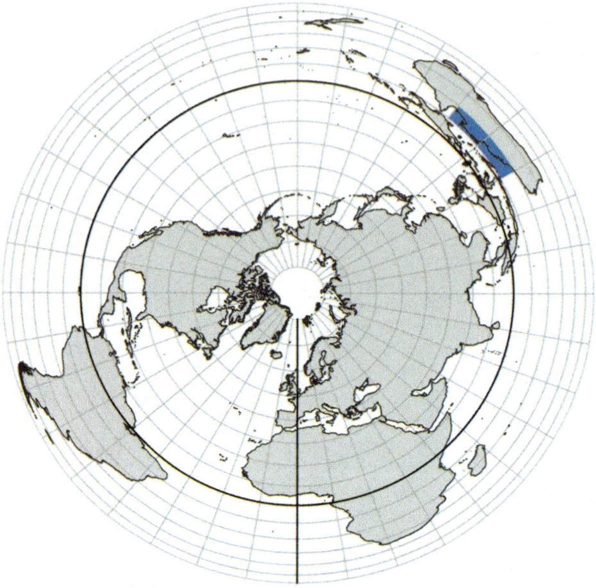

Utricularia floridana **Nash**
Synonyms:

Range area (grAH km^2): NEA: 13000; NEO: 10200
Ecozone origin-status: NEA: N; NEO: N
World rarity status: VR
Ecozone-endemic macroregion occurrence: not endemic to any world ecozone
Life form(s): SU, FF
Ploidy state: NA
Speciation age (Ma): NA
Populations also occur in terrestrial, wetland and/ or riparian habitats: NO
Long distance movement vectors: HYDROCHORY (NEO to NEA; NEA to NEO)
Nuisance status and uses: no nuisance issues; phytoremediation potential uses
Endangered status (IUCN): LC

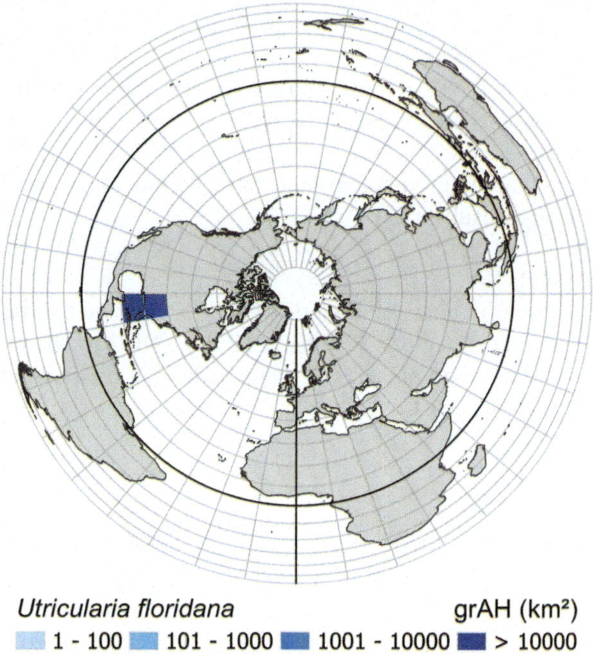

Utricularia floridana Nash

Utricularia foliosa **L.**
Synonyms:

Range area (grAH km²): NEA: 13000; NEO: 327900; AFR: 89800
Ecozone origin-status: NEA: N; NEO: N; AFR: N
World rarity status: C
Ecozone-endemic macroregion occurrence: not endemic to any world ecozone
Life form(s): FF (hydrosubshrub)
Ploidy state: P
Speciation age (Ma): NA
Populations also occur in terrestrial, wetland and/ or riparian habitats: NO
Long distance movement vectors: HYDROCHORY (NEO to NEA)
Nuisance status and uses: no nuisance issues; medicinal, ornamental uses
Endangered status (IUCN): LC

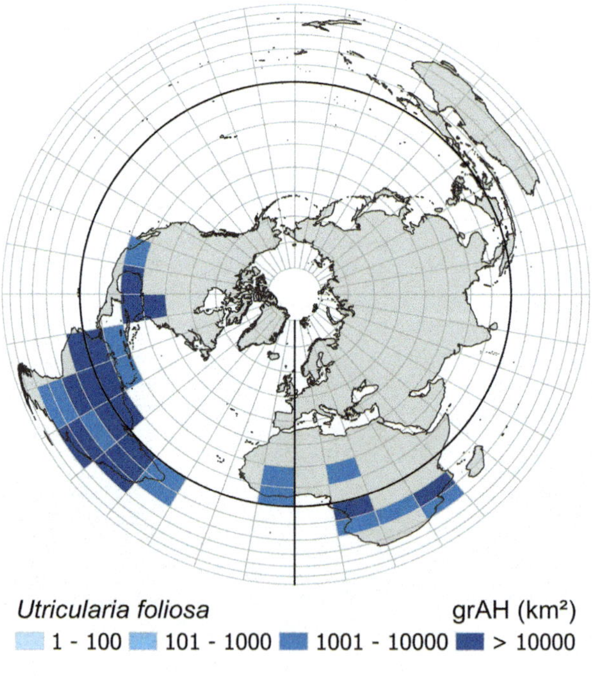

Utricularia foliosa L.
Image copyright: G. Ervin, with permission

Utricularia foveolata **Edgew.**
Synonyms:

Range area (grAH km^2): OR: 48800; AUS: 12900; AFR: 31700
Ecozone origin-status: OR: N; AUS: N; AFR: N
World rarity status: R
Ecozone-endemic macroregion occurrence: not endemic to any world ecozone
Life form(s): EM
Ploidy state: NA
Speciation age (Ma): NA
Populations also occur in terrestrial, wetland and/ or riparian habitats: YES
Long distance movement vectors: NA
Nuisance status and uses: no nuisance issues; no known uses
Endangered status (IUCN): LC

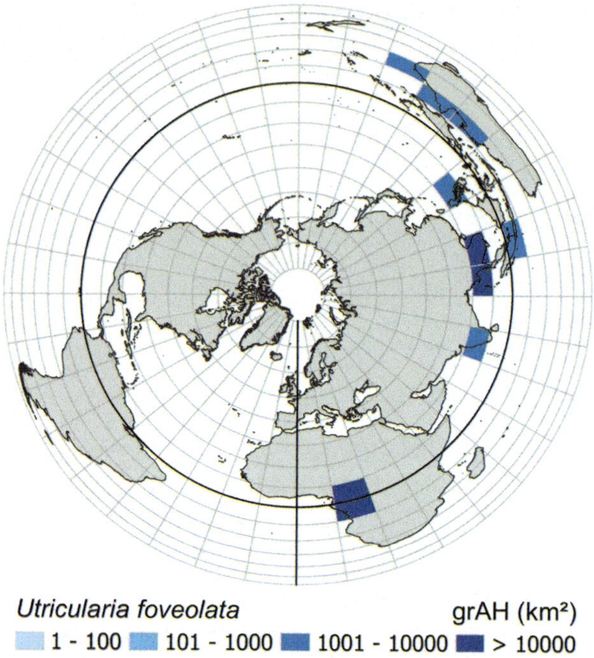

Utricularia foveolata Edgew.

***Utricularia fulva* F. Muell.**
Synonyms:

Range area (grAH km^2): AUS: 13500
Ecozone origin-status: AUS: E
World rarity status: VR
Ecozone-endemic macroregion occurrence: AUS: NA
Life form(s): SU, EM
Ploidy state: NA

Speciation age (Ma): NA
Populations also occur in terrestrial, wetland and/ or riparian habitats: YES
Long distance movement vectors: NA
Nuisance status and uses: no nuisance issues; carnivorous plant hobby species
Endangered status (IUCN): LC

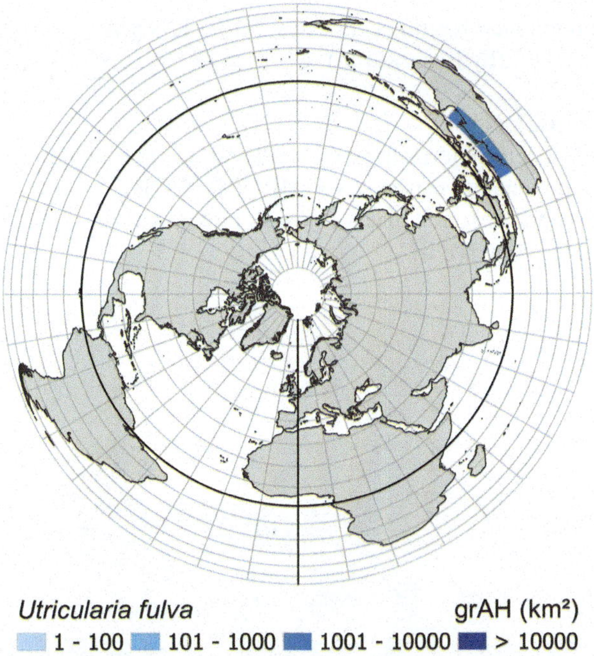

Utricularia fulva F. Muell.

***Utricularia geminiscapa* Benj.**
Synonyms:

Range area (grAH km²): NEA: 352900
Ecozone origin-status: NEA: E; (recent reports suggest an introduced population in South Island, New Zealand)
World rarity status: C
Ecozone-endemic macroregion occurrence: NEA: E NAM
Life form(s): SU, FF (hydrosubshrub)
Ploidy state: NA
Speciation age (Ma): NA
Populations also occur in terrestrial, wetland and/ or riparian habitats: NO
Long distance movement vectors: ?ANTHROPOCHORY
Nuisance status and uses: no nuisance issues; no known use
Endangered status (IUCN): LC

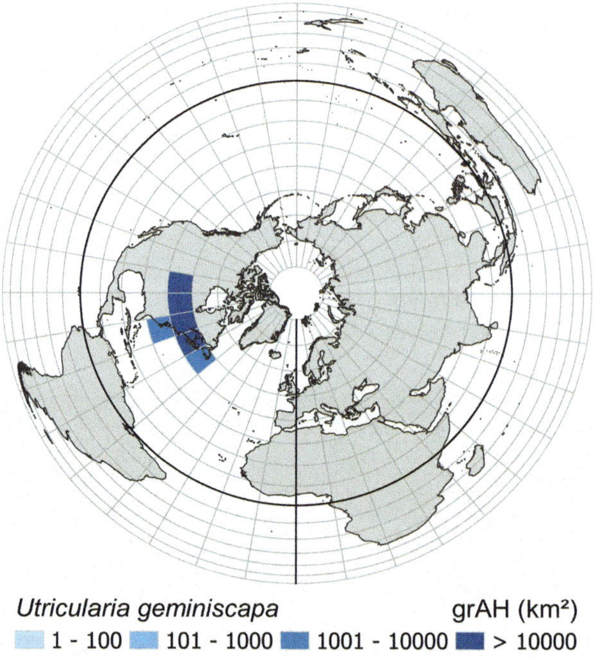

***Utricularia gibba* L.**
Synonyms: *Vesiculina gibba* (L.) Raf.

Range area (grAH km²): PAL: 83800; NEA: 382000; NEO: 388300; OR: 126000; AUS: 77000; AFR: 307300; present in OCE
Ecozone origin-status: PAL: N; NEA: N; NEO: N; OR: N; AUS: N; AFR: N; OCE: I
World rarity status: C

Ecozone-endemic macroregion occurrence: not endemic to any world ecozone
Life form(s): SU, FF (hydrosubshrub)
Ploidy state: P
Speciation age (Ma): NA
Populations also occur in terrestrial, wetland and/ or riparian habitats: NO
Long distance movement vectors: BIRD (waterfowl appear to be a major vector for the plant's spread through North Island, New Zealand, and possibly also initial entry to New Zealand via birds flying from Australia, where the plant is native [34, 135]; ANTHROPOCHORY (within AUS, introduction to New Zealand via the aquarium trade is also suggested as a possible vector: [164]).
Nuisance status and uses: no or generally low (mainly in New Zealand) nuisance issues; water garden plant and carnivorous plant hobby species, phytoremediation potential
Endangered status (IUCN): LC

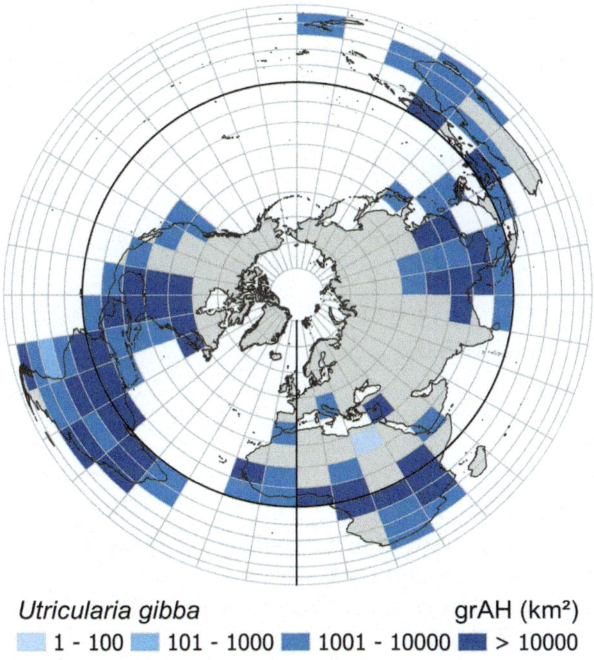

Utricularia gibba L.
Image copyright: A. Efremov, with permission

Utricularia hamiltonii F.E. Lloyd
Synonyms:

Range area (grAH km^2): AUS: 6300
Ecozone origin-status: AUS: E
World rarity status: VR
Ecozone-endemic macroregion occurrence: AUS: NA
Life form(s): EM
Ploidy state: NA
Speciation age (Ma): NA
Populations also occur in terrestrial, wetland and/ or riparian habitats: YES
Long distance movement vectors: NA
Nuisance status and uses: no nuisance issues; no known uses
Endangered status (IUCN): LC

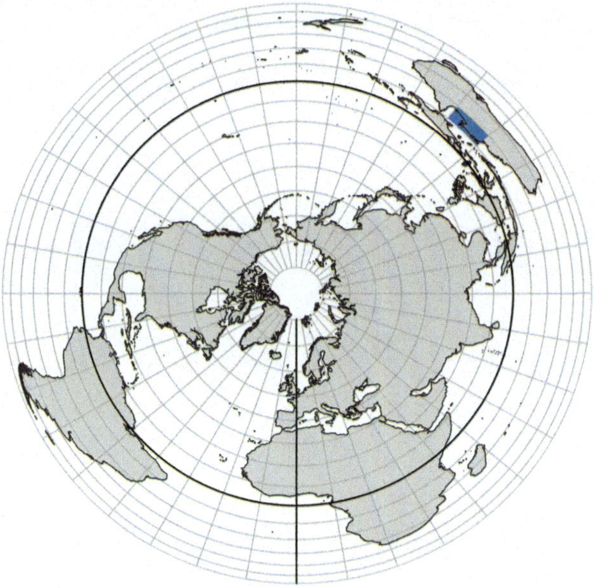

***Utricularia helix* P.Taylor**
Synonyms:

Range area (grAH km^2): AUS: 600
Ecozone origin-status: AUS: E
World rarity status: VR
Ecozone-endemic macroregion occurrence: AUS: WA
Life form(s): EM
Ploidy state: NA
Speciation age (Ma): NA
Populations also occur in terrestrial, wetland and/ or riparian habitats: YES
Long distance movement vectors: NA
Nuisance status and uses: no nuisance issues; no known uses
Endangered status (IUCN): NA

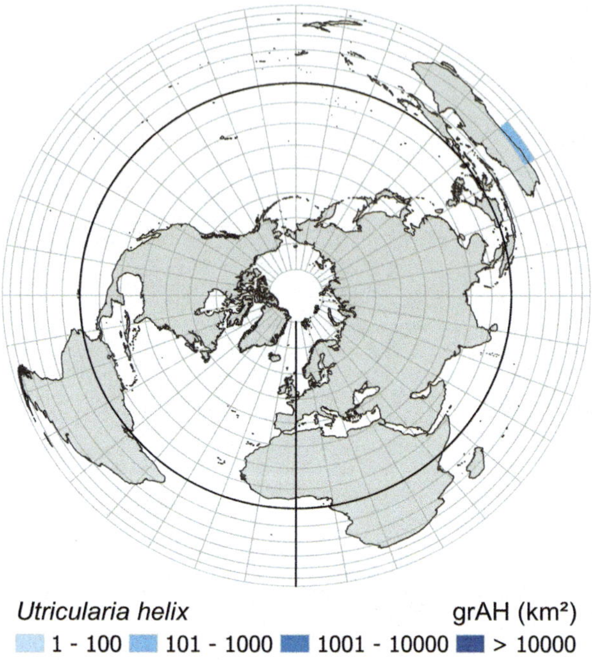

Utricularia heterosepala **Benj.**
Synonyms:

Range area (grAH km^2): OR: 17900
Ecozone origin-status: OR: E
World rarity status: VR
Ecozone-endemic macroregion occurrence: OR: SUP
Life form(s): SU, FR, EM
Ploidy state: NA
Speciation age (Ma): NA
Populations also occur in terrestrial, wetland and/ or riparian habitats: YES
Long distance movement vectors: NA
Nuisance status and uses: no nuisance issues; carnivorous plant hobby species and aquarium trade
Endangered status (IUCN): LC

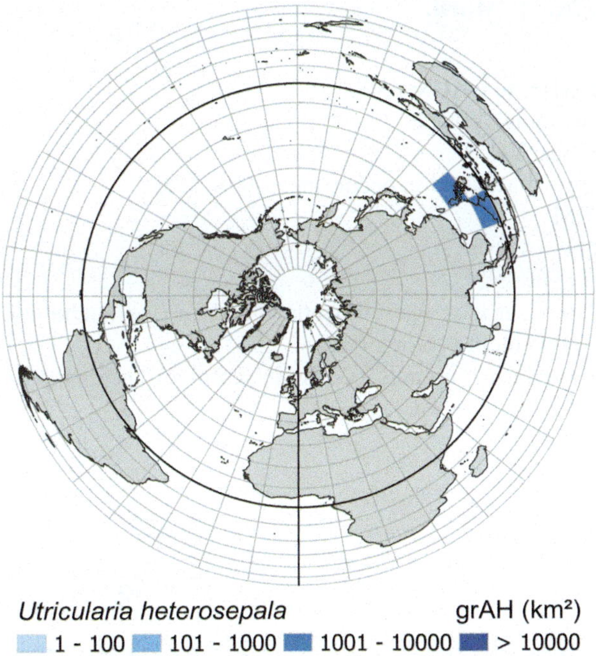

Utricularia heterosepala Benj.
Image copyright: iNaturalist CC0 @leaf0605

***Utricularia holtzei* F. Muell.**
Synonyms:

Range area (grAH km^2): AUS: 6300
Ecozone origin-status: AUS: E
World rarity status: VR
Ecozone-endemic macroregion occurrence: AUS: NA
Life form(s): EM
Ploidy state: NA
Speciation age (Ma): NA
Populations also occur in terrestrial, wetland and/ or riparian habitats: YES
Long distance movement vectors: NA
Nuisance status and uses: no nuisance issues; carnivorous plant hobby species
Endangered status (IUCN): LC

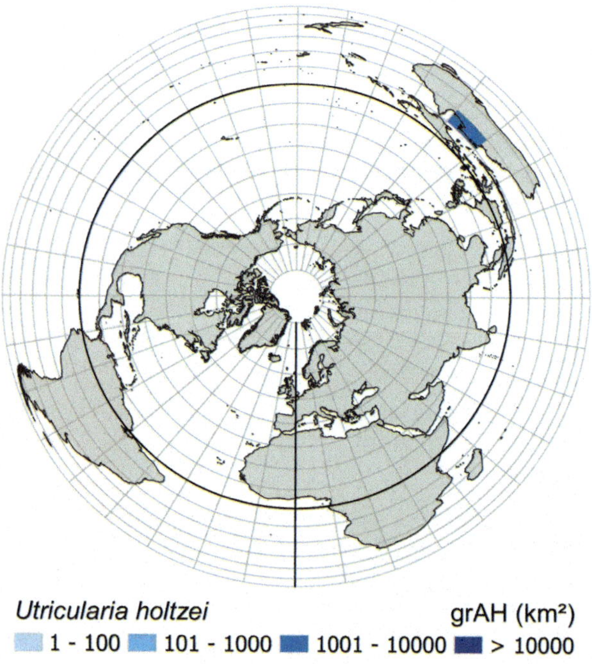

***Utricularia humboldtii* R.H.Schomb.**

Synonyms: *Calpidisca humboldtii* (R.H.Schomb.) Gleason, *Orchyllium humboldtii* (R.H.Schomb.) Barnhart

Range area (grAH km^2): NEO: 102000
Ecozone origin-status: NEO: E
World rarity status: R
Ecozone-endemic macroregion occurrence: NEO: NWSA, AMZ, CHA
Life form(s): SU, EM (also epiphytic, growing in the water-filled leaf axils of some species of bromeliad, including *Brocchinia micrantha* and *Orectanthe* spp.)
Ploidy state: NA
Speciation age (Ma): NA
Populations also occur in terrestrial, wetland and/ or riparian habitats: YES
Long distance movement vectors: NA
Nuisance status and uses: no nuisance issues; carnivorous plant hobby species
Endangered status (IUCN): LC

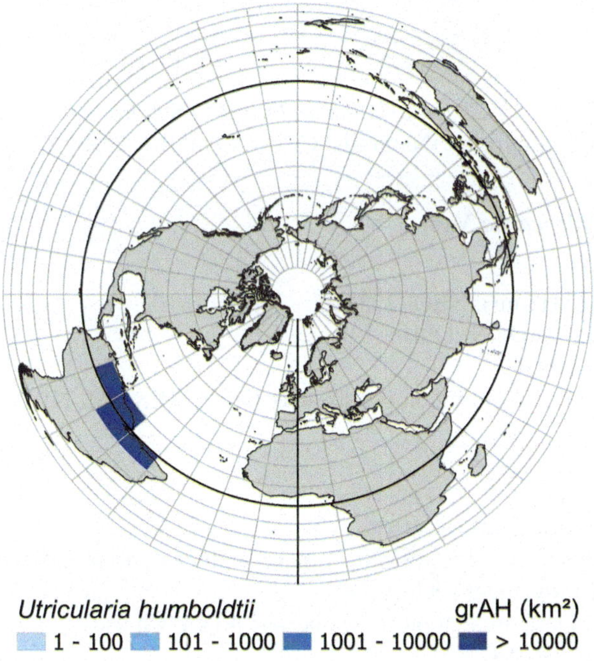

Utricularia humboldtii R.H.Schomb.

***Utricularia hydrocarpa* Vahl**
Synonyms:

Range area (grAH km^2): NEO: 223400
Ecozone origin-status: NEO: E
World rarity status: C
Ecozone-endemic macroregion occurrence: NEO: MES AM, NWSA, AMZ, CHA, PAR
Life form(s): FF
Ploidy state: NA
Speciation age (Ma): NA
Populations also occur in terrestrial, wetland and/ or riparian habitats: NO
Long distance movement vectors: NA
Nuisance status and uses: no nuisance issues; no known uses
Endangered status (IUCN): LC

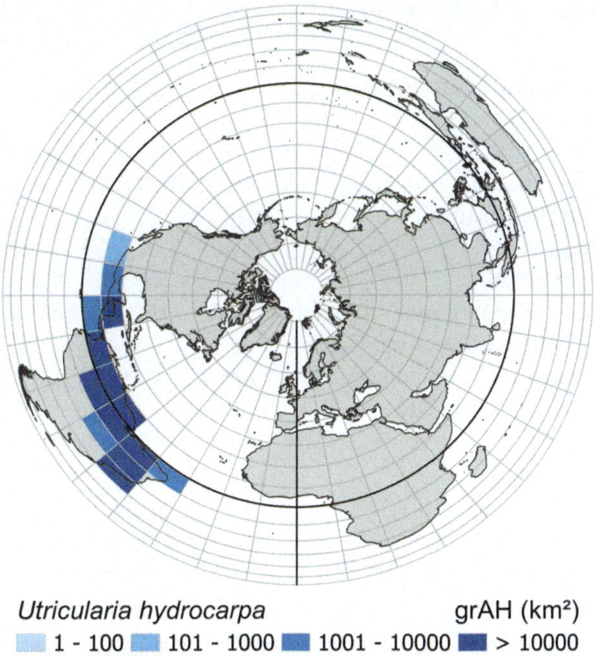

Utricularia hydrocarpa Vahl
Image copyright: iNaturalist CC0 @tsssss

Utricularia incisa **(A.Rich.) Alain**
Synonyms:

Range area (grAH km^2): NEO: 15100
Ecozone origin-status: NEO: E
World rarity status: VR
Ecozone-endemic macroregion occurrence: NEO: MES AM
Life form(s): FF (hydrosubshrub)
Ploidy state: NA
Speciation age (Ma): NA
Populations also occur in terrestrial, wetland and/ or riparian habitats: NO
Long distance movement vectors: NA
Nuisance status and uses: no nuisance issues; no known uses
Endangered status (IUCN): DD

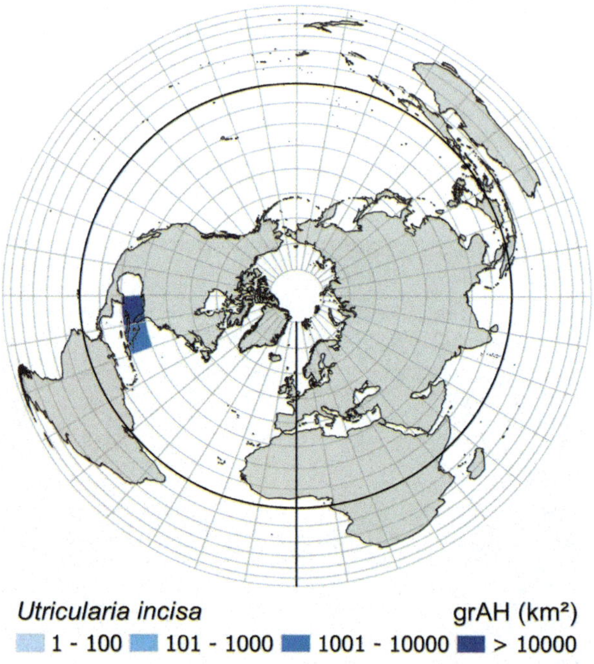

Utricularia inflexa **Forssk.**
Synonyms:

Range area (grAH km^2): PAL: 18000; OR: 8100; AUS: 2900; AFR: 285900
Ecozone origin-status: PAL: N; OR: N; AUS: N; AFR: N
World rarity status: C
Ecozone-endemic macroregion occurrence: not endemic to any world ecozone
Life form(s): FF (hydrosubshrub)

Ploidy state: NA
Speciation age (Ma): NA
Populations also occur in terrestrial, wetland and/ or riparian habitats: NO
Long distance movement vectors: NA
Nuisance status and uses: no nuisance issues; no known uses
Endangered status (IUCN): NA

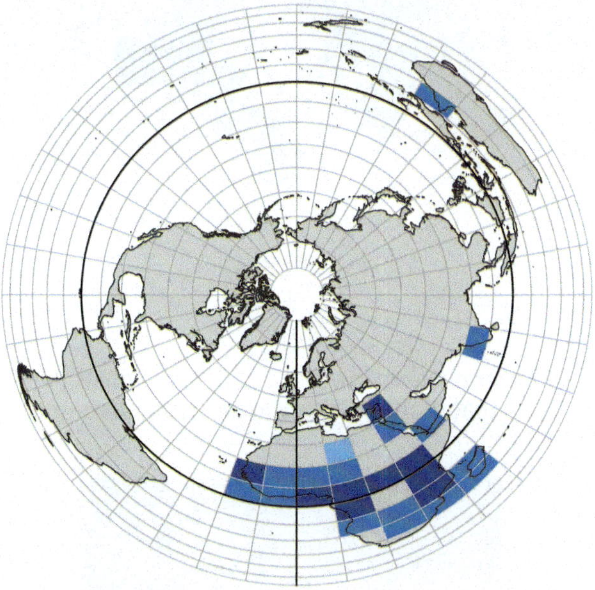

Utricularia intermedia **Hayne**
Synonyms: *Lentibularia intermedia* (Hayne) Nieuwl. & Lunell

Range area (grAH km^2): PAL: 735400; NEA: 918600; OR: 38000
Ecozone origin-status: PAL: N; NEA: N; OR: N
World rarity status: C
Ecozone-endemic macroregion occurrence: not endemic to any world ecozone
Life form(s): FF
Ploidy state: DP
Speciation age (Ma): 8.0
Populations also occur in terrestrial, wetland and/ or riparian habitats: NO
Long distance movement vectors: ?AV; HYDROCHORY (PAL to OR)
Nuisance status and uses: no nuisance issues; no known uses
Endangered status (IUCN): LC

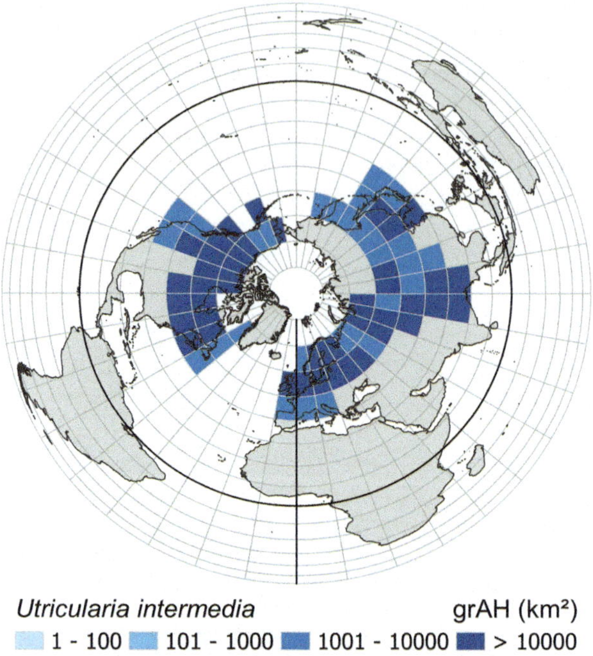

Utricularia intermedia Hayne
Image copyright: J.C. Schou, with permission

***Utricularia juncea* Vahl**

Synonyms: *Stomoisia juncea* (Vahl) Barnhart

Range area (grAH km^2): NEA: 82100; NEO: 135400; AFR: 7800
Ecozone origin-status: NEA: N; NEO: N; AFR: I
World rarity status: C
Ecozone-endemic macroregion occurrence: not endemic to any world ecozone
Life form(s): EM
Ploidy state: D
Speciation age (Ma): NA
Populations also occur in terrestrial, wetland and/ or riparian habitats: YES
Long distance movement vectors: ANTHROPOCHORY
Nuisance status and uses: no nuisance issues; carnivorous plant hobby species
Endangered status (IUCN): LC

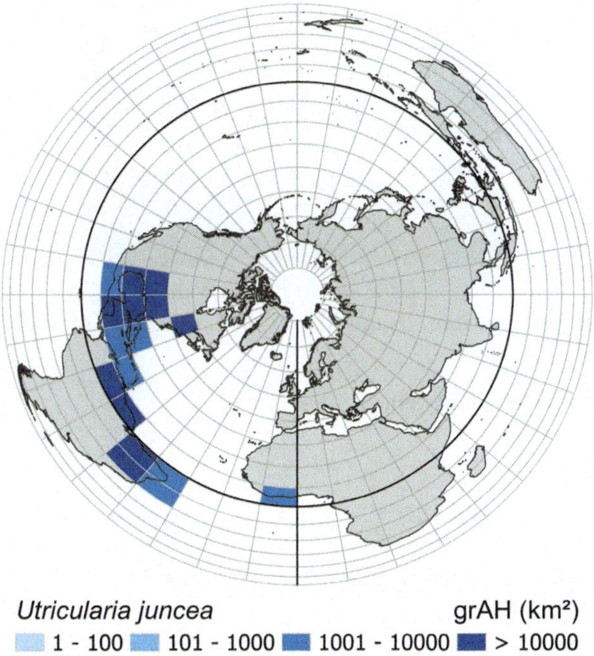

Utricularia juncea Vahl

***Utricularia kamienskii* F.Muell.**
Synonyms:

Range area (grAH km^2): AUS: 6300
Ecozone origin-status: AUS: E
World rarity status: VR
Ecozone-endemic macroregion occurrence: AUS: NA
Life form(s): EM
Ploidy state: NA
Speciation age (Ma): NA
Populations also occur in terrestrial, wetland and/ or riparian habitats: YES
Long distance movement vectors: NA
Nuisance status and uses: no nuisance issues; no known uses
Endangered status (IUCN): LC

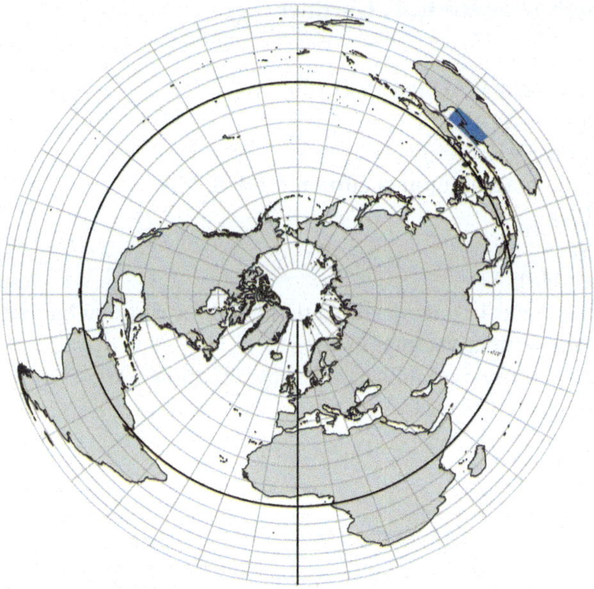

Utricularia kamienskii F.Muell.
Image copyright: iNaturalist CC-BY @hugo_innes (Hugo Innes)

***Utricularia kimberleyensis* C.A. Gardner**
Synonyms:

Range area (grAH km^2): AUS: 13500
Ecozone origin-status: AUS: E
World rarity status: VR
Ecozone-endemic macroregion occurrence: AUS: NA
Life form(s): EM
Ploidy state: NA
Speciation age (Ma): NA
Populations also occur in terrestrial, wetland and/ or riparian habitats: YES
Long distance movement vectors: NA
Nuisance status and uses: no nuisance issues; no known uses
Endangered status (IUCN): LC

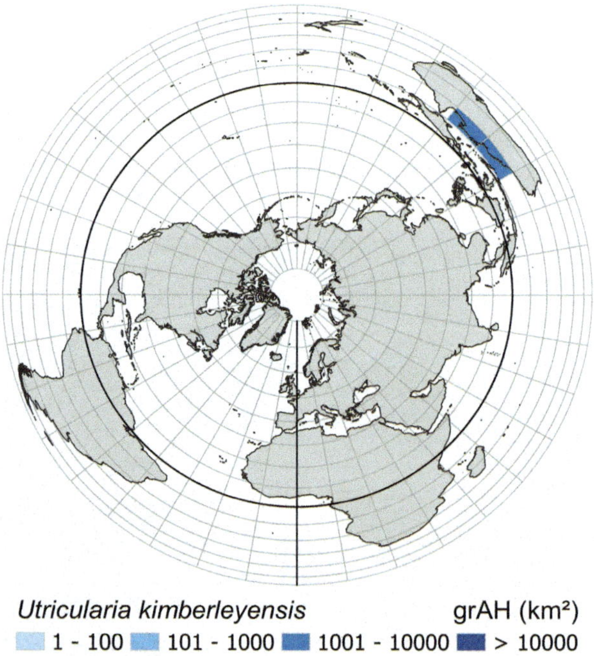

Utricularia kimberleyensis C.A. Gardner
Image copyright: iNaturalist CC-BY @thilokrueger (Thilo Krueger)

***Utricularia lazulina* P.Taylor**
Synonyms:

Range area (grAH km^2):OR: 8100
Ecozone origin-status: OR: E
World rarity status: VR
Ecozone-endemic macroregion occurrence: OR: IND
Life form(s): EM
Ploidy state: NA
Speciation age (Ma): NA
Populations also occur in terrestrial, wetland and/ or riparian habitats: YES
Long distance movement vectors: NA
Nuisance status and uses: no nuisance issues; no known uses
Endangered status (IUCN): LC

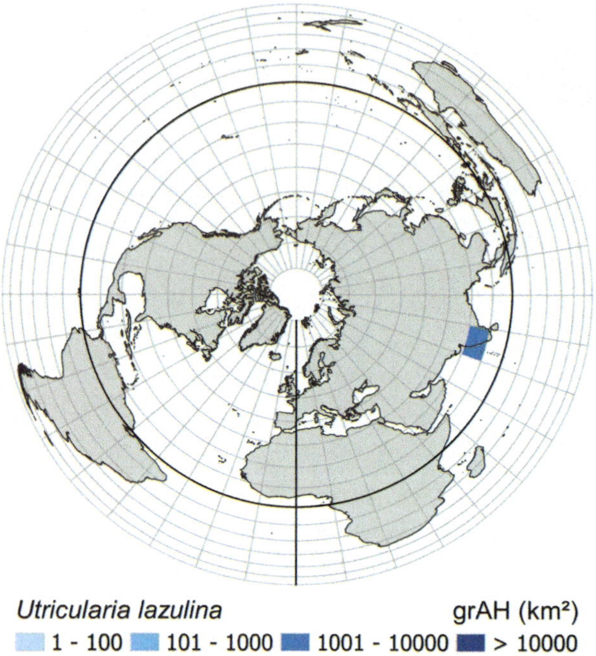

Utricularia lazulina P.Taylor
Image copyright: iNaturalist CC0 @ajit (Ajit Ampalakkad)

Utricularia leptoplectra **F. Muell.**
Synonyms:

Range area (grAH km^2): AUS: 13500
Ecozone origin-status: AUS: E
World rarity status: VR
Ecozone-endemic macroregion occurrence: AUS: NA
Life form(s): EM
Ploidy state: NA
Speciation age (Ma): NA
Populations also occur in terrestrial, wetland and/ or riparian habitats: YES
Long distance movement vectors: NA
Nuisance status and uses: no nuisance issues; no known uses
Endangered status (IUCN): NA

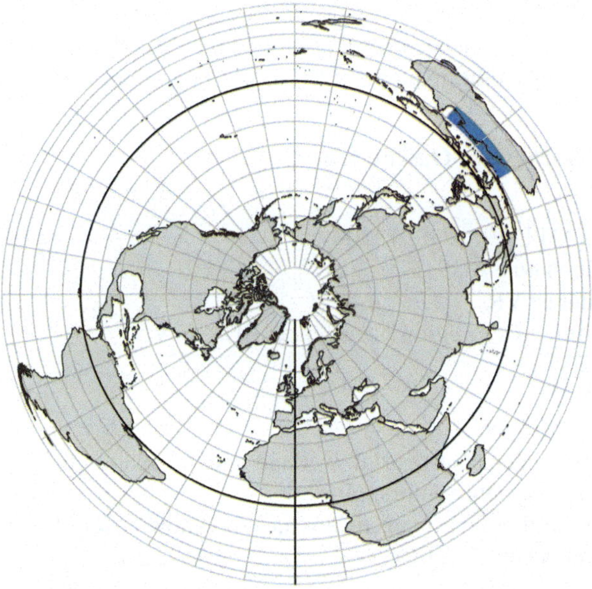

Utricularia leptoplectra grAH (km²)
1 - 100 101 - 1000 1001 - 10000 > 10000

Utricularia leptoplectra F. Muell.

***Utricularia leptorhyncha* O.Schwarz**
Synonyms:

Range area (grAH km^2): AUS: 13500
Ecozone origin-status: AUS: E
World rarity status: VR
Ecozone-endemic macroregion occurrence: AUS: NA
Life form(s): EM
Ploidy state: NA
Speciation age (Ma): NA
Populations also occur in terrestrial, wetland and/ or riparian habitats: YES
Long distance movement vectors: NA
Nuisance status and uses: no nuisance issues; no known uses
Endangered status (IUCN): LC

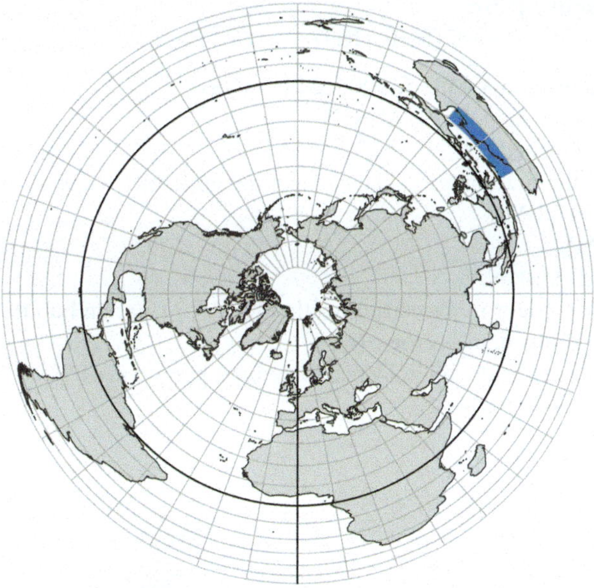

Utricularia leptorhyncha O.Schwarz

***Utricularia limosa* R.Br.**
Synonyms: *Nelipus limosa* (R.Br.) Raf.

Range area (grAH km^2): OR: 53900; AUS: 34600
Ecozone origin-status: OR: N; AUS: N
World rarity status: R
Ecozone-endemic macroregion occurrence: not endemic to any world ecozone
Life form(s): SU, EM
Ploidy state: NA
Speciation age (Ma): NA
Populations also occur in terrestrial, wetland and/ or riparian habitats: YES
Long distance movement vectors: NA
Nuisance status and uses: no nuisance issues; no known uses
Endangered status (IUCN): LC

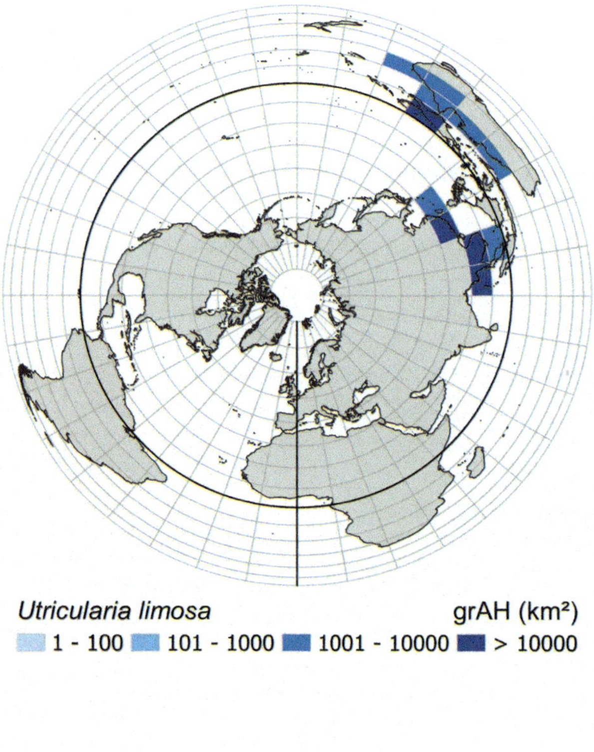

Utricularia limosa R.Br.
Image copyright: iNaturalist CC-BY @hugo_innes (Hugo Innes)

Utricularia lowriei **R.W.Jobson**
Synonyms:

Range area (grAH km^2): AUS: 2900
Ecozone origin-status: AUS: E
World rarity status: VR
Ecozone-endemic macroregion occurrence: AUS: NA
Life form(s): EM
Ploidy state: NA
Speciation age (Ma): NA
Populations also occur in terrestrial, wetland and/ or riparian habitats: YES
Long distance movement vectors: NA
Nuisance status and uses: no nuisance issues; no known uses
Endangered status (IUCN): LC

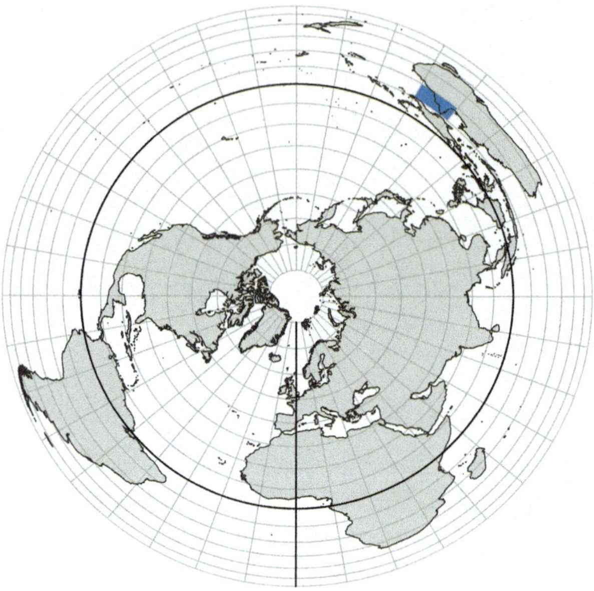

Utricularia minor **L.**
Synonyms: *Lentibularia minor* (L.) Raf.

Range area (grAH km^2): PAL: 735000; NEA: 848700; OR: 47600; AUS: 12800; AFR: 7200
Ecozone origin-status: PAL: N; NEA: N; OR: N; AUS: N; AFR: N
World rarity status: C
Ecozone-endemic macroregion occurrence: not endemic to any world ecozone
Life form(s): SU, FF (hydrosubshrub)
Ploidy state: DP
Speciation age (Ma): 8.0
Populations also occur in terrestrial, wetland and/ or riparian habitats: NO
Long distance movement vectors: ?AV
Nuisance status and uses: no nuisance issues; aquarium trade
Endangered status (IUCN): LC

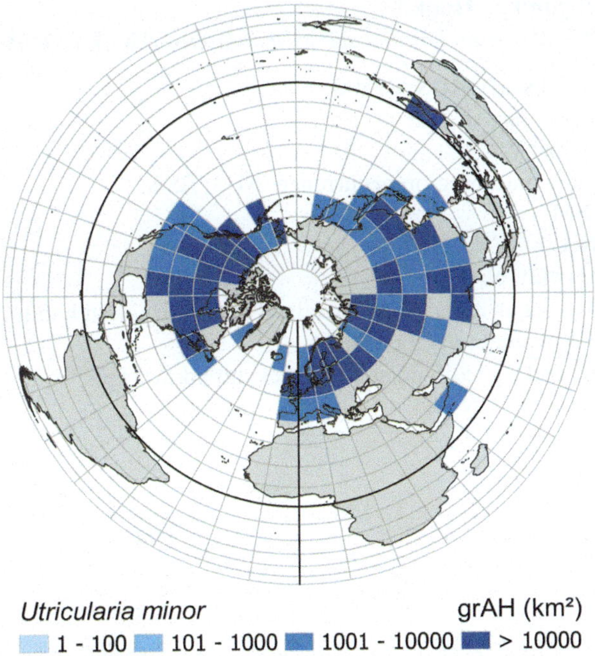

Utricularia minor L.
Image copyright: J.C. Schou, with permission

Utricularia monanthos **Hook.f.**
Synonyms: *Utricularia dichotoma* subsp. *monanthos* (Hook.f.) R.W.Jobson

Range area (grAH km^2): AUS: 27500
Ecozone origin-status: AUS: E
World rarity status: VR
Ecozone-endemic macroregion occurrence: AUS: ESA, NZ
Life form(s): EM
Ploidy state: NA
Speciation age (Ma): NA
Populations also occur in terrestrial, wetland and/ or riparian habitats: YES
Long distance movement vectors: NA
Nuisance status and uses: no nuisance issues; no known uses
Endangered status (IUCN): LC

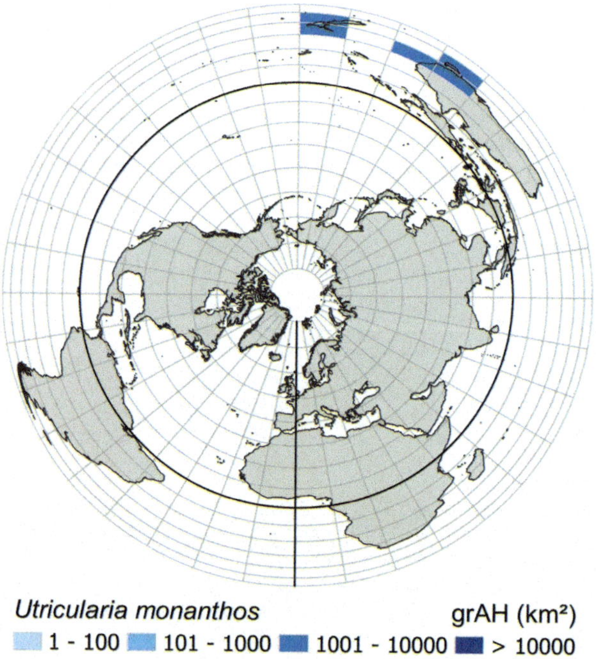

Utricularia monanthos Hook.f.
Image copyright: iNaturalist CC0 @cowirrie

***Utricularia muelleri* Kamienski**
Synonyms:

Range area (grAH km^2): AUS: 29300
Ecozone origin-status: AUS: E
World rarity status: VR
Ecozone-endemic macroregion occurrence: AUS: NG, NA
Life form(s): FF (hydrosubshrub)
Ploidy state: NA
Speciation age (Ma): NA
Populations also occur in terrestrial, wetland and/ or riparian habitats: NO
Long distance movement vectors: NA
Nuisance status and uses: no nuisance issues; no known uses
Endangered status (IUCN): LC

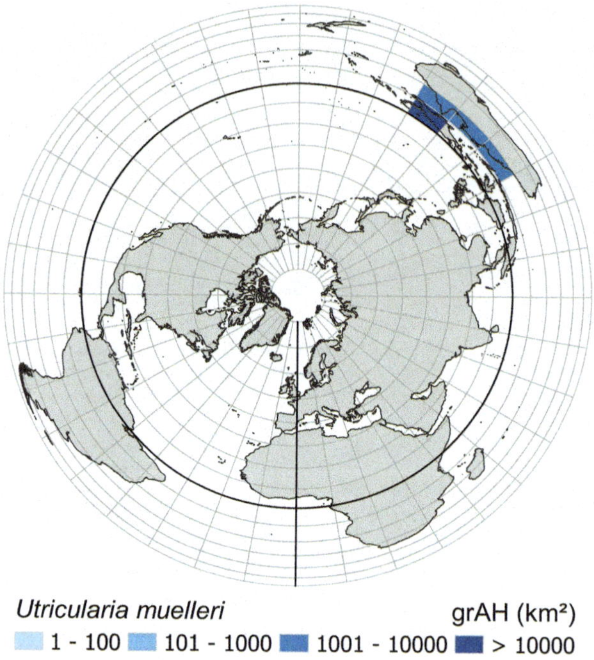

Utricularia myriocista A.St.-Hil. & Girard
Synonyms:

Range area (grAH km^2): NEO: 205700
Ecozone origin-status: NEO: E
World rarity status: C
Ecozone-endemic macroregion occurrence: NEO: MES AM, NWSA, AMZ, CHA, PAR
Life form(s): FF (hydrosubshrub)
Ploidy state: NA
Speciation age (Ma): NA
Populations also occur in terrestrial, wetland and/ or riparian habitats: NO
Long distance movement vectors: NA
Nuisance status and uses: no nuisance issues; ornamental
Endangered status (IUCN): NA

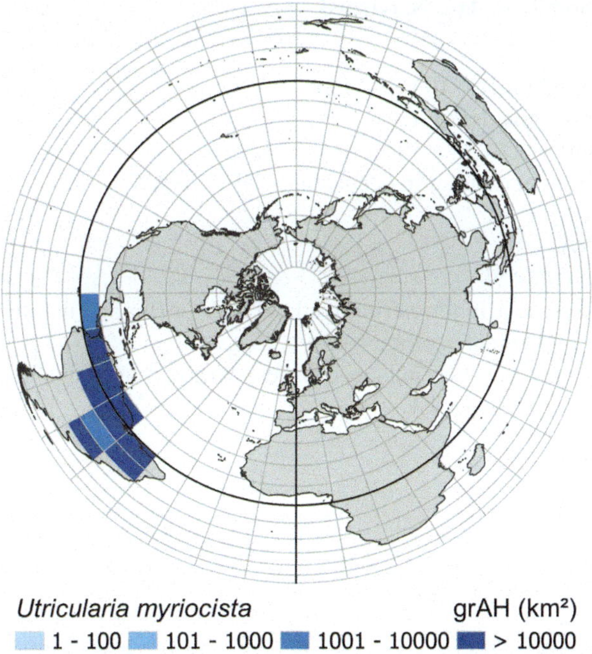

Utricularia myriocista A.St.-Hil. & Girard

***Utricularia nana* A.St.-Hil. & Girard**
Synonyms:

Range area (grAH km^2): NEO: 162300
Ecozone origin-status: NEO: E
World rarity status: C
Ecozone-endemic macroregion occurrence: NEO: NWSA, AMZ, CHA, PAR
Life form(s): EM
Ploidy state: NA
Speciation age (Ma): NA
Populations also occur in terrestrial, wetland and/ or riparian habitats: YES
Long distance movement vectors: NA
Nuisance status and uses: no nuisance issues; carnivorous plants hobby trade
Endangered status (IUCN): LC

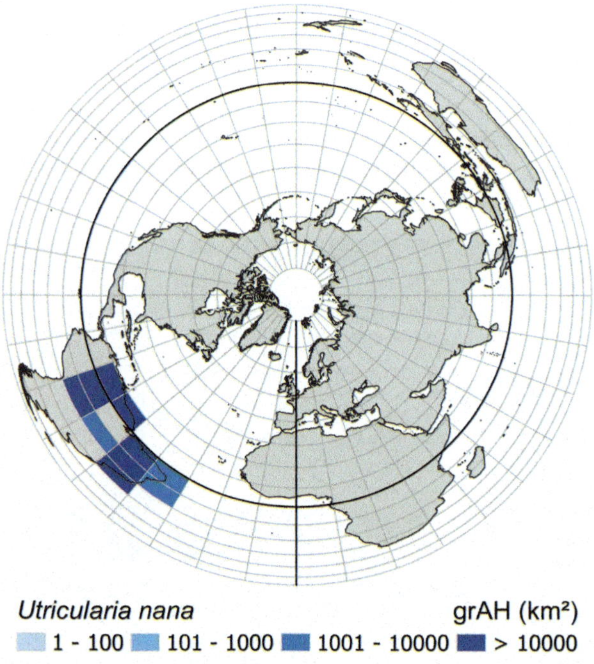

***Utricularia naviculata* P.Taylor**
Synonyms:

Range area (grAH km^2): NEO: 65600
Ecozone origin-status: NEO: E
World rarity status: R
Ecozone-endemic macroregion occurrence: NEO: AMZ
Life form(s): FF

Ploidy state: NA
Speciation age (Ma): NA
Populations also occur in terrestrial, wetland and/ or riparian habitats: NO
Long distance movement vectors: NA
Nuisance status and uses: no nuisance issues; no known uses
Endangered status (IUCN): DD

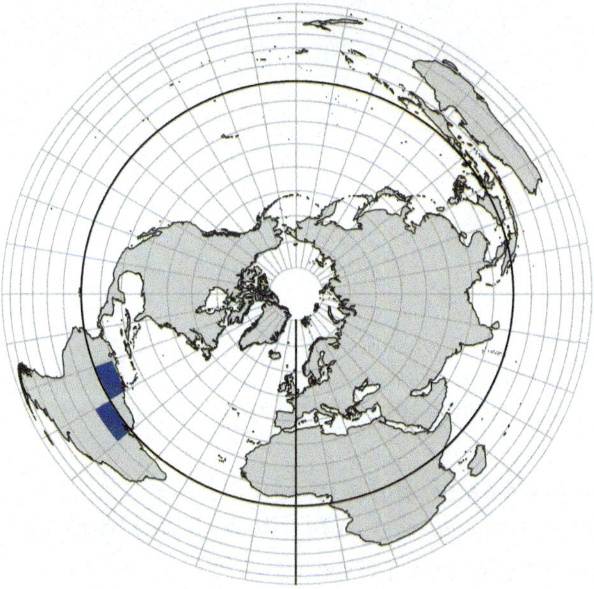

Utricularia naviculata grAH (km²)
1 - 100 101 - 1000 1001 - 10000 > 10000

Utricularia nelumbifolia **Gardner**
Synonyms:

Range area (grAH km^2): NEO: 23500
Ecozone origin-status: NEO: E
World rarity status: VR
Ecozone-endemic macroregion occurrence: NEO: CHA, PAR
Life form(s): EM (epiphytic in bromeliads)
Ploidy state: NA
Speciation age (Ma): NA
Populations also occur in terrestrial, wetland and/ or riparian habitats: YES
Long distance movement vectors: NA
Nuisance status and uses: no nuisance issues; carnivorous plants hobby trade
Endangered status (IUCN): LC

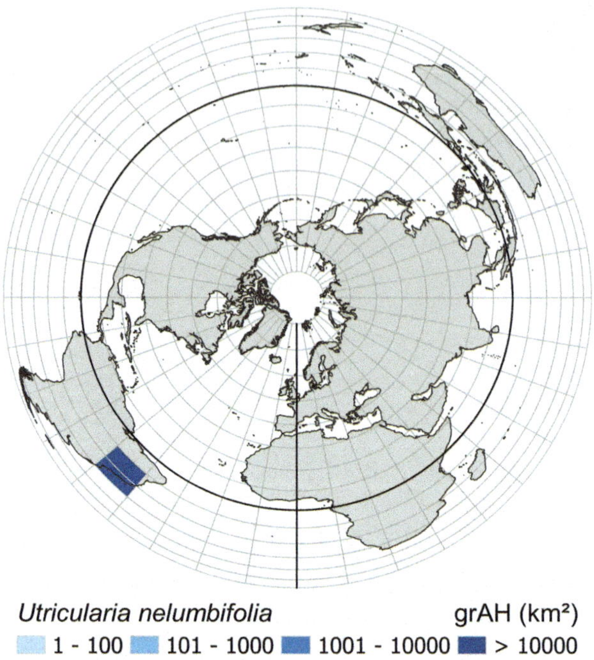

Utricularia neottioides **A.St.-Hil. & Girard**
Synonyms: *Avesicaria neottioides* (A.St.-Hil. & Girard) Barnhart

Range area (grAH km^2): NEO: 151700
Ecozone origin-status: NEO: E
World rarity status: R
Ecozone-endemic macroregion occurrence: NEO: MES AM, NWSA, AMZ, CHA
Life form(s): SU (rheophyte, attached to rocks)
Ploidy state: NA
Speciation age (Ma): NA
Populations also occur in terrestrial, wetland and/ or riparian habitats: NO
Long distance movement vectors: NA
Nuisance status and uses: no nuisance issues; partially-successful attempts have been made to grow the species in cultivation, primarily for the carnivorous plants hobby trade
Endangered status (IUCN): LC

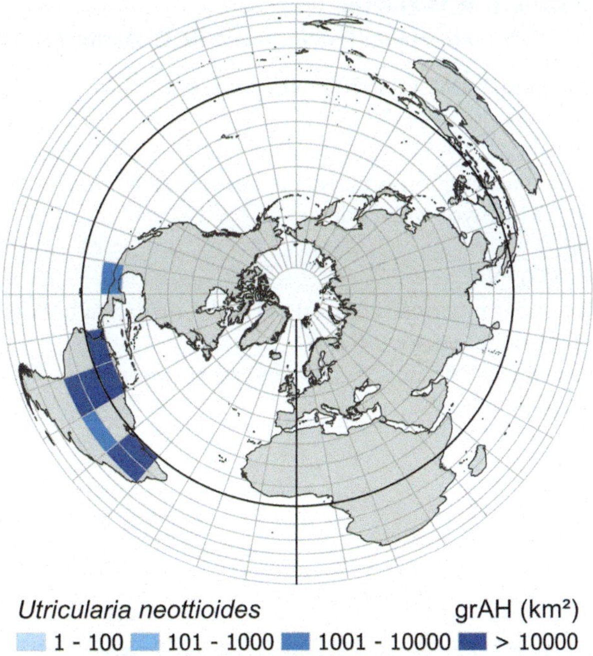

Utricularia neottioides A.St.-Hil. & Girard

***Utricularia ochroleuca* R.W.Hartm.**

Synonyms: *Utricularia intermedia* f. *ochroleuca* (R.W.Hartm.) Komiya

Range area (grAH km²): PAL: 236700; NEA: 187600
Ecozone origin-status: PAL: N; NEA: N
World rarity status: C
Ecozone-endemic macroregion occurrence: not endemic to any world ecozone
Life form(s): SU, EM (hydrosubshrub)
Ploidy state: DP
Speciation age (Ma): NA
Populations also occur in terrestrial, wetland and/ or riparian habitats: YES
Long distance movement vectors: NA
Nuisance status and uses: no nuisance issues; no known uses
Endangered status (IUCN): LC

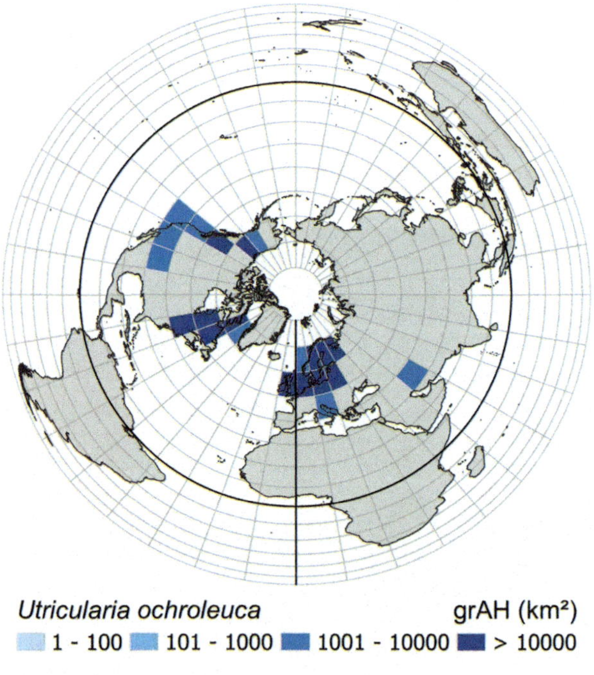

Utricularia ochroleuca R.W.Hartm.
Image copyright: J.C. Schou, with permission

***Utricularia olivacea* C.Wright ex Griseb.**
Synonyms: *Biovularia olivacea* (C.Wright ex Griseb.) Kamienski

Range area (grAH km^2): NEA: 8200; NEO: 209100
Ecozone origin-status: NEA: N; NEO: N
World rarity status: C
Ecozone-endemic macroregion occurrence: not endemic to any world ecozone
Life form(s): FF
Ploidy state: NA
Speciation age (Ma): NA
Populations also occur in terrestrial, wetland and/ or riparian habitats: NO
Long distance movement vectors: NA
Nuisance status and uses: no nuisance issues; no known uses
Endangered status (IUCN): LC

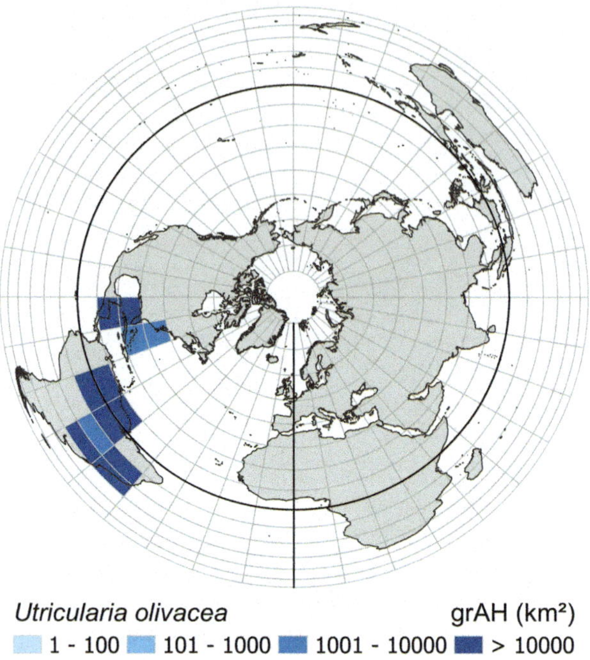

Utricularia oliveriana **Steyerm.**
Synonyms:

Range area (grAH km^2): NEO: 193400
Ecozone origin-status: NEO: E
World rarity status: C
Ecozone-endemic macroregion occurrence: NEO: NWSA, AMZ, CHA
Life form(s): SU, EM (rheophyte, attached to rocks; hydrosubshrub)
Ploidy state: NA
Speciation age (Ma): NA
Populations also occur in terrestrial, wetland and/ or riparian habitats: NO
Long distance movement vectors: NA
Nuisance status and uses: no nuisance issues; no known uses
Endangered status (IUCN): LC

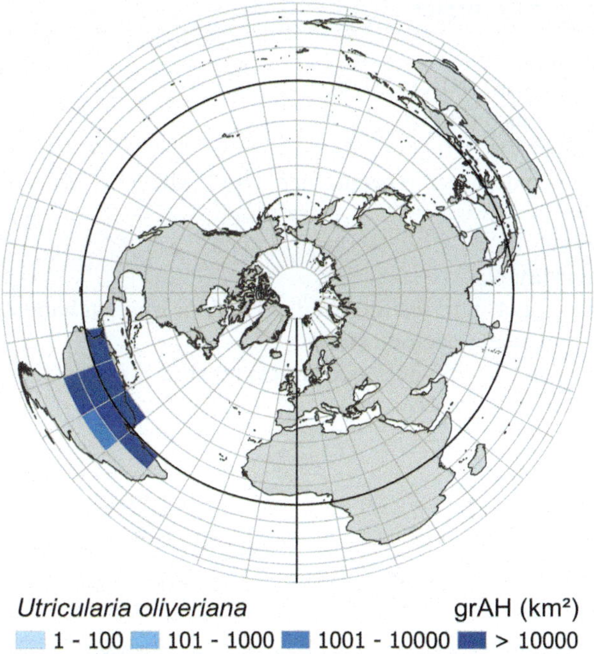

Utricularia oliveriana Steyerm.
Image copyright: iNaturalist CC-BY @nbareschu (Nicolás Baresch Uribe)

Utricularia perversa **P.Taylor**
Synonyms:

Range area (grAH km^2): NEO: 25300
Ecozone origin-status: NEO: E
World rarity status: VR
Ecozone-endemic macroregion occurrence: NEO: MG MX, MES AM
Life form(s): FF
Ploidy state: NA
Speciation age (Ma): NA
Populations also occur in terrestrial, wetland and/ or riparian habitats: NO
Long distance movement vectors: NA
Nuisance status and uses: no nuisance issues; no known uses
Endangered status (IUCN): NA

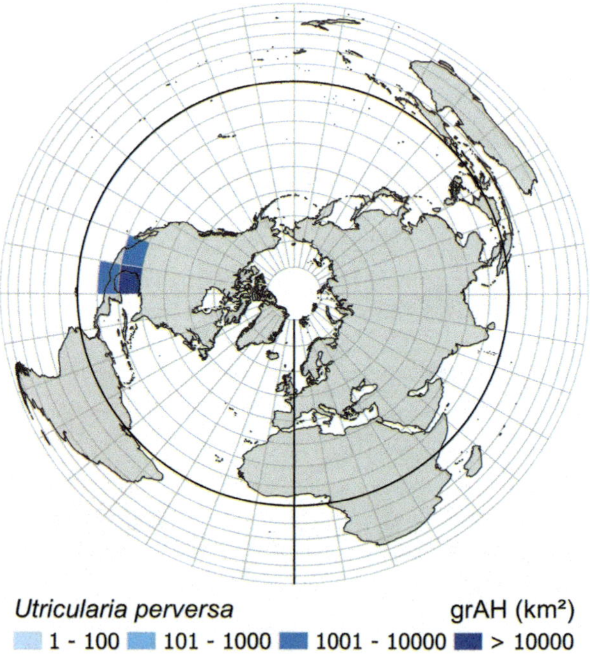

Utricularia perversa P.Taylor
Image copyright: T. Lobato-de Magalhães

***Utricularia platensis* Speg.**
Synonyms:

Range area (grAH km^2): NEO: 68400
Ecozone origin-status: NEO: E
World rarity status: R
Ecozone-endemic macroregion occurrence: NEO: AMZ, CHA, PAR
Life form(s): FF (hydrosubshrub)
Ploidy state: NA
Speciation age (Ma): NA
Populations also occur in terrestrial, wetland and/ or riparian habitats: NO
Long distance movement vectors: NA
Nuisance status and uses: no nuisance issues; no known uses
Endangered status (IUCN): LC

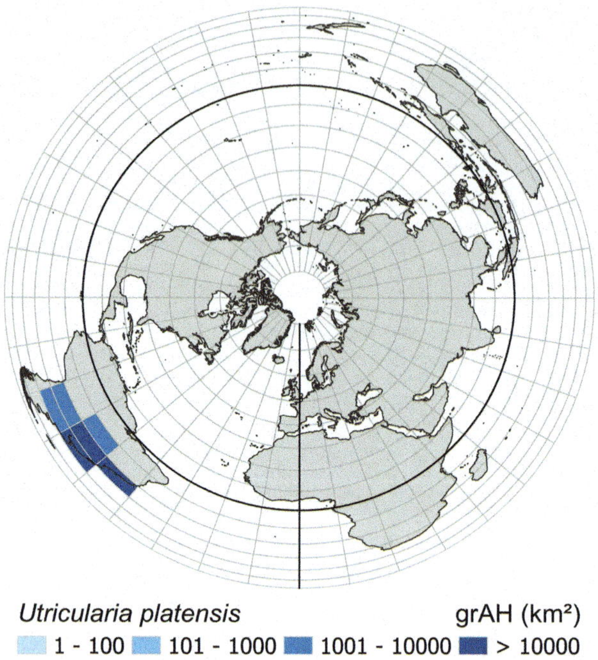

Utricularia poconensis **Fromm.**
Synonyms:

Range area (grAH km^2): NEO: 58400
Ecozone origin-status: NEO: E
World rarity status: R
Ecozone-endemic macroregion occurrence: NEO: AMZ, CHA, PAR
Life form(s): FF (hydrosubshrub)
Ploidy state: NA
Speciation age (Ma): NA
Populations also occur in terrestrial, wetland and/ or riparian habitats: NO
Long distance movement vectors: NA
Nuisance status and uses: no nuisance issues; no known uses
Endangered status (IUCN): LC

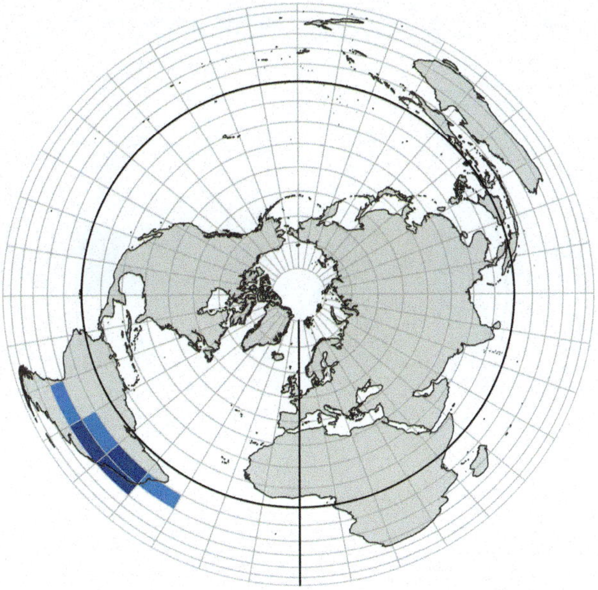

Utricularia punctata **Wall. ex A. DC.**
Synonyms:

Range area (grAH km^2): OR: 66200; AUS: 11800
Ecozone origin-status: OR: N; AUS: N
World rarity status: R
Ecozone-endemic macroregion occurrence: not endemic to any world ecozone
Life form(s): FF (hydrosubshrub)
Ploidy state: NA
Speciation age (Ma): NA
Populations also occur in terrestrial, wetland and/ or riparian habitats: NO
Long distance movement vectors: NA
Nuisance status and uses: no nuisance issues; no known uses
Endangered status (IUCN): LC

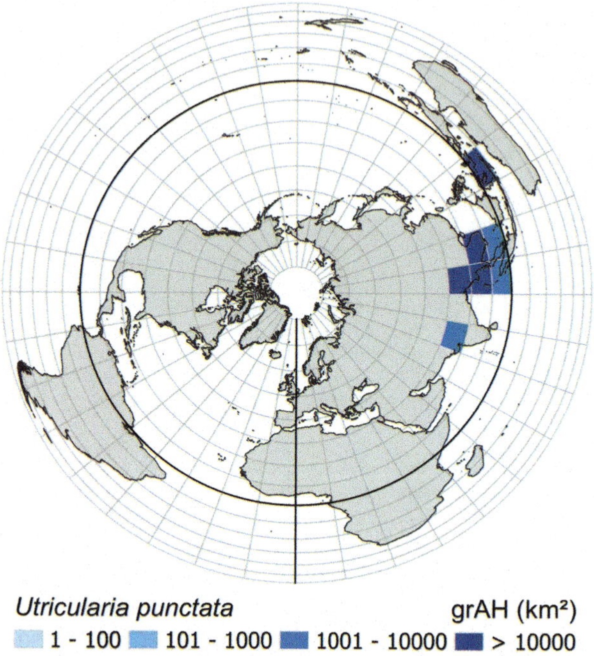

Utricularia punctata Wall. ex A. DC.

Utricularia purpurea **Walter**
Synonyms: *Vesiculina purpurea* (Walter) Raf.

Range area (grAH km^2): NEA: 378700; NEO: 59600
Ecozone origin-status: NEA: N; NEO: N
World rarity status: C
Ecozone-endemic macroregion occurrence: not endemic to any world ecozone
Life form(s): FF
Ploidy state: NA
Speciation age (Ma): NA
Populations also occur in terrestrial, wetland and/ or riparian habitats: NO
Long distance movement vectors: NA
Nuisance status and uses: no nuisance issues; phytoremediation potential use
Endangered status (IUCN): LC

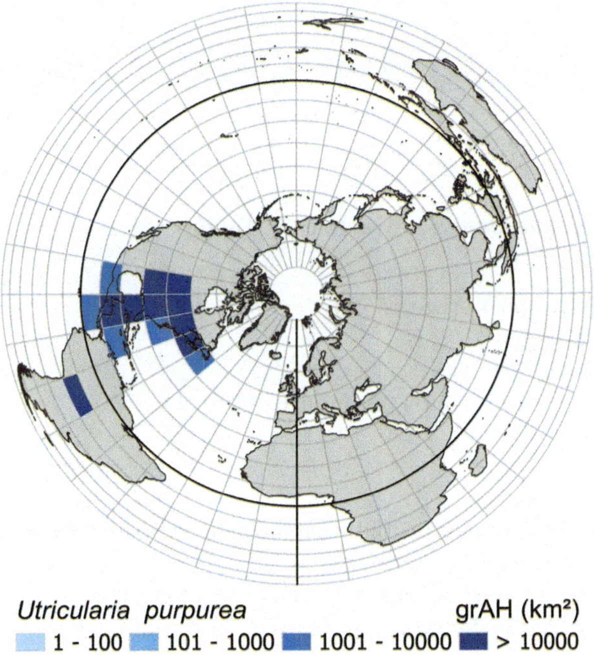

Utricularia purpurea Walter

***Utricularia quinquedentata* F. Muell. ex P. Taylor**
Synonyms:

Range area (grAH km^2): AUS: 16500
Ecozone origin-status: AUS: E
World rarity status: VR
Ecozone-endemic macroregion occurrence: AUS: NA
Life form(s): EM
Ploidy state: NA
Speciation age (Ma): NA
Populations also occur in terrestrial, wetland and/ or riparian habitats: YES
Long distance movement vectors: NA
Nuisance status and uses: no nuisance issues; no known uses
Endangered status (IUCN): LC

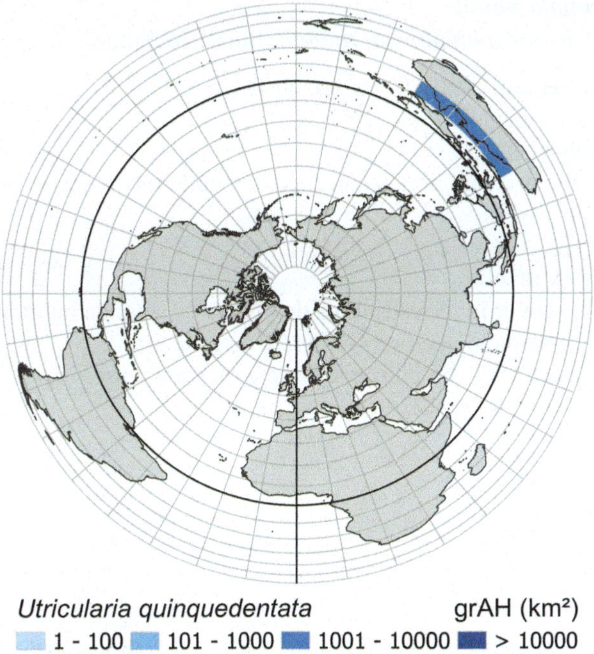

Utricularia quinquedentata F. Muell. ex P. Taylor

Utricularia radiata **Small**

Synonyms: *Utricularia inflata* var. *radiata* (Small) W.Stone

Range area (grAH km²): NEA: 120100; NEO: 24100
Ecozone origin-status: NEA: N; NEO: N
World rarity status: R
Ecozone-endemic macroregion occurrence: not endemic to any world ecozone
Life form(s): FF
Ploidy state: P
Speciation age (Ma): NA
Populations also occur in terrestrial, wetland and/ or riparian habitats: NO
Long distance movement vectors: HYD (NEA to NEO)
Nuisance status and uses: no nuisance issues; no known uses
Endangered status (IUCN): LC

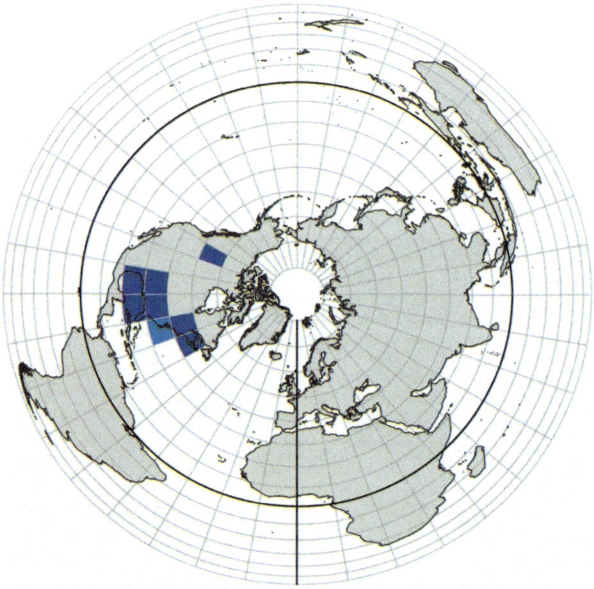

Utricularia radiata Small

***Utricularia raynalii* P.Taylor**
Synonyms:

Range area (grAH km^2): AFR: 139500
Ecozone origin-status: AFR: E
World rarity status: R
Ecozone-endemic macroregion occurrence: AFR: SAH/ AR, FOR, EA, ZAM
Life form(s): FF
Ploidy state: NA
Speciation age (Ma): NA
Populations also occur in terrestrial, wetland and/ or riparian habitats: NO
Long distance movement vectors: BIRD (probably waterfowl: [106])
Nuisance status and uses: no nuisance issues; no known uses
Endangered status (IUCN): LC

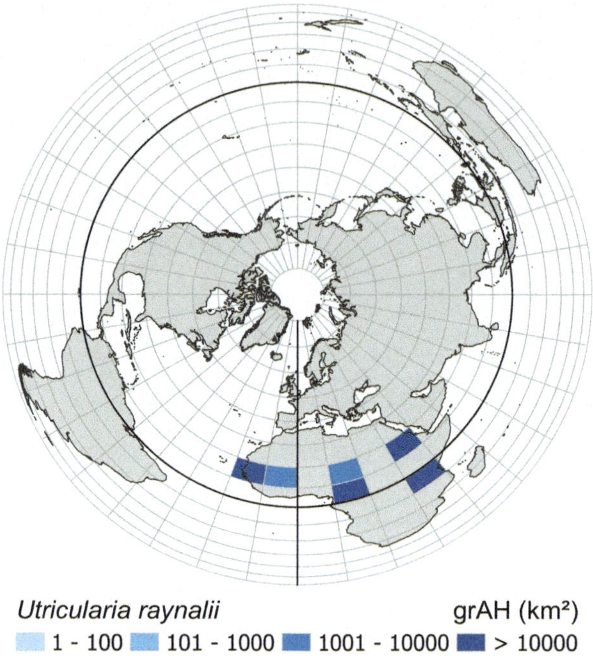

Utricularia raynalii P.Taylor

***Utricularia reflexa* Oliv.**
Synonyms: *Utricularia bangweolensis* R.E.Fr.

Range area (grAH km^2): AFR: 227500
Ecozone origin-status: AFR: E
World rarity status: C
Ecozone-endemic macroregion occurrence: AFR: SAH/ AR, FOR, EA, ZAM, SDES, MAD
Life form(s): FF
Ploidy state: NA
Speciation age (Ma): NA
Populations also occur in terrestrial, wetland and/ or riparian habitats: NO
Long distance movement vectors: NA
Nuisance status and uses: no nuisance issues; carnivorous plant hobby trade
Endangered status (IUCN): LC

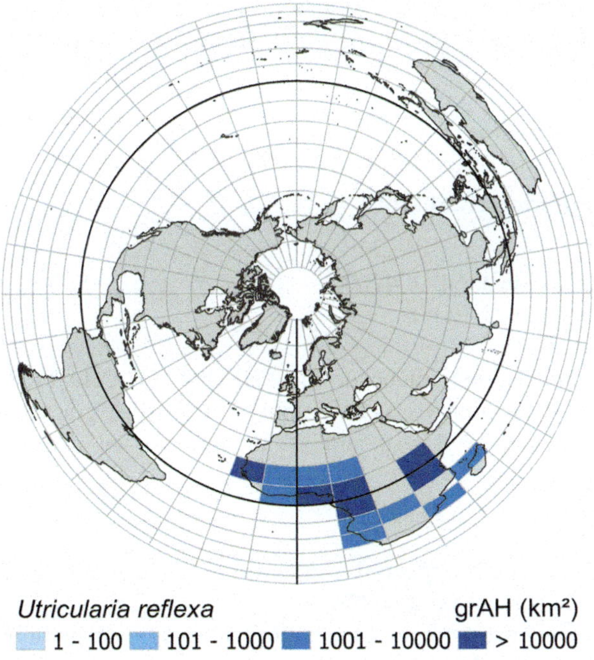

Utricularia reflexa Oliv.

***Utricularia reniformis* A. St.-Hil.**
Synonyms:

Range area (grAH km^2): NEO: 47600
Ecozone origin-status: NEO: E
World rarity status: R
Ecozone-endemic macroregion occurrence: NEO: AMZ, CHA,PAR
Life form(s): EM (also grows as an epiphyte on tussock grasses in the southern coastal mountains of Brazil, and as an epiphyte on other epiphytes, e.g., bromeliads)
Ploidy state: NA
Speciation age (Ma): NA
Populations also occur in terrestrial, wetland and/ or riparian habitats: YES
Long distance movement vectors: NA
Nuisance status and uses: no nuisance issues; carnivorous plant hobby trade
Endangered status (IUCN): LC

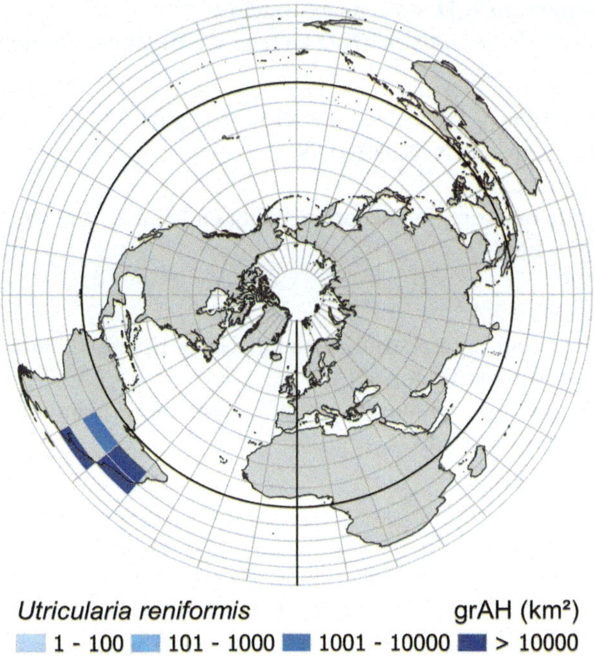

Utricularia reniformis A. St.-Hil.
Image copyright: Y. Domínguez, with permission

***Utricularia resupinata* B.D. Greene ex Bigelow**
Synonyms: *Lecticula resupinata* (B.D.Greene ex Bigelow) Barnhart, *Utricularia greenei* Oakes

Range area (grAH km^2): NEA: 358200; NEO: 103200
Ecozone origin-status: NEA: N; NEO: N
World rarity status: C
Ecozone-endemic macroregion occurrence: not endemic to any world ecozone
Life form(s): SU, EM
Ploidy state: NA
Speciation age (Ma): NA
Populations also occur in terrestrial, wetland and/ or riparian habitats: NO
Long distance movement vectors: NA
Nuisance status and uses: no nuisance issues; carnivorous plant hobby trade
Endangered status (IUCN): LC

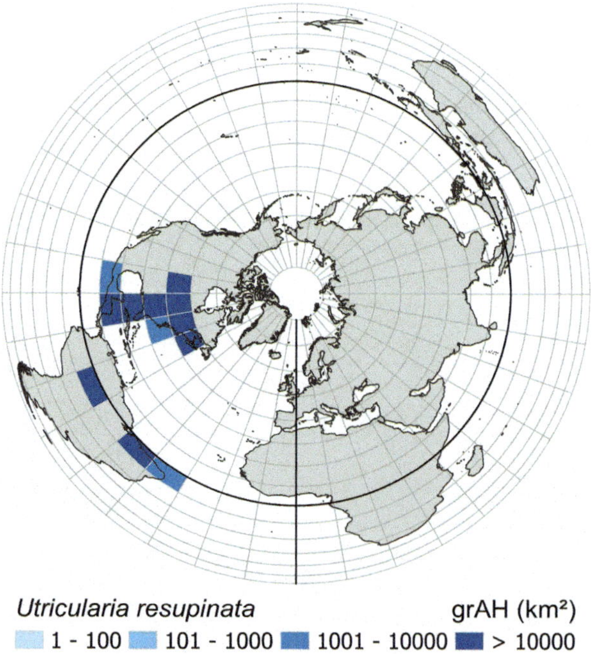

Utricularia resupinata B.D. Greene ex Bigelow
Image copyright: iNaturalist CC-BY @mcburly

***Utricularia rhododactylos* P.Taylor**
Synonyms:

Range area (grAH km^2): AUS: 6300
Ecozone origin-status: AUS: E
World rarity status: VR
Ecozone-endemic macroregion occurrence: AUS: NA
Life form(s): SU, EM
Ploidy state: NA
Speciation age (Ma): NA
Populations also occur in terrestrial, wetland and/ or riparian habitats: YES
Long distance movement vectors: NA
Nuisance status and uses: no nuisance issues; no known uses
Endangered status (IUCN): LC

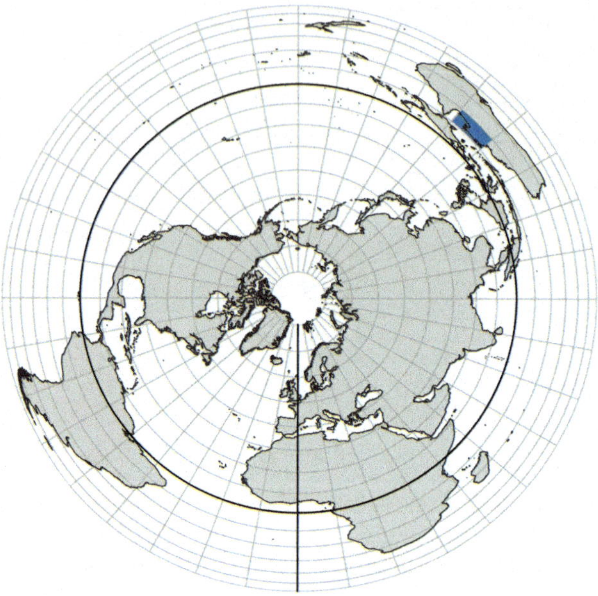

Utricularia rigida **Benj.**
Synonyms:

Range area (grAH km^2): AFR: 46900
Ecozone origin-status: AFR: E
World rarity status: R
Ecozone-endemic macroregion occurrence: AFR: SAH/ AR, FOR
Life form(s): SU, EM (rheophyte, attached to rocks in fast-flowing rivers and streams)
Ploidy state: NA
Speciation age (Ma): NA
Populations also occur in terrestrial, wetland and/ or riparian habitats: NO
Long distance movement vectors: NA
Nuisance status and uses: no nuisance issues; no known uses
Endangered status (IUCN): LC

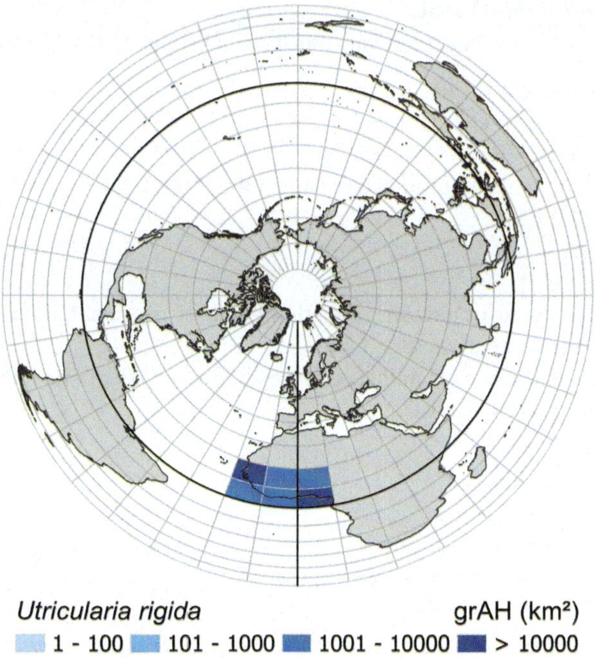

Utricularia rigida Benj.
Image copyright: iNaturalist CC-BY @rob_palmer (Rob Palmer)

Utricularia smithiana **Wight**
Synonyms: *Utricularia caerulea* var. *smithiana* (Wight) C.B.Clarke

Range area (grAH km^2): OR: 12200
Ecozone origin-status: OR: E
World rarity status: VR
Ecozone-endemic macroregion occurrence: OR: IND, ICH
Life form(s): SU, EM
Ploidy state: NA
Speciation age (Ma): NA
Populations also occur in terrestrial, wetland and/ or riparian habitats: YES
Long distance movement vectors: NA
Nuisance status and uses: no nuisance issues; carnivorous plants hobby trade
Endangered status (IUCN): LC

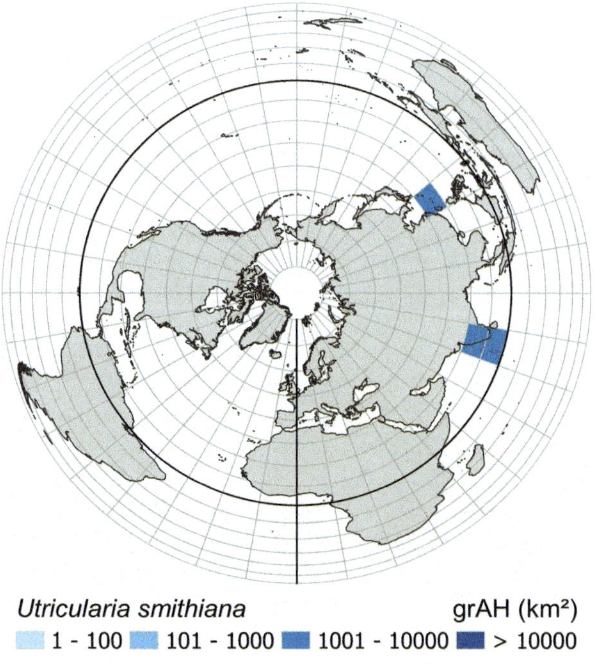

Utricularia smithiana Wight
Image copyright: iNaturalist CC-BY @maxhsieh (Shen ching chen)

***Utricularia spruceana* Benth. ex Oliv.**
Synonyms: *Lecticula spruceana* (Benth. ex Oliv.) Barnhart

Range area (grAH km^2): NEO: 98400
Ecozone origin-status: NEO: E
World rarity status: R
Ecozone-endemic macroregion occurrence: NEO: AMZ
Life form(s): SU, EM
Ploidy state: NA
Speciation age (Ma): NA
Populations also occur in terrestrial, wetland and/ or riparian habitats: YES
Long distance movement vectors: NA
Nuisance status and uses: no nuisance issues; no known uses
Endangered status (IUCN): NA

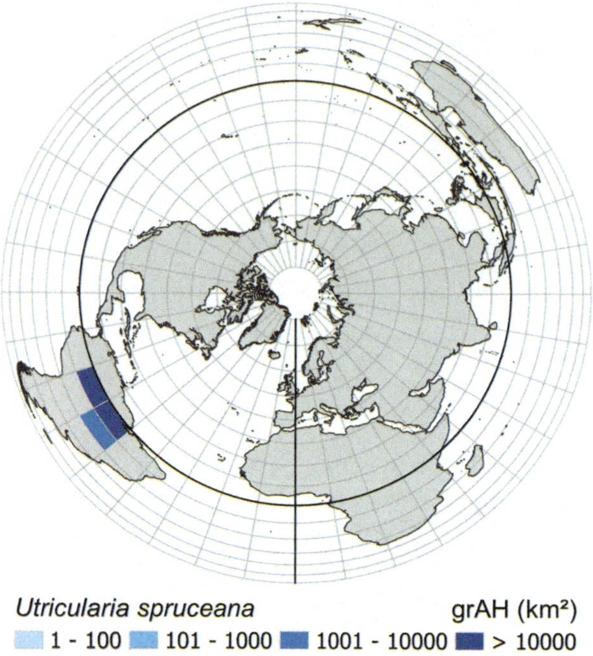

Utricularia spruceana Benth. ex Oliv.

***Utricularia stellaris* L.f.**
Synonyms: *Lepiactis stel*laris (L.f.) Raf.

Range area (grAH km^2): OR: 29100; AUS: 29600; AFR: 234700
Ecozone origin-status: OR: N; AUS: N; AFR: N
World rarity status: C
Ecozone-endemic macroregion occurrence: not endemic to any world ecozone
Life form(s): FF (hydrosubshrub)
Ploidy state: D
Speciation age (Ma): NA
Populations also occur in terrestrial, wetland and/ or riparian habitats: NO
Long distance movement vectors: NA
Nuisance status and uses: no nuisance issues; folk medicine
Endangered status (IUCN): NA

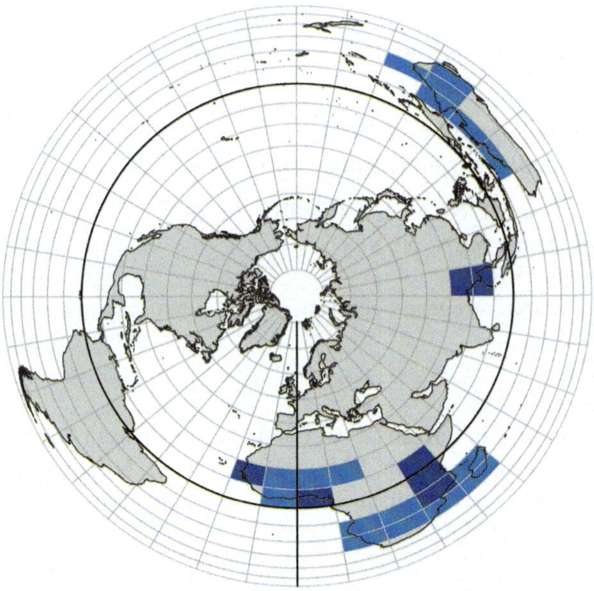

Utricularia stellaris L.f.

***Utricularia striata* Leconte ex Torr.**
Synonyms: *Trilobulina striata* (Leconte ex Torr.) Raf.

Range area (grAH km^2): NEA: 13000; NEO: 10300
Ecozone origin-status: NEA: N; NEO: N
World rarity status: VR
Ecozone-endemic macroregion occurrence: not endemic to any world ecozone
Life form(s): SU
Ploidy state: NA
Speciation age (Ma): NA
Populations also occur in terrestrial, wetland and/ or riparian habitats: YES
Long distance movement vectors: HYD (NEA to NEO; NEO to NEA)
Nuisance status and uses: no nuisance issues; no known uses
Endangered status (IUCN): LC

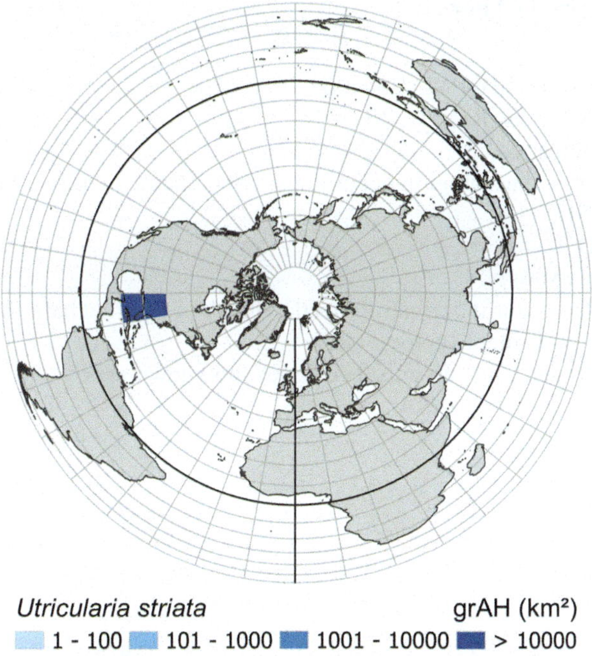

Utricularia striata Leconte ex Torr.
Image copyright: iNaturalist CC-BY @er-birds

Utricularia stygia **G.Thor**
Synonyms:

Range area (grAH km^2): PAL: 296400; NEA: 5200
Ecozone origin-status: PAL: N; NEA: N
World rarity status: C
Ecozone-endemic macroregion occurrence: not endemic to any world ecozone
Life form(s): SU
Ploidy state: D
Speciation age (Ma): NA
Populations also occur in terrestrial, wetland and/ or riparian habitats: NO
Long distance movement vectors: NA
Nuisance status and uses: no nuisance issues; no known uses
Endangered status (IUCN): NA

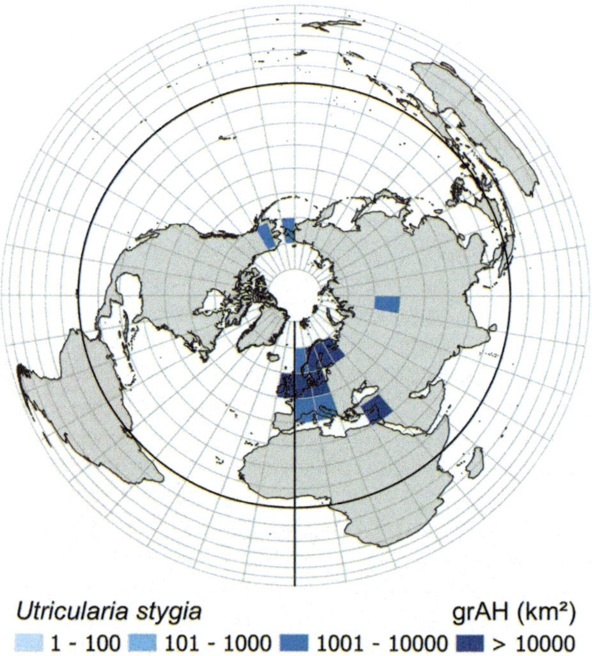

Utricularia stygia G.Thor
Image copyright: J.C. Schou, with permission

Utricularia subulata **L.**
Synonyms: *Setiscapella subulata* (L.) Barnhart

Range area (grAH km^2): NEA: 339500; NEO: 338500; OR: 26600; AUS: 16000; AFR: 180500
Ecozone origin-status: NEA: N; NEO: N; OR: N; AUS: N; AFR: N
World rarity status: C
Ecozone-endemic macroregion occurrence: not endemic to any world ecozone
Life form(s): EM
Ploidy state: D
Speciation age (Ma): NA
Populations also occur in terrestrial, wetland and/ or riparian habitats: YES
Long distance movement vectors: NA
Nuisance status and uses: no nuisance issues; carnivorous plant hobby trade
Endangered status (IUCN): LC

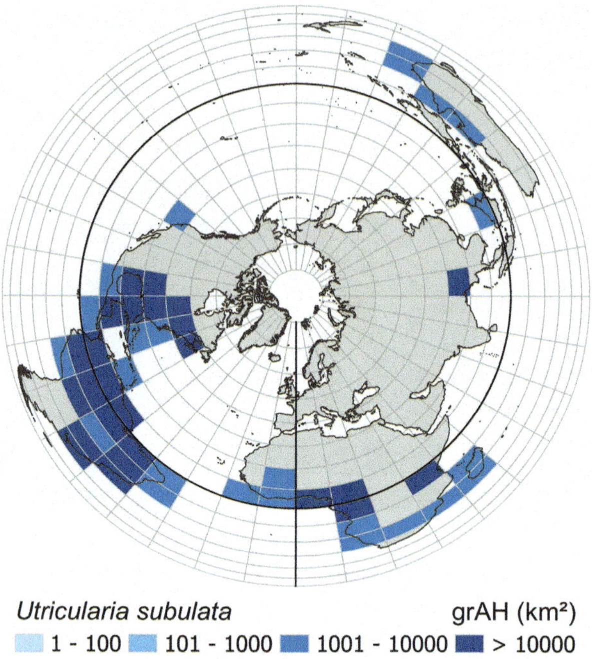

Utricularia subulata L.
Image copyright: iNaturalist CC-BY @er-birds

Utricularia tenella **R.Br.**
Synonyms: *Polypompholyx tenella* (R.Br.) Lehm., *Vesiculina tenella* (R.Br.) Raf.

Range area (grAH km^2): AUS: 22000
Ecozone origin-status: AUS: E
World rarity status: VR
Ecozone-endemic macroregion occurrence*:* AUS: WA, ESA
Life form(s): EM
Ploidy state: NA
Speciation age (Ma): NA
Populations also occur in terrestrial, wetland and/ or riparian habitats: YES
Long distance movement vectors: NA
Nuisance status and uses: no nuisance issues; no known uses
Endangered status (IUCN): LC

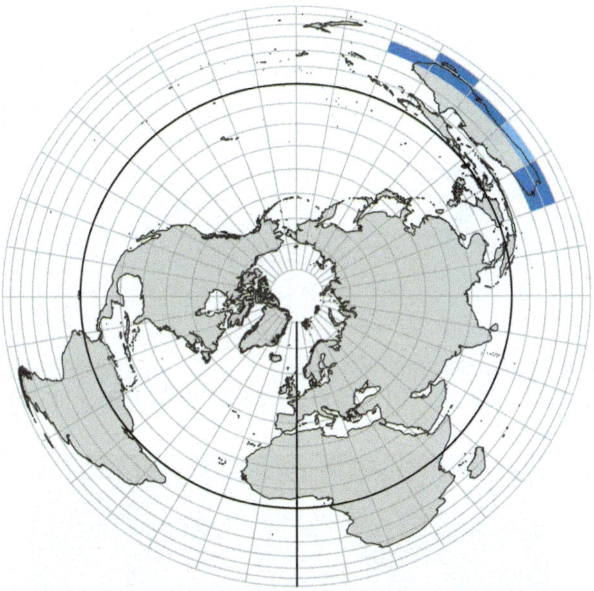

Utricularia tenuissima **Tutin**
Synonyms:

Range area (grAH km^2): NEO: 163200
Ecozone origin-status: NEO: E
World rarity status: C
Ecozone-endemic macroregion occurrence*:* NEO: NWSA, AMZ, CHA
Life form(s): EM

Ploidy state: NA
Speciation age (Ma): NA
Populations also occur in terrestrial, wetland and/ or riparian habitats: YES
Long distance movement vectors: NA
Nuisance status and uses: no nuisance issues; no known uses
Endangered status (IUCN): LC

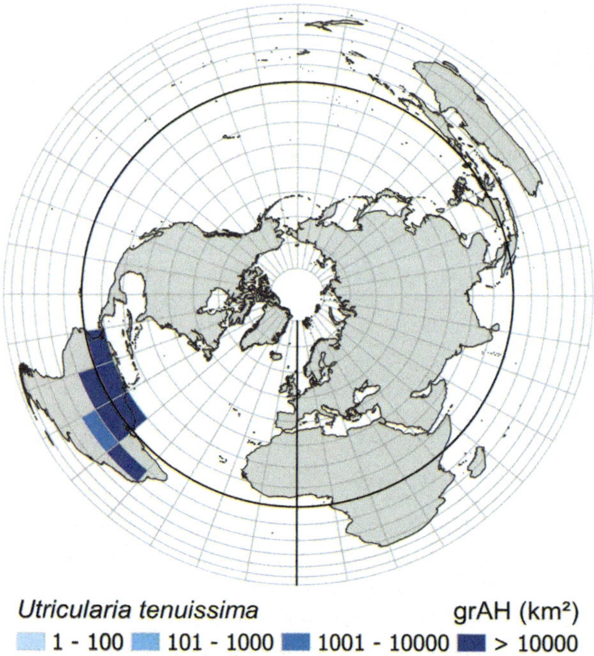

Utricularia tenuissima Tutin
Image copyright: iNaturalist CC-BY @rvgalli (Hervé Galliffet)

Utricularia terrae-reginae **P.Taylor**
Synonyms:

Range area (grAH km^2): AUS: 2900
Ecozone origin-status: AUS: E
World rarity status: VR
Ecozone-endemic macroregion occurrence: AUS: NA
Life form(s): EM
Ploidy state: NA
Speciation age (Ma): NA
Populations also occur in terrestrial, wetland and/ or riparian habitats: YES
Long distance movement vectors: NA
Nuisance status and uses: no nuisance issues; no known uses
Endangered status (IUCN): LC

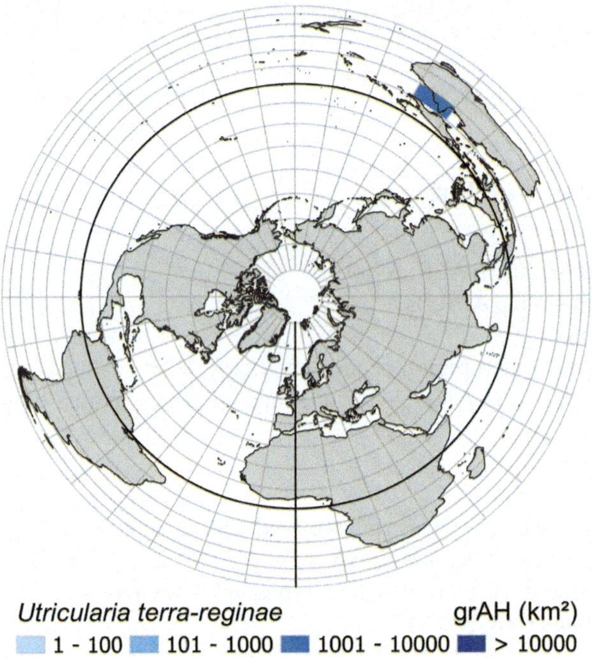

Utricularia terrae-reginae P.Taylor
Image copyright: iNaturalist CC-BY @hugo_innes (Hugo Innes)

***Utricularia tetraloba* P.Taylor**
Synonyms:

Range area (grAH km^2): AFR: 3400
Ecozone origin-status: AFR: E
World rarity status: VR
Ecozone-endemic macroregion occurrence: AFR: FOR
Life form(s): SU, EM (rheophyte, attached to rocks in fast-flowing rivers and streams)
Ploidy state: NA
Speciation age (Ma): NA
Populations also occur in terrestrial, wetland and/ or riparian habitats: NO
Long distance movement vectors: NA
Nuisance status and uses: no nuisance issues; no known uses
Endangered status (IUCN): VU

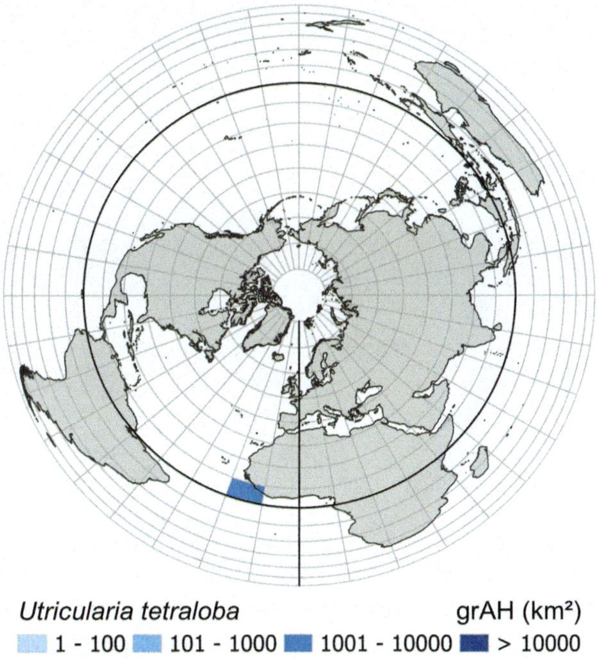

***Utricularia trichophylla* Spruce ex Oliv.**
Synonyms:

Range area (grAH km^2): NEO: 239400
Ecozone origin-status: NEO: E
World rarity status: C
Ecozone-endemic macroregion occurrence: NEO: MES AM, NWSA, AMZ, CHA, PAR
Life form(s): SU, EM
Ploidy state: NA
Speciation age (Ma): NA
Populations also occur in terrestrial, wetland and/ or riparian habitats: YES
Long distance movement vectors: NA
Nuisance status and uses: no nuisance issues; ornamental
Endangered status (IUCN): LC

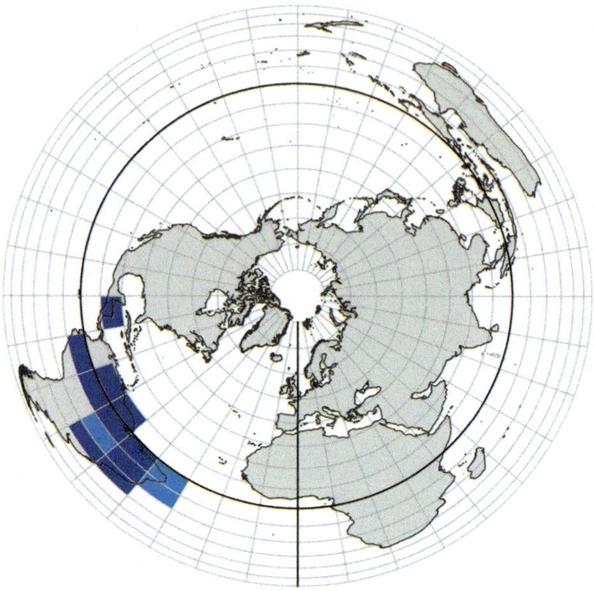

Utricularia triflora **P.Taylor**
Synonyms:

Range area (grAH km^2): AUS: 6300
Ecozone origin-status: AUS: E
World rarity status: VR
Ecozone-endemic macroregion occurrence*:* AUS: NA
Life form(s): EM
Ploidy state: NA
Speciation age (Ma): NA
Populations also occur in terrestrial, wetland and/ or riparian habitats: YES
Long distance movement vectors: NA
Nuisance status and uses: no nuisance issues; no known uses
Endangered status (IUCN): LC

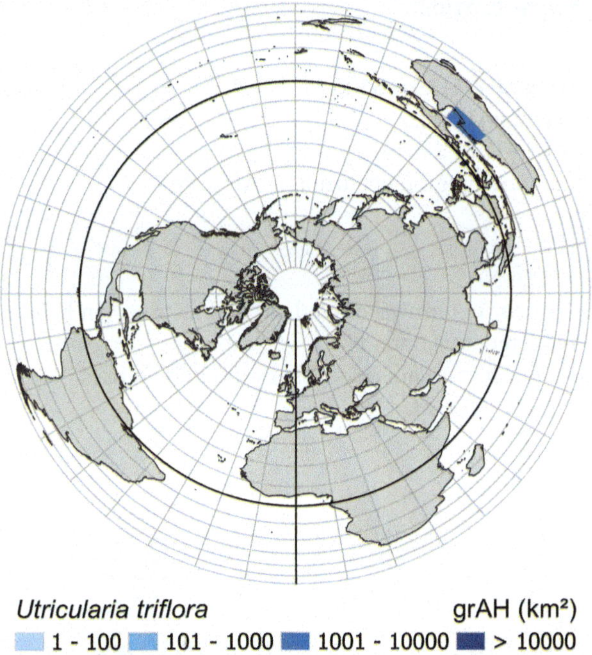

Utricularia triflora P.Taylor
Image copyright: Z. Madycki, with permission

Utricularia tubulata **F. Muell.**
Synonyms:

Range area (grAH km^2): AUS: 16500
Ecozone origin-status:AUS: E
World rarity status: VR
Ecozone-endemic macroregion occurrence*:* AUS: NA
Life form(s): FF
Ploidy state: NA
Speciation age (Ma): NA
Populations also occur in terrestrial, wetland and/ or riparian habitats: NO
Long distance movement vectors: NA
Nuisance status and uses: no nuisance issues; no known uses
Endangered status (IUCN): LC

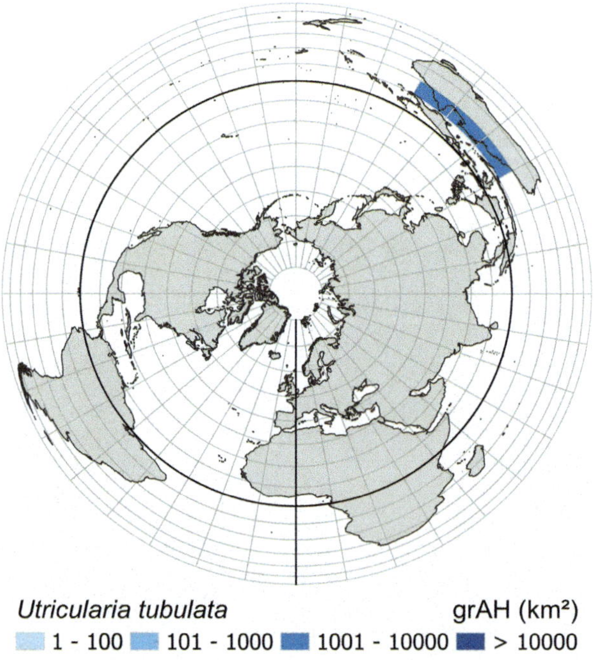

Utricularia tubulata F. Muell.
Image copyright: iNaturalist CC-BY @hthilokrueger (Thilo Krueger)

Utricularia uliginosa **Vahl**
Synonyms:

Range area (grAH km^2): PAL: 14800; OR: 89500; AUS: 63400; present in OCE
Ecozone origin-status: PAL: N; OR: N; AUS: N; OCE: N
World rarity status: C
Ecozone-endemic macroregion occurrence*:* not endemic to any world ecozone
Life form(s): SU, EM
Ploidy state: NA
Speciation age (Ma): NA
Populations also occur in terrestrial, wetland and/ or riparian habitats: YES
Long distance movement vectors: NA
Nuisance status and uses: no nuisance issues; no known uses
Endangered status (IUCN): LC

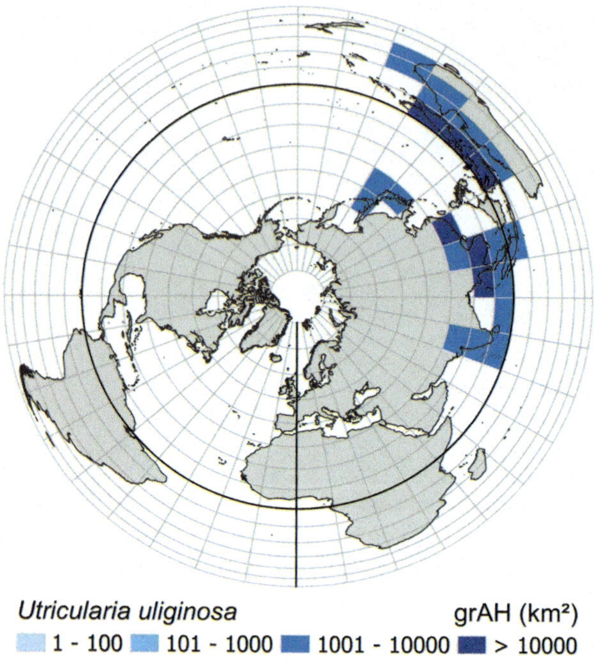

Utricularia viscosa **Spruce ex Oliv.**
Synonyms:

Range area (grAH km^2): NEO: 175000
Ecozone origin-status: NEO: E
World rarity status: C
Ecozone-endemic macroregion occurrence: NEO: MES AM, NWSA, AMZ, CHA
Life form(s): SU, EM
Ploidy state: NA
Speciation age (Ma): NA
Populations also occur in terrestrial, wetland and/ or riparian habitats: YES
Long distance movement vectors: NA
Nuisance status and uses: no nuisance issues; no known uses
Endangered status (IUCN): LC

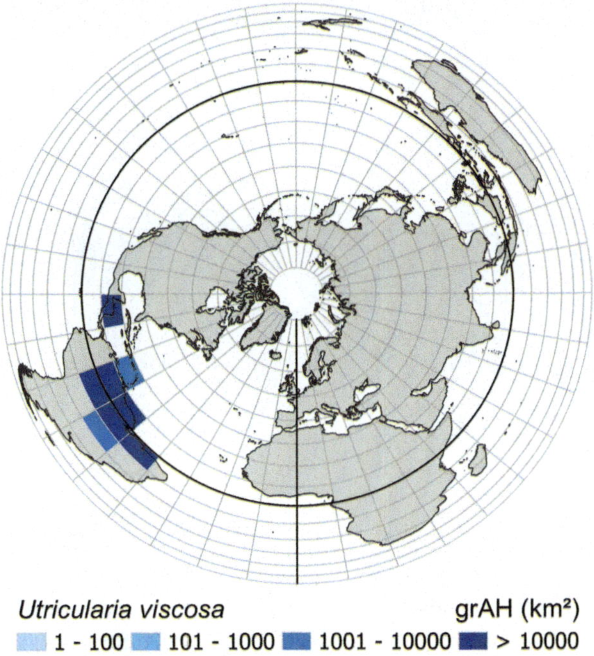

***Utricularia vitellina* Ridl.**
Synonyms: *Utricularia aurea* Ridl.

Range area (grAH km^2): OR: 6700
Ecozone origin-status: OR: E
World rarity status: VR
Ecozone-endemic macroregion occurrence: OR: SUP
Life form(s): EM
Ploidy state: NA
Speciation age (Ma): NA
Populations also occur in terrestrial, wetland and/ or riparian habitats: YES
Long distance movement vectors: NA
Nuisance status and uses: no nuisance issues; no known uses
Endangered status (IUCN): LC

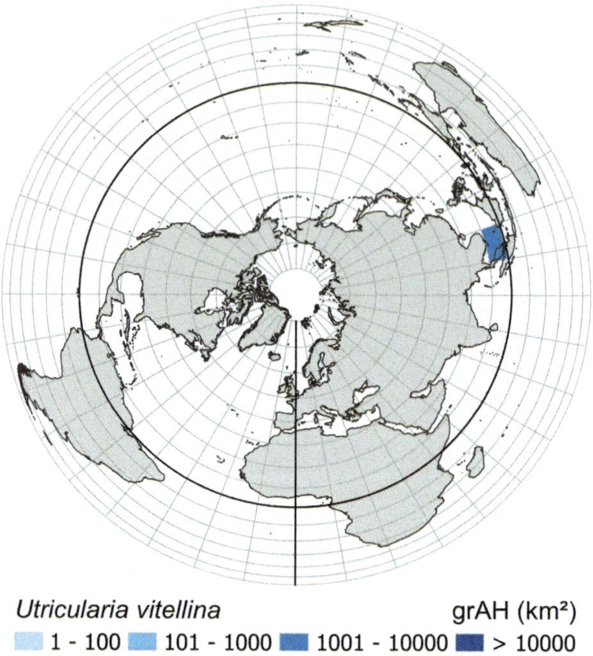

***Utricularia volubilis* R.Br.**
Synonyms:

Range area (grAH km^2): AUS: 2200
Ecozone origin-status: AUS: E
World rarity status: VR
Ecozone-endemic macroregion occurrence: AUS: WA
Life form(s): SU, EM
Ploidy state: NA
Speciation age (Ma): NA
Populations also occur in terrestrial, wetland and/ or riparian habitats: YES
Long distance movement vectors: NA
Nuisance status and uses: no nuisance issues;carnivorous plant hobby trade
Endangered status (IUCN): LC

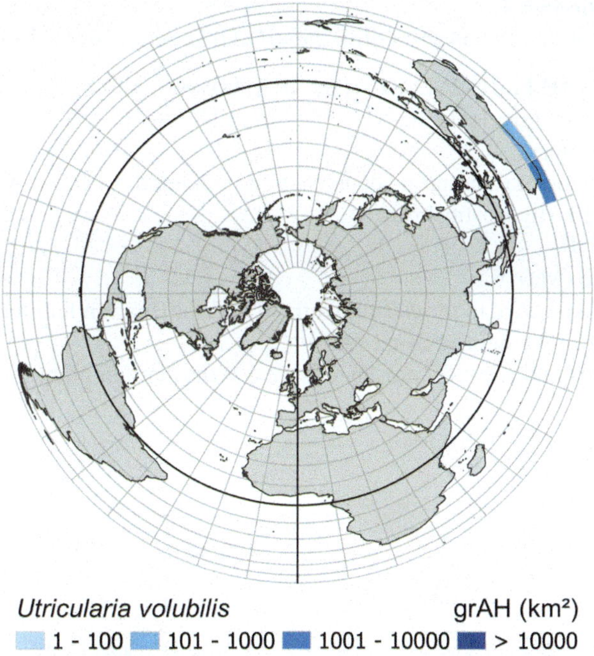

Utricularia volubilis R.Br.

***Utricularia vulgaris* L.**
Synonyms: *Lentibularia major* Gilib., *Lentibularia vulgaris* (L.) Moench

Range area (grAH km^2): PAL: 846100; NEA: 904900; NEO: 36100; OR: 18200
Ecozone origin-status: PAL: N; NEA: N; NEO: N; OR: N
World rarity status: C
Ecozone-endemic macroregion occurrence: not endemic to any world ecozone
Life form(s): FF (hydrosubshrub)
Ploidy state: DP
Speciation age (Ma): 6.5
Populations also occur in terrestrial, wetland and/ or riparian habitats: NO
Long distance movement vectors: ?AV
Nuisance status and uses: no nuisance issues; water garden plant, medicinal potential uses
Endangered status (IUCN): LC

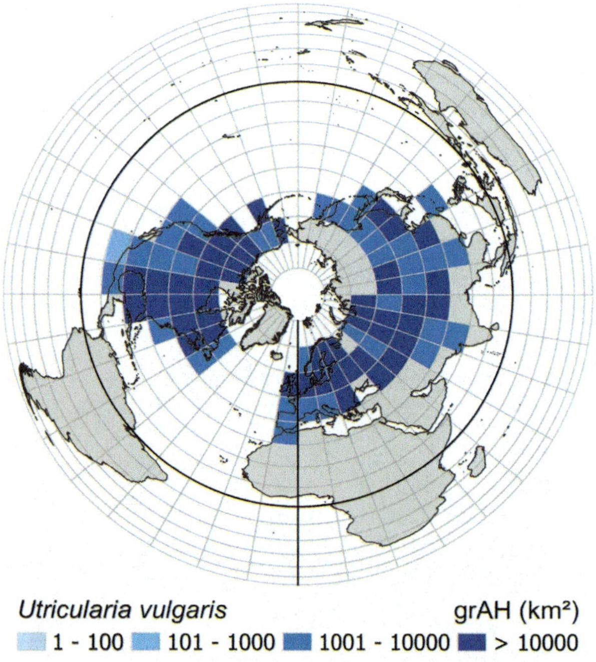

Utricularia vulgaris L.
Image copyright: J.C. Schou, with permission

Utricularia warmingii **Kamienski**
Synonyms:

Range area (grAH km^2): NEO: 35600
Ecozone origin-status: NEO: E
World rarity status: R
Ecozone-endemic macroregion occurrence: NEO: NWSA, AMZ, PAR
Life form(s): FF
Ploidy state: NA
Speciation age (Ma): NA
Populations also occur in terrestrial, wetland and/ or riparian habitats: NO
Long distance movement vectors: NA
Nuisance status and uses: no nuisance issues; no known uses
Endangered status (IUCN): LC

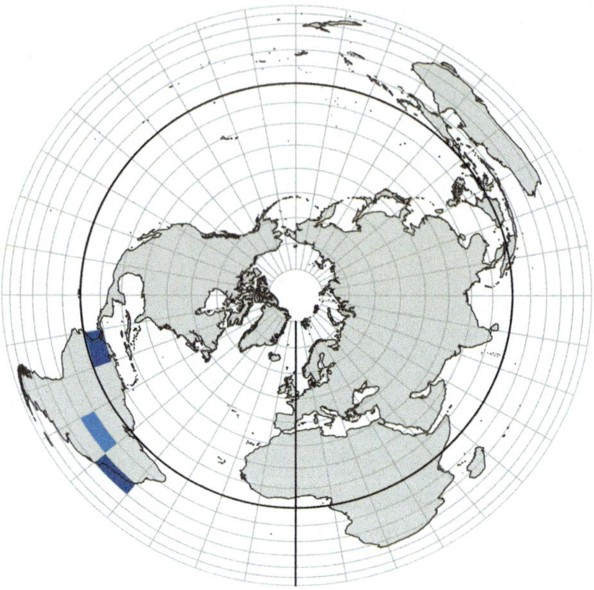

3.28 LIMNANTHACEAE (Brassicales)

The Limnanthaceae contain four macrophyte species, all in the genus *Limnanthes*, and endemic to the Western North America macroregion of NEA. *Limnanthes douglasii* has also been introduced to PAL and AUS. None causes any nuisance issues. All four species are world-rare, with three VR and one R 25, none assessed for endangered status.

***Limnanthes douglasii* R.Br.**
Synonyms:

Range area (grAH km^2): PAL: 119800; NEA: 18800; AUS: 12800
Ecozone origin-status: PAL: I; NEA: E(I); AUS: I
World rarity status: R
Ecozone-endemic macroregion occurrence: NEA: W NAM
Life form(s): EM
Ploidy state: D
Speciation age (Ma): NA
Populations also occur in terrestrial, wetland and/ or riparian habitats: YES
Long distance movement vectors: ANTHROPOCHORY
Nuisance status and uses: no nuisance issues; garden plant
Endangered status (IUCN): NA

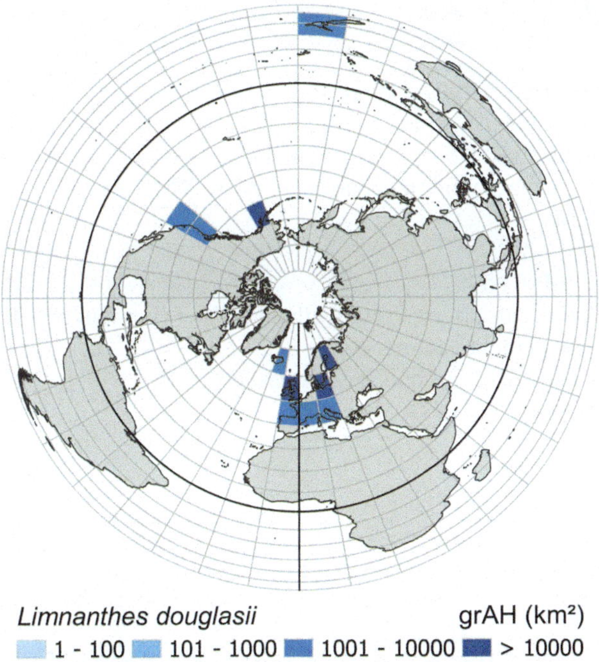

Limnanthes douglasii R.Br.
Image copyright: iNaturalist CC-BY @icosahedron

Limnanthes floccosa **Howell**
Synonyms:

Range area (grAH km²): NEA: 7500
Ecozone origin-status: NEA: E
World rarity status: VR
Ecozone-endemic macroregion occurrence: NEA: W NAM
Life form(s): EM
Ploidy state: D
Speciation age (Ma): NA
Populations also occur in terrestrial, wetland and/ or riparian habitats: YES
Long distance movement vectors: NA
Nuisance status and uses: no nuisance issues; cosmetic
Endangered status (IUCN): NA

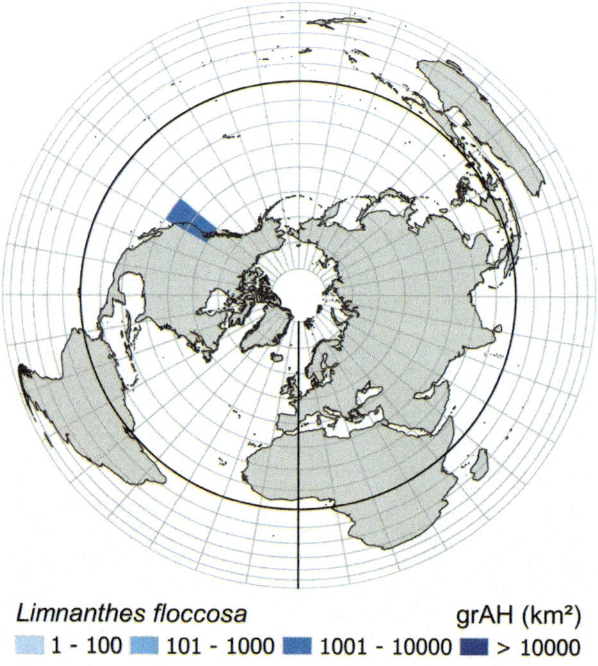

Limnanthes floccosa Howell
Image copyright: iNaturalist CC-BY @sheriff_woody_pct (Matt Berger)

***Limnanthes macounii* Trel.**
Synonyms:

Range area (grAH km^2): NEA: 7500
Ecozone origin-status: NEA: E
World rarity status: VR
Ecozone-endemic macroregion occurrence: NEA: W NAM
Life form(s): EM
Ploidy state: D
Speciation age (Ma): NA
Populations also occur in terrestrial, wetland and/ or riparian habitats: YES
Long distance movement vectors: NA
Nuisance status and uses: no nuisance issues; no known uses
Endangered status (IUCN): NA

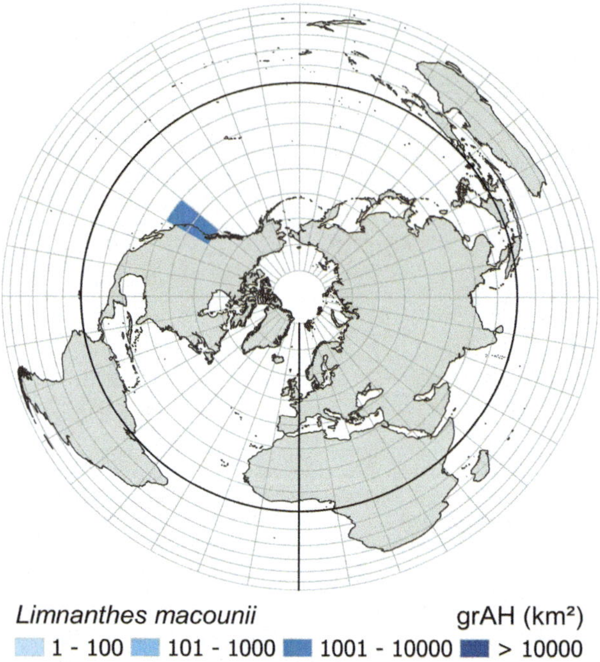

Limnanthes macounii Trel.

***Limnanthes vinculans* Ornduff**
Synonyms:

Range area (grAH km^2): NEA: 1000
Ecozone origin-status: NEA: E
World rarity status: VR
Ecozone-endemic macroregion occurrence: NEA: W NAM
Life form(s): EM
Ploidy state: D
Speciation age (Ma): NA
Populations also occur in terrestrial, wetland and/ or riparian habitats: YES
Long distance movement vectors: NA
Nuisance status and uses: no nuisance issues; no known uses
Endangered status (IUCN): NA

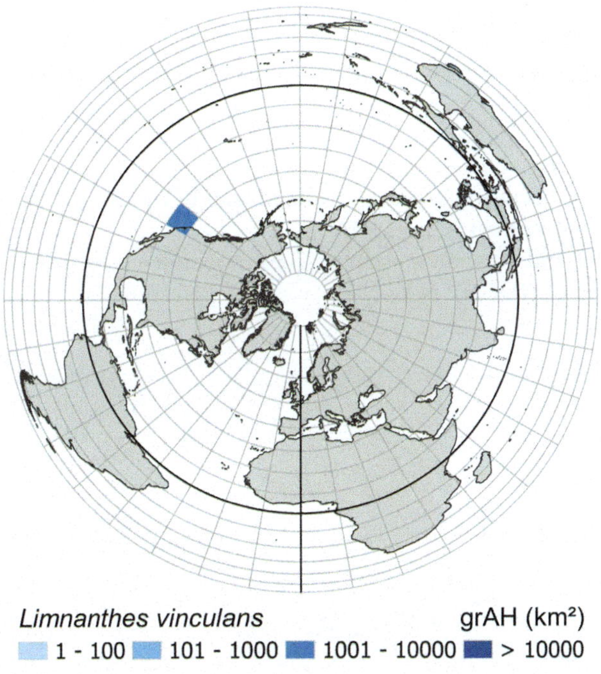

3.29 LINDERNIACEAE (Lamiales)

The Linderniaceae contain 10 macrophyte species in two genera, mainly found at low latitudes, and present in all ecozones except ANT. There are seven species in NEO (two endemic and two introduced), six in AFR (two endemic, one introduced) and six in OR (one endemic, one introduced), four in AUS (three being introduced), three in OCE (two introduced) and three in PAL (one introduced), and one species native in NEA. Only one species, *Lindernia dubia* causes occasional nuisance, as a weed of rice in Europe and Asia. They are mainly common at world scale, with one species VR, two R and seven C. Three species have not been assessed for endangered status, and seven are LC.

Quick Links to Genera

1. *Lindernia*
2. *Torenia*

Lindernia brachyphylla Pennell

Synonyms:

Range area (grAH km^2): NEO: 121700
Ecozone origin-status: NEO: E
World rarity status: R
Ecozone-endemic macroregion occurrence: NEO: NWSA, AMZ, CHA
Life form(s): EM
Ploidy state: NA
Speciation age (Ma): NA
Populations also occur in terrestrial, wetland and/ or riparian habitats: YES
Long distance movement vectors: NA
Nuisance status and uses: no nuisance issues; no known uses
Endangered status (IUCN): NA

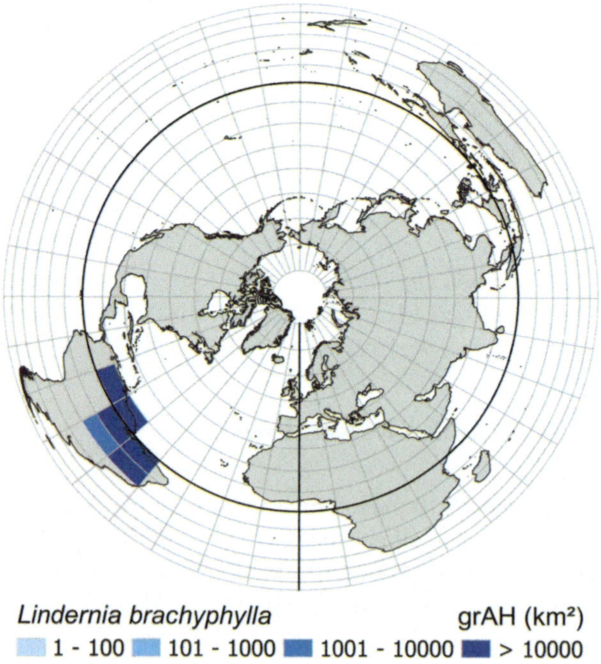

***Lindernia crustacea* (L.) F.Muell.**

Synonyms: *Torenia crustacea* (L.) Cham. & Schltdl.

Range area (grAH km²): PAL: 107500; NEO: 297200; OR: 128000; AUS: 57400; AFR: 164000; present in OCE
Ecozone origin-status: PAL: N ; NEO: N; OR: N; AUS: N; AFR: N; OCE: N
World rarity status: C
Ecozone-endemic macroregion occurrence: not endemic to any world ecozone
Life form(s): EM
Ploidy state: D
Speciation age (Ma): NA
Populations also occur in terrestrial, wetland and/ or riparian habitats: YES
Long distance movement vectors: NA
Nuisance status and uses: no nuisance issues; used as a medicine and for food
Endangered status (IUCN): LC

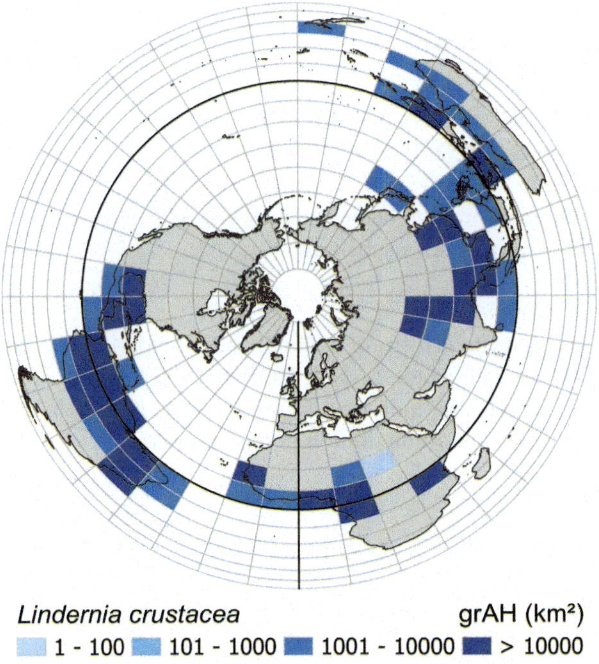

Lindernia crustacea (L.) F.Muell.
Image copyright: A.T. Dung-Le, with permission

***Lindernia diffusa* (L.) Wettst.**
Synonyms: *Vandellia diffusa* L.

Range area (grAH km^2): NEO: 291500; AFR: 249500
Ecozone origin-status: NEO: I; AFR: E(I)
World rarity status: C
Ecozone-endemic macroregion occurrence: SAH/ AR, FOR, EA
Life form(s): EM (sometimes a subshrub)
Ploidy state: NA
Speciation age (Ma): NA
Populations also occur in terrestrial, wetland and/ or riparian habitats: YES
Long distance movement vectors: ANTHROPOCHORY
Nuisance status and uses: no nuisance issues; harvested from the wild for local use as a medicine, also phytoremediation potential uses
Endangered status (IUCN): LC

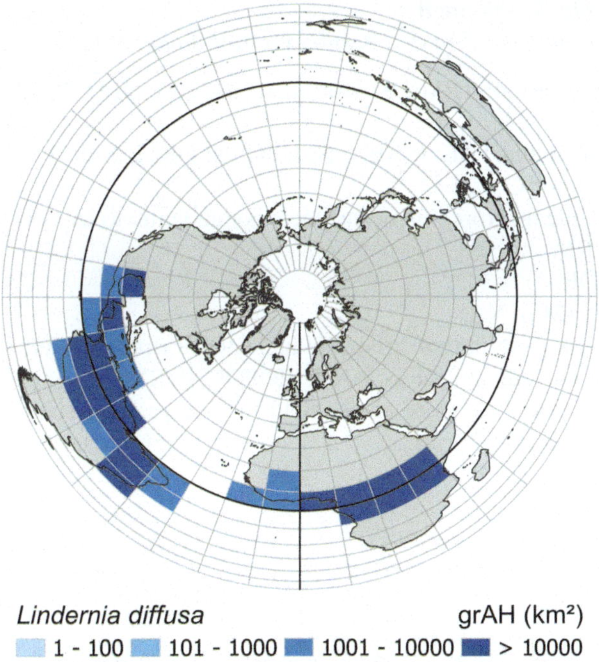

Lindernia diffusa (L.) Wettst.

Lindernia dubia **(L.) Pennell**
Synonyms: *Ilysanthes dubia* (L.) Barnhart; *Gratiola dubia* L.

Range area (grAH km^2): PAL: 92300; NEA: 392100; NEO: 216200; OR: 98600; AFR: 6000
Ecozone origin-status: PAL: I; NEA: N; NEO: N; OR: I; AFR: I
World rarity status: C
Ecozone-endemic macroregion occurrence: not endemic to any world ecozone
Life form(s): EM
Ploidy state: DP
Speciation age (Ma): NA
Populations also occur in terrestrial, wetland and/ or riparian habitats: YES
Long distance movement vectors: ANTHROPOCHORY
Nuisance status and uses: weed of rice fields in Europe and Asia; no known uses
Endangered status (IUCN): LC

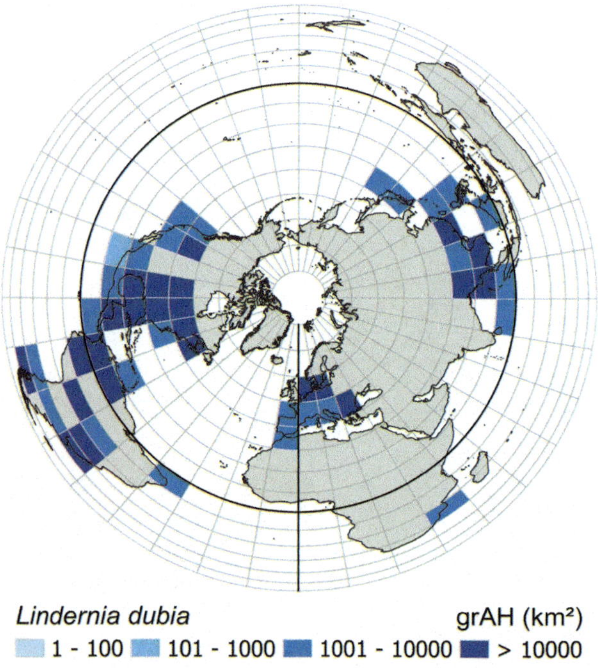

***Lindernia hyssopoides* (L.) Haines**
Synonyms:

Range area (grAH km²): OR: 43100; AUS: 8300
Ecozone origin-status: OR: E(I); AUS: I
World rarity status: R
Ecozone-endemic macroregion occurrence: OR: ICH, IND
Life form(s): EM
Ploidy state: D
Speciation age (Ma): NA
Populations also occur in terrestrial, wetland and/ or riparian habitats: YES
Long distance movement vectors: ANTHROPOCHORY
Nuisance status and uses: no nuisance issues; harvested from the wild for local use as a medicine
Endangered status (IUCN): LC

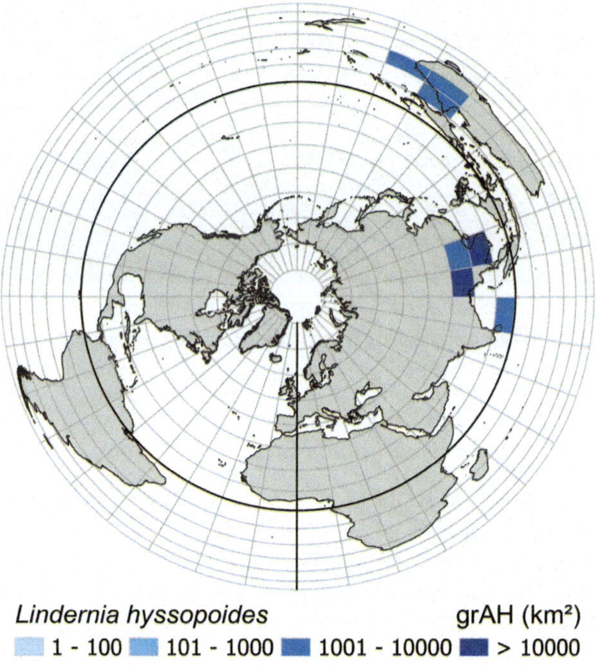

Lindernia hyssopoides (L.) Haines

***Lindernia linearifolia* (Engl.) Eb.Fisch.**
Synonyms: *Craterostigma linearifolium* Engl.

Range area (grAH km^2): AFR: 2900
Ecozone origin-status: AFR: E
World rarity status: VR
Ecozone-endemic macroregion occurrence: AFR: ZAM
Life form(s): EM
Ploidy state: NA
Speciation age (Ma): NA
Populations also occur in terrestrial, wetland and/ or riparian habitats: YES
Long distance movement vectors: NA
Nuisance status and uses: no nuisance issues; no known uses
Endangered status (IUCN): NA

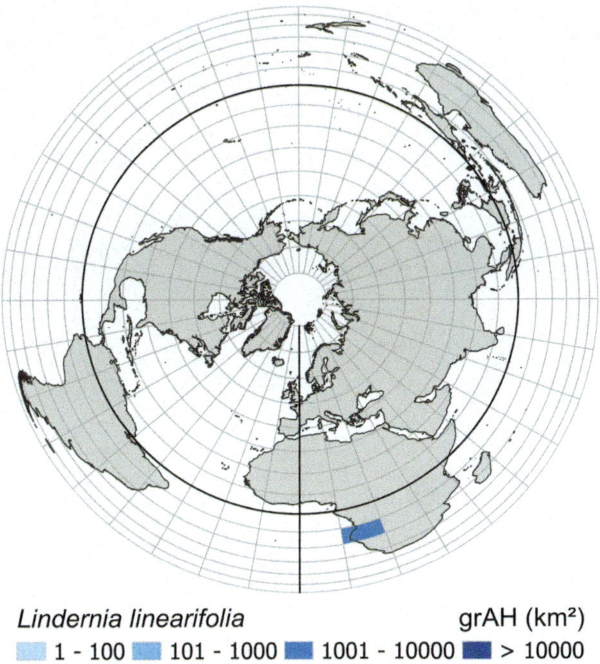

Lindernia microcalyx **Pennell & Stehlé**
Synonyms:

Range area (grAH km^2): NEO: 183200
Ecozone origin-status: NEO: E
World rarity status: C
Ecozone-endemic macroregion occurrence: NEO: NWSA, AMZ, CHA, PAR
Life form(s): EM
Ploidy state: NA
Speciation age (Ma): NA
Populations also occur in terrestrial, wetland and/ or riparian habitats: YES
Long distance movement vectors: NA
Nuisance status and uses: no nuisance issues; no known uses
Endangered status (IUCN): NA

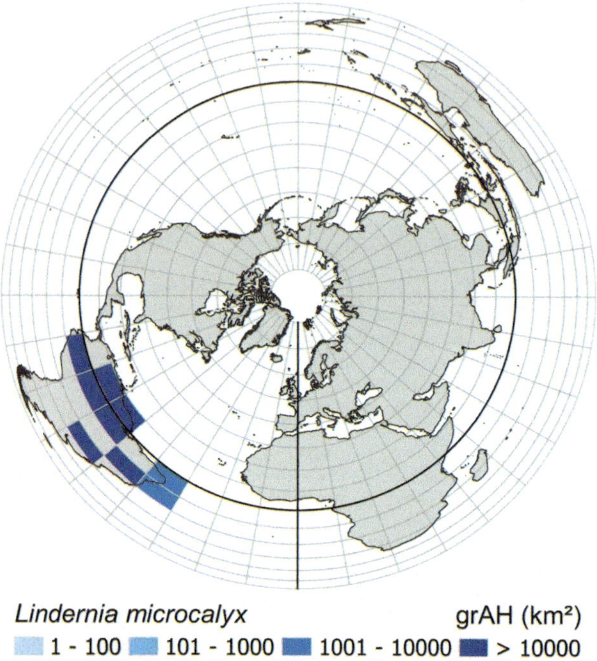

Lindernia microcalyx Pennell & Stehlé

***Lindernia procumbens* (Krock.) Borbás**

Synonyms: *Anagalloides procumbens* Krock.; *Pyxidaria procumbens* (Krock.) Borbás

Range area (grAH km^2): PAL: 224600; OR: 81200; AUS: 3600; present in OCE
Ecozone origin-status: PAL: N; OR: N; AUS: I; OCE: N
World rarity status: C
Ecozone-endemic macroregion occurrence: not endemic to any world ecozone
Life form(s): EM (sometimes a subshrub)
Ploidy state: D
Speciation age (Ma): NA
Populations also occur in terrestrial, wetland and/ or riparian habitats: YES
Long distance movement vectors: ANTHROPOCHORY
Nuisance status and uses: no nuisance issues; harvested from the wild for local use as a medicine
Endangered status (IUCN): LC

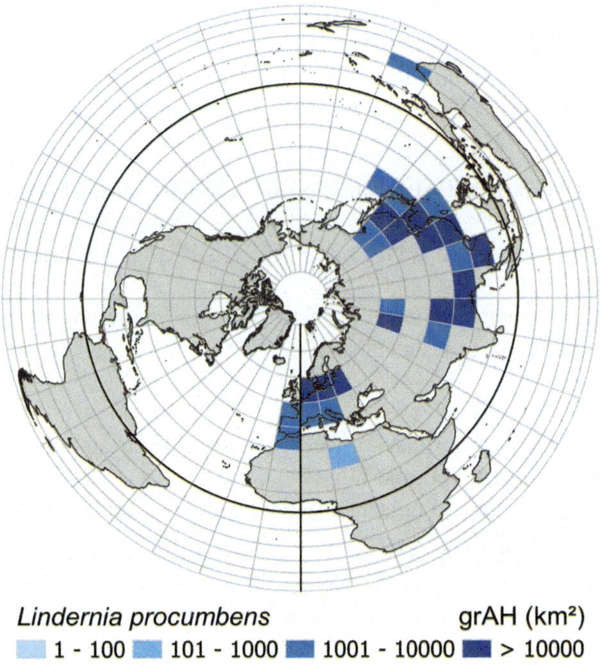

Lindernia procumbens (Krock.) Borbás
Image copyright: iNaturalist CC-BY @chiuluan

Lindernia rotundifolia (L.) Alston
Synonyms: *Bonnaya rotundifolia* (L.) Benth.; *Ilysanthes rotundifolia* (L.) Benth.; *Torenia rotundifolia* (L.) Cordem.; *Gratiola rotundifolia* L.; *Morgania rotundifolia* (L.) Spreng.

Range area (grAH km^2): PAL: 1300; NEO: 230900; OR: 91700; AFR: 158500; AUS: 3600; present in OCE
Ecozone origin-status: PAL: I; NEO: N; OR: N; AFR: N; AUS: I; OCE: I
World rarity status: C
Ecozone-endemic macroregion occurrence: not endemic to any world ecozone
Life form(s): EM (sometimes a subshrub)
Ploidy state: NA
Speciation age (Ma): NA
Populations also occur in terrestrial, wetland and/ or riparian habitats: YES
Long distance movement vectors: NA
Nuisance status and uses: no nuisance issues; aquarium trade, medicinal uses
Endangered status (IUCN): LC

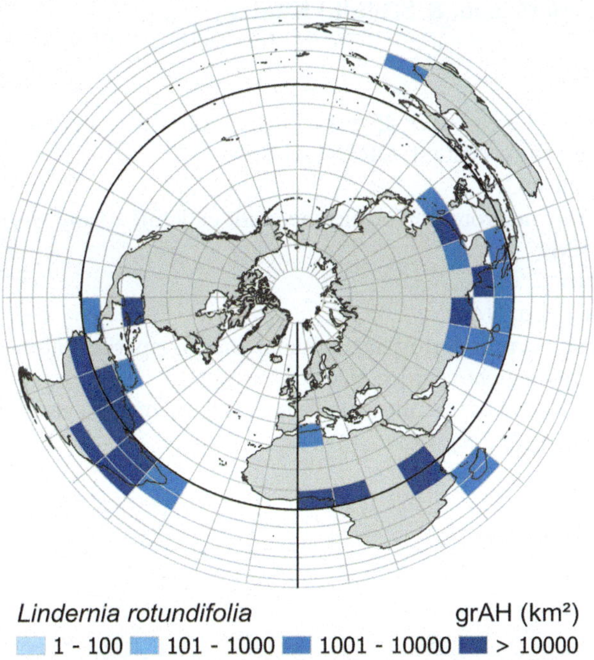

Lindernia rotundifolia (L.) Alston
Image copyright: A. Efremov, with permission

***Torenia thouarsii* (Cham. & Schltdl.) Kuntze**

Synonyms: *Lindernia thouarsii* (Cham. & Schltdl.) Edwin; *Nortenia thouarsii* Cham. & Schltdl.

Range area (grAH km^2): NEO: 257600; OR: 44100; AFR: 265100
Ecozone origin-status: NEO: I; OR: N; AFR: N
World rarity status: C
Ecozone-endemic macroregion occurrence: not endemic to any world ecozone
Life form(s): EM
Ploidy state: NA
Speciation age (Ma): NA
Populations also occur in terrestrial, wetland and/ or riparian habitats: YES
Long distance movement vectors: NA
Nuisance status and uses: no nuisance issues; folk medicine uses
Endangered status (IUCN): LC

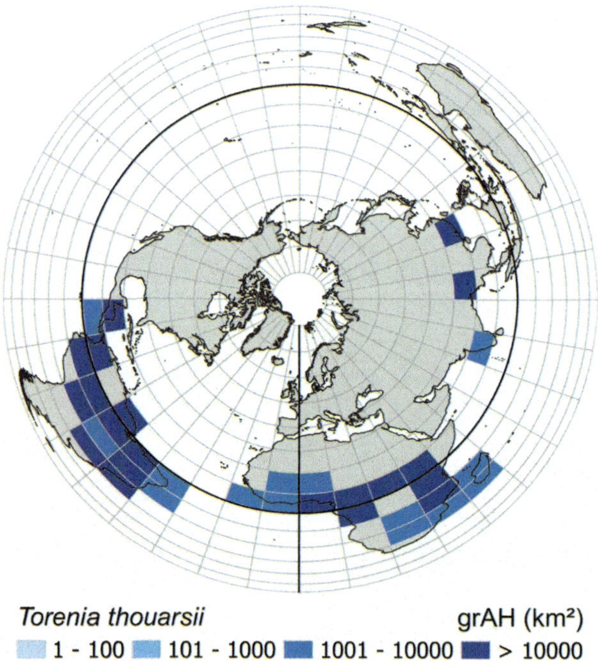

3.30 LYTHRACEAE (Myrtales)

The Lythraceae contain 103 freshwater macrophyte species in eight genera, which are mainly world-rare, with values of VR 29%, R 44% and C 27%. Species of this family occur in all world ecozones except ANT, with particularly high richness in sub-tropical to tropical regions. There are 40 species in AFR (29 of them endemic to the ecozone), 38 in NEO (25 endemic), 36 in OR (17 endemic), 28 in PAL (10 endemic),19 species in NEA (one endemic), 16 in AUS and four in OCE (no endemics). Anthropochory varies in importance between world ecozones in the global distribution of these species, with I-origin species particularly prominent in NEA (10 species) and AUS (six species), and five or fewer introduced species in the other ecozones. All the species of OCE are introduced. Several species of this family cause nuisance issues, especially *Lythrum salicaria, Trapa natans* and *Rotala rotundifolia*, which are major invasive aquatic weeds in NEA. Others mainly cause weed problems in rice, in Asia (*Rotala densa* and *Rotala indica*), or in both Asia and Europe (*Rotala ramosior*). Of the 103 species 48% are NA/DD for endangered status, 39% are LC, and 13% are classed in the higher at-risk categories, with 3% being CR.

Quick Links to Genera

1. *Ammannia*
2. *Cuphea*
3. *Didiplis*
4. *Lythrum*
5. *Rotala*
6. *Trapa*

Ammannia alternifolia **H.Perrier**
Synonyms:

Range area (grAH km^2): AFR: 8000
Ecozone origin-status: AFR: E
World rarity status: VR
Ecozone-endemic macroregion occurrence: AFR: MAD
Life form(s): EM
Ploidy state: NA
Speciation age (Ma): NA
Populations also occur in terrestrial, wetland and/ or riparian habitats: YES
Long distance movement vectors: NA
Nuisance status and uses: no nuisance issues; no known uses
Endangered status (IUCN): CR

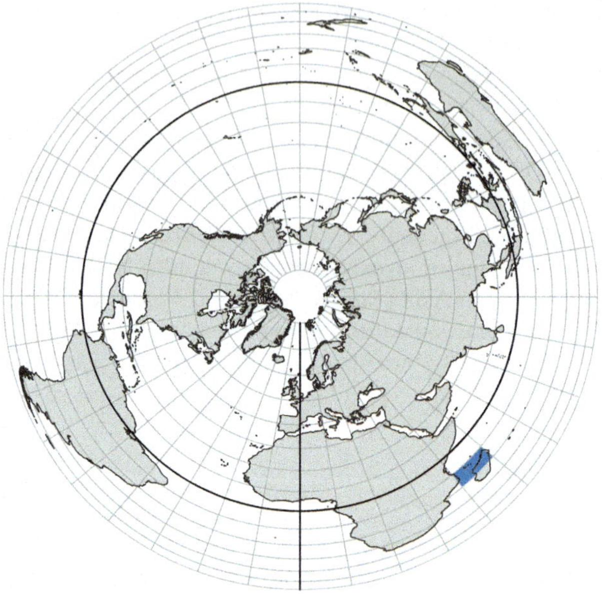

Ammannia auriculata **Willd.**

Synonyms: *Ammannia senegalensis* var. *auriculata* (Willd.) Hiern

Range area (grAH km^2): PAL: 80400; NEA: 14300; NEO: 201600; OR: 43300; AUS: 20300; AFR: 236800; present in OCE

Ecozone origin-status: PAL: N; NEA: N; NEO: N; OR: N; AUS: N; AFR: N; OCE: I

World rarity status: C

Ecozone-endemic macroregion occurrence: not endemic to any world ecozone

Life form(s): EM

Ploidy state: DP

Speciation age (Ma): NA

Populations also occur in terrestrial, wetland and/ or riparian habitats: YES

Long distance movement vectors: ANTHROPOCHORY

Nuisance status and uses: no nuisance issues; folk medicine

Endangered status (IUCN): LC

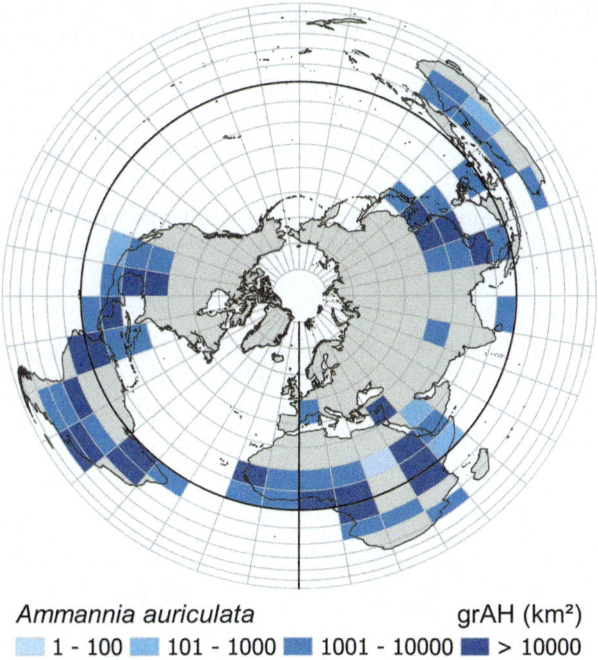

Ammannia auriculata Willd.

Ammannia baccifera **L.**
Synonyms:

Range area (grAH km^2): PAL: 234700; NEO: 8100; OR: 142600 ; AUS: 58900; AFR: 301100; present in OCE
Ecozone origin-status: PAL: N; NEO: I; OR: N; AUS: N; AFR: N; OCE: I
World rarity status: C
Ecozone-endemic macroregion occurrence: not endemic to any world ecozone
Life form(s): EM (sometimes a shrub)
Ploidy state: D
Speciation age (Ma): NA
Populations also occur in terrestrial, wetland and/ or riparian habitats: YES
Long distance movement vectors: ANTHROPOCHORY
Nuisance status and uses: no nuisance issues; medicinal uses
Endangered status (IUCN): LC

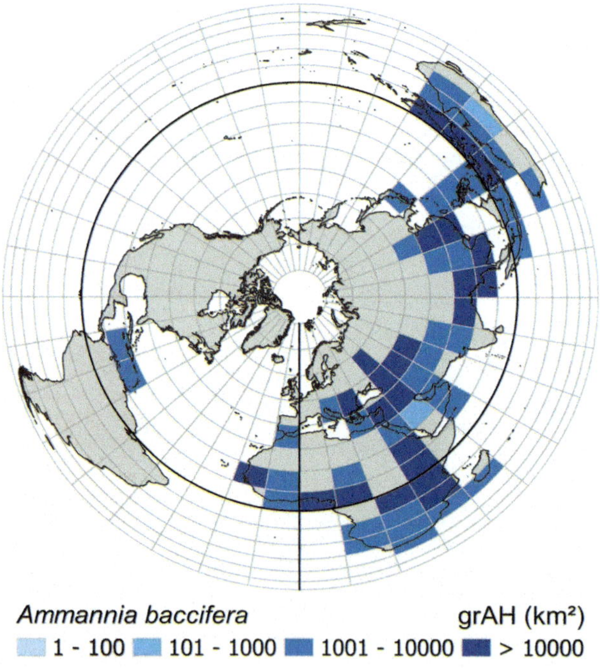

Ammannia baccifera L.
Image copyright: A.T. Dung-Le, with permission

***Ammannia coccinea* Rottb.**
Synonyms:

Range area (grAH km^2): PAL: 47900; NEA: 311700; NEO: 159800; OR: 21500; AUS: 5600; AFR: 10500; present in OCE
Ecozone origin-status: PAL: N; NEA: N; NEO: N; OR: N; AUS: N; AFR: N; OCE: I
World rarity status: C
Ecozone-endemic macroregion occurrence: not endemic to any world ecozone
Life form(s): EM
Ploidy state: P
Speciation age (Ma): NA
Populations also occur in terrestrial, wetland and/ or riparian habitats: YES
Long distance movement vectors: ANTHROPOCHORY
Nuisance status and uses: no nuisance issues; medicinal uses
Endangered status (IUCN): NA

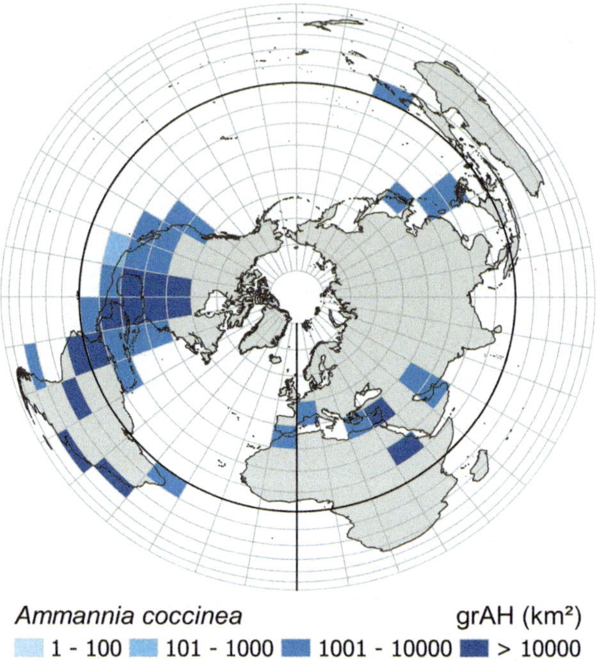

Ammannia coccinea Rottb.
Image copyright: iNaturalist CC-BY @botanygirl

***Ammannia crassicaulis* Guill. & Perr.**
Synonyms: *Nesaea crassicaulis* (Guill. & Perr.) Koehne

Range area (grAH km^2): AFR: 41700
Ecozone origin-status: AFR: E
World rarity status: R
Ecozone-endemic macroregion occurrence: AFR: SAH/ AR, FOR, ZAM, SDES
Life form(s): EM (hydrosubshrub)
Ploidy state: NA
Speciation age (Ma): NA
Populations also occur in terrestrial, wetland and/ or riparian habitats: YES
Long distance movement vectors: NA
Nuisance status and uses: no nuisance issues; aquarium trade
Endangered status (IUCN): LC

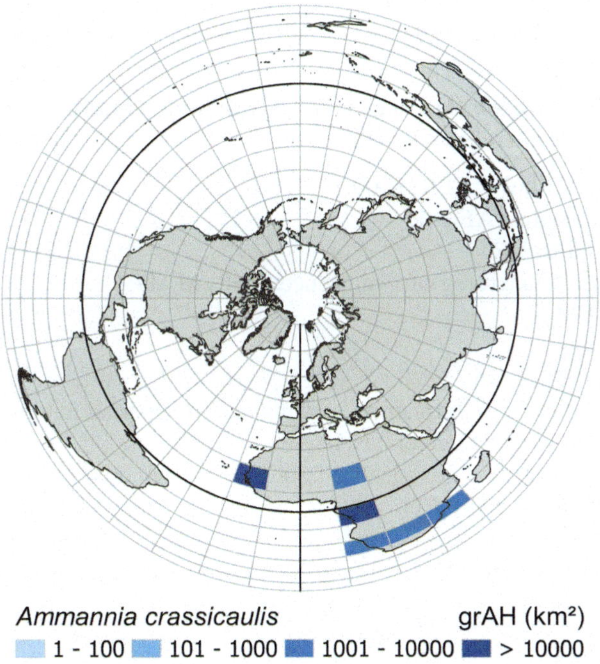

Ammannia crassicaulis Guill. & Perr.
Image copyright: iNaturalist CC0 @aztekium (Carlos G Velazco-Macias)

Ammannia heterophylla (H.Perrier) S.A.Graham & Gandhi
Synonyms: *Nesaea heterophylla* H.Perrier

Range area (grAH km^2): AFR: 8000
Ecozone origin-status: AFR: E
World rarity status: VR
Ecozone-endemic macroregion occurrence: AFR: MAD
Life form(s): EM
Ploidy state: NA
Speciation age (Ma): NA
Populations also occur in terrestrial, wetland and/ or riparian habitats: YES
Long distance movement vectors: NA
Nuisance status and uses: no nuisance issues; no known uses
Endangered status (IUCN): EN

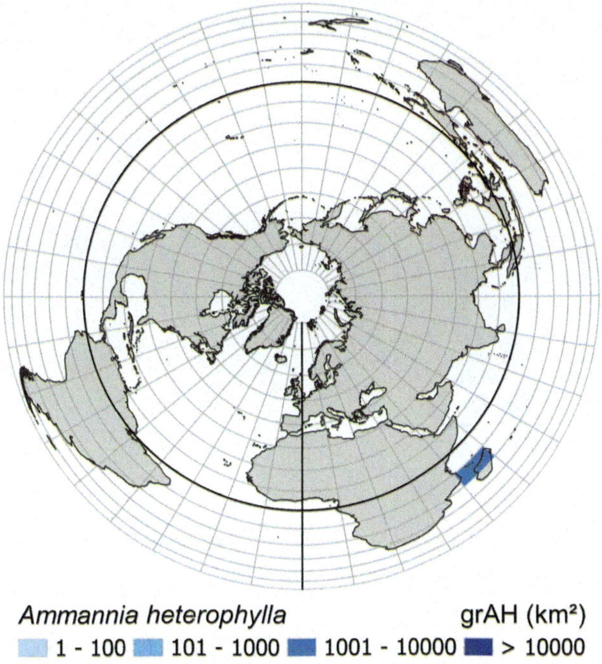

Ammannia latifolia **L.**
Synonyms: *Ammannia lythrifolia* Salisb.

Range area (grAH km^2): NEA: 8200; NEO: 196300
Ecozone origin-status: NEA: N; NEO: N
World rarity status: C
Ecozone-endemic macroregion occurrence: not endemic to any world ecozone
Life form(s): EM
Ploidy state: P
Speciation age (Ma): NA
Populations also occur in terrestrial, wetland and/ or riparian habitats: YES
Long distance movement vectors: NA
Nuisance status and uses: no nuisance issues; aquarium trade
Endangered status (IUCN): NA

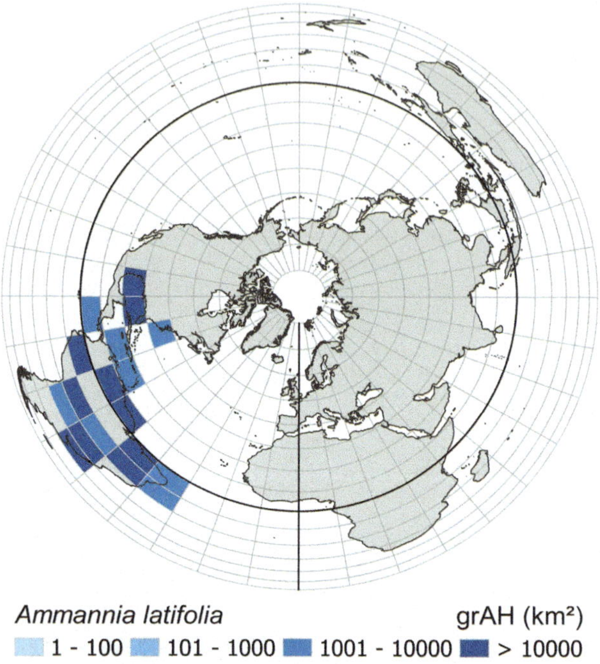

Ammannia latifolia L.

***Ammannia mossambicensis* (A.Fern. & Diniz) S.A.Graham & Gandhi**
Synonyms: *Hionanthera mossambicensis* A.Fern. & Diniz

Range area (grAH km^2): AFR: 34700
Ecozone origin-status: AFR: E
World rarity status: R
Ecozone-endemic macroregion occurrence: AFR: ZAM
Life form(s): EM (sometimes a hydrosubshrub)
Ploidy state: NA
Speciation age (Ma): NA
Populations also occur in terrestrial, wetland and/ or riparian habitats: YES
Long distance movement vectors: NA
Nuisance status and uses: no nuisance issues; no known uses
Endangered status (IUCN): NA

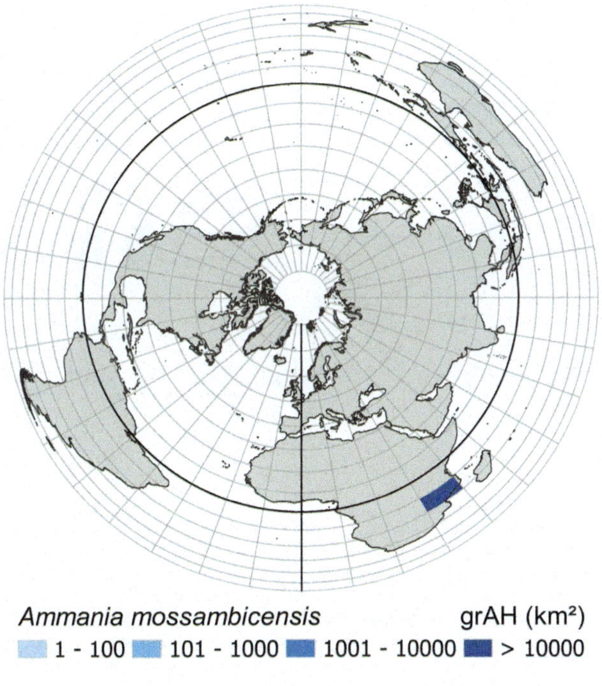

***Ammannia myriophylloides* Dunn**
Synonyms: *Rotala wallichii* (Hook.f.) Koehne

Range area (grAH km^2): OR: 24900
Ecozone origin-status: OR: E
World rarity status: VR
Ecozone-endemic macroregion occurrence: OR: ICH
Life form(s): SU, EM
Ploidy state: NA
Speciation age (Ma): NA
Populations also occur in terrestrial, wetland and/ or riparian habitats: NO
Long distance movement vectors: NA
Nuisance status and uses: no nuisance issues; no known uses
Endangered status (IUCN): NA

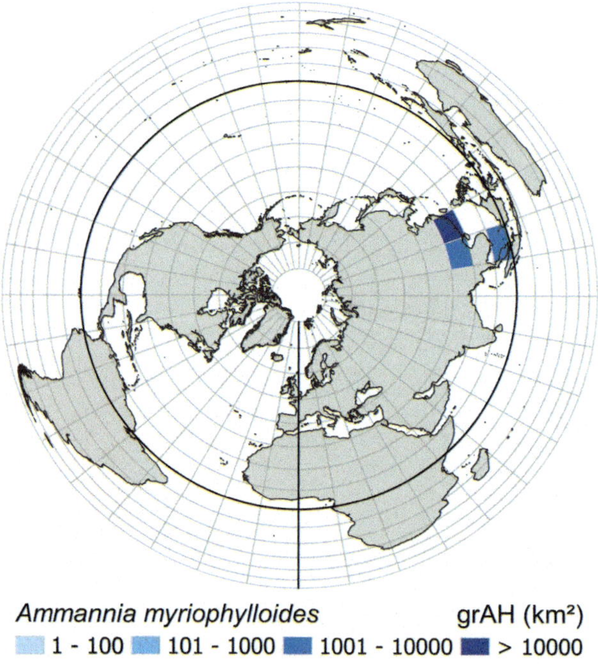

Ammannia petrensis **(M.G.Gilbert & Thulin) S.A.Graham & Gandhi**
Synonyms: *Nesaea petrensis* M.G.Gilbert & Thulin

Range area (grAH km^2): AFR: 900
Ecozone origin-status: AFR: E
World rarity status: VR
Ecozone-endemic macroregion occurrence: AFR: SAH/ AR
Life form(s): EM
Ploidy state: NA
Speciation age (Ma): NA
Populations also occur in terrestrial, wetland and/ or riparian habitats: YES
Long distance movement vectors: NA
Nuisance status and uses: no nuisance issues; no known uses
Endangered status (IUCN): NA

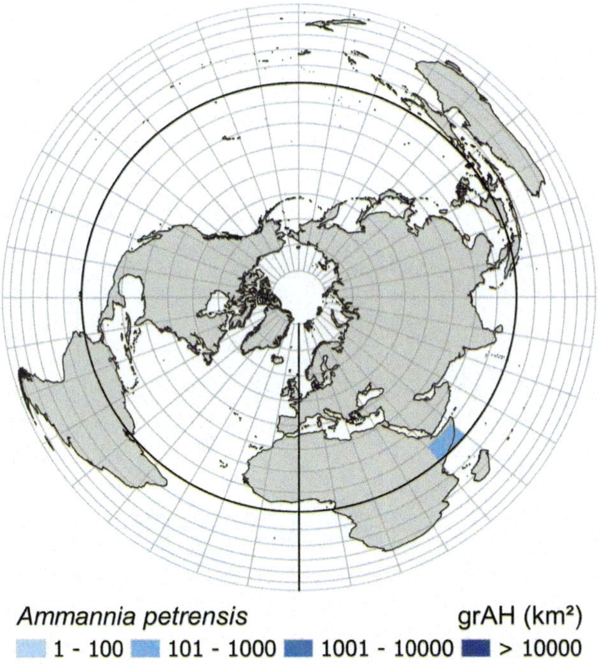

***Ammannia prieuriana* Guill. & Perr.**
Synonyms:

Range area (grAH km^2): PAL: 11200; AFR: 304000
Ecozone origin-status: PAL: I; AFR: E(I)
World rarity status: C
Ecozone-endemic macroregion occurrence: AFR: SAH/ AR, FOR, EA, ZAM, SDES
Life form(s): EM
Ploidy state: D
Speciation age (Ma): NA
Populations also occur in terrestrial, wetland and/ or riparian habitats: YES
Long distance movement vectors: ANTHROPOCHORY
Nuisance status and uses: no nuisance issues; no known uses
Endangered status (IUCN): LC

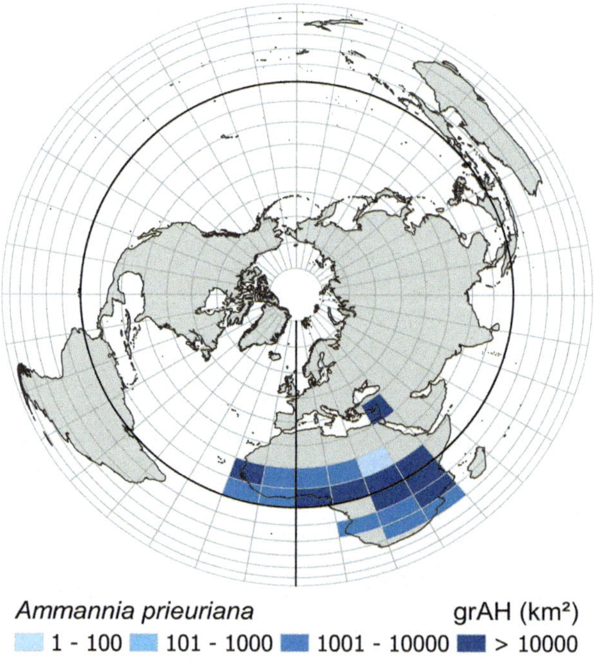

Ammannia prieuriana Guill. & Perr.

Ammannia robusta **Heer & Regel**
Synonyms:

Range area (grAH km^2): PAL: 19300; NEA: 315900; NEO: 125000
Ecozone origin-status: PAL: I; NEA: N; NEO: N
World rarity status: C
Ecozone-endemic macroregion occurrence: not endemic to any world ecozone
Life form(s): EM
Ploidy state: DP

Speciation age (Ma): NA
Populations also occur in terrestrial, wetland and/ or riparian habitats: YES
Long distance movement vectors: ANTHROPOCHORY
Nuisance status and uses: no nuisance issues; no known uses
Endangered status (IUCN): LC

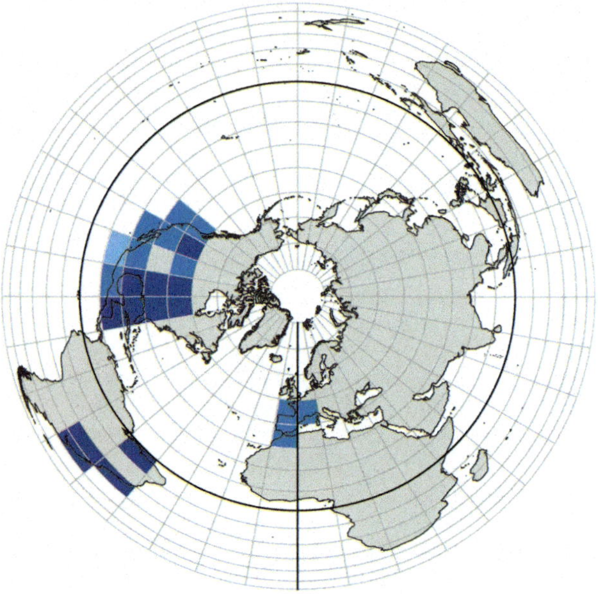

Ammannia robusta Heer & Regel
Image copyright: iNaturalist CC-BY @sheriff_woody_pct (Matt Berger)

Ammannia senegalensis **Lam.**
Synonyms:

Range area (grAH km^2): AFR: 112200
Ecozone origin-status: AFR: E
World rarity status: R
Ecozone-endemic macroregion occurrence: AFR: SAH/ AR, FOR, ZAM, SDES
Life form(s): SU, EM
Ploidy state: NA
Speciation age (Ma): NA
Populations also occur in terrestrial, wetland and/ or riparian habitats: YES
Long distance movement vectors: NA
Nuisance status and uses: no nuisance issues; aquarium trade
Endangered status (IUCN): LC

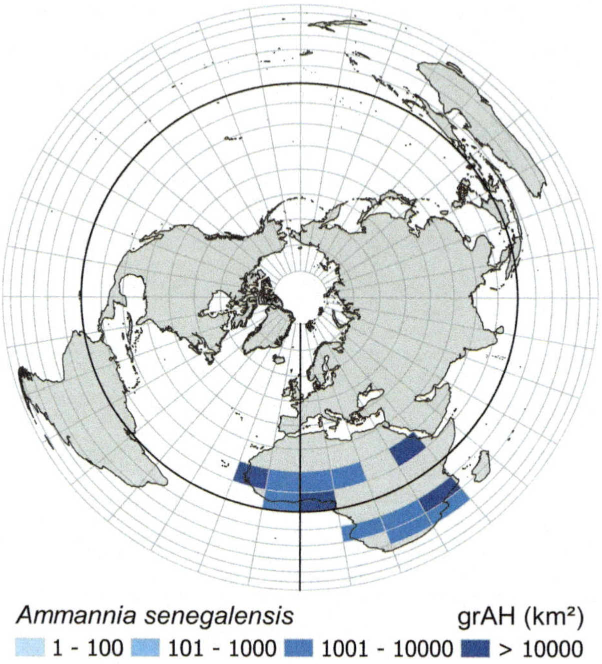

***Ammannia verticillata* (Ard.) Lam.**
Synonyms: *Cornelia verticillata* Ard.

Range area (grAH km^2): PAL: 104000; OR: 60900
Ecozone origin-status: PAL: E(I); OR: I
World rarity status: R
Ecozone-endemic macroregion occurrence: PAL: W EUR, MED, E EUR, C ASIA
Life form(s): EM
Ploidy state: NA
Speciation age (Ma): NA
Populations also occur in terrestrial, wetland and/ or riparian habitats: YES
Long distance movement vectors: ANTHROPOCHORY
Nuisance status and uses: no nuisance issues; no known uses
Endangered status (IUCN): LC

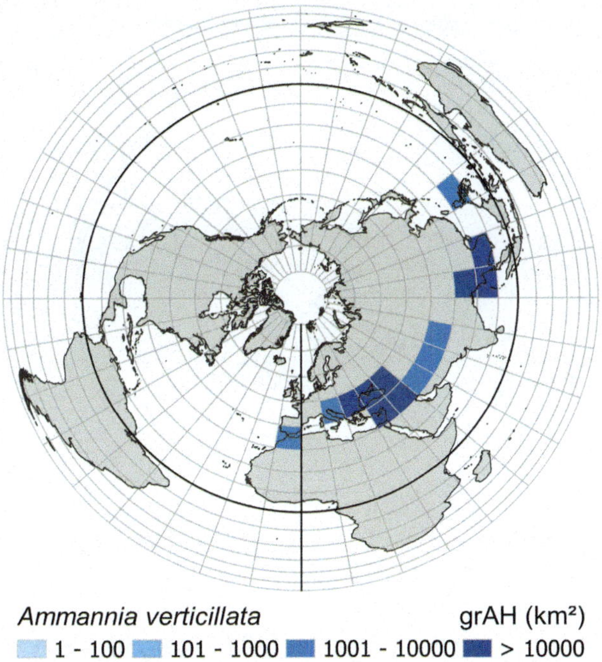

***Cuphea alatosperma* T.B.Cavalc. & S.A.Graham**
Synonyms:

Range area (grAH km^2): NEO: 46000
Ecozone origin-status: NEO: E
World rarity status: R
Ecozone-endemic macroregion occurrence: NEO: AMZ
Life form(s): EM (subshrub)
Ploidy state: NA
Speciation age (Ma): NA
Populations also occur in terrestrial, wetland and/ or riparian habitats: YES
Long distance movement vectors: NA
Nuisance status and uses: no nuisance issues; no known uses
Endangered status (IUCN): NA

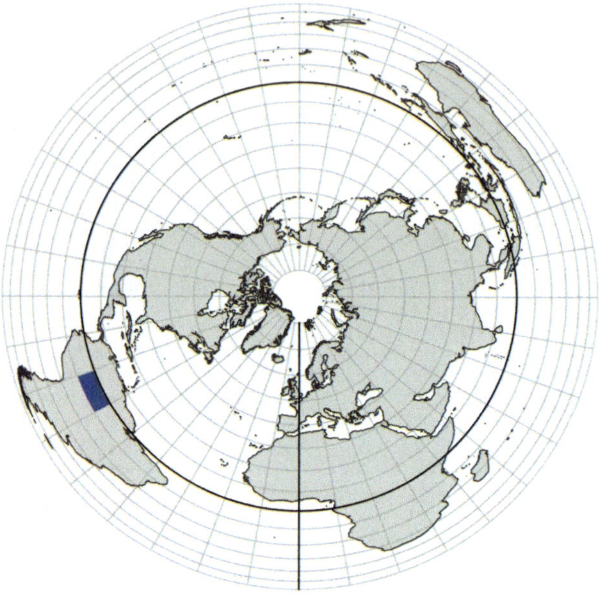

Cuphea anagalloidea **A.St.-Hil.**
Synonyms:

Range area (grAH km^2): NEO: 100300
Ecozone origin-status: NEO: E
World rarity status: R
Ecozone-endemic macroregion occurrence: NEO: AMZ, CHA, PAR
Life form(s): SU, EM
Ploidy state: NA
Speciation age (Ma): NA
Populations also occur in terrestrial, wetland and/ or riparian habitats: YES
Long distance movement vectors: NA
Nuisance status and uses: no nuisance issues; aquarium trade
Endangered status (IUCN): NA

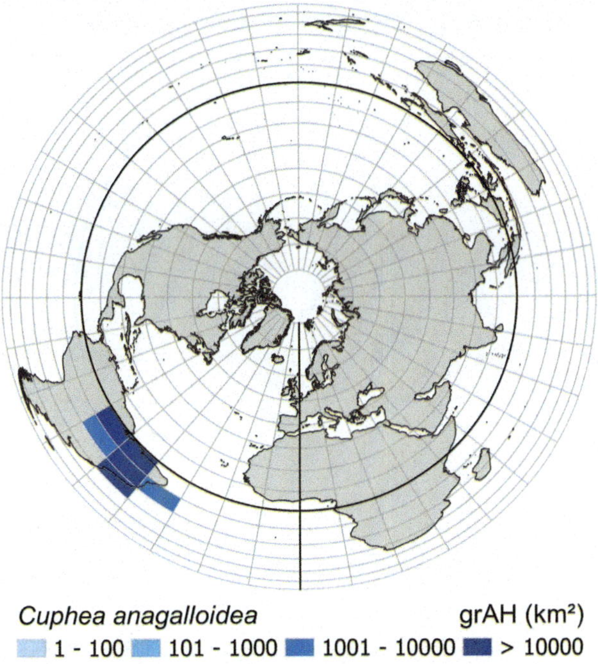

***Cuphea anamariae* T.B.Cavalc. & S.A.Graham**
Synonyms:

Range area (grAH km^2): NEO: 12100
Ecozone origin-status: NEO: E
World rarity status: VR
Ecozone-endemic macroregion occurrence: NEO: CHA
Life form(s): EM (subshrub)
Ploidy state: NA
Speciation age (Ma): NA
Populations also occur in terrestrial, wetland and/ or riparian habitats: YES
Long distance movement vectors: NA
Nuisance status and uses: no nuisance issues; no known uses
Endangered status (IUCN): NA

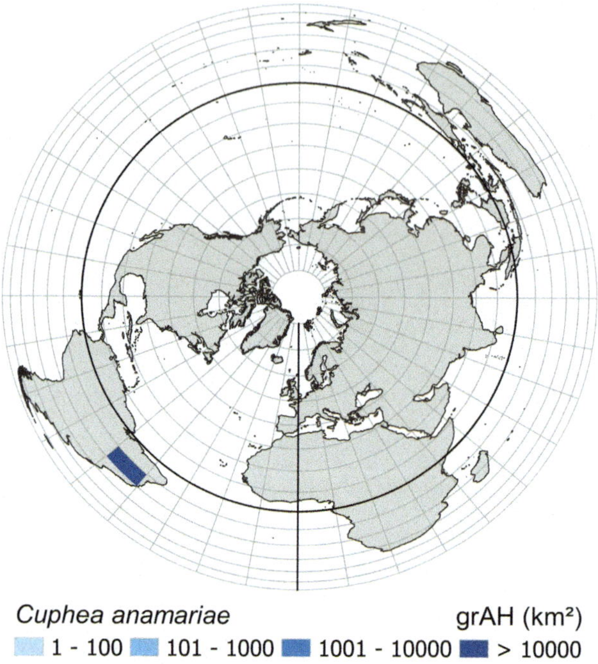

Cuphea antisyphilitica **Kunth**
Synonyms:

Range area (grAH km^2): NEO: 246200
Ecozone origin-status: NEO: E
World rarity status: C
Ecozone-endemic macroregion occurrence: NEO: NWSA, AMZ, CHA, PAR
Life form(s): EM
Ploidy state: P
Speciation age (Ma): NA
Populations also occur in terrestrial, wetland and/ or riparian habitats: YES
Long distance movement vectors: NA
Nuisance status and uses: no nuisance issues; medicinal potential
Endangered status (IUCN): NA

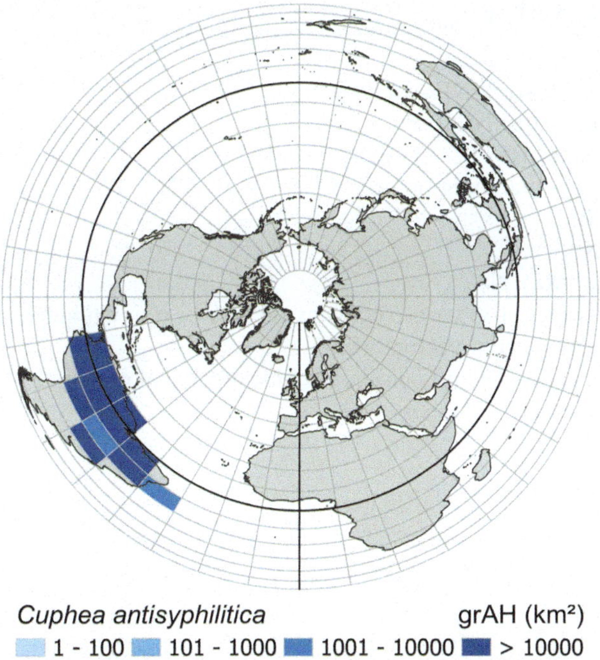

Cuphea antisyphilitica Kunth
Image copyright: iNaturalist CC0 @tsssss

Cuphea aperta Koehne
Synonyms:

Range area (grAH km^2): NEO: 67900
Ecozone origin-status: NEO: E
World rarity status: R
Ecozone-endemic macroregion occurrence: NEO: CHA, PAR
Life form(s): EM
Ploidy state: NA
Speciation age (Ma): NA
Populations also occur in terrestrial, wetland and/ or riparian habitats: YES
Long distance movement vectors: NA
Nuisance status and uses: no nuisance issues; medicinal potential
Endangered status (IUCN): NA

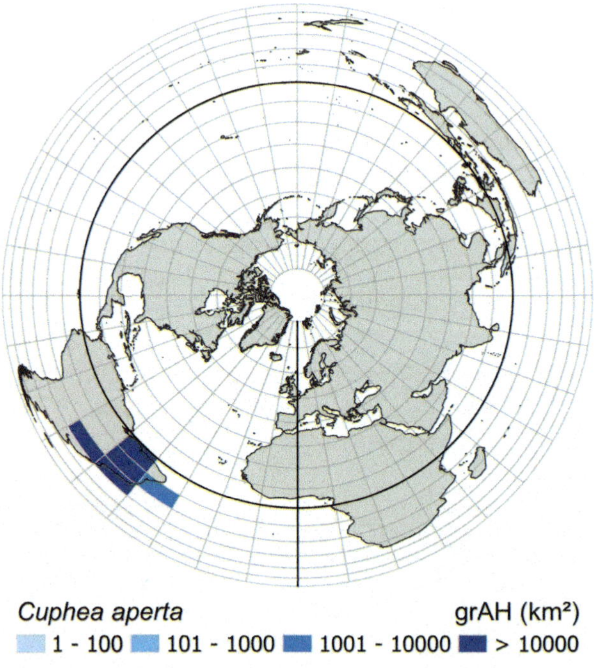

Cuphea arenarioides A.St.-Hil.
Synonyms:

Range area (grAH km^2): NEO: 52200
Ecozone origin-status: NEO: E
World rarity status: R
Ecozone-endemic macroregion occurrence: NEO: AMZ, CHA, PAR
Life form(s): EM

Ploidy state: NA
Speciation age (Ma): NA
Populations also occur in terrestrial, wetland and/ or riparian habitats: YES
Long distance movement vectors: NA
Nuisance status and uses: no nuisance issues; no known uses
Endangered status (IUCN): NA

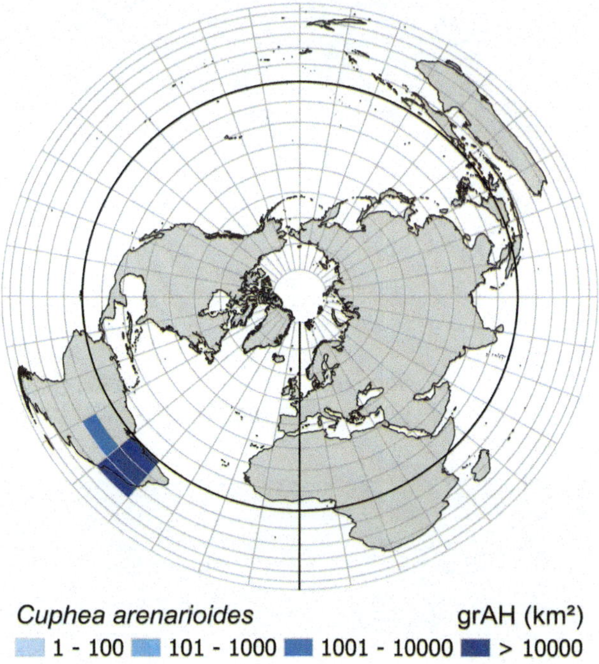

Cuphea campylocentra **Griseb.**
Synonyms:

Range area (grAH km^2): NEO: 55500
Ecozone origin-status: NEO: E
World rarity status: R
Ecozone-endemic macroregion occurrence: NEO: AMZ, CHA, PAR
Life form(s): EM
Ploidy state: D
Speciation age (Ma): NA
Populations also occur in terrestrial, wetland and/ or riparian habitats: YES
Long distance movement vectors: NA
Nuisance status and uses: no nuisance issues; medicinal potential
Endangered status (IUCN): NA

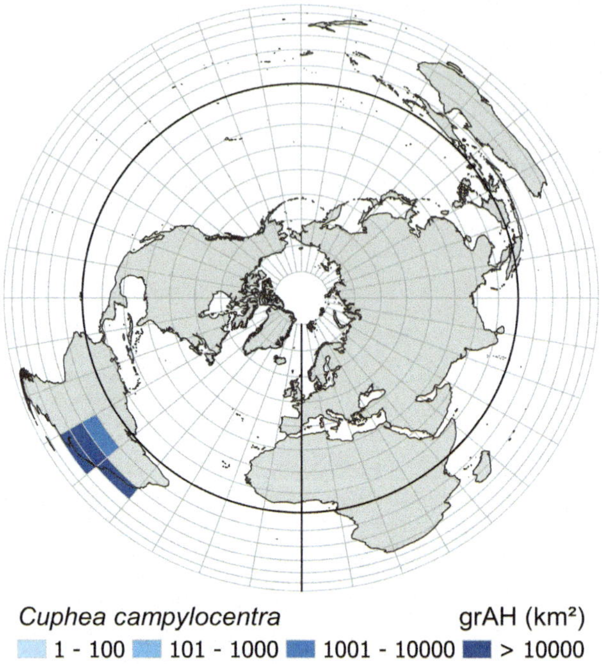

Cuphea campylocentra Griseb.
Image copyright: iNaturalist CC-BY @ttamagno (Tomás Tamagno)

***Cuphea carthagenensis* (Jacq.) J.F.Macbr.**
Synonyms: *Lythrum carthagenense* Jacq.

Range area (grAH km²): PAL: 9000; NEA: 34000; NEO: 55500; OR: 91500; AUS: 46800; AFR: 21400; present in OCE
Ecozone origin-status: PAL: I; NEA: I; NEO: E(I); OR: I; AUS: I; AFR: I; OCE: I
World rarity status: C
Ecozone-endemic macroregion occurrence: NEO: MG MX, MES AM, NWSA, AND, AMZ, CHA, PAR
Life form(s): EM (sometimes a subshrub)
Ploidy state: D
Speciation age (Ma): NA
Populations also occur in terrestrial, wetland and/ or riparian habitats: YES
Long distance movement vectors: ANTHROPOCHORY
Nuisance status and uses: a nuisance weed in many sub-tropical and tropical areas outside its native range; medicinal potential and food uses
Endangered status (IUCN): NA

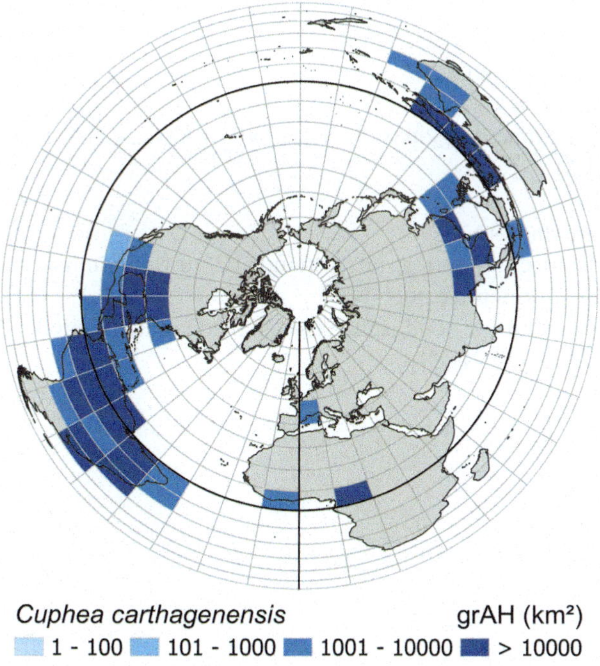

Cuphea carthagenensis (Jacq.) J.F.Macbr.

***Cuphea circaeoides* Sm. ex Sims**
Synonyms:

Range area (grAH km^2): NEO: 84800
Ecozone origin-status: NEO: E
World rarity status: R
Ecozone-endemic macroregion occurrence: NEO: AMZ, CHA, PAR
Life form(s): EM
Ploidy state: P
Speciation age (Ma): NA
Populations also occur in terrestrial, wetland and/ or riparian habitats: YES
Long distance movement vectors: NA
Nuisance status and uses: no nuisance issues; medicinal potential
Endangered status (IUCN): NA

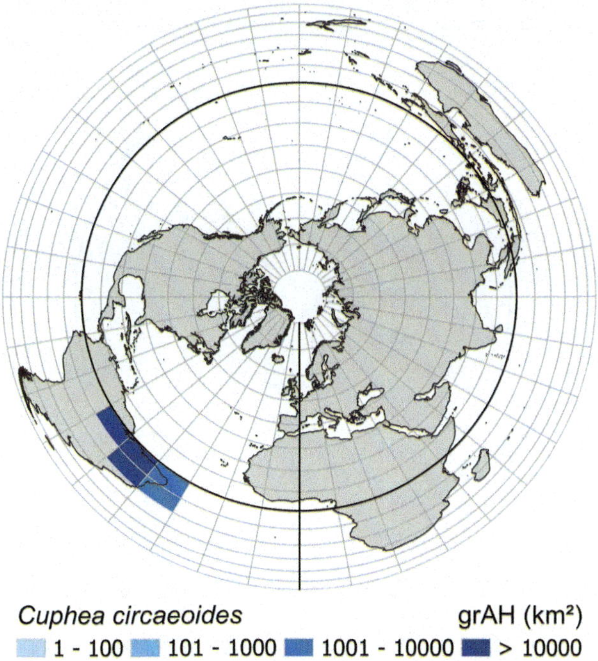

Cuphea fruticosa **Spreng.**
Synonyms:

Range area (grAH km^2): NEO: 72200
Ecozone origin-status: NEO: E
World rarity status: R
Ecozone-endemic macroregion occurrence: NEO: CHA, PAR
Life form(s): EM (subshrub)
Ploidy state: NA
Speciation age (Ma): NA
Populations also occur in terrestrial, wetland and/ or riparian habitats: YES
Long distance movement vectors: NA
Nuisance status and uses: no nuisance issues; medicinal, ornamental uses
Endangered status (IUCN): NA

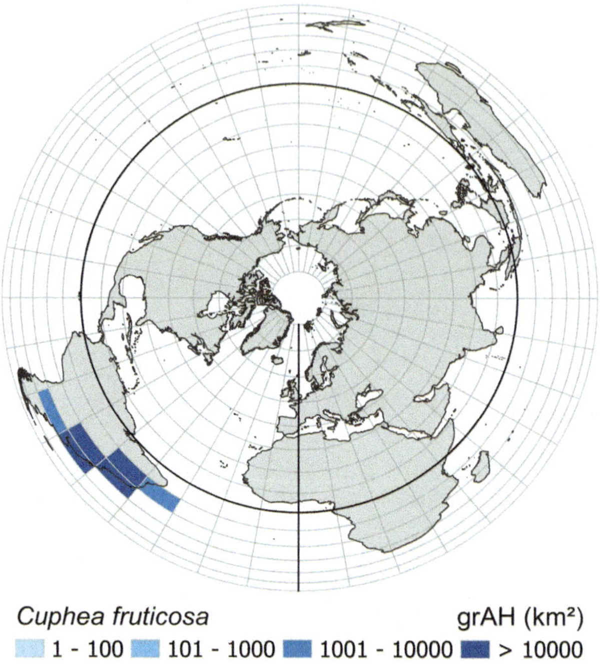

Cuphea fruticosa Spreng.
Image copyright: iNaturalist CC-BY @candela_castro

***Cuphea lindmaniana* Koehne ex Bacig.**
Synonyms:

Range area (grAH km^2): NEO: 47900
Ecozone origin-status: NEO: E
World rarity status: R

Ecozone-endemic macroregion occurrence: NEO: CHA, PAR
Life form(s): EM
Ploidy state: NA
Speciation age (Ma): NA
Populations also occur in terrestrial, wetland and/ or riparian habitats: YES
Long distance movement vectors: NA
Nuisance status and uses: no nuisance issues; medicinal potential
Endangered status (IUCN): NA

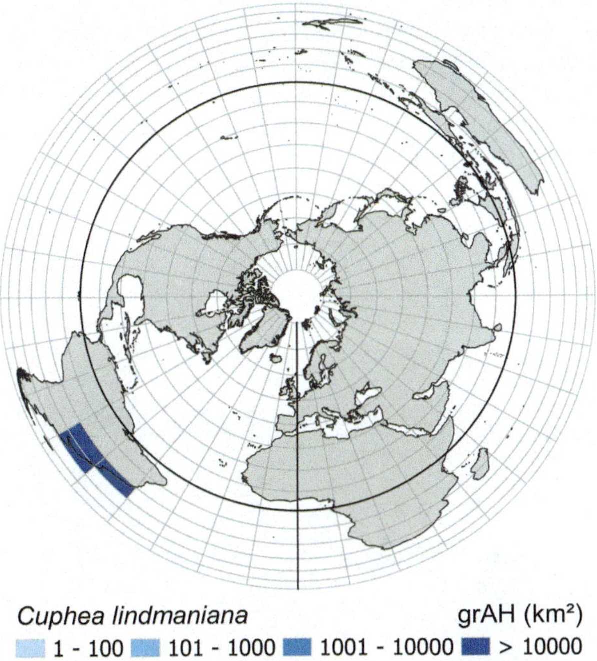

Cuphea lindmaniana Koehne ex Bacig.

***Cuphea melvilla* Lindl.**
Synonyms: *Melvilla speciosa* A.Anderson ex Raf.

Range area (grAH km^2): NEO: 244900
Ecozone origin-status: NEO: E
World rarity status: C
Ecozone-endemic macroregion occurrence: NEO: NWSA, AMZ, AND, CHA, PAR
Life form(s): EM (shrub)
Ploidy state: P
Speciation age (Ma): NA
Populations also occur in terrestrial, wetland and/ or riparian habitats: YES
Long distance movement vectors: NA
Nuisance status and uses: no nuisance issues; harvested from the wild for local use as a medicine, and often grown as an ornamental in the tropics
Endangered status (IUCN): LC

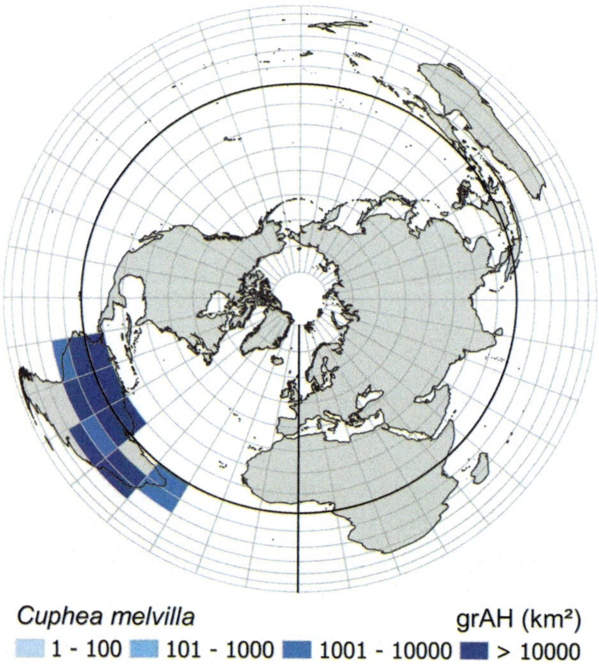

Cuphea melvilla Lindl.
Image copyright: iNaturalist CC-BY @marcia_martins (Márcia Cristina Martins da Silva)

Cuphea micrantha Kunth
Synonyms:

Range area (grAH km^2): NEO: 179800
Ecozone origin-status: NEO: E
World rarity status: C
Ecozone-endemic macroregion occurrence: NEO: MES AM, NWSA, AMZ, CHA, PAR
Life form(s): EM (sometimes a subshrub)
Ploidy state: D
Speciation age (Ma): NA
Populations also occur in terrestrial, wetland and/ or riparian habitats: YES
Long distance movement vectors: NA
Nuisance status and uses: no nuisance issues; medicinal potential uses
Endangered status (IUCN): NA

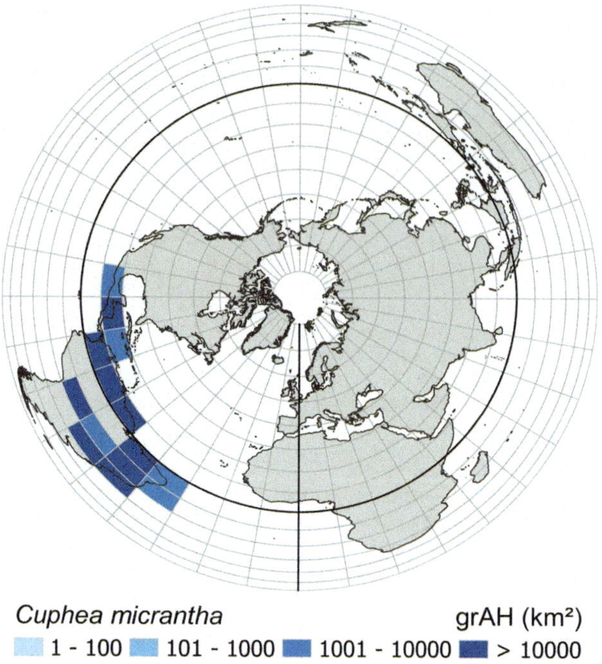

Cuphea mimuloides **Schltdl. & Cham.**
Synonyms:

Range area (grAH km^2): NEO: 65700
Ecozone origin-status: NEO: E
World rarity status: R
Ecozone-endemic macroregion occurrence: NEO: MG MX, MES AM, CHA, PAR
Life form(s): EM (subshrub)
Ploidy state: P
Speciation age (Ma): NA
Populations also occur in terrestrial, wetland and/ or riparian habitats: YES
Long distance movement vectors: NA
Nuisance status and uses: no nuisance issues; medicinal potential uses
Endangered status (IUCN): NA

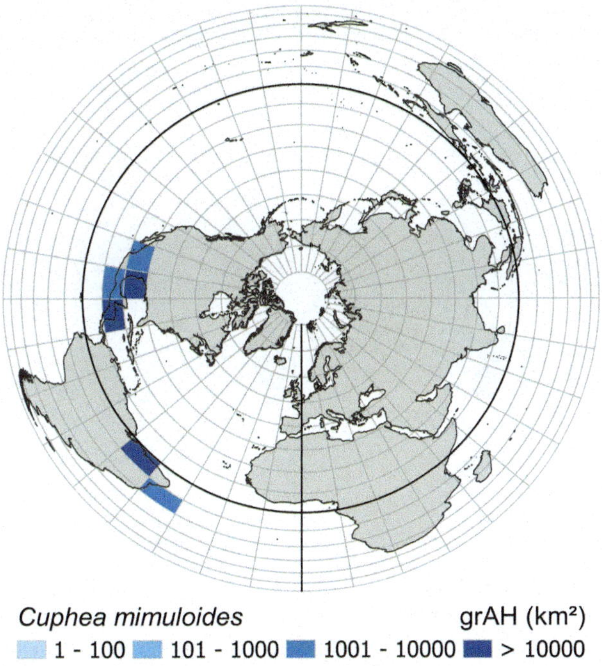

***Cuphea odonelli* Lourteig**
Synonyms:

Range area (grAH km^2): NEO: 129100
Ecozone origin-status: NEO: E
World rarity status: R
Ecozone-endemic macroregion occurrence: NEO: NWSA, AMZ, AND, CHA
Life form(s): EM
Ploidy state: D
Speciation age (Ma): NA
Populations also occur in terrestrial, wetland and/ or riparian habitats: YES
Long distance movement vectors: NA
Nuisance status and uses: no nuisance issues; no known uses
Endangered status (IUCN): NA

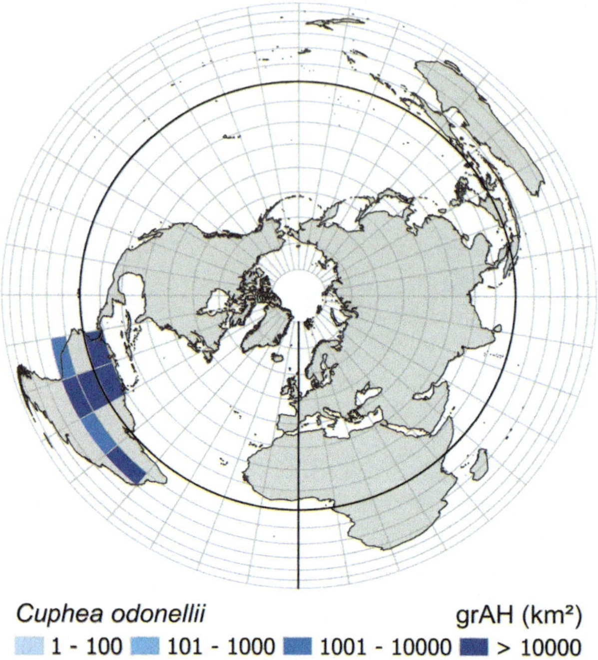

Cuphea odonelli Lourteig

***Cuphea pascuorum* Mart. ex Koehne**
Synonyms:

Range area (grAH km^2): NEO: 22900
Ecozone origin-status: NEO: E
World rarity status: VR
Ecozone-endemic macroregion occurrence: NEO: AMZ, CHA, PAR
Life form(s): EM
Ploidy state: P
Speciation age (Ma): NA
Populations also occur in terrestrial, wetland and/ or riparian habitats: YES
Long distance movement vectors: NA
Nuisance status and uses: no nuisance issues; medicinal potential uses
Endangered status (IUCN): NA

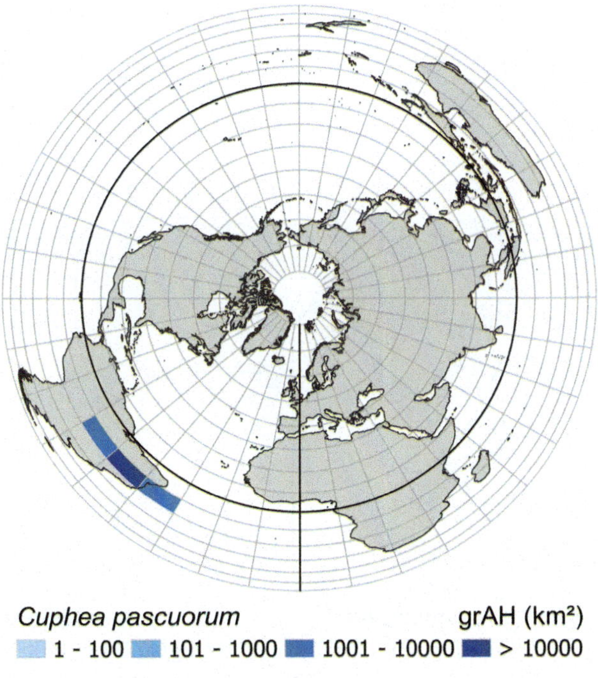

***Cuphea polymorpha* A.St.-Hil.**
Synonyms:

Range area (grAH km²): NEO: 54400
Ecozone origin-status: NEO: E
World rarity status: R
Ecozone-endemic macroregion occurrence: NEO: AMZ, CHA, PAR
Life form(s): EM
Ploidy state: D
Speciation age (Ma): NA
Populations also occur in terrestrial, wetland and/ or riparian habitats: YES
Long distance movement vectors: NA
Nuisance status and uses: no nuisance issues; no known uses
Endangered status (IUCN): NA

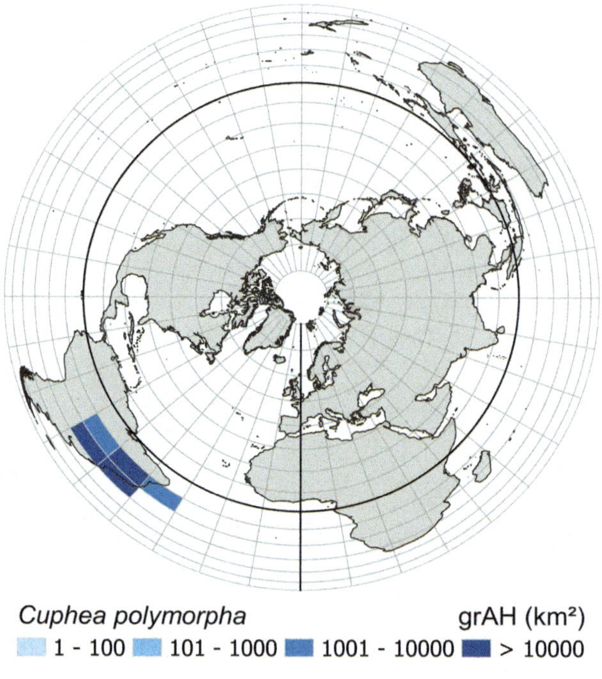

Cuphea polymorpha A.St.-Hil.
Image copyright: Y Domínguez, with permission

Cuphea racemosa **A.St.-Hil.**
Synonyms:

Range area (grAH km^2): NEO: 215700
Ecozone origin-status: NEO: E
World rarity status: C
Ecozone-endemic macroregion occurrence: NEO: MG MX, MES AM, NWSA, AND, AMZ, CHA, PAR
Life form(s): EM (subshrub)
Ploidy state: P
Speciation age (Ma): NA
Populations also occur in terrestrial, wetland and/ or riparian habitats: YES
Long distance movement vectors: NA
Nuisance status and uses: no nuisance issues; medicinal uses
Endangered status (IUCN): NA

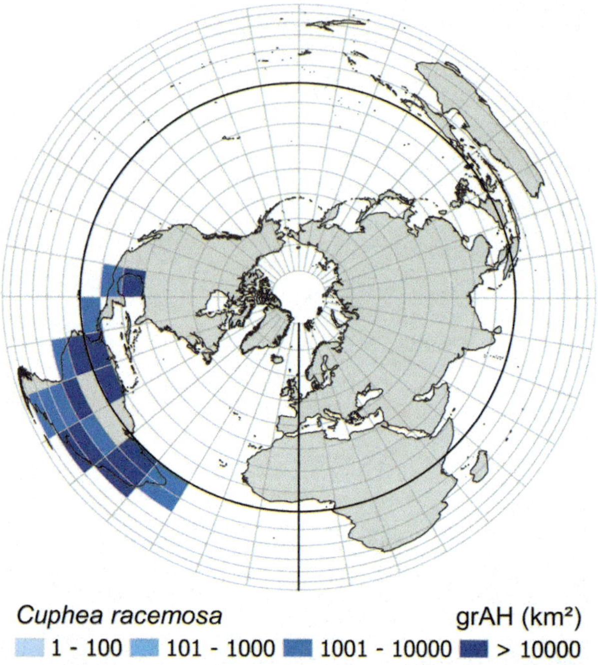

Cuphea racemosa A.St.-Hil.
Image copyright: iNaturalist CC-BY @romigaleota (Romi Galeota Lencina)

***Cuphea repens* Koehne**
Synonyms:

Range area (grAH km^2): NEO: 150300
Ecozone origin-status: NEO: E
World rarity status: R
Ecozone-endemic macroregion occurrence: NEO: NWSA, AND, AMZ, CHA

Life form(s): EM (subshrub)
Ploidy state: NA
Speciation age (Ma): NA
Populations also occur in terrestrial, wetland and/ or riparian habitats: YES
Long distance movement vectors: NA
Nuisance status and uses: no nuisance issues; no known uses
Endangered status (IUCN): NA

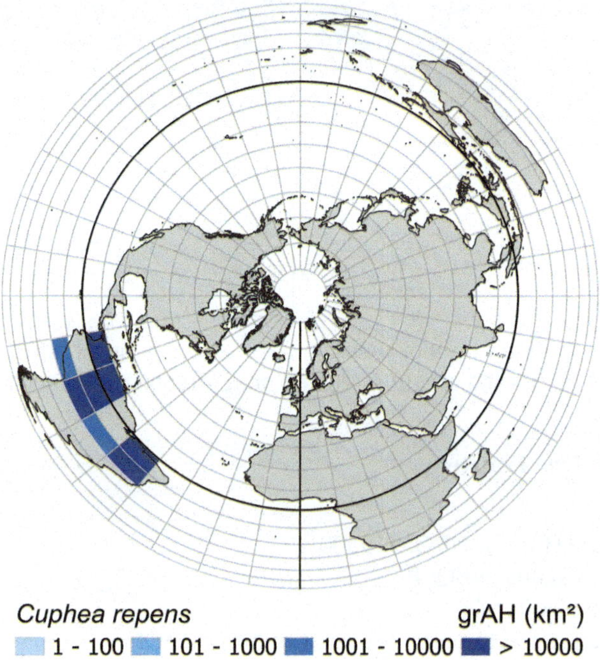

***Cuphea rionegrensis* Lourteig**
Synonyms:

Range area (grAH km^2): NEO: 53600
Ecozone origin-status: NEO: E
World rarity status: R
Ecozone-endemic macroregion occurrence: NEO: AMZ
Life form(s): EM
Ploidy state: NA
Speciation age (Ma): NA
Populations also occur in terrestrial, wetland and/ or riparian habitats: YES
Long distance movement vectors: NA
Nuisance status and uses: no nuisance issues; no known uses
Endangered status (IUCN): NA

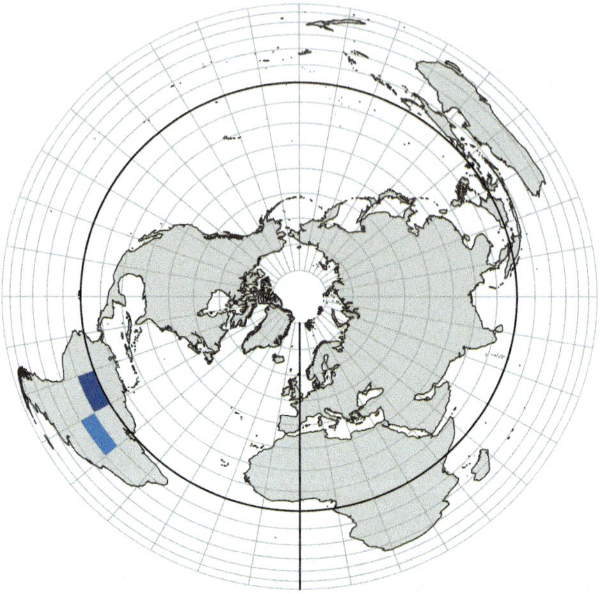

Cuphea scolnikiae **Lourteig**
Synonyms:

Range area (grAH km^2): NEO: 119500
Ecozone origin-status: NEO: E
World rarity status: R
Ecozone-endemic macroregion occurrence: NEO: AMZ
Life form(s): EM (subshrub)
Ploidy state: NA
Speciation age (Ma): NA
Populations also occur in terrestrial, wetland and/ or riparian habitats: YES
Long distance movement vectors: NA
Nuisance status and uses: no nuisance issues; no known uses
Endangered status (IUCN): NA

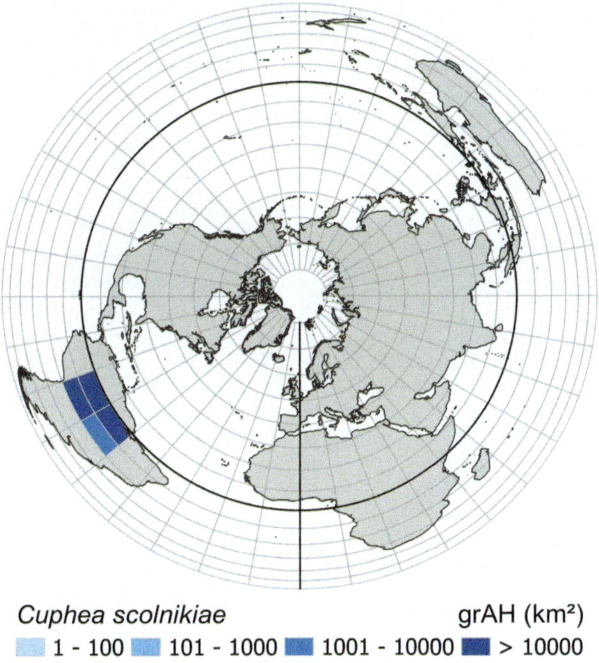

***Cuphea sessilifolia* Mart.**
Synonyms:

Range area (grAH km^2): NEO: 166300
Ecozone origin-status: NEO: E
World rarity status: C
Ecozone-endemic macroregion occurrence: NEO: AMZ, CHA, PAR
Life form(s): EM (subshrub)
Ploidy state: P
Speciation age (Ma): NA
Populations also occur in terrestrial, wetland and/ or riparian habitats: YES
Long distance movement vectors: NA
Nuisance status and uses: no nuisance issues; medicinal potential
Endangered status (IUCN): NA

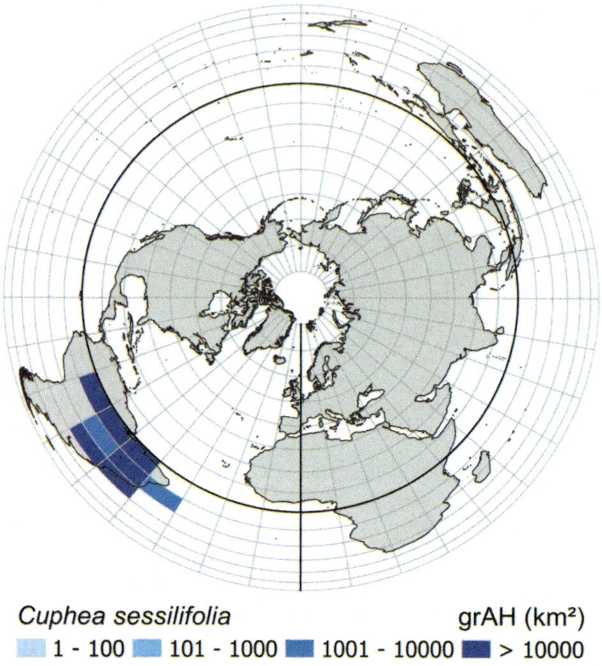

***Cuphea setosa* Koehne**
Synonyms:

Range area (grAH km^2): NEO: 107600
Ecozone origin-status: NEO: E
World rarity status: R
Ecozone-endemic macroregion occurrence: NEO: MES AM, NWSA, AMZ
Life form(s): EM (subshrub)
Ploidy state: P
Speciation age (Ma): NA
Populations also occur in terrestrial, wetland and/ or riparian habitats: YES
Long distance movement vectors: NA
Nuisance status and uses: no nuisance issues; medicinal potential
Endangered status (IUCN): NA

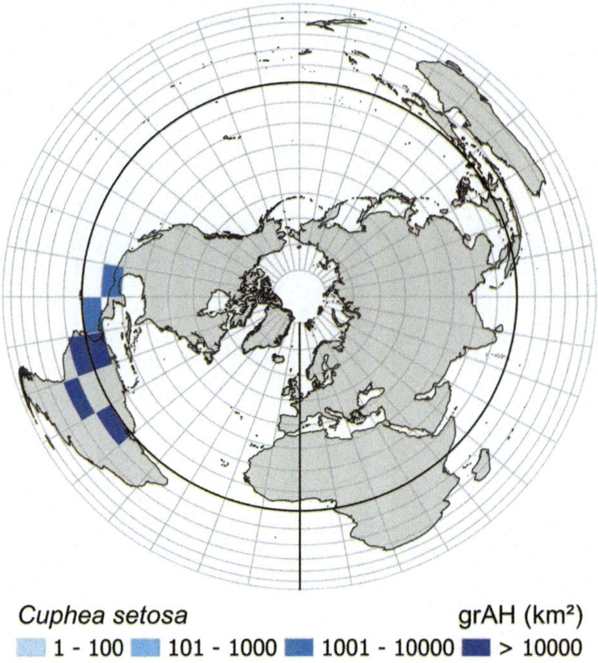

Cuphea tuberosa **Koehne**
Synonyms:

Range area (grAH km^2): NEO: 55500
Ecozone origin-status: NEO: E
World rarity status: R
Ecozone-endemic macroregion occurrence: NEO: AMZ, CHA, PAR
Life form(s): EM
Ploidy state: D
Speciation age (Ma): NA
Populations also occur in terrestrial, wetland and/ or riparian habitats: YES
Long distance movement vectors: NA
Nuisance status and uses: no nuisance issues; medicinal potential uses
Endangered status (IUCN): NA

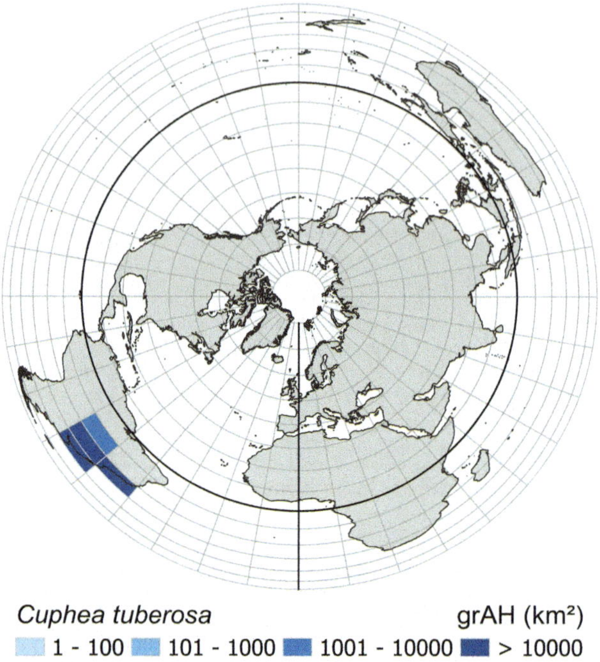

***Decodon verticillatus* (L.) Elliott**
Synonyms: *Lythrum verticillatum* L.

Range area (grAH km²): NEA: 338500; NEO: 24100
Ecozone origin-status: NEA: N; NEO: N
World rarity status: C
Ecozone-endemic macroregion occurrence: not endemic to any world ecozone
Life form(s): EM (sometimes subshrub)
Ploidy state: NA
Speciation age (Ma): NA
Populations also occur in terrestrial, wetland and/ or riparian habitats: YES
Long distance movement vectors: HYD (NEA to NEO)
Nuisance status and uses: no nuisance issues; medicinal potential
Endangered status (IUCN): LC

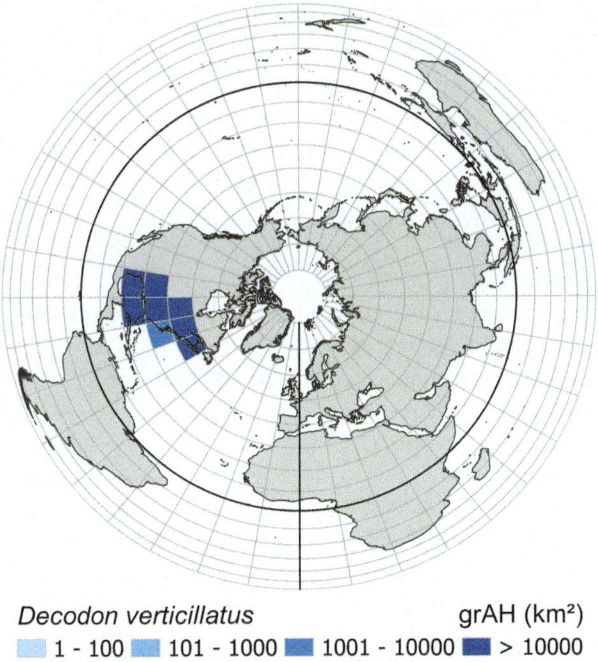

Decodon verticillatus (L.) Elliott

Didiplis diandra **(Nutt. ex DC.) Alph.Wood**
Synonyms: *Peplis diandra* Nutt. ex DC.

Range area (grAH km^2): NEA: 314800
Ecozone origin-status: NEA: E
World rarity status: C
Ecozone-endemic macroregion occurrence: NEA: E NAM
Life form(s): SU, EM (hydrosubshrub)
Ploidy state: NA
Speciation age (Ma): NA
Populations also occur in terrestrial, wetland and/ or riparian habitats: YES
Long distance movement vectors: NA
Nuisance status and uses: no nuisance issues; aquarium trade
Endangered status (IUCN): NA

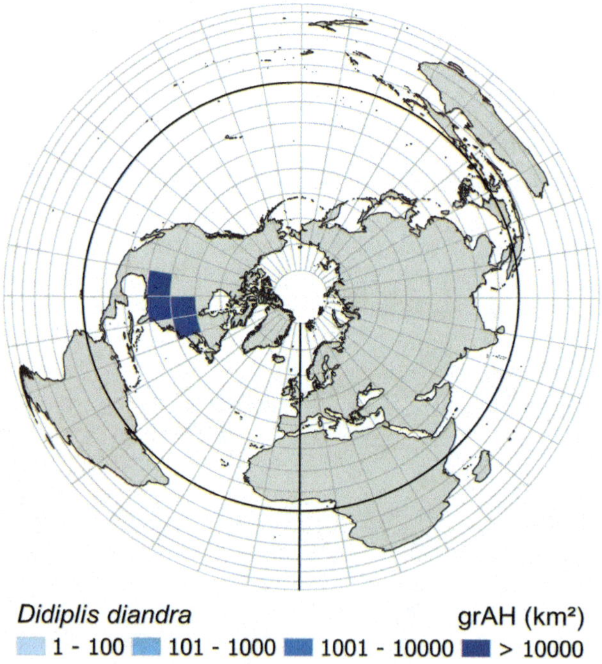

Didiplis diandra (Nutt. ex DC.) Alph.Wood
Image copyright: iNaturalist CC-BY @jim_keesling (Jim Keesling)

Lythrum alatum Pursh
Synonyms: *Pythagorea alata* (Pursh) Raf; *Salicaria alata* (Pursh) Lunell

Range area (grAH km^2): NEA: 357000; NEO: 56200
Ecozone origin-status: NEA: N; NEO: N
World rarity status: C
Ecozone-endemic macroregion occurrence: not endemic to any world ecozone
Life form(s): EM (sometimes a subshrub)
Ploidy state: P
Speciation age (Ma): NA
Populations also occur in terrestrial, wetland and/ or riparian habitats: YES
Long distance movement vectors: NA
Nuisance status and uses: no nuisance issues; medicinal potential
Endangered status (IUCN): LC

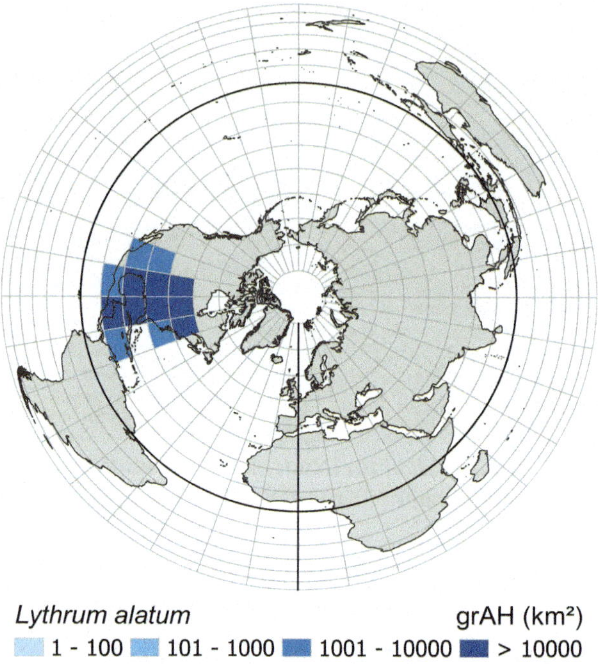

Lythrum alatum Pursh
Image copyright: iNaturalist CC-BY @brybrysciguy

***Lythrum baeticum* Gonz.Albo**
Synonyms:

Range area (grAH km^2): PAL: 19300
Ecozone origin-status: PAL: E
World rarity status: VR
Ecozone-endemic macroregion occurrence: PAL: W EUR, MED
Life form(s): EM

Ploidy state: NA
Speciation age (Ma): NA
Populations also occur in terrestrial, wetland and/ or riparian habitats: YES
Long distance movement vectors: NA
Nuisance status and uses: no nuisance issues; no known uses
Endangered status (IUCN): NT

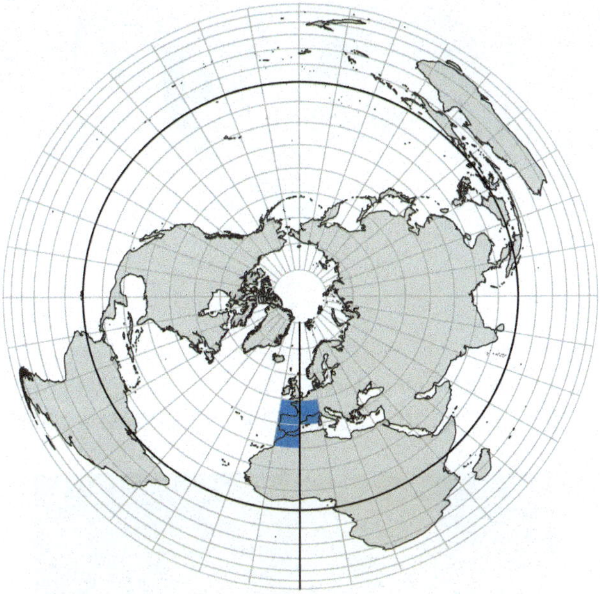

Lythrum borysthenicum **(M.Bieb. ex Schrank) Litv.**
Synonyms: *Ammannia borysthenica* (M.Bieb. ex Schrank) DC.; *Middendorfia borysthenica* (M.Bieb. ex Schrank) Trautv.; *Peplis borysthenica* M.Bieb. ex Schrank

Range area (grAH km^2): PAL: 212400
Ecozone origin-status: PAL: E
World rarity status: C
Ecozone-endemic macroregion occurrence: PAL: W EUR, MED, E EUR, UR/ W SIB, C ASIA, CH
Life form(s): EM
Ploidy state: P
Speciation age (Ma): NA
Populations also occur in terrestrial, wetland and/ or riparian habitats: YES
Long distance movement vectors: NA
Nuisance status and uses: no nuisance issues; no known uses
Endangered status (IUCN): LC

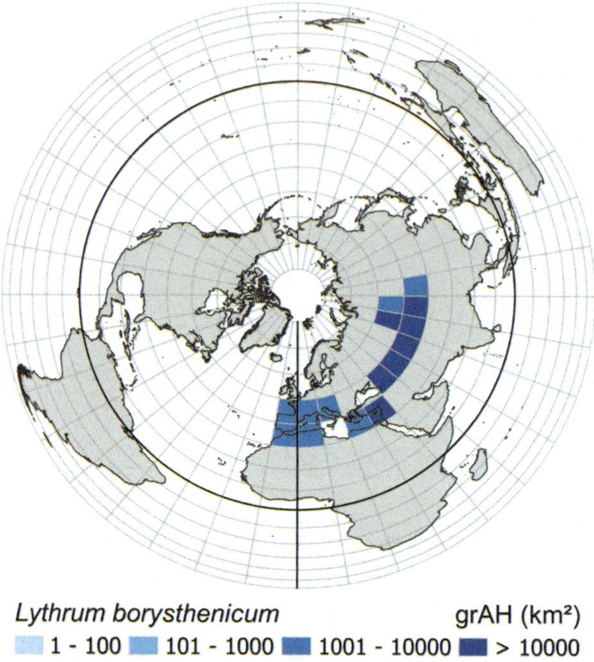

Lythrum borysthenicum (M.Bieb. ex Schrank) Litv.
Image copyright: iNaturalist CC-BY @davidclp (David Pereira)

Lythrum hyssopifolia **(M.Bieb. ex Schrank) Litv.**

Synonyms: *Chabraea hyssopifolia* (L.) Bubani; *Hyssopifolia parviflora* Opiz; *Salicaria hyssopifolia* (L.) Lam.

Range area (grAH km^2): PAL: 327100; NEA: 92900; NEO: 94800; AUS: 41400; AFR: 50800
Ecozone origin-status: PAL: N; NEA: I; NEO: I; AUS: N; AFR: N
World rarity status: C
Ecozone-endemic macroregion occurrence: not endemic to any world ecozone
Life form(s): EM
Ploidy state: P
Speciation age (Ma): NA
Populations also occur in terrestrial, wetland and/ or riparian habitats: YES
Long distance movement vectors: ANTHROPOCHORY
Nuisance status and uses: no nuisance issues; medicinal uses
Endangered status (IUCN): LC

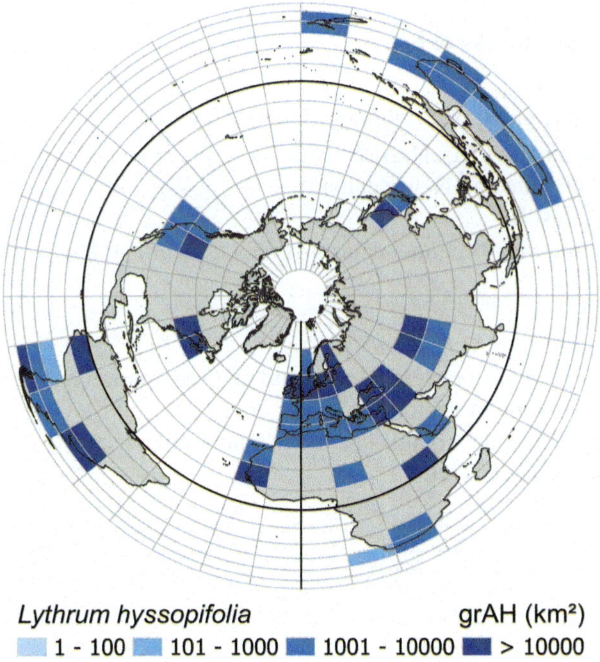

Lythrum hyssopifolia (M.Bieb. ex Schrank) Litv.
Image copyright: iNaturalist CC-BY @jacqui-nz (Jacqui Geux)

Lythrum junceum **Banks & Sol.**
Synonyms:

Range area (grAH km^2): PAL: 79000; AUS: 29300
Ecozone origin-status: PAL: N; AUS: N (some sources consider it introduced/ naturalised in Australia and New Zealand)
World rarity status: R
Ecozone-endemic macroregion occurrence: not endemic to any world ecozone
Life form(s): EM
Ploidy state: D
Speciation age (Ma): NA
Populations also occur in terrestrial, wetland and/ or riparian habitats: YES
Long distance movement vectors: NA
Nuisance status and uses: no nuisance issues; no known uses
Endangered status (IUCN): LC

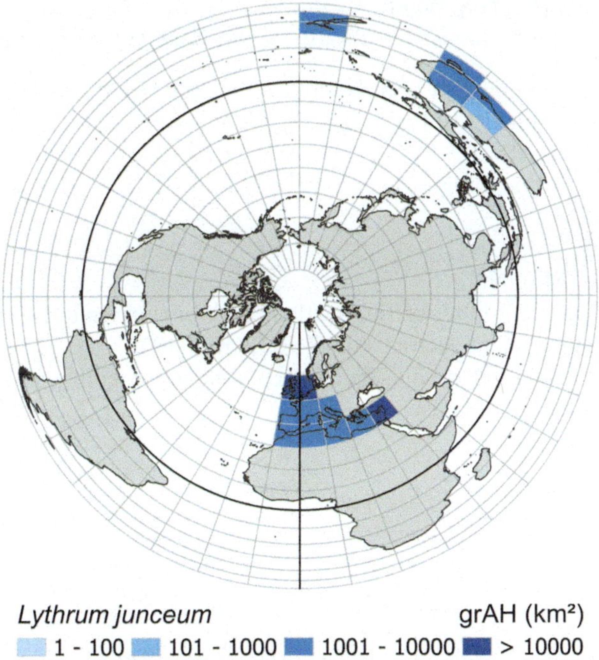

Lythrum junceum Banks & Sol.
Image copyright: J.C. Schou, with permission

***Lythrum portula* (L.) D.A.Webb**
Synonyms: *Ammannia portula* (L.) Baill.; *Peplis portula* L.

Range area (grAH km^2): PAL: 306200; NEA: 11000; NEO: 23800; AUS: 11500
Ecozone origin-status: PAL: E(I); NEA: I; NEO: I; AUS: I
World rarity status: R
Ecozone-endemic macroregion occurrence: PAL: N EUR, W EUR, E EUR, C ASIA
Life form(s): EM
Ploidy state: D
Speciation age (Ma): NA
Populations also occur in terrestrial, wetland and/ or riparian habitats: YES
Long distance movement vectors: ANTHROPOCHORY
Nuisance status and uses: no nuisance issues; forage use, edible
Endangered status (IUCN): LC

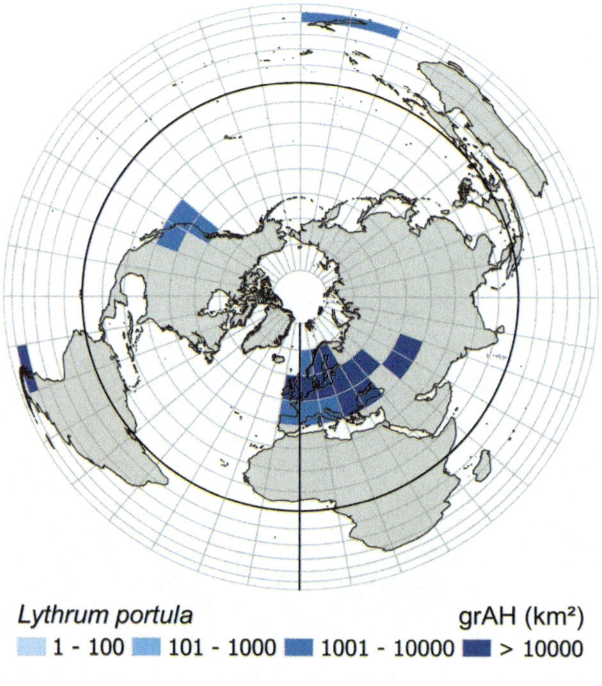

Lythrum portula (L.) D.A.Webb
Image copyright: J.C. Schou, with permission

***Lythrum salicaria* L.**
Synonyms:

Range area (grAH km^2): PAL: 751400; NEA: 505300; NEO: 57200; OR: 45600; AUS: 40100; AFR: 30800
Ecozone origin-status: PAL: N; NEA: I; NEO: I; OR: N; AUS: N; AFR: N
World rarity status: C
Ecozone-endemic macroregion occurrence: not endemic to any world ecozone
Life form(s): EM
Ploidy state: DP
Speciation age (Ma): NA
Populations also occur in terrestrial, wetland and/ or riparian habitats: YES
Long distance movement vectors: ANTHROPOCHORY; BIRD: *Anas crecca* (Eurasian Teal)
Nuisance status and uses: invasive aquatic weed in North America; medicinal and ornamental uses
Endangered status (IUCN): LC

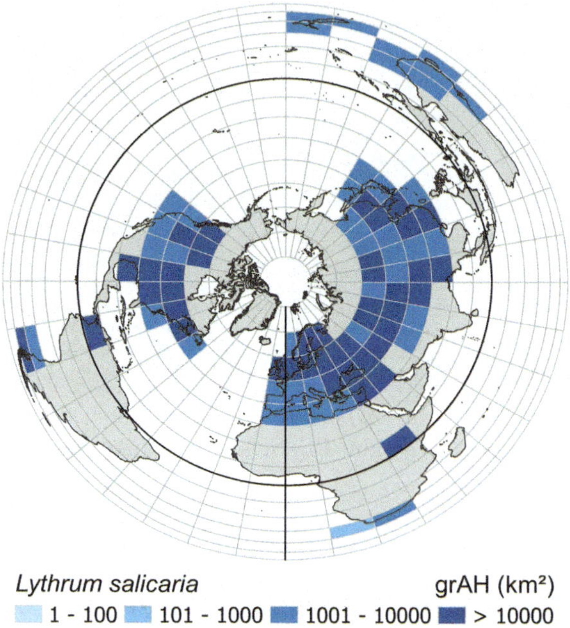

Lythrum salicaria L.
Image copyright: J.C. Schou, with permission

***Lythrum thymifolia* L.**

Synonyms: *Editeles thymifolia* (L.) Raf.; *Lythrum hyssopifolia* var. *thymifolia* (L.) Ball; *Lythrum hyssopifolia* subsp. *thymifolia* (L.) Batt.; *Middendorfia thymifolia* (L.) Tacik; *Salicaria thymifolia* (L.) Lam.

Range area (grAH km^2): PAL: 101500
Ecozone origin-status: PAL: E
World rarity status: R
Ecozone-endemic macroregion occurrence: PAL: MED, W EUR, E EUR, C ASIA
Life form(s): EM
Ploidy state: D
Speciation age (Ma): NA
Populations also occur in terrestrial, wetland and/ or riparian habitats: YES
Long distance movement vectors: NA
Nuisance status and uses: no nuisance issues; no known uses
Endangered status (IUCN): LC

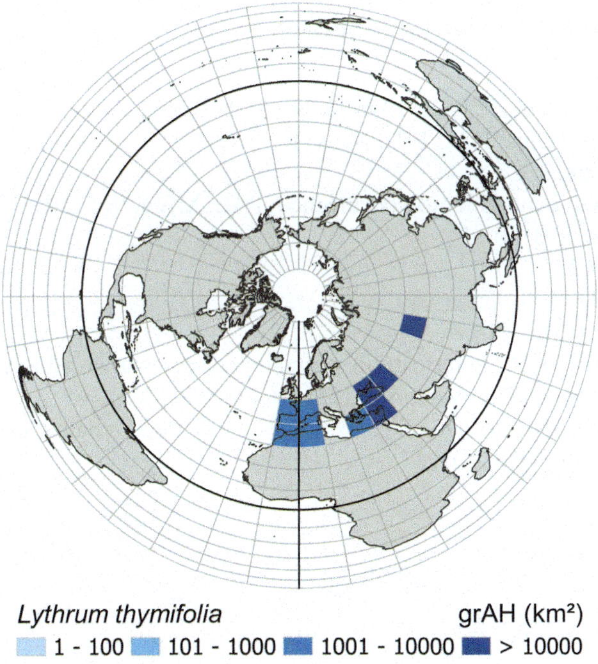

Lythrum thymifolia L.
Image copyright: A. Efremov, with permission

***Lythrum tribracteatum* Salzm. ex Spreng.**
Synonyms:

Range area (grAH km^2): PAL: 234600; NEA: 21100
Ecozone origin-status: PAL: E(I); NEA: I
World rarity status: C
Ecozone-endemic macroregion occurrence: PAL: MED, W EUR, E EUR, C ASIA, CH, UR/ W SIB
Life form(s): EM
Ploidy state: D
Speciation age (Ma): NA
Populations also occur in terrestrial, wetland and/ or riparian habitats: YES
Long distance movement vectors: ANTHROPOCHORY
Nuisance status and uses: no nuisance issues; occasional ornamental use
Endangered status (IUCN): LC

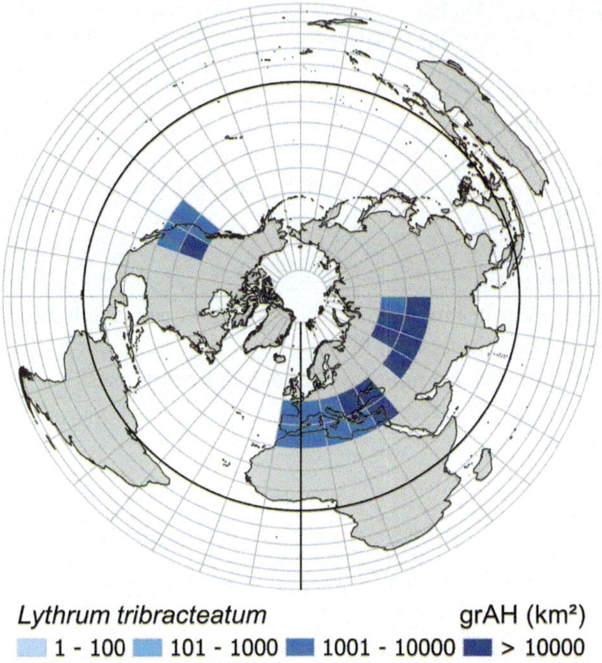

Lythrum tribracteatum Salzm. ex Spreng.

***Lythrum volgense* D.A.Webb**

Synonyms: *Peplis alternifolia* M.Bieb.

Range area (grAH km^2): PAL: 191800
Ecozone origin-status: PAL: E
World rarity status: C
Ecozone-endemic macroregion occurrence: PAL: E EUR, C ASIA, CH, UR/ W SIB
Life form(s): EM
Ploidy state: D
Speciation age (Ma): NA
Populations also occur in terrestrial, wetland and/ or riparian habitats: YES
Long distance movement vectors: NA
Nuisance status and uses: no nuisance issues; no known uses
Endangered status (IUCN): NA

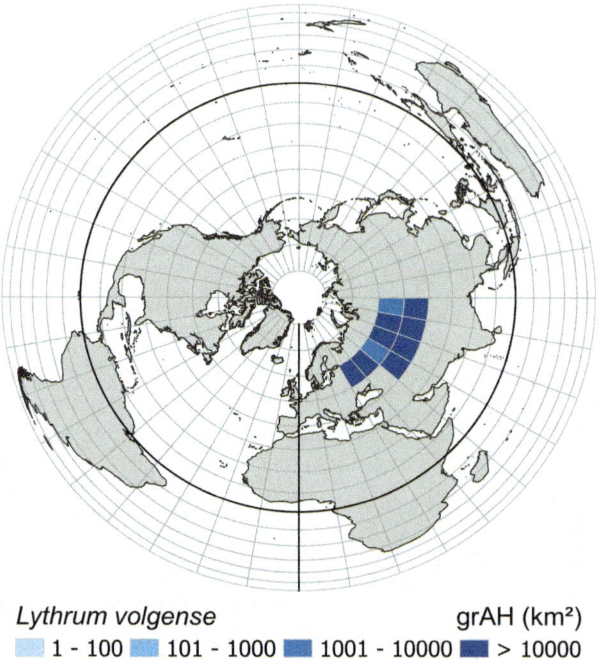

Nesaea pedicellata **Hiern**
Synonyms: *Ammannia pedicellata* (Hiern) S.A.Graham & Gandhi

Range area (grAH km^2): AFR: 93700
Ecozone origin-status: AFR: E
World rarity status: R
Ecozone-endemic macroregion occurrence: AFR: EA
Life form(s): SU, EM (subshrub)
Ploidy state: NA
Speciation age (Ma): NA
Populations also occur in terrestrial, wetland and/ or riparian habitats: YES
Long distance movement vectors: NA
Nuisance status and uses: no nuisance issues; aquarium trade
Endangered status (IUCN): VU

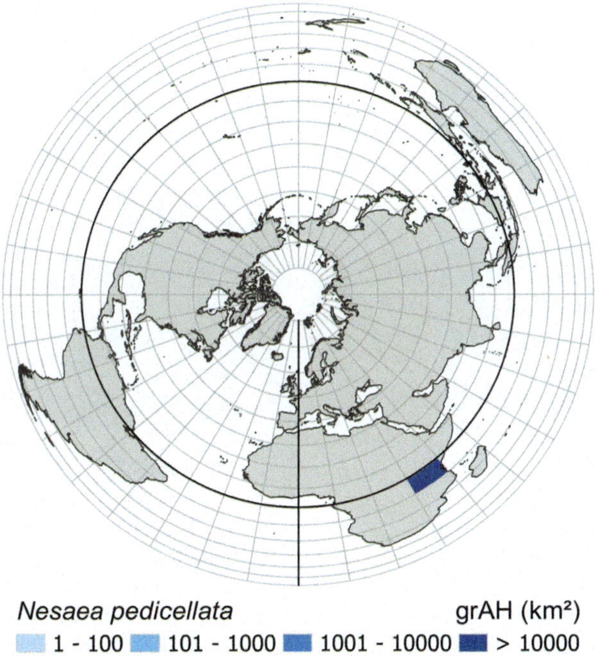

***Nesaea triflora* (L.f.) Kunth**
Synonyms:

Range area (grAH km^2): OR: 10300; AUS: 6300; AFR: 103800
Ecozone origin-status: OR: N; AUS: N; AFR: N
World rarity status: R
Ecozone-endemic macroregion occurrence: not endemic to any world ecozone
Life form(s): SU, EM (sometimes a subshrub)
Ploidy state: P
Speciation age (Ma): NA
Populations also occur in terrestrial, wetland and/ or riparian habitats: YES
Long distance movement vectors: NA
Nuisance status and uses: no nuisance issues; aquarium trade
Endangered status (IUCN): EN

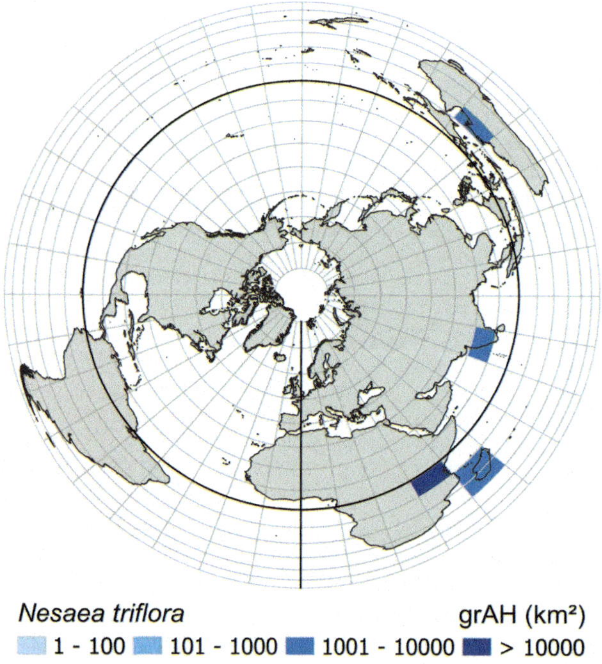

Rotala capensis **(Harv.) A.Fern. & Diniz**
Synonyms: *Suffrenia capensis* Harv.

Range area (grAH km^2): AFR: 65100
Ecozone origin-status: AFR: E
World rarity status: R
Ecozone-endemic macroregion occurrence: AFR: FOR, ZAM, SDES
Life form(s): SU, EM
Ploidy state: NA
Speciation age (Ma): NA
Populations also occur in terrestrial, wetland and/ or riparian habitats: YES
Long distance movement vectors: NA
Nuisance status and uses: no nuisance issues; no known uses
Endangered status (IUCN): NA (classed as LC in South Africa, where most of its records occur)

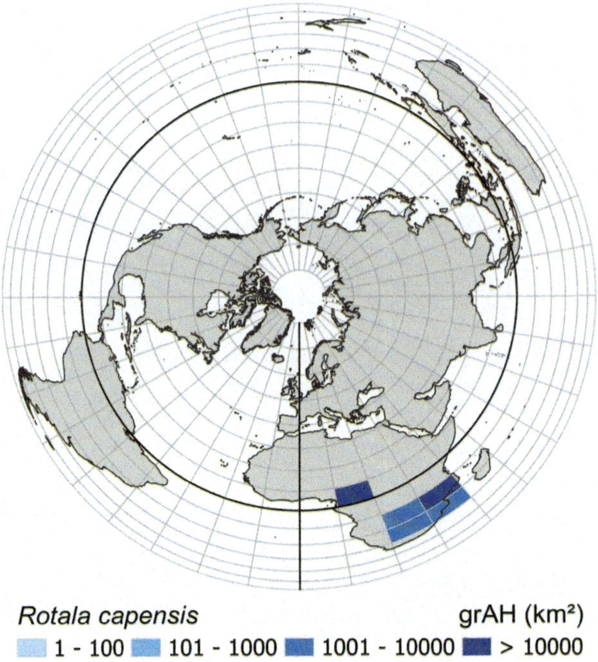

Rotala cookii **K.T.Joseph & Sivar.**
Synonyms:

Range area (grAH km^2): OR: 8100
Ecozone origin-status: OR: E
World rarity status: VR
Ecozone-endemic macroregion occurrence: OR: IND
Life form(s): SU, EM
Ploidy state: NA
Speciation age (Ma): NA
Populations also occur in terrestrial, wetland and/ or riparian habitats: NO
Long distance movement vectors: NA
Nuisance status and uses: no nuisance issues; no known uses
Endangered status (IUCN): EN

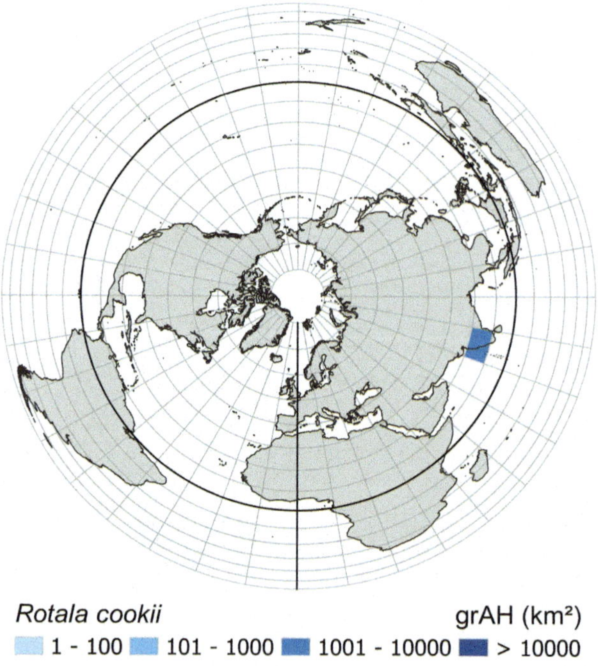

***Rotala cordata* Koehne**
Synonyms:

Range area (grAH km²): OR: 53700
Ecozone origin-status: OR: E
World rarity status: R
Ecozone-endemic macroregion occurrence: OR: IND, ICH
Life form(s): SU, EM
Ploidy state: NA
Speciation age (Ma): NA
Populations also occur in terrestrial, wetland and/ or riparian habitats: YES
Long distance movement vectors: NA
Nuisance status and uses: no nuisance issues; no known uses
Endangered status (IUCN): NA

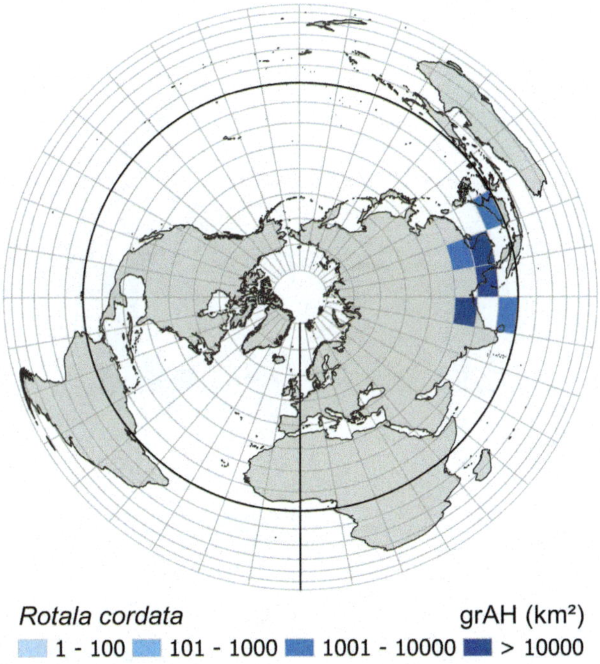

Rotala densiflora **(Roth) Koehne**
Synonyms: *Ammannia densiflora* Roth; *Ditheca densiflora* (Roth) Miq.

Range area (grAH km^2): PAL: 49300; OR: 112200; AUS: 12900; AFR: 24000
Ecozone origin-status: PAL: N; OR: N; AUS: N; AFR: I
World rarity status: C
Ecozone-endemic macroregion occurrence: not endemic to any world ecozone
Life form(s): SU, EM
Ploidy state: NA
Speciation age (Ma): NA
Populations also occur in terrestrial, wetland and/ or riparian habitats: YES
Long distance movement vectors: ANTHROPOCHORY
Nuisance status and uses: some reported aquatic weed issues, mainly in rice paddy fields in Asia; folk medicine
Endangered status (IUCN): LC

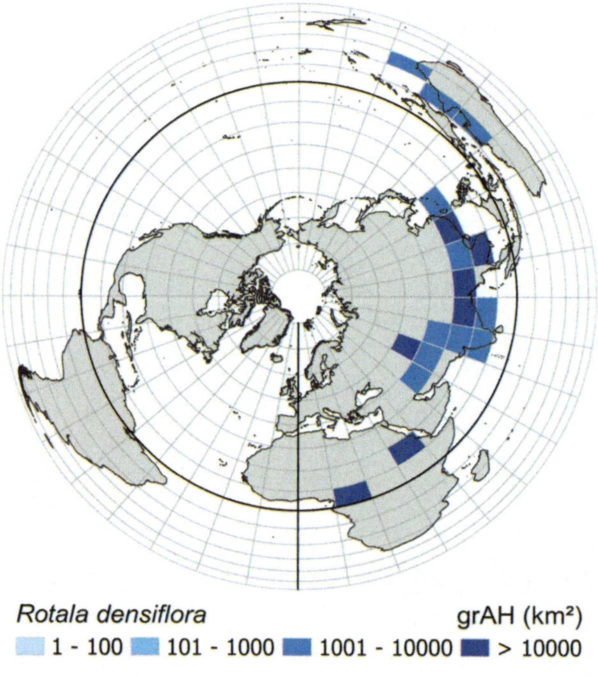

Rotala diandra **(F.Muell.) Koehne**
Synonyms: *Ameletia diandra* F.Muell.; *Ammannia diandra* (F.Muell.) Benth.

Range area (grAH km²): AUS: 21500
Ecozone origin-status: AUS: E
World rarity status: VR
Ecozone-endemic macroregion occurrence: AUS: NA, WA
Life form(s): SU, EM
Ploidy state: NA
Speciation age (Ma): NA
Populations also occur in terrestrial, wetland and/ or riparian habitats: YES
Long distance movement vectors: NA
Nuisance status and uses: no nuisance issues; no known uses
Endangered status (IUCN): NA

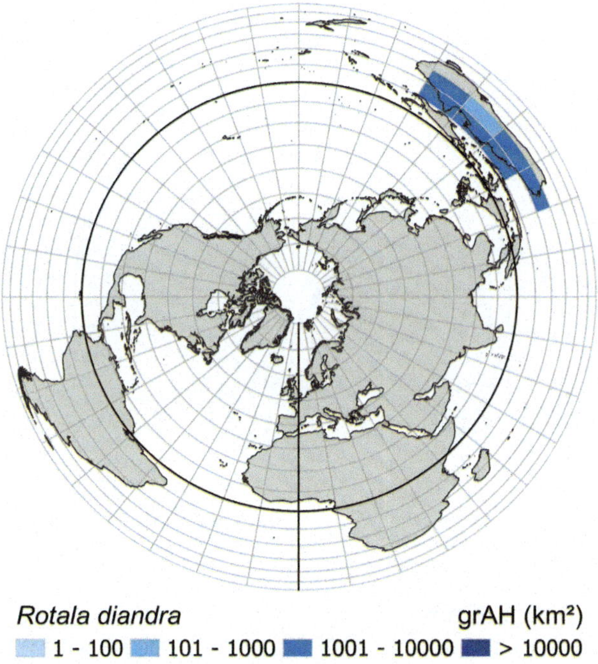

***Rotala dinteri* Koehne**
Synonyms:

Range area (grAH km^2): AFR: 11700
Ecozone origin-status: AFR: E
World rarity status: VR
Ecozone-endemic macroregion occurrence: AFR: ZAM
Life form(s): SU, ?EM (sometimes a subshrub)
Ploidy state: NA
Speciation age (Ma): NA
Populations also occur in terrestrial, wetland and/ or riparian habitats: YES
Long distance movement vectors: NA
Nuisance status and uses: no nuisance issues; no known uses
Endangered status (IUCN): NA

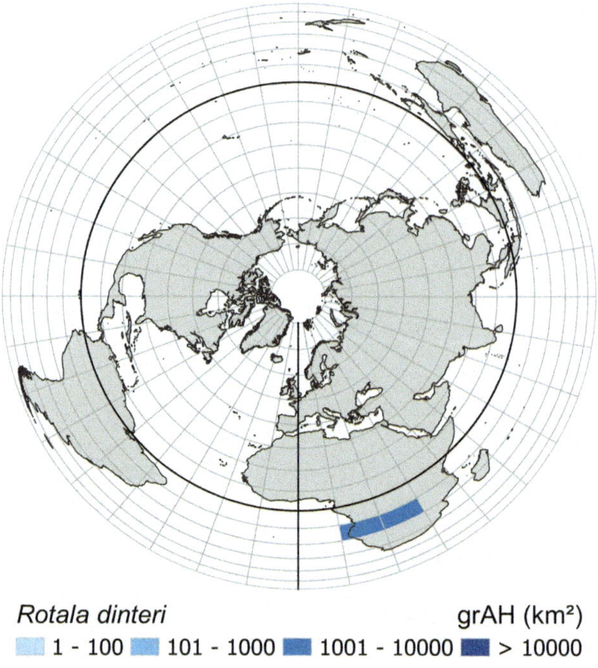

***Rotala elatinoides* Hiern**
Synonyms:

Range area (grAH km²): AFR: 36800
Ecozone origin-status: AFR: E
World rarity status: R
Ecozone-endemic macroregion occurrence: AFR: SAH/ AR, FOR
Life form(s): SU, EM
Ploidy state: NA
Speciation age (Ma): NA
Populations also occur in terrestrial, wetland and/ or riparian habitats: YES
Long distance movement vectors: NA
Nuisance status and uses: no nuisance issues; no known uses
Endangered status (IUCN): LC

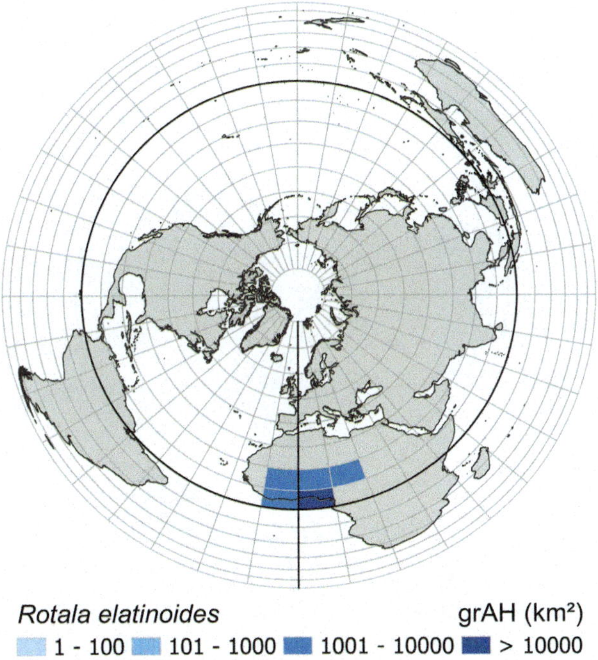

***Rotala filiformis* (Bellardi) Hiern**
Synonyms: *Suffrenia filiformis* Bellardi

Range area (grAH km^2): AFR: 195500
Ecozone origin-status: AFR: E
World rarity status: C
Ecozone-endemic macroregion occurrence: AFR: FOR, EA, ZAM, SDES
Life form(s): SU, EM
Ploidy state: NA
Speciation age (Ma): NA
Populations also occur in terrestrial, wetland and/ or riparian habitats: YES
Long distance movement vectors: NA
Nuisance status and uses: no nuisance issues; no known uses
Endangered status (IUCN): NA

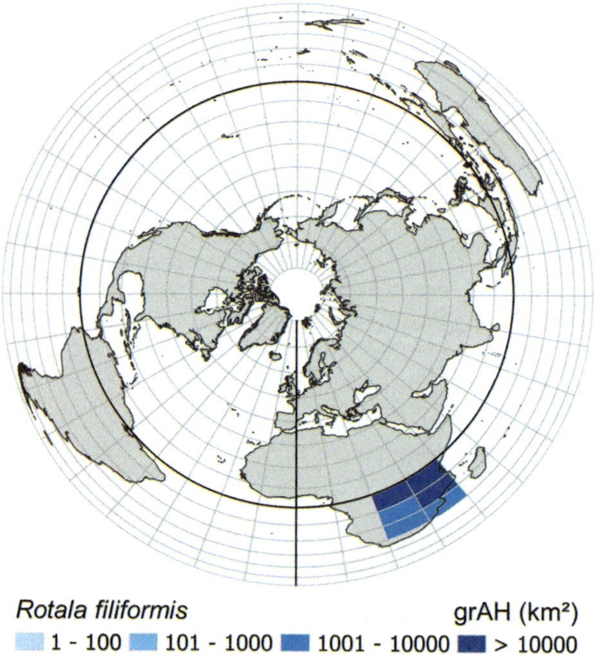

Rotala filiformis (Bellardi) Hiern
Image copyright: R. Fourie, with permission

Rotala fimbriata Wight
Synonyms: *Ammannia pentandra* var. *fimbriata* (Wight) C.B.Clarke

Range area (grAH km^2): OR: 11900
Ecozone origin-status: OR: E
World rarity status: VR
Ecozone-endemic macroregion occurrence: OR: IND
Life form(s): SU, EM
Ploidy state: NA
Speciation age (Ma): NA
Populations also occur in terrestrial, wetland and/ or riparian habitats: YES
Long distance movement vectors: NA
Nuisance status and uses: no nuisance issues; no known uses
Endangered status (IUCN): LC

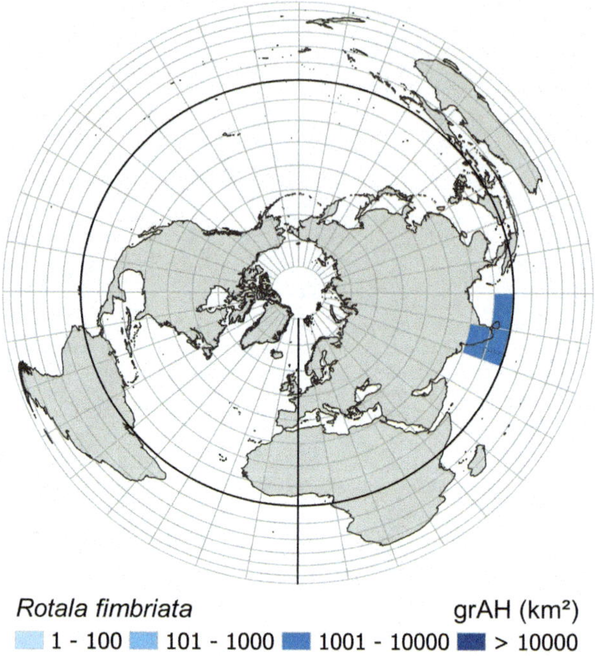

***Rotala floribunda* (Wight) Koehne**
Synonyms: *Ameletia floribunda* (Wight) Wight; *Nimmoia floribunda* Wight

Range area (grAH km^2): OR: 8100
Ecozone origin-status: OR: E
World rarity status: VR
Ecozone-endemic macroregion occurrence: OR: IND
Life form(s): EM (sometimes rheophytic on steep sloping rocks and cliff faces on which water is trickling down; as well as lake shores)
Ploidy state: NA
Speciation age (Ma): NA
Populations also occur in terrestrial, wetland and/ or riparian habitats: YES
Long distance movement vectors: NA
Nuisance status and uses: no nuisance issues; folk medicine
Endangered status (IUCN): VU

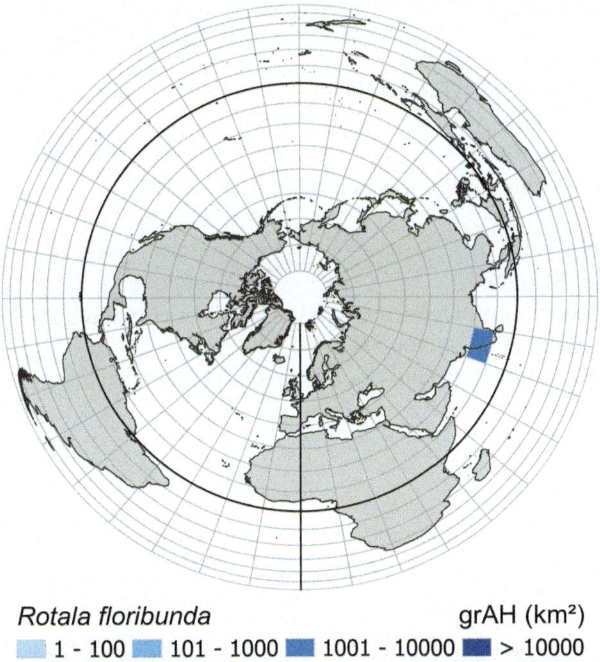

Rotala floribunda (Wight) Koehne

***Rotala fluitans* Pohnert**
Synonyms:

Range area (grAH km^2): AFR: 11700
Ecozone origin-status: AFR: E
World rarity status: VR
Ecozone-endemic macroregion occurrence: AFR: ZAM
Life form(s): ?SU, EM
Ploidy state: NA
Speciation age (Ma): NA
Populations also occur in terrestrial, wetland and/ or riparian habitats: YES
Long distance movement vectors: NA
Nuisance status and uses: no nuisance issues; no known uses
Endangered status (IUCN): LC

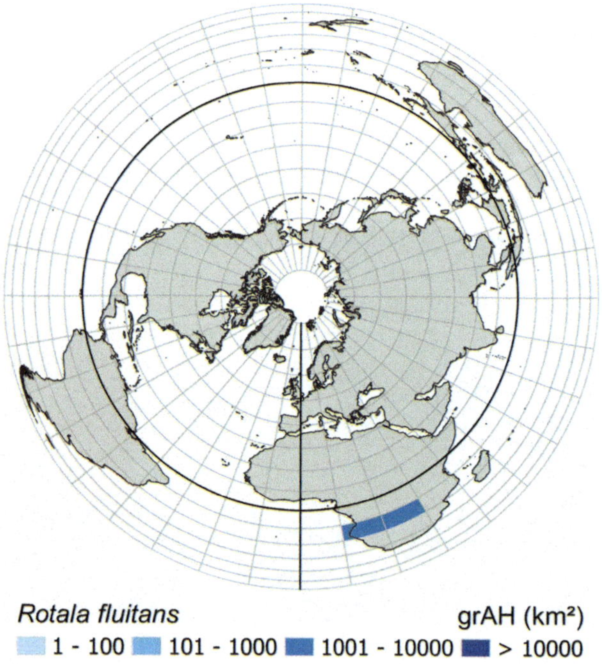

Rotala fontinalis **Hiern**
Synonyms:

Range area (grAH km^2): AFR: 54200
Ecozone origin-status: AFR: E
World rarity status: R
Ecozone-endemic macroregion occurrence: AFR: SAH/ AR, ZAM
Life form(s): EM (subshrub)
Ploidy state: NA
Speciation age (Ma): NA
Populations also occur in terrestrial, wetland and/ or riparian habitats: YES
Long distance movement vectors: NA
Nuisance status and uses: no nuisance issues; no known uses
Endangered status (IUCN): DD

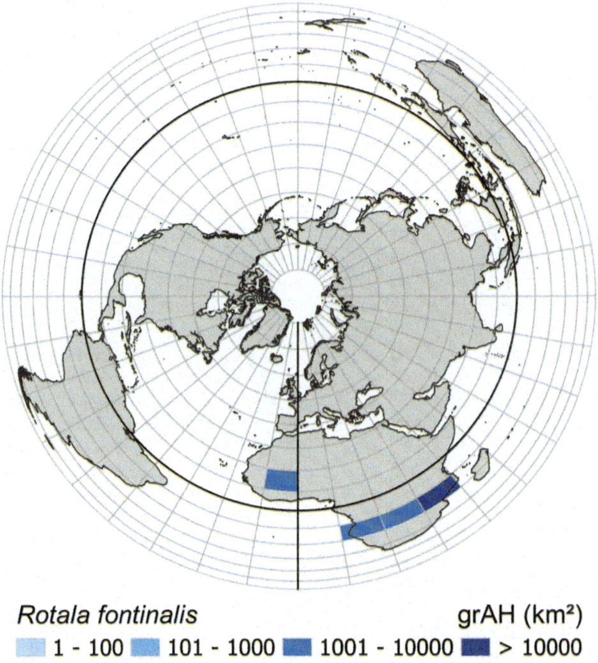

***Rotala gerardii* Boutique**

Synonyms:

Range area (grAH km²): AFR: 27200
Ecozone origin-status: AFR: E
World rarity status: VR
Ecozone-endemic macroregion occurrence: AFR: FOR
Life form(s): ?EM (subshrub)
Ploidy state: NA
Speciation age (Ma): NA
Populations also occur in terrestrial, wetland and/ or riparian habitats: YES
Long distance movement vectors: NA
Nuisance status and uses: no nuisance issues; no known uses
Endangered status (IUCN): NT

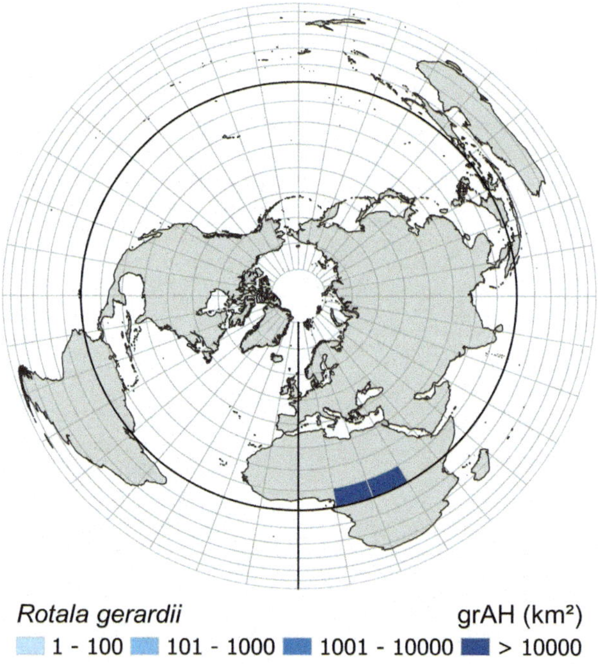

***Rotala gossweileri* Koehne**
Synonyms:

Range area (grAH km^2): AFR: 165000
Ecozone origin-status: AFR: E
World rarity status: C
Ecozone-endemic macroregion occurrence: AFR: SAH/ AR, FOR, EA, ZAM
Life form(s): EM
Ploidy state: NA
Speciation age (Ma): NA
Populations also occur in terrestrial, wetland and/ or riparian habitats: YES
Long distance movement vectors: NA
Nuisance status and uses: no nuisance issues; no known uses
Endangered status (IUCN): LC

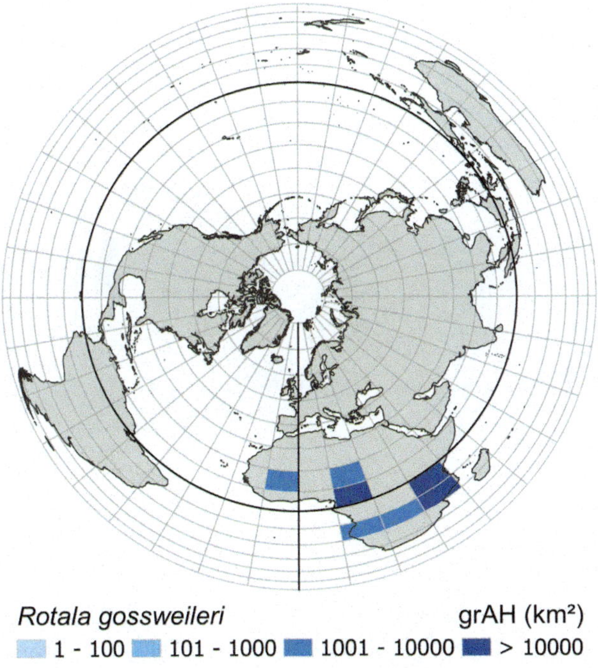

Rotala hexandra **Wall. ex Koehne**
Synonyms:

Range area (grAH km^2): OR: 36100
Ecozone origin-status: OR: E
World rarity status: R
Ecozone-endemic macroregion occurrence: OR: ICH, SUP
Life form(s): EM
Ploidy state: NA
Speciation age (Ma): NA
Populations also occur in terrestrial, wetland and/ or riparian habitats: YES
Long distance movement vectors: NA
Nuisance status and uses: no nuisance issues; no known uses
Endangered status (IUCN): NA

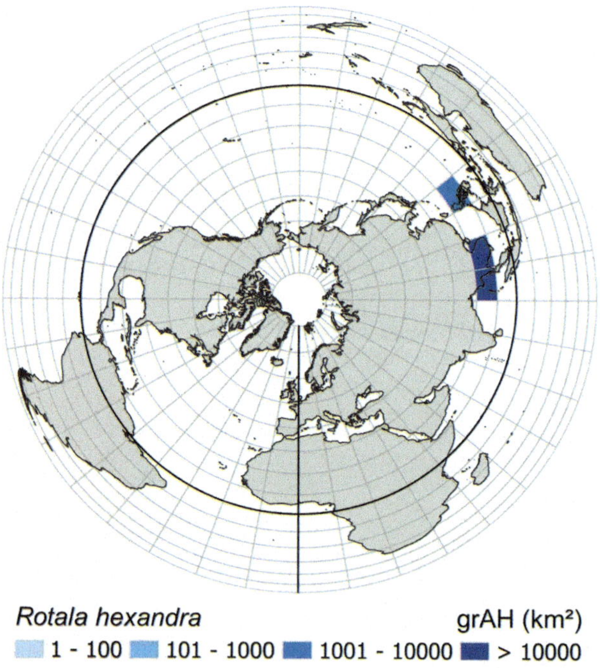

Rotala hippuris **Makino**
Synonyms:

Range area (grAH km^2): PAL: 7500
Ecozone origin-status: PAL: E
World rarity status: VR
Ecozone-endemic macroregion occurrence: PAL: FER/ J
Life form(s): SU, EM
Ploidy state: NA
Speciation age (Ma): NA
Populations also occur in terrestrial, wetland and/ or riparian habitats: YES

Long distance movement vectors: NA
Nuisance status and uses: no nuisance issues; aquarium trade
Endangered status (IUCN): NA

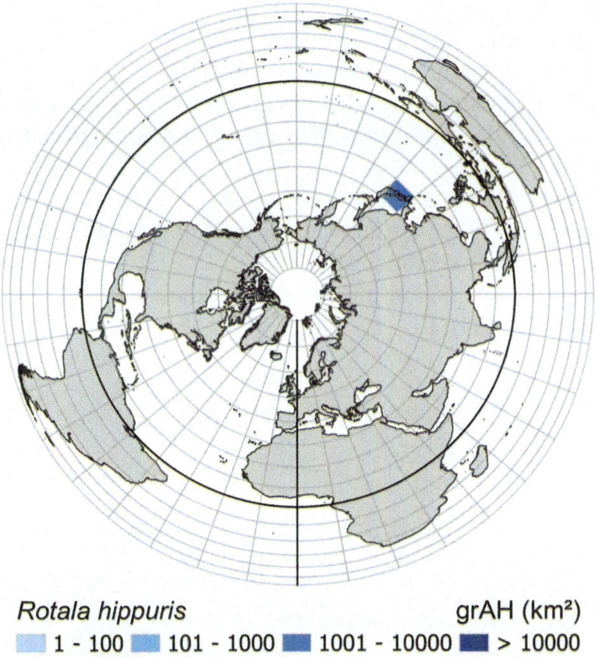

Rotala hippuris Makino
Image copyright: Tian-Chuan Hsu, with permission

***Rotala illecebroides* (Arn.) Koehne**
Synonyms: *Ammannia illecebroides* Arn.

Range area (grAH km^2): OR: 27700
Ecozone origin-status: OR: E
World rarity status: VR
Ecozone-endemic macroregion occurrence: OR: IND
Life form(s): SU, EM (sometimes rheophytic, on wet rocks beside streams, and on dripping rocks in hilly areas; as well as in marshes, springs and wet flushe)
Ploidy state: NA
Speciation age (Ma): NA
Populations also occur in terrestrial, wetland and/ or riparian habitats: YES
Long distance movement vectors: NA
Nuisance status and uses: no nuisance issues; no known uses
Endangered status (IUCN): LC

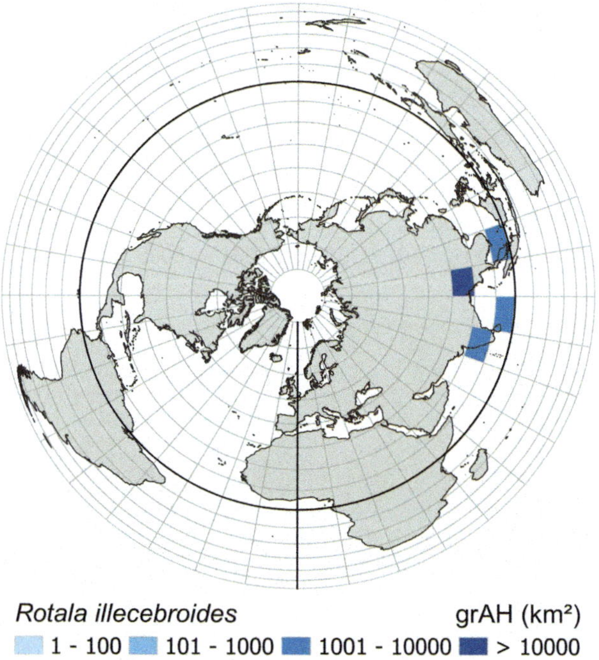

***Rotala indica* (Willd.) Koehne**

Synonyms: *Ameletia indica* (Willd.) DC.; *Ammannia peploides* Spreng.; *Peplis indica* Willd.

Range area (grAH km²): PAL: 108600; NEA: 1000; OR: 115100; AUS: 5400; AFR: 18200
Ecozone origin-status: PAL: N; NEA: I; OR: N; AUS: I; AFR: I
World rarity status: C
Ecozone-endemic macroregion occurrence: not endemic to any world ecozone
Life form(s): SU, EM
Ploidy state: P
Speciation age (Ma): NA
Populations also occur in terrestrial, wetland and/ or riparian habitats: YES
Long distance movement vectors: ANTHROPOCHORY
Nuisance status and uses: aquatic weed in rice paddies in Asia; aquarium trade
Endangered status (IUCN): LC

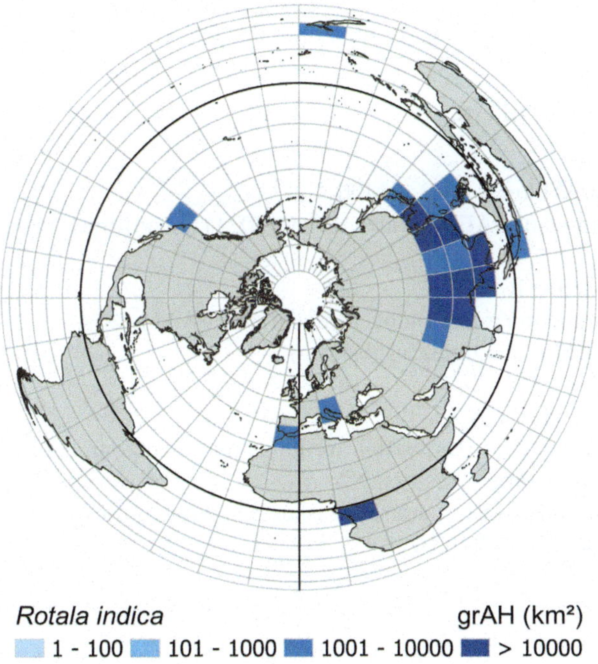

Rotala indica (Willd.) Koehne

***Rotala juniperina* A.Fern.**
Synonyms:

Range area (grAH km^2): AFR: 43400
Ecozone origin-status: AFR: E (also recent (2021) unconfirmed report from western India)
World rarity status: R
Ecozone-endemic macroregion occurrence: AFR: ZAM
Life form(s): EM (sometimes a subshrub)
Ploidy state: NA
Speciation age (Ma): NA
Populations also occur in terrestrial, wetland and/ or riparian habitats: YES
Long distance movement vectors: NA
Nuisance status and uses: no nuisance issues; no known uses
Endangered status (IUCN): NA

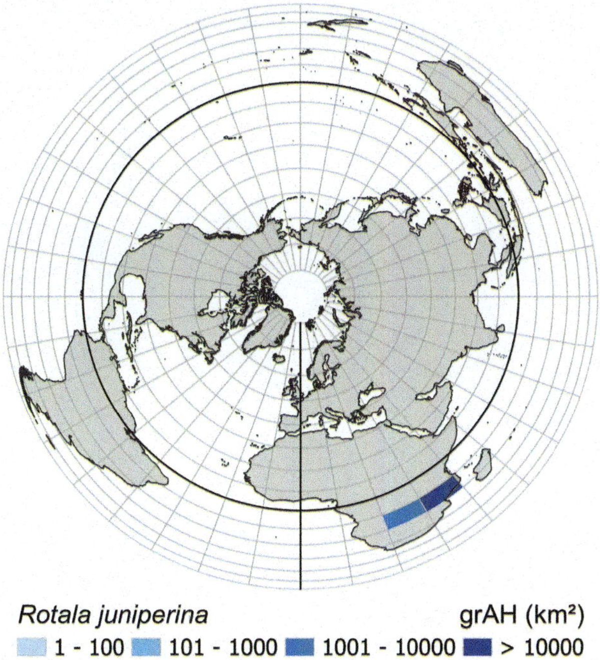

Rotala lucalensis **A.Fern. & Diniz**
Synonyms:

Range area (grAH km^2): OR: 8100; AFR: 46400
Ecozone origin-status: OR: I; AFR: E(I)
World rarity status: R
Ecozone-endemic macroregion occurrence: AFR: ZAM
Life form(s): EM
Ploidy state: NA
Speciation age (Ma): NA
Populations also occur in terrestrial, wetland and/ or riparian habitats: YES
Long distance movement vectors: ANTHROPOCHORY
Nuisance status and uses: no nuisance issues; no known uses
Endangered status (IUCN): NA

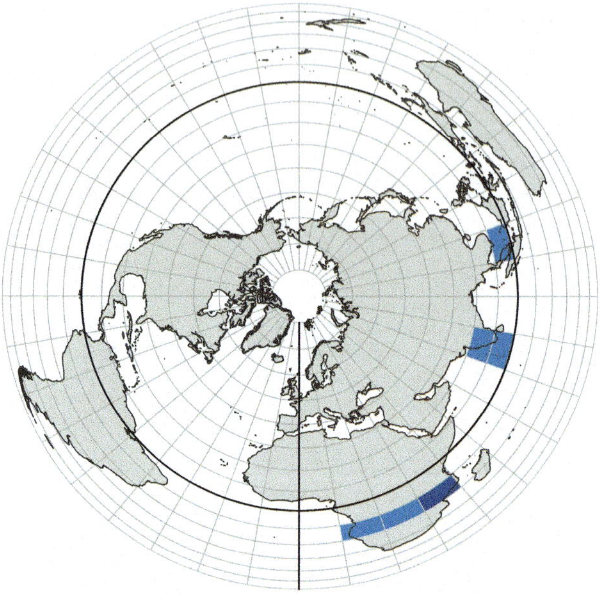

Rotala macrandra **Koehne**

Synonyms: *Rotala rotundifolia* var. *macrandra* (Koehne) M.R.Almeida

Range area (grAH km^2): OR: 36200
Ecozone origin-status: OR: E
World rarity status: R
Ecozone-endemic macroregion occurrence: OR: IND, SUP
Life form(s): SU, EM
Ploidy state: NA
Speciation age (Ma): NA
Populations also occur in terrestrial, wetland and/ or riparian habitats: YES
Long distance movement vectors: NA
Nuisance status and uses: no nuisance issues; aquarium trade
Endangered status (IUCN): LC

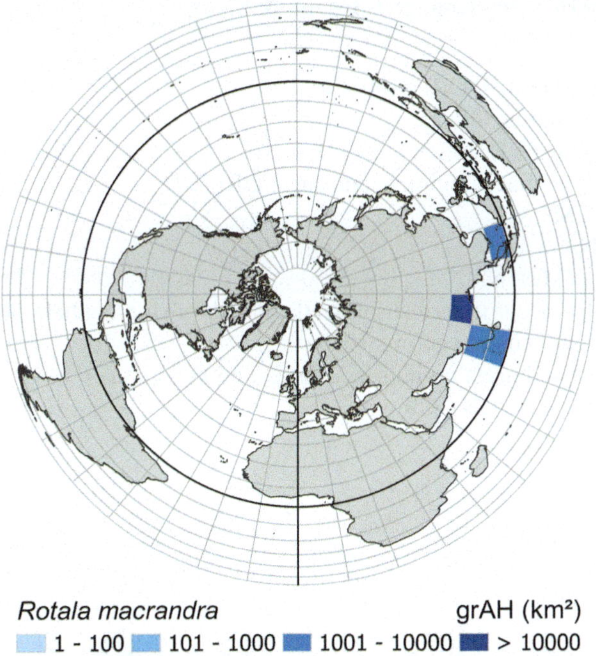

Rotala macrandra Koehne

Rotala malabarica **Pradeep, K.T.Joseph & Sivar.**
Synonyms:

Range area (grAH km^2): OR: 8100
Ecozone origin-status: OR: E
World rarity status: VR
Ecozone-endemic macroregion occurrence: OR: IND
Life form(s): EM
Ploidy state: NA
Speciation age (Ma): NA
Populations also occur in terrestrial, wetland and/ or riparian habitats: YES
Long distance movement vectors: NA
Nuisance status and uses: no nuisance issues; no known uses
Endangered status (IUCN): CR

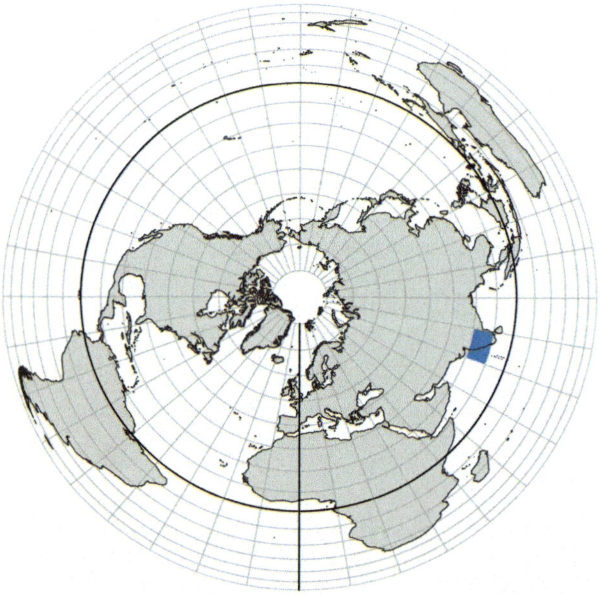

***Rotala malampuzhensis* R.V.Nair ex C.D.K.Cook**

Synonyms:

Range area (grAH km^2): World: OR: 8100
Ecozone origin-status: OR: E
World rarity status: VR
Ecozone-endemic macroregion occurrence: OR: IND
Life form(s): EM
Ploidy state: NA
Speciation age (Ma): NA
Populations also occur in terrestrial, wetland and/ or riparian habitats: YES
Long distance movement vectors: NA
Nuisance status and uses: no nuisance issues; no known uses
Endangered status (IUCN): LC

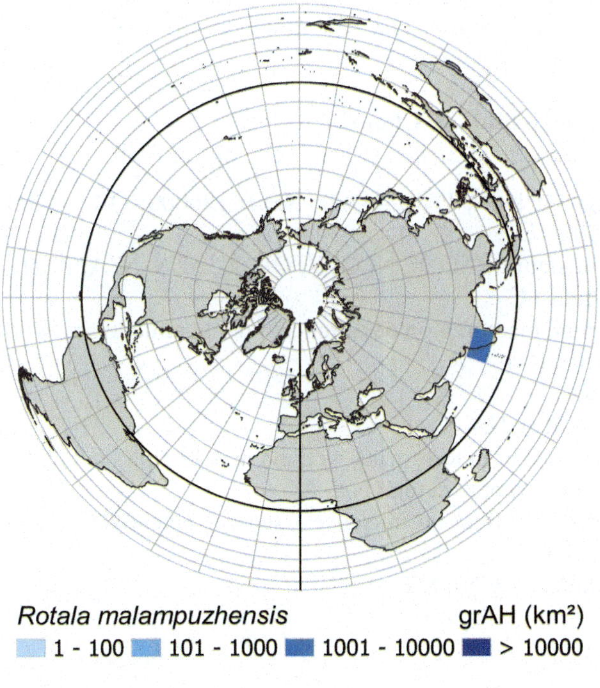

Rotala malampuzhensis R.V.Nair ex C.D.K.Cook

***Rotala mexicana* Schltdl. & Cham.**
Synonyms: *Ammannia mexicana* (Cham. & Schltdl.) Baill.

Range area (grAH km^2): PAL: 44700; NEO: 186600; OR: 31700; AUS: 51800; AFR: 125500
Ecozone origin-status: PAL: N; NEO: N; OR: N; AUS: N; AFR: N
World rarity status: C
Ecozone-endemic macroregion occurrence: not endemic to any world ecozone
Life form(s): SU, EM
Ploidy state: NA
Speciation age (Ma): NA
Populations also occur in terrestrial, wetland and/ or riparian habitats: YES
Long distance movement vectors: NA
Nuisance status and uses: no nuisance issues; a potential source of insecticides, aquarium trade
Endangered status (IUCN): LC

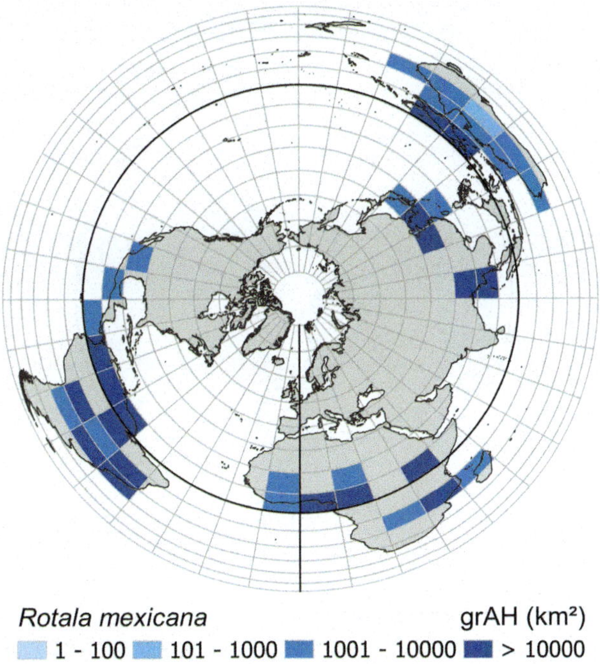

Rotala milne-redheadii **A.Fern. & Diniz**
Synonyms:

Range area (grAH km²): AFR: 8700
Ecozone origin-status: AFR: E
World rarity status: VR
Ecozone-endemic macroregion occurrence: AFR: ZAM
Life form(s): EM
Ploidy state: NA
Speciation age (Ma): NA
Populations also occur in terrestrial, wetland and/ or riparian habitats: YES
Long distance movement vectors: NA
Nuisance status and uses: no nuisance issues; no known uses
Endangered status (IUCN): NA

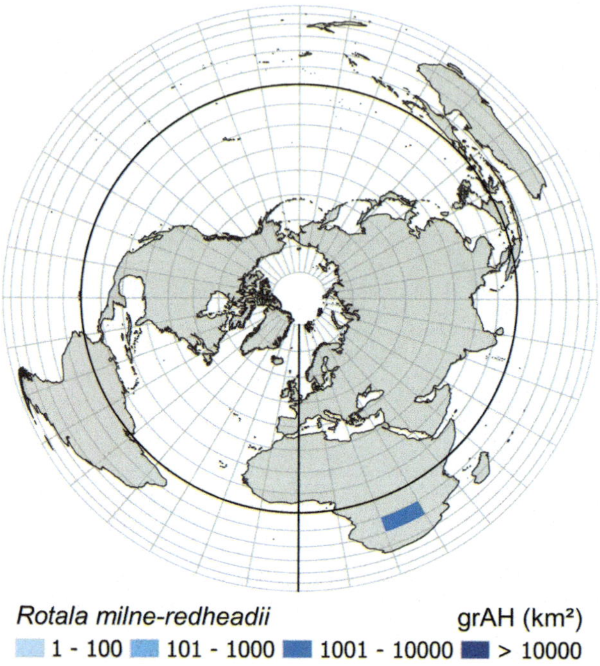

Rotala myriophylloides **Welw. ex Hiern**
Synonyms:

Range area (grAH km^2): AFR: 150100
Ecozone origin-status: AFR: E
World rarity status: R
Ecozone-endemic macroregion occurrence: AFR: EA, ZAM, MAD
Life form(s): SU, EM
Ploidy state: NA
Speciation age (Ma): NA
Populations also occur in terrestrial, wetland and/ or riparian habitats: YES
Long distance movement vectors: NA
Nuisance status and uses: no nuisance issues; some medicinal uses
Endangered status (IUCN): NA

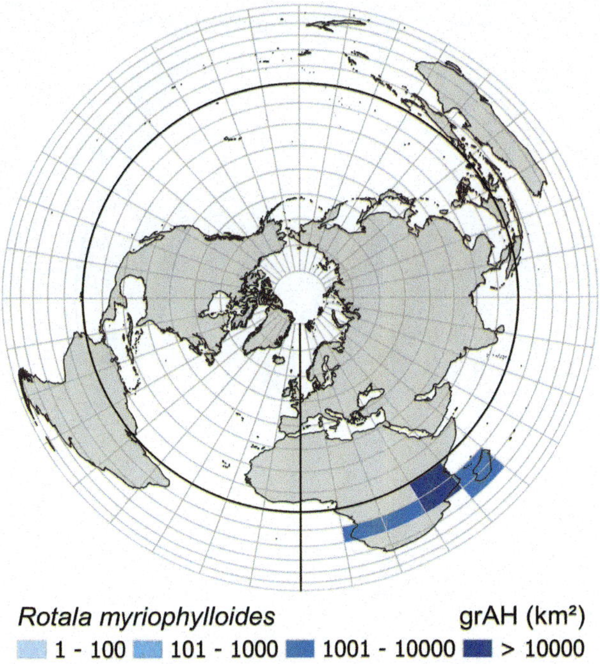

Rotala occultiflora **Koehne**
Synonyms:

Range area (grAH km^2): OR: 8100; AUS: 26100
Ecozone origin-status: OR: N; AUS: N
World rarity status: R
Ecozone-endemic macroregion occurrence: not endemic to any world ecozone
Life form(s): SU, EM
Ploidy state: NA
Speciation age (Ma): NA
Populations also occur in terrestrial, wetland and/ or riparian habitats: YES
Long distance movement vectors: NA
Nuisance status and uses: no nuisance issues; aquarium trade
Endangered status (IUCN): LC

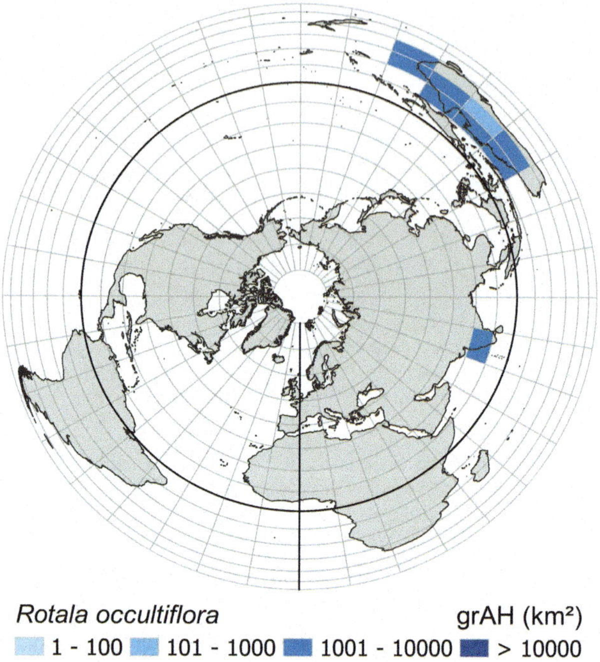

***Rotala pterocalyx* A.Raynal**
Synonyms:

Range area (grAH km²): AFR: 27500
Ecozone origin-status: AFR: E
World rarity status: R
Ecozone-endemic macroregion occurrence: AFR: SAH/ AR, FOR
Life form(s): EM
Ploidy state: NA
Speciation age (Ma): NA
Populations also occur in terrestrial, wetland and/ or riparian habitats: YES
Long distance movement vectors: NA
Nuisance status and uses: no nuisance issues; no known uses
Endangered status (IUCN): NA

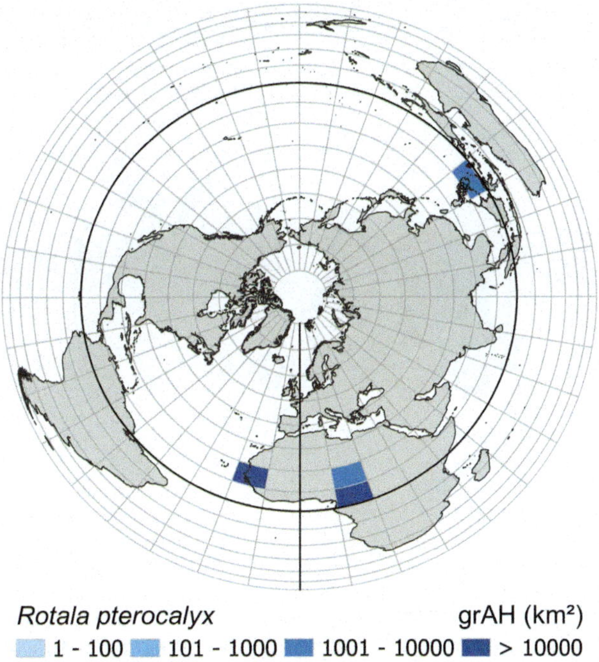

Rotala ramosior **(L.) Koehne**
Synonyms: *Ammannia ramosior* L.

Range area (grAH km^2): PAL: 26300; NEA: 390400; NEO: 188000; OR: 21500
Ecozone origin-status: PAL: I; NEA: N; NEO: N; OR: I
World rarity status: C
Ecozone-endemic macroregion occurrence: not endemic to any world ecosystem
Life form(s): SU, EM
Ploidy state: DP
Speciation age (Ma): NA
Populations also occur in terrestrial, wetland and/ or riparian habitats: YES
Long distance movement vectors: ANTHROPOCHORY
Nuisance status and uses: occasional nuisance issues reported when growing outside its native range, notably in rice fields in Europe, Taiwan and the Philippines; aquarium trade, medicinal potential
Endangered status (IUCN): LC

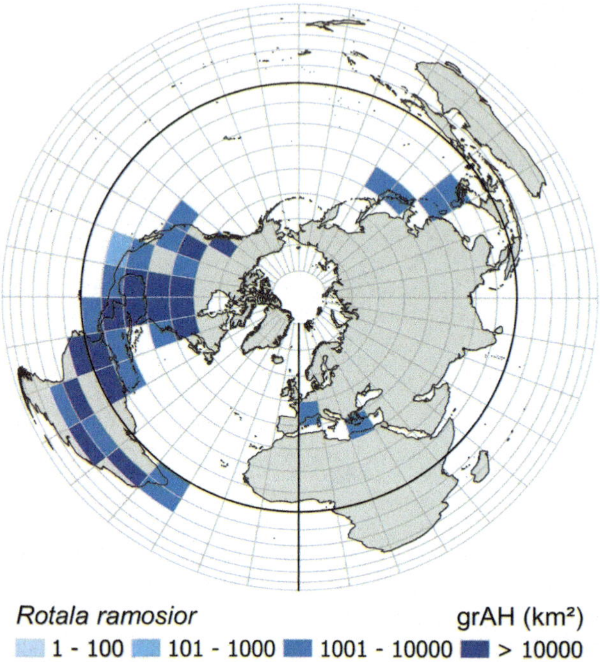

Rotala ramosior (L.) Koehne
Image copyright: iNaturalist CC-BY @calloftheloon

***Rotala ritchiei* Koehne**

Synonyms:

Range area (grAH km²): OR: 8100
Ecozone origin-status: OR: E
World rarity status: VR
Ecozone-endemic macroregion occurrence: OR: IND
Life form(s): EM
Ploidy state: NA
Speciation age (Ma): NA
Populations also occur in terrestrial, wetland and/ or riparian habitats: NO
Long distance movement vectors: NA
Nuisance status and uses: no nuisance issues; no known uses
Endangered status (IUCN): EN

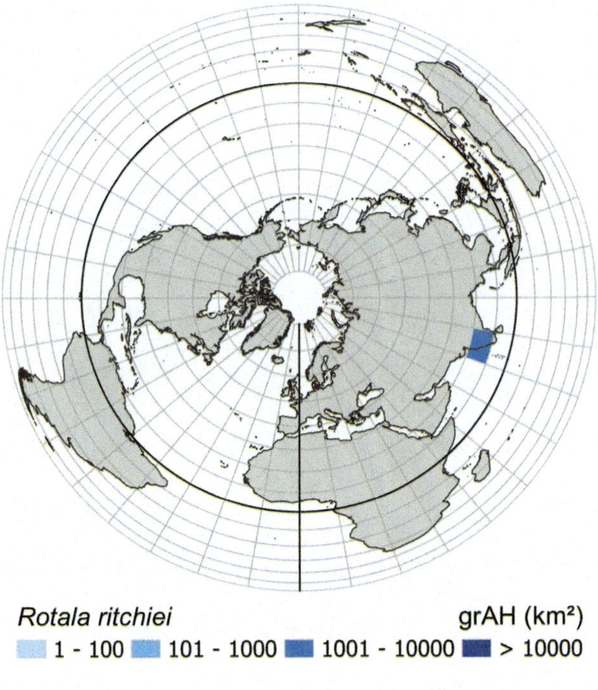

***Rotala rotundifolia* (Buch.-Ham. ex Roxb.) Koehne**

Synonyms: *Ameletia rotundifolia* (Buch.-Ham. ex Roxb.) Wight; *Ammannia rotundifolia* Buch.-Ham. ex Roxb.; *Mirkooa rotundifolia* (Buch.-Ham. ex Roxb.) Wight

Range area (grAH km^2): PAL: 44700; NEA: 13000; NEO: 21700; OR: 103900; AUS: 18900
Ecozone origin-status: PAL: N; NEA: I; NEO: I; OR:N; AUS: I
World rarity status: C
Ecozone-endemic macroregion occurrence: not endemic to any world ecozone
Life form(s): SU, EM
Ploidy state: NA
Speciation age (Ma): NA
Populations also occur in terrestrial, wetland and/ or riparian habitats: YES
Long distance movement vectors: ANTHROPOCHORY
Nuisance status and uses: invasive aquatic weed in North America, also a common weed in rice paddies and wet places in India, China, Taiwan, Thailand, Laos, and Vietnam; aquarium trade, medicinal and phytoremediation potential uses
Endangered status (IUCN): ELC

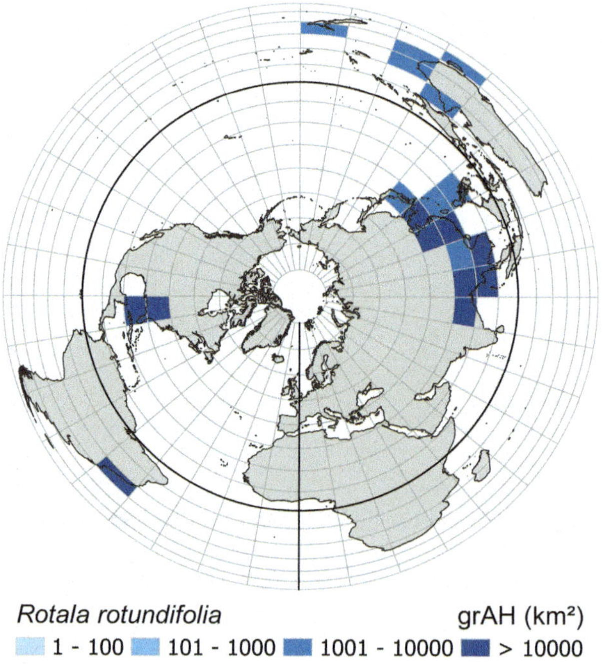

Rotala rotundifolia (Buch.-Ham. ex Roxb.) Koehne
Image copyright: A.T. Dung-Le, with permission

Rotala rubra **(Buch.-Ham. ex D.Don) H.Hara**
Synonyms: *Ammania rubra* Buch.-Ham. ex D. Don

Range area (grAH km^2): OR: 19900
Ecozone origin-status: OR: E
World rarity status: VR
Ecozone-endemic macroregion occurrence: OR: IND
Life form(s): SU, EM
Ploidy state: NA
Speciation age (Ma): NA
Populations also occur in terrestrial, wetland and/ or riparian habitats: YES
Long distance movement vectors: NA
Nuisance status and uses: no nuisance issues; no known uses
Endangered status (IUCN): NA

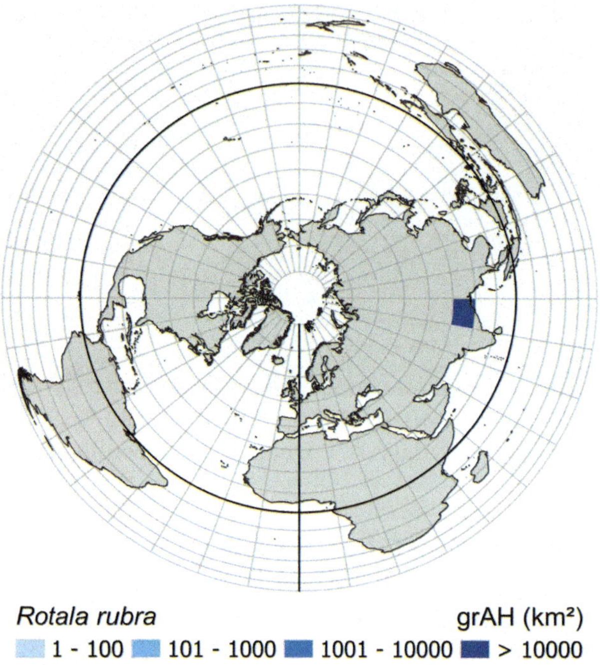

Rotala serpiculoides **Welw. ex Hiern**
Synonyms:

Range area (grAH km^2): AFR: 96600
Ecozone origin-status: AFR: E
World rarity status: R
Ecozone-endemic macroregion occurrence: AFR: EA, ZAM
Life form(s): EM
Ploidy state: NA
Speciation age (Ma): NA
Populations also occur in terrestrial, wetland and/ or riparian habitats: YES
Long distance movement vectors: NA
Nuisance status and uses: no nuisance issues; no known uses
Endangered status (IUCN): NA

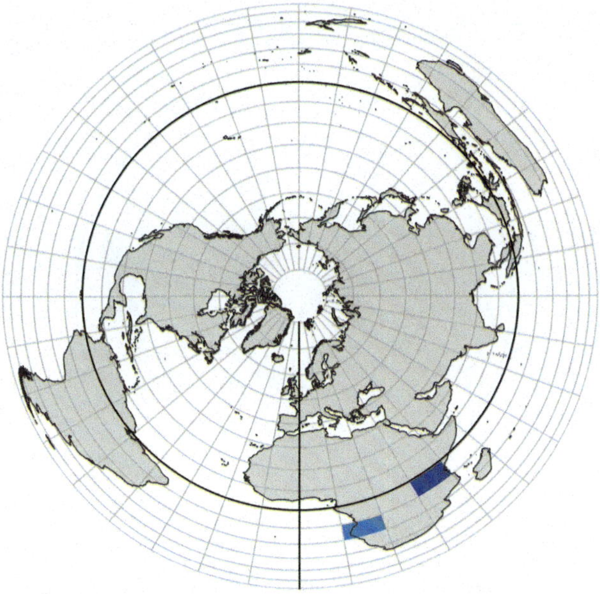

***Rotala serpyllifolia* (Roth) Bremek.**
Synonyms: *Micranthus serpyllifolius* Roth

Range area (grAH km^2): OR: 62300
Ecozone origin-status: OR: E
World rarity status: R
Ecozone-endemic macroregion occurrence: OR: IND, ICH
Life form(s): SU, EM
Ploidy state: NA
Speciation age (Ma): NA
Populations also occur in terrestrial, wetland and/ or riparian habitats: YES
Long distance movement vectors: NA
Nuisance status and uses: no nuisance issues; aquarium hobby plant
Endangered status (IUCN): LC

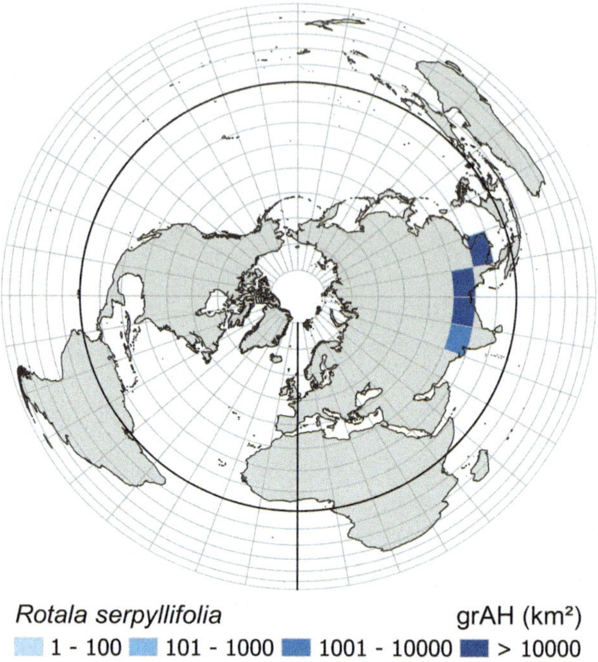

Rotala serpyllifolia (Roth) Bremek.
Image copyright: iNaturalist CC0 @radhaveach (Radha Veach)

Rotala simpliciuscula **Koehne**
Synonyms: *Ammannia simpliciuscula* Kurz

Range area (grAH km^2): OR: 38000
Ecozone origin-status: OR: E
World rarity status: R
Ecozone-endemic macroregion occurrence: OR: IND
Life form(s): SU, EM
Ploidy state: NA
Speciation age (Ma): NA
Populations also occur in terrestrial, wetland and/ or riparian habitats: YES
Long distance movement vectors: NA
Nuisance status and uses: no nuisance issues; no known uses
Endangered status (IUCN): NA (possibly extinct, last recorded 1925)

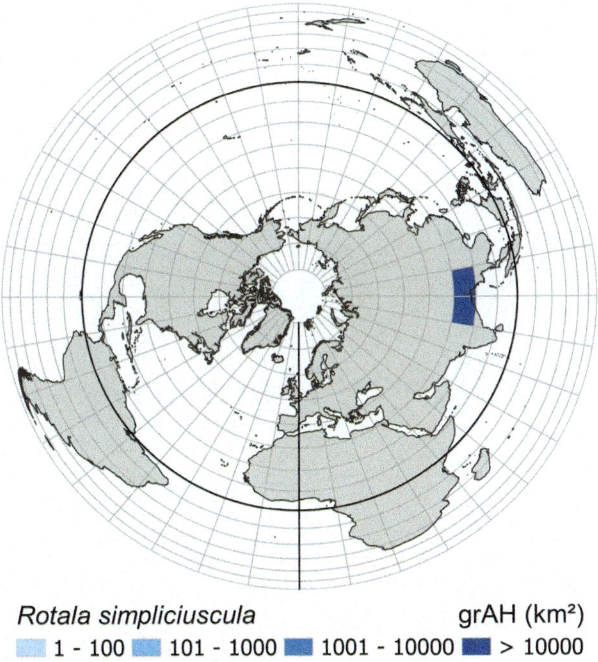

***Rotala smithii* A.Fern. & Diniz**
Synonyms:

Range area (grAH km^2): AFR: 18200
Ecozone origin-status: AFR: E
World rarity status: VR
Ecozone-endemic macroregion occurrence: AFR: FOR
Life form(s): EM (subshrub)
Ploidy state: NA
Speciation age (Ma): NA
Populations also occur in terrestrial, wetland and/ or riparian habitats: YES
Long distance movement vectors: NA
Nuisance status and uses: no nuisance issues; no known uses
Endangered status (IUCN): VU

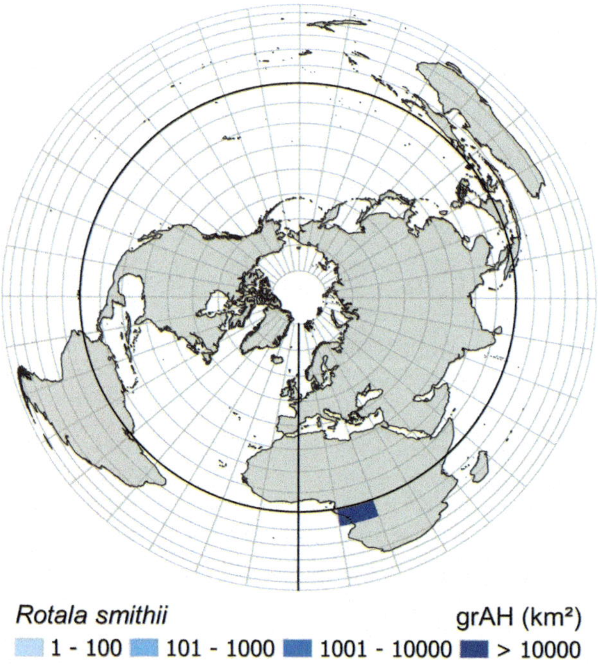

***Rotala stagnina* Hiern**
Synonyms: *Ammannia elatinoides* G.W.Schimp. ex Hiern

Range area (grAH km^2): AFR: 81800
Ecozone origin-status: AFR: E
World rarity status: R
Ecozone-endemic macroregion occurrence: AFR: SAH/ AR, EA, FOR
Life form(s): EM

Ploidy state: NA
Speciation age (Ma): NA
Populations also occur in terrestrial, wetland and/ or riparian habitats: YES
Long distance movement vectors: NA
Nuisance status and uses: no nuisance issues; no known uses
Endangered status (IUCN): LC

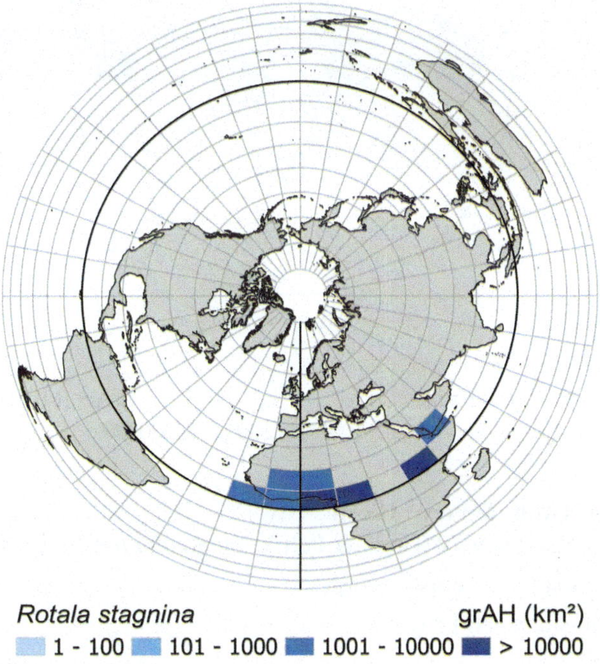

Rotala subrotunda **Koehne**
Synonyms: *Ammannia subrotunda* Wall. ex Kurz

Range area (grAH km^2): OR: 29100
Ecozone origin-status: OR: E
World rarity status: VR
Ecozone-endemic macroregion occurrence: OR: IND, ICH
Life form(s): EM
Ploidy state: NA
Speciation age (Ma): NA
Populations also occur in terrestrial, wetland and/ or riparian habitats: YES
Long distance movement vectors: NA
Nuisance status and uses: no nuisance issues; no known uses
Endangered status (IUCN): NA

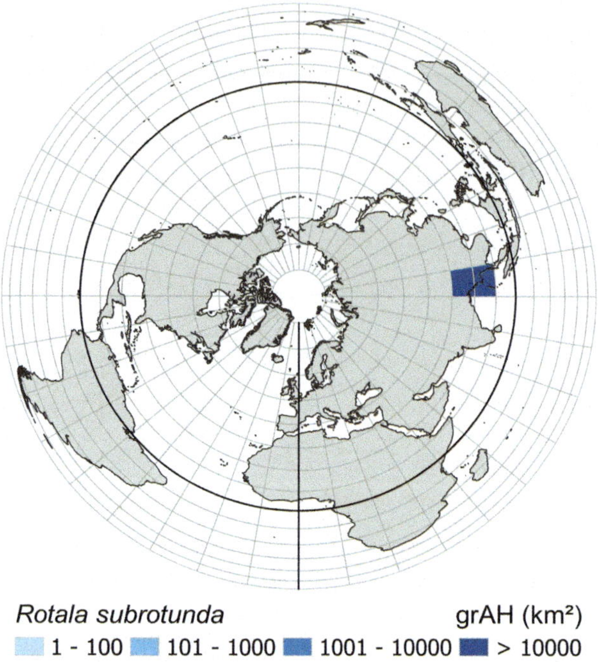

***Rotala tenella* (Guill. & Perr.) Hiern**

Synonyms: *Ameletia tenella* (Guill. & Perr.); *Ammannia tenella* Guill. & Perr.

Range area (grAH km^2): AFR: 251700
Ecozone origin-status: AFR: E
World rarity status: C
Ecozone-endemic macroregion occurrence: AFR: SAH/ AR, FOR, EA, ZAM, SDES
Life form(s): EM
Ploidy state: NA
Speciation age (Ma): NA
Populations also occur in terrestrial, wetland and/ or riparian habitats: YES
Long distance movement vectors: NA
Nuisance status and uses: no nuisance issues; folk medicine
Endangered status (IUCN): LC

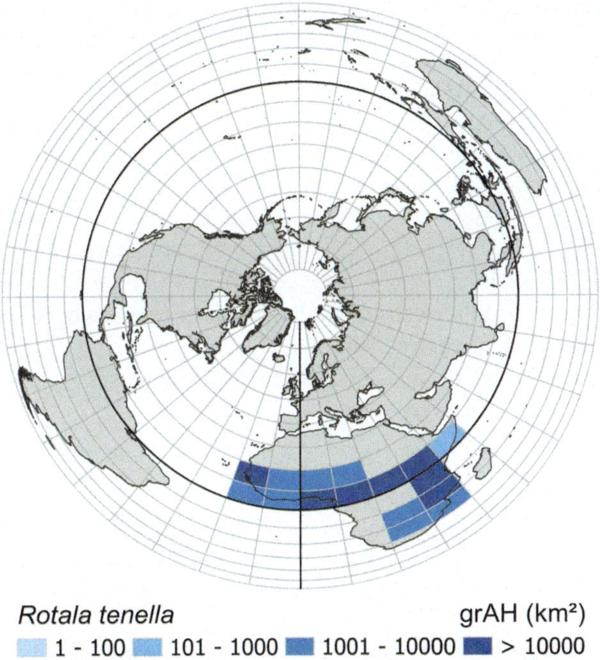

Rotala tenella (Guill. & Perr.) Hiern

***Rotala thymoides* Exell**
Synonyms:

Range area (grAH km^2): AFR: 2900
Ecozone origin-status: AFR: E
World rarity status: VR
Ecozone-endemic macroregion occurrence: AFR: ZAM
Life form(s): EM
Ploidy state: NA
Speciation age (Ma): NA
Populations also occur in terrestrial, wetland and/ or riparian habitats: ?YES
Long distance movement vectors: NA
Nuisance status and uses: no nuisance issues; no known uses
Endangered status (IUCN): NA

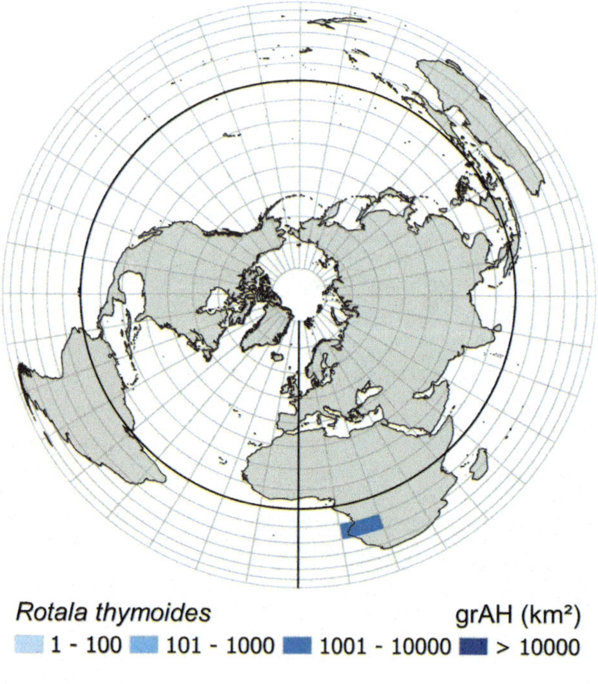

Rotala tulunadensis **Prasad, Biju, Ravi & Bhat**
Synonyms:

Range area (grAH km^2): OR: 8100
Ecozone origin-status: OR: E
World rarity status: VR
Ecozone-endemic macroregion occurrence: OR: IND
Life form(s): SU, EM
Ploidy state: NA
Speciation age (Ma): NA
Populations also occur in terrestrial, wetland and/ or riparian habitats: YES
Long distance movement vectors: NA
Nuisance status and uses: no nuisance issues; aquarium trade
Endangered status (IUCN): NA

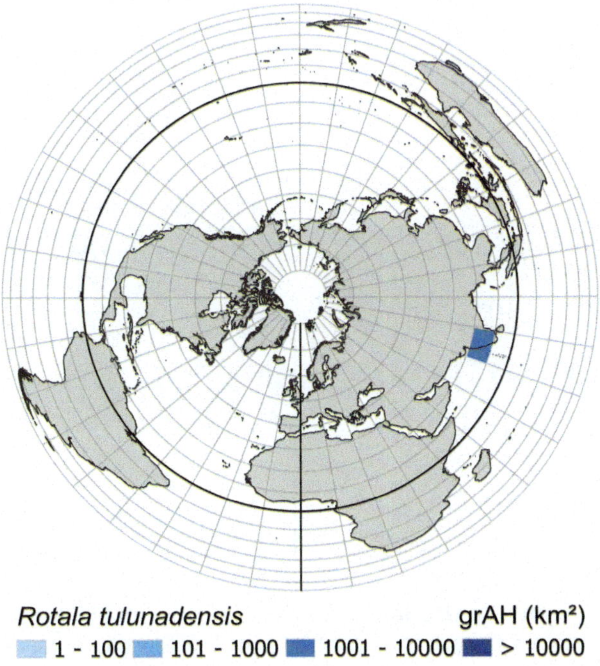

***Rotala verdcourtii* A.Fern.**
Synonyms:

Range area (grAH km^2): AFR: 93700
Ecozone origin-status: AFR: E
World rarity status: R
Ecozone-endemic macroregion occurrence: AFR: EA
Life form(s): SU, EM
Ploidy state: NA
Speciation age (Ma): NA
Populations also occur in terrestrial, wetland and/ or riparian habitats: YES
Long distance movement vectors: NA
Nuisance status and uses: no nuisance issues; no known uses
Endangered status (IUCN): NA

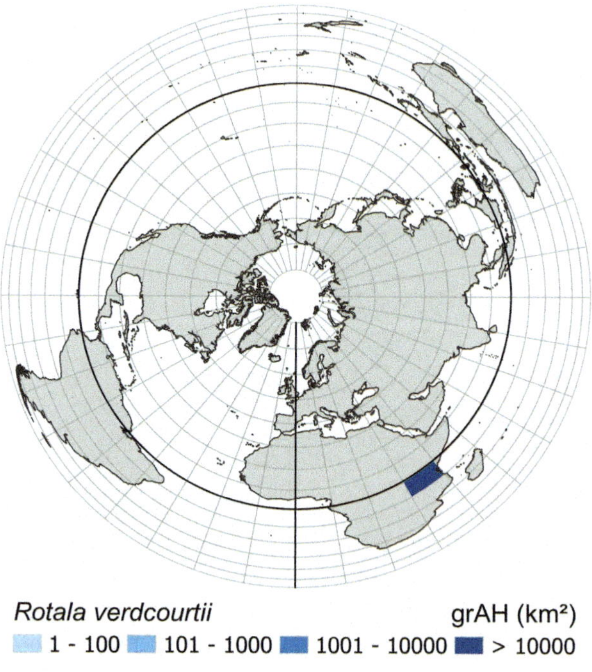

***Rotala verticillaris* L.**
Synonyms:

Range area (grAH km²): OR: 10300; AUS: 2900; AFR: 13500
Ecozone origin-status: OR: N; AUS: I; AFR: N
World rarity status: VR
Ecozone-endemic macroregion occurrence: not endemic to any world ecozone
Life form(s): SU, EM (sometimes a subshrub)
Ploidy state: NA
Speciation age (Ma): NA
Populations also occur in terrestrial, wetland and/ or riparian habitats: YES
Long distance movement vectors: ANTHROPOCHORY
Nuisance status and uses: no nuisance issues; aquarium trade
Endangered status (IUCN): LC

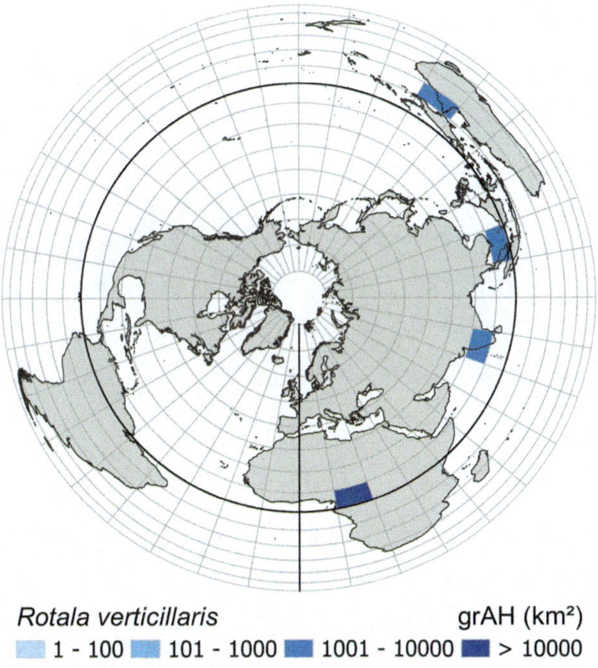

***Rotala wallichii* (Hook. f.) Koehne**

Synonyms: *Hydrolythrum wallichii* Hook.f.; *Ammannia wallichii* (Hook.f.) Kurz

Range area (grAH km^2): OR: 53900
Ecozone origin-status: OR: E
World rarity status: R
Ecozone-endemic macroregion occurrence: OR: ICH, SUP
Life form(s): SU, EM
Ploidy state: NA
Speciation age (Ma): NA
Populations also occur in terrestrial, wetland and/ or riparian habitats: NO
Long distance movement vectors: NA
Nuisance status and uses: no nuisance issues; aquarium trade
Endangered status (IUCN): NA

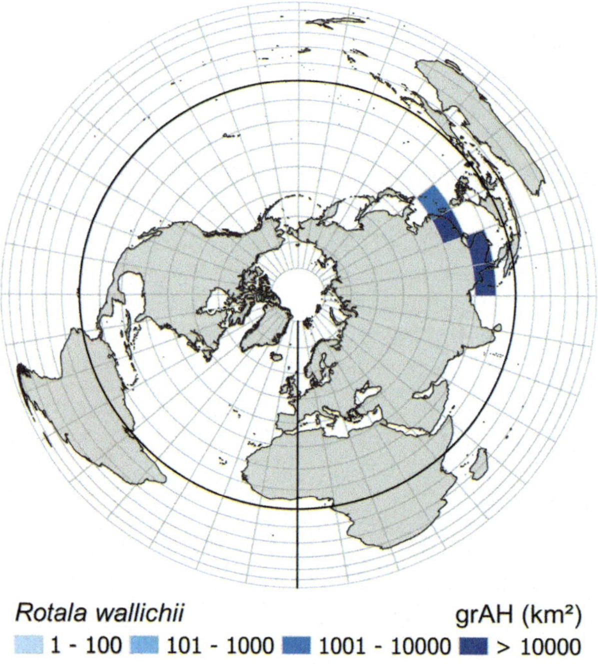

Rotala wallichii (Hook. f.) Koehne
Image copyright: T. Malchinski, with permission

Rotala welwitschii Exell
Synonyms:

Range area (grAH km^2): AFR: 31200
Ecozone origin-status: AFR: E
World rarity status: VR
Ecozone-endemic macroregion occurrence: AFR: SAH/ AR, FOR, ZAM
Life form(s): EM
Ploidy state: NA
Speciation age (Ma): NA

Populations also occur in terrestrial, wetland and/ or riparian habitats: YES
Long distance movement vectors: NA
Nuisance status and uses: no nuisance issues; no known uses
Endangered status (IUCN): LC

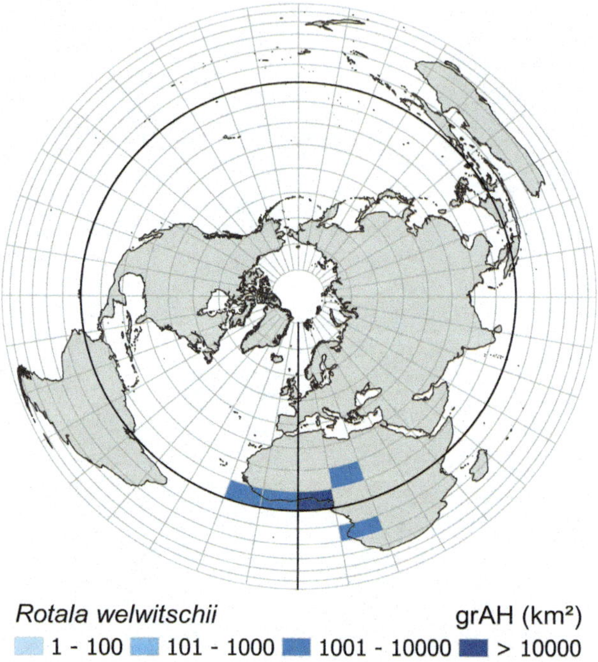

Rotala welwitschii Exell

***Trapa colchica* Albov**
Synonyms: *Trapa natans* L.

Range area (grAH km^2): PAL: 15300
Ecozone origin-status: PAL: E
World rarity status: VR
Ecozone-endemic macroregion occurrence: PAL: E EUR
Life form(s): FR
Ploidy state: NA
Speciation age (Ma): NA
Populations also occur in terrestrial, wetland and/ or riparian habitats: NO
Long distance movement vectors: NA
Nuisance status and uses: no nuisance issues; no known uses
Endangered status (IUCN): CR

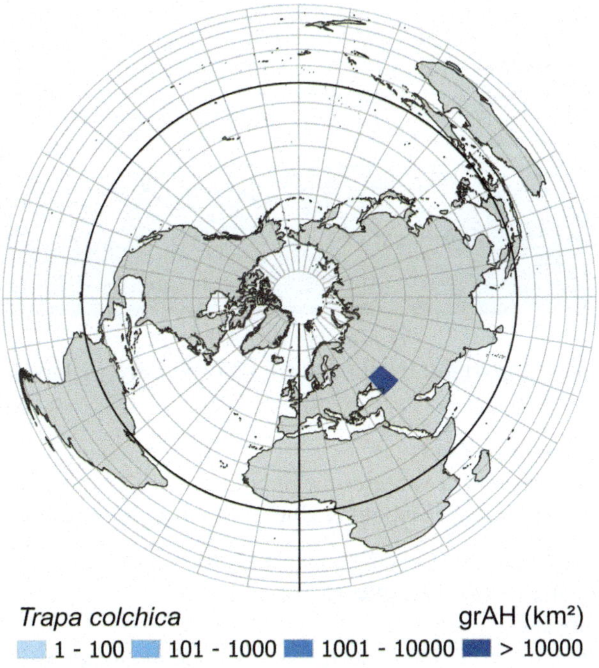

Trapa colchica Albov

Trapa incisa **Siebold & Zucc.**
Synonyms: *Trapa bispinosa* var. *incisa* (Siebold & Zucc.) Franch. & Sav.; *Trapa natans* var. *incisa* (Siebold & Zucc.) Makino

Range area (grAH km^2): PAL: 84700; OR: 81200
Ecozone origin-status: PAL: N; OR: N
World rarity status: C
Ecozone-endemic macroregion occurrence: not endemic to any world ecozone
Life form(s): FR (hydrosubshrub)
Ploidy state: DP
Speciation age (Ma): NA
Populations also occur in terrestrial, wetland and/ or riparian habitats: NO
Long distance movement vectors: NA
Nuisance status and uses: no nuisance issues; sometimes cultivated for its edible seeds
Endangered status (IUCN): LC

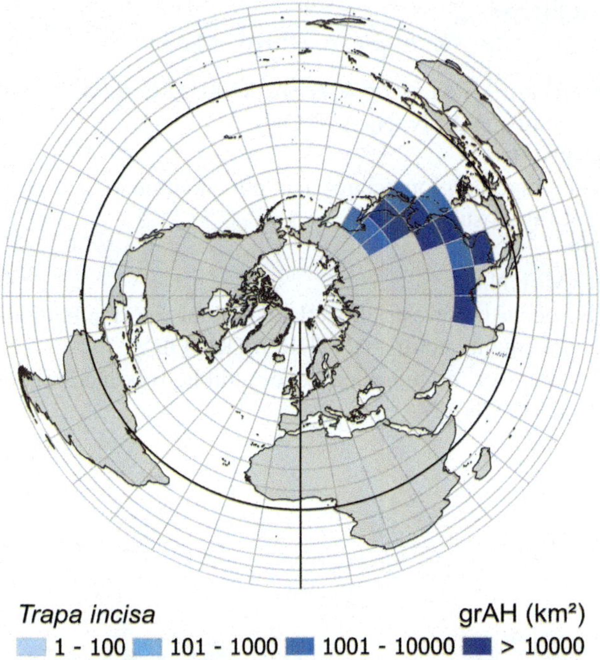

Trapa incisa Siebold & Zucc.
Image copyright: A. Efremov, with permission

***Trapa japonica* Flerow**
Synonyms: *Trapa natans* var. *bispinosa* (Roxb.) Makino

Range area (grAH km²): PAL: 72800; NEA: 8200; OR: 2600
Ecozone origin-status: PAL: N; NEA: I; OR: N
World rarity status: R
Ecozone-endemic macroregion occurrence: not endemic to any world ecozone
Life form(s): FR (hydrosubshrub)
Ploidy state: DP
Speciation age (Ma): NA
Populations also occur in terrestrial, wetland and/ or riparian habitats: NO
Long distance movement vectors: ANTHROPOCHORY
Nuisance status and uses: no nuisance issues; edible, folk medicine
Endangered status (IUCN): NA

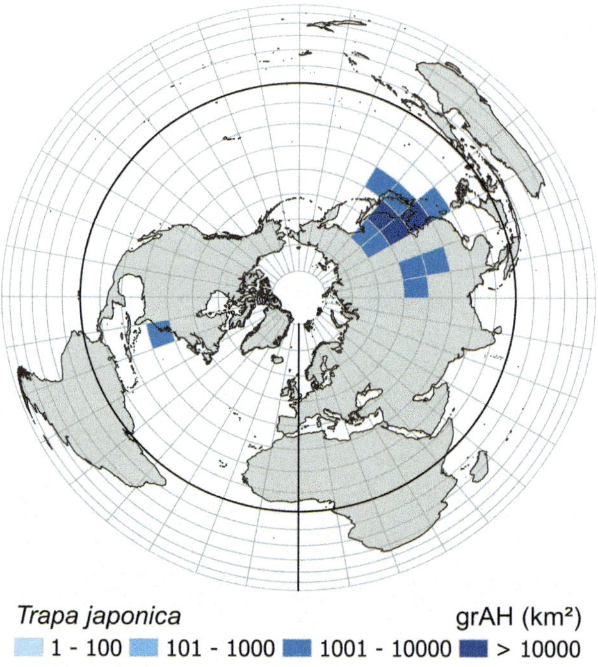

Trapa japonica Flerow
Image copyright: A. Efremov, with permission

Trapa maleevii **V.N.Vassil.**
Synonyms: *Trapa natans* var. *natans* L.

Range area (grAH km^2): PAL: 31300
Ecozone origin-status: PAL: E
World rarity status: VR
Ecozone-endemic macroregion occurrence: PAL: E EUR
Life form(s): FR
Ploidy state: NA
Speciation age (Ma): NA
Populations also occur in terrestrial, wetland and/ or riparian habitats: NO
Long distance movement vectors: NA
Nuisance status and uses: no nuisance issues; no known uses
Endangered status (IUCN): VU

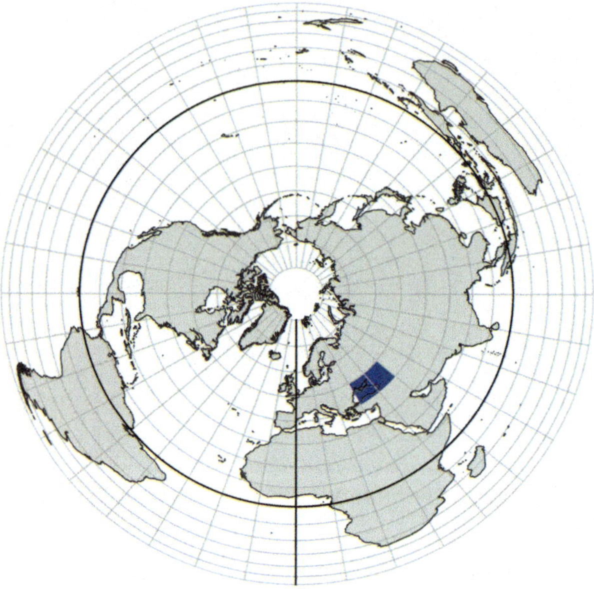

***Trapa natans* L.**
Synonyms:

Range area (grAH km^2): PAL: 449300; NEA: 77300; NEO: 7600; OR: 116900; AFR: 255700
Ecozone origin-status: PAL: N; NEA: I; NEO: I; OR: N; AFR: N
World rarity status: C
Ecozone-endemic macroregion occurrence: not endemic to any world ecozone
Life form(s): FR
Ploidy state: DP
Speciation age (Ma): NA
Populations also occur in terrestrial, wetland and/ or riparian habitats: NO
Long distance movement vectors: ANTHROPOCHORY
Nuisance status and uses: invasive aquatic weed in North America; medicinal, food, water garden and phytoremediation uses
Endangered status (IUCN): LC

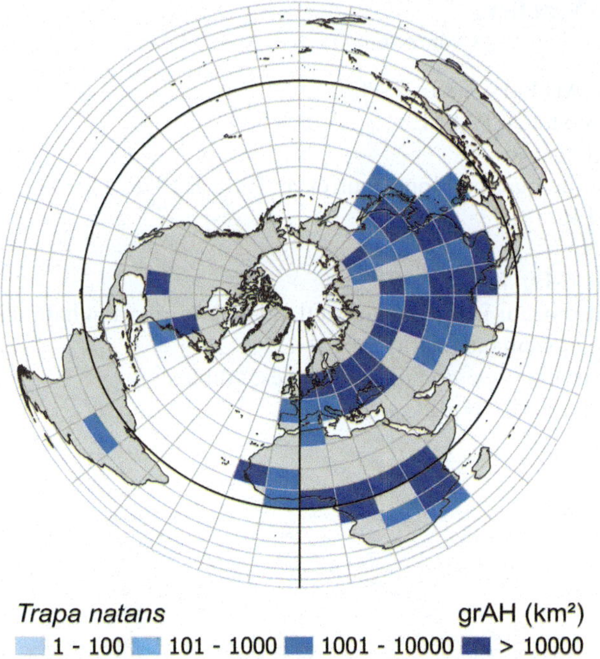

Trapa natans L.
Image copyright: J.C. Schou, with permission

***Trapa rossica* Vassiliev**

Synonyms: *Trapa natans* var. *natans* L.

Range area (grAH km^2): PAL: 78900
Ecozone origin-status: PAL: E
World rarity status: R
Ecozone-endemic macroregion occurrence: PAL: E EUR, UR/ W SIB, CH, FER/ J
Life form(s): FR
Ploidy state: NA
Speciation age (Ma): NA
Populations also occur in terrestrial, wetland and/ or riparian habitats: NO
Long distance movement vectors: NA
Nuisance status and uses: no nuisance issues; edible
Endangered status (IUCN): LC

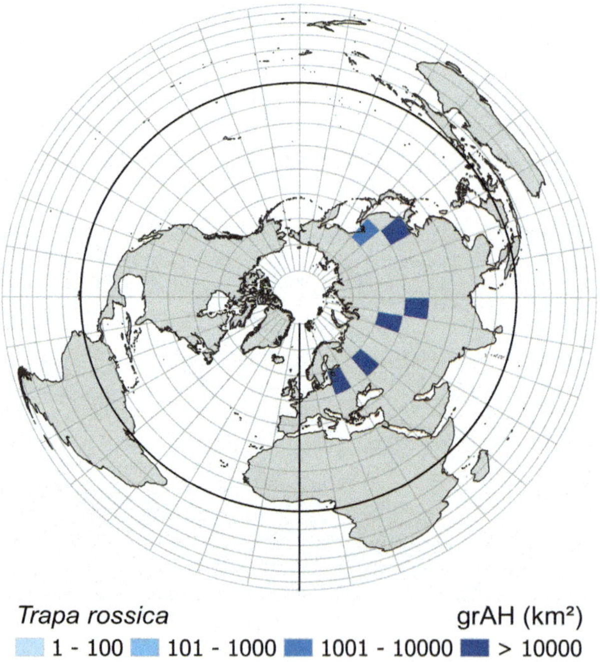

Trapa rossica Vassiliev
Image copyright: A. Efremov, with permission

3.31 MALVACEAE (Malvales)

Three species in Malvaceae are freshwater macrophytes, all in the genus *Hibiscus*. They occur primarily in low-latitude regions. All three species occur in NEO (one endemic to that ecozone), two in NEA, and one in OR, one in OCE and one introduced in PAL. One species was introduced in OR and in NEA. None of these species causes aquatic weed problems. Two of the three macrophyte species of Malvaceae are rare at world scale and one is common, giving values of R 67% and C 33%. In terms of endangered status, two of the three macrophyte species of Malvaceae are unassessed, and the other one is LC.

***Hibiscus coccineus* Walter**
Synonyms:

Range area (grAH km^2): PAL: 40800; NEA: 34000; NEO: 24100; OR: 17900
Ecozone origin-status: PAL: I; NEA: I; NEO: E(I); OR: I
World rarity status: R
Ecozone-endemic macroregion occurrence: NEO: MG MX
Life form(s): EM
Ploidy state: D
Speciation age (Ma): NA
Populations also occur in terrestrial, wetland and/ or riparian habitats: YES
Long distance movement vectors: ANTHROPOCHORY
Nuisance status and uses: no nuisance issues; garden plant use, medicinal potential
Endangered status (IUCN): NA

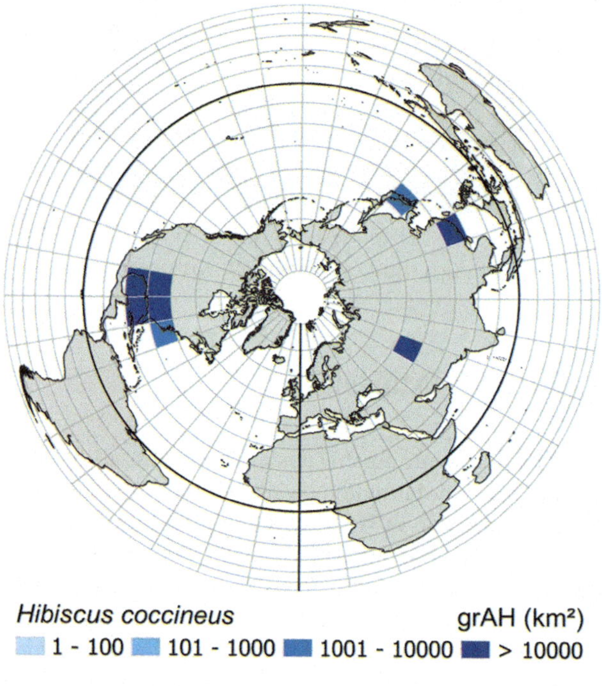

Hibiscus coccineus Walter
Image copyright: iNaturalist CC-BY @bgsmith (Brooke Smith)

Hibiscus furcellatus **Desr.**
Synonyms:

Range area (grAH km^2): NEO: 276900; present in OCE
Ecozone origin-status: NEO: N; OCE: N
World rarity status: C
Ecozone-endemic macroregion occurrence: not endemic to any world ecozone
Life form(s): EM (shrub)
Ploidy state: D
Speciation age (Ma): NA
Populations also occur in terrestrial, wetland and/ or riparian habitats: YES
Long distance movement vectors: NA
Nuisance status and uses: no nuisance issues; medicinal, forage, ornamental uses
Endangered status (IUCN): LC

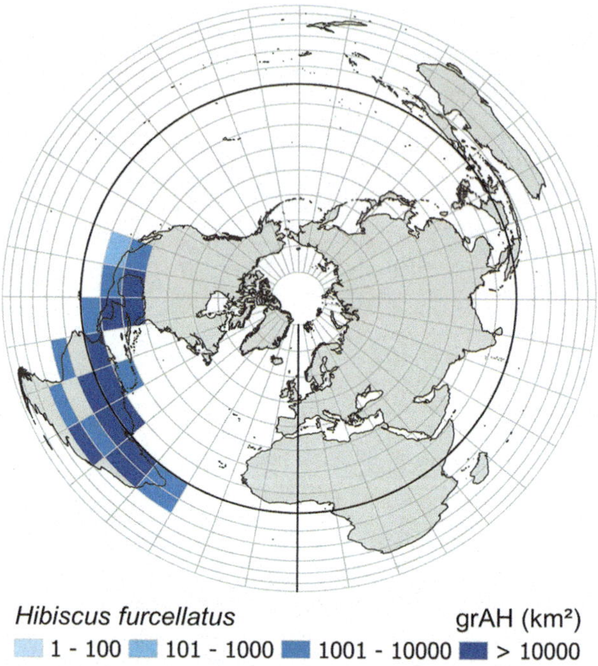

Hibiscus furcellatus Desr.

***Hibiscus grandiflorus* Michx.**
Synonyms:

Range area (grAH km^2): NEA: 13000; NEO: 24100
Ecozone origin-status: NEA: N; NEO: N
World rarity status: R
Ecozone-endemic macroregion occurrence: not endemic to any world ecozone
Life form(s): EM
Ploidy state: DP
Speciation age (Ma): NA
Populations also occur in terrestrial, wetland and/ or riparian habitats: YES
Long distance movement vectors: HYD (NEA to NEO, and NEO to NEA)
Nuisance status and uses: no nuisance issues; medicinal potential and garden use
Endangered status (IUCN): NA

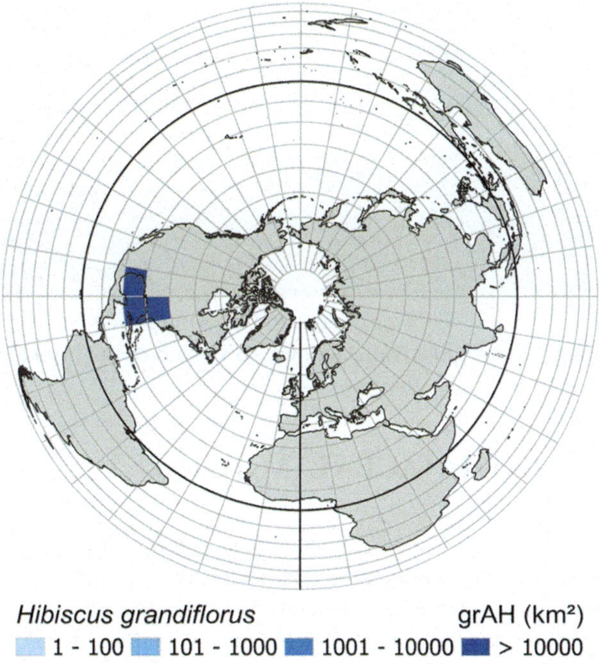

Hibiscus grandiflorus Michx.
Image copyright: iNaturalist CC-BY @jayhorn (Jay Horn)

3.32 MELASTOMATACEAE (Myrtales)

The Melastomataceae have six freshwater macrophyte species, five in the genus *Acisanthera*, and *Nepsera aquatica* the only species in its genus. All are Neotropical endemics, which have not been introduced anywhere else globally, and none cause aquatic weed problems. One species is VR, three R and two C. In terms of endangered status, none of the six macrophyte species of Melastomataceae has been assessed.

Quick Links to Genera

1. *Acisanthera*
2. *Nepsera*

***Acisanthera bivalvis* (Aubl.) Cogn.**

Synonyms: *Noterophila bivalvis* (Aubl.) Kriebel & M.J.Rocha

Range area (grAH km^2): NEO: 136500
Ecozone origin-status: NEO: E
World rarity status: R
Ecozone-endemic macroregion occurrence: NEO: MES AM, AMZ, CHA, PAR
Life form(s): EM
Ploidy state: NA
Speciation age (Ma): NA
Populations also occur in terrestrial, wetland and/ or riparian habitats: YES
Long distance movement vectors: NA
Nuisance status and uses: no nuisance issues; no known uses
Endangered status (IUCN): NA

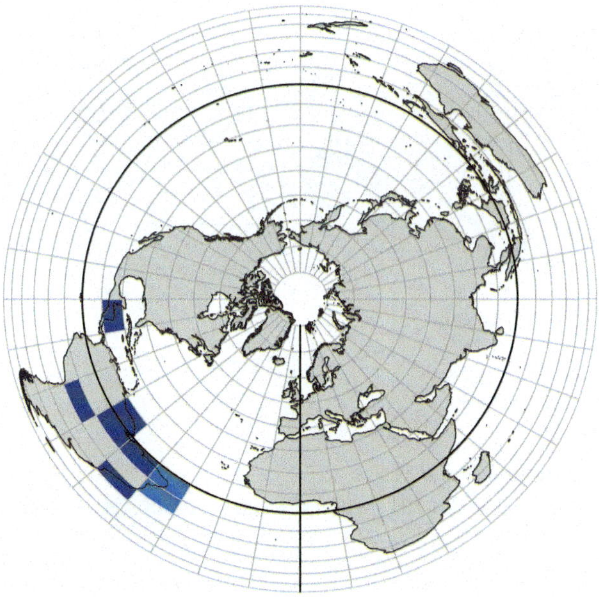

Acisanthera fluitans **Cogn.**
Synonyms: *Noterophila limnobios* (DC.) Mart.

Range area (grAH km^2): NEO: 65800
Ecozone origin-status: NEO: E
World rarity status: R
Ecozone-endemic macroregion occurrence: NEO: MES AM, AMZ, CHA, PAR
Life form(s): EM
Ploidy state: NA
Speciation age (Ma): NA
Populations also occur in terrestrial, wetland and/ or riparian habitats: YES
Long distance movement vectors: NA
Nuisance status and uses: no nuisance issues; no known uses
Endangered status (IUCN): NA

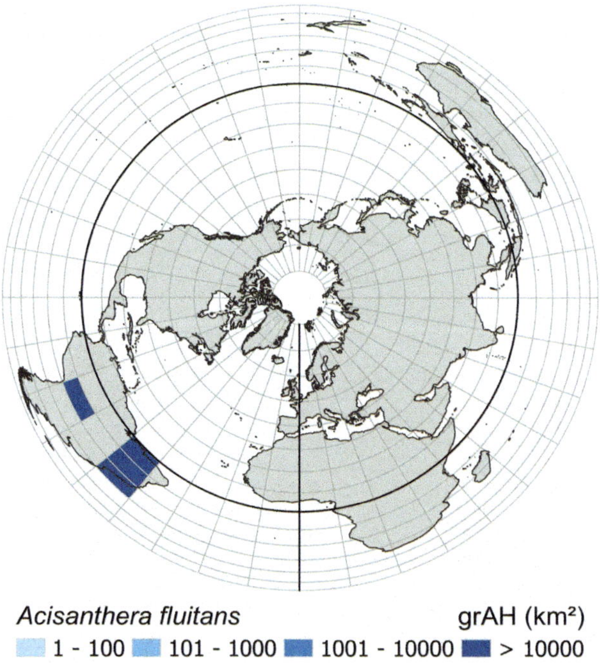

***Acisanthera inundata* (DC.) Triana**
Synonyms: *Noterophila inundata* (DC.) Mart.; *Melastoma inundatum* (DC.) Steud.; *Microlicia inundata* DC.

Range area (grAH km^2): NEO: 27600
Ecozone origin-status: NEO: E
World rarity status: VR
Ecozone-endemic macroregion occurrence: NEO: AMZ, PAR
Life form(s): ?EM
Ploidy state: NA
Speciation age (Ma): NA
Populations also occur in terrestrial, wetland and/ or riparian habitats: ?YES
Long distance movement vectors: NA
Nuisance status and uses: no nuisance issues; no known uses
Endangered status (IUCN): NA

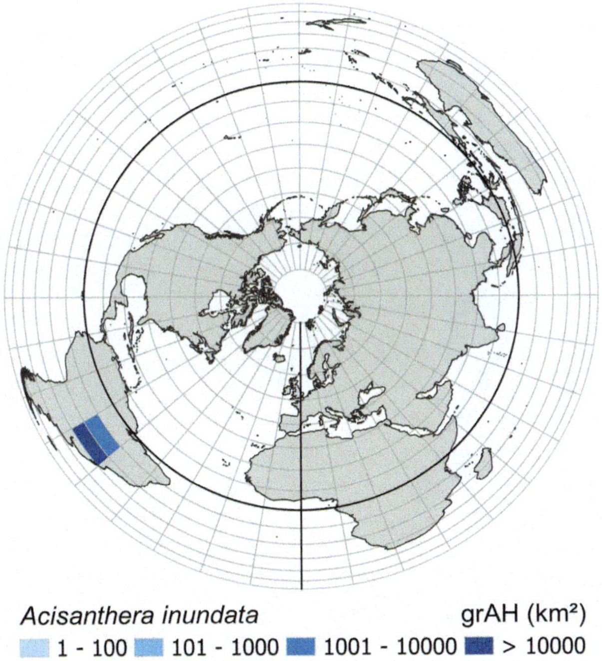

Acisanthera limnobios **(DC.) Triana**
Synonyms: *Noterophila limnobios* (DC.) Mart.

Range area (grAH km^2): NEO: 191600
Ecozone origin-status: NEO: E
World rarity status: C
Ecozone-endemic macroregion occurrence: NEO: MG MX, MES AM, NWSA, CHA, AMZ, PAR
Life form(s): EM
Ploidy state: NA
Speciation age (Ma): NA
Populations also occur in terrestrial, wetland and/ or riparian habitats: YES
Long distance movement vectors: NA
Nuisance status and uses: no nuisance issues; no known uses
Endangered status (IUCN): NA

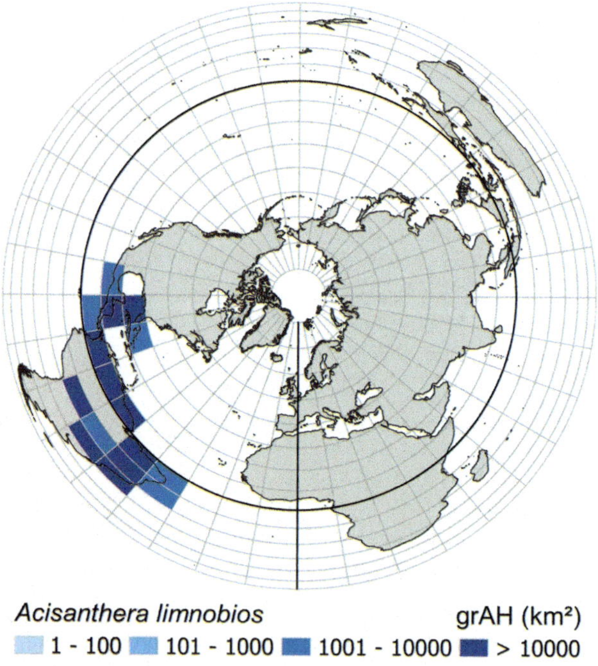

Acisanthera nana **Ule**
Synonyms: *Noterophila nana* (Ule) Kriebel & M.J.Rocha

Range area (grAH km^2): NEO: 105700
Ecozone origin-status: NEO: E
World rarity status: R
Ecozone-endemic macroregion occurrence: NEO: MES AM, NWSA, AMZ,
Life form(s): EM
Ploidy state: NA
Speciation age (Ma): NA
Populations also occur in terrestrial, wetland and/ or riparian habitats: YES
Long distance movement vectors: NA
Nuisance status and uses: no nuisance issues; no known uses
Endangered status (IUCN): NA

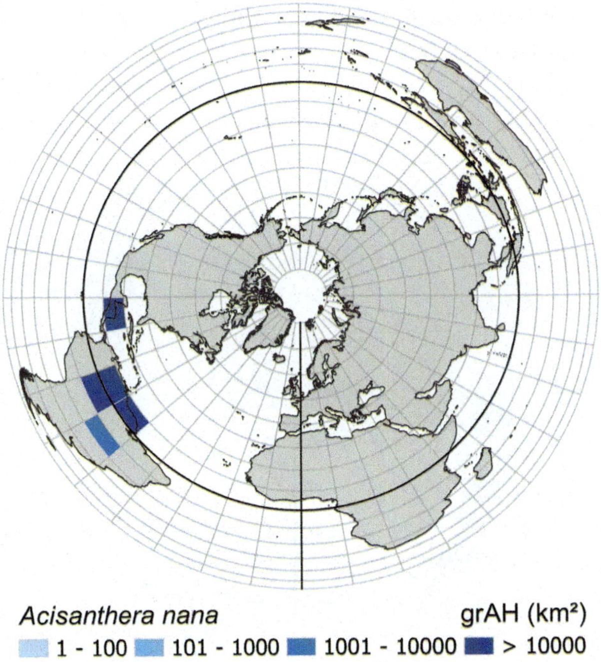

Nepsera aquatica **(Aubl.) Naudin**

Synonyms: *Aciotis aquatica* (Aubl.) G.Don; *Melastoma aquaticum* Aubl.; *Rhexia aquatica* (Aubl.) Sw., Spennera aquatica (Aubl.) Mart. ex DC.; *Tibouchina aquatica* (Aubl.) M.Gómez; *Xeracina aquatica* (Aubl.) Raf.

Range area (grAH km^2): NEO: 226400
Ecozone origin-status: NEO: E
World rarity status: C
Ecozone-endemic macroregion occurrence: NEO: MES AM, NWSA, AMZ, CHA, PAR
Life form(s): EM
Ploidy state: NA
Speciation age (Ma): NA
Populations also occur in terrestrial, wetland and/ or riparian habitats: YES
Long distance movement vectors: NA
Nuisance status and uses: no nuisance issues; folk medicine
Endangered status (IUCN): NA

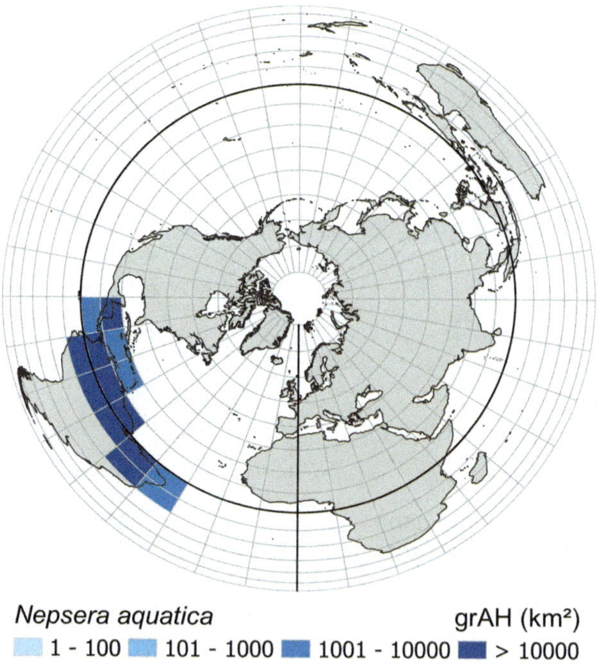

3.33 MENYANTHACEAE (Asterales)

With 69 macrophyte species in six genera, the Menyanthaceae are a widespread family, with at least one of the species occurring in every world ecozone. Macrophyte species of this family tend to show strong ecozone-endemism, with no less than 31 of the 39 species found in AUS being endemics, 13 of the 16 Afrotropical species being endemic to AFR, and eight endemics among the 17 species found in OR. Elsewhere the numbers are lower, with five endemic species among the 11 occurring in NEO, and just one among the nine Menyanthaceae macrophyte species of NEA. There are no endemics among the six species in PAL, and three in OCE (two of which were introduced). The single species found in ANT, *Liparophyllum gunnii*, is native to the sub-Antarctic islands. Elsewhere there are four I-origin species in each of NEA and NEO, three in AFR, two in AUS, and one in OR. There are no introduced macrophyte species of this family in PAL. Two species, *Nymphoides cristata* and *Nymphoides peltata* are invasive, both in NEA, while *N. peltata* also causes aquatic weed problems in AUS. The 69 species of are dominated by world-rare species, with VR 74%, R 17% and C 9%. In terms of endangered status, 68% of species are NA/DD, 20% are LC, and 12% are classed in the higher at-risk categories, with 2 species (3%) CR.

Quick Links to Genera

1. *Liparophyllum*
2. *Menyanthes*
3. *Nephrophyllidium*
4. *Nymphoides*
5. *Ornduffia*
6. *Villarsia*

Liparophyllum capitatum **(Nees) Tippery & Les**
Synonyms: *Villarsia capitata* Nees

Range area (grAH km^2): AUS: 11700
Ecozone origin-status: AUS: E
World rarity status: VR
Ecozone-endemic macroregion occurrence: AUS: WA, NA, ESA
Life form(s): EM
Ploidy state: D
Speciation age (Ma): NA
Populations also occur in terrestrial, wetland and/ or riparian habitats: YES
Long distance movement vectors: NA
Nuisance status and uses: NA
Endangered status (IUCN): NA

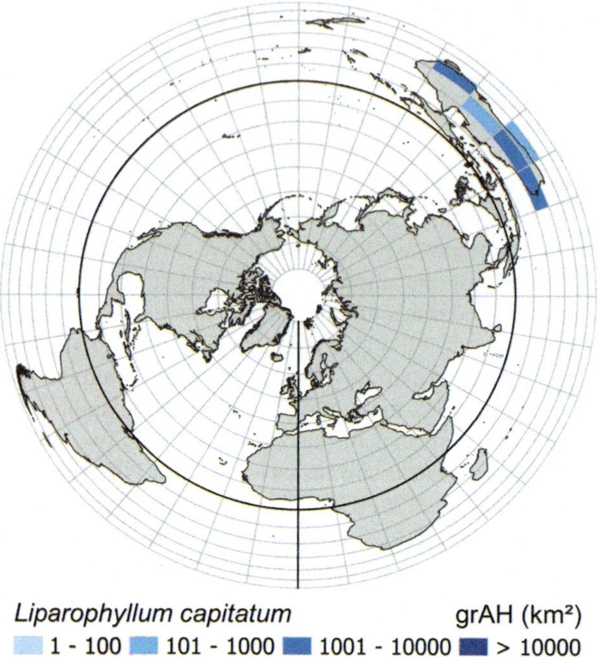

Liparophyllum congestiflorum (F. Muell.) Tippery & Les
Synonyms: *Villarsia congestiflora* F. Muell.

Range area (grAH km^2): AUS: 3600
Ecozone origin-status: AUS: E
World rarity status : VR
Ecozone-endemic macroregion occurrence: AUS: WA
Life form(s): EM
Ploidy state: D
Speciation age (Ma): NA
Populations also occur in terrestrial, wetland and/ or riparian habitats: YES
Long distance movement vectors: NA
Nuisance status and uses: NA
Endangered status (IUCN): NA

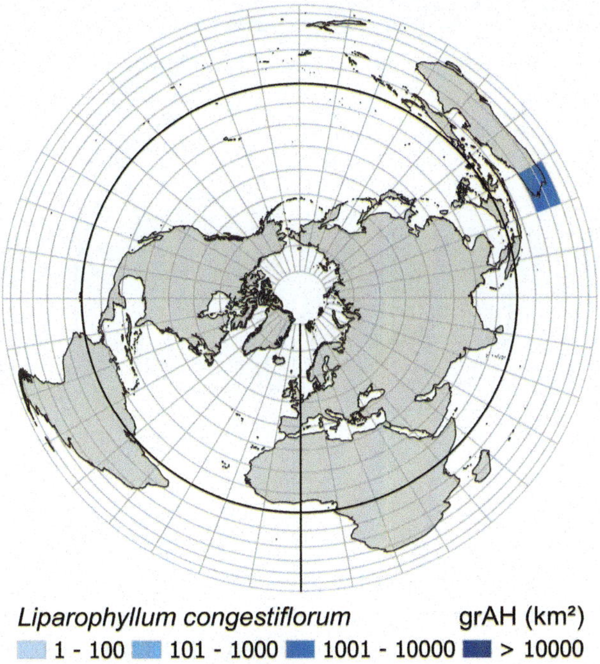

***Liparophyllum exaltatum* (Sol. ex Sims) Tippery & Les**
Synonyms: *Villarsia exaltata* G. Don

Range area (grAH km^2): AUS: 21500
Ecozone origin-status: AUS: E
World rarity status: VR
Ecozone-endemic macroregion occurrence: AUS: NA, ESA
Life form(s): EM
Ploidy state: NA
Speciation age (Ma): NA
Populations also occur in terrestrial, wetland and/ or riparian habitats: YES
Long distance movement vectors: NA
Nuisance status and uses: NA
Endangered status (IUCN): NA

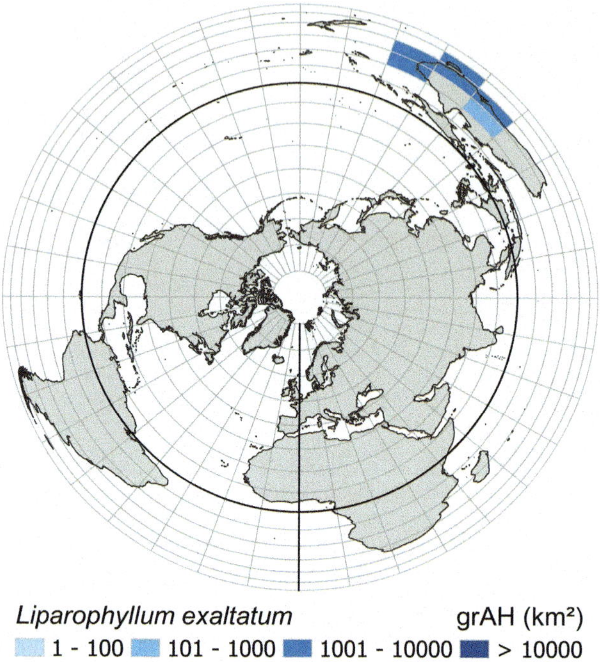
Liparophyllum exaltatum
grAH (km²)
1 - 100
101 - 1000
1001 - 10000
> 10000

Liparophyllum exaltatum (Sol. ex Sims) Tippery & Les

***Liparophyllum exiguum* (F. Muell.) Tippery & Les**
Synonyms: *Nymphoides exigua* (F. Muell.) Kuntze

Range area (grAH km^2): AUS: 3800
Ecozone origin-status: AUS: E
World rarity status: VR
Ecozone-endemic macroregion occurrence: AUS: ESA
Life form(s): EM
Ploidy state: NA
Speciation age (Ma): 6.4
Populations also occur in terrestrial, wetland and/ or riparian habitats: YES
Long distance movement vectors: ?AV
Nuisance status and uses: NA
Endangered status (IUCN): NA

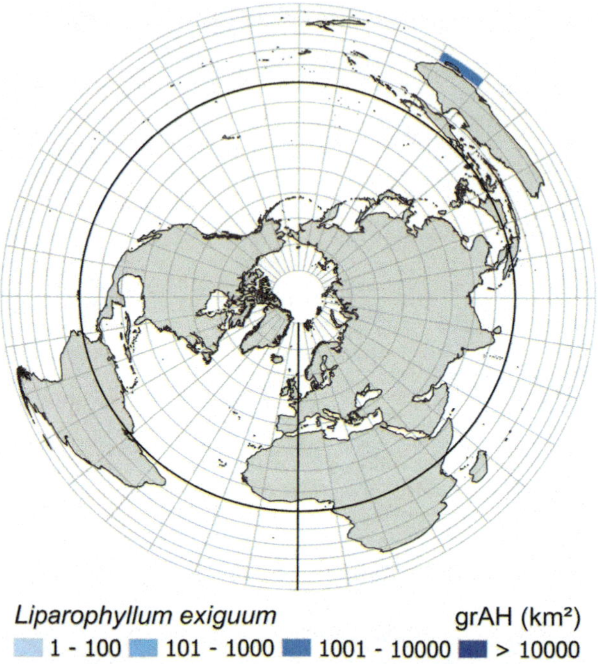

Liparophyllum exiguum (F. Muell.) Tippery & Les

Liparophyllum gunnii **Hook. f.**
Synonyms: *Limnanthemum gunnii* (Hook.f.) Hook.f., *Villarsia gunnii* (Hook.f.) Hook.f.

Range area (grAH km2): AUS: 20700; present in ANT
Ecozone origin-status: AUS: N; ANT: N
World rarity status: VR

Ecozone-endemic macroregion occurrence: not endemic to any world ecozone
Life form(s): EM
Ploidy state: NA
Speciation age (Ma): 6.4
Populations also occur in terrestrial, wetland and/ or riparian habitats: YES
Long distance movement vectors: ?AV
Nuisance status and uses: NA
Endangered status (IUCN): NA

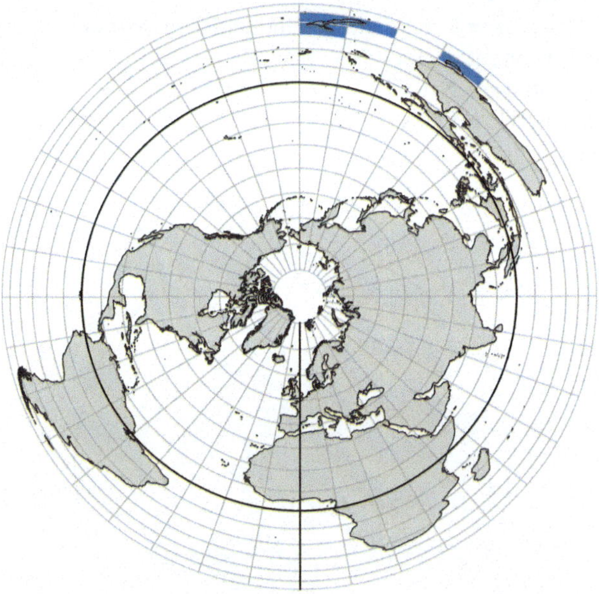

Liparophyllum gunnii Hook. f.
Image copyright: iNaturalist CC-BY @murray_dawson (Murray Dawson)

***Liparophyllum lasiospermum* (F. Muell.) Tippery & Les**
Synonyms: *Villarsia lasiosperma* F. Muell.

Range area (grAH km^2): AUS: 2200
Ecozone origin-status: AUS: E
World rarity status: VR
Ecozone-endemic macroregion occurrence: AUS: WA
Life form(s): EM
Ploidy state: D
Speciation age (Ma): NA
Populations also occur in terrestrial, wetland and/ or riparian habitats: YES
Long distance movement vectors: NA
Nuisance status and uses: NA
Endangered status (IUCN): NA

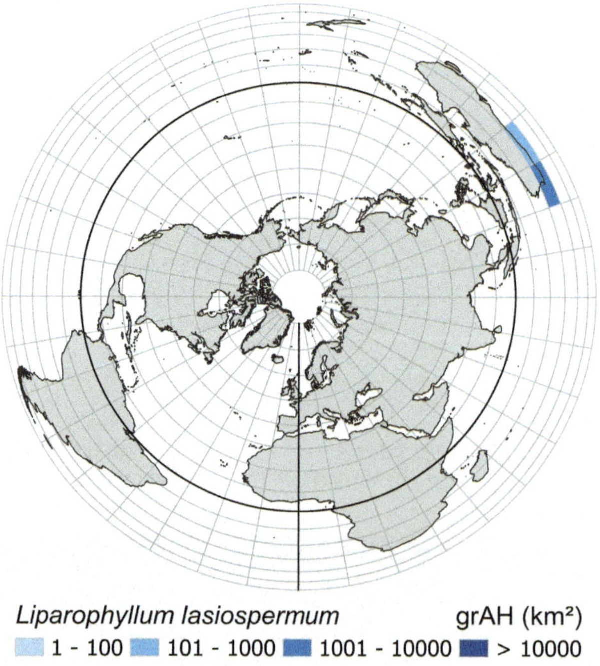

***Liparophyllum latifolium* (Benth.) Tippery & Les**
Synonyms: *Villarsia latifolia* Benth.

Range area (grAH km^2): AUS: 1500
Ecozone origin-status: AUS: E
World rarity status: VR
Ecozone-endemic macroregion occurrence: AUS: WA
Life form(s): EM

Ploidy state: D
Speciation age (Ma): NA
Populations also occur in terrestrial, wetland and/ or riparian habitats: YES
Long distance movement vectors: NA
Nuisance status and uses: no nuisance issues; used for revegetation work and as garden plant locally in West Australia
Endangered status (IUCN): NA

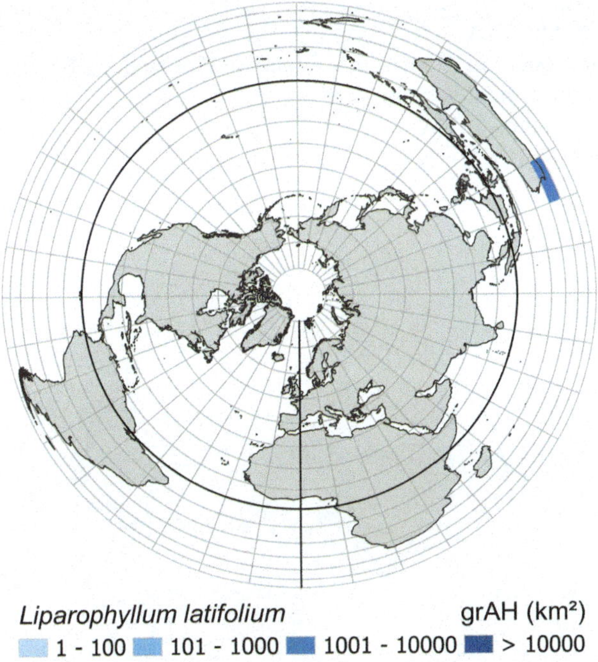

Liparophyllum latifolium (Benth.) Tippery & Les
Image copyright: R. Davis, with permission

***Liparophyllum violifolium* (F. Muell.) Tippery & Les**
Synonyms: *Villarsia violifolia* F. Muell.

Range area (grAH km^2): AUS: 8700
Ecozone origin-status: AUS: E
World rarity status: VR
Ecozone-endemic macroregion occurrence: AUS: NA, WA
Life form(s): EM
Ploidy state: D
Speciation age (Ma): NA
Populations also occur in terrestrial, wetland and/ or riparian habitats: YES
Long distance movement vectors: NA
Nuisance status and uses: NA
Endangered status (IUCN): NA

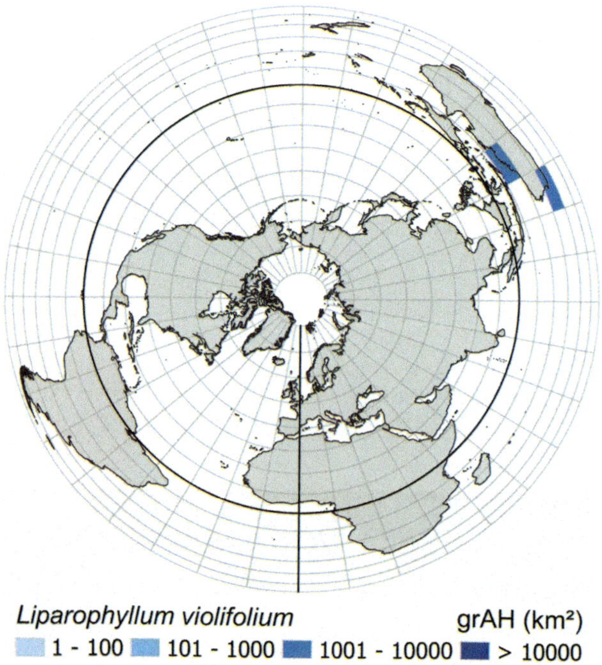

***Menyanthes trifoliata* L.**
Synonyms: *Menyanthes americana* Sweet, *Menyanthes latifolia* Raf., *Menyanthes palustris* Gray, *Menyanthes paradoxa* Fr., *Menyanthes tridentata* Raf.

Range area (grAH km^2): PAL: 822700; NEA: 896700; OR: 45000; AUS: 5400
Ecozone origin-status: PAL: N; NEA: N; OR: N; AUS: I
World rarity status: C
Ecozone-endemic macroregion occurrence: not endemic to any world ecozone
Life form(s): EM, FR
Ploidy state: DP
Speciation age (Ma): NA
Populations also occur in terrestrial, wetland and/ or riparian habitats: YES
Long distance movement vectors: BIRD: *Spatula querquedula* (Garganey); *Anas platyrhynchos* (Northern Mallard); *Anas crecca*: (Eurasian Teal)
Nuisance status and uses: no nuisance issues; medicinal uses to treat leukaemia and anorexia, and as a laxative, plants used in horticulture (as water garden plants)
Endangered status (IUCN): LC

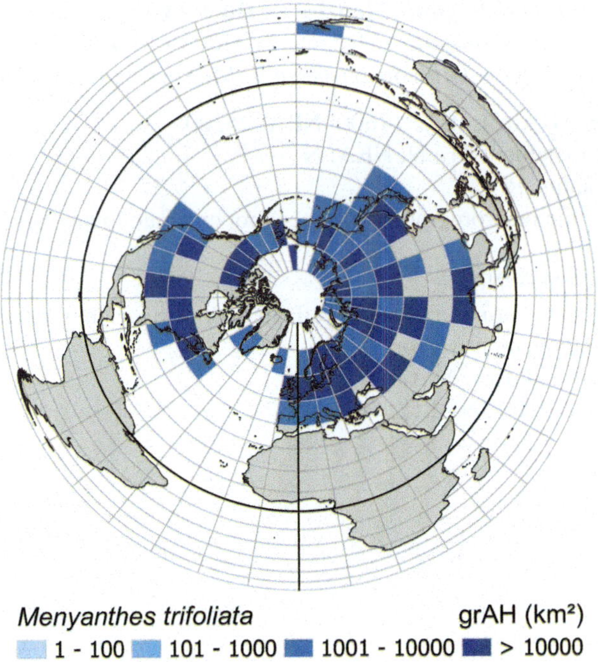

Menyanthes trifoliata L.
Image copyright: J. C. Schou, with permission

***Nephrophyllidium crista-galli* (Menzies ex Hook.) Gilg**
Synonyms: *Fauria crista-galli* (Menzies ex Hook.) Makino, *Menyanthes crista-galli* Menzies ex Hook., *Villarsia crista-galli* (Menzies ex Hook.) Hook.

Range area (grAH km^2): PAL: 17600; NEA: 25200
Ecozone origin-status: PAL: N; NEA: N
World rarity status: R
Ecozone-endemic macroregion occurrence: not endemic to any world ecozone
Life form(s): EM
Speciation age (Ma): NA
Populations also occur in terrestrial, wetland and/ or riparian habitats: YES
Long distance movement vectors: NA
Nuisance status and uses: NA
Endangered status (IUCN): LC

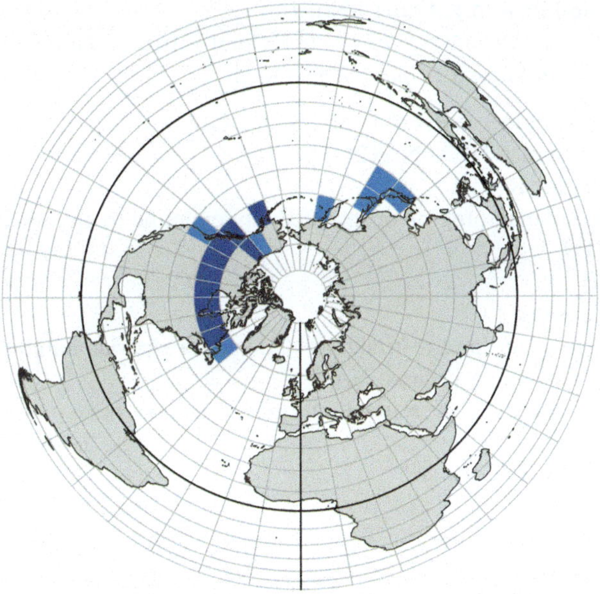

Nephrophyllidium crista-galli grAH (km²)
1 - 100 101 - 1000 1001 - 10000 > 10000

Nephrophyllidium crista-galli (Menzies ex Hook.) Gilg

***Nymphoides aquatica* (J.F. Gmel.) Kuntze**

Synonyms: *Limnanthemum aquaticum* (J.F.Gmel.) Britton, *Trachysperma aquatica* (J.F.Gmel.) House, *Villarsia aquatica* J.F.Gmel.

Range area (grAH km^2): NEA: 90300; NEO: 24100
Ecozone origin-status: NEA: N; NEO: N
World rarity status: R
Ecozone-endemic macroregion occurrence: not endemic to any world ecozone
Life form(s): FR
Ploidy state: P
Speciation age (Ma): 1.0
Populations also occur in terrestrial, wetland and/ or riparian habitats: NO
Long distance movement vectors: NA
Nuisance status and uses: occasional nuisance issues but not a major aquatic weed; aquarium trade.
Endangered status (IUCN): LC

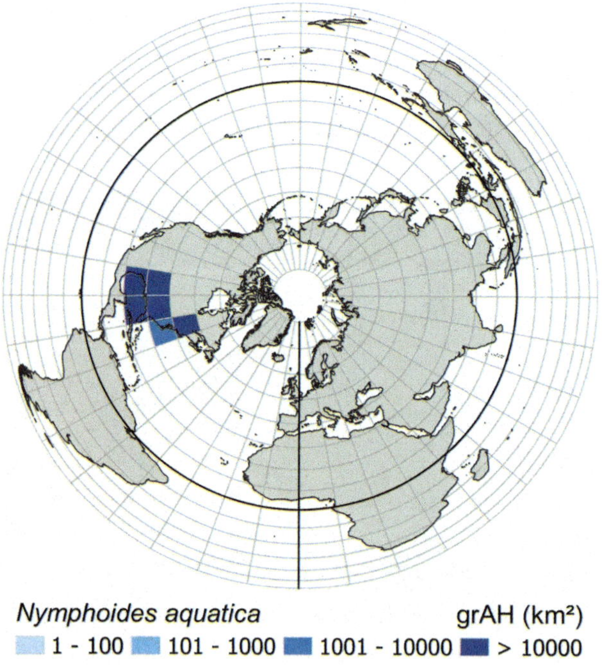

Nymphoides aquatica (J.F. Gmel.) Kuntze
Image copyright: Z. Li, with permission

***Nymphoides aurantiaca* (Dalzell) Kuntze**
Synonyms: *Limnanthemum aurantiacum* Dalzell

Range area (grAH km²): OR: 45700; AUS: 32900
cozone origin-status: OR: N; AUS: N
World rarity status: R
Ecozone-endemic macroregion occurrence: not endemic to any world ecozone
Life form(s): FR
Ploidy state: NA
Speciation age (Ma): 1.9
Populations also occur in terrestrial, wetland and/ or riparian habitats: NO
Long distance movement vectors: NA
Nuisance status and uses: NA
Endangered status (IUCN): LC

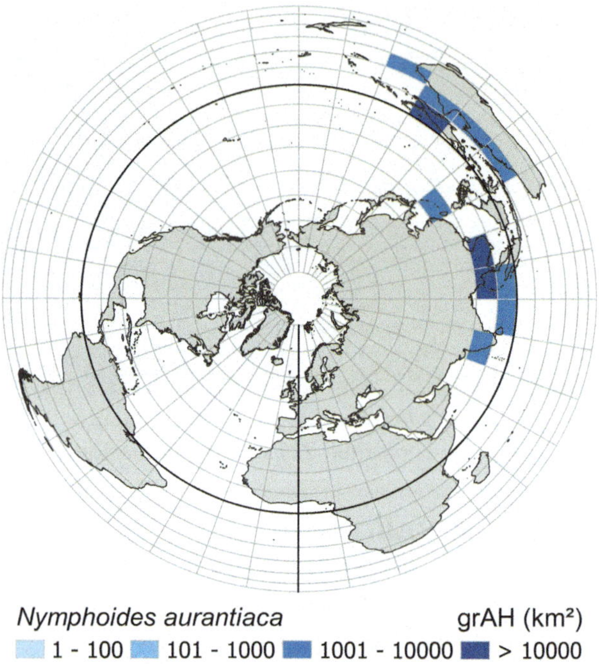

Nymphoides aurantiaca (Dalzell) Kuntze
Image copyright: N. Tippery, with permission

***Nymphoides beaglensis* Aston**
Synonyms:

Range area (grAH km²): AUS: 7200
Ecozone origin-status: AUS: E
World rarity status: VR
Ecozone-endemic macroregion occurrence: AUS: NA
Life form(s): FR
Ploidy state: NA
Speciation age (Ma): 2.6
Populations also occur in terrestrial, wetland and/ or riparian habitats: NO
Long distance movement vectors: NA
Nuisance status and uses: NA
Endangered status (IUCN): NA

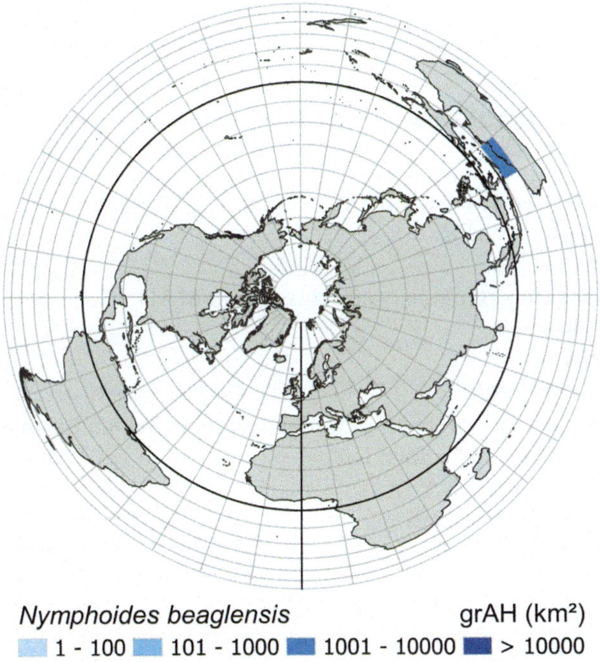

Nymphoides beaglensis Aston
Image copyright: N. Tippery, with permission

Nymphoides bosseri A.Raynal
Synonyms:

Range area (grAH km^2): AFR: 10100
Ecozone origin-status: AFR: E
World rarity status: VR
Ecozone-endemic macroregion occurrence: AFR: MAD
Life form(s): FR
Ploidy state: NA
Speciation age (Ma): NA
Populations also occur in terrestrial, wetland and/ or riparian habitats: NO
Long distance movement vectors: NA
Nuisance status and uses: NA
Endangered status (IUCN): EN

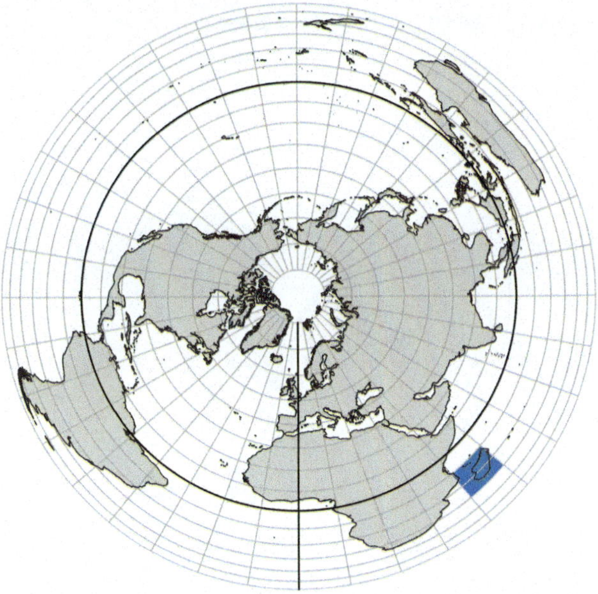

***Nymphoides brevipedicellata* (Vatke) A.Raynal**
Synonyms: *Limnanthemum brevipedicellatum* Vatke

Range area (grAH km^2): AFR: 154300
Ecozone origin-status: AFR: E
World rarity status: R
Ecozone-endemic macroregion occurrence: AFR: SAH/ AR , FOR, EA, ZAM
Life form(s): FR
Ploidy state: NA
Populations also occur in terrestrial, wetland and/ or riparian habitats: NO
Speciation age (Ma): NA
Long distance movement vectors: NA
Nuisance status and uses: NA
Endangered status (IUCN): LC

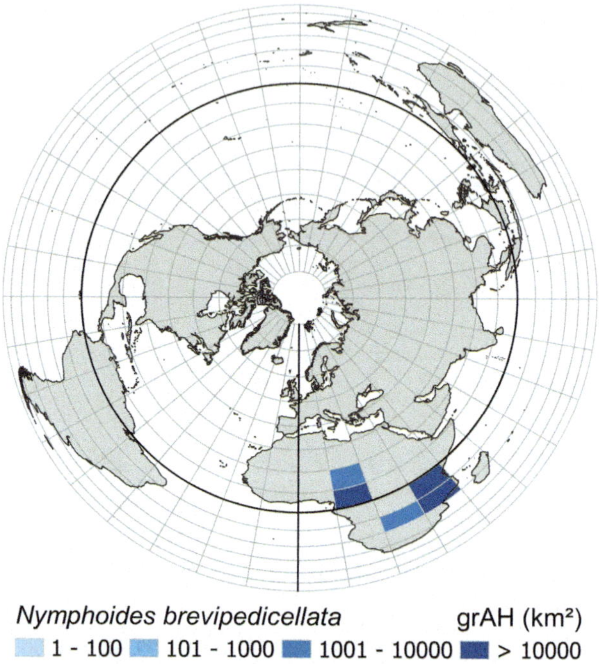

Nymphoides brevipedicellata (Vatke) A.Raynal
Image copyright: Z. Li, with permission

***Nymphoides cambodiana* (Hance) Tippery**
Synonyms: *Villarsia cambodiana* Hance

Range area (grAH km^2): OR: 15700
Ecozone origin-status: OR: E
World rarity status: VR
Ecozone-endemic macroregion occurrence: OR: ICH
Life form(s): FR
Ploidy state: NA
Speciation age (Ma): 1.8
Populations also occur in terrestrial, wetland and/ or riparian habitats: NO
Long distance movement vectors: NA
Nuisance status and uses: NA
Endangered status (IUCN): NA

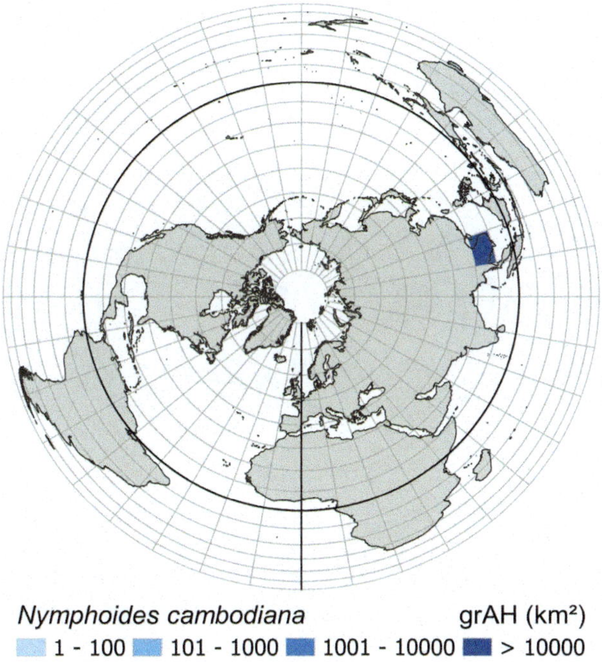

***Nymphoides cordata* (Elliott) Fernald**
Synonyms: *Nymphoides lacunosum* Fernald

Range area (grAH km^2): NEA: 119800; present in OCE
Ecozone origin-status: NEA: E(I); OCE: I
World rarity status: R
Ecozone-endemic macroregion occurrence: NEA: E NAM, W NAM
Life form(s): FR
Ploidy state: P
Speciation age (Ma): 0.6
Populations also occur in terrestrial, wetland and/ or riparian habitats: NO
Long distance movement vectors: ANTHROPOCHORY
Nuisance status and uses: NA
Endangered status (IUCN): LC

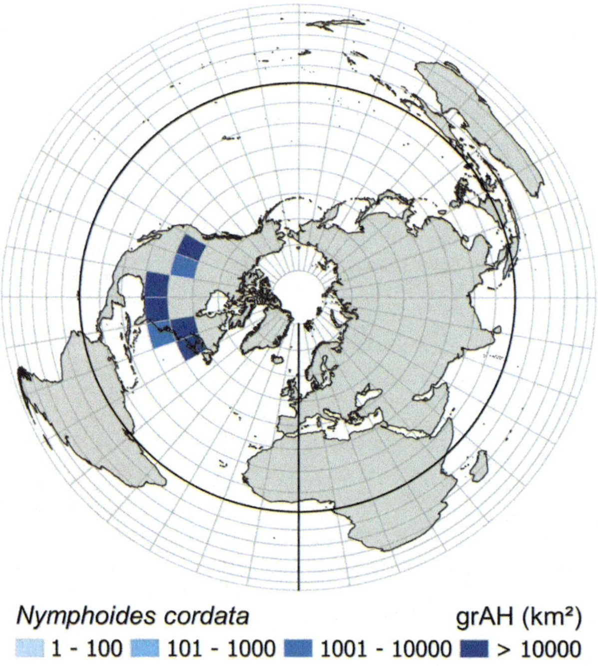

Nymphoides cordata (Elliott) Fernald

***Nymphoides coreana* (H. Lév.) H. Hara**
Synonyms: *Limnanthemum coreanum* H.Lév.

Range area (grAH km^2): PAL: 50900; OR: 2600
Ecozone origin-status: PAL: N; OR: N
World rarity status: R
Ecozone-endemic macroregion occurrence: not endemic to any world ecozone
Life form(s): FR
Ploidy state: P
Speciation age (Ma): 0.9
Populations also occur in terrestrial, wetland and/ or riparian habitats: NO
Long distance movement vectors: NA
Nuisance status and uses: NA
Endangered status (IUCN): NA

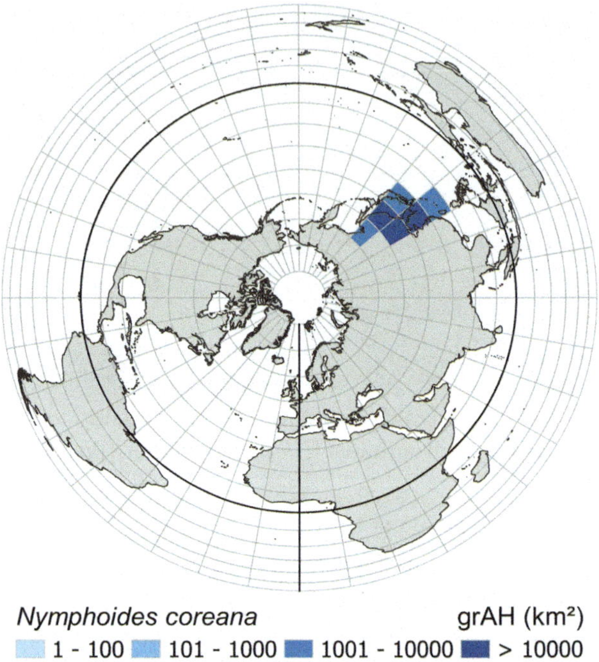

Nymphoides coreana (H. Lév.) H. Hara
Image copyright: Z. Li, with permission

Nymphoides crenata **(F. Muell.) Kuntze**
Synonyms: *Limnanthemum crenatum* F.Muell., *Villarsia crenata* (F.Muell.) F.Muell.

Range area (grAH km^2): AUS: 37000
Ecozone origin-status: AUS: E
World rarity status: R
Ecozone-endemic macroregion occurrence: AUS: NA, ESA, WA
Life form(s): FR
Ploidy state: D
Speciation age (Ma): 2.4
Populations also occur in terrestrial, wetland and/ or riparian habitats: NO
Long distance movement vectors: NA
Nuisance status and uses: NA
Endangered status (IUCN): NA

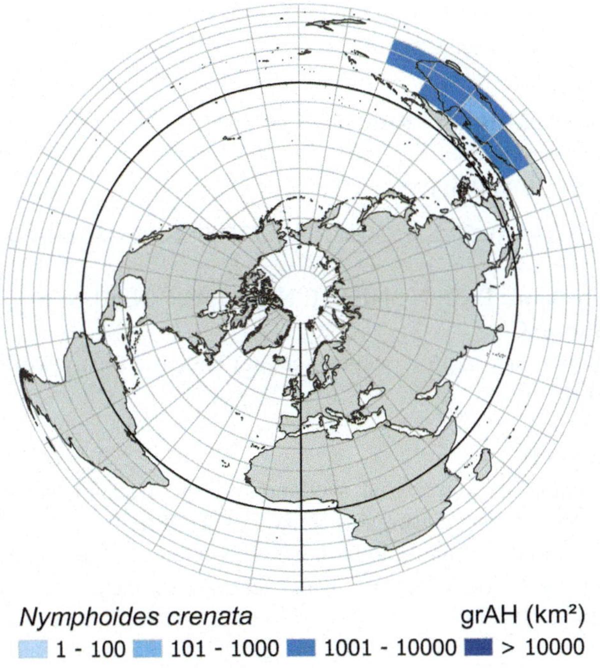

Nymphoides crenata (F. Muell.) Kuntze
Image copyright: K. Ward, with permission

Nymphoides cristata **(Roxb.) Kuntze**
Synonyms: *Limnanthemum cristatum* (Roxb.) Griseb., *Menyanthes cristata* Roxb., *Villarsia cristata* (Roxb.) Spreng.

Range area (grAH km^2): PAL: 58500; NEA: 77500; NEO: 24100; OR: 65400; AUS: 11800
Ecozone origin-status: PAL: N ; NEA: I; NEO: I; OR: N; AUS: N
World rarity status: C
Ecozone-endemic macroregion occurrence: not endemic to any world ecozone
Life form(s): FR
Ploidy state: D
Speciation age (Ma): 1.4
Populations also occur in terrestrial, wetland and/ or riparian habitats: NO
Long distance movement vectors: ANTHROPOCHORY
Nuisance status and uses: major aquatic weed in at least one world ecozone outside home range [54, 80]; horticulture, water garden ornamental, medical
Endangered status (IUCN): NA

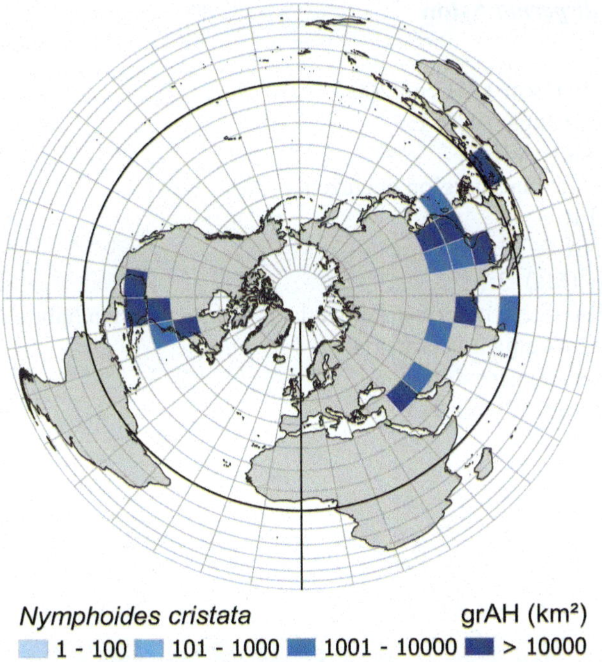

Nymphoides cristata (Roxb.) Kuntze
Image copyright: A. Efremov, with permission

***Nymphoides disperma* Aston**
Synonyms:

Range area (grAH km^2): AUS: 7200
Ecozone origin-status: AUS: E
World rarity status: VR
Ecozone-endemic macroregion occurrence: AUS: NA
Life form(s): FR
Ploidy state: NA
Speciation age (Ma): 3.1
Populations also occur in terrestrial, wetland and/ or riparian habitats: NO
Long distance movement vectors: NA
Nuisance status and uses: NA
Endangered status (IUCN): NA

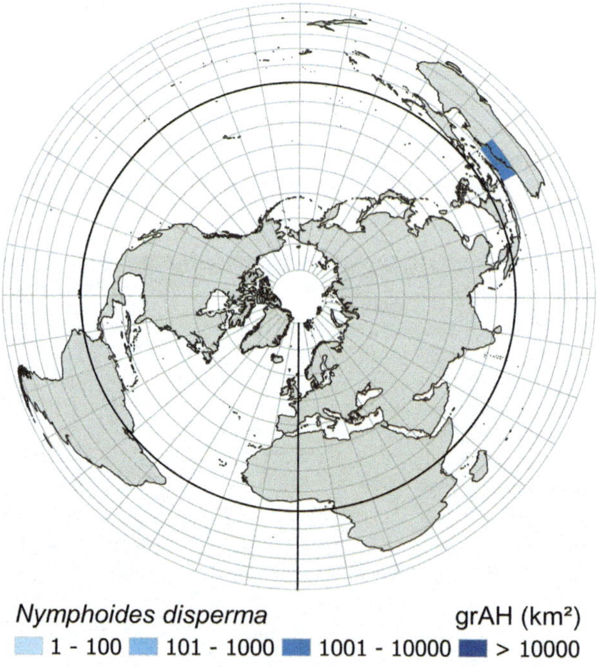

***Nymphoides elegans* A. Raynal**
Synonyms: *Nymphoides moratiana* A. Raynal

Range area (grAH km^2): AFR: 10100
Ecozone origin-status: AFR: E
World rarity status: VR
Ecozone-endemic macroregion occurrence: AFR: MAD
Life form(s): FR
Ploidy state: NA
Speciation age (Ma): NA
Populations also occur in terrestrial, wetland and/ or riparian habitats: NO
Long distance movement vectors: NA
Nuisance status and uses: NA
Endangered status (IUCN): EN

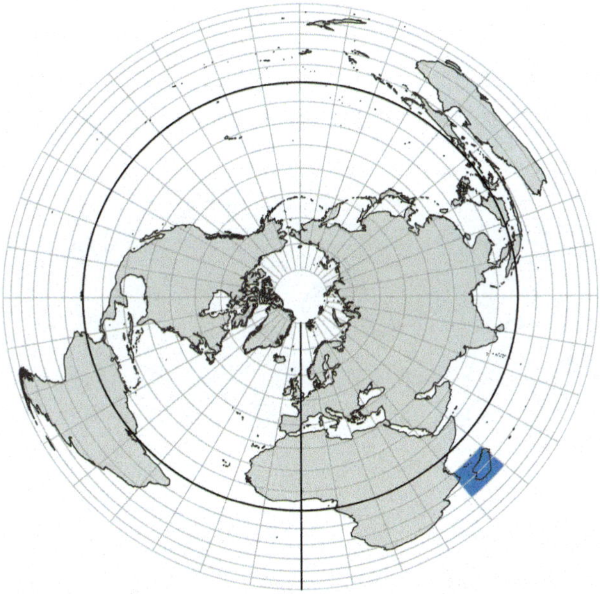

Nymphoides elliptica **Aston**
Synonyms:

Range area (grAH km^2): AUS: 2900
Ecozone origin-status: AUS: E
World rarity status: VR
Ecozone-endemic macroregion occurrence: AUS: NA
Life form(s): FR
Ploidy state: NA
Speciation age (Ma): 2.5
Populations also occur in terrestrial, wetland and/ or riparian habitats: NO
Long distance movement vectors: NA
Nuisance status and uses: NA
Endangered status (IUCN): NA

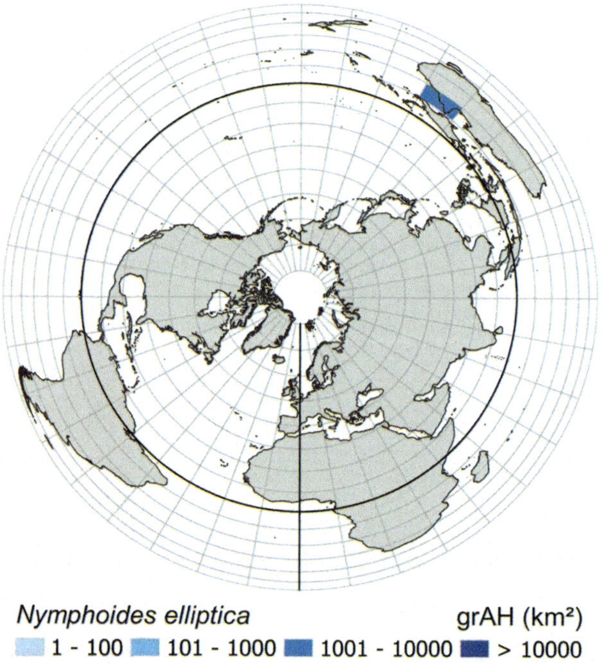

***Nymphoides exiliflora* (F. Muell.) Kuntze**

Synonyms: *Limnanthemum exiliflorum* (F.Muell.) Benth., *Villarsia exiliflora* F.Muell.

Range area (grAH km^2): AUS: 27400
Ecozone origin-status: AUS: E
World rarity status: VR
Ecozone-endemic macroregion occurrence: AUS: NG, NA, ESA
Life form(s): FR
Ploidy state: NA
Speciation age (Ma): 2.5
Populations also occur in terrestrial, wetland and/ or riparian habitats: NO
Long distance movement vectors: NA
Nuisance status and uses: no nuisance issues; used in water garden and aquarium trade in Australia
Endangered status (IUCN): NA

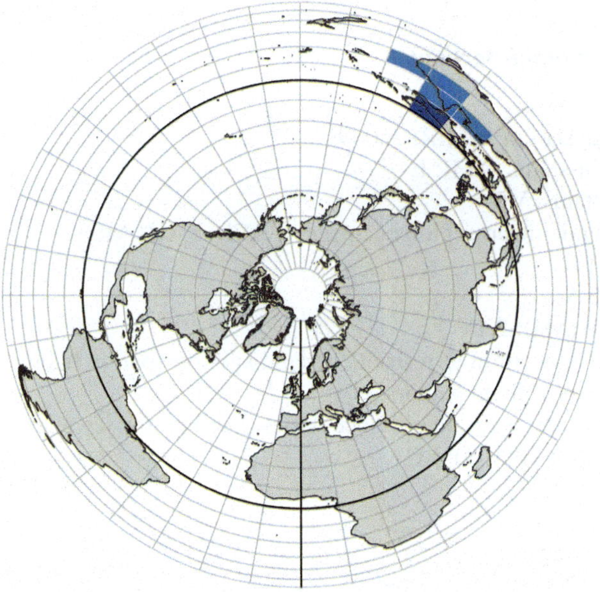

Nymphoides exiliflora (F. Muell.) Kuntze
Image copyright: N. Tippery, with permission

Nymphoides ezannoi **Berhaut**
Synonyms:

Range area (grAH km^2): AFR: 26000
Ecozone origin-status: AFR: E
World rarity status: VR
Ecozone-endemic macroregion occurrence: AFR: SAH/ AR
Life form(s): FR
Ploidy state: NA
Speciation age (Ma): 2.1
Populations also occur in terrestrial, wetland and/ or riparian habitats: NO
Long distance movement vectors: NA
Nuisance status and uses: no nuisance issues; handicrafts, jewellery, etc.
Endangered status (IUCN): LC

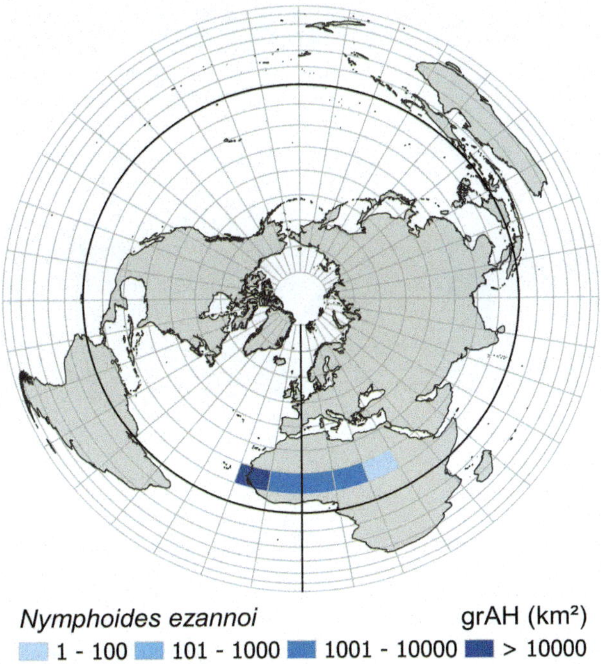

***Nymphoides fallax* Ornduff**
Synonyms:

Range area (grAH km^2): NEO: 41800
Ecozone origin-status: NEO: E
World rarity status: R
Ecozone-endemic macroregion occurrence: NEO: MG MX, MES AM
Life form(s): FR
Ploidy state: P
Speciation age (Ma): 0.6
Populations also occur in terrestrial, wetland and/ or riparian habitats: NO
Long distance movement vectors: NA
Nuisance status and uses: no nuisance issues; medicinal use
Endangered status (IUCN): NA

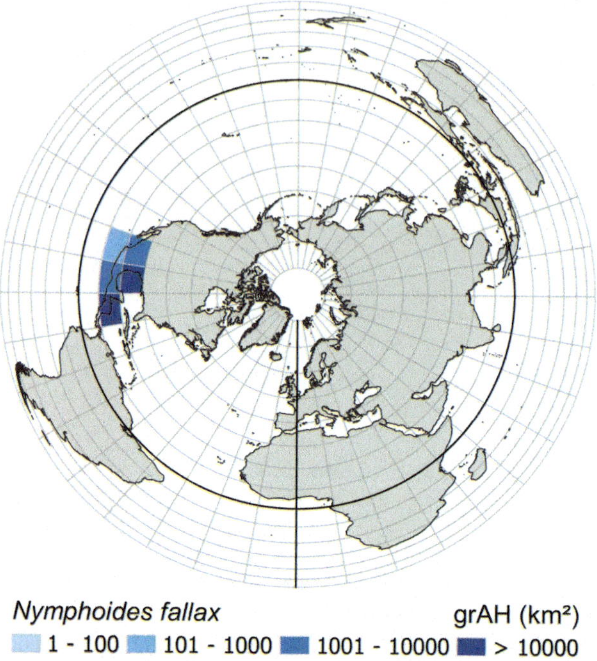

Nymphoides fallax Ornduff
Image copyright: T. Lobato-de Magalhães

***Nymphoides flaccida* L.B.Sm.**

Synonyms:

Range area (grAH km^2): NEO: 16600
Ecozone origin-status: NEO: E
World rarity status: VR
Ecozone-endemic macroregion occurrence: NEO: NWSA
Life form(s): FR
Ploidy state: NA
Speciation age (Ma): NA
Populations also occur in terrestrial, wetland and/ or riparian habitats: NO
Long distance movement vectors: NA
Nuisance status and uses: no nuisance issues; aquarium trade
Endangered status (IUCN): NA

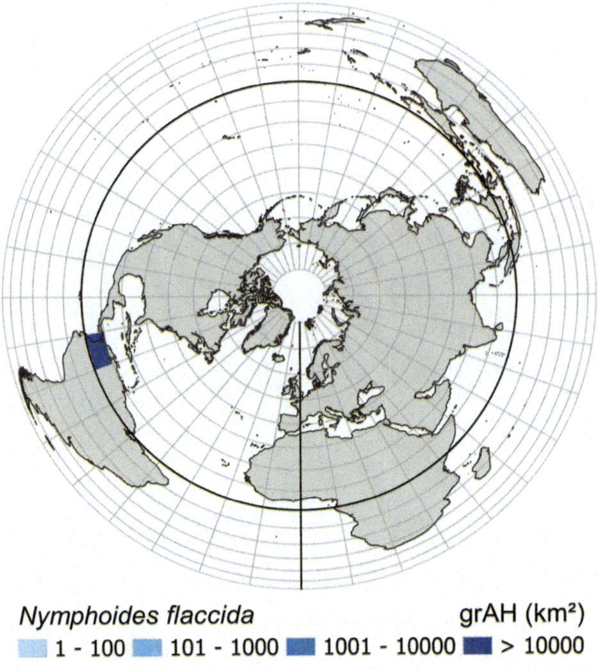

***Nymphoides forbesiana* (Griseb.) Kuntze**
Synonyms: *Limnanthemum forbesianum* Griseb.

Range area (grAH km^2): OR: 2200; AFR: 196400
Ecozone origin-status: OR: I; AFR: E(I)
World rarity status: C
Ecozone-endemic macroregion occurrence: AFR: FOR, EA, ZAM
Life form(s): FR
Ploidy state: NA
Speciation age (Ma): NA
Populations also occur in terrestrial, wetland and/ or riparian habitats: NO
Long distance movement vectors: ANTHROPOCHORY
Nuisance status and uses: no nuisance issues; no known uses
Endangered status (IUCN): LC

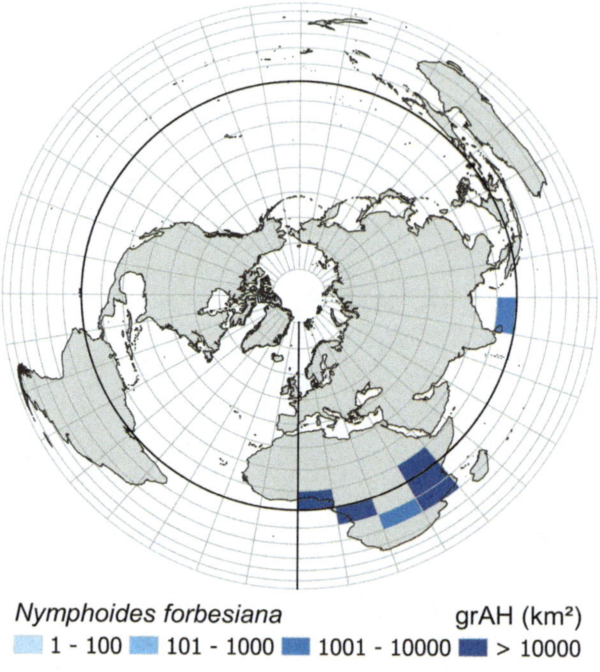

Nymphoides forbesiana (Griseb.) Kuntze

Nymphoides furculifolia **Specht**
Synonyms:

Range area (grAH km^2): AUS: 13500
Ecozone origin-status: AUS: E
World rarity status: VR
Ecozone-endemic macroregion occurrence: AUS: NA
Life form(s): FR
Ploidy state: NA
Speciation age (Ma): 2.4
Populations also occur in terrestrial, wetland and/ or riparian habitats: NO
Long distance movement vectors: NA
Nuisance status and uses: no nuisance issues; no known uses
Endangered status (IUCN): NA

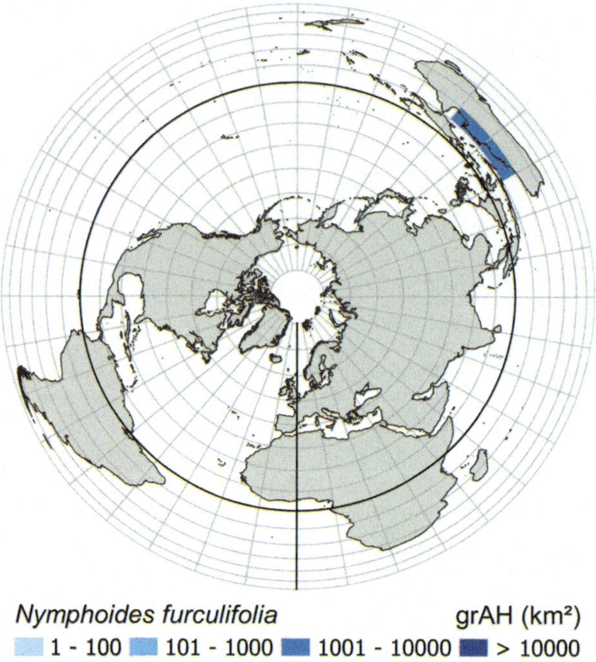

Nymphoides furculifolia Specht
Image copyright: N. Tippery, with permission

***Nymphoides geminata* (R.Br.) Kuntze**

Synonyms: *Limnanthemum geminatum* (R.Br.) Griseb., *Villarsia geminata* R.Br.

Range area (grAH km^2): AUS: 44800
Ecozone origin-status: AUS: E
World rarity status: R
Ecozone-endemic macroregion occurrence: AUS: NG, NA, ESA, NZ
Life form(s): FR
Ploidy state: P
Speciation age (Ma): 1.5
Populations also occur in terrestrial, wetland and/ or riparian habitats: NO
Long distance movement vectors: NA
Nuisance status and uses: no nuisance issues; no known uses
Endangered status (IUCN): NA

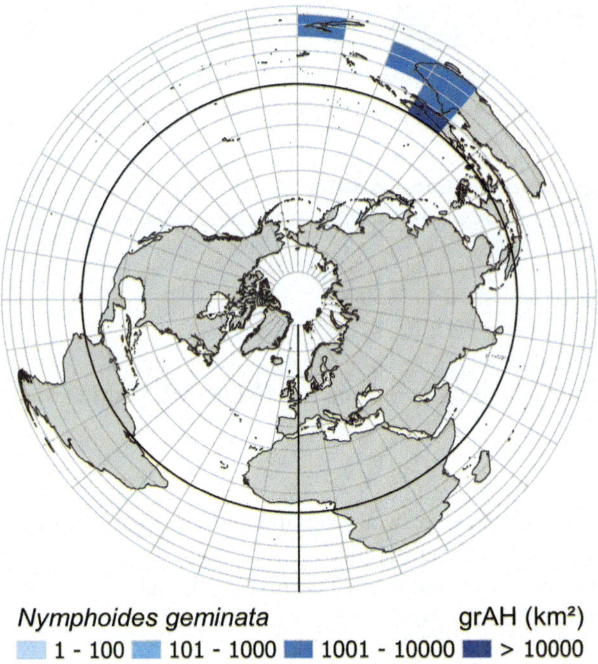

Nymphoides geminata (R.Br.) Kuntze

***Nymphoides grayana* (Griseb.) Kuntze**
Synonyms: *Limnanthemum grayanum* Griseb., *Trachysperma grayana* (Griseb.) House

Range area (grAH km^2): NEO: 55900
Ecozone origin-status: NEO: E
World rarity status: R
Ecozone-endemic macroregion occurrence: NEO: MES AM, AMZ, CHA
Life form(s): FR
Ploidy state: P
Speciation age (Ma): 0.5
Populations also occur in terrestrial, wetland and/ or riparian habitats: NO
Long distance movement vectors: NA
Nuisance status and uses: no nuisance issues; medicinal
Endangered status (IUCN): NA

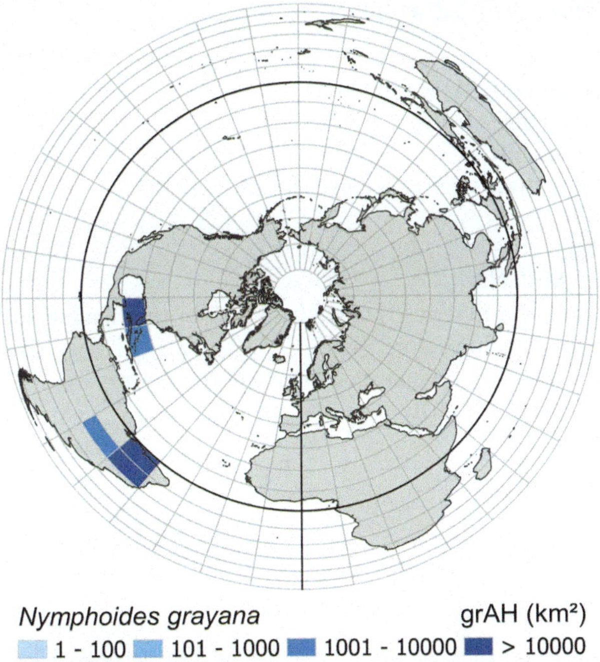

Nymphoides grayana (Griseb.) Kuntze
Image copyright: Z. Li, with permission

***Nymphoides guineensis* A.Raynal**
Synonyms:

Range area (grAH km^2): AFR: 3400
Ecozone origin-status: AFR: E
World rarity status: VR
Ecozone-endemic macroregion occurrence: AFR: FOR
Life form(s): FR
Ploidy state: NA
Speciation age (Ma): NA
Populations also occur in terrestrial, wetland and/ or riparian habitats: NO
Long distance movement vectors: NA
Nuisance status and uses: no nuisance issues; horticulture
Endangered status (IUCN): DD

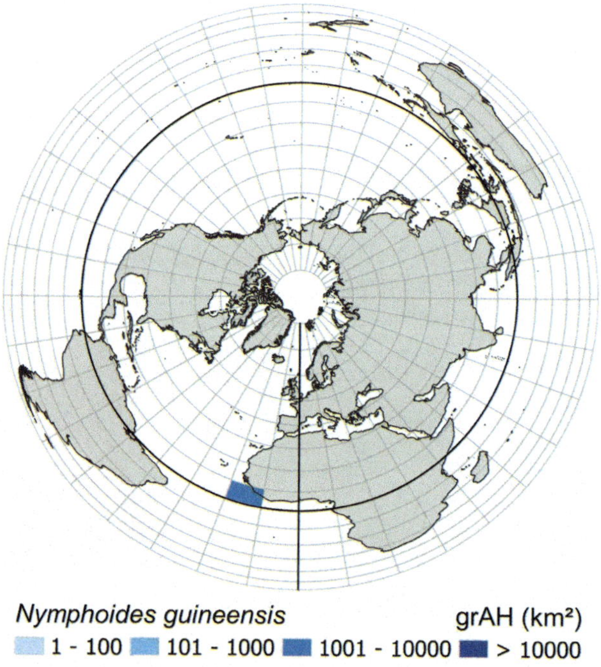

Nymphoides hastata **(Dop) Kerr**
Synonyms: *Limnanthemum hastatum* Dop

Range area (grAH km^2): OR: 15700
Ecozone origin-status: OR: E
World rarity status: VR
Ecozone-endemic macroregion occurrence: OR: ICH
Life form(s): FR
Ploidy state: NA
Speciation age (Ma): NA
Populations also occur in terrestrial, wetland and/ or riparian habitats: NO
Long distance movement vectors: NA
Nuisance status and uses: no nuisance issues; no known uses
Endangered status (IUCN): NA

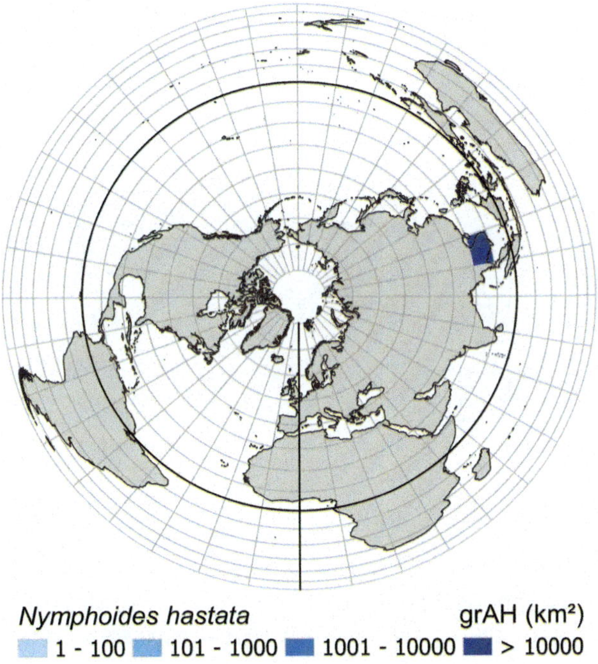

***Nymphoides herzogii* A. Galán & G. Navarro**
Synonyms:

Range area (grAH km^2): NEO: 21100
Ecozone origin-status: NEO: E
World rarity status: VR
Ecozone-endemic macroregion occurrence: NEO: AND
Life form(s): FR
Ploidy state: NA
Speciation age (Ma): NA
Populations also occur in terrestrial, wetland and/ or riparian habitats: NO
Long distance movement vectors: NA
Nuisance status and uses: no nuisance issues; no known uses
Endangered status (IUCN): EN

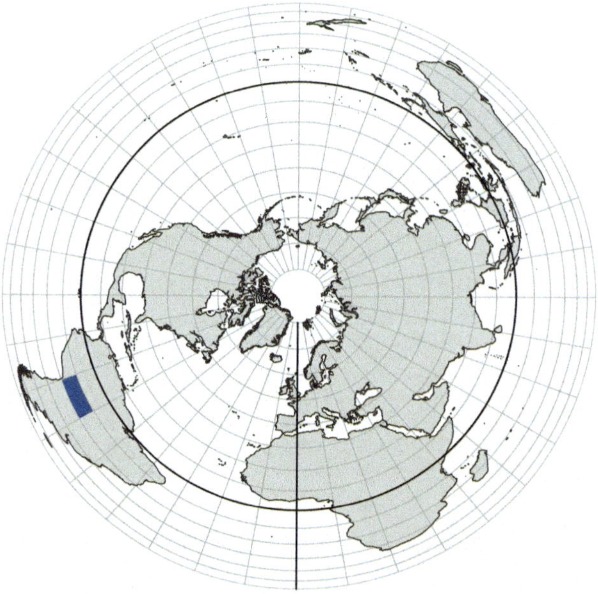

Nymphoides herzogii grAH (km²)
1 - 100 101 - 1000 1001 - 10000 > 10000

Nymphoides herzogii A. Galán & G. Navarro
Image Copyright: L. Fung, with permission

***Nymphoides humilis* A. Raynal**
Synonyms:

Range area (grAH km^2): AFR: 17700
Ecozone origin-status: AFR: E
World rarity status: VR
Ecozone-endemic macroregion occurrence: AFR: SAH/ AR, FOR
Life form(s): FR (hydrosubshrub)
Ploidy state: NA
Speciation age (Ma): NA
Populations also occur in terrestrial, wetland and/ or riparian habitats: NO
Long distance movement vectors: NA
Nuisance status and uses: no nuisance issues; no known uses
Endangered status (IUCN): DD

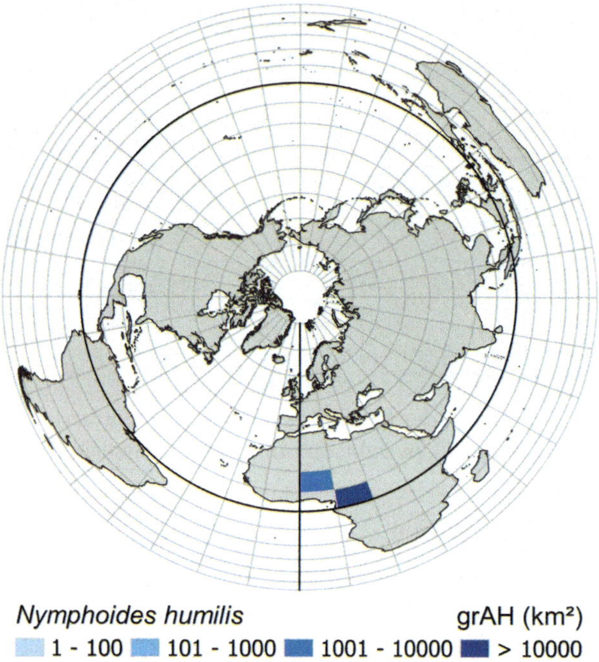

***Nymphoides hydrocharoides* (F. Muell.) Kuntze**
Synonyms: *Nymphoides aurantiaca* (Dalzell) Kuntze

Range area (grAH km^2): AUS: 29300
Ecozone origin-status: AUS: E
World rarity status: VR
Ecozone-endemic macroregion occurrence: AUS: NG, NA
Life form(s): FR
Ploidy state: NA
Speciation age (Ma): NA
Populations also occur in terrestrial, wetland and/ or riparian habitats: NO
Long distance movement vectors: NA
Nuisance status and uses: no nuisance issues; no known uses
Endangered status (IUCN): NA

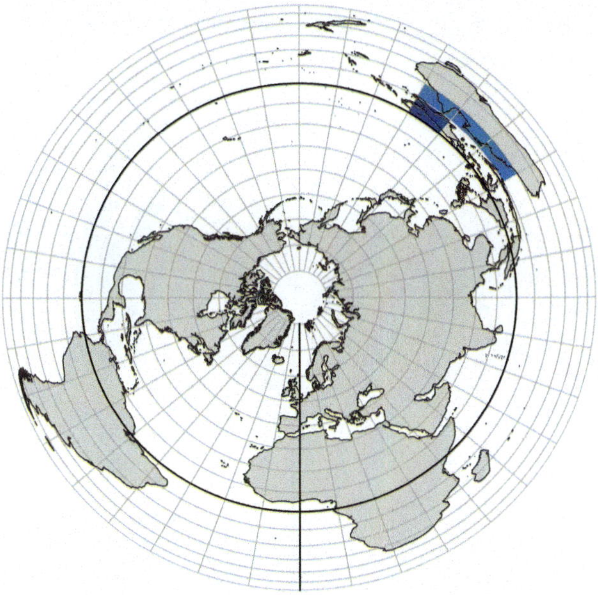

Nymphoides hydrophylla **(Lour.) Kuntze**

Synonyms: *Limnanthemum hydrophyllum* (Lour.) Griseb., *Menyanthes hydrophyllum* Lour., *Villarsia hydrophyllum* (Lour.) Roem. & Schult.

Range area (grAH km^2): PAL: 19400; NEA: 34000; NEO: 24100; OR: 130500; AUS: 24500; AFR: 8000

Ecozone origin-status: PAL: N; NEA: I; NEO: I; OR: N; AUS: N; AFR: I

World rarity status: C

Ecozone-endemic macroregion occurrence: not endemic to any world ecozone

Life form(s): FR

Ploidy state: D

Speciation age (Ma): NA

Populations also occur in terrestrial, wetland and/ or riparian habitats: NO

Long distance movement vectors: ANTHROPOCHORY

Nuisance status and uses: no nuisance issues; used as a medicine for ulcers and bites and as antihelmenthic, aquarium trade

Endangered status (IUCN): LC

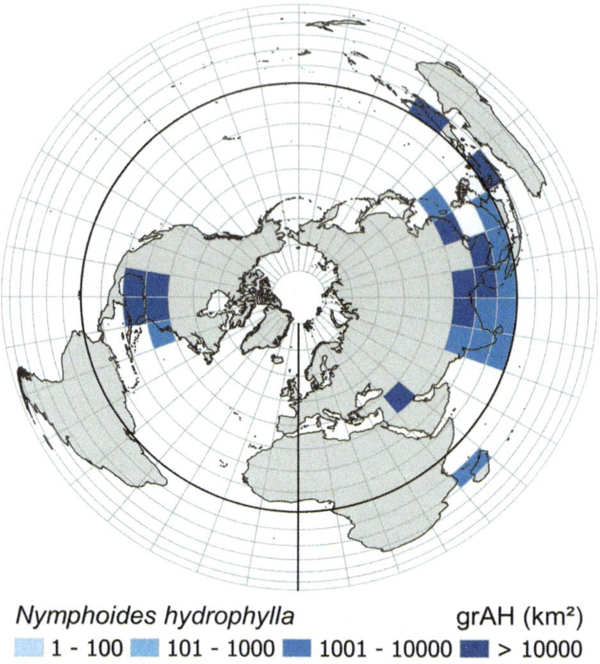

Nymphoides hydrophylla (Lour.) Kuntze
Image copyright: Z. Li, with permission

***Nymphoides indica* (L.) Kuntze**
Synonyms: *Limnanthemum indicum* (L.) Griseb., *Menyanthes indica* L., *Villarsia indica (*L.) Vent.

Range area (grAH km^2): PAL: 104800; NEA: 13000; NEO: 348000; OR: 128600; AUS: 63600; AFR: 170800; present in OCE
Ecozone origin-status: PAL: N; NEA: I; NEO: I; OR: N; AUS: N; AFR: I; OCE: N
World rarity status: C
Ecozone-endemic macroregion occurrence: not endemic to any world ecozone

Life form(s): FR
Ploidy state: D
Speciation age (Ma): 0.7
Populations also occur in terrestrial, wetland and/ or riparian habitats: NO
Long distance movement vectors: ANTHROPOCHORY
Nuisance status and uses: low or no nuisance issues; horticulture (water gardens), ornamental, folk medicine, phytoremediation uses
Endangered status (IUCN): LC

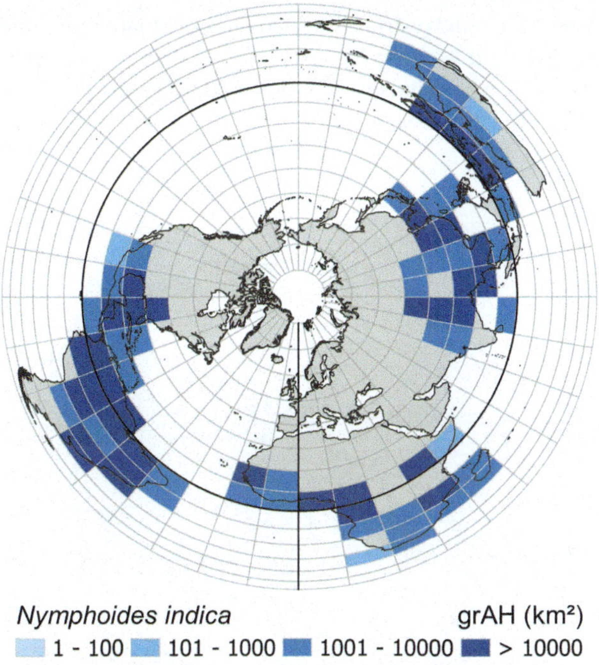

Nymphoides indica (L.) Kuntze
Image copyright: N. Tippery, with permission

Nymphoides krishnakesara **K.T.Joseph & Sivar.**
Synonyms:

Range area (grAH km^2): OR: 8100
Ecozone origin-status: OR: E
World rarity status: VR
Ecozone-endemic macroregion occurrence: OR: IND
Life form(s): FR
Ploidy state: NA
Speciation age (Ma): NA
Populations also occur in terrestrial, wetland and/ or riparian habitats: NO
Long distance movement vectors: NA
Nuisance status and uses: no nuisance issues; no known uses
Endangered status (IUCN): EN

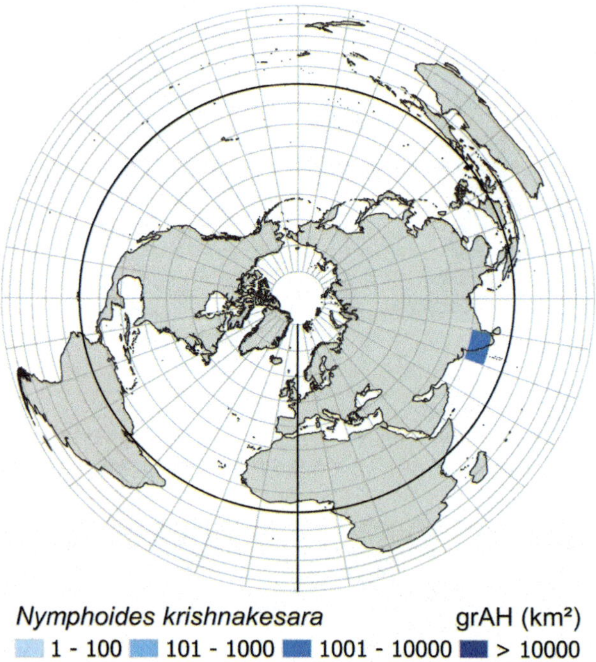

Nymphoides lungtanensis **S.P. Li, T.H. Hsieh & C.C. Lin**
Synonyms:

Range area (grAH km^2): OR: 2600
Ecozone origin-status: OR: E
World rarity status: VR
Ecozone-endemic macroregion occurrence: OR: ICH
Life form(s): FR
Ploidy state: NA
Speciation age (Ma): NA
Long distance movement vectors: NA
Populations also occur in terrestrial, wetland and/ or riparian habitats: NO
Nuisance status and uses: no nuisance issues; no known uses
Endangered status (IUCN): NA

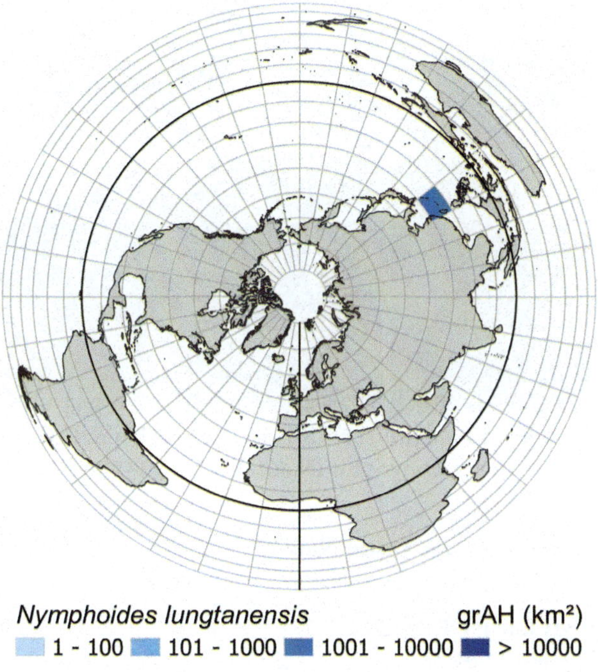

Nymphoides lungtanensis S.P. Li, T.H. Hsieh & C.C. Lin
Image copyright: iNaturalist CC0 @gbno

Nymphoides macrospermum **Vasud.Nair**
Synonyms:

Range area (grAH km^2): OR: 8100
Ecozone origin-status: OR: E
World rarity status: VR
Ecozone-endemic macroregion occurrence: OR: IND
Life form(s): FR
Ploidy state: P
Speciation age (Ma): NA
Populations also occur in terrestrial, wetland and/ or riparian habitats: NO
Long distance movement vectors: NA
Nuisance status and uses: no nuisance issues; no known uses
Endangered status (IUCN): CR

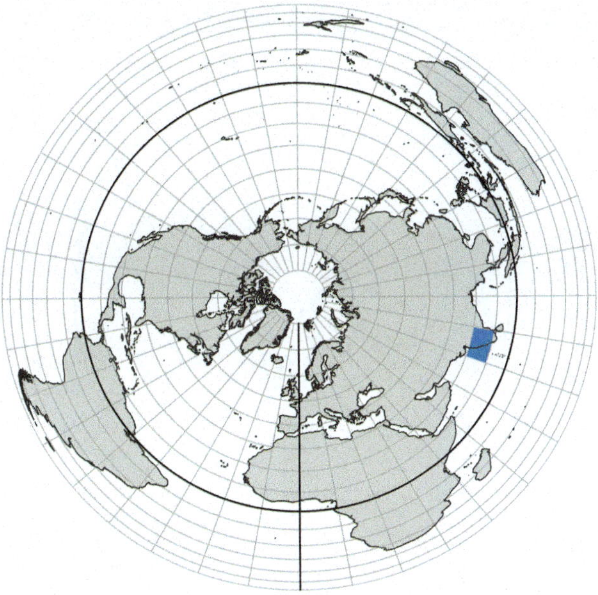

Nymphoides macrospermum grAH (km²)
1 - 100 101 - 1000 1001 - 10000 > 10000

Nymphoides milnei **A. Raynal**
Synonyms:

Range area (grAH km²): AFR: 8700
Ecozone origin-status: AFR: E
World rarity status: VR
Ecozone-endemic macroregion occurrence: AFR: ZAM
Life form(s): FR
Ploidy state: NA
Speciation age (Ma): NA
Populations also occur in terrestrial, wetland and/ or riparian habitats: NO
Long distance movement vectors: NA
Nuisance status and uses: no nuisance issues; no known uses
Endangered status (IUCN): NA

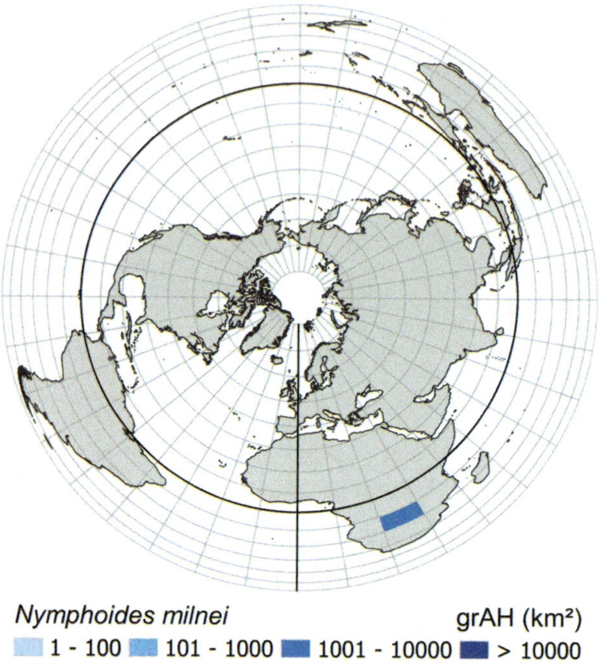

***Nymphoides minima* (F. Muell.) Kuntze**
Synonyms: *Limnanthemum minimum* F.Muell., *Villarsia minima* (F.Muell.) F.Muell.

Range area (grAH km^2): AUS: 16500
Ecozone origin-status: AUS: E
World rarity status: VR
Ecozone-endemic macroregion occurrence: AUS: NA
Life form(s): FR
Ploidy state: NA
Speciation age (Ma): 1.7
Populations also occur in terrestrial, wetland and/ or riparian habitats: NO
Long distance movement vectors: NA
Nuisance status and uses: no nuisance issues; aquarium trade
Endangered status (IUCN): NA

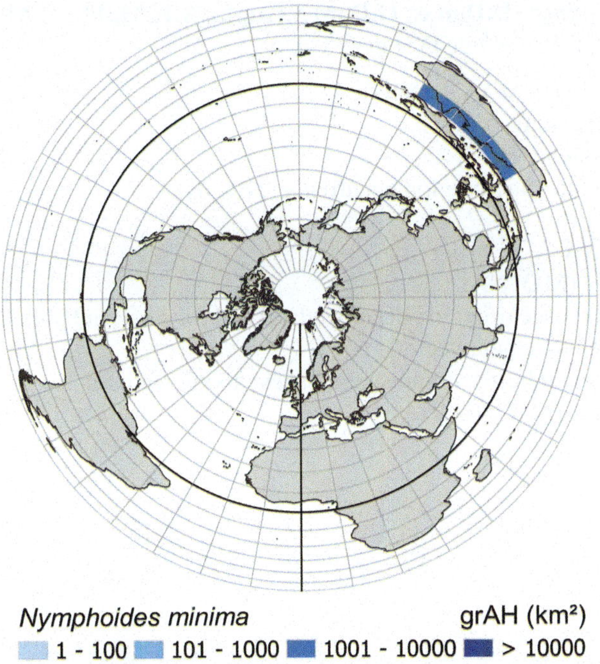

Nymphoides minima (F. Muell.) Kuntze
Image copyright: N. Tippery, with permission

Nymphoides minor **(D.Don ex G.Don) S.Gupta, A.K.Mukh. & M.Mondal**
Synonyms: *Villarsia minor* D.Don ex G.Don

Range area (grAH km^2): OR: 8100
Ecozone origin-status: OR: E
World rarity status: VR
Ecozone-endemic macroregion occurrence: OR: IND
Life form(s): FR
Ploidy state: NA
Speciation age (Ma): NA
Populations also occur in terrestrial, wetland and/ or riparian habitats: NO
Long distance movement vectors: NA
Nuisance status and uses: no nuisance issues; no known uses
Endangered status (IUCN): NA

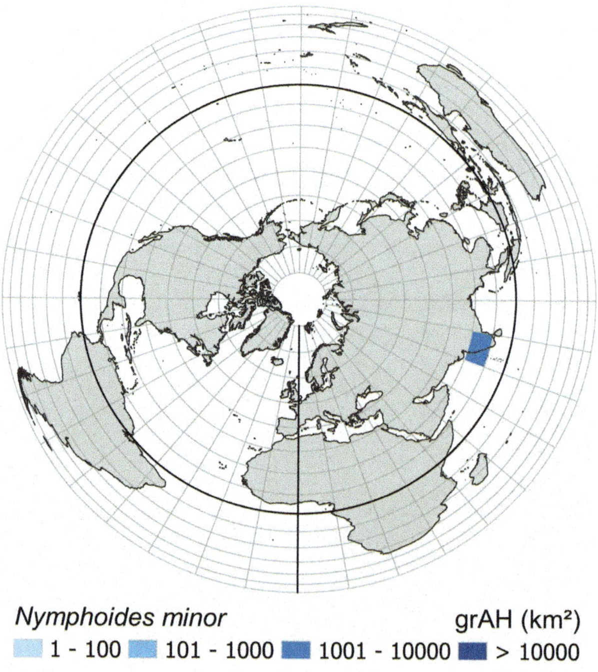

Nymphoides montana **Aston**

Synonyms: (considered a hybrid, *Nymphoides* x *montana* Aston by some sources: e.g., [118])

Range area (grAH km^2): AUS: 14300
Ecozone origin-status: AUS: E
World rarity status: VR
Ecozone-endemic macroregion occurrence: AUS: ESA, NZ
Life form(s): FR
Ploidy state: P
Speciation age (Ma): 2.2
Populations also occur in terrestrial, wetland and/ or riparian habitats: NO
Long distance movement vectors: NA
Nuisance status and uses: no nuisance issues; no known uses
Endangered status (IUCN): NA

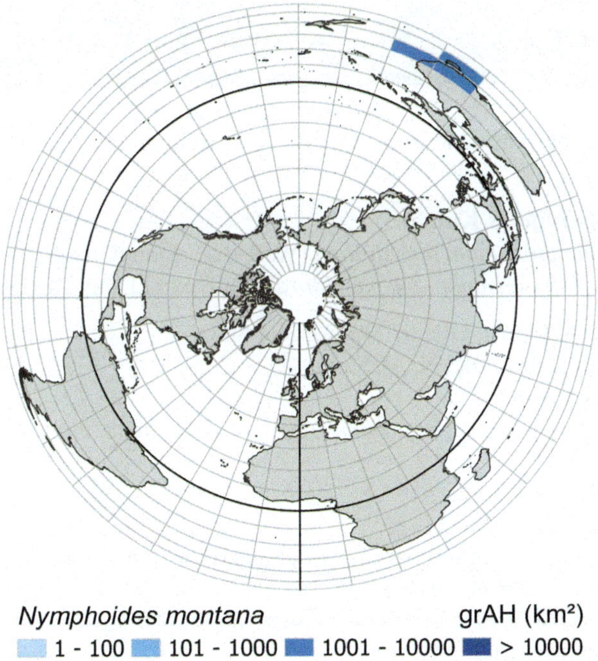

Nymphoides montana Aston

Nymphoides parvifolia **Kuntze**
Synonyms: *Limnanthemum parvifolium* (Wall. ex G.Don) Griseb., *Villarsia parvifolia* Wall. ex G.Don, *Nymphoides parviflora* (Wall. ex G.Don) Tippery

Range area (grAH km^2): OR: 67000; AUS: 44600
Ecozone origin-status: OR: N; AUS: N
World rarity status: R
Ecozone-endemic macroregion occurrence: not endemic to any world ecozone
Life form(s): FR
Ploidy state: NA
Speciation age (Ma): 0.4
Populations also occur in terrestrial, wetland and/ or riparian habitats: NO
Long distance movement vectors: NA
Nuisance status and uses: no nuisance issues; no known uses
Endangered status (IUCN): LC

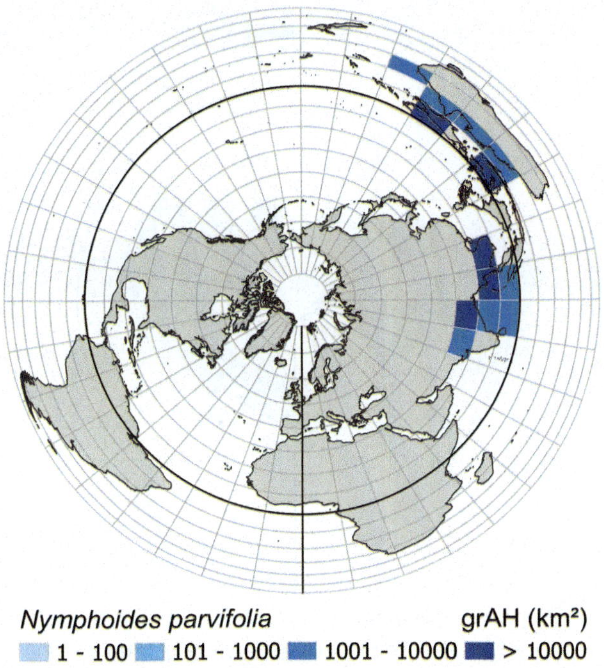

***Nymphoides peltata* (S.G.Gmel.) Kuntze**

Synonyms: *Limnanthemum peltatum* S.G.Gmel., *Villarsia peltata (S.G.Gmel.)* Roem. & Schult.

Range area (grAH km^2): PAL: 623500; NEA: 359600; NEO: 16500; OR: 45600; AUS: 5400; AFR: 6000; present in OCE

Ecozone origin-status: PAL: N; NEA: I; NEO: I; OR: N; AUS: I; N; AFR: I; OCE: I

World rarity status: C

Ecozone-endemic macroregion occurrence: not endemic to any world ecozone

Life form(s): FR

Ploidy state: P

Speciation age (Ma): 2.5

Populations also occur in terrestrial, wetland and/ or riparian habitats: NO

Long distance movement vectors: ANTHROPOCHORY; BIRD: *Anas platyrhynchos* (Northern Mallard)

Nuisance status and uses: major aquatic weed in at least one world ecozone outside home range [54, 80]; horticulture, water gardens,

medicinal, ornamental

Endangered status (IUCN): LC

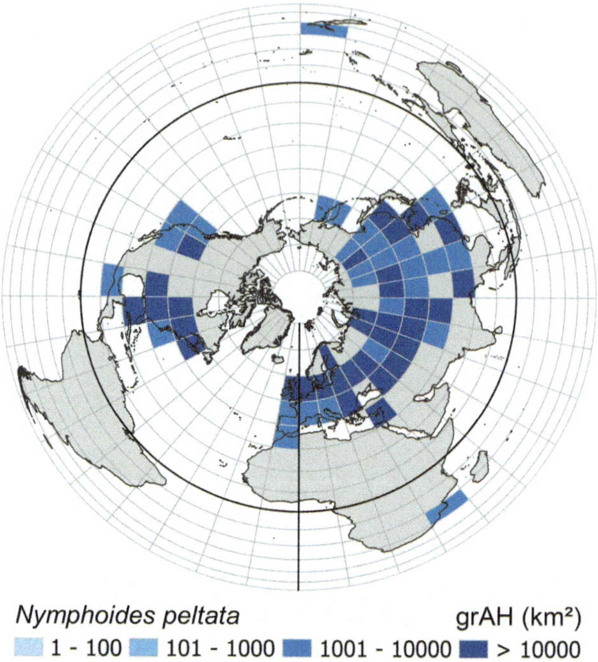

Nymphoides peltata (S.G.Gmel.) Kuntze
Image copyright: J. C.Schou, with permission

Nymphoides planosperma **Aston**
Synonyms:

Range area (grAH km^2): AUS: 6300
Ecozone origin-status: AUS: E
World rarity status: VR
Ecozone-endemic macroregion occurrence: AUS: NA
Life form(s): FR
Ploidy state: NA
Speciation age (Ma): 0.8
Populations also occur in terrestrial, wetland and/ or riparian habitats: NO
Long distance movement vectors: NA
Nuisance status and uses: no nuisance issues; no known uses
Endangered status (IUCN): NA

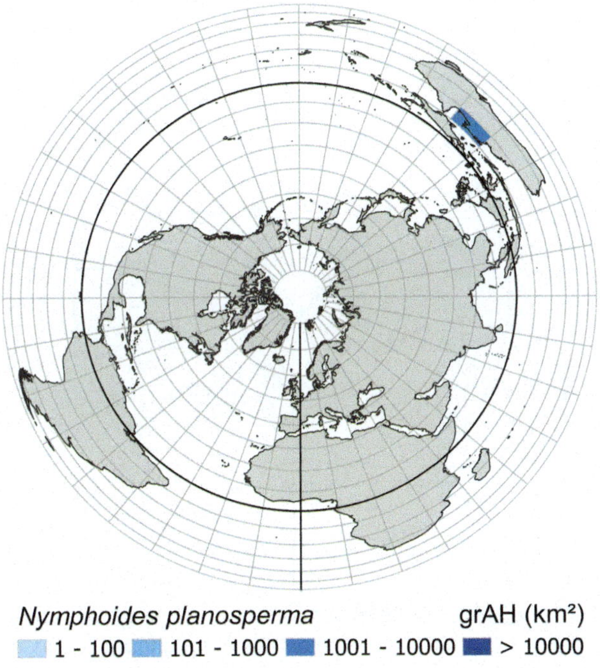

Nymphoides quadriloba **Aston**
Synonyms:

Range area (grAH km^2): AUS: 16500
Ecozone origin-status: AUS: E
World rarity status: VR
Ecozone-endemic macroregion occurrence: AUS: NA
Life form(s): FR
Ploidy state: NA
Speciation age (Ma): 0.3
Populations also occur in terrestrial, wetland and/ or riparian habitats: NO
Long distance movement vectors: NA
Nuisance status and uses: no nuisance issues; no known uses
Endangered status (IUCN): NA

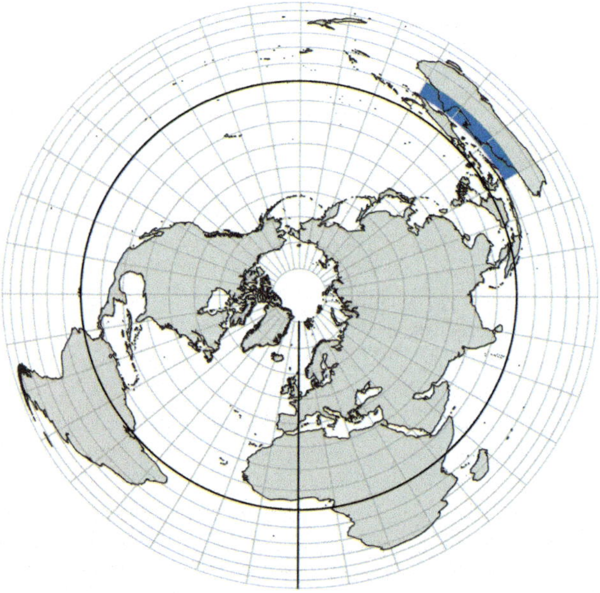

Nymphoides rautanenii **(N.E.Br.) A.Raynal**
Synonyms: *Limnanthemum rautanenii* N.E.Br.

Range area (grAH km^2): AFR: 17700
Ecozone origin-status: AFR: E
World rarity status: VR
Ecozone-endemic macroregion occurrence: AFR: ZAM
Life form(s): FR
Ploidy state: NA
Speciation age (Ma): NA
Populations also occur in terrestrial, wetland and/ or riparian habitats: NO
Long distance movement vectors: NA
Nuisance status and uses: no nuisance issues; no known uses
Endangered status (IUCN): LC

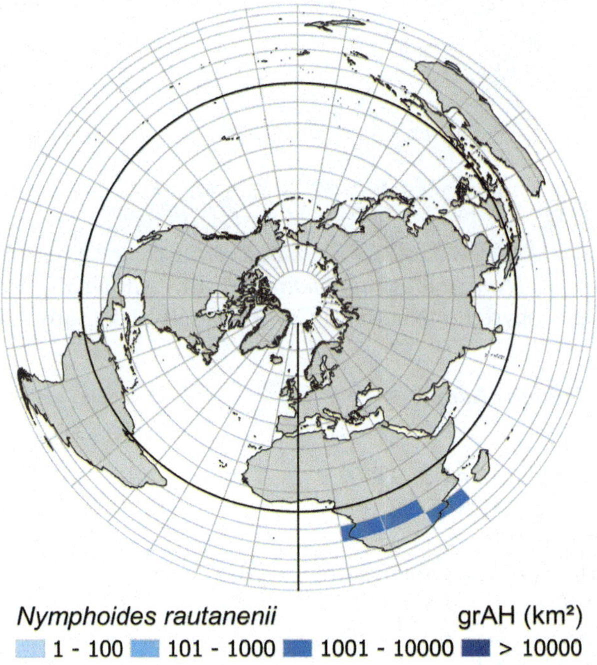

Nymphoides rautanenii (N.E.Br.) A.Raynal

***Nymphoides simulans* Aston**
Synonyms:

Range area (grAH km^2): AUS: 9300
cozone origin-status: AUS: E
World rarity status: VR
Ecozone-endemic macroregion occurrence: AUS: NA
Life form(s): FR
Ploidy state: NA
Speciation age (Ma): 1.1
Populations also occur in terrestrial, wetland and/ or riparian habitats: NO
Long distance movement vectors: NA
Nuisance status and uses: no nuisance issues; no known uses
Endangered status (IUCN): NA

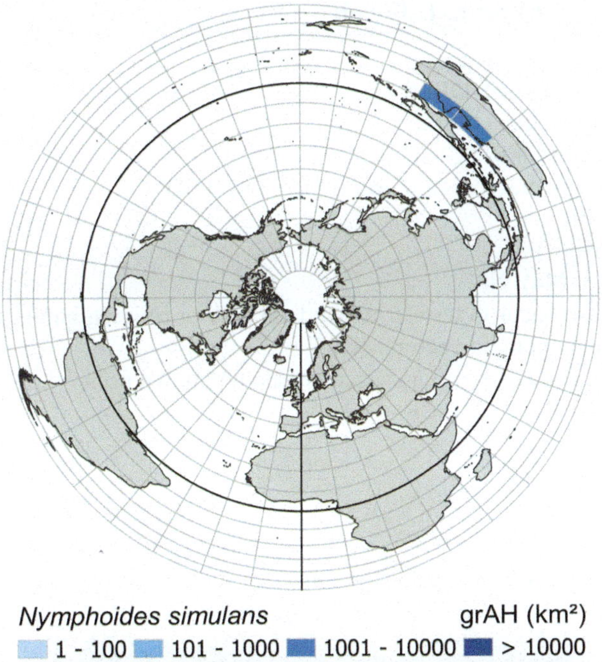

Nymphoides simulans Aston
Image copyright: N. Tippery, with permission

***Nymphoides sivarajanii* K.T. Joseph**
Synonyms:

Range area (grAH km^2): OR: 8100
Ecozone origin-status: OR: E
World rarity status: VR
Ecozone-endemic macroregion occurrence: OR: IND
Life form(s): FR
Ploidy state: NA
Speciation age (Ma): NA
Populations also occur in terrestrial, wetland and/ or riparian habitats: NO
Long distance movement vectors: NA
Nuisance status and uses: no nuisance issues; no known uses
Endangered status (IUCN): CR

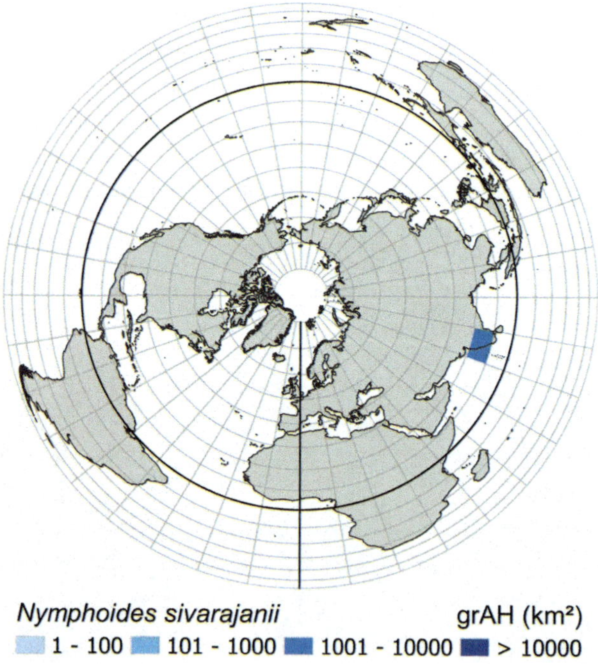

Nymphoides spinulosperma **Aston**
Synonyms:

Range area (grAH km^2): AUS: 9500
Ecozone origin-status: AUS: E
World rarity status: VR
Ecozone-endemic macroregion occurrence: AUS: ESA
Life form(s): FR
Ploidy state: NA
Speciation age (Ma): 0.6
Populations also occur in terrestrial, wetland and/ or riparian habitats: NO
Long distance movement vectors: NA
Nuisance status and uses: no nuisance issues; horticulture, water garden plant in Australia
Endangered status (IUCN): NA

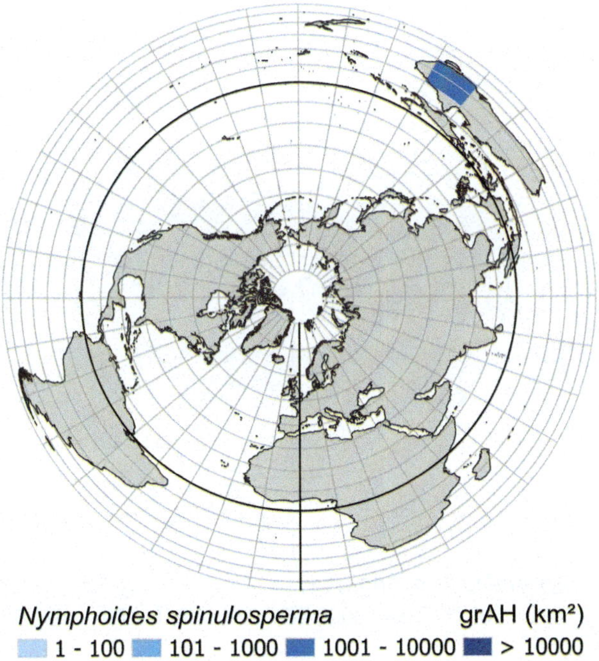

***Nymphoides spongiosa* Aston**
Synonyms:

Range area (grAH km^2): AUS: 6300
Ecozone origin-status: AUS: E
World rarity status: VR
Ecozone-endemic macroregion occurrence: AUS: NA
Life form(s): FR
Ploidy state: NA
Speciation age (Ma): 1.0
Populations also occur in terrestrial, wetland and/ or riparian habitats: NO
Long distance movement vectors: NA
Nuisance status and uses: no nuisance issues; no known uses
Endangered status (IUCN): NA

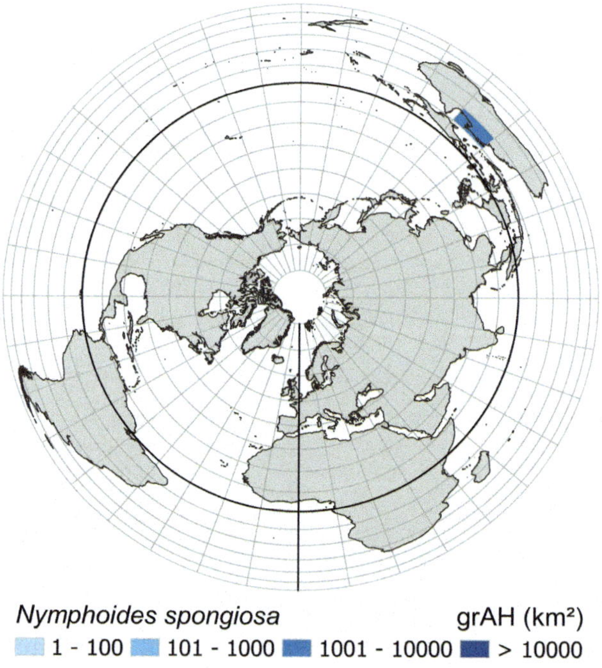

Nymphoides subacuta **Aston**
Synonyms:

Range area (grAH km²): AUS: 6300
Ecozone origin-status: AUS: E
World rarity status: VR
Ecozone-endemic macroregion occurrence: AUS: NA
Life form(s): FR, EM
Ploidy state: NA
Speciation age (Ma): NA
Populations also occur in terrestrial, wetland and/ or riparian habitats: YES
Long distance movement vectors: NA
Nuisance status and uses: no nuisance issues; no known uses
Endangered status (IUCN): NA

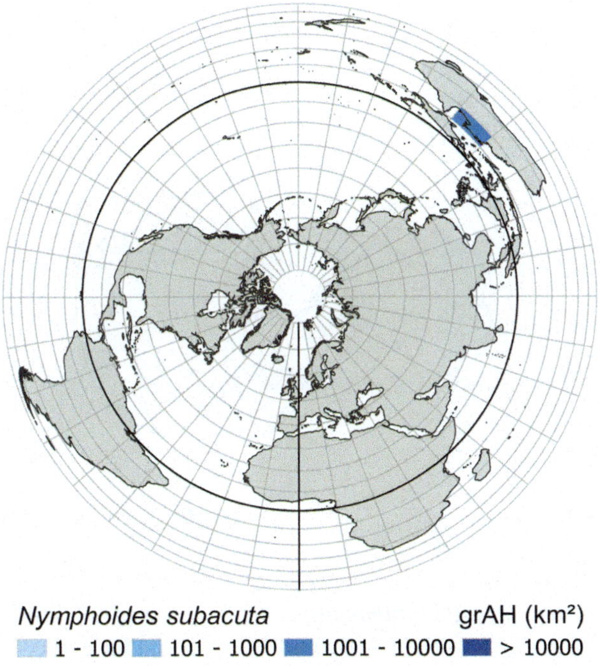

***Nymphoides tenuissima* A. Raynal**
Synonyms:

Range area (grAH km²): AFR: 8700
Ecozone origin-status: AFR: E
World rarity status: VR
Ecozone-endemic macroregion occurrence: AFR: ZAM
Life form(s): FR
Ploidy state: NA
Speciation age (Ma): NA
Populations also occur in terrestrial, wetland and/ or riparian habitats: NO
Long distance movement vectors: NA
Nuisance status and uses: no nuisance issues; no known uses
Endangered status (IUCN): EN

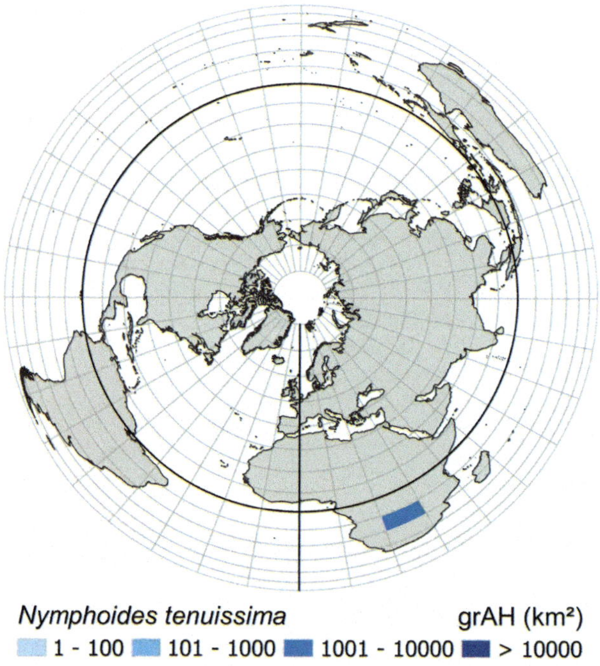

***Nymphoides tonkinensis* (Dop) P.H. Hô**

Synonyms: *Limnanthemum tonkinense* Dop, *Nymphoides aurantiaca* (Dalzell) Kuntze

Range area (grAH km^2): OR: 15700
Ecozone origin-status: OR: E
World rarity status: VR
Ecozone-endemic macroregion occurrence: OR: ICH
Life form(s): FR
Ploidy state: NA
Speciation age (Ma): NA
Populations also occur in terrestrial, wetland and/ or riparian habitats: NO
Long distance movement vectors: NA
Nuisance status and uses: no nuisance issues; no known uses
Endangered status (IUCN): NA

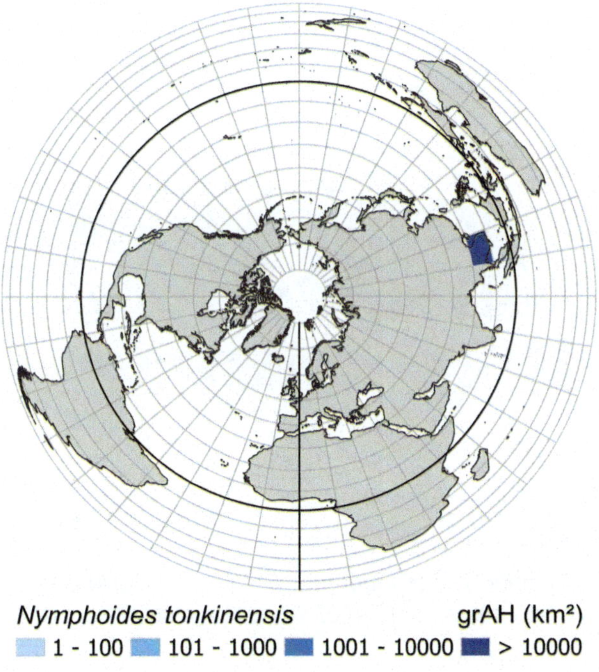

Nymphoides triangularis **Aston**
Synonyms:

Range area (grAH km²): AUS: 2900
Ecozone origin-status: AUS: E
World rarity status: VR
Ecozone-endemic macroregion occurrence: AUS: NA
Life form(s): FR
Ploidy state: NA
Speciation age (Ma): 2.9
Populations also occur in terrestrial, wetland and/ or riparian habitats: NO
Long distance movement vectors: NA
Nuisance status and uses: no nuisance issues; no known uses
Endangered status (IUCN): NA

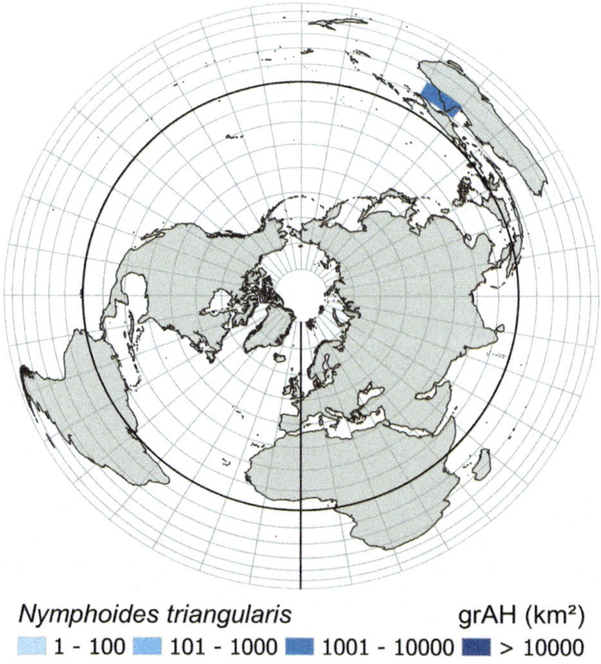

Nymphoides triangularis Aston
Image copyright: N. Tippery, with permission

Nymphoides verrucosa (R.E. Fr.) A. Galán & G. Navarro
Synonyms: *Limnanthemum verrucosum* R.E.Fr.

Range area (grAH km^2): NEO: 41100
Ecozone origin-status: NEO: E
World rarity status: R
Ecozone-endemic macroregion occurrence: NEO: AND, PAR
Life form(s): FR
Ploidy state: NA
Speciation age (Ma): NA
Populations also occur in terrestrial, wetland and/ or riparian habitats: NO
Long distance movement vectors: NA
Nuisance status and uses: no nuisance issues; medicinal uses
Endangered status (IUCN): NA

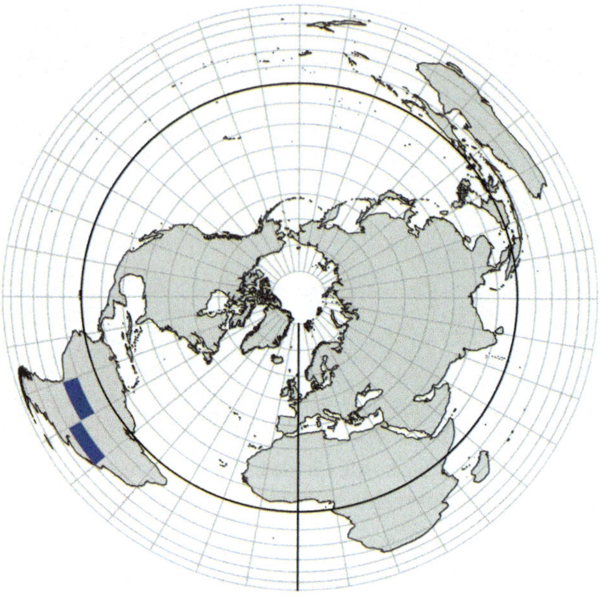

Ornduffia albiflora **(F. Muell.) Tippery & Les**
Synonyms: *Villarsia albiflora* F. Muell., *Limnanthemum albiflorum* (F.Muell.) F.Muell., *Renealmia albiflora* (F.Muell.) Kuntze

Range area (grAH km^2): AUS: 1500
Ecozone origin-status: AUS: E
World rarity status: VR
Ecozone-endemic macroregion occurrence: AUS: WA
Life form(s): EM, FR
Ploidy state: NA
Speciation age (Ma): NA
Populations also occur in terrestrial, wetland and/ or riparian habitats: YES
Long distance movement vectors: NA
Nuisance status and uses: no nuisance issues; horticulture, water garden plant
Endangered status (IUCN): NA

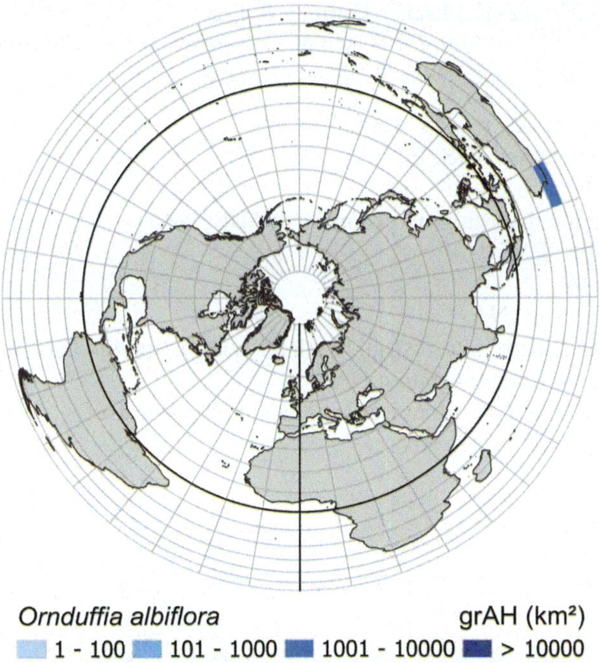

Ornduffia albiflora (F. Muell.) Tippery & Les

***Ornduffia calthifolia* (F. Muell.) Tippery & Les**
Synonyms: *Villarsia calthifolia* F. Muell., *Limnanthemum calthifolium* (F.Muell.) F.Muell., *Renealmia calthifolia* (F.Muell.) Kuntze

Range area (grAH km^2): AUS: 1500
Ecozone origin-status: AUS: E
World rarity status: VR
Ecozone-endemic macroregion occurrence: AUS: WA
Life form(s): FR, EM
Ploidy state: NA
Speciation age (Ma): NA
Populations also occur in terrestrial, wetland and/ or riparian habitats: YES
Long distance movement vectors: NA
Nuisance status and uses: no nuisance issues; no known uses
Endangered status (IUCN): NA

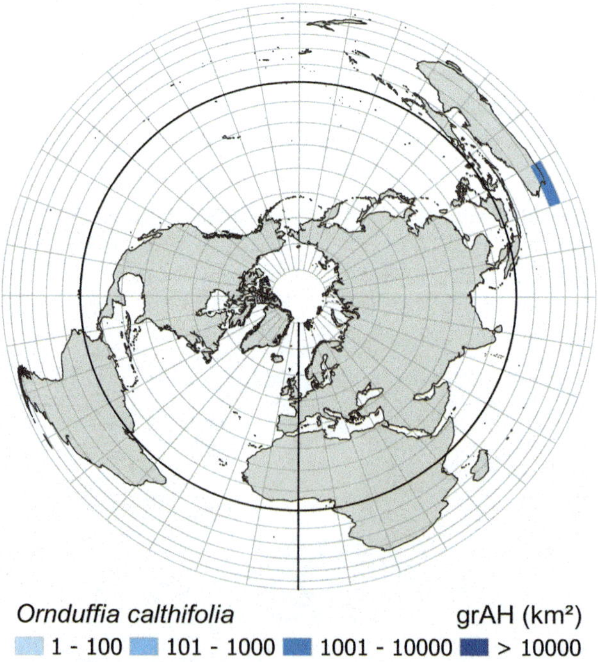

Ornduffia calthifolia (F. Muell.) Tippery & Les

***Ornduffia marchantii* (Ornduff) Tippery & Les**
Synonyms: *Villarsia marchantii* Ornduff

Range area (grAH km^2): AUS: 1500
Ecozone origin-status: AUS: E
World rarity status: VR
Ecozone-endemic macroregion occurrence: AUS: WA
Life form(s): FR, EM
Ploidy state: NA
Speciation age (Ma): NA
Populations also occur in terrestrial, wetland and/ or riparian habitats: YES
Long distance movement vectors: NA
Nuisance status and uses: no nuisance issues; no known uses
Endangered status (IUCN): NA

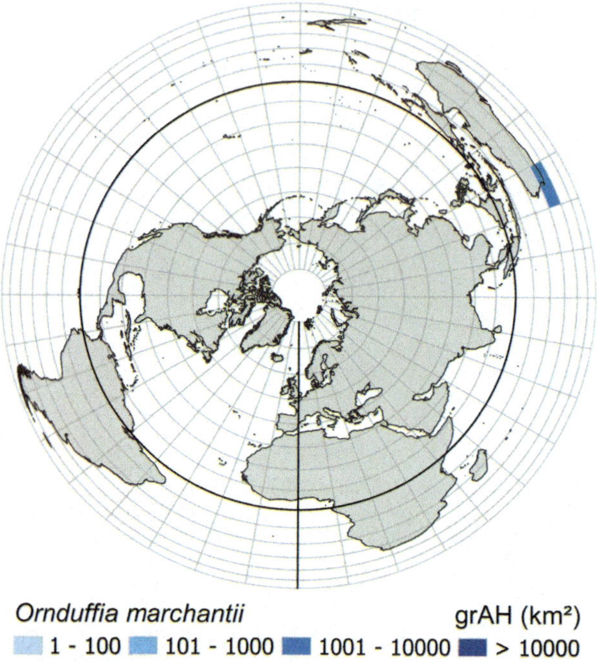

Ornduffia marchantii (Ornduff) Tippery & Les

***Ornduffia parnassiifolia* (Labill.) Tippery & Les**
Synonyms: *Villarsia parnassiifolia* (Labill.) R.Br.

Range area (grAH km^2): AUS: 5200
Ecozone origin-status: AUS: E
World rarity status: VR
Ecozone-endemic macroregion occurrence: AUS: WA, ESA
Life form(s): FR, EM
Ploidy state: NA
Speciation age (Ma): NA
Populations also occur in terrestrial, wetland and/ or riparian habitats: YES
Long distance movement vectors: NA
Nuisance status and uses: no nuisance issues; no known uses
Endangered status (IUCN): NA

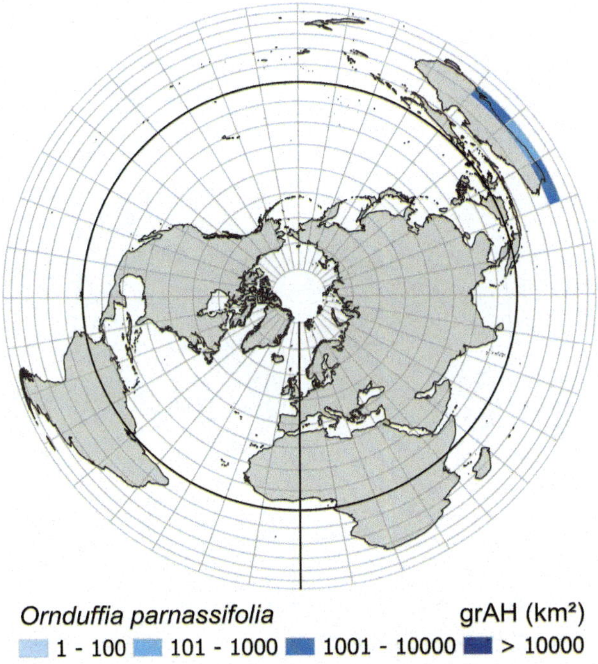

***Ornduffia reniformis* (R. Br.) Tippery & Les**

Synonyms: *Villarsia reniformis* R. Br.

Range area (grAH km^2): AUS: 24400
Ecozone origin-status: AUS: E
World rarity status: VR
Ecozone-endemic macroregion occurrence: AUS: NA, ESA
Life form(s): FR, EM
Ploidy state: P
Speciation age (Ma): NA
Populations also occur in terrestrial, wetland and/ or riparian habitats: YES
Long distance movement vectors: NA
Nuisance status and uses: no nuisance issues; used in Australia as water garden plant
Endangered status (IUCN): NA

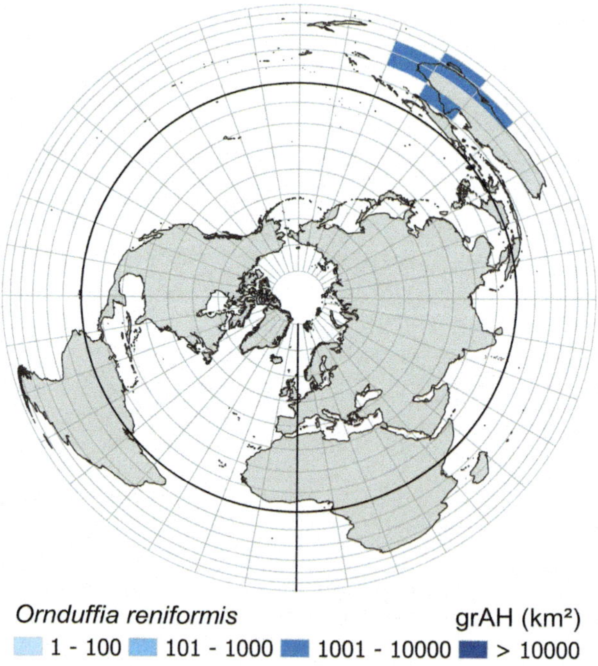

Ornduffia reniformis (R. Br.) Tippery & Les

***Ornduffia submersa* (Aston) Tippery & Les**
Synonyms: *Villarsia submersa* Aston

Range area (grAH km^2): AUS: 1500
Ecozone origin-status: AUS: E
World rarity status: VR
Ecozone-endemic macroregion occurrence: AUS: WA
Life form(s): SU, FR, EM
Ploidy state: NA
Speciation age (Ma): NA
Populations also occur in terrestrial, wetland and/ or riparian habitats: YES
Long distance movement vectors: NA
Nuisance status and uses: no nuisance issues; no known uses
Endangered status (IUCN): NA

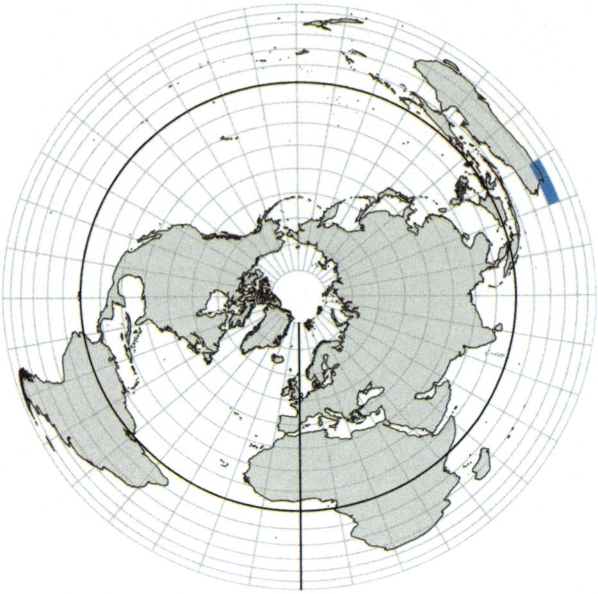

Villarsia capensis **Merr.**
Synonyms: *Menyanthes capensis* (Houtt.) Thunb., *Renealmia capensis* Houtt.

Range area (grAH km^2): AFR: 9100
Ecozone origin-status: AFR: E
World rarity status: VR
Ecozone-endemic macroregion occurrence: AFR: ZAM, CAPE
Life form(s):, EM, FR
Ploidy state: P
Speciation age (Ma): NA
Populations also occur in terrestrial, wetland and/ or riparian habitats: YES
Long distance movement vectors: NA
Nuisance status and uses: no nuisance issues; horticulture, water garden plant
Endangered status (IUCN): LC

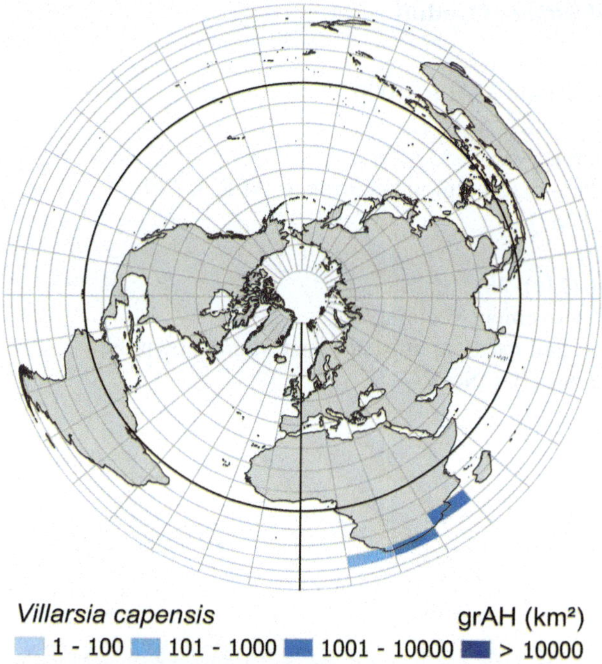

Villarsia capensis Merr.
Image copyright: iNaturalist CC-BY @linkie

Villarsia goldblattiana **Ornduff**
Synonyms:

Range area (grAH km^2): AFR: 1000
Ecozone origin-status: AFR: E
World rarity status: VR
Ecozone-endemic macroregion occurrence: AFR: CAPE
Life form(s):, EM, FR
Ploidy state: P
Speciation age (Ma): NA
Long distance movement vectors: NA
Nuisance status and uses: no nuisance issues; no known uses
Endangered status (IUCN): VU

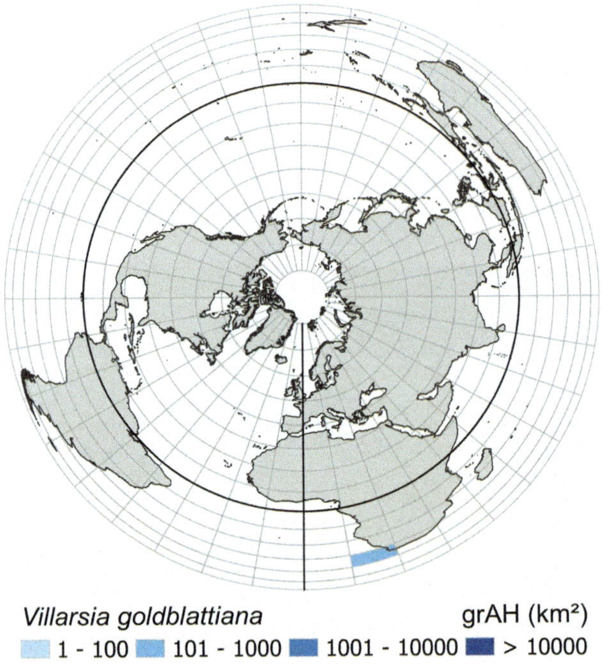

Villarsia goldblattiana Ornduff
Image copyright: iNaturalist CC0 @flamelily

***Villarsia manningiana* Ornduff**
Synonyms:

Range area (grAH km^2): AFR: 1000
Ecozone origin-status: AFR: E
World rarity status: VR
Ecozone-endemic macroregion occurrence: AFR: CAPE
Life form(s):, EM, FR
Ploidy state: P
Speciation age (Ma): NA
Long distance movement vectors: NA
Nuisance status and uses: no nuisance issues; no known uses
Endangered status (IUCN): VU

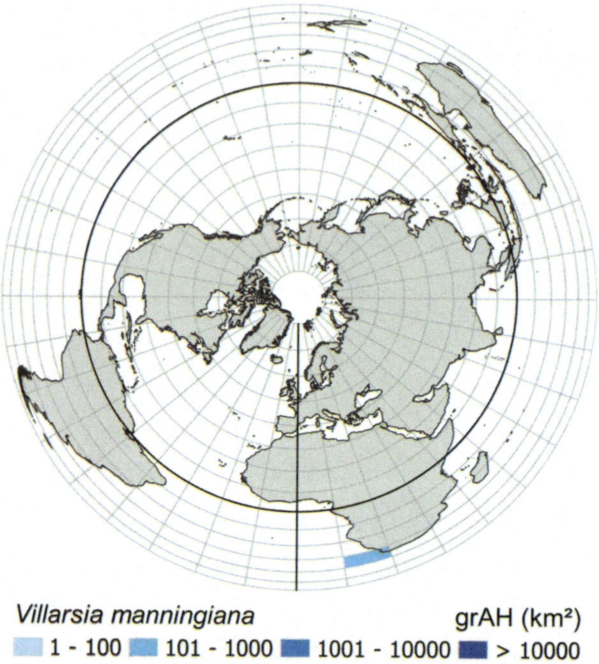

Villarsia manningiana Ornduff
Image copyright: iNaturalist CC0

***Villarsia umbricola* Aston**

Synonyms: *Ornduffia umbricola* (Aston) Tippery & Les

Range area (grAH km^2): AUS: 10900
Ecozone origin-status: AUS: E
World rarity status: VR
Ecozone-endemic macroregion occurrence: AUS: ESA
Life form(s):, EM, FR
Ploidy state: P
Speciation age (Ma): NA
Long distance movement vectors: NA
Nuisance status and uses: no nuisance issues; horticulture, used in Australia as water garden plant
Endangered status (IUCN): NA

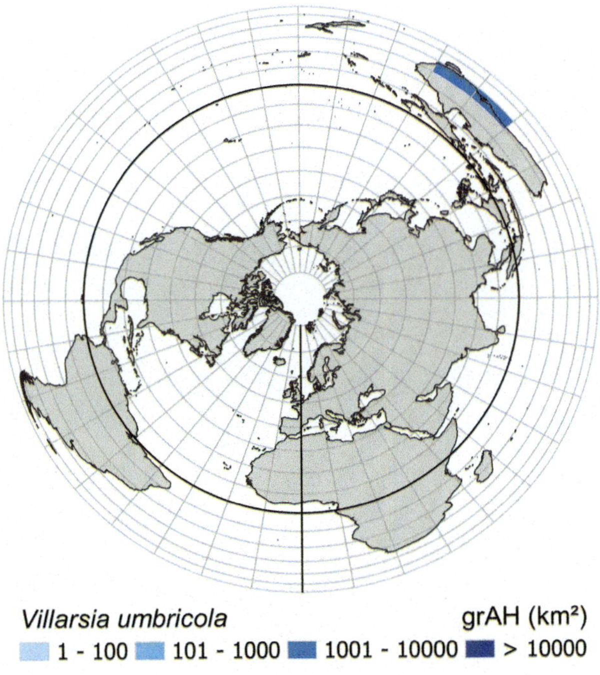

Grey Literature, Web and DOI Sources for Species Entries

Except where indicated, all sources refer to the "Nuisance status and uses" category in species entries of Chapter 3.

Acanthaceae

https://aquatickart.in/hygrophila-balsamica-hygrophila-balsamica.html
http://www.asianjournalofchemistry.co.in/(X(1))/User/ViewFreeArticle.aspx?ArticleID=22_3_110
https://doi.org/10.1016/S0378-8741(01)00346-4
https://doi.org/10.1155/2013/680425
https://rprcbbsr.in/View/Downloads/news/Hygrophila%20auriculata%2013%207%2015.pdf
https://tropica.com/en/plants/plantdetails/Hygrophiladifformis(051BDT)/4485
https://www.aquariumgardens.co.uk/hygrophila-costata-647-p.asp
https://www.aquariumgardens.co.uk/hygrophila-polysperma-248-p.asp
https://www.aquasabi.com/Hygrophila-triflora
https://www.fishlaboratory.com/fish/hygrophila-corymbosa/
https://www.ijpcbs.com/articles/a-details-study-on-hygrophila-difformis.pdf
https://www.iucnredlist.org/search?query=Hygrophila%20polysperma
https://www.lilieswatergardens.co.uk/justicia-americana-barerooted-p-2279.html
https://doi.org/10.1201/9780429429095
https://www.lume.ufrgs.br/bitstream/handle/10183/235617/001138071.pdf?sequence=1
http://repository-tnmgrmu.ac.in/18094/1/261710421deepika.pdf

T. Lobato-de Magalhães et al., *World Atlas of Freshwater Macrophytes*, Wetlands: Ecology, Conservation and Management 10,
https://doi.org/10.1007/978-3-031-52749-4

https://www.researchgate.net/publication/301635022_Studies_on_the_potential_therapeutic_effects_on_theaquatic_macrophytes_namely_Cabomba_aquatica_Ceratophyllum_demersum_and_Hygrophila_corymbosa
https://www.researchgate.net/publication/347716933_isolation_and_characterization_of_secondary_metabolites_and_evaluation_of_biological_activity_of_Hygrophila_erecta_Burmf_Hochr
https://www.wettropicsplan.org.au/wp-content/uploads/2020/11/Wetland-Plants-of-the-Wet-Tropics-FINAL-WEB-compressed-b.pdf

Amaranthaceae

http://www.iucngisd.org/gisd/species.php?sc=763
http://www.iucngisd.org/gisd/speciesname/Alternanthera+sessilis
https://botanicaargentina.org.ar/wp-content/uploads/2019/06/978-987-688-332-0-baja-resoluci%C3%B3n.pdf
https://doi.org/10.1007/978-0-387-70638-2_94
https://doi.org/10.1007/s11157-020-09552-y
https://doi.org/10.1016/j.jep.2022.115608
https://doi.org/10.1163/20021975-99990285
https://doi.org/10.1201/9780429429095
https://doi.org/10.33448/rsd-v10i4.13714
https://naturalhistory.si.edu/sites/default/files/media/file/medicinalplantsmaster.pdf
https://phytochem.nal.usda.gov/phytochem/ethnoPlants/show/2362?qlookup=Alternanthera+tetramera&offset=0&max=20&et=
https://pubmed.ncbi.nlm.nih.gov/17353090/
https://ri.conicet.gov.ar/handle/11336/24328
https://www.flowgrow.de/db/aquaticplants/alternanthera-aquatica
https://www.nparks.gov.sg/florafaunaweb/flora/3/3/3307#description
https://www.pro-shrimp.co.uk/aquarium-plants/4572-alternanthera-sessilis-aquadip-8717344711357.html
https://www.researchgate.net/publication/327900020_Plantas_como_Ferramentas_para_a_Remediacao_Ambiental_uma_Revisao_da_Literatura
https://link.springer.com/article/10.1007/s13762-018-1705-z
https://ctujs.ctu.edu.vn/index.php/ctujs/article/view/382

Apiaceae

https://doi.org/10.30972/bon.2813572;
http://www.habitas.org.uk/flora/species.asp?item=3678
https://botanicaargentina.org.ar/wp-content/uploads/2019/06/978-987-688-332-0-baja-resoluci%C3%B3n.pdf

https://docs.bvsalud.org/biblioref/2018/10/916684/plantas-medicinales-de-los-andes-y-la-amazonia-la-flora-magica-_Qa3dgqr.pdf
https://doi.org/10.1007/978-0-387-70638-2_137 https://phytochem.nal.usda.gov/phytochem/plants/show/160?part=&_ubiq=&ubiq=on
https://doi.org/10.1007/978-0-387-70638-2_322
https://doi.org/10.1007/978-0-387-70638-2_578
https://doi.org/10.1007/s11157-020-09552-y
https://doi.org/10.1007/s11270-016-3150-2
https://doi.org/10.1016/j.fitote.2010.11.010
https://doi.org/10.1016/j.jep.2006.10.032
https://doi.org/10.1016/j.scitotenv.2016.10.171
https://doi.org/10.1021/np50048a028
https://doi.org/10.1080/10412905.2010.9700290
https://doi.org/10.1201/9780429429095
https://doi.org/10.3109/13880209.2015.1102949
https://doi.org/10.4103/phrev.phrev_35_16
https://gobotany.nativeplanttrust.org/species/sium/suave/
https://journal.unisza.edu.my/agrobiotechnology/index.php/agrobiotechnology/article/view/202
https://mailorder.crug-farm.co.uk/?pid=10262
https://naturalhistory.si.edu/sites/default/files/media/file/medicinalplantsmaster.pdf
https://phytochem.nal.usda.gov
https://powo.science.kew.org/taxon/urn:lsid:ipni.org:names:140092-2
https://repositorio.ufsc.br/bitstream/handle/123456789/129388/329004.pdf?sequence=1&isAllowed=y
https://repositorio.unesp.br/bitstream/handle/11449/108725/000766039.pdf
https://ri.conicet.gov.ar/handle/11336/24328
https://temperate.theferns.info/plant/Oenanthe+sarmentosa
https://tropica.com/en/plants/plantdetails/Lilaeopsisbrasiliensis(040TC)/4460
https://vestnik-bio-msu.elpub.ru/jour/article/view/93/94
https://wildseed.co.uk/product/species/wild-flowers/thyselium-palustre-peucedanum-palustre/
https://www.agfonds.lv/herbs/herbs-g-h/greater-water-parsnip-sium-latifolium/
https://www.aphis.usda.gov/plant_health/plant_pest_info/weeds/downloads/wra/Lilaeopsis-brasiliensis.pdf
https://www.aquariumgardens.co.uk/lilaeopsis-novae-zelandiae-412-p.asp
https://www.aquasabi.com/Lilaeopsis-carolinensis
https://www.aquasabi.com/Lilaeopsis-macloviana
https://www.aquasabi.com/Lilaeopsis-mauritiana
https://www.deacua.com/tienda/lilaeopsis-chinensis/
https://www.forager.org.uk/component/opencart/c-391-fool-039-s-watercress?Itemid=112
https://www.gardenersworld.com/plants/oenanthe-javanica-flamingo/
https://www.gardenia.net/plant/eryngium-variifolium

https://www.iucnredlist.org/search?query=Apium%20graveolens%20&searchType=species
https://www.iucnredlist.org/search?query=berula%20erecta&searchType=species
https://www.iucnredlist.org/search?query=Centella%20asiatica%20&searchType=species
https://www.iucnredlist.org/search?query=Cicuta%20bulbifera&searchType=species
https://www.iucnredlist.org/species/164030/13575513
https://www.iucnredlist.org/species/164203/13575099
https://www.iucnredlist.org/species/168749/6532868
https://www.iucnredlist.org/species/175289/212680128
https://www.misin.msu.edu/facts/detail/?project=misin&id=355&cname=Water%20hemlock
https://www.ncbi.nlm.nih.gov/pmc/articles/PMC3116297/
https://www.ncbi.nlm.nih.gov/pmc/articles/PMC6463588/
https://www.nickys-nursery.co.uk/product_info.php/culantro-mexican-coriander-eryngium-foetidum-200-seeds-h302-p-20716
https://www.rhs.org.uk/plants/11724/oenanthe-fistulosa/details
https://www.shootgardening.co.uk/plant/eryngium-aquaticum
https://www.udesc.br/arquivos/cav/id_cpmenu/1349/dissertacao_karina_monti-beller_15676053075669_1349.pdf
https://www.watershednursery.com/nursery/plant-finder/eryngium-aristulatum/
http://icm.fch.lisboa.ucp.pt/resources/Documentos/CEPCEP/Dicionario,%20Letra%20C.pdf
https://pubmed.ncbi.nlm.nih.gov/37241835/
https://www.ncbi.nlm.nih.gov/pmc/articles/PMC10221068/ http://www.polibotan-ica.mx/pdf/pb12/PlantasMedicinalesQro.pdf
https://www.suncoasttropicals.com.au/shop/aquatic-plants/oxygenating-plants/lilaeopsis-polyantha/
https://www.thekrib.com/Plants/Plants/NZ/
https://www.ncbi.nlm.nih.gov/pmc/articles/PMC8779323/
https://homeomart.com/products/oenanthe-crocata-homeopathy-dilution-6c-30c-200c-1m-10m https://pubmed.ncbi.nlm.nih.gov/24012643/ https://econtent.hogrefe.com/doi/abs/10.1027/0227-5910/a000200?journalCode=cri
https://en.wikipedia.org/wiki/Peucedanum_palustre
https://pubmed.ncbi.nlm.nih.gov/17345511/
https://temperate.theferns.info/plant/Peucedanum+palustre
https://temperate.theferns.info/plant/Sium+latifolium https://www.tandfonline.com/doi/abs/10.1080/10412905.2005.9698940
https://www.ncbi.nlm.nih.gov/pmc/articles/PMC10367252/ https://pfaf.org/user/plant.aspx?LatinName=Sium+sisarum

Apocynaceae

https://en.wikipedia.org/wiki/Oxystelma_esculentum

Araliaceae

http://yschang.cmu.edu.tw/old/Journal.files/2008/Antioxidant%20and%20antiproliferative%20activities%20of%20the%20four%20Hydrocotyle%20species%20from%20Taiwan.pdf
https://aquaessentials.co.uk/products/hydrocotyle-vulgaris-marsh-pennywort
https://doi.org/10.1007/978-0-387-70638-2_764
https://doi.org/10.1007/978-3-030-35691-0_11
https://doi.org/10.1016/j.indcrop.2014.02.013
https://doi.org/10.2306/scienceasia1513-1874.2006.32.323
https://doi.org/10.5539/jsd.v14n2p66
https://hayloft.co.uk/product/hydrocotyle-vulgaris/pt50hyvul1-in22
https://hortodidatico.ufsc.br/erva-capitao/
https://keyserver.lucidcentral.org/weeds/data/media/Html/hydrocotyle_bonariensis.htm
https://phytochem.nal.usda.gov
https://ri.conicet.gov.ar/handle/11336/24328
https://tropica.com/en/plants/plantdetails/Hydrocotyletripartita(039B)/4458
https://tropica.com/en/plants/plantdetails/Hydrocotyleverticillata(039)/4457
https://www.ajol.info/index.php/ajb/article/view/59417
https://www.flowgrow.de/db/aquaticplants/hydrocotyle-sibthorpioides
https://www.gardenia.net/plant/hydrocotyle-umbellata
https://www.gardenia.net/plant/hydrocotyle-verticillata
https://www.invasive-species.org/species/floating-pennywort/
https://www.rhs.org.uk/plants/239153/hydrocotyle-sibthorpioides-crystal-confetti-(v)/details

Asteraceae

https://doi.org/10.2174/1389201021666200922161529
https://doi.org/10.4314/tjpr.v16i1.23
https://doi.org/10.22037/ijpr.2017.2027
http://redlist.sanbi.org/species.php?species=3162-30
http://scielo.sld.cu/scielo.php?script=sci_arttext&pid=S1028-47962012000200005&lng=es&tlng=en
http://www.knowledgebank.irri.org/training/fact-sheets/item/eclipta-prostrata
http://www.plantsoftheworldonline.org/taxon/urn:lsid:ipni.org:names:32497-2

https://alienplantsbelgium.myspecies.info/content/bidens-radiata https://ann-clinmicrob.biomedcentral.com/articles/10.1186/s12941-015-0074-4
https://citeseerx.ist.psu.edu/viewdoc/download?doi=10.1.1.908.2235&rep=rep1&type=pdf
https://docs.bvsalud.org/biblioref/2018/10/916684/plantas-medicinales-de-los-andes-y-la-amazonia-la-flora-magica-_Qa3dgqr.pdf
https://doi.org/10.1002/pca.2555
https://doi.org/10.1007/s10600-012-0112-9
https://doi.org/10.1007/s11157-020-09552-y
https://doi.org/10.1016/j.jep.2006.10.032
https://doi.org/10.1080/10412905.2009.9700103
https://doi.org/10.1111/j.1365-2621.2011.02615.x
https://doi.org/10.1186/s12906-016-1014-3
https://doi.org/10.1590/S1519-69842012000300013
https://doi.org/10.2478/s11756-009-0061-3
https://doi.org/10.2478/v10042-008-0082-8
https://doi.org/10.25135/rnp.59.16.09.067
https://doi.org/10.29328/JOURNAL.JPSP.1001050
https://doi.org/10.4103/pm.pm_71_19
https://gobotany.nativeplanttrust.org/species/bidens/frondosa/
https://gobotany.nativeplanttrust.org/species/bidens/laevis/
https://gobotany.nativeplanttrust.org/species/bidens/tripartita/
https://gobotany.nativeplanttrust.org/species/petasites/frigidus/
https://lewiscottageplants.co.uk/index.php/contact-2/
https://link.springer.com/content/pdf/10.1007%2F978-3-319-77093-2_38-1.pdf
https://patents.google.com/patent/KR20160143838A/en
https://phytochem.nal.usda.gov
https://pubmed.ncbi.nlm.nih.gov/16447448/
https://pubmed.ncbi.nlm.nih.gov/24892666/
https://pubmed.ncbi.nlm.nih.gov/31136898/
https://pubmed.ncbi.nlm.nih.gov/32297807/
https://pubmed.ncbi.nlm.nih.gov/33221497/
https://pubmed.ncbi.nlm.nih.gov/33087028/
https://pubmed.ncbi.nlm.nih.gov/23506988/
https://repositorio.ufsc.br/bitstream/handle/123456789/129388/329004.pdf?sequence=1&isAllowed=y
https://repositorio.unb.br/bitstream/10482/31437/1/2017_JanaeLyonMillion.pdf
https://repositoriodigital.uns.edu.ar/handle/123456789/507
https://ri.conicet.gov.ar/handle/11336/24328
https://rjptonline.org/HTMLPaper.aspx?Journal=Research%20Journal%20of%20Pharmacy%20and%20Technology;PID=2013-6-9-26
https://temperate.theferns.info/viewtropical.php?id=Bidens+parviflora
https://tropical.theferns.info/viewtropical.php?id=Acmella+uliginosa
https://tropical.theferns.info/viewtropical.php?id=Centipeda+minima
https://tropical.theferns.info/viewtropical.php?id=Enydra+fluctuans

https://tropical.theferns.info/viewtropical.php?id=Grangea+maderaspatana
https://tropical.theferns.info/viewtropical.php?id=Tagetes+filifolia
https://www.arca.fiocruz.br/bitstream/icict/48188/2/m_et_all.pdf
https://www.cabidigitallibrary.org/doi/10.1079/cabicompendium.26246
https://www.cal-ipc.org/plants/profile/cotula-coronopifolia-profile/
https://www.efloraofgandhinagar.in/plant-details.php?cateUrl=herb&plantUrl=acmella-paniculata
https://www.fishkeeper.co.uk/help-and-advice/plant/aquarium-plants/gymnocoronis-spilanthoides-
https://www.fishkeeper.co.uk/help-and-advice/plant/aquarium-plants/shinnersia-rivularis-
https://www.floridaaquatic.com/plant-blank-2
https://www.frontiersin.org/articles/10.3389/fphar.2021.708618/full
https://www.gardenia.net/plant/bidens-laevis
https://www.icmbio.gov.br/portal/images/stories/imgs-unidades-coservacao/parna_matogrossensee.pdf
https://www.invasive.org/browse/subinfo.cfm?sub=5563
https://www.iucnredlist.org/species/168968/6556955
https://www.jocpr.com/abstract/the-effect-antioxidant-aqueous-crude-extract-in-acmella-ciliata-kunth-asteraceae-5774.html
https://www.mdpi.com/1420-3049/26/12/3695
https://www.ncbi.nlm.nih.gov/pmc/articles/PMC3059449/
https://www.ncbi.nlm.nih.gov/pmc/articles/PMC3581292/
https://www.ncbi.nlm.nih.gov/pmc/articles/PMC5041393/
https://www.ncbi.nlm.nih.gov/pmc/articles/PMC5110660/
https://www.ncbi.nlm.nih.gov/pmc/articles/PMC6244959/
https://www.ncbi.nlm.nih.gov/pmc/articles/PMC8615741/
https://www.phytojournal.com/archives/2019/vol8issue2/PartP/8-1-433-163.pdf
https://www.researchgate.net/publication/287047257_Sphaeranthus_amaranthoides_-_A_review
https://www.rhs.org.uk/plants/4689/cotula-coronopifolia/details
https://www.scielo.br/j/rbcf/a/xFPS8HmL5sngrf6jLgpBHBK/?lang=en&format=pdf
https://www.sciencedirect.com/science/article/abs/pii/S0882401017304072?via%3Dihub
https://www.sciencedirect.com/science/article/pii/S2405665018300593
https://www.specialplants.net/shop/seeds/tagetes_filifolia_dropshot/
https://www.tandfonline.com/doi/abs/10.1080/10412905.2004.9698767
https://www.sciencedirect.com/science/article/abs/pii/S0031942200826370
https://www.sciencedirect.com/science/article/abs/pii/S1567576917303405 https://www.indianetzone.com/38/cyathocline_purpurea_plant.htm https://www.sciencedirect.com/science/article/abs/pii/S0378874112008227
https://www.researchgate.net/publication/333829800_Amino_Acid_Profiles_Antimicrobial_Activity_and_Anti-nutritional_Contents_of_Two_Wild_Edible_Plants_Sphenoclea_zeylanica_Gaertn_and_Sphaerantus_peguensis_Kurz_ex_CB_Clarke
https://pfaf.org/user/Plant.aspx?LatinName=Tephroseris+palustris

Balsaminaceae

https://tropical.theferns.info/viewtropical.php?id=Hydrocera+triflora

Boraginaceae

https://doi.org/10.1691/ph.2015.4873
https://tropical.theferns.info/viewtropical.php?id=Coldenia+procumbens
https://tropical.theferns.info/viewtropical.php?id=Heliotropium+supinum#:~:text=Heliotropium%20supinum%20is%20a%20prostrate,308
https://uses.plantnet-project.org/en/Heliotropium_ovalifolium_(PROTA)#Heliotropium_supinum
https://www.gardenersworld.com/plants/myosotis-scorpioides/
https://www.researchgate.net/publication/325042138_Rotula_aquatica_Lour-A_Review_on_Medicinal_uses_Phytochemistry_and_Pharmacological_actions
https://www.rhs.org.uk/plants/142245/myosotis-laxa/details
https://www.researchgate.net/publication/290302132_Anti-inflammatory_and_hypoglycemic_effects_of_heliotropium_keralense_in_albino_rats https://www.semanticscholar.org/paper/Micropropagation-of-Heliotropium-keralense-a-and-Sebastian-Hariharan/0889ed3b91869402a40c8d0b22c246eaeddd55f1
https://www.ingentaconnect.com/content/govi/pharmaz/2015/00000070/00000006/art00001 https://pfaf.org/USER/Plant.aspx?LatinName=Myosotis+scorpioides
https://doi.org/10.1079/IVP2004598

Brassicaceae

http://abs.bibl.u-szeged.hu/index.php/abs/article/view/2426
http://jehe.abzums.ac.ir/article-1-476-en.html
http://www.naturalmedicinalherbs.net/herbs/c/cardamine-prorepens.php
https://doi.org/10.1007/978-0-387-70638-2_1059
https://doi.org/10.1007/s11104-013-1783-0
https://doi.org/10.1007/s11157-020-09552-y
https://doi.org/10.1016/j.apgeochem.2019.104377
https://doi.org/10.1016/j.jep.2006.10.032
https://doi.org/10.1016/j.scitotenv.2016.10.171
https://doi.org/10.1016/j.yrtph.2019.104443
https://doi.org/10.1186/s13002-016-0100-4
https://doi.org/10.1201/9780429429095
https://doi.org/10.3390/su12051927

https://doi.org/10.4141/cjps2012-006
https://explorer.natureserve.org/Taxon/ELEMENT_GLOBAL.2.143570/Rorippa_calycina
https://gobotany.nativeplanttrust.org/species/cardamine/pensylvanica/
https://keyserver.lucidcentral.org/weeds/data/media/Html/rorippa_palustris.htm
https://link.springer.com/article/10.1007/s10750-006-0106-3
https://phytochem.nal.usda.gov
https://southernseedexchange.com/products/watercress-200-seeds-culinary-medicinal-herb-seeds-gmo-free-nasturtium-officinale
https://temperate.theferns.info/plant/Cardamine+bonariensis
https://totallywilduk.co.uk/2021/04/14/identify-horseradish/
https://tropica.com/en/plants/plantdetails/Cardaminelyrata(024)/4441
https://www.aquasabi.com/Rorippa-aquatica;
https://www.sciencedirect.com/science/article/abs/pii/S0926669017305873
https://temperate.theferns.info/plant/Cardamine+lyrata
https://pfaf.org/User/Plant.aspx?LatinName=Rorippa+amphibia https://www.picturethisai.com/wiki/Rorippa_amphibia.html
https://swbiodiversity.org/seinet/taxa/index.php?taxon=44671&clid=2692
http://sinavef.senasica.gob.mx/CNRF/AreaDiagnostico/DocumentosReferencia/Documentos/MaterialDivulgativo/Infografias/Catalogo%20de%20malezas%20de%20Mexico%20Familia%20Brassicaceae%20(Cruciferae).pdf

Cabombaceae

http://dnr.alaska.gov/ag/Index/IntroductiontoCommonNativeInvasiveFreshwaterPlantsinAlaska.pdf
https://cjs.sljol.info/articles/abstract/10.4038/cjs.v51i2.8010/
https://doi.org/10.1201/9780429429095
https://link.springer.com/article/10.1007/BF02868922
https://nt.gov.au/environment/weeds/weeds-in-the-nt/A-Z-list-of-weeds-in-the-NT/cabomba
https://onlinelibrary.wiley.com/doi/abs/10.1111/wre.12518
https://phytochem.nal.usda.gov
https://repositorio.ufscar.br/bitstream/handle/ufscar/2060/DissKST.pdf?sequence=1&isAllowed=y
https://ri.conicet.gov.ar/handle/11336/24328
https://tropica.com/en/plants/plantdetails/Cabombaaquatica(015BDT)/4431
https://tropical.theferns.info/viewtropical.php?id=Brasenia+schreberi
https://www.aquasabi.com/Cabomba-furcata-piauhyensis-Pot
https://www.gardenia.net/plant/brasenia-schreberi
https://www.gardenia.net/plant/cabomba-caroliniana

https://www.jocpr.com/articles/evaluation-of-bioactive-compounds-from-the-fresh-water-weed-cabomba-caroliniana-a-gray.pdf
https://www.nparks.gov.sg/florafaunaweb/flora/5/9/5991
https://www.earth.com/plant-encyclopedia/Angiosperms/Cabombaceae/cabomba-palaeformis/en/ https://www.aquasabi.com/Cabomba-palaeformis-Rio-Dulce-Bunch

Campanulaceae

http://redlist.sanbi.org/species.php?species=1720-139
http://redlist.sanbi.org/species.php?species=1723-4
http://redlist.sanbi.org/species.php?species=1727-1
https://botanicaargentina.org.ar/wp-content/uploads/2019/06/978-987-688-332-0-baja-resoluci%C3%B3n.pdf
https://darcyeverest.co.uk/product/hypsela-reniformis-lobelia-oligophylla/
https://doi.org/10.1007/978-981-10-8022-7_47
https://doi.org/10.1016/j.fitote.2016.04.013
https://doi.org/10.1016/j.fitote.2018.12.021
https://doi.org/10.1023/A:1025630301162
https://doi.org/10.3389/fphar.2021.638210
https://inis.iaea.org/collection/NCLCollectionStore/_Public/16/052/16052679.pdf
https://inis.iaea.org/search/searchsinglerecord.aspx?recordsFor=SingleRecord&RN=14760593
https://northwestmeadowscapes.com/products/elegant-calicoflower-seeds-downingia-elegans?variant=15897179226202
https://pfaf.org/user/Plant.aspx?LatinName=Lobelia+siphilitica&utm_content=cmp-true
https://phytochem.nal.usda.gov
https://tropical.theferns.info/viewtropical.php?id=Lobelia+chinensis
https://tropical.theferns.info/viewtropical.php?id=Lobelia+zeylanica
https://www.backyardgardener.com/plantname/lobelia-paludosa-swamp-lobelia/
https://www.gardenia.net/plant/lobelia-cardinalis-cardinal-flower
https://www.gardenia.net/plant/lobelia-siphilitica-great-blue-lobelia
https://www.gardeningknowhow.com/lawn-care/lawn-substitutes/blue-star-creeper-lawn/blue-star-creeper-plant.html
https://www.jstor.org/stable/27802092
https://www.lilieswatergardens.co.uk/lobelia-chinensis-7-cm-pot-p-771.html
https://www.perennials.com/plants/isotoma-fluviatilis.html?utm_content=cmp-true
https://pubmed.ncbi.nlm.nih.gov/31277907/
https://pubmed.ncbi.nlm.nih.gov/32860351/
https://pfaf.org/user/Plant.aspx?LatinName=Lobelia+dortmanna

Caryophyllaceae

https://decs.bvsalud.org/es/ths/resource/?id=26699&filter=&q=&lang_another=pt-br https://conecte.es/index.php/es/plantas/1140-illecebrum-verticillatum/usos-tradicionales http://www.wildflowerweb.co.uk/plant/2838/coral-necklace

http://www.epharmacognosy.com/2022/02/stellaria-aquatica.html https://link.springer.com/article/10.1007/s11130-008-0092-x

Ceratophyllaceae

https://doi.org/10.1007/978-0-387-70638-2_327
https://doi.org/10.1007/978-981-32-9664-0_4
https://doi.org/10.1007/s11157-020-09552-y
https://doi.org/10.1201/9780429429095
https://doi.org/10.3390/su12051927
https://jst-ud.vn/jst-ud/article/view/2493
https://www.flowgrow.de/db/aquaticplants/ceratophyllum-submersum
https://www.nzpcn.org.nz/flora/species/ceratophyllum-demersum/
https://www.rhsplants.co.uk/plants/_/ceratophyllum-demersum/classid.2000022455/

Convolvulaceae

http://old.cnpgc.embrapa.br/publicacoes/divulga/GCD44.html
http://tropical.theferns.info/viewtropical.php?id=Ipomoea+asarifolia
http://www.iepa.ap.gov.br/metadados/instituicoes/iepa/projetos/ressacas/documentos/1finalVEGETACAO.pdf
http://www.iucngisd.org/gisd/species.php?sc=477
https://doi.org/10.1007/978-3-030-00099-8_7
https://doi.org/10.1007/978-981-32-9664-0_4
https://doi.org/10.1007/BF02860761
https://doi.org/10.1007/s11157-020-09552-y
https://doi.org/10.1007/s13659-012-0015-2
https://doi.org/10.1039/C29690000709
https://phytochem.nal.usda.gov
https://pubmed.ncbi.nlm.nih.gov/27448455/
https://repositorio.ufsc.br/bitstream/handle/123456789/129388/329004.pdf?sequence=1&isAllowed=y
https://revistas.cientifica.edu.pe/index.php/cientifica/article/view/159/178
https://ri.conicet.gov.ar/handle/11336/24328

https://tropical.theferns.info/viewtropical.php?id=Ipomoea+aquatica
https://tropical.theferns.info/viewtropical.php?id=Ipomoea+carnea
https://uses.plantnet-project.org/en/Ipomoea_asarifolia_(PROTA)
https://worldcrops.org/crops/water-spinach
https://www.alice.cnptia.embrapa.br/alice/handle/doc/783791
https://www.cabidigitallibrary.org/doi/10.1079/cabicompendium.28785
https://www.gardenia.net/plant/ipomoea-alba-moonflower
https://www.semanticscholar.org/paper/Angiospermas-acu%C3%A1ticas-en-el-Per%C3%BA%3A-usos-y-estado-en-Aponte-P%C3%A9rez-Irigoyen/ecf5df365e4c63832bd7440018d3bfd870eb4e22

Crassulaceae

https://freshwaterhabitats.org.uk/projects/other-projects/controlling-crassula-helmsii/
https://www.invasive-species.org/species/australian-swamp-stonecrop/
https://www.misin.msu.edu/facts/detail/?id=171
http://www.wildflowerweb.co.uk/plant/4000/pigmyweed

Droseraceae

https://carnivorousplantnursery.com/products/aldrovanda-vesiculosa

Elatinaceae

https://explorer.natureserve.org/Taxon/ELEMENT_GLOBAL.2.815519/Elatine_ojibwayensis
https://www.cabidigitallibrary.org/doi/10.1079/cabicompendium.83102500
https://www.imedpub.com/articles/antifungal-potentiality-of-some-medicinal-plant-extracts-against-bipolaris-oryzae-breda-de-haan.pdf
https://pubmed.ncbi.nlm.nih.gov/25853974/ https://asianjournalofchemistry.co.in/User/ViewFreeArticle.aspx?ArticleID=25_15_6
https://doi.org/10.3109/13880209.2015.1028079
https://www.picturethisai.com/wiki/Elatine.html
http://beta.floranorthamerica.org/Elatine_ambigua
http://www.watergardenparadise.com.au/submergedplants.php https://www.nzpcn.org.nz/flora/species/elatine-gratioloides/?download=pdf
https://www.botanikks.com/plants/elatine-heterandra-mason/541807/1
https://shop.glassaqua.com/products/elatine-hydropiper?variant=19609623986262

Euphorbiaceae

https://doi.org/10.1590/S1984-29612013000400013
https://naturalhistory.si.edu/sites/default/files/media/file/medicinalplantsmaster.pdf
https://www.academia.edu/18398609/Notes_on_Guyanas_medical_ethnobotany
https://www.essencejournal.com/vol2/issue1/pdf/13a.1.pdf

Fabaceae

https://academica-e.unavarra.es/handle/2454/914
http://static.sites.sbq.org.br/rvq.sbq.org.br/pdf/v10n3a18.pdf
http://www.asianjournalofchemistry.co.in/User/ViewFreeArticle.aspx?ArticleID=24_4_15
http://www.hear.org/pier/species/neptunia_plena.htm
http://www.journal-aquaticscience.com/article_73311_6f1b85b191a85f268c100f560a723aa9.pdf
https://botanicaargentina.org.ar/wp-content/uploads/2019/06/978-987-688-332-0-baja-resoluci%C3%B3n.pdf
https://docplayer.com.br/134941626-Estudo-quimico-de-aeschynomene-filosa-leguminosae-alene-cortes-de-queiroz-e-silva.html
https://doi.org/10.1007/978-0-387-70638-2_1063
https://doi.org/10.1007/s11157-020-09552-y
https://doi.org/10.1080/10934520701629781
https://doi.org/10.1590/S1984-29612013000400013
https://doi.org/10.2174/187152406778226725
https://doi.org/10.7717/peerj.6736
https://jbiochemtech.com/storage/models/article/RT5gXdFkUwxhuDVj7nqyr9mAqLHUBxNvf3qNapGPjqeaaPJnxF82nF8aLAb6/synthesis-of-silver-nanoparticles-using-aeschynomene-indica-l-aqueous-leaf-extract-and-evaluation-.pdf
https://keyserver.lucidcentral.org/weeds/data/media/Html/neptunia_oleracea_and_neptunia_plena.htm
https://nas.er.usgs.gov/queries/GreatLakes/FactSheet.aspx?Species_ID=2730
https://phytochem.nal.usda.gov
https://repository.unam.edu.na/bitstream/handle/11070/1653/Dolly_2016.pdf?sequence=1&isAllowed=y
https://tropical.theferns.info/viewtropical.php?id=Aeschynomene+americana
https://tropical.theferns.info/viewtropical.php?id=Aeschynomene+aspera
https://tropical.theferns.info/viewtropical.php?id=Aeschynomene+indica
https://tropical.theferns.info/viewtropical.php?id=Neptunia+oleracea
https://web.archive.org/web/20180413023418id_/http://ijppsjournal.com/Vol4Suppl4/4508.pdf

https://www.embrapa.br/busca-de-publicacoes/-/publicacao/1073295/especies-nativas-da-flora-brasileira-de-valor-economico-atual-ou-potencial-plantas-para-o-futuro-regiao-centro-oeste
https://www.embrapa.br/busca-de-publicacoes/-/publicacao/132611/influencia-do-arsenio-e-fosforo-sobre-o-crescimento-de-sesbania-exasperata-kunth
https://www.gardenia.net/plant/aeschynomene-fluitans
https://www.gardenia.net/plant/neptunia-oleracea
https://www.inaturalist.org/taxa/75294-Aeschynomene-rudis
https://www.nature.com/articles/s41467-021-21094-7
https://www.plantsjournal.com/archives/2019/vol7issue4/PartC/7-4-14-165.pdf
https://www.rarepalmseeds.com/neptunia-plena
https://www.researchgate.net/publication/262619951_Phytochemical_study_and_evaluation_of_anti-inflammatory_activity_of_Aeschynomene_fluminensis_vell_Fabaceae
https://www.semanticscholar.org/paper/Angiospermas-acu%C3%A1ticas-en-el-Per%C3%BA%3A-usos-y-estado-en-Aponte-P%C3%A9rez-Irigoyen/ecf5df365e4c63832bd7440018d3bfd870eb4e22
https://www.tropicalforages.info/text/entities/aeschynomene_americana.htm
https://pubag.nal.usda.gov/catalog/818876
https://www.researchgate.net/publication/342244935_Potential_of_Leaves_of_Water_Mimosa_Neptunia_plena_in_Increasing_Spermatogenesis_Activity_of_Male_Rat

Haloragaceae

http://www.eppo.int/INVASIVE_PLANTS/iap_list/Myriophyllum_heterophyllum.htm
http://www.hanaquatics.com/myriophyllum-mattogrossense/
http://www.scielo.org.mx/scielo.php?script=sci_arttext&pid=S0188-88972018000100109&lng=pt&nrm=iso
https://agris.fao.org/agris-search/search.do?recordID=US201301748371
https://doi.org/10.1007/978-3-030-00099-8_7
https://doi.org/10.1007/s00244-007-9070-9 https://www.pondsplantsandmore.com/Foxtail_Myriophyllum_heterophyllum_p/psh_foxtail.htm
https://doi.org/10.1007/978-981-32-9664-0_4
https://doi.org/10.1016/j.chemosphere.2005.02.079
https://doi.org/10.1016/S0273-1223(96)00749-4
https://doi.org/10.1034/j.1600-0706.2003.12666.x
https://doi.org/10.1201/9780429429095
https://doi.org/10.3390/su12051927
https://doi.org/10.4172/2155-6199.1000404
https://en.aqua-fish.net/plants/myriophyllum-hippuroides
https://en.hortipedia.com/Myriophyllum_pinnatum

https://en.hortipedia.com/Myriophyllum_quitense
https://phytochem.nal.usda.gov/phytochem/ethnoPlants/show/11348?qlookup=Myriophyllum+braziliense&offset=0&max=20&et=
https://plants.ifas.ufl.edu/plant-directory/myriophyllum-laxum/
https://plantsforponds.co.uk/products/upright-water-milfoil-myriophyllum-crispatum
https://pubmed.ncbi.nlm.nih.gov/27255321/
https://repositorio.ufscar.br/bitstream/handle/ufscar/2060/DissKST.pdf?sequence=1&isAllowed=y
https://tropica.com/en/plants/plantdetails/Proserpinacapalustris%27Cuba%27(037CTC)/4766
https://vicflora.rbg.vic.gov.au/flora/taxon/7567bbdc-fa64-42c9-a1f6-cb8b04f28d7f
https://vicflora.rbg.vic.gov.au/flora/taxon/b7962d70-4147-4893-b818-d8b9fac2d432
https://www.aquagreen.com.au/plant_data/Myriophyllum_dicoccum.html
https://www.aquagreen.com.au/plant_data/Myriophyllum_muricatum.html
https://www.aquasabi.com/Myriophyllum-pinnatum
https://www.aquasabi.com/Myriophyllum-simulans https://recentlyextinctspecies.com/?view=article&id=10000:myriophyllum-sparsiflorum&catid=159
https://www.aquasabi.com/Myriophyllum-tetrandrum-Bunch
https://www.cabidigitallibrary.org/doi/10.1079/cabicompendium.34940
https://www.corma.cl/wp-content/uploads/2020/12/Guia-Campo-Plantas-Vasculares-Acuaticas-WEB.pdf
https://www.fishkeeper.co.uk/help-and-advice/plant/pond-plants/myriophyllum-propinquum
https://www.fishkeeper.co.uk/pond-plant-myriophyllum-crispatum-upright-water-milfoil
https://www.floraphyta.com.au/plants/myriophyllum-gracile
https://www.flowgrow.de/db/aquaticplants/myriophyllum-mezianum/
https://www.flowgrow.de/db/aquaticplants/myriophyllum-quitense
https://www.gbif.org/tools/zoom/simple.html?src=//api.gbif.org/v1/image/unsafe/http%3A%2F%2Fn2t.net%2Fark%3A%2F65665%2Fm38d386c77-f75c-477c-bd64-96be7304c67b
https://www.hindawi.com/journals/bmri/2013/890803/
https://www.invasive.org/alien/pubs/midatlantic/mysp.htm
https://www.invasivespeciesinfo.gov/aquatic/plants/eurasian-watermilfoil)
https://www.ncbi.nlm.nih.gov/pmc/articles/PMC5148709/
https://www.nzpcn.org.nz/flora/species/myriophyllum-robustum/
https://www.plantlife.org.uk/uk/discover-wild-plants-nature/plant-fungi-species/parrots-feather
https://www.researchgate.net/publication/281060403_Biology_Biotechnology_and_Sustainable_Development
https://www.rhs.org.uk/plants/69328/myriophyllum-spicatum/details
https://www.semanticscholar.org/paper/Angiospermas-acu%C3%A1ticas-en-el-Per%C3%BA%3A-usos-y-estado-en-Aponte-P%C3%A9rez-Irigoyen/ecf5df365e4c63832bd7440018d3bfd870eb4e22
https://www.shootgardening.com/plants/myriophyllum-propinquum

https://www.watergardenplants.co.uk/products/myriophyllum-spicatum
https://www.watersidenursery.co.uk/myriophyllum-verticillatum-whorled-milfoil
https://www.yarraranges.vic.gov.au/PlantDirectory/Aquatic-Semi-aquatic/Myriophyllum-amphibium
https://www.yarraranges.vic.gov.au/PlantDirectory/Aquatic-Semi-aquatic/Myriophyllum-pedunculatum-ssp.-pedunculatum
https://doi.org/10.1080/15226514.2012.723068
https://www.aqualifeaquarium.com.au/myriophyllum-papillosum-milfoil-bunch https://ppnn.org.au/plantlist/myriophyllum-papillosum/
https://niwa.co.nz/media-gallery/detail/109727/27816 https://www.nzpcn.org.nz/flora/species/myriophyllum-robustum/
https://gobotany.nativeplanttrust.org/species/myriophyllum/sibiricum/
https://www.picturethisai.com/wiki/Myriophyllum_tenellum.html
https://www.picturethisai.com/wiki/Myriophyllum_triphyllum.html https://www.nzpcn.org.nz/flora/species/myriophyllum-triphyllum/
https://www.picturethisai.com/wiki/Myriophyllum_votschii.html

Hydatellaceae

https://doi.org/10.1002/tax.12026

Hydroleaceae

https://doi.org/10.35818/acta.v5i1.20
http://publish.plantnet-project.org/project/riceweeds_en/collection/collection/information/details/HYMZE
http://www.iepa.ap.gov.br/metadados/instituicoes/iepa/projetos/ressacas/documentos/1finalVEGETACAO.pdf
https://phytochem.nal.usda.gov
https://tropical.theferns.info/viewtropical.php?id=Hydrolea+zeylanica
https://www.fnps.org/plant/hydrolea-corymbosa

Hydrostachyaceae

https://phytokeys.pensoft.net/article/58538/
https://www.researchgate.net/publication/280876903_Etude_ethnobotanique_et_screening_phytochimique_d%27Hydrostachys_plumosa_A_Juss_ex_Tul_Hydrostachyaceae_espece_aquatique_endemique_de_Mandritsara_Madagascar
https://pza.sanbi.org/hydrostachys-polymorpha

Hypericaceae

https://doi.org/10.4141/cjps2012-006
https://www.devonpondplants.co.uk/product/hypericum-elodes/
http://www.wildflowerweb.co.uk/plant/4123/marsh-st-johns-wort

Lamiaceae

http://www.sech.info/ACTAS/Acta%20n%C2%BA%2052.%20IV%20Jornadas%20Ib%C3%A9ricas%20de%20Horticultura%20Ornamental/Ponencias%20invitadas/Plantas%20Ibero-Macaron%C3%A9sicas%20con%20valor%20ornamental.pdf
https://aquaessentials.co.uk/products/pogostemon-stellatus
https://botanicaargentina.org.ar/wp-content/uploads/2019/06/978-987-688-332-0-baja-resoluci%C3%B3n.pdf
https://dennerle.com/en/plants-database/pogostemon-quadrifolius/
https://docs.bvsalud.org/biblioref/2018/10/916684/plantas-medicinales-de-los-andes-y-la-amazonia-la-flora-magica-_Qa3dgqr.pdf
https://doi.org/10.1007/978-0-387-70638-2_1000
https://doi.org/10.1007/978-0-387-70638-2_1003
https://doi.org/10.1007/978-0-387-70638-2_957
https://doi.org/10.1007/978-0-387-70638-2_999
https://doi.org/10.1007/s11157-020-09552-y
https://doi.org/10.1016/j.jep.2006.10.032
https://doi.org/10.1016/S0160-4120(03)00091-6
https://doi.org/10.1155/2019/6087509
https://doi.org/10.1186/s13002-016-0100-4
https://doi.org/10.1201/9780429429095
https://doi.org/10.29328/JOURNAL.JPSP.1001050
https://explorer.natureserve.org/Taxon/ELEMENT_GLOBAL.2.145565/Pogogyne_abramsii
https://lincspplants.co.uk/product/lycopus-europaeus/
https://phytochem.nal.usda.gov
https://powo.science.kew.org/taxon/urn:lsid:ipni.org:names:450735-1#synonyms
https://repositorio.ufsc.br/bitstream/handle/123456789/129388/329004.pdf?sequence=1&isAllowed=y
https://temperate.theferns.info/plant/Lycopus+asper
https://temperate.theferns.info/plant/Teucrium+scordium
https://tropica.com/en/plants/plantdetails/Pogostemonhelferi(053H)/4499
https://www.aquasabi.com/Pogostemon-yatabeanus
https://www.arca.fiocruz.br/bitstream/icict/48188/2/m_et_all.pdf
https://www.drugs.com/npc/bugleweed.html

https://www.gardenersworld.com/plants/mentha-aquatica/
https://www.gardenersworld.com/plants/mentha-pulegium/
https://www.gardenersworld.com/plants/mentha-spicata/
https://www.healthbenefitstimes.com/water-mint/
https://www.infoteca.cnptia.embrapa.br/infoteca/bitstream/doc/884520/1/livro-plantastropicais-2.pdf
https://www.ncbi.nlm.nih.gov/pmc/articles/PMC4171855/
https://www.norfolkherbs.co.uk/product/mint-thai-mentha-arvensis/
https://www.oand3.com/product/peppermint-mentha-arvensis-cornmint-oil-essential-oil-organic/
https://www.scielo.br/j/rbfar/a/nkwKBHG6VyF9QySBSqPS9TJ/?lang=pt&format=pdf
https://doi.org/10.3390/molecules25061422
https://pfaf.org/user/Plant.aspx?LatinName=Pogogyne+douglasii

Lentibulariaceae

http://www.bestcarnivorousplants.net/index.php?node=product&id=1202-utricularia-arnhemica-prince-regent-river-kimberley-nt-australia
http://www.iepa.ap.gov.br/metadados/instituicoes/iepa/projetos/ressacas/documentos/1finalVEGETACAO.pdf
https://carnivorousplantnursery.com/blogs/carnivorous-plants-by-genus/genlisea
https://carnivorousplantnursery.com/products/utricularia-gibba
https://carnivorousplantnursery.com/products/utricularia-vulgaris
https://carnivorousplantresource.com/product/utricularia-heterosepala/
https://doi.org/10.1007/s00244-017-0498-2
https://doi.org/10.1007/s11157-020-09552-y
https://doi.org/10.1201/9780429429095
https://doi.org/10.22059/poll.2019.288505.676
https://doi.org/10.4172/2155-6199.1000404
https://herbaria.plants.ox.ac.uk/bol/plants400/Profiles/UV/Utricularia
https://johnscarnivorousplants.com/products/utricularia-fulva
https://peerj.com/articles/4914.pdf
https://shop.aquaria-usa.com/product/utricularia-heterosepala/
https://uk.inaturalist.org/taxa/840996-Utricularia-humboldtii
https://vosserareplants.com/product/utricularia-biloba/?lang=en
https://vosserareplants.com/product/utricularia-chrysantha-kings-cascades/?lang=en
https://vosserareplants.com/product/utricularia-nana-red-leaf/?lang=en
https://www.araflora.com/p695/bladderwort_utricularia_humboldtii_serra_do_neblinai

https://www.araflora.com/p704/bladderwort_utricularia_nelumbifolia
https://www.cabidigitallibrary.org/doi/10.1079/cabicompendium.117747
https://www.earthandjungle.com/variety/2rtRSuXv/utricularia-benjaminiana
https://www.earthandjungle.com/variety/zmg9GKe2/utricularia-holtzei
https://www.floraphyta.com.au/plants/utricularia-aurea
https://www.flytrapcare.com/phpBB3/utricularia-smithiana-and-others-flowers-t43806.html
https://www.hccarnivorousplants.co.uk/products/genlisea-hispidula
https://www.hccarnivorousplants.co.uk/products/utricularia-dichotoma
https://www.icmbio.gov.br/portal/images/stories/imgs-unidades-coservacao/parna_matogrossensee.pdf
https://www.littleshopofhorrors.co.uk/product/utricularia-reniformis/
https://www.nzpcn.org.nz/flora/species/utricularia-geminiscapa/
https://www.rarexoticseeds.com/en/utricularia-juncea-seeds-southern-bladderwort.html
https://www.researchgate.net/profile/Sanjeet-Kumar-16/publication/340886563_Ecological_Mapping_Pharmacological_Activity_of_Utricularia_aurea_Lour_A_Carnivorous_Plant_of_Odisha/links/5ea2af87458515ec3a02f7bd/Ecological-Mapping-Pharmacological-Activity-of-Utricularia-aurea-Lour-A-Carnivorous-Plant-of-Odisha.pdf
https://www.sarracenia.com/faq/faq5330.html
https://www.sarracenia.com/faq/faq5340.html
https://www.sarracenia.com/faq/faq5616.html
https://www.thaiscience.info/journals/Article/SONG/10987031.pdf
https://www.thecps.org.uk/seedbank/genlisea-africana/
https://www.fishfish.fr/plante/utricularia-minor
http://www.bestcarnivorousplants.net/index.php?node=product&id=1008-utricularia-reflexa-huge-traps-zambia-5-plants
https://carnivores.zone/en/produit/utricularia-resupinata/
https://indiabiodiversity.org/species/show/226354
https://www.floriculturapaiefilho.com.br/utricularia-subulata https://www.carnivorasonline.com.br/utricularia-subulata.html
https://www.exotic-plants.de/seeds/carnivorous-plants/Utricularia-volubilis.php https://www.amazon.ca/Utricularia-Volubilis-bladderwort-Carnivorous-Seeds/dp/B00MUDKJMO

Limnanthaceae

https://www.gardenersworld.com/plants/limnanthes-douglasii/
https://calscape.org/Limnanthes-floccosa-()
http://beta.floranorthamerica.org/Limnanthes

Linderniaceae

https://doi.org/10.25081/jmhe.2020.v6.6240
https://gobotany.nativeplanttrust.org/species/lindernia/dubia/
https://jtsl.ub.ac.id/index.php/jtsl/article/view/137
https://naturalhistory.si.edu/sites/default/files/media/file/medicinalplantsmaster.pdf
https://powo.science.kew.org/taxon/urn:lsid:ipni.org:names:1061345-2
https://powo.science.kew.org/taxon/urn:lsid:ipni.org:names:1189750-2
https://powo.science.kew.org/taxon/urn:lsid:ipni.org:names:255156-2
https://tropica.com/en/plants/plantdetails/Linderniarotundifolia(045)/4468
https://tropical.theferns.info/viewtropical.php?id=Lindernia+diffusa
https://tropical.theferns.info/viewtropical.php?id=Lindernia+hyssopoides
https://tropical.theferns.info/viewtropical.php?id=Lindernia+procumbens
https://www.cabidigitallibrary.org/doi/10.1079/cabicompendium.115307
https://doi.org/10.1016/j.jep.2019.112493

Lythraceae

http://dnr.alaska.gov/ag/Index/IntroductiontoCommonNativeInvasiveFreshwaterPlantsinAlaska.pdf
http://redlist.sanbi.org/species.php?species=3692-1
http://temperate.theferns.info/plant/Lythrum+portula
http://tropical.theferns.info/viewtropical.php?id=Cuphea+aperta
http://tropical.theferns.info/viewtropical.php?id=Cuphea+carthagenensis
http://tropical.theferns.info/viewtropical.php?id=Cuphea+racemosa
http://www.bdtd.ufrr.br/tde_busca/processaPesquisa.php?pesqExecutada=1&id=327
http://www.guitarfish.org/2008/11/24/rotala-verticillaris
http://www.ijstr.org/final-print/july2014/Laboratory-Study-Of-Arsenic-Uptake-And-Phytoremediation-Potential-Of-Three-Aquatic-Macrophytes-Of-Meghalaya-India.pdf
http://www.sech.info/ACTAS/Acta%20n%C2%BA%2052.%20IV%20Jornadas%20Ib%C3%A9ricas%20de%20Horticultura%20Ornamental/Ponencias%20invitadas/Plantas%20Ibero-Macaron%C3%A9sicas%20con%20valor%20ornamental.pdf
https://academicjournals.org/journal/JMPR/article-abstract/17F061D15070
https://aquaticarts.com/products/golden-nesaea
https://botanicaargentina.org.ar/wp-content/uploads/2019/06/978-987-688-332-0-baja-resoluci%C3%B3n.pdf
https://buceplant.com/products/ammania-senegalensis
https://doi.org/10.1002/cbdv.202100041
https://doi.org/10.1007/978-0-387-70638-2_103

https://doi.org/10.1007/978-3-662-08614-8_5
https://doi.org/10.1007/s00044-013-0682-5
https://doi.org/10.1016/0305-1978(94)00071-N
https://doi.org/10.1016/j.indcrop.2016.04.008
https://doi.org/10.1016/j.jep.2006.10.032
https://doi.org/10.1016/j.jep.2012.01.001
https://doi.org/10.1016/j.jep.2015.05.017
https://doi.org/10.1016/j.jep.2015.12.054
https://doi.org/10.1016/S2225-4110(16)30057-8
https://doi.org/10.1080/10408398909527495
https://doi.org/10.11646/phytotaxa.514.2.8
https://doi.org/10.1201/9780429429095
https://doi.org/10.1590/S0102-33062003000400009
https://doi.org/10.1691/ph.2014.4635
https://doi.org/10.2174/1568026619666190412101047
https://doi.org/10.3329/jujbs.v6i2.36587
https://doi.org/10.4103/pr.pr_42_17
https://doi.org/10.4172/2155-6199.1000404
https://doi.org/10.5772/intechopen.106742
https://gd.eppo.int/reporting/article-6984
https://indiabiodiversity.org/species/show/226213
https://kingcounty.gov/services/environment/animals-and-plants/noxious-weeds/weed-identification/purple-loosestrife.aspx
https://naturalhistory.si.edu/sites/default/files/media/file/medicinalplantsmaster.pdf
https://phytochem.nal.usda.gov
https://plantsforponds.co.uk/products/water-chestnut-trapa-natans
https://pubs.acs.org/doi/abs/10.1021/jo01056a005
https://recentlyextinctspecies.com/myrtales-evening-primrose-fuschias-myrtles-pomegranates-etc/rotala-simpliciuscula
https://repositorio.ufmg.br/handle/1843/FARD-8S6TT3
https://repositorio.ufpe.br/bitstream/123456789/3480/1/arquivo6099_1.pdf
https://repositorio.ufsc.br/bitstream/handle/123456789/129388/329004.pdf?sequence=1&isAllowed=y
https://ri.conicet.gov.ar/handle/11336/24328
https://ria.ua.pt/bitstream/10773/7941/1/5149.pdf
https://shop.aquaria-usa.com/product/rotala-tulunadensis/
https://tropica.com/en/plants/plantdetails/Rotalamacrandra(032TC)/4445
https://tropica.com/en/plants/plantdetails/Rotalawallichii(032ATC)/18748
https://tropical.theferns.info/viewtropical.php?id=Cuphea+melvilla
https://tropical.theferns.info/viewtropical.php?id=Trapa+incisa
https://www.aphis.usda.gov/plant_health/plant_pest_info/weeds/downloads/wra/Rotala-%20rotundifolia.pdf
https://www.aquagreen.com.au/plant_data/Rotala_occultiflora.html
https://www.aquariumgardens.co.uk/rotala-rotundifolia-443-p.asp
https://www.aquariumsource.com/rotala-indica/

https://www.aquasabi.com/Ammannia-crassicaulis
https://www.aquasabi.com/Rotala-mexicana-Goias
https://www.aquasabi.com/Rotala-ramosior-Florida
https://www.aquaticplantcentral.com/threads/rotala-serpyllifolia.143181/
https://www.cabidigitallibrary.org/doi/10.1079/cabicompendium.113690
https://www.flowgrow.de/db/aquaticplants/ammannia-latifolia
https://www.flowgrow.de/db/aquaticplants/cuphea-anagalloidea
https://www.flowgrow.de/db/aquaticplants/didiplis-diandra
https://www.flowgrow.de/db/aquaticplants/rotala-hippuris
https://www.flowgrow.de/db/aquaticplants/rotala-indica
https://www.gardenersworld.com/plants/lythrum-salicaria/
https://www.herbal-organic.com/en/herb/21912
https://www.lume.ufrgs.br/bitstream/handle/10183/131911/000981101.pdf?sequence=1
https://www.picturethisai.com/wiki/Lythrum_tribracteatum.html
https://www.researchgate.net/publication/287848428_Screening_of_aquatic_plant_extracts_for_herbicidal_fungicidal_and_insecticidal_activity
https://www.researchgate.net/publication/311064628_An_Ethnomedical_Pharmacological_and_Phytochemical_Review_of_Ammannia_baccifera_L
https://www.researchgate.net/publication/323178350_LYTHRUM_SALICARIA_L_IN_THE_HUMID_AREA_VEGETATION
https://botanicaargentina.org.ar/wp-content/uploads/2019/06/978-987-688-332-0-baja-resoluci%C3%B3n.pdf
https://www.researchgate.net/publication/332405686_Medicinal_Chemistry_of_Alternative_Therapeutics_Novelty_and_Hopes_with_Genus_Ammannia
https://www.scielo.br/j/rbfar/a/nkwKBHG6VyF9QySBSqPS9TJ/?lang=pt&format=pdf
https://www.semanticscholar.org/paper/Advances-in-Study-on-Chemical-Constituents-of-in-Luo-Chen/9dc67334b62738cd8f78dca86c9a4b7b9c970d38?pdf
https://www.udesc.br/arquivos/cav/id_cpmenu/1349/dissertacao_karina_montibeller_15676053075669_1349.pdf
https://temperate.theferns.info/plant/Lythrum+portula
https://buceplant.com/products/nesaea-triflora https://shop.glassaqua.com/products/nesaea-triflora?variant=33392260186198
https://indiabiodiversity.org/species/show/251641
https://www.mdpi.com/2076-3417/12/1/290
https://www.eattheweeds.com/tag/trapa-rossica/

Malvaceae

http://citeseerx.ist.psu.edu/viewdoc/download?doi=10.1.1.560.8735&rep=rep1&type=pdf
http://nativeplants.hawaii.edu/plant/view/Hibiscus_furcellatus

http://www.wildsouthflorida.com/lindenleaf.rosemallow.html
https://patents.google.com/patent/US20030072820
https://www.alice.cnptia.embrapa.br/alice/handle/doc/783791
https://www.gardenia.net/plant/hibiscus-coccineus
https://www.gardenia.net/plant/hibiscus-grandiflorus

Melastomataceae

http://mbm.urca.br/pdf/dissertacoes/MARCOS%20AUR%C3%89LIO%20FIGUEIREDO%20DOS%20SANTOS.pdf
https://floram.org/article/588e2208e710ab87018b460c/pdf/floram-11-1-47.pdf

Menyanthaceae

http://dnr.alaska.gov/ag/Index/IntroductiontoCommonNativeInvasiveFreshwaterPlantsinAlaska.pdf
http://www.rareplants.es/shop/product.asp?P_ID=13451
http://www.walliscreekwatergarden.com.au/product/villarsia-umbricola/
https://doi.org/10.1201/9780429429095
https://doi.org/10.2134/jeq1999.00472425002800050009x
https://doi.org/10.2307/2444173
https://doi.org/10.4103/2221-1691.225615
https://gardenworld.com.au/shop/water-gardens/water-plants/zone-3-plants/marbled-marshwort/
https://phytochem.nal.usda.gov
https://plants.ifas.ufl.edu/plant-directory/nymphoides-cristata/
https://powo.science.kew.org/taxon/urn:lsid:ipni.org:names:900365-1
https://ri.conicet.gov.ar/handle/11336/24328
https://ria.ua.pt/bitstream/10773/7941/1/5149.pdf
https://silverhillseeds.co.za/product/villarsia-capensis/
https://www.abcplants.com/product-page/nymphoides-hydrophylla-taiwan-1
https://www.apacewa.org.au/plant/villarsia-latifolia/
https://www.aquagreen.com.au/plant_data/Nymphoides_minima.html
https://www.aquariumcoop.com/blogs/aquarium/banana-plant-an-unusual-aquarium-plant-for-beginners
https://www.floraphyta.com.au/plants/nymphoides-exiliflora
https://www.gardenia.net/plant/nymphoides-indica
https://www.instagram.com/p/Ce5UQYyolVF/
https://www.iucnredlist.org/search?query=Nymphoides%20peltata
https://www.iucnredlist.org/search/grid?query=Nymphoides%20ezannoi
https://www.iucnredlist.org/search/grid?query=Nymphoides%20guineensis

https://www.iucnredlist.org/species/194158/8884939

https://www.rhs.org.uk/plants/11057/menyanthes-trifoliata/details

https://www.semanticscholar.org/paper/Angiospermas-acu%C3%A1ticas-en-el-Per%C3%BA%3A-usos-y-estado-en-Aponte-P%C3%A9rez-Irigoyen/ecf5df365e4c63832bd7440018d3bfd870eb4e22

https://www.watergardenplants.co.uk/products/nymphoides-peltata?variant=8943395143723

https://www.wetland-plants.co.uk/shop/british-native/menyanthes-trifoliata-bog-bean-native/

https://www.yarraranges.vic.gov.au/PlantDirectory/Aquatic-Semi-aquatic/Ornduffia-reniformis

www.iucnredlist.org/species/163993/42397943#use-trade

References

1. Agami M, Waisel Y (1986) The role of mallard ducks (*Anas platyrhynchos*) in distribution and germination of seeds of the submerged hydrophyte *Najas marina* L. Oecologia 68:473–475
2. Akasaka M, Takamura N, Mitsuhashi H, Kadono Y (2010) Effects of land use on aquatic macrophyte diversity and water quality of ponds. Freshw Biol 55:909–922
3. Alahuhta J (2015) Geographic patterns of lake macrophyte communities and species-richness at regional scales. J Veg Sci 26:564–575
4. Alahuhta J, Kanninen A, Hellsten S, Vuorif K, Kuoppala M, Hämäläinen H (2013) Environmental and spatial correlates of community composition, richness and status of boreal lake macrophytes. Ecol Indic 32:172–181
5. Alahuhta J, Kosten S, Akasaka M, Auderset D, Azzella MM, Bolpagni R, Bove CP, Chambers PA, Chappuis E, Clayton J, de Winton M, Ecke F, Gacia E, Gecheva G, Grillas P, Hauxwell J, Hellsten S, Hjort J, Hoyer MV, Ilg C, Kolada A, Kuoppala M, Lauridsen T, Li E-H, Lukács BA, Mjelde M, Mikulyuk A, Mormul RP, Nishihiro J, Oertli B, Rhazi L, Rhazi M, Sass L, Schranz C, Søndergaard M, Yamanouchi T, Yu Q, Wang H, Willby N, Zhang X-K, Heino J (2017a) Global variation in the beta diversity of lake macrophytes is driven by environmental heterogeneity rather than latitude. J Biogeogr 44:1758–1769
6. Alahuhta J, Virtala A, Hjort J, Eck F, Johnson LB, Sass L, Heino J (2017b) Average niche breadths of species in lake macrophyte communities respond to ecological gradients variably in four regions on two continents. Oecologia 184:219–235
7. Alahuhta J, Lindholm M, Baastrup-Spohr L, García-Girón J, Toivanen M, Heino J, Murphy K (2021) Macroecology of macrophytes in the freshwater realm: patterns, mechanisms and implications. Aquat Bot 168:103325. https://doi.org/10.1016/j.aquabot.2020.103325
8. Ali MM, Hamad A, Springuel IV, Murphy KJ (1995) Environmental factors affecting submerged macrophyte communities in regulated waterbodies in Egypt. Arch Hydrobiol 133:107–128
9. Ali MM, Murphy KJ, Abernethy VJ (1999) Macrophyte functional variables v. species assemblages as predictors of trophic status in flowing waters. Hydrobiologia 415:131–138
10. Beck MW, Alahuhta J (2017) Ecological determinants of *Potamogeton* taxa in glacial lakes: assemblage composition, species-richness, and species-level approach. Aquat Sci 79:427–441
11. Bhattacharjee S, Islam MT, Kabir ME, Kabir MM (2021) Land-use and land-cover change detection in a north-eastern wetland ecosystem of Bangladesh using remote sensing and GIS techniques. Earth Syst Environ 5:319–340. https://doi.org/10.1007/s41748-021-00228-3

T. Lobato-de Magalhães et al., *World Atlas of Freshwater Macrophytes*, Wetlands: Ecology, Conservation and Management 10,
https://doi.org/10.1007/978-3-031-52749-4

12. Bini LM, Thomaz SM, Murphy KJ, Camargo AFM (1999) Aquatic macrophyte distribution in relation to water and sediment conditions in the Itaipu Reservoir, Brazil. Hydrobiologia 415:147–154
13. Bobrov A, Gałosz J, Jopek M, Movergoz E (2015) *Ranunculus schmalhausenii* (section Batrachium, Ranunculaceae), a neglected water crowfoot endemic to Fennoscandia – a case of rapid hybrid speciation in postglacial environment of North Europe. Phytotaxa 233:101–138. https://doi.org/10.11646/phytotaxa.233.2.1
14. Bohm BA, Nicholls KW, Ornduff R (1986) Flavonoids of the Menyanthaceae: intra- and interfamilial relationships. Am J Bot 73:204–213. https://doi.org/10.1002/j.1537-2197.1986.tb08521.x
15. Borsch T, Wiersema JH, Hellquist CB, Löhne C, Govers K (2014) Speciation in North American water lilies: evidence for the hybrid origin of the newly discovered Canadian endemic *Nymphaea loriana* sp. nov. (Nymphaeaceae) in a past contact zone. Botany 92:867–882. https://doi.org/10.1139/cjb-2014-0060
16. Bottino F, Calijuri M, Murphy KJ (2014) Temporal and spatial variation of limnological variables and biomass of different macrophyte species in a Neotropical reservoir (São Paulo – Brazil). Acta Limn Bras 25(4):387–397
17. Boulangeat I, Lavergne S, Van Es J, Garraud L, Thuiller W (2012) Niche-breadth, rarity and ecological characteristics within a regional flora spanning large environmental gradients. J Biogeogr 39:204–214
18. Bussmann RW, Sharon D (2018) Medicinal plants of the Andes and the Amazon - the magic and medicinal flora of Northern Peru. Ethnobot Res Appl 15:1–295
19. Butcher RW (1933) Studies on the ecology of rivers: I. On the distribution of macrophytic vegetation in the rivers of Britain. J Ecol 21:58–91
20. Calhoun AJ, Mushet DM, Bell KP, Boix D, Fitzsimons JA, Isselin-Nondedeu F (2017a) Temporary wetlands: challenges and solutions to conserving a 'disappearing' ecosystem. Biol Conserv 211(1):3–11. https://doi.org/10.1016/j.biocon.2016.11.024
21. Calhoun AJ, Mushet DM, Alexander LC, DeKeyser ES, Fowler L, Lane CR, Lang MW, Rains MC, Richter SC, Walls SC (2017b) The significant surface-water connectivity of "geographically isolated wetlands". Wetlands 37:801–806. https://doi.org/10.1007/s13157-017-0887-3
22. Capers RS, Selsky R, Bugbee GJ (2010) The relative importance of local conditions and regional processes in structuring aquatic plant communities. Freshw Biol 55:952–966
23. Cardina J (2021) Lives of weeds: opportunism, resistance and folly. Comstock Pub Ass, Ithaca
24. Carrillo-Angeles IG, Suzán-Azpiri H, Mandujano MC, Golubov J, Martínez-Ávalos JG (2016) Niche-breadth and the implications of climate change in the conservation of the genus *Astrophytum* (Cactaceae). J Arid Environ 124:310–317
25. Carvalho P, Thomaz SM, Kobayashi JT, Bini LM (2013) Species-richness increases the resilience of wetland plant communities in a tropical floodplain. Aust Ecol 38:592–598
26. Casanova MT, Karol KG (2023) Charophytes of Australia's Northern Territory – II. Tribe Nitelleae Aust Syst Bot 36:322–353. https://doi.org/10.1071/SB22029
27. Ceballos G, Ehrlich PR, Barnosky AD, García A, Pringle RM, Palmer TM (2015) Accelerated modern human–induced species losses: entering the sixth mass extinction. Sci Adv 1:e1400253. https://doi.org/10.1126/sciadv.1400253
28. Chambers PA, Lacoul P, Murphy KJ, Thomaz SM (2008) Global diversity of aquatic macrophytes in freshwater. In: Balian EV, Lévêque C, Segers H, Martens K (eds) Freshwater animal diversity assessment. Developments in hydrobiology, vol 198. Springer, Dordrecht. https://doi.org/10.1007/978-1-4020-8259-7_2
29. Chappuis E, Ballesteros E, Gacia E (2012) Distribution and richness of aquatic plants across Europe and Mediterranean countries: patterns, environmental driving factors and comparison with total plant richness. J Veg Sci 23:985–997
30. Chappuis E, Gacia E, Ballesteros E (2014) Environmental factors explaining the distribution and diversity of vascular aquatic macrophytes in a highly heterogeneous Mediterranean region. Aquat Bot 113:72–82

31. Chen LY, Chen JM, Gituru RW, Temam TD, Wang QF (2012a) Generic phylogeny and historical biogeography of Alismataceae, inferred from multiple DNA sequences. Mol Phylogen Evol 63:407–416
32. Chen LY, Chen JM, Gituru RW, Wang QF (2012b) Generic phylogeny, historical biogeography and character evolution of the cosmopolitan aquatic plant family Hydrocharitaceae. BMC Evol Biol 12:30–41
33. Clausen P, Nolet BA, Fox A, Klaassen M (2002) Long-distance endozoochorous dispersal of submerged macrophyte seeds by migratory waterbirds in northern Europe – a critical review of possibilities and limitations. Acta Oecol 23:191–203
34. Compton TJ, Winton M de, Leathwick JR, Wadhwa S (2012) Predicting spread of invasive macrophytes in New Zealand lakes using indirect measures of human accessibility. Freshw Biol 57:938–948. https://doi.org/10.1111/j.1365-2427.2012.02754.x
35. Coughlan NE, Kelly TC, Davenport J, Jansen MAK (2017a) Up, up and away: bird-mediated ectozoochorous dispersal between aquatic environments. Freshw Biol 62:631–648
36. Coughlan NE, Kelly TC, Jansen MAK (2017b) "Step by step": high frequency short-distance epizoochorous dispersal of aquatic macrophytes. Biol Invasions 19:625–634
37. Cowardin LM, Carter V, Golet VFC, LaRoe ET (1979) Classification of wetlands and Deepwater habitats of the United States. U.S. Fish and Wildlife Service. FWS/OBS-79/31. Washington, DC
38. Cowie RH, Bouchet P, Fontaine B (2022) The sixth mass extinction: fact, fiction or speculation? Biol Rev 97:640–663. https://doi.org/10.1111/brv.12816
39. Crow GE (1993) Species diversity in aquatic angiosperms: latitudinal patterns. Aquat Bot 44:229–258
40. Culumber ZW, Anaya-Rojas JM, Booker WW, Hooks AP, Lange EC, Pluer B, Ramírez-Bullón N, Travis J (2019) Widespread biases in ecological and evolutionary studies. Bioscience 69:631–640. https://doi.org/10.1093/biosci/biz063
41. Dar MA, Wani GA, Reshi ZA, Al-Qarawi AA, Abd Allah EF, Shah MA (2020) Stage-specific ploidy level variations in invasive species in comparison to rare endemics in Kashmir Himalaya. Flora 262:151525. https://doi.org/10.1016/j.flora.2019.151525
42. Davidson TA, Mackay AW, Wolski P, Mazebedi R, Murray-Hudson M, Todd M (2012) Seasonal and spatial hydrological variability drives aquatic biodiversity in a flood-pulsed, sub-tropical wetland. Freshw Biol 57:1253–1265
43. Escalante T, Rodríguez-Tapia G, Szumik C, Morrone JJ, Rivas M (2010) Delimitation of the Nearctic region according to mammalian distributional patterns. J Mammal 91:1381–1388. https://doi.org/10.1644/10-MAMM-A-136.1
44. Federal Geographic Data Committee (2013) Classification of wetlands and deepwater habitats of the United States. FGDC-STD-004-2013. 2° Ed Wetlands Subcommittee, Federal Geographic Data Committee and U.S. Fish and Wildlife Service, Washington, DC
45. Figuerola J, Green AJ (2002) Dispersal of aquatic organisms by waterbirds: a review of past research and priorities for future studies. Freshw Biol 47:483–494
46. Figuerola J, Charalambidou I, Santamaría L, Green AJ (2010) Internal dispersal of seeds by waterfowl: effect of seed size on gut passage time and germination patterns. Naturwissenschaften 97:555–565
47. GBIF (2023) The global biodiversity information facility. https://www.gbif.org/citation-guidelines. Accessed 11 Apr 2023
48. Gessner F (1955) Hydrobotanik: Die physiologischen grundlagen der pflanzenverbreitung im wasser. Deutscher Verlag der Wissenschaften, Berlin I:160–161
49. Global Economy (2020) Percent agricultural land: country rankings. https://www.theglobaleconomy.com/rankings/Percent_agricultural_land/. Accessed 11 Jan 2023
50. Grime JP (1979) Plant strategies and vegetation processes. Wiley, Chichester
51. Grime JP (2006) Plant strategies, vegetation processes, and ecosystem properties. Wiley
52. Haslam SM (1978) River plants. Cambridge University Press, Cambridge, UK

53. Hickey RJ (1981) A new *Isoetes* from Jamaica. Am Fern J 71:69–74
54. Hofstra D, Schoelynck J, Ferrell J, Coetzee J, de Winton M, Bickel TO, Champion P, Madsen J, Bakker ES, Hilt S, Matheson F, Netherland, Gross, EM (2020) On the move: new insights on the ecology and management of native and alien macrophytes. Aquat Bot 162:103190. https://doi.org/10.1016/j.aquabot.2019.103190
55. Horvitz N, Wang R, Wan FH, Nathan R (2017) Pervasive human-mediated large-scale invasion: analysis of spread patterns and their underlying mechanisms in 17 of China's worst invasive plants. J Ecol 105:85–94. https://doi.org/10.1111/1365-2745.12692
56. Hughes AC, Orr MC, Ma K, Costello MJ, Waller J, Provoost P, Yang Q, Zhu C, Qiao H (2021) Sampling biases shape our view of the natural world. Ecography 44:1259–1269. https://doi.org/10.1111/ecog.05926
57. Huntley B (1993) Species-richness in north-temperate zone forests. J Biogeogr 20:163–180. https://doi.org/10.2307/2845669
58. iNaturalist (2023). https://www.inaturalist.org/. Accessed Jul 2023
59. Iseli E, Chisholm C, Lenoir J, Haider S, Seipel T, Barros A, Hargreaves AL, Kardol P, Lembrechts JJ, McDougall K, Rashid I, Rumpf SB, Arévalo JR, Cavieres L, Daehler C, Dar PA, Endress B, Jakobs G, Jiménez A, Küffer C, Mihoc M, Milbau A, Morgan JW, Naylor BJ, Pauchard A, Backes AR, Reshi ZA, Rew LJ, Righetti D, Shannon JM, Valencia G, Walsh N, Wright GT, Alexander JM (2023) Rapid upwards spread of non-native plants in mountains across continents. Nature Ecol Evolut 7:405–413. https://doi.org/10.1038/s41559-022-01979-6
60. IUCN (2023) International Union for conservation of nature and natural resources. https://www.iucnredlist.org/. Accessed Mar 2023
61. Jacob MCM, Albuquerque UP (eds) (2021) Local food plants of Brazil. Ethnobiology. Springer
62. Janauer GA, Gaberscik A, Kvet J, Germ M, Exler N (eds) (2018) Macrophytes of the river Danube Basin. Academia Praha, Czech Republic
63. Johnson WC, Millett BV, Gilmanov T, Voldseth RA, Guntenspergen GR, Naugle DE (2005) Vulnerability of northern prairie wetlands to climate change. Bioscience 55:863–872. https://doi.org/10.1641/0006-3568(2005)055[0863:VONPWT]2.0.CO;2
64. Jones JI, Li W, Maberly SC (2003) Area, altitude and aquatic plant diversity. Ecography 26:411–420
65. Judd WS, Campbell CS, Kellogg EA, Stevens PF, Donoghue MJ (2008) Plant systematics: a phylogenetic approach, 3rd edn. Sinauer Associates, Sunderland
66. Kennedy MP, Lang P, Tapia-Grimaldo J, Varandas-Martins S, Bruce A, Hastie A, Lowe S, Ali MM, Briggs J, Sichingabula H, Murphy KJ (2015) Environmental drivers of aquatic macrophyte communities in southern tropical African rivers: Zambia as a case study. Aquat Bot 124:19–28
67. Kennedy MP, Lang P, Tapia-Grimaldo J, Varandas-Martins S, Bruce A, Moore I, Taubert R, Macleod-Nolan C, McWaters S, Briggs J, Lowe S, Saili K, Sichingabula H, Willems F, Dallas H, Morrison S, Franceschini C, Bottino F, Murphy KJ (2017) Niche-breadth of freshwater macrophytes occurring in tropical southern African rivers predicts species global latitudinal range. Aquat Bot 136:121–130
68. Khan ZR, Chowdhury NS, Sharmin S, Sohrab MH (2018) Medicinal values of aquatic plant genus *Nymphoides* grown in Asia: a review. Asian Pac J Trop Biomed 8(2):113–119. https://doi.org/10.4103/2221-1691.225615
69. Klink CA, Machado RB (2005) Conservation of the Brasilian cerrado. Conserv Biol 19:707–713
70. Kusumoto B, Chao A, Eiserhardt W, Svenning J-C, Shiono T, Kubota Y (2023) Occurrence-based diversity estimation reveals macroecological and conservation knowledge gaps for global woody plants. bioRxiv 2023.04.09.536180. https://doi.org/10.1101/2023.04.09.536180

71. Laird MG, Bradshaw JD (2004) The break-up of a long-term relationship: the cretaceous separation of New Zealand from Gondwana. Gondwana Res 7:273–286. https://doi.org/10.1016/S1342-937X(05)70325-7
72. Lall N (ed) (2020) Aquatic plants: pharmaceutical and cosmetic applications. CRC Press, Baton Rouge. https://doi.org/10.1201/9780429429095
73. Les DH (2017) Aquatic dicotyledons of North America: ecology, life history, and systematics. CRC Press, Taylor & Francis. https://doi.org/10.1201/9781315118116
74. Les DH (2022) Aquatic monocotyledons of North America: ecology, life history, and systematics. CRC Press, Baton Rouge, p 568
75. Les DH, Crawford DJ, Kimball RT, Moody ML, Landolt KE (2003) Biogeography of discontinuously distributed hydrophytes: a molecular appraisal of intercontinental disjunctions. Int J Plant Sci 164:917–932
76. Lewis N, Murphy KJ, da Costa V, Filho L, Campos JB, Thomaz SM, Dickinson G, Dibble E, Macleod H, Brown M, Burns F, Campbell E, Downie L, Home-Robertson P, Livingstone P, Mackintosh F, McArthur N, McBean M, O'Halloran K, Padial A, Palmar J, Rickard I (2008) Estrutura funcional e diversidade da vegetação ripária de três rios do Brasil. Cadernos de Biodiversidade 5:10–18
77. Lobato-de Magalhães T, Otte ML (eds) (2023) Wetlands for remediation in the tropics: wet ecosystems for nature-based solutions, vol 9. Springer Nature. https://doi.org/10.1007/978-3-031-23665-5
78. Lobato-de Magalhães T, Rico Y, Cabrera-Toledo D, Martinez M (2020) Plant functional connectivity of *Nymphoides fallax* in geographically isolated temporary wetlands in Mexican highlands. Aquat Bot 164:103215. https://doi.org/10.1016/j.aquabot.2020.103215
79. Lobato-de Magalhães T, Murphy K, Efremov A, Chepinoga V, Davidson T, Molina-Navarro E (2021) Ploidy state of aquatic macrophytes: global distribution and drivers. Aquat Bot 173:103417. https://doi.org/10.1016/j.aquabot.2021.103417
80. Lobato-de Magalhães T, Murphy K, Efremov A, Davidson TA, Molina-Navarro E, Wood KA, Tapia Grimaldo J, Hofstra D, Fu H, Ortegón-Aznar I (2023) How on earth did that get there? Natural and human vectors of aquatic macrophyte global distribution. Hydrobiologia 850:1515–1542. https://doi.org/10.1007/s10750-022-05107-0
81. Lobato-de Magalhães T, Murphy K, Tapia Grimaldo J, Davidson TA, Molina-Navarro E, De-Nova JA, Efremov A (2024 In review) Global hotspots of endemicity, rarity and speciation of aquatic macrophytes. Mar Freshw Res
82. Löffler H (1984) The importance of mountains for animal distribution, species speciation, and faunistic evolution (with special attention to inland waters). In: Proceedings of a workshop of the Arbeitsgemeinschaft für Vergleichende Hochgebirgsforschung, Munich, Dec 1982, pp 299–304. https://doi.org/10.2307/3673232
83. Luna B, Moreno JM (2010) Range-size, local abundance and germination niche-breadth in Mediterranean plants of two life-forms. Plant Ecol 210:85–95
84. Maitner B, Gallagher R, Svenning JC, Tietje M, Wenk EH, Eiserhardt WL (2023) A global assessment of the Raunkiæran shortfall in plants: geographic biases in our knowledge of plant traits. New Phytol 240:1345–1354. https://doi.org/10.1111/nph.18999
85. Marchese MR, Saigo M, Zilli FL, Capello S, Devercelli M, Montalto L, Paporello G, Wantzen KM (2014) Food webs of the Paraná River floodplain: assessing basal sources using stable carbon and nitrogen isotopes. Limnologica 46:22–30. https://doi.org/10.1016/j.limno.2013.11.004
86. Martin G, Coetzee JA (2011) Pet stores, aquarists and the internet trade as modes of introduction and spread of invasive macrophytes in South Africa. Water SA 37:371–380
87. McCutcheon SC, Schnoor JL (eds) (2003) Phytoremediation: transformation and control of contaminants. Wiley, New York
88. McNyset KM (2005) Use of ecological niche modelling to predict distributions of freshwater fish species in Kansas. Ecol Freshw Fish 14:243–255

89. Menge BA, Chan F, Dudas S, Eerkes-Medrano D, Grorud-Colvert K, Heiman K, Hessing-Lewis N, Iles A, Milston-Clements R, Noble M, Page-Albins K, Richmond E, Rilov G, Rose J, Tyburczy J, Vinueza L, Zarnetske P (2009) Terrestrial ecologists ignore aquatic literature: asymmetry in citation breadth in ecological publications and implications for generality and progress in ecology. J Exper Mar Bio Ecol 377(2):93–100. https://doi.org/10.1016/j.jembe.2009.06.024
90. Mesters CML (1995) Shifts in macrophyte species composition as a result of eutrophication and pollution in Dutch transboundary streams over the past decades. J Aquat Ecosyst Health 4:295–305
91. Mikulyuk A, Sharma S, van Egeren S, Erdmann E, Nault ME, Hauxwell J (2011) The relative role of environmental, spatial, and land-use patterns in explaining aquatic macrophyte community composition. Can J Fish Aquat Sci 68:1778–1789
92. Milne JM, Murphy KJ, Thomaz SM (2005) Comunidades de plantas aquáticas do alto rio Paraná: respostas às alterações do estresse ambiental. Cadernos de Biodiversidade 5:11–15
93. Mitsch WJ, Gosselink JG, Anderson CJ, Siobhan Fennessy MS (2023). Wetlands. 6th Ed.. Wiley. Pp- 672
94. Molina-Navarro E, Martínez-Pérez S, Sastre-Merlín A (2017) Addressing uncertainties in the environmental sustainability of limno-reservoirs: insights from the synthesis of multidisciplinary research. Lakes Reser: Res Manag 22(2):134–147. https://doi.org/10.1111/lre.12171
95. Morandeira NS, Kandus P (2015) Multi-scale analysis of environmental constraints on macrophyte distribution, floristic groups and plant diversity in the lower Paraná River floodplain. Aquat Bot 123:13–25
96. Moraru M, Chelariu EL, Brînză M, Goanță M, Draghia L (2017) Aspects regarding the ornamental value of plants from *Eryngium* genus. Lucrări Stiințifice USAMV Iași Seria Horticultură 60:389–394. https://repository.uaiasi.ro/xmlui/handle/20.500.12811/1241
97. Morin X, Lechowicz MJ (2013) Niche breadth and range area in North American trees. Ecography 36:300–312
98. Morueta-Holme N, Enquist BJ, McGill BJ, Boyle B, Jørgensen PM, Ott JE, Peet RK, Símová I, Sloat LL, Thiers B, Violle C, Wiser SK, Dolins S, Donoghue JC, Kraft NJB, Regetz J, Schildhauer M, Spencer N, Svenning J-C (2013) Habitat area and climate stability determine geographical variation in plant species range sizes. Ecol Lett 16:1446–1454
99. Murphy KJ (2002) Plant communities and plant diversity in softwater lakes of Northern Europe. Aquat Bot 73(4):287–324
100. Murphy KJ, Eaton JW (1983) The effects of pleasure-boat traffic on macrophyte growth in canals. J Appl Ecol 20:713–729
101. Murphy KJ, Rørslett B, Springuel I (1990) Strategy analysis of submerged lake macrophyte communities: an international example. Aquat Bot 36:303–323
102. Murphy KJ, Tippett R, Beveridge M (eds) (1994) The ecology of Loch Lomond. Developments in hydrobiology series, Dordrecht: Kluwer. Hydrobiologia 290:1–170
103. Murphy KJ, Dickinson G, Thomaz SM, Bini LM, Dick K, Greaves K, Kennedy M, Livingstone S, McFerran H, Milne J, Oldroyd J, Wingfield R (2003) Aquatic plant communities and predictors of diversity in a sub-tropical river floodplain: the Upper Rio Paraná, Brazil. Aquat Bot 77:257–276
104. Murphy K, Efremov A, Davidson T, Molina-Navarro E, Fidanza K, Crivelari Betiol TC, Chambers P, Tapia Grimaldo J, Varandas Martins S, Springuel I, Kennedy M, Mormul R, Dibble E, Hofstra D, Lukács BA, Gebler D, Båstrup-Spohr L, Urrutia Estrada J (2019) World distribution, diversity and endemism of aquatic macrophytes. Aquat Bot 158:103127. https://doi.org/10.1016/j.aquabot.2019.06.006
105. Murphy K, Carvalho P, Efremov A, Tapia Grimaldo J, Molina-Navarro E, Davidson TA, Thomaz SM (2020) Latitudinal variation in global range-size of aquatic macrophyte species shows evidence for a Rapoport effect. Freshwat Biol 65:1622–1640. https://doi.org/10.1111/fwb.13528

106. Murray-Hudson F, Cheek M, Taylor RW (2018) *Utricularia raynalii* PTaylor (Lentibulariaceae), newly discovered S of The Congo Basin. Africa
107. Ngangbam RD, Devi NP, Devi MH, Singh PK (2019) Rediscovery of *Aldrovanda vesiculosa* L. (Droseraceae), an endangered plant, after six decades in India from Manipur with studies on micromorphology and physico-chemistry of water. Reinwardtia 18(2). https://doi.org/10.14203/reinwardtia.v18i2.3749
108. Nies G, Reusch T (2005) Evolutionary divergence and possible incipient speciation in post-glacial populations of a cosmopolitan aquatic plant. J Evol Biol 18:19–26
109. O'Hare MT, Murphy KJ (1999) Invertebrate hydraulic microhabitat and community structure in *Callitriche stagnalis* Scop. Patches. Hydrobiologia 415:169–176
110. Orchard AE (1981) A revision of South American *Myriophyllum* (Haloragaceae) and its repercussions on some Australian and North American species. Brunonia 4:27–65. https://doi.org/10.1071/BRU9810027
111. Otte ML, Fang W-T, Jiang M (2021) A framework for identifying reference wetland conditions in highly altered landscapes. Wetlands 41:40. https://doi.org/10.1007/s13157-021-01439-0
112. Parsons WT, Cuthbertson EG (1992) Noxious weeds of Australia. Inkata Press, Melbourne
113. Peres CK, Lambrecht RW, Tavares DA, Chiba WA (2018) Alien express: the threat of aquarium e-commerce introducing invasive aquatic plants in Brazil. Perspect Ecol Conserv 16:221–227
114. Pezeshki SR (2001) Wetland plant responses to soil flooding. Environ Exp Bot 46:299–312. https://doi.org/10.1016/S0098-8472(01)00107-1
115. Pip E (1988) Niche congruency of aquatic macrophytes in Central North America with respect to 5 water chemistry parameters. Hydrobiologia 162:173–182
116. Pollux BJA, Santamaría L, Ouborg NJ (2005) Differences in endozoochorous dispersal between aquatic plant species, with reference to plant population persistence in rivers. Freshw Biol 50:232–242
117. Pott VJ, Pott A (2000) Plantas aquáticas do Pantanal. Embrapa, Brasília
118. POWO (2023) Plants of the world online. The Royal Botanic Gardens, Kew. http://www-plantsoftheworldonlineorg/. Accessed 8 Jul 2023
119. Pulido C, Riera JL, Ballesteros E, Chappuis E, Gacia E (2015) Predicting aquatic macrophyte occurrence in soft-water oligotrophic lakes (Pyrenees mountain range). J Limnol 74:143–154
120. Ramsar (1971) What are wetlands? Ramsar Information Paper no 1 https://www.ramsar.org/document/ramsar-information-paper-no-1-what-are-wetlands/. Accessed 6 Jul 2023
121. Ramsey J, Ramsey TS (2014) Ecological studies of polyploidy in the 100 years following its discovery. Philos Trans R Soc Lond B 369:20130352. https://doi.org/10.1098/rstb.2013.0352
122. Record S, Strecker A, Tuanmu M-N, Beaudrot L, Zarnetske P, Belmaker J, Gerstner B (2018) Does scale matter? A systematic review of incorporating biological realism when predicting changes in species distributions. PLoS One 13(4):e0194650. https://doi.org/10.1371/journal.pone.0194650
123. Renner SS, Zhang L (2004) Biogeography of the *Pistia* clade (Araceae): based on chloroplast and mitochondrial DNA sequences and Bayesian divergence time inference. Syst Biol 53:422–432
124. Rice A, Glick L, Abadi S, Einhorn M, Kopelman NM, Salman-Minkov A, Mayzel J, Chay IO, Mayrose I (2015) The chromosome counts database (CCDB)–a community resource of plant chromosome numbers. New Phytol 206(1): 19–26. https://doi.org/10.1111/nph.13191
125. Rice A, Šmarda P, Novosolov M, Drori M, Glick L, Sabath N, Meiri S, Belmaker J, Mayrose I (2019) The global biogeography of polyploid plants. Nat Ecol Evol 3:265–273. https://doi.org/10.1038/s41559-018-0787-9
126. Ritchie H (2018) How urban is the world? OurWorldInData. https://ourworldindataorg/how-urban-is-the-world/. Accessed 4 Nov 2022
127. Rolon AS, Maltchik L (2006) Environmental factors as predictors of aquatic macrophyte richness and composition in wetlands of southern Brazil. Hydrobiologia 556:221–231

128. Rolon AS, Lacerda T, Maltchik L, Guadagnin DL (2008) Influence of area, habitat and water chemistry on richness and composition of macrophyte assemblages in southern Brasilian wetlands. J Veg Sci 19:221–228
129. Rørslett B (1987a) A generalized spatial niche model for aquatic macrophytes. Aquat Bot 29:63–81
130. Rørslett B (1987b) Niche statistics of submerged macrophytes in Tyrifjord, a large oligotrophic Norwegian lake. Arch Hydrobiol 111:283–308
131. Rørslett B (1988) Niche extension of aquatic macrophytes in hydrolakes: predictive assessment of environmental impacts. Int Rev Gesamten Hydrobiol Hydrogr 73:129–143
132. Rørslett B (1991) Principal determinants of aquatic macrophyte richness in northern European lakes. Aquat Bot 39:173–193
133. Rørslett B, Agami M (1987) Downslope limits of aquatic macrophytes, a test of the transient niche hypothesis. Aquat Bot 29:83–95
134. Sabbatini MR, Murphy KJ (1996) Submerged plant survival strategy in relation to management and environmental pressures in drainage channel habitats. Hydrobiologia 340:191–195
135. Salmon B (2001) Carnivorous plants of New Zealand. Ecosphere Publications, Auckland
136. Sandel B, Arge L, Dalsgaard B, Davies RG, Gaston KJ, Sutherland WJ, Svenning J-C (2011) The influence of late quaternary climate-change velocity on species endemism. Science 334(5056):660–664. https://doi.org/10.1126/science.1210173
137. Santamaría L (2002) Why are most aquatic plants widely distributed? Dispersal, clonal growth and small-scale heterogeneity in a stressful environment. Acta Oecol 23:137–154
138. Santos AM, Thomaz SM (2007) Aquatic macrophyte diversity in lagoons of a tropical floodplain: the role of connectivity and water level. Austr Ecol 32:177–190
139. Schou JC, Moeslund B, van de Weyer K, Wiegleb G, Lansdown RV, Holm P, Baastrup-Spohr L, Sand-Jensen K (2023) Key to the Aquatic Plants of Northern and Central Europe including Britain and Ireland. Princeton University Press
140. Sculthorpe CD (1967) The biology of aquatic vascular plants. Edward Arnold, London
141. Simpson MG (2019) Plant systematics, 3rd edn. Elsevier, Amsterdam
142. Siquiera T, Bini LM, Cianciaruso MV, Roque FO, Trivinho-Strixino S (2009) The role of niche measures in explaining the abundance-distribution relationship in tropical lotic chironomids. Hydrobiologia 636:163–172
143. Slatyer RA, Hirst M, Sexton JP (2013) Niche breadth predicts geographical range size: a general ecological pattern. Ecol Lett 16:1104–1114
144. Soininen J, Bartels P, Heino J, Luoto M, Hillebrand H (2015) Toward more integrated ecosystem research in aquatic and terrestrial environments. Bioscience 65:174–182. https://doi.org/10.1093/biosci/biu216
145. Sokoloff DD, Remizowa MV, Macfarlane TD, Rudall PJ (2008) Classification of the early-divergent angiosperm family Hydatellaceae: one genus instead of two, four new species and sexual dimorphism in dioecious taxa. Taxon 57:179–200
146. Sokoloff DD, Marques I, Macfarlane TD, Remizowa MV, Lam VKY, Pellicer J, Hidalgo O, Rudall PJ, Graham SW (2019) Cryptic species in an ancient flowering-plant lineage (Hydatellaceae, Nymphaeales) revealed by molecular and micromorphological data. Taxon 68(1):1–19. https://doi.org/10.1002/tax.12026
147. Soons MB, Brochet AL, Kleyheeg E, Green AJ (2016) Seed dispersal by dabbling ducks: an overlooked dispersal pathway for a broad spectrum of plant species. J Ecol 104:443–455
148. Stevens OA (1963) Handbook of North Dakota plants. North Dakota Institute for Regional Studies, Fargo
149. Tapia Grimaldo J, Bini LM, Landeiro V, O'Hare MT, Kennedy MP, Caffrey J, Spink A, Varandas Martins S, Kennedy M, Murphy KJ (2016) Spatial and environmental drivers of macrophyte diversity and community composition in temperate and tropical calcareous rivers. Aquat Bot 132:49–61

150. Tapia Grimaldo J, O'Hare M, Kennedy M, Davidson TA, Bonilla-Barbosa J, Santamaría B, Gettys L, Varandas Martins S, ThomazS MK (2017) Environmental drivers of freshwater macrophyte diversity and community composition in calcareous warm-water rivers of America and Africa. Freshw Biol 62:1511–1527
151. Trindade CRT, Landeiro VL, Schneck F (2018) Macrophyte functional groups elucidate the relative role of environmental and spatial factors on species-richness and assemblage structure. Hydrobiologia 823:217–230. https://doi.org/10.1007/s10750-018-3709-6
152. USACE (1987) Corps of Engineers wetland delineation manual, Technical report Y87-1 https://usacecontentdmoclcorg/digital/collection/p266001coll1/id/4532/. Accessed 23 Jul 2023
153. Van de Peer Y, Ashman TL, Soltis PS, Soltis DE (2021) Polyploidy: an evolutionary and ecological force in stressful times. Plant Cell 33(1):11–26. https://doi.org/10.1093/plcell/koaa015
154. van Geest GJ, Wolters H, Roozen FCJM, Coops H, Roijackers RMM, Buijse AD, Scheffer M (2005) Water-level fluctuations affect macrophyte richness in floodplain lakes. Hydrobiologia 539:239–248
155. Van Meter KJ, Basu NB (2015) Signatures of human impact: size distributions and spatial organization of wetlands in the Prairie Pothole landscape. Ecol Appl 25: 451–465. https://doi.org/10.1890/14-0662.1
156. Varandas Martins S, Milne J, Thomaz SM, McWaters S, Mormul RP, Kennedy M, Murphy K (2013) Human and natural drivers of changing macrophyte community dynamics over twelve years in a Neotropical riverine floodplain system. Aquat Conserv Mar Freshwat Ecosyst 23:678–697
157. Vestergaard O, Sand-Jensen K (2000) Aquatic macrophyte richness in Danish lakes in relation to alkalinity, transparency, and lake area. Can J Fish Aquat Sci 57:2022–2031
158. Viana DS, Santamaría L, Schwenk K, Manca M, Hobæk A, Mjelde M, Preston CD, Gornall RJ, Croft JM, King RA, Green AJ, Figuerola J (2014) Environment and biogeography drive aquatic plant and cladoceran species-richness across Europe. Freshw Biol 59:2096–2106
159. Viana DS, Figuerola J, Schwenk K, Manca M, Hobæk A, Mjelde M, Preston CD, Gornall RJ, Croft JM, King RA, Green AJ, Santamaría L (2015) Assembly mechanisms determining high species turnover in aquatic communities over regional and continental scales. Ecography 38:1–8
160. Vieira LS (1992) Fitoterapia da Amazônia: manual das plantas medicinais. Ed Agr Ceres, São Paulo
161. Volkova PA, Arutynyan NG, Shanzer IA, Chemeris EV, Bobrov AA (2018) Genetic variability of Eurasian *Nuphar* species unravels possible routes in which freshwater plants could fill their wide areas. Aquat Bot 145:49–57
162. Wani GA, Shah MA, Reshi ZA, Dar MA (2018) Polyploidy determines the stage of invasion: clues from Kashmir Himalayan aquatic flora. Acta Physiol Plant 40(3). https://doi.org/10.1007/s11738-018-2629-4
163. Wasowicz P, Przedpelska-Wasowicz EM, GuÐmundsdóttir L, Tamayo M (2014) *Vallisneria spiralis* and *Egeria densa* (Hydrocharitaceae) in arctic and subarctic Iceland. New J Bot 4:85–89
164. Webb CJ, Sykes WR, Garnock-Jones PJ (1988) Flora of New Zealand. Vol IV: naturalised pteridophytes, gymnosperms, dicotyledons. Department of Scientific and Industrial Research: Botany Division, Christchurch, New Zealand
165. World Bank (2022) Urban Development, understanding poverty, topics. https://www.worldbank.org/en/topic/urbandevelopment/. Accessed 11 Jan 2023
166. Xu Z, Tian J, Rapanarivo SHJV, Letsara R, Rakotonasolo RA, Onjalalaina GE, Hu GW, Wang QF (2020) *Hydrostachys flabellifera* (Hydrostachyaceae), a new species from Madagascar. PhytoKeys 167:45–56. https://doi.org/10.3897/phytokeys.167.58538

167. Yılmaz G, Çolak MA, Özgencil IK, Metin M, Korkmaz M, Ertuğrul S, Soyluer M, Bucak T, Tavşanoğlu NÜ, Özkan K, Akyürek Z, Beklioğlu M, Jeppesen E (2021) Decadal changes in size, salinity, waterbirds, and fish in lakes of the Konya Closed Basin, Turkey, associated with climate change and increasing water abstraction for agriculture. Inland Waters 11:538–555. https://doi.org/10.1080/20442041.2021.1924034
168. Yuan Y, Zhu X, Mushet DM, Otte ML (2018) Multi-element fingerprinting of waters to evaluate connectivity among depressional wetlands. Ecol Indic 97:398–409. https://doi.org/10.1016/j.ecolind.2018.10.033
169. Zhu J, Yu D, Xu X (2015) The phylogeographic structure of *Hydrilla verticillata* (Hydrocharitaceae) in China and its implications for the biogeographic history of this worldwide-distributed submerged macrophyte. BMC Evol Biol 15(95):95. https://doi.org/10.1186/s12862-015-0381-6

The manufacturer's authorised representative in the EU is Springer Nature Customer Service Centre GmbH, Europaplatz 3, 69115 Heidelberg, Germany. If you have any concerns regarding our products, please contact ProductSafety@springernature.com

Printed and bound by CPI Group (UK) Ltd, Croydon, CR0 4YY
01/10/2026
02225907-0019